Longhorned Woodboring Beetles

(Coleoptera: Cerambycidae and Disteniidae)

Longhorned Woodboring Beetles
(Coleoptera: Cerambycidae and Disteniidae)
Primary Types of the Smithsonian Institution

Steven W. Lingafelter
Eugenio H. Nearns
Gérard L. Tavakilian
Miguel A. Monné
Michael Biondi

A Smithsonian Contribution to Knowledge

Smithsonian Institution
Scholarly Press

Washington, D.C.
2014

Abstract. The primary types of longhorned woodboring beetles (Coleoptera: Cerambycidae, Disteniidae) of the National Museum of Natural History (Smithsonian Institution) are catalogued and figured, current through 2012 (but also including many 2013 holotypes). Data on the original combination, current combination, current tribal classification, and type locality are verified and presented. A history of the collection, with biographies of the major systematists who contributed to its growth, is presented. This collection now contains 2,156 primary types from 82 countries, making it among the most important collections of longhorned woodborers in the world in terms of quantity and breadth. The collection includes 1,815 holotypes, 280 lectotypes (including 260 designated herein), and 6 neotypes (including 4 designated herein). A total of 113 proposed names are determined to be infrasubspecific and therefore unavailable, following the provisions of ICZN 45.6, and are excluded from the formal catalog. At least 38 primary types believed to have been present in the Smithsonian Institution collection could not be found and are presumed lost or deposited in other collections. Ninety-five specimens were labeled and segregated by others as primary types, but no evidence of their formal descriptions could be found and they are determined to be *nomina nuda*.

Published by
SMITHSONIAN INSTITUTION SCHOLARLY PRESS
P.O. Box 37012, MRC 957
Washington, D.C. 20013-7012
www.scholarlypress.si.edu

Cover images: Derancistrus hovorei *Lingafelter & Woodley, 2007;* Calliclytus macoris *Lingafelter, 2011 (drawing by Taina Litwak).*

Library of Congress Cataloging-in-Publication Data

Lingafelter, Steven W.
Longhorned woodboring beetles (Coleoptera: Cerambycidae and Disteniidae) : primary types of the Smithsonian Institution / Steven W. Lingafelter, Eugenio H. Nearns, Gérard L. Tavakilian, Miguel A. Monné, Michael Biondi.
pages cm. — (A Smithsonian contribution to knowledge)
Includes bibliographical references.
ISBN 978-1-935623-40-3 (alk. paper)
1. Smithsonian Institution. 2. Cerambycidae—Classification. 3. Cerambycidae—Catalogs and collections. I. Nearns, Eugenio H. (Eugenio Hernán) II. Tavakilian, Gérard L. (Gérard Luc) III. Monné, Miguel A. IV. Biondi, Michael, 1982– V. Title. VI. Series: Smithsonian contribution to knowledge.
QL596.C4L55 2014
595.76'48—dc23

2014005388

ISBN: 978-1-935623-40-3 (print)
ISBN: 978-1-935623-41-0 (ebook)

Printed in Canada

♾ The paper used in this publication meets the minimum requirements of the American National Standard for Permanence of Paper for Printed Library Materials Z39.48–1992.

Contents

Preface

The history of the National Insect Collection of the Smithsonian Institution—and especially that of the longhorned woodboring beetle collection—cannot be discussed without an understanding of its dependence upon the United States Department of Agriculture (USDA) for its foundation and growth.†

The Smithsonian Institution was founded in 1846 but had no dedicated entomology staff until the Department of Entomology was established in 1881. From 1879 to 1881, Professor John Henry Comstock served as Chief Entomologist of the USDA. Although he had no formal appointment with the Smithsonian, Comstock played an active role in seeking to establish a national collection of insects. For example, he reported in 1879 that he purchased 500 Schmitt boxes to house the collection nucleus at the Smithsonian, and other reports show that he acquired 1,600 entomological slides and 40 cases of pinned material (mostly Coleoptera and Lepidoptera) during his tenure as Chief Entomologist with the USDA (Sabrosky, 1964).

In 1881 Charles Valentine Riley (Figure 1a) was selected as Chief Entomologist of the USDA, replacing Comstock, and concurrently served as the first curator of the Smithsonian's Department of Entomology, although this was an unsalaried position. Comstock never formally held this title (Sabrosky, 1964; Flint et al., 1976; Gurney, 1976; Spilman, 1989). Riley probably was initially the most instrumental force in founding and developing the Smithsonian insect collections. He had established himself as a preeminent entomologist, working for nearly 20 years on agriculturally important insects in Missouri and elsewhere before being designated USDA Chief Entomologist (Gurney, 1976; Spilman, 1989). After his designation as Honorary Curator at the Smithsonian Institution, he formally transferred the USDA collections to the Institution in 1881, along with his private collection of 115,000 pinned specimens in 1885 (Sabrosky, 1964; Spilman, 1989). Prior to these transfers—and the formal establishment of a Department of Entomology with full time staff—the Smithsonian had an agreement with its collaborators (e.g., John Lawrence LeConte) who worked on loaned material collected by "the Institution" that the "type specimens of insects . . . [can] be reclaimed at any time when required by the Institution" (Sabrosky, 1964). The appointment and determination of Riley enhanced the ability of the Smithsonian to properly house and care for collections and stabilized the environment for the secure guardianship of primary types. As part of the agreement to make the formal transfer of

† The following publications as well as correspondence held by the National Museum of Natural History (NMNH) and the Smithsonian Institution Archives website (siarchives.si.edu) served as valuable resources: Howard et al. (1928), Sabrosky (1964), Mallis (1971), Flint et al. (1976), Gurney (1976), Spilman (1989), and Miller et al. (1998).

Figure 1. Early Smithsonian-based entomologists: **(a)** Charles V. Riley; **(b)** Colonel Thomas L. Casey; **(c)** Eugene A. Schwarz; **(d)** Warren S. Fisher (a – courtesy Special Collections, National Agricultural Library; b – courtesy Smithsonian Institution Archives; c – National Museum of Natural History files; d – courtesy Joe and Catherine Hubert).

his private collection in 1885, Riley lobbied successfully to the Secretary of the Smithsonian, Spencer Fullerton Baird, for permanent, paid staff to care for the collections (Sabrosky, 1964).

Martin L. Linell was the first Coleoptera specialist officially hired and also the first to describe Cerambycidae at the Smithsonian. Serving from 1889 until his sudden death in 1897 at the young age of 48, Linell was only the second salaried entomologist at the museum; John Smith, a lepidopterist, was the first. Linell's position was Museum Aid (Flint et al., 1976), and his broad job duties and relatively short career before his death meant he had a relatively sparse publication record; he nevertheless described 16 species of Cerambycidae in his eight years at the Smithsonian, starting with two published in 1896, representing some of the earliest described material in the collections and, therefore, the foundation for the Cerambycidae type collection. Linell was educated at Lund University in Sweden and his papers clearly demonstrated his classical training in languages and science. Chittenden (1897) writes: ". . . had his life been spared his name would have been placed in the first rank of the systematists of his day."

The Coleoptera collections surged in breadth and volume when Eugene A. Schwarz (Figures 1c; 2a), a USDA employee hired by Riley in 1878, became formally associated with the Smithsonian Institution in 1895. Schwarz had developed a great reputation and gained the respect of Riley, LeConte, and others for his superior, broad knowledge of Coleoptera, dedication to collections, and academic generosity (Mallis, 1971). While receiving a USDA salary, he was designated as Honorary Custodian of Coleoptera and was based at the Smithsonian for much of the time from 1895 through 1928 (Gurney, 1976; Mallis, 1971). Although Schwarz was not specifically interested in Cerambycidae, he was a prolific general collector, and many of the primary types at the Smithsonian were collected and curated by him. His collaborative, generous spirit fostered close associations with many other coleopterists, especially Henry G. Hubbard, with whom he collected for many years, and Herbert Barber (Figure 2c), a subsequent Custodian of Coleoptera. Schwarz made over 25 major expeditions throughout his career to all parts of the United States—especially Michigan, Florida, Texas, and Arizona—Cuba, Mexico, Guatemala, and Panama (Howard et al., 1928). In addition to his own specimens, the museum received many important collections during his 30 years of association due to the professional relationships (like with Hubbard and Barber) he cultivated (Mallis, 1971).

As mentioned, and due in large part to the collections that Schwarz donated, the Coleoptera collection grew substantially during this time, and Schwarz was instrumental in introducing measures to continually improve its care and organization (Schwarz, 1891; Mallis, 1971). Schwarz wrote a report in 1906 (published in 1926 at the request of S. A. Rohwer, former head of the USDA Bureau of Entomology) that gave the status of the collection of that year, approximately 10 years after his appointment as Custodian of Coleoptera. In that report he estimated that the number of species included about 22,000 determined and 13,000 undetermined. Remarkably, this number included only about 500 primary types. Fewer than 100 drawers were available for the Coleoptera collection at the time, and most beetles were preserved in Schmitt boxes.

The Cerambycidae collection grew tremendously with the association of Warren S. Fisher (Figures 1d; 2b, d, e), a woodboring beetle specialist. Before his affiliation with the Smithsonian, Fisher was hired in 1912 as an assistant in the Forest Insect Investigations unit of the USDA, headed by A. D. Hopkins (Cushman et al., 1972). Other well-known entomologists in this unit included A. G. Boving and F. C. Craighead. This laboratory was based in Falls Church, Virginia, not far from the Smithsonian Institution, and Fisher was therefore able to receive additional mentoring from Schwarz, even before being based at the Smithsonian. In this capacity Fisher learned an immense amount on the biology and identification of many species in several families of woodboring beetles, and in 1915 his knowledge gained him an appointment as Specialist on Forest Coleoptera under Schwarz (Cushman et al., 1972). Thus began his 35-year career at the National Museum of Natural

Figure 2. Early Smithsonian-based entomologists: **(a)** Eugene A. Schwarz; **(b)** Warren S. Fisher; **(c)** Herbert S. Barber; **(d, e)** Warren S. Fisher (a, b, c – courtesy Smithsonian Institution Archives, Image #84-3567; d, e – courtesy Joe and Catherine Hubert).

History (NMNH), the new building of the Smithsonian Institution that had been completed just three years prior. Although he did not engage in much fieldwork, his contacts as a forest entomologist brought him many important Indian, Philippine, and Caribbean collections from A. D. Hopkins, C. F. Baker, H. M. Pendlebury, F. C. Drescher, and others, which he methodically studied and described (Cushman, 1929; Cushman et al., 1972). Like Schwarz, Fisher cultivated relationships with other international cerambycid researchers who studied material (and deposited holotypes) at the Smithsonian. Fisher ultimately published more than 135 papers, mostly on Cerambycidae and Buprestidae. His legacy of 373 holotypes of Cerambycidae at the Smithsonian is the largest number for any researcher at the Institution in its history. After Fisher's retirement in 1948, it was nearly 50 years before another specialist on Cerambycidae was based at the Smithsonian.

Before and during part of Fisher's tenure, Colonel Thomas L. Casey (Figure 1b) was a well-known and prolific coleopterist, who spent much of his time in Washington, DC. While never formally associated with the Smithsonian, Casey made a permanent mark on the beetle collection. Casey, a military engineer, was often stationed in biologically rich parts of the world from the 1890s through the 1910s. During his free time, he would enthusiastically collect beetles, ultimately developing a collection of more than 116,000 specimens (Buchanan, 1935; Blackwelder, 1950). Casey was very active in researching several families of beetles and described 671 species of Cerambycidae. When he died in 1925, he bequeathed his entire collection to the Smithsonian. This represented the single greatest boost to the number of primary types of Cerambycidae at the Institution, eclipsing the number described by Fisher, although many of Casey's species have since been determined to be variants of previously described species. Casey's collection is maintained separately from the other Coleoptera, according to stipulations of his bequest, and this has also facilitated its study. Casey's peculiar use of obscure locality codes, aversion to placing type labels on specimens, and discrepancies with localities and original series counts based on the literature and correspondence required careful attention by a curator. The Smithsonian designated L. L. Buchanan to spend half of his workday for nearly five years preparing the Casey Collection—by labeling all the specimens and verifying the type series with the original literature and Casey correspondence—so it could be studied by other researchers (Buchanan, 1935).

During Fisher's tenure, another important event occurred that greatly boosted the cerambycid holdings of the Smithsonian. Fisher had collaborated with Professor Charles Baker, dean of the College of Agriculture at the University of the Philippines in Los Baños, Philippines, in the early and mid-1920s. Baker was a highly dedicated collector of all insects, especially Cerambycidae, throughout the Philippines as well as from Borneo (Malaysia). He was very active in getting this material out to specialists all over the world, including Fisher. When Baker died in 1927, he willed his entire collection to the Smithsonian Institution. The Cerambycidae were the largest represented family, filling 122 Schmitt boxes and numbering more than 20,000 specimens (Rohwer and Cushman correspondence, 1928; Cushman, 1929; Mallis, 1971). The collection includes many primary types described by Fisher, but also parts of syntypic series described by other researchers such as Christopher Aurivillius (Naturhistoriska riksmuseet, Stockholm, Sweden) and Karl Heller (Staatliches Museum für Tierkunde, Dresden, Germany).

The next event that catapulted the Smithsonian's Cerambycidae collection to among the best in the world, rivaling those of the Muséum national d'Histoire naturelle in Paris and the Natural History Museum in London, was the purchase of the Freidrich F. Tippmann Cerambycidae Collection in 1957. Tippmann (Figures 3a, b) was born in 1894 in Hungary and subsequently lived in Vienna, Austria. He studied electrical and chemical engineering, and developed an expertise in cement production. This became a lucrative career that took him all over the world as a consultant on major industrial projects. Wherever he traveled, he acquired from local collectors or museums large collections of Cerambycidae, and thus his collection grew tremendously, ultimately becoming the largest private collection of Cerambycidae in the world at the time (Mandl, 1979). When it was purchased by the Smithsonian Institution, it included over 400 holotypes, including 140 described by Tippmann and 232 described by Stephan Breuning, who studied Tippmann's material and was the most prolific cerambycid taxonomist in the world. Tippmann's collection nearly doubled the general holdings of Smithsonian Cerambycidae, adding almost 100,000 specimens (United States National Museum Accession Memorandum #213112, Cat. No. 18080, 12 November 1957). Because the Tippmann collection was worldwide in scope (especially strong in Asian, African, European, and Australian species), it immediately elevated the Smithsonian holdings to worldwide coverage, replacing the previous geographical emphasis on North America.

Figure 3. (a, b) Friedrich F. Tippmann (courtesy Christian Mandl, grandson of F. Tippmann).

The Smithsonian Cerambycidae collection now held superior international stature and was regularly studied and contributed to by other researchers on Cerambycidae from around the world, such as Frederico Lane, Ubirajara Martins (Figure 4e), J. Linsley Gressitt, E. Gorton Linsley, and John Chemsak. While rich, the collection languished in a very poor curatorial state, still having no dedicated curator, but it continued to grow through the collecting efforts of some of the expanding Coleoptera staff that included collections-oriented researchers George Vogt and Theodore "Ted" Spilman (each of whom published a few papers on Cerambycidae), Terry Erwin, and Paul Spangler (Flint et al., 1976; Miller et al., 1998). In 1996 Steven Lingafelter was hired by the USDA and based at the Smithsonian as the curator dedicated to the group, almost 50 years after Warren Fisher retired and 40 years since the acquisition of the Tippmann Collection.

The Cerambycidae collection of approximately 1,300 drawers had, just a few years earlier, been moved from the NMNH building to the Smithsonian Museum Support Center, a storage facility in Suitland, Maryland. This was done for many "orphaned" insect groups for which there was no dedicated specialist. One of the top priorities as curator was to return the collection to the museum, where it could be curated and enhanced for research. After 1998, when it returned to NMNH, and through 2013, the Cerambycidae collection experienced renewed growth in numbers of holotypes and non-types, primarily due to expeditions led by Lingafelter and contacts fostered in the international community of Cerambycidae researchers (Figure 4). Major expeditions focusing on Cerambycidae were made in this period to Japan, China, South Korea, Russia, Turkey, Armenia, Puerto Rico, Bolivia, Dominican Republic, Fiji, Nicaragua, and Vietnam. These trips resulted in accessions of nearly 10,000 Cerambycidae specimens. A major effort has been made to enhance curation of the long neglected Cerambycidae, and specialists such as Mikhail Danilevsky, Eduard Vives, Miguel Monné (Figure 4b), Marcela Monné (Figure 4c), Frank Hovore (Figure 4c), James Wappes (Figure 4a), Charyn Micheli, Eugenio Nearns (Figure 4d), Ian Swift (Figure 4c), Daniel Heffern, Robert Turnbow, Francesco Vitali, Hilda Waqa-Sakiti, Elisabeth Roberts, and Greg Setliff have worked in the collection, identifying and curating many tribes. Through collaboration with the above researchers

Figure 4. Contemporary Smithsonian Institution longhorned beetle workers and collaborators: **(a)** James Wappes; **(b)** Steve Lingafelter (left) and Miguel Monné (right); **(c)** (left to right) Marcela Monné, Miguel Monné, Ian Swift, Frank Hovore; **(d)** Gérard Tavakilian (left) and Eugenio Nearns (right); **(e)** (left to right) Sonia Casari, Ubirajara Martins, Antonio Santos-Silva.

Figure 5. The Smithsonian Institution's state-of-the-art collection of Cerambycidae: **(a)** fireproof and pest-proof cabinets and drawers; **(b)** example of a curated drawer of Cerambycidae; **(c)** example of a drawer of holotypes.

and the voluminous identifications provided by Ubirajara Martins (Figure 4e), Antonio Santos-Silva (Figure 4e), Maria Helena Galileo, Gérard Tavakilian (Figure 4d), Meiying Lin, and others, nearly 50% of the tribes have had their addenda identified and incorporated, the nomenclature updated, and the species databased. As of this writing, the Cerambycidae collection fills almost 2,000 drawers (Figures 5a, b, c), and 8,000 identified species have been databased. It is estimated that the general collection, when fully curated and databased, will include more than 16,000 identified species.

Lingafelter (Figure 4b) envisioned the need to establish a website of the primary types of Cerambycidae and closely related families in 2004. Specimens were routinely shipped worldwide to facilitate the research of systematists in other countries, but this expended significant resources and introduced risk to the specimens. An online image database, along with the data in searchable fields, would obviate sending the material much of the time. Through collaboration with Eugenio Nearns and funding through the USDA, Lingafelter developed "Cerambycoid Primary Types of the Smithsonian Institution" (smithsoniancerambycidae.com), an online database for primary type specimens of longhorned woodboring beetles at the NMNH (Lingafelter et al., 2013), including more than 5,000 images taken since 2004 by Rick Stanley (Bethesda, Maryland), Alexander Boyson (Rockville, Maryland), and Elisabeth Roberts (USDA). It is now a model that is in use for collections in Brazil (Galileo et al., 2012; Martins et al., 2012; Monné et al., 2012), Bolivia (Ledesma Arias et al., 2012), and Costa Rica (Solís et al., 2012).

Acknowledgments

This book is the culmination of an enormous effort and collaboration by many individuals over the past 10 years. Funding has come via contracts from the USDA and the Smithsonian Institution, and we thank both organizations for recognizing the value and urgency of this research. More than 5,000 photographs were taken in the course of this project. Rick Stanley (Bethesda, Maryland) took many of the dorsal habitus photographs as a volunteer in the summer of 2004. Additional photographs were taken by Alexander Boyson (Rockville, Maryland), who interned in 2008. Elisabeth Roberts (Systematic Entomology Laboratory) took many of the label images as well as dorsal habitus images of recently acquired holotypes. She, along with Lucrecia Rodriguez and Michele Touchet (both of SEL), also prepared the new lectotype labels and their images.

We thank Andrés Garzón-Moreno for his work on verifying the status of many of the Casey types and those from China through careful examination of the literature and the collection. For access to the fossil collections in the Department of Paleobiology at the Smithsonian Institution where the four fossil Cerambycidae holotypes are housed, we thank Conrad Labandeira and Finnegan Marsh. For assistance providing much of the literature, we thank the University of New Mexico Interlibrary Loan Service. We thank Alicia M. Hodson (Gaithersburg, Maryland) for formatting many of the plates.

We thank James Wappes (San Antonio, Texas) for being the most active proponent of building the Smithsonian Institution holotype collection. Jim has been incredibly active by continually sending material to researchers (especially Ubirajara Martins and Maria Helena Galileo in Brazil) so that it can described. In the past 10 years, Jim has been responsible for enhancing the Smithsonian collection by many dozens of holotypes. We also want to thank all of our other collaborators and fellow taxonomists, many of whom are mentioned in the Preface, who have worked with our material and deposited their holotypes here. In particular, Antonio Santos-Silva, Ubirajara Martins, Maria Helena Galileo, Marcela Monné, Jim Wappes, and Michael Ivie deserve special acknowledgment.

We give special thanks to Terrence Walters, Identification Technology Program, USDA, APHIS-PPQ, who provided funding and support for photography and travel throughout Europe by E. Nearns. This was critical in resolving many syntype deposition questions and generating thousands of images. Special thanks are given also to Kelly B. Miller, University of New Mexico, who provided the Visionary Digital Passport equipment that was crucial for photographing much material, especially abroad. We thank Christian Mandl, grandson of Friedrich Tippmann, and Joe and Catherine Hubert, relatives of Warren Fisher, for providing photos and biographical information. We thank Norman Woodley for his consultation regarding the format of this work and for providing a critical review. We also thank Ted MacRae (Monsanto) who gave this work a final review and provided useful comments.

Introduction

Although Linell was the first staff member to describe Cerambycidae at the Smithsonian Institution, the oldest primary type is one described by Shimer (1868), *Liopus xanthoxyli*—now *Sternidius alpha* (Say)—a mere 22 years after the founding of the Institution. It wasn't until the 1890s that the next Cerambycidae types were described and deposited in the United States National Museum (USNM), the original name of the Smithsonian's National Museum of Natural History (NMNH). Figure 6 shows the growth of the primary type collection since. About two-thirds of all the Cerambycidae primary types were added to the collection between 1910 and 1949. These dates correspond with a period of beetle research that included Schwarz, Casey, Tippmann, Breuning, and Fisher, among others. The 1970s through 1990s show a rapid decline in numbers, corresponding with the retirement of Fisher and lack of acquisition of other cerambycid collections. The 2000s through the time of this publication reveal a strong resurgence in the growth of the cerambycid primary type collection, due to many cerambycid collecting expeditions by Lingafelter.

By the end of 2013, the Smithsonian Institution contained 2,156 primary types of Cerambycidae and Disteniidae. There are no primary types of the other cerambycoid families, Vesperidae and Oxypeltidae (Figure 5c). Not all primary types from 2013 were available for this book; however, it is complete through 2012. Nearly all of the primary types are holotypes, but the types also include 280 lectotypes (260 designated herein) and 6 neotypes (4 designated herein). Four fossil holotypes from the Florissant, Colorado, formation described by Wickham (1913, 1917) are also included, although these are maintained in the paleoentomology collections of the NMNH's Department of Paleobiology. The Smithsonian Institution (Figure 7) holds the third largest collection in the world based on primary types and is the leading institution outside of Europe. The Muséum national d'Histoire naturelle, Paris, France, holds more than 10,000 primary types (Tavakilian and Chevillotte, 2013) and the Natural History Museum, London follows with more than 6,700 primary types (Max Barclay, pers. comm.). It should be noted that the latter two collections overlap somewhat, because they include shared syntypes of the same species in their totals.

More than 30 authors have described Cerambycidae and had their primary types deposited in the Smithsonian collections. Some of these authors were stationed at the Smithsonian or otherwise had a strong connection. Most, however, were not affiliated with the Smithsonian yet had a major impact. The 10 most active taxonomists who had their longhorned beetle holotypes and lectotypes deposited at the Smithsonian include Casey (671), Fisher (373), Breuning (203), Gressitt (146), Tippmann (138), Martins (119 [including 84 as co-author with M. H. M. Galileo]), Aurivillius (109), Galileo (84, as co-author with U. R. Martins), L.& S. Dillon (53), and Schaeffer (43). They represent

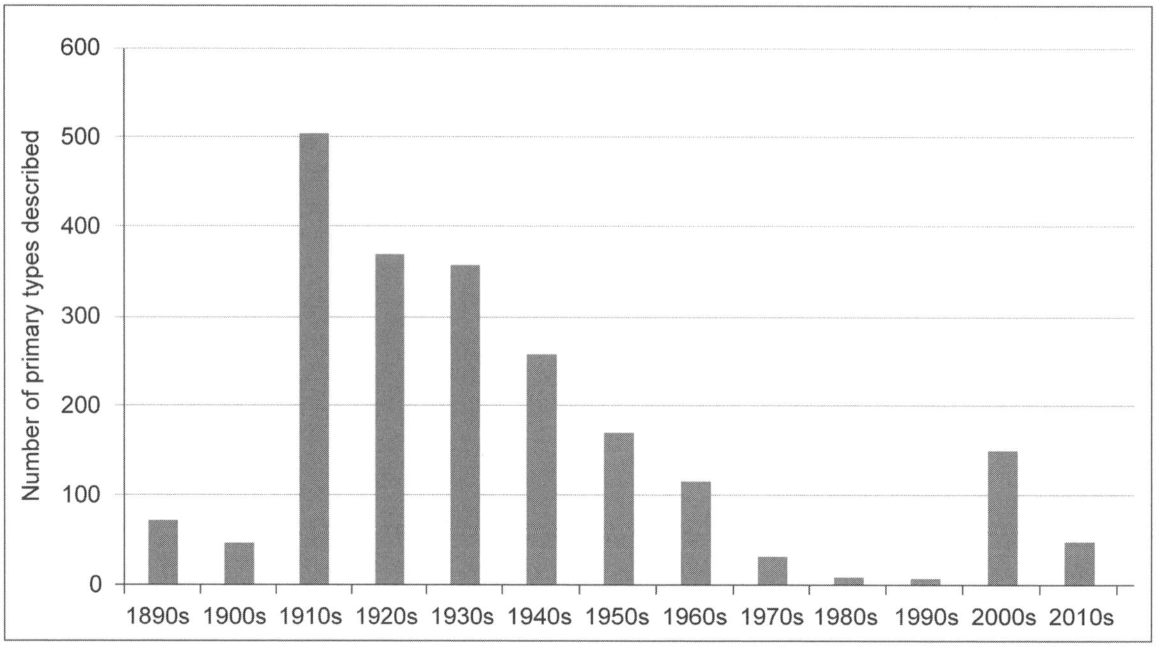

Figure 6. Graph showing Smithsonian Institution cerambycoid relative primary type collection growth by decade.

Figure 7. Facade of the Smithsonian Institution's National Museum of Natural History.

86% of the primary types of Cerambycidae and Disteniidae in the Smithsonian collection, with a total of 1,855 primary types among them. Other important authors of material in the collections include Lingafelter (41 primary types), Chemsak (36), Linsley (34), Santos-Silva (31), Melzer (25), Julio Micheli and Charyn Micheli (23, sometimes publishing separately), Hovore (20), Linell (18), Miguel Monné and Marcela Monné (17, sometimes publishing separately), Lane (15), Kriesche (12), Nearns (11), and Tavakilian (6). It should be noted that many of these researchers have coauthored with one another; therefore, there is some overlap in the representation of the numbers of types for each one.

The primary types originated in 82 countries (43% of all countries), making the Smithsonian Institution among the very best in terms of worldwide representation. The top 10 most highly represented countries account for 68% of the collection (1,461 types): United States (excluding Puerto Rico & Guam) (694), Philippines (145), Mexico (91), Malaysia (90), Indonesia (88), Brazil (84), Panama (77), China (including Tibet, excluding Taiwan) (67), Costa Rica (63) and Bolivia (62).

Other countries with primary types represented in the Smithsonian collections include, in alphabetical order:

Afghanistan (1)
Angola (2)
Anguila (1)
Antigua & Barbuda (2)
Argentina (5)
Australia (4)
Austria (2)
Bahamas (2)
Barbados (3)
Belize (6)
British Virgin Islands (4)
Cameroon (4)
Canada (16)
Chile (1)
Colombia (24)
Congo (7)
Croatia (2)
Cuba (62)
Dominica (5)
Dominican Republic (45)
Ecuador (28)
Egypt (1)
El Salvador (5)
Ethiopia (2)
Fiji (1)
French Guiana (4)
Gabon (3)
Guadeloupe (2)
Guam (3)
Guatemala (17)
Guinea-Bissau (1)
Guyana (3)
Haiti (20)
Honduras (13)
India (14)
Iran (1)
Italy (1)
Ivory Coast (3)
Jamaica (15)
Japan (7)
Kenya (11)
Laos (5)
Madagascar (3)
Marshall Islands (5)
Mauritius (1)
Micronesia (21)
Montserrat (1)
Myanmar (2)
Namibia (1)
Navassa Island (1)
New Caledonia (1)
Nicaragua (2)
Northern Mariana Islands (3)
Palau (12)
Papua New Guinea (36)
Paraguay (9)
Peru (50)
Puerto Rico (34)
Russia (2)
Saint Lucia (2)
Seychelles (1)
Singapore (2)
South Korea (1)
Sri Lanka (1)
Sudan (2)
Taiwan (29)
Tanzania (26)
Thailand (7)
Trinidad & Tobago (5)
Turkey (2)
Venezuela (36)
Vietnam (10)

METHODS

Determination of Subspecific and Infrasubspecific Names

As discussed in Lingafelter and Nearns (2013), the determination of whether a species group name is infrasubspecific or subspecific is sometimes difficult. Much of the Smithsonian material had to be carefully researched to make these determinations, and we discovered that many of the "types," although labeled and segregated by earlier researchers, were not, in fact, primary types, because their trinomials were determined to be infrasubspecific. The terms "aberration", "ab." and "morph" *always* denote infrasubspecific status. The terms "variety" and "form" (including use of the terms "var.", "forma", "v." and "f.") denote infrasubspecific status if published *after* 1960. Before 1961, the terms denote infrasubspecific status *only if* the author expressly gave it that status, based on Article 45.6.4 (ICZN, 1999).

Since all of Tippmann's—and most of Breuning's—types were published before 1961, we had to consider all the exceptions and qualifications laid out in Article 45.6.4, and that required familiarity with all the literature through 1985 that might have used the names for a species or subspecies or senior homonym. In Appendix 1, we include a list of names that we have determined to be of infrasubspecific rank and therefore unavailable and excluded from this work. Appendix 2 lists the most complex cases where names were originally published with the qualifier "var." but that we determined are available species-group names. Lingafelter and Nearns (2013) developed a dichotomous key as a tool to make determinations of subspecific or infrasubspecific rank for trinomial names.

Determination of Holotype or Syntypic Series and Lectotype Designations

Another problematic issue with this project is the determination of whether a specimen is a holotype or part of a syntypic series. The label, whether it says "type", "cotype", or any of a number of variants, is irrelevant to this determination. What is critical is if the original literature makes an indication or reference to only one or to more than one specimen. We have followed Article 73.1.3 (ICZN, 1999), which, paraphrased, says that if there is an indication that there is only one specimen (for example, explicit ". . .based on a single male" or implicit, such as giving a single measurement, rather than a range, "measuring 6.2 mm long, 1.4 mm wide,") then that specimen can be regarded as the holotype, provided the other label data (locality, collector, etc.) agree with the published record. Following Article 74.6, if what we called a holotype is later determined to be part of a syntypic series, then that assumption of holotype becomes a lectotype designation.

If, however, a measurement range is given in the original species account, or the species account otherwise implies or states explicitly that there is more than one specimen, then we have followed Recommendation 73F of the ICZN to avoid assuming a holotype in these cases. In those cases, we have designated a lectotype or recorded one of the syntypes, particularly if there is evidence that a lectotype designation would be better made in the future from material at another institution. Some may argue that this goes against Recommendation 74G in that it is being done for curatorial purposes. However, the only other option in a comprehensive catalog of the primary types of the third largest cerambycid collection in the world would be to have photos of all the syntypes, if lectotypes were not designated, since all the syntypes together represent the "primary type" of that particular taxon. That would add approximately 2,000 images and hundreds of pages to this volume, since each syntype would have to be treated equally with photos of the labels and habitus. In some cases, we would have an entire page (or more) of images devoted to the syntypic series of just one species. Therefore, for reasons of practicality and the taxonomic need to promote stability by linking the name to one specimen, we have determined that these lectotype designations are necessary.

In the Smithsonian Institution collection, there are 223 Casey names of Cerambycidae that were described from a series of specimens without his selection of a holotype. There are others described by Aurivillius, Heller, Hamilton, and a few other authors. To further complicate things, Casey did not label the type specimens, as such, in his collection, but indicated verbally to J. C. Crawford (who then conveyed that intention in a letter to then Assistant Secretary of the Smithsonian Institution

Alexander Wetmore) that the specimen that had the epithet name label affixed was to be considered as the holotype (Buchanan, 1935). Buchanan, who curated the Casey collection after its bequest, carefully labeled these specimens with labels bearing "TYPE" and the others in the series as "paratypes" in order to facilitate future work in the collection. Buchanan (1935) indicated that Marie Siebrecht prepared a catalog of the entire collection that indicated the types; however, that has remained an unpublished manuscript. Since Casey did not designate holotypes in the original publications and Buchanan's work was curatorial rather than a formal fixation of types, it is prudent to designate lectotypes for these species. Therefore, following the criteria laid out in Article 74.7, we formally designate lectotypes for the 223 Casey species and subspecies that exist as a syntypic series.

For each of these new lectotype designations (listed in Appendix 7), we include a photo of the lectotype and its labels, including the "TYPE USNM" label as affixed by Buchanan and our lectotype label that indicates the original combination, followed by "Lectotype designated by Lingafelter, Nearns, Biondi, 2012/2013" or "Lectotype designated by Lingafelter et al., 2013." We include the following verbiage in the remarks with each one: "This specimen is herein designated as the lectotype in order to stabilize the taxonomy and facilitate further identifications of this species."

We have tried to resolve the type status of all of this material in the NMNH, but realize that there may be additional primary types present in the NMNH (for example, as partial syntypic series with the remainder scattered in one or more institutions, or unlabeled/mislabeled specimens). In some cases, as with Aurivillius labeled material, it is preferable to designate lectotypes for specimens residing in the Naturhistoriska riksmuseet, Stockholm, since he served as director there and that is where the bulk of his material was deposited (following Recommendation 74D [ICZN, 1999]). Likewise, with much of Karl Heller's material, it would be appropriate to designate lectotypes at the Museum für Tierkunde, Dresden, where he served as a staff researcher for many years.

Misplaced Types, *nomina nuda*, and Neotypes

The database of Tavakilian and Chevillotte (2013) and the catalogues of Western Hemisphere Cerambycidae by Monné (2005a, 2005b, 2006, 2013a, 2013b, 2013c) were consulted regularly for help in making these determinations. We include a list of species that were labeled as types and segregated in the Smithsonian type collection, but for which we were unable to find any publication that confirmed that they were described. Therefore, they are listed as *nomina nuda* (Appendix 3). We also include a list of specimens that had labels indicating they were types, but for which our literature and museum research indicated that they are not, in fact, primary types (Appendix 4). We further include a list of species for which there is some evidence that suggests the Smithsonian Institution should have the primary type(s), but that we were unable to locate (Appendix 5). Hopefully this list will stimulate a more thorough search by curators of other collections to locate them or determine their proper depository. It is probable that there are others in this category, but it would be a Herculean task to find all such records in the literature and contact all museums for potential NMNH types.

Likewise, it is possible that some of the primary types included in this work actually belong to another institution. In some cases, the literature indicates a different depository but since it is impossible to know the entire history of each type and any agreements that may have been made leading to their current disposition, they are considered as NMNH types for the time being. This work may stimulate proper repatriation based on individual case examination.

Finally, there are four instances of new neotype designations made from NMNH material in cases where the original type is confirmed or believed to be lost. All NMNH neotypes are summarized in Appendix 6.

ARRANGEMENT OF TEXT AND PLATES

We have made the determination that the best way to present the data for the taxa is to sort them alphabetically by specific epithet, subspecific epithet (if applicable), author, and year. In this way, each species can be located rapidly without regard to different subsequent or future combinations and without needing an index. The negative aspect of this arrangement is when a common specific epithet is used for many species, it requires looking at the author and in some cases the year and original genus for further determination. For example, Breuning used the epithet "*strandi*" for 18 different species of longhorned beetles in the Smithsonian collection. Since they will usually appear on the same page, we think that the inconvenience is minimized.

The text for each primary type is arranged as follows: The first line contains the original specific epithet, original subspecific epithet (if applicable), author, year: page number. This is followed by the figure number of the dorsal habitus and label images. This is followed by the kind of type (either holotype, lectotype, neotype, or syntype). The next line is the complete species or subspecies name in the original combination. The third line is the current name which can differ from the original combination if the species is placed in a different genus or is regarded as a synonym of an older name. The fourth line is the current classification listing the family, subfamily, and tribe based on modern references including Bezark and Monné (2013), Bouchard et al. (2011), Bousquet (2007), Bousquet et al. (2009), Löbl and Smetana (2009, 2011), Danilevsky (2010, 2011, 2012a–c), Kasatkin and Miroshnikov (2011), Miroshnikov (2011a–d), and Özdikmen (2011). The fifth line is the type locality to the most specific level possible based on the label data, literature, and other data. Country, province (or department or state) are listed in most cases, even if these data are not present on the label or in the original literature. There is a "Remarks" section where additional information such as lectotype designation justification, inconsistencies with the label(s), or other applicable historical information is presented.

The scientific names are spelled exactly as they are in the original publications, with the following exceptions corrected. Following Article 32.5.2 (ICZN, 1999), specific and subspecific epithets that contain capital letters, diacritics, hyphens, or other non-letter marks are corrected to use only the current Latin alphabet. For example, all hyphens (with one exception, noted below) and apostrophes are removed and the formerly separate letters are united. Ligature letters are converted to their original elements. Any numerals are converted to a Latin word and joined without any hyphen. Any spaces within an epithet are removed. The only time a hyphen is retained is, following Article 32.5.2.4.3, "If the first element is a Latin letter used to denote descriptively a character of the taxon, it must be retained and connected to the remainder of the name by a hyphen." An example of this is *Trachyderes MM-ornata* Tippmann, 1960. In this case, the "*MM*" refers to the pattern of the elytra that looks like two Ms. Therefore, the name is corrected to all lowercase letters; however, the hyphen is retained: "*mm-ornata*". Most of the examples provided in Article 32.5.2 are represented among the Smithsonian primary Cerambycoid types. In all cases where names were corrected, an indication of the original error in name formation is made in the remarks.

An elaboration on the type locality section is warranted here. According to the ICZN (1999, Article 76.1 [see Recommendation 76A]), the type locality is the actual location at which the primary type was collected. Most of the time the label provides the most accurate and specific information available and we provide photos of all the labels. There are many instances, however, where the published type locality differs in specificity from the label data. There are also instances where the specimens are unlabeled or erroneously labeled. In those situations, an examination of the primary literature and/or correspondence was used to clarify these data. Most of the time, the locality data is somewhat vague or ambiguous, such that an exact location cannot be determined. Taking information from the labels, primary literature, coden definitions (in the case of Casey's types) and correspondence, we provide the most accurate possible type locality for each primary type.

The habitus and label images referred to above are presented in plates in the center of this book in the same order as the text. Larger images, and in some cases additional views, can be found in the online searchable database at smithsoniancerambycidae.com (Lingafelter et al., 2014).

The Smithsonian Institution's Primary Type Collection of Cerambycidae and Disteniidae

***abbreviatus* Casey, 1893**: 593 (Fig. 8a, b), Holotype
Original combination: *Ptychodes abbreviatus* Casey
Current name: *Neoptychodes trilineatus* (Linnaeus, 1771)
Current placement: Cerambycidae: Lamiinae: Monochamini
Type locality: USA: Arizona

***aberrans* Aurivillius, 1927b**: 560 (Fig. 8c, d), Holotype
Original combination: *Pterolophia aberrans* Aurivillius
Current name: same
Current placement: Cerambycidae: Lamiinae: Pteropliini
Type locality: Philippines: Samar

***aberrans* Fisher, 1935c**: 618 (Fig. 8e, f), Holotype
Original combination: *Heteropalpoides aberrans* Fisher
Current name: same
Current placement: Cerambycidae: Lamiinae: Apomecynini
Type locality: Malaysia: Borneo, Sabah, Mt. Kinabalu, Pakka, 10,000 ft.

***abnormalis* Tippmann, 1953d**: 317 (Fig. 8g, h), Holotype
Original combination: *Phaedinus abnormalis* Tippmann
Current name: same
Current placement: Cerambycidae: Cerambycinae: Trachyderini
Type locality: Brazil: Santa Catarina, São Bento do Sul, 850 m.

***abruptus* Casey, 1912**: 301 (Fig. 8i, j), Holotype
Original combination: *Hypermallus abruptus* Casey
Current name: *Anelaphus villosus* (Fabricius, 1793)
Current placement: Cerambycidae: Cerambycinae: Elaphidiini
Type locality: No locality data

***acangauna* Martins & Galileo, 2004**: 538 (Fig. 8k, l), Holotype
Original combination: *Adesmus acangauna* Martins & Galileo
Current name: same
Current placement: Cerambycidae: Lamiinae: Hemilophini
Type locality: Panama: Bocas del Toro: 4 km north of Continental Divide, Gualaca-Chiriquí Grande Highway

***aceris* Fisher, 1917**: 215 (Fig. 8m, n), Holotype
Original combination: *Xylotrechus aceris* Fisher
Current name: same
Current placement: Cerambycidae: Cerambycinae: Clytini
Type locality: USA: Washington, D.C.

***acomana* Casey, 1912**: 328 (Fig. 8o, p), Lectotype
Original combination: *Batyle acomana* Casey
Current name: *Batyle ignicollis* (Say, 1824)
Current placement: Cerambycidae: Cerambycinae: Trachyderini
Type locality: USA: New Mexico
Remarks: This specimen is herein designated as the lectotype in order to stabilize the taxonomy and facilitate further identifications of this species.

***acomanus* Casey, 1913**: 337 (Fig. 8q, r), Holotype
Original combination: *Graphisurus (Graphisurus) acomanus* Casey
Current name: *Acanthocinus obliquus* (LeConte, 1862)
Current placement: Cerambycidae: Lamiinae: Acanthocinini
Type locality: USA: New Mexico

actinophorae **Fisher, 1934b**: 36 (Fig. 8s, t), Holotype
Original combination: *Exocentrus actinophorae* Fisher
Current name: *Exocentrus (Pseudocentrus) actinophorae* Fisher, 1934b
Current placement: Cerambycidae: Lamiinae: Acanthocinini
Type locality: Indonesia: Java, Semarang, teak forest

acunai **Fisher, 1935b**: 206 (Fig. 8u, v), Holotype
Original combination: *Cyrtinus acunai* Fisher
Current name: *Cyrtinus acunai* Fisher, 1935b
Current placement: Cerambycidae: Lamiinae: Cyrtinini
Type locality: Cuba: Buenos Aires, Trinidad Mountains

acunai **Fisher, 1936d**: 344 (Fig. 8w, x), Holotype
Original combination: *Pentomacrus acunai* Fisher
Current name: *Plectromerus acunai* (Fisher, 1936d)
Current placement: Cerambycidae: Cerambycinae: Plectromerini
Type locality: Cuba: Santiago de Cuba, Loma del Gato, Sierra del Cobre, Oriente

acuticauda **Casey, 1913**: 274 (Fig. 9a, b), Lectotype
Original combination: *Typocerus acuticauda* Casey
Current name: same
Current placement: Cerambycidae: Lepturinae: Lepturini
Type locality: USA: Massachusetts
Remarks: This specimen is herein designated as the lectotype in order to stabilize the taxonomy and facilitate further identifications of this species.

acutipennis **Fisher, 1935c**: 586 (Fig. 9c, d), Holotype
Original combination: *Dymasius acutipennis* Fisher
Current name: *Imbrius acutipennis* (Fisher, 1935c)
Current placement: Cerambycidae: Cerambycinae: Cerambycini
Type locality: Malaysia: Borneo, Sabah, Mt. Kinabalu, 5500 ft., Lumu Lumu

acutipennis **Fisher, 1947a**: 54 (Fig. 9e, f), Holotype
Original combination: *Epimelitta acutipennis* Fisher
Current name: same
Current placement: Cerambycidae: Cerambycinae: Rhinotragini
Type locality: Guyana

adamsi adamsi **Gressitt, 1956**: 90 (Fig. 9g, h), Holotype
Original combination: *Ceresium (Ceresium) adamsi adamsi* Gressitt
Current name: *Ceresium adamsi* Gressitt, 1956
Current placement: Cerambycidae: Cerambycinae: Callidiopini
Type locality: Micronesia: Pohnpei (formerly Ponape Island), Mantalanim Plantation

adamsi kusiana **Gressitt, 1956**: 91 (Fig. 9i, j), Holotype
Original combination: *Ceresium (Ceresium) adamsi kusiana* Gressitt
Current name: same
Current placement: Cerambycidae: Cerambycinae: Callidiopini
Type locality: Micronesia: Kusaie Hill 541, 165 m.

adiectus **Galileo & Martins, 2010**: 35 (Fig. 9k, l), Holotype
Original combination: *Hovorelus adiectus* Galileo & Martins
Current name: same
Current placement: Cerambycidae: Prioninae: Anacolini
Type locality: Costa Rica: Heredia, Estación El Ceibo, 10 km SE, La Virgen, 450–550 m, 10°20′N, 84°05′W
Remarks: The full locality data was published in Galileo & Martins (2012).

admirala admirala **Dillon & Dillon, 1951**: 33 (Fig. 9m, n), Holotype
Original combination: *Gnoma admirala admirala* Dillon & Dillon
Current name: *Gnoma admirala* Dillon & Dillon, 1951
Current placement: Cerambycidae: Lamiinae: Gnomini
Type locality: Papua New Guinea: Manus Province (Admiralty Islands)

admirala pallida **Dillon & Dillon, 1951**: 34 (Fig. 9o, p), Holotype
Original combination: *Gnoma admirala pallida* Dillon & Dillon
Current name: *Gnoma admirala* Dillon & Dillon, 1951
Current placement: Cerambycidae: Lamiinae: Gnomini
Type locality: Papua New Guinea: Manus Province, Los Negros Island (Admiralty Islands)

aeneonigrum **Fisher, 1937**: 147 (Fig. 9q, r), Holotype
Original combination: *Gnomidolon aeneonigrum* Fisher
Current name: *Tetraibidion aurivillii* (Gounelle, 1909)
Current placement: Cerambycidae: Cerambycinae: Hexoplini
Type locality: Brazil: Santa Catarina, Nova Teutonia

aenigma **Galileo & Martins, 2006c**: 181 (Fig. 9s, t), Holotype
Original combination: *Eburodacrys aenigma* Galileo & Martins
Current name: same
Current placement: Cerambycidae: Cerambycinae: Eburiini
Type locality: This specimen lacked a locality label or published locality data.

***aequicollis* Casey, 1912**: 306 (Fig. 9u, v), Holotype
Original combination: *Geropa aequicollis* Casey
Current name: *Geropa concolor* (LeConte, 1873)
Current placement: Cerambycidae: Cerambycinae: Achrysonini
Type locality: USA: Texas, Brownsville

***aereiventris* Tippmann, 1960**: 208 (Fig. 9w, x), Holotype
Original combination: *Antodice aereiventris* Tippmann
Current name: *Apagomera aereiventris* (Tippmann, 1960)
Current placement: Cerambycidae: Lamiinae: Aerenicini
Type locality: Bolivia: Cochabamba, Reg. Chapare 400 m.

***aethiops bergmanni* Tippmann, 1958**: 151 (Fig. 10a, b), Holotype
Original combination: *Dorcadion (Carinatorcadion) aethiops bergmanni* Tippmann
Current name: same
Current placement: Cerambycidae: Lamiinae: Dorcadionini
Type locality: Austria: Mannersdorf am Leithagebirge

***affinis* Breuning, 1936b**: 319 (Fig. 10c, d), Holotype
Original combination: *Olenecamptus affinis* Breuning
Current name: same
Current placement: Cerambycidae: Lamiinae: Dorcaschematini
Type locality: Indonesia: West Kalimantan, Borneo, Pontianak
Remarks: The locality of this type does not match the published type locality. Further research may show it to not be a primary type.

***affinis* Breuning, 1940d**: 435 (Fig. 10e, f), Holotype
Original combination: *Anandra affinis* Breuning
Current name: *Anandra bilineaticeps* Pic, 1939
Current placement: Cerambycidae: Lamiinae: Agapanthiini
Type locality: Indonesia: Java

***agave* Swift & Ray, 2008**: 17 (Fig. 10g, h), Neotype
Original combination: *Tragidion agave* Swift & Ray
Current name: same
Current placement: Cerambycidae: Cerambycinae: Trachyderini
Type locality: USA: California, Santa Rosa Mountains, 2000 ft.
Remarks: Although the authors indicated that they returned the holotype to the Smithsonian Institution, it could not be located. The neotype is here designated.

***agnatus* Melzer, 1934**: 71 (Fig. 10i, j), Holotype
Original combination: *Haruspex agnatus* Melzer
Current name: *Haruspex bivittis* (White, 1855)
Current placement: Cerambycidae: Cerambycinae: Piezocerini
Type locality: Costa Rica: Puntarenas, Golfo Dulce Sándalo

***alaskanus* Schaeffer, 1909**: 385 (Fig. 10k, l), Holotype
Original combination: *Pogonocherus alaskanus* Schaeffer
Current name: *Pogonocherus penicillatus* LeConte, 1850
Current placement: Cerambycidae: Lamiinae: Pogonocherini
Type locality: USA: Alaska

***albisterna* Martins & Galileo, 2004**: 539 (Fig. 10m, n), Holotype
Original combination: *Iareonycha albisterna* Martins & Galileo
Current name: same
Current placement: Cerambycidae: Lamiinae: Hemilophini
Type locality: Panama: Nusagandi area (Cordillera Igar trail)

***albofasciatum* Linell, 1897a**: 393 (Fig. 10o, p), Holotype
Original combination: *Elaphidion albofasciatum* Linell
Current name: *Anelaphus albofasciatus* (Linell, 1897a)
Current placement: Cerambycidae: Cerambycinae: Elaphidiini
Type locality: USA: California, Los Angeles Co.

***albofasciatus* Breuning, 1938a**: 359 (Fig. 10q, r), Holotype
Original combination: *Prosoplus albofasciatus* Breuning
Current name: same
Current placement: Cerambycidae: Lamiinae: Pterolophini
Type locality: Papua New Guinea
Remarks: This may not be the actual holotype since Tavakilian and Chevillotte (2013) suggests the holotype is present in the Paris Museum and the locality data is somewhat incomplete compared to the published description.

***albofasciatus* Fisher, 1926**: 16 (Fig. 10s, t), Holotype
Original combination: *Leptostylus albofasciatus* Fisher
Current name: *Leptostylopsis albofasciatus* (Fisher, 1926)
Current placement: Cerambycidae: Lamiinae: Acanthocinini
Type locality: Cuba: Cayamas

***albofasciatus* Fisher, 1932**: 64 (Fig. 10u, v), Holotype
Original combination: *Pentanodes albofasciatus* Fisher
Current name: same
Current placement: Cerambycidae: Cerambycinae: Tillomorphini
Type locality: Cuba: Cardenas

***albofemorata* Aurivillius, 1927b**: 580 (Fig. 10w, x), Holotype
Original combination: *Serixia albofemorata* Aurivillius
Current name: same
Current placement: Cerambycidae: Lamiinae: Saperdini
Type locality: Malaysia: Borneo, Sandakan

***alboguttata* Aurivillius, 1922**: 433 (Fig. 11a, b), Holotype
Original combination: *Hemicryllis alboguttata* Aurivillius
Current name: same
Current placement: Cerambycidae: Lamiinae: Saperdini
Type locality: Singapore

***alboguttatus* Breuning, 1957a**: 13 (Fig. 11c, d), Holotype
Original combination: *Enes alboguttatus* Breuning
Current name: same
Current placement: Cerambycidae: Lamiinae: Acanthocinini
Type locality: Indonesia: Maluku, Halmahera

***alboguttatus* Fisher, 1925a**: 240 (Fig. 11e, f), Holotype
Original combination: *Exocentrus alboguttatus* Fisher
Current name: *Exocentrus (Oligopsis) alboguttatus* Fisher, 1925a
Current placement: Cerambycidae: Lamiinae: Acanthocinini
Type locality: Philippines: Mindanao, Davao

***alboguttatus* Fisher, 1935c**: 619 (Fig. 11g, h), Holotype
Original combination: *Emeopedus alboguttatus* Fisher
Current name: *Emeopedus (Longicornemeopedus) alboguttatus* Fisher, 1935c
Current placement: Cerambycidae: Lamiinae: Acanthocinini
Type locality: Malaysia: Borneo, Sabah, Mt. Kinabalu, Kamborangah, 7000 ft.

***albolateralis* Breuning, 1940d**: 414 (Fig. 11i, j), Holotype
Original combination: *Tetraulax albolateralis* Breuning
Current name: same
Current placement: Cerambycidae: Lamiinae: Petrognathini
Type locality: Ivory Coast: Dimbokro

***albolineata* Fisher, 1944a**: 4 (Fig. 11k, l), Holotype
Original combination: *Eburia albolineata* Fisher
Current name: same
Current placement: Cerambycidae: Cerambycinae: Eburiini
Type locality: Venezuela: Monagas, Caripito

***albolineatum* Monné & Monné, 2009**: 362 (Fig. 11m, n), Holotype
Original combination: *Nealcidion albolineatum* Monné & Monné
Current name: same
Current placement: Cerambycidae: Lamiinae: Acanthocinini
Type locality: Venezuela: Lara, Parque Nacional Yacambú (9.5 km east Sanaro, 1650 m.

***albomaculata* Fisher, 1925a**: 220 (Fig. 11o, p), Holotype
Original combination: *Neoegesina albomaculata* Fisher
Current name: *Egesina albomaculata* (Fisher, 1925a)
Current placement: Cerambycidae: Lamiinae: Pteropliini
Type locality: Malaysia: Borneo, Sandakan

***albomaculata* Martins, 1962a**: 282 (Fig. 11q, r), Holotype
Original combination: *Compsa albomaculata* Martins
Current name: same
Current placement: Cerambycidae: Cerambycinae: Neoibidionini
Type locality: Colombia: Fusagasugá

***albomarmorata* Breuning, 1939b**: 517 (Fig. 11s, t), Holotype
Original combination: *Mesosa (Saimia) albomarmorata* Breuning
Current name: same
Current placement: Cerambycidae: Lamiinae: Mesosini
Type locality: Indonesia: Sumatra, Medan

***albonotatus* Casey, 1912**: 361 (Fig. 11u, v), Holotype
Original combination: *Xylotrechus albonotatus* Casey
Current name: *Xylotrechus albonotatus* Casey, 1912
Current placement: Cerambycidae: Cerambycinae: Clytini
Type locality: USA: California, Siskiyou Co.

***albopicta* Fisher, 1925a**: 229 (Fig. 11w, x), Holotype
Original combination: *Ebaeides albopicta* Fisher
Current name: same
Current placement: Cerambycidae: Lamiinae: Acanthocinini
Type locality: Philippines: Basilan

***albopunctulata* Heller, 1923a**: 424 (Fig. 12a, b), Syntype
Original combination: *Mispila albopunctulata* Heller
Current name: *Mispila albopunctulata* Heller, 1923a
Current placement: Cerambycidae: Lamiinae: Pteropliini
Type locality: Philippines: Mindanao, Butuan
Remarks: The USNM has 9 other syntypes. Other syntypes are present in the Staatliches Museum für Tierkunde, Dresden, Germany where Heller served as a staff researcher for many years.

***alboscutellaris* Breuning, 1938b**: 227 (Fig. 12c, d), Holotype
Original combination: *Anauxesis alboscutellaris* Breuning
Current name: same
Current placement: Cerambycidae: Lamiinae: Agapanthiini
Type locality: Tanzania: Namupa
Remarks: Tavakilian and Chevillotte (2013) indicate the holotype is in the Zoologische Staatssammlung des Bayerischen Staates, Munich, Germany, however, this specimen and its labels agree with the original publication.

***alboscutellata* Fisher, 1926**: 3 (Fig. 12e, f), Holotype
Original combination: *Ataxia alboscutellata* Fisher
Current name: same
Current placement: Cerambycidae: Lamiinae: Pteropliini
Type locality: Cuba: Cayamas

***albosignata* Breuning, 1940d**: 425 (Fig. 12g, h), Holotype
Original combination: *Mispila albosignata* Breuning
Current name: same
Current placement: Cerambycidae: Lamiinae: Pteropliini
Type locality: Sri Lanka
Remarks: Since the original description states the type locality as Halbinsel Malakka, Singapore, further research may show this to not be the holotype.

***albosignata* Breuning, 1950b**: 319 (Fig. 12i, j), Holotype
Original combination: *Homonoea (Homonoea) albosignata* Breuning
Current name: *Homonoea albosignata* Breuning, 1950b
Current placement: Cerambycidae: Lamiinae: Homonoeini
Type locality: Philippines, Samar

***albosignatus* Fisher, 1935a**: 53 (Fig. 12k, l), Holotype
Original combination: *Leptostylus albosignatus* Fisher
Current name: *Leptostylopsis antillarum* (Fisher, 1925d)
Current placement: Cerambycidae: Lamiinae: Acanthocinini
Type locality: USA: Ponce, Puerto Rico

***albostictica* Breuning, 1942a**: 158 (Fig. 12m, n), Holotype
Original combination: *Ischnolea albostictica* Breuning
Current name: same
Current placement: Cerambycidae: Lamiinae: Desmiphorini
Type locality: Brazil: Santa Catarina, Mafra

***albostictica* Breuning, 1942a**: 161 (Fig. 12o, p), Holotype
Original combination: *Sophronica albostictica* Breuning
Current name: same
Current placement: Cerambycidae: Lamiinae: Desmiphorini
Type locality: Congo: Kasaï-Oriental

***albovarius* Fisher, 1925a**: 243 (Fig. 12q, r), Holotype
Original combination: *Exocentrus albovarius* Fisher
Current name: same
Current placement: Cerambycidae: Lamiinae: Acanthocinini
Type locality: Philippines: Mindanao, Davao

***albovittata* Breuning, 1943c**: 45 (Fig. 12s, t), Holotype
Original combination: *Sybra albovittata* Breuning
Current name: same
Current placement: Cerambycidae: Lamiinae: Apomecynini
Type locality: Papua New Guinea: Bongu

***aldabrensis* Linell, 1897b**: 701 (Fig. 12u, v), Holotype
Original combination: *Leptocera aldabrensis* Linell
Current name: *Iresioides aldabrensis* (Linell, 1897b)
Current placement: Cerambycidae: Cerambycinae: Glaucytini
Type locality: Seychelles: Aldabra Island

***alicei* Lane, 1977**: 233 (Fig. 12w, x), Holotype
Original combination: *Oncideres alicei* Lane
Current name: same
Current placement: Cerambycidae: Lamiinae: Onciderini
Type locality: Brazil: Bahia
Remarks: A second "holotype" is present in MZSP. Further study needs to be made to clarify which specimen was the intended holotype since both specimens match the original description.

***alpha* var. *floridanus* Hamilton, 1896**: 125 (Fig. 13a, b), Lectotype
Original combination: *Liopus alpha* var. *floridanus* Hamilton
Current name: *Sternidius alpha* (Say, 1827)
Current placement: Cerambycidae: Lamiinae: Acanthocinini
Type locality: USA: Florida, Biscayne
Remarks: The name following the binomen was not expressly given infrasubspecific rank and the work did not indicate unambiguously that the name was infrasubspecific, therefore, it is deemed a subspecies and available. This specimen is herein designated as the lectotype in order to stabilize the taxonomy and facilitate further identifications of this species.

***alternata* Gressitt, 1938a**: 51 (Fig. 13c, d), Holotype
Original combination: *Pterolophia alternata* Gressitt
Current name: same
Current placement: Cerambycidae: Lamiinae: Pteropliini
Type locality: China: Sichuan Province, Suifu

***alticola* Casey, 1913**: 239 (Fig. 13e, f), Holotype
Original combination: *Leptacmaeops (Leptacmaeops) alticola* Casey
Current name: *Cortodera longicornis* (Kirby, 1837)
Current placement: Cerambycidae: Lepturinae: Rhagiini
Type locality: USA: Colorado, Veta Pass

***alticola* Casey, 1913**: 342 (Fig. 13g, h), Lectotype
Original combination: *Dectes alticola* Casey
Current name: *Dectes texanus* LeConte, 1862
Current placement: Cerambycidae: Lamiinae: Acanthocinini
Type locality: USA: Colorado
Remarks: This specimen is herein designated as the lectotype in order to stabilize the taxonomy and facilitate further identifications of this species.

***alticola* Gressitt, 1948a**: 43 (Fig. 13i, j), Holotype
Original combination: *Leptura alticola* Gressitt
Current name: same
Current placement: Cerambycidae: Lepturinae: Lepturini
Type locality: China: Sichuan Province (probably), near Tibet border, Wahu Pass, 14,000–16,000 ft.

***alutaceus* Casey, 1912**: 237 (Fig. 13k, l), Holotype
Original combination: *Prionus (Prionus) alutaceus* Casey
Current name: *Prionus californicus* Motschulsky, 1845
Current placement: Cerambycidae: Prioninae: Prionini
Type locality: USA: Arizona

***amabilis* Tippmann, 1960**: 101 (Fig. 13m, n), Holotype
Original combination: *Udeteromorphus amabilis* Tippmann
Current name: *Myzomorphus amabilis* (Tippmann, 1960)
Current placement: Cerambycidae: Prioninae: Anacolini
Type locality: Bolivia: Cochabamba, Reg. Chapare

***amaliae* Galileo & Martins, 2011**:184 (Fig. 13o, p), Holotype
Original combination: *Esthlogena amaliae* Galileo & Martins
Current name: same
Current placement: Cerambycidae: Lamiinae: Pteropliini
Type locality: Guatemala: 5 km. south of Santiago Atitlan, Suchitepéquez, 1500 m.

***amazonica* Tippmann, 1960**: 116 (Fig. 13q, r), Holotype
Original combination: *Eburia (Eburostola) amazonica* Tippmann
Current name: *Eburostola amazonica* (Tippmann, 1960)
Current placement: Cerambycidae: Cerambycinae: Eburiini
Type locality: Brazil: Pará, Obidos

***amboinicus* Breuning, 1957a**: 11 (Fig. 13s, t), Holotype
Original combination: *Pithomictus amboinicus* Breuning
Current name: *Pithomyctus (Parapithomyctus) amboinicus* Breuning, 1957a
Current placement: Cerambycidae: Lamiinae: Rhodopini
Type locality: Indonesia: Maluku, Ambon Island (Amboina)
Remarks: Tavakilian and Chevillotte (2013) suggest the holotype is present in Berlin, however, this specimen and its labels match the original description.

***amnicola* Casey, 1913**: 326 (Fig. 13u, v), Holotype
Original combination: *Hyperplatys amnicola* Casey
Current name: *Hyperplatys maculata* Haldeman, 1847
Current placement: Cerambycidae: Lamiinae: Acanthocinini
Type locality: USA: Iowa

***amoena* Fisher, 1937d**: 148 (Fig. 13w, x), Holotype
Original combination: *Compsa amoena* Fisher
Current name: same
Current placement: Cerambycidae: Cerambycinae: Neoibidionini
Type locality: Brazil: Santa Catarina, Nova Teutonia

***amoena* Hamilton, 1896**: 161 (Fig. 14a, b), Holotype
Original combination: *Amphionycha amoena* Hamilton
Current name: *Cathetopteron amoena* Hamilton, 1896
Current placement: Cerambycidae: Lamiinae: Hemilophini
Type locality: USA: Texas, near Brownsville

***amoenoides* Fisher, 1938**: 151 (Fig. 14c, d), Holotype
Original combination: *Adesmus amoenoides* Fisher
Current name: *Cuiciuna amoenoides* (Fisher, 1938)
Current placement: Cerambycidae: Lamiinae: Hemilophini
Type locality: Brazil: Santa Catarina, Nova Teutonia

***amoenus hudsonicus* Casey, 1912**: 284 (Fig. 14e, f), Lectotype
Original combination: *Phymatodes (Microcallidium) amoenus hudsonicus* Casey
Current name: *Phymatodes amoenus* (Say, 1824)
Current placement: Cerambycidae: Cerambycinae: Callidiini
Type locality: USA: New York
Remarks: This specimen is herein designated as the lectotype in order to stabilize the taxonomy and facilitate further identifications of this species.

***ampla* Casey, 1912**: 273 (Fig. 14g, h), Lectotype
Original combination: *Anocomis ampla* Casey
Current name: *Semanotus ligneus amplus* (Casey, 1912)
Current placement: Cerambycidae: Cerambycinae: Callidiini
Type locality: USA: California
Remarks: This specimen is herein designated as the lectotype in order to stabilize the taxonomy and facilitate further identifications of this species

***ampla angusta* Casey, 1912**: 273 (Fig. 14i, j), Holotype
Original combination: *Anocomis ampla angusta* Casey
Current name: *Semanotus ligneus amplus* (Casey, 1912)
Current placement: Cerambycidae: Cerambycinae: Callidiini
Type locality: USA: California

***ampla arizonae* Casey, 1924**: 236 (Fig. 14k, l), Holotype
Original combination: *Anocomis ampla arizonae* Casey
Current name: *Semanotus ligneus parvicollis* (Casey, 1912)
Current placement: Cerambycidae: Cerambycinae: Callidiini
Type locality: USA: Arizona

ampla basalis **Casey, 1924**: 237 (Fig. 14m, n), Holotype
Original combination: *Anocomis ampla basalis* Casey
Current name: *Semanotus ligneus basalis* (Casey, 1924)
Current placement: Cerambycidae: Cerambycinae: Callidiini
Type locality: Canada: British Columbia, Terrace

ampla conformis **Casey, 1924**: 236 (Fig. 14o, p), Lectotype
Original combination: *Anocomis ampla conformis* Casey
Current name: *Semanotus ligneus parvicollis* (Casey, 1912)
Current placement: Cerambycidae: Cerambycinae: Callidiini
Type locality: USA: Utah, Vineyard
Remarks: This specimen is herein designated as the lectotype in order to stabilize the taxonomy and facilitate further identifications of this species.

ampla obliviosa **Casey, 1924**: 237 (Fig. 14q, r), Holotype
Original combination: *Anocomis ampla obliviosa* Casey
Current name: *Semanotus ligneus amplus* (Casey, 1912)
Current placement: Cerambycidae: Cerambycinae: Callidiini
Type locality: USA: California

ampla terracensis **Casey, 1924**: 237 (Fig. 14s, t), Lectotype
Original combination: *Anocomis ampla terracensis* Casey
Current name: *Semanotus ligneus amplus* (Casey, 1912)
Current placement: Cerambycidae: Cerambycinae: Callidiini
Type locality: Canada: British Columbia, Terrace
Remarks: This specimen is herein designated as the lectotype in order to stabilize the taxonomy and facilitate further identifications of this species.

amplians **Casey, 1912**: 231 (Fig. 14u, v), Holotype
Original combination: *Orthosoma amplians* Casey
Current name: *Orthosoma brunneum* (Forster, 1771)
Current placement: Cerambycidae: Prioninae: Macrotomini
Type locality: USA: Indiana
Remarks: Although the original description suggests there may be more than one specimen in the original series, there is only one in the USNM collection.

ampliceps **Casey, 1912**: 216 (Fig. 14w, x), Lectotype
Original combination: *Parandra ampliceps* Casey
Current name: *Neandra brunnea* (Fabricius, 1798)
Current placement: Cerambycidae: Parandrinae: Parandrini
Type locality: USA: Indiana
Remarks: This specimen is herein designated as the lectotype in order to stabilize the taxonomy and facilitate further identifications of this species.

amplipennis **Casey, 1913**: 295 (Fig. 15a, b), Holotype
Original combination: *Hammoderus amplipennis* Casey
Current name: *Goes tesselatus* (Haldeman, 1847)
Current placement: Cerambycidae: Lamiinae: Monochamini
Type locality: USA: Colorado

amputatum **Casey, 1912**: 259 (Fig. 15c, d), Holotype
Original combination: *Asemum amputatum* Casey
Current name: *Asemum striatum* (Linnaeus, 1758)
Current placement: Cerambycidae: Spondylidinae: Asemini
Type locality: USA: Massachusetts, Wellesley

anatahana **Gressitt, 1956**: 142 (Fig. 15e, f), Holotype
Original combination: *Sybra anatahana* Gressitt
Current name: same
Current placement: Cerambycidae: Lamiinae: Apomecynini
Type locality: Northern Mariana Islands: Anatahan Island

anatolicus **Daniel & Daniel, 1898**: 76 (Fig. 15g, h), Holotype
Original combination: *Pogonochaerus anatolicus* Daniel & Daniel
Current name: same
Current placement: Cerambycidae: Lamiinae: Pogonocherini
Type locality: Turkey: Antalya, Adalia
Remarks: Tavakilian and Chevillotte (2013) suggest the holotype is present in the Deutsches Entomologisches Institut, Eberswalde, however, this specimen and its labels match the original description.

andesiana **Casey, 1912**: 352 (Fig. 15i, j), Holotype
Original combination: *Cyllene andesiana* Casey
Current name: *Megacyllene andesiana* (Casey, 1912)
Current placement: Cerambycidae: Cerambycinae: Clytini
Type locality: Ecuador: Babahoyo, foot of Andes

andicola **Breuning, 1940d**: 409 (Fig. 15k, l), Holotype
Original combination: *Tuberolamia andicola* Breuning
Current name: same
Current placement: Cerambycidae: Lamiinae: Morimopsini
Type locality: Peru: Huancabamba, 3000 m.

andinus **Tippmann, 1960**: 185 (Fig. 15m, n), Holotype
Original combination: *Tropanisopodus andinus* Tippmann
Current name: same
Current placement: Cerambycidae: Lamiinae: Acanthocinini
Type locality: Bolivia: Yungas de Corani, 2500 m.

anemum **Martins & Galileo, 2004**: 540 (Fig. 15o, p), Holotype
Original combination: *Sybaguasu anemum* Martins & Galileo
Current name: same
Current placement: Cerambycidae: Lamiinae: Hemilophini
Type locality: Panama: Chiriquí Province, Continental Divide Trail

***angaturama* Galileo & Martins, 2008a**: 2 (Fig. 15q, r), Holotype
Original combination: *Oncideres angaturama* Galileo & Martins
Current name: same
Current placement: Cerambycidae: Lamiinae: Onciderini
Type locality: Peru: San Martin Province, Cordillera Escalera, Tarapoto-Iquitos Road at USM Centro De Biodiversidad, S 06°27.526′, W76°17.275′, El. 1000 m.

***angolana* Aurivillius, 1913**: 239 (Fig. 15s, t), Lectotype
Original combination: *Haploparmena angolana* Aurivillius
Current name: same
Current placement: Cerambycidae: Lamiinae: Parmenini
Type locality: Angola: Bailunde
Remarks: This specimen is herein designated as the lectotype in order to stabilize the taxonomy and facilitate further identifications of this species. One additional paralectotype is present in the USNM and at least one is present in Zoologische Staatssammlung des Bayerischen Staates, Munich, Germany.

***angulatum* Casey, 1924**: 239 (Fig. 15u, v), Holotype
Original combination: *Physocnemum angulatum* Casey
Current name: *Physocnemum brevilineum* (Say, 1824)
Current placement: Cerambycidae: Cerambycinae: Callidiini
Type locality: USA: Michigan, Marquette

***angulicollis* Chemsak, 1969**: 307 (Fig. 15w, x), Holotype
Original combination: *Tetranodus angulicollis* Chemsak
Current name: same
Current placement: Cerambycidae: Cerambycinae: Tillomorphini
Type locality: Honduras: Tegucigalpa

***angulifascia* Aurivillius, 1922**: 415 (Fig. 16a, b), Holotype
Original combination: *Demonax angulifascia* Aurivillius
Current name: same
Current placement: Cerambycidae: Cerambycinae: Clytini
Type locality: Philippines: Luzon, Mt. Makiling

***angulifera* Casey, 1912**: 354 (Fig. 16c, d), Holotype
Original combination: *Cyllene angulifera* Casey
Current name: *Megacyllene angulifera* (Casey, 1912)
Current placement: Cerambycidae: Cerambycinae: Clytini
Type locality: USA: Kansas

***angulosus* Casey, 1913**: 338 (Fig. 16e, f), Holotype
Original combination: *Graphisurus (Graphisurus) angulosus* Casey
Current name: *Acanthocinus angulosus* (Casey, 1913)
Current placement: Cerambycidae: Lamiinae: Acanthocinini
Type locality: USA: New Mexico

***angustata* Schaeffer, 1905b**: 162 (Fig. 16g, h), Syntype
Original combination: *Rhopalophora angustata* Schaeffer
Current name: same
Current placement: Cerambycidae: Cerambycinae: Rhopalophorini
Type locality: USA: Texas, Brownsville, Esperanza Ranch

***angusticollis* Aurivillius, 1928**: 318 (Fig. 16i, j), Holotype
Original combination: *Demonax angusticollis* Aurivillius
Current name: same
Current placement: Cerambycidae: Cerambycinae: Clytini
Type locality: Philippines: Negros Oriental, Cuernos Mountains

***angusticollis* Casey, 1913**: 292 (Fig. 16k, l), Holotype
Original combination: *Monochamus angusticollis* Casey
Current name: *Monochamus titillator* (Fabricius, 1775)
Current placement: Cerambycidae: Lamiinae: Monochamini
Type locality: USA: Texas

***angusticollis* Fisher, 1925a**: 248 (Fig. 16m, n), Holotype
Original combination: *Exocentrus angusticollis* Fisher
Current name: *Exocentrus (Camptomyne) angusticollis* Fisher, 1925a
Current placement: Cerambycidae: Lamiinae: Acanthocinini
Type locality: Philippines: Mindanao, Dapitan

***angusticollis* var. *sibuyanus* Aurivillius, 1928**: 319 (Fig. 16o, p), Lectotype
Original combination: *Demonax angusticollis* var. *sibuyanus* Aurivillius
Current name: *Demonax angusticollis* Aurivillius, 1928
Current placement: Cerambycidae: Cerambycinae: Clytini
Type locality: Philippines: Sibuyan
Remarks: The name following the binomen was not expressly given infrasubspecific rank and the work did not indicate unambiguously that the name was infrasubspecific, therefore, it is deemed a subspecies and available. This specimen is herein designated as the lectotype in order to stabilize the taxonomy and facilitate further identifications of this species.

***angustolineata* Gressitt, 1935g**: 277 (Fig. 16q, r), Holotype
Original combination: *Ocalemia angustolineata* Gressitt
Current name: *Leptostrangalia angustolineata* (Gressitt, 1935g)
Current placement: Cerambycidae: Lepturinae: Lepturini
Type locality: Philippines: Luzon, Lamao

***angustulus* Casey, 1912**: 241 (Fig. 16s, t), Lectotype
Original combination: *Prionus (Prionus) angustulus* Casey
Current name: *Prionus californicus* Motschulsky, 1845
Current placement: Cerambycidae: Prioninae: Prionini
Type locality: USA: Arizona, Canyon of Colorado River
Remarks: This specimen is herein designated as the lectotype in order to stabilize the taxonomy and facilitate further identifications of this species.

***annulata* Casey, 1913**: 300 (Fig. 16u, v), Holotype
Original combination: *Synaphaeta annulata* Casey
Current name: *Synaphaeta guexi* (LeConte, 1852)
Current placement: Cerambycidae: Lamiinae: Mesosini
Type locality: USA: California, Tulare Co.

***annulatum* var. *peninsulare* Schaeffer, 1908**: 339 (Fig. 16w, x), Holotype
Original combination: *Tragidion annulatum* var. *peninsulare* Schaeffer
Current name: *Tragidion annulatum* LeConte, 1858
Current placement: Cerambycidae: Cerambycinae: Trachyderini
Type locality: USA: California, Santa Rosa (San Felipe)
Remarks: The name following the binomen was not expressly given infrasubspecific rank and the work did not indicate unambiguously that the name was infrasubspecific, therefore, it is deemed a subspecies and available.

***annulicornis* Breuning, 1942a**: 145 (Fig. 17a, b), Holotype
Original combination: *Frea (Crossotofrea) annulicornis* Breuning
Current name: same
Current placement: Cerambycidae: Lamiinae: Ceroplesini
Type locality: Tanzania: Dar es Salaam

***annulicornis* Breuning, 1942a**: 153 (Fig. 17c, d), Holotype
Original combination: *Dolichosybra annulicornis* Breuning
Current name: same
Current placement: Cerambycidae: Lamiinae: Apomecynini
Type locality: Brazil: Santa Catarina, Corupá (formerly Hansa-Humboldt)

***annulicornis* Breuning, 1948b**: 32 (Fig. 17e, f), Holotype
Original combination: *Pothyne annulicornis* Breuning
Current name: same
Current placement: Cerambycidae: Lamiinae: Agapanthiini
Type locality: Philippines: Mindanao

***annulicornis* Fisher, 1926**: 9 (Fig. 17g, h), Holotype
Original combination: *Eupogonius annulicornis* Fisher
Current name: same
Current placement: Cerambycidae: Lamiinae: Desmiphorini
Type locality: Cuba: Cayamas

***antecurrens* Wickham, 1913**: 295 (Fig. 17i, j), Holotype
Original combination: *Leptura antecurrens* Wickham
Current name: *Anoplodera antecurrens* (Wickham, 1913)
Current placement: Cerambycidae: Lepturinae: Lepturini
Type locality: USA: Colorado, Florissant
Remarks: This is one of four fossil specimens at the USNM and is housed in the Paleobiology (Paleoentomology) section.

***antennata* Martins, Chemsak & Linsley, 1966**: 211 (Fig. 17k, l), Holotype
Original combination: *Apomethia antennata* Martins, Chemsak & Linsley
Current name: *Niophis antennata* (Martins, Chemsak, & Linsley, 1966)
Current placement: Cerambycidae: Cerambycinae: Ectenessini
Type locality: Panama: Pearl Island, San Jose

***antennatum* Monné & Monné, 2009**: 361 (Fig. 17m, n), Holotype
Original combination: *Nealcidion antennatum* Monné & Monné
Current name: same
Current placement: Cerambycidae: Cerambycinae: Acanthocinini
Type locality: Panama: Chiriquí Province, 21 km. northwest Volcán

***antennatum hesperum* Casey, 1912**: 288 (Fig. 17o, p), Lectotype
Original combination: *Callidium antennatum hesperum* Casey
Current name: same
Current placement: Cerambycidae: Cerambycinae: Callidiini
Type locality: USA: Nevada
Remarks: This specimen is herein designated as the lectotype in order to stabilize the taxonomy and facilitate further identifications of this species.

***antennatus* Galileo & Martins, 2007c**: 709 (Fig. 17q, r), Holotype
Original combination: *Piruapsis antennatus* Galileo & Martins
Current name: same
Current placement: Cerambycidae: Cerambycinae: Necydalopsini
Type locality: Trinidad and Tobago: Trinidad, north Range, Arina-Blanchisseuse Rd., Textel, near Morne Bleu, 2300 ft.

***antillarum* Fisher, 1925b**: 5 (Fig. 17s, t), Holotype
Original combination: *Leptostylus antillarum* Fisher
Current name: *Leptostylopsis antillarum* (Fisher, 1925b)
Current placement: Cerambycidae: Lamiinae: Acanthocinini
Type locality: USA: Puerto Rico, Culebra

***antillarum* Fisher, 1932**: 42 (Fig. 17u, v), Holotype
Original combination: *Elaphidion antillarum* Fisher
Current name: *Anopliomorpha antillarum* (Fisher, 1932)
Current placement: Cerambycidae: Cerambycinae: Elaphidiini
Type locality: Haiti: Source Matelas

***anulatus* Lingafelter, 2013**: 61 (Fig. 17w, x), Holotype
Original combination: *Oedudes anulatus* Lingafelter
Current name: same
Current placement: Cerambycidae: Lamiinae: Hemilophini
Type locality: Dominican Republic: Peravia Province, 5 km W of road to El Rio, S. of Pedregal, 19°05.092′N, 70°35.864′W, 52 m.

***apicalis* Casey, 1914**: 371 (Fig. 18a, b), Holotype
Original combination: *Tessaropa apicalis* Casey
Current name: *Tessaropa tenuipes* (Haldeman, 1848)
Current placement: Cerambycidae: Cerambycinae: Methiini
Type locality: USA: Pennsylvania, Rockville (?)
Remarks: The label is not in exact agreement with the published data.

***apicalis* Fisher, 1936a**: 171 (Fig. 18c, d), Holotype
Original combination: *Trypogeus apicalis* Fisher
Current name: same
Current placement: Cerambycidae: Lepturinae: Lepturini
Type locality: Indonesia: Java, Mt. Tangkoeban Prahoe, Preanger 4000–5000 ft.

***apicalis* Melzer, 1933**: 376 (Fig. 18e, f), Holotype
Original combination: *Amphicnaeia apicalis* Melzer
Current name: same
Current placement: Cerambycidae: Lamiinae: Apomecynini
Type locality: Costa Rica: Cartago, Turrialba, 800 m.

***apicalis* Tippmann, 1960**: 102 (Fig. 18g, h), Holotype
Original combination: *Temnopis apicalis* Tippmann
Current name: *Temnopis oculata* Zajciw, 1960
Current placement: Cerambycidae: Cerambycinae: Oemini
Type locality: Bolivia: Santa Cruz, 500 m.

***apicata* Casey, 1924**: 280 (Fig. 18i, j), Lectotype
Original combination: *Strangalia (Anastrangalia) apicata* Casey
Current name: *Anastrangalia sanguinea* (LeConte, 1859)
Current placement: Cerambycidae: Lepturinae: Lepturini
Type locality: USA: Colorado
Remarks: This specimen is herein designated as the lectotype in order to stabilize the taxonomy and facilitate further identifications of this species.

***apicatum* Casey, 1912**: 323 (Fig. 18k, l), Holotype
Original combination: *Tragidion apicatum* Casey
Current name: *Tragidion coquus* (Linnaeus, 1758)
Current placement: Cerambycidae: Cerambycinae: Trachyderini
Type locality: USA: Texas

***apicatus* Linsley, 1935a**: 140 (Fig. 18m, n), Holotype
Original combination: *Nephalius apicatus* Linsley
Current name: *Mephritus apicatus* (Linsley, 1935a)
Current placement: Cerambycidae: Cerambycinae: Elaphidiini
Type locality: Panama: La Chorrera

***apicicornis* Gressitt, 1937a**: 92 (Fig. 18o, p), Holotype
Original combination: *Aglaophis apicicornis* Gressitt
Current name: *Paraclytus apicicornis* (Gressitt, 1937a)
Current placement: Cerambycidae: Cerambycinae: Anaglyptini
Type locality: China: Sichuan Province, near Muping 1000–3000 ft.

***apiciventris* Casey, 1913**: 211 (Fig. 18q, r), Holotype
Original combination: *Stenocorus (Stenocorus) apiciventris* Casey
Current name: *Stenocorus nubifer* (LeConte, 1859)
Current placement: Cerambycidae: Lepturinae: Rhagiini
Type locality: USA: California, Santa Cruz Mts.

***approximata* Dillon & Dillon, 1945**: 123 (Fig. 18s, t), Holotype
Original combination: *Sternycha approximata* Dillon & Dillon
Current name: same
Current placement: Cerambycidae: Lamiinae: Onciderini
Type locality: Costa Rica: Limón, Hamburg Farm, Reventazón, Ebene

***aragua* Martins & Galileo, 1990**: 93 (Fig. 18u, v), Holotype
Original combination: *Oncideres aragua* Martins & Galileo
Current name: same
Current placement: Cerambycidae: Lamiinae: Onciderini
Type locality: Venezuela: Aragua, Rancho Grande

***arapahoe* Casey, 1924**: 278 (Fig. 18w, x), Lectotype
Original combination: *Typocerus arapahoe* Casey
Current name: *Typocerus octonotatus* (Haldeman, 1847)
Current placement: Cerambycidae: Lepturinae: Lepturini
Type locality: USA: Colorado, Wray
Remarks: This specimen is herein designated as the lectotype in order to stabilize the taxonomy and facilitate further identifications of this species.

***archboldi* Lingafelter & Chemsak, 2002**: 571 (Fig. 19a, b), Holotype
Original combination: *Enaphalodes archboldi* Lingafelter & Chemsak
Current name: same
Current placement: Cerambycidae: Cerambycinae: Elaphidiini
Type locality: USA: Florida, Highlands Co., Archbold Biological Station

***arctofasciata* Gressitt, 1940**: 147 (Fig. 19c, d), Holotype
Original combination: *Pterolophia arctofasciata* Gressitt
Current name: *Pterolophia (Hylobrotus) postfasciculata* Pic, 1934
Current placement: Cerambycidae: Lamiinae: Pteropliini
Type locality: China: Hainan Province, Hainan Island, Ta Hian

***argentea* Gressitt, 1956**: 109 (Fig. 19e, f), Holotype
Original combination: *Glaucytes argentea* Gressitt
Current name: *Scituloglaucytes argentea* (Gressitt, 1956)
Current placement: Cerambycidae: Lamiinae: Glaucytini
Type locality: Palau: Babelthuap Island, E. Ngatpang, 65 m.

***argenteofasciata* Tippmann, 1960**: 178 (Fig. 19g, h), Holotype
Original combination: *Acanthoderes (Trichacanthoderes) argenteofasciata* Tippmann
Current name: *Zikanita argenteofasciata* (Tippmann, 1960)
Current placement: Cerambycidae: Lamiinae: Acanthoderini
Type locality: Peru

***argentina* Lane, 1978**: 72 (Fig. 19i, j), Holotype
Original combination: *Polyrhaphis argentina* Lane
Current name: same
Current placement: Cerambycidae: Lamiinae: Polyrhaphidini
Type locality: Argentina: Salta, General Ballivián

***argentipleura* Gressitt, 1956**: 158 (Fig. 19k, l), Holotype
Original combination: *Sciadella (Sciadella) argentipleura* Gressitt
Current name: *Sciades (Estoliops) argentipleura* (Gressitt, 1956)
Current placement: Cerambycidae: Lamiinae: Acanthocinini
Type locality: Micronesia: Caroline Islands, Truk, Moen

***argus* Casey, 1912**: 342 (Fig. 19m, n), Holotype
Original combination: *Tylosis argus* Casey
Current name: *Tylosis maculatus* LeConte, 1850
Current placement: Cerambycidae: Cerambycinae: Trachyderini
Type locality: USA: New Mexico

***aridus* Casey, 1913**: 343 (Fig. 19o, p), Lectotype
Original combination: *Dectes aridus* Casey
Current name: *Dectes texanus* LeConte, 1862
Current placement: Cerambycidae: Lamiinae: Acanthocinini
Type locality: Mexico: Durango, Tepehuanes
Remarks: This specimen is herein designated as the lectotype in order to stabilize the taxonomy and facilitate further identifications of this species.

***aridus* Casey, 1924**: 224 (Fig. 19q, r), Holotype
Original combination: *Archodontes aridus* Casey
Current name: *Archodontes melanopus aridus* Casey, 1924
Current placement: Cerambycidae: Prioninae: Macrotomini
Type locality: USA: Arizona

***arizonense* Casey, 1891**: 28 (Fig. 19s, t), Lectotype
Original combination: *Elaphidion arizonense* Casey
Current name: *Orwellion gibbulum arizonense* (Casey, 1891)
Current placement: Cerambycidae: Cerambycinae: Elaphidiini
Type locality: USA: Arizona
Remarks: This specimen is herein designated as the lectotype in order to stabilize the taxonomy and facilitate further identifications of this species.

***arizonica* Fisher, 1920**: 158 (Fig. 19u, v), Holotype
Original combination: *Ataxia arizonica* Fisher
Current name: same
Current placement: Cerambycidae: Lamiinae: Pteropliini
Type locality: USA: Arizona, Sabino Canyon

***arizonica* Schaeffer, 1908**: 352 (Fig. 19w, x), Syntype
Original combination: *Methia arizonica* Schaeffer
Current name: same
Current placement: Cerambycidae: Cerambycinae: Methiini
Type locality: USA: Arizona, Rincon Mtns.

***arizonicus* Casey, 1912**: 228 (Fig. 20a, b), Lectotype
Original combination: *Paramallus arizonicus* Casey
Current name: *Neomallodon arizonicus* (Casey, 1912)
Current placement: Cerambycidae: Prioninae: Mallodonini
Type locality: USA: Arizona
Remarks: This specimen is herein designated as the lectotype in order to stabilize the taxonomy and facilitate further identifications of this species.

***arizonicus* Casey, 1912**: 318 (Fig. 20c, d), Holotype
Original combination: *Stenaspis arizonicus* Casey
Current name: *Stenaspis verticalis arizonicus* Casey, 1912
Current placement: Cerambycidae: Cerambycinae: Trachyderini
Type locality: USA: Arizona, Baboquivari Mts.

***arizonicus* Schaeffer, 1908**: 346 (Fig. 20e, f), Holotype
Original combination: *Pogonocherus arizonicus* Schaeffer
Current name: *Pogonocherus arizonicus* Schaeffer, 1908
Current placement: Cerambycidae: Lamiinae: Pogonocherini
Type locality: USA: Arizona, Huachuca Mts., Carrs Peak, 9000 ft.

***armatipes* Martins, 1962a**: 280 (Fig. 20g, h), Holotype
Original combination: *Octoplon armatipes* Martins
Current name: *Epacroplon armatipes* (Martins, 1962a)
Current placement: Cerambycidae: Cerambycinae: Hexoplini
Type locality: Bolivia: Santa Cruz, 500 m.

***armatum* var. *integrum* Tippmann, 1960**: 118 (Fig. 20i, j), Holotype
Original combination: *Hexoplon armatum* var. *integrum* Tippmann
Current name: *Hexoplon integrum* Tippmann, 1960
Current placement: Cerambycidae: Cerambycinae: Hexoplini
Type locality: Bolivia: Santa Cruz, 500 m.
Remarks: The name following the binomen was not expressly given infrasubspecific rank and the work did not indicate unambiguously that the name was infrasubspecific, therefore, it is deemed a subspecies and available.

***artocarpi* Fisher, 1934b**: 36 (Fig. 20k, l), Holotype
Original combination: *Exocentrus artocarpi* Fisher
Current name: *Exocentrus (Centenexocentrus) artocarpi* Fisher, 1934b
Current placement: Cerambycidae: Lamiinae: Acanthocinini
Type locality: Indonesia: Java, Bogor, ex. Buitenzorg, 250 m.

***aspera* Knull, 1962**: 105 (Fig. 20m, n), Holotype
Original combination: *Gymnopsyra aspera* Knull
Current name: same
Current placement: Cerambycidae: Cerambycinae: Elaphidiini
Type locality: USA: Texas, Big Bend National Park, 4000 ft.

***aspera impressa* Casey, 1912**: 264 (Fig. 20o, p), Lectotype
Original combination: *Nothorhina aspera impressa* Casey
Current name: *Megasemum asperum* (LeConte, 1854)
Current placement: Cerambycidae: Spondylidinae: Asemini
Type locality: USA: Colorado, Boulder Co.
Remarks: This specimen is herein designated as the lectotype in order to stabilize the taxonomy and facilitate further identifications of this species.

***aspera rectipennis* Casey, 1912**: 264 (Fig. 20q, r), Holotype
Original combination: *Nothorhina aspera rectipennis* Casey
Current name: *Megasemum asperum* (LeConte, 1854)
Current placement: Cerambycidae: Spondylidinae: Asemini
Type locality: USA: New Mexico, Cloudcroft, 9000 ft.

***ataxiiformis* Linsley, 1935a**: 148 (Fig. 20s, t), Holotype
Original combination: *Aulacoscapus ataxiiformis* Linsley
Current name: *Miltesthus marginatus* Bates, 1872
Current placement: Cerambycidae: Cerambycinae: Elaphidiini
Type locality: Guatemala: El Peten, Uaxactum

***ater* Aurivillius, 1922**: 422 (Fig. 20u, v), Holotype
Original combination: *Demonax ater* Aurivillius
Current name: same
Current placement: Cerambycidae: Cerambycinae: Clytini
Type locality: Philippines: Mindanao, Dapitan

***ater* Fisher, 1935c**: 603 (Fig. 20w, x), Holotype
Original combination: *Dihammus ater* Fisher
Current name: *Acalolepta atra* (Fisher, 1935c)
Current placement: Cerambycidae: Lamiinae: Monochamini
Type locality: Malaysia: Borneo, Sabah, Mt. Kinabalu, 7000 ft., Kamborangah

***aterrima* Casey, 1924**: 276 (Fig. 21a, b), Holotype
Original combination: *Brachysomida aterrima* Casey
Current name: *Brachysomida atra* (LeConte, 1850)
Current placement: Cerambycidae: Lepturinae: Rhagiini
Type locality: Canada: Alberta, Calgary

aterrima **Fisher, 1931c**: 198 (Fig. 21c, d), Holotype
Original combination: *Moneilema (Collapteryx) aterrima* Fisher
Current name: *Moneilema aterrimum* Fisher, 1931c
Current placement: Cerambycidae: Lamiinae: Moneilemini
Type locality: Mexico: San Luis Potosí

atlanticus kusaiecus **Gressitt, 1956**: 134 (Fig. 21e, f), Holotype
Original combination: *Prosoplus atlanticus kusaiecus* Gressitt
Current name: same
Current placement: Cerambycidae: Lamiinae: Pterolophini
Type locality: Micronesia: Kusaie, Lelu Island, 100 m.

atollorum **Gressitt, 1956**: 158 (Fig. 21g, h), Holotype
Original combination: *Sciadella (Micronesiella) atollorum* Gressitt
Current name: *Sciades (Micronesiella) atollorum* (Gressitt, 1956)
Current placement: Cerambycidae: Lamiinae: Acanthocinini
Type locality: Micronesia: Bukaru, Falalop Atoll, Fuasua

atrata **Fisher, 1935c**: 590 (Fig. 21i, j), Holotype
Original combination: *Rhaphuma atrata* Fisher
Current name: same
Current placement: Cerambycidae: Cerambycinae: Clytini
Type locality: Malaysia: Borneo, Sabah, Mt. Kinabalu, 3000 ft., Kiau

atratus **Fisher, 1932**: 3 (Fig. 21k, l), Holotype
Original combination: *Monodesmus atratus* Fisher
Current name: same
Current placement: Cerambycidae: Prioninae: Meroscelisini
Type locality: Cuba: Pinar del Río, Sierra Rangel

atricollis **Heller, 1924a**: 31 (Fig. 21m, n), Lectotype
Original combination: *Merionoeda atricollis* Heller
Current name: same
Current placement: Cerambycidae: Cerambycinae: Stenopterini
Type locality: Malaysia: Borneo, Sandakan
Remarks: This specimen is herein designated as the lectotype in order to stabilize the taxonomy and facilitate further identifications of this species. Other paralectotype(s) are present in the Staatliches Museum für Tierkunde, Dresden, Germany.

atriplicis **Linsley, 1957b**: 160 (Fig. 21o, p), Holotype
Original combination: *Amannus atriplicis* Linsley
Current name: same
Current placement: Cerambycidae: Cerambycinae: Trachyderini
Type locality: USA: Texas, Presidio

atrisetosus **Casey, 1913**: 383 (Fig. 21q, r), Lectotype
Original combination: *Tetraopes atrisetosus* Casey
Current name: *Tetraopes femoratus* LeConte, 1847
Current placement: Cerambycidae: Lamiinae: Tetraopini
Type locality: USA: New Mexico, Fort Wingate
Remarks: This specimen is herein designated as the lectotype in order to stabilize the taxonomy and facilitate further identifications of this species.

atromaculatus **Fisher, 1926**: 18 (Fig. 21s, t), Holotype
Original combination: *Leptostylus atromaculatus* Fisher
Current name: *Leptostylopsis atromaculatus* (Fisher, 1926)
Current placement: Cerambycidae: Lamiinae: Acanthocinini
Type locality: Cuba: Central Constancia

atropos **Dillon & Dillon, 1948**: 257 (Fig. 21u, v), Holotype
Original combination: *Cylindrepomus atropos* Dillon & Dillon
Current name: same
Current placement: Cerambycidae: Lamiinae: Gnomiini
Type locality: Philippines: Samar

atrovittatus **Fisher, 1925b**: 9 (Fig. 21w, x), Holotype
Original combination: *Leiopus atrovittatus* Fisher
Current name: *Styloleptus atrovittatus* (Fisher, 1925b)
Current placement: Cerambycidae: Lamiinae: Acanthocinini
Type locality: Cuba: Yara

attenuata **Fisher, 1926**: 12 (Fig. 22a, b), Holotype
Original combination: *Estola attenuata* Fisher
Current name: same
Current placement: Cerambycidae: Lamiinae: Desmiphorini
Type locality: Jamaica

attenuata **Gressitt, 1956**: 158 (Fig. 22c, d), Holotype
Original combination: *Sciadella (Micronesiella) attenuata* Gressitt
Current name: *Sciades (Micronesiella) attenuatus* (Gressitt, 1956)
Current placement: Cerambycidae: Lamiinae: Acanthocinini
Type locality: Micronesia: Yap Island, Dugor-Rumu, 10 m.

aurantia **Martins & Galileo, 2007c**: 1146 (Fig. 22e, f), Holotype
Original combination: *Tyrinthia aurantia* Martins & Galileo
Current name: same
Current placement: Cerambycidae: Lamiinae: Hemilophini
Type locality: Ecuador: Pichincha, Rio Palenque Res. Sta., 45 km. south of Santo Domingo, 220 m.

***aurantius* Martins & Galileo, 2010**: 87 (Fig. 22g, h), Holotype
Original combination: *Estoloides (Estoloides) aurantius* Martins & Galileo
Current name: same
Current placement: Cerambycidae: Lamiinae: Desmiphorini
Type locality: Mexico: Quintana Roo, Bacalar, elevation 90 ft., 18°44.52′N, 88°21.59′W

***auratoides* Breuning, 1943c**: 37 (Fig. 22i, j), Holotype
Original combination: *Adetus auratoides* Breuning
Current name: *Adetus subcostatus* Aurivillius, 1900
Current placement: Cerambycidae: Lamiinae: Apomecynini
Type locality: Venezuela: Maracay, Aragua

***aureatus* Casey, 1913**: 213 (Fig. 22k, l), Holotype
Original combination: *Stenocorus (Stenocorus) aureatus* Casey
Current name: *Stenocorus vestitus* (Haldeman, 1847)
Current placement: Cerambycidae: Lepturinae: Rhagiini
Type locality: USA: California

***aureatus gilvicornis* Casey, 1913**: 213 (Fig. 22m, n), Holotype
Original combination: *Stenocorus (Stenocorus) aureatus gilvicornis* Casey
Current name: *Stenocorus vestitus* (Haldeman, 1847)
Current placement: Cerambycidae: Lepturinae: Rhagiini
Type locality: USA: California

***aureola* Casey, 1913**: 268 (Fig. 22o, p), Lectotype
Original combination: *Leptura (Cosmosalia) aureola* Casey
Current name: *Cosmosalia chrysocoma* (Kirby, 1837)
Current placement: Cerambycidae: Lepturinae: Lepturini
Type locality: USA: Colorado
Remarks: This specimen is herein designated as the lectotype in order to stabilize the taxonomy and facilitate further identifications of this species.

***aureopleura* Gressitt, 1956**: 158 (Fig. 22q, r), Holotype
Original combination: *Sciadella (Sciadella) aureopleura* Gressitt
Current name: *Sciades (Estoliops) aureopleura* (Gressitt, 1956)
Current placement: Cerambycidae: Lamiinae: Acanthocinini
Type locality: Micronesia: Yap Island, Hill behind Yaptown, 50 m.

***aureovittatus* Aurivillius, 1903**: 316 (Fig. 22s, t), Lectotype
Original combination: *Hypocrites aureovittatus* Aurivillius
Current name: *Psilacestes aureovittatus* (Aurivillius, 1903)
Current placement: Cerambycidae: Cerambycinae: Callichromatini
Type locality: Tanzania: West Usambara (Tanganyika)
Remarks: This specimen is herein designated as the lectotype in order to stabilize the taxonomy and facilitate further identifications of this species. Other paralectotype(s) are present in the Naturhistoriska Riksmuseet, Stockholm, Sweden.

***aureovridis* Gressitt, 1941**: 332 (Fig. 22u, v), Holotype
Original combination: *Ocalemia aureovridis* Gressitt
Current name: *Metalloleptura gahani* (Plavilstshikov, 1921)
Current placement: Cerambycidae: Lepturinae: Lepturini
Type locality: Thailand: Khun Tan Mts., 3000 ft.

***auricapilla* var. *vittulata* Melzer, 1935a**: 240 (Fig. 22w, x), Holotype
Original combination: *Phaea auricapilla* var. *vittulata* Melzer
Current name: *Phaea phthisica* Bates, 1881
Current placement: Cerambycidae: Lamiinae: Tetraopini
Type locality: Costa Rica: Limón, Hamburg Farm, Reventazón, Ebene
Remarks: The name following the binomen was not expressly given infrasubspecific rank and the work did not indicate unambiguously that the name was infrasubspecific, therefore, it is deemed a subspecies and available.

***auricollis* Casey, 1913**: 232 (Fig. 23a, b), Lectotype
Original combination: *Leptacmaeops (Leptacmaeops) auricollis* Casey
Current name: *Cortodera cubitalis* (LeConte, 1861)
Current placement: Cerambycidae: Lepturinae: Rhagiini
Type locality: USA: California, Mokelumne Hill
Remarks: This specimen is herein designated as the lectotype in order to stabilize the taxonomy and facilitate further identifications of this species.

***auricoma* Lingafelter, 2008**: 366 (Fig. 23c, d), Holotype
Original combination: *Elaphidion auricoma* Lingafelter
Current name: same
Current placement: Cerambycidae: Cerambycinae: Elaphidiini
Type locality: Dominican Republic: La Altagracia Province, PN del Este, Guaraguao, 0–5 meters, 18°19.568′N, 68°48.500′W

***auripenne* Casey, 1893**: 587 (Fig. 23e, f), Lectotype
Original combination: *Tragidion auripenne* Casey
Current name: same
Current placement: Cerambycidae: Cerambycinae: Trachyderini
Type locality: USA: Utah
Remarks: This specimen is herein designated as the lectotype in order to stabilize the taxonomy and facilitate further identifications of this species.

***auripilis densepilosa* Casey, 1924**: 281 (Fig. 23g, h), Holotype
Original combination: *Cosmosalia auripilis densepilosa* Casey
Current name: *Cosmosalia chrysocoma* (Kirby, 1837)
Current placement: Cerambycidae: Lepturinae: Lepturini
Type locality: USA: Utah, North Fork, Provo Canyon

***aurivillii* Fisher, 1935c**: 620 (Fig. 23i, j), Holotype
Original combination: *Glenea aurivillii* Fisher
Current name: same
Current placement: Cerambycidae: Lamiinae: Saperdini
Type locality: Malaysia: Borneo, Sabah, Mt. Kinabalu, 3300 ft., Kenokok

***aurocincta nigroapicalis* Fisher, 1947a**: 52 (Fig. 23k, l), Holotype
Original combination: *Odontocera aurocincta nigroapicalis* Fisher
Current name: *Odontocera aurocincta aurocincta* Bates, 1873
Current placement: Cerambycidae: Cerambycinae: Rhinotragini
Type locality: Mexico: Veracruz, La Gloria Cardel

***auropilosa* Tippmann, 1953c**: 212 (Fig. 23m, n), Holotype
Original combination: *Odontocera auropilosa* Tippmann
Current name: same
Current placement: Cerambycidae: Cerambycinae: Rhinotragini
Type locality: Bolivia: Cochabamba, Reg. Chapare 400 m.

***aurora* Casey, 1913**: 244 (Fig. 23o, p), Holotype
Original combination: *Acmaeops aurora* Casey
Current name: *Acmaeops proteus* (Kirby, 1837)
Current placement: Cerambycidae: Lepturinae: Rhagiini
Type locality: USA: Colorado, Boulder Co.

***aurovittata* var. *debilis* Melzer, 1935b**: 184 (Fig. 23q, r), Holotype
Original combination: *Dihammaphora aurovittata* var. *debilis* Melzer
Current name: *Dihammaphora aurovittata* Bates, 1880
Current placement: Cerambycidae: Cerambycinae: Rhopalophorini
Type locality. Costa Rica. San José, 1000–1200 m.
Remarks: The name following the binomen was not expressly given infrasubspecific rank and the work did not indicate unambiguously that the name was infrasubspecific, therefore, it is deemed a subspecies and available.

***ayri* Galileo & Martins, 2006c**: 184 (Fig. 23s, t), Holotype
Original combination: *Eburodacrys ayri* Galileo & Martins
Current name: same
Current placement: Cerambycidae: Cerambycinae: Eburiini
Type locality: Colombia: Río Magdalena

***ayri* Martins & Galileo, 2012**: 111 (Fig. 23u, v), Holotype
Original combination: *Pseudestola ayri* Martins & Galileo
Current name: same
Current placement: Cerambycidae: Lamiinae: Desmiphorini
Type locality: Costa Rica: Limón, Hamburg Farm, Santa Clara

***aztecus* Casey, 1912**: 246 (Fig. 23w, x), Lectotype
Original combination: *Prionus (Riponus) aztecus* Casey
Current name: *Prionus (Neopolyarthron) aztecus* Casey, 1912
Current placement: Cerambycidae: Prioninae: Prionini
Type locality: Mexico: Durango, Sierra de Durango
Remarks: This specimen is herein designated as the lectotype in order to stabilize the taxonomy and facilitate further identifications of this species.

***azurea* Demets, 1976**: 136 (Fig. 24a, b), Holotype
Original combination: *Xenochroma azurea* Demets
Current name: *Monnechroma azurea* (Demets, 1976)
Current placement: Cerambycidae: Cerambycinae: Callichromatini
Type locality: Costa Rica: Limón, Hamburg Farm, Reventazón, Ebene

***azureoopalescens* Tippmann, 1953c**: 210 (Fig. 24c, d), Holotype
Original combination: *Acyphoderes azureoopalescens* Tippmann
Current name: *Acyphoderes auricapilla* Fisher, 1947a
Current placement: Cerambycidae: Cerambycinae: Rhinotragini
Type locality: Brazil: Paraná, Arapongas

***badius* Fisher, 1925a**: 242 (Fig. 24e, f), Holotype
Original combination: *Exocentrus badius* Fisher
Current name: same
Current placement: Cerambycidae: Lamiinae: Acanthocinini
Type locality: Philippines: Mindanao, Davao

***bakeri* Aurivillius, 1922**: 408 (Fig. 24g, h), Holotype
Original combination: *Chlorophorus bakeri* Aurivillius
Current name: same
Current placement: Cerambycidae: Cerambycinae: Clytini
Type locality: Philippines: Luzon, Mt. Banahao
Remarks: The original description is somewhat unclear as to specifying if there was more than one primary type. The USNM specimen matches the description and data and is presumed to be the holotype. It is possible that an additional specimen examined by Aurivillius is in the Naturhistoriska Riksmuseet, Stockholm, Sweden.

***bakeri* Aurivillius, 1922**: 421 (Fig. 24i, j), Holotype
Original combination: *Demonax bakeri* Aurivillius
Current name: same
Current placement: Cerambycidae: Cerambycinae: Clytini
Type locality: Malaysia: Borneo, Sandakan

***bakeri* Aurivillius, 1926**: 113 (Fig. 24k, l), Lectotype
Original combination: *Daphisia bakeri* Aurivillius
Current name: *Menesia makilingi* (Heller, 1924c)
Current placement: Cerambycidae: Lamiinae: Saperdini
Type locality: Philippines: Negros Oriental, Cuernos Mountains
Remarks: This specimen is herein designated as the lectotype in order to stabilize the taxonomy and facilitate further identifications of this species. Other paralectotype(s) are present in the Naturhistoriska Riksmuseet, Stockholm, Sweden.

***bakeri* Aurivillius, 1927b**: 555 (Fig. 24m, n), Holotype
Original combination: *Agnia bakeri* Aurivillius
Current name: same
Current placement: Cerambycidae: Lamiinae: Lamiini
Type locality: Philippines: Sibuyan, Romblon

***bakeri* Aurivillius, 1927b**: 558 (Fig. 24o, p), Holotype
Original combination: *Homonaeomorpha bakeri* Aurivillius
Current name: *Bumetopia bakeri* (Aurivillius, 1927b)
Current placement: Cerambycidae: Lamiinae: Homonoeini
Type locality: Philippines: Sibuyan, Romblon
Remarks: The genus name misspelled on label.

***bakeri* Fisher, 1925a**: 214 (Fig. 24q, r), Holotype
Original combination: *Platyzeargyra bakeri* Fisher
Current name: *Egesina bakeri* (Fisher, 1925a)
Current placement: Cerambycidae: Lamiinae: Pteropliini
Type locality: Malaysia: Borneo, Sandakan

***bakeri* Fisher, 1925a**: 258 (Fig. 24s, t), Holotype
Original combination: *Enes bakeri* Fisher
Current name: *Enes (Sibuyenes) bakeri* Fisher, 1925a
Current placement: Cerambycidae: Lamiinae: Acanthocinini
Type locality: Philippines: Sibuyan, Romblon

***bakeri* Fisher, 1925a**: 265 (Fig. 24u, v), Holotype
Original combination: *Xaenapta bakeri* Fisher
Current name: same
Current placement: Cerambycidae: Lamiinae: Acanthocinini
Type locality: Malaysia: Borneo, Sandakan

***bakeri* var. *immaculata* Aurivillius, 1926**: 114 (Fig. 24w, x), Lectotype
Original combination: *Daphisia bakeri* var. *immaculata* Aurivillius
Current name: *Menesia makilingi* (Heller, 1924c)
Current placement: Cerambycidae: Lamiinae: Saperdini
Type locality: Philippines: Negros Oriental, Cuernos Mountains
Remarks: The name following the binomen was not expressly given infrasubspecific rank and the work did not indicate unambiguously that the name was infrasubspecific, therefore, it is deemed a subspecies and available. This specimen is herein designated as the lectotype in order to stabilize the taxonomy and facilitate further identifications of this species.

***bakeri* var. *semisignata* Aurivillius, 1926**: 114 (Fig. 25a, b), Lectotype
Original combination: *Daphisia bakeri* var. *semisignata* Aurivillius
Current name: *Menesia makilingi* (Heller, 1924c)
Current placement: Cerambycidae: Lamiinae: Saperdini
Type locality: Philippines: Mindanao, Surigao
Remarks: The name following the binomen was not expressly given infrasubspecific rank and the work did not indicate unambiguously that the name was infrasubspecific, therefore, it is deemed a subspecies and available. This specimen is herein designated as the lectotype in order to stabilize the taxonomy and facilitate further identifications of this species.

***bakeri* var. *vittulata* Aurivillius, 1926**: 114 (Fig. 25c, d), Lectotype
Original combination: *Daphisia bakeri* var. *vittulata* Aurivillius
Current name: *Menesia makilingi* (Heller, 1924c)
Current placement: Cerambycidae: Lamiinae: Saperdini
Type locality: Philippines: Basilan
Remarks: The name following the binomen was not expressly given infrasubspecific rank and the work did not indicate unambiguously that the name was infrasubspecific, therefore, it is deemed a subspecies and available. This specimen is herein designated as the lectotype in order to stabilize the taxonomy and facilitate further identifications of this species.

***balineae* Heller, 1915**: 43 (Fig. 25e, f), Lectotype
Original combination: *Oberea balineae* Heller
Current name: *Oberea balineae* Heller, 1915
Current placement: Cerambycidae: Lamiinae: Obereini
Type locality: Philippines: Luzon, Mt. Makiling
Remarks: This specimen is herein designated as the lectotype in order to stabilize the taxonomy and facilitate further identifications of this species. Other paralectotype(s) are present in the Staatliches Museum für Tierkunde, Dresden, Germany.

***balloui* Fisher, 1926**: 2 (Fig. 25g, h), Holotype
Original combination: *Liosynaphaeta balloui* Fisher
Current name: same
Current placement: Cerambycidae: Lamiinae: Mesosini
Type locality: Jamaica

***balteata* Lane, 1972**: 368 (Fig. 25i, j), Holotype
Original combination: *Natagaima balteata* Lane
Current name: *Irundisaua balteata* (Lane, 1972)
Current placement: Cerambycidae: Lamiinae: Acanthoderini
Type locality: Colombia: Natagaima

***balteata* var. *inapicalis* Tippmann, 1960**: 190 (Fig. 25k, l), Holotype
Original combination: *Dectes (Canidia) balteata* var. *inapicalis* Tippmann
Current name: *Canidia cincticornis balteata* (Lacordaire, 1872)
Current placement: Cerambycidae: Lamiinae: Acanthocinini
Type locality: Colombia: Bogota
Remarks: The name following the binomen was not expressly given infrasubspecific rank and the work did not indicate unambiguously that the name was infrasubspecific, therefore, it is deemed a subspecies and available.

***baluensis* Fisher, 1935c**: 587 (Fig. 25m, n), Holotype
Original combination: *Strangalia baluensis* Fisher
Current name: same
Current placement: Cerambycidae: Lepturinae: Lepturini
Type locality: Malaysia: Borneo, Sabah, Mt. Kinabalu, 5500 ft., Lumu Lumu

***bangueyensis nigripes* Aurivillius, 1926**: 101 (Fig. 25o, p), Holotype
Original combination: *Glenea bangueyensis nigripes* Aurivillius
Current name: *Glenea bangueyensis* Aurivillius, 1920
Current placement: Cerambycidae: Lamiinae: Saperdini
Type locality: Philippines: Negros Oriental, Cuernos Mountains

***baoruco* Lingafelter, 2011**: 66 (Fig. 25q, r), Holotype
Original combination: *Tilloclytus baoruco* Lingafelter
Current name: same
Current placement: Cerambycidae: Cerambycinae: Anaglyptini
Type locality: Dominican Republic: Pedernales, PN Sierra de Baoruco, Las Abejas, 1150 m.

***barbicornis itaitubensis* Tippmann, 1953c**: 227 (Fig. 25s, t), Lectotype
Original combination: *Batus barbicornis itaitubensis* Tippmann
Current name: *Batus barbicornis* (Linnaeus, 1764)
Current placement: Cerambycidae: Cerambycinae: Trachyderini
Type locality: Brazil: Pará, Itaituba
Remarks: This specimen is herein designated as the lectotype in order to stabilize the taxonomy and facilitate further identifications of this species.

***baroni* Casey, 1912**: 223 (Fig. 25u, v), Holotype
Original combination: *Mallodon baroni* Casey
Current name: *Mallodon dasystomus* (Say, 1824)
Current placement: Cerambycidae: Prioninae: Macrotomini
Type locality: Mexico: Guerrero

***basale* var. *inbasale* Tippmann, 1960**: 133 (Fig. 25w, x), Holotype
Original combination: *Ethemon basale* var. *inbasale* Tippmann
Current name: *Ethemon imbasale* Tippmann, 1960
Current placement: Cerambycidae: Cerambycinae: Compsocerini
Type locality: Bolivia: Cochabamba, 2800 m
Remarks: The name following the binomen was not expressly given infrasubspecific rank and the work did not indicate unambiguously that the name was infrasubspecific, therefore, it is deemed a subspecies and available.

***basalis* Casey, 1912**: 220 (Fig. 26a, b), Holotype
Original combination: *Spondylis basalis* Casey
Current name: *Neospondylis upiformis* (Mannerheim, 1843)
Current placement: Cerambycidae: Spondylidinae: Spondylidini
Type locality: USA: Washington

***basalis* Fisher, 1925a**: 227 (Fig. 26c, d), Holotype
Original combination: *Ebaeides basalis* Fisher
Current name: same
Current placement: Cerambycidae: Lamiinae: Acanthocinini
Type locality: Philippines: Negros Oriental, Cuernos Mountains

***basalis* Schwarzer, 1925**: 25 (Fig. 26e, f), Syntype
Original combination: *Xylotrechus basalis* Schwarzer
Current name: *Xylotrechus atronotatus* Pic, 1917
Current placement: Cerambycidae: Cerambycinae: Clytini
Type locality: China: Taiwan, Formosa, Kankau (Koshun)

***basalis dichroma* Casey, 1913**: 231 (Fig. 26g, h), Lectotype
Original combination: *Leptacmaeops (Leptacmaeops) basalis dichroma* Casey
Current name: *Cortodera cubitalis* (LeConte, 1861)
Current placement: Cerambycidae: Lepturinae: Rhagiini
Type locality: USA: California, Santa Barbara
Remarks: This specimen is herein designated as the lectotype in order to stabilize the taxonomy and facilitate further identifications of this species.

***basicristata* Breuning, 1938a**: 337 (Fig. 26i, j), Holotype
Original combination: *Pterolophia basicristata* Breuning
Current name: *Pterolophia (Ale) basicristata* Breuning, 1938a
Current placement: Cerambycidae: Lamiinae: Pteropliini
Type locality: Indonesia: Sumatra, Medan
Remarks: It is unclear if this is actually the holotype since Tavakilian and Chevillotte (2013) indicate that it is present in in the Naturhistoriska Riksmuseet, Stockholm, Sweden.

***basiflava* Breuning, 1943c**: 57 (Fig. 26k, l), Holotype
Original combination: *Estola basiflava* Breuning
Current name: same
Current placement: Cerambycidae: Lamiinae: Desmiphorini
Type locality: Venezuela: Aragua, Maracay

***basifulvus* Lingafelter & Micheli, 2009**: 31 (Fig. 26m, n), Holotype
Original combination: *Leptostylopsis basifulvus* Lingafelter & Micheli
Current name: same
Current placement: Cerambycidae: Lamiinae: Acanthocinini
Type locality: Dominican Republic: La Altagracia Province, Punta Cana, near Ecological Reserve, 0–5 m, 18°30.477′N, 68°22.499′W.

***basilaris* Galileo & Martins, 2012c**: 299 (Fig. 26o, p), Holotype
Original combination: *Capillicornis basilaris* Galileo & Martins
Current name: same
Current placement: Cerambycidae: Lamiinae: Desmiphoriini
Type locality: Ecuador: Pichincha, Puerto Quito

***basipenicillata* Tippmann, 1960**: 159 (Fig. 26q, r), Holotype
Original combination: *Oreodera basipenicillata* Tippmann
Current name: same
Current placement: Cerambycidae: Lamiinae: Acanthoderini
Type locality: Bolivia: Cochabamba, Reg. Chapare 400 m.

***basiradiata* Tippmann, 1960**: 161 (Fig. 26s, t), Holotype
Original combination: *Oreodera basiradiata* Tippmann
Current name: same
Current placement: Cerambycidae: Lamiinae: Acanthoderini
Type locality: Brazil: Amapa, Macapa

***basirufus* Gressitt, 1940**: 184 (Fig. 26u, v), Holotype
Original combination: *Exocentrus basirufus* Gressitt
Current name: *Exocentrus (Camptomyne) basirufus* Gressitt, 1940
Current placement: Cerambycidae: Lamiinae: Acanthocinini
Type locality: China: Hainan Province, Hainan Island, Chung Kon

***basisuturale* Gressitt, 1935c**: 573 (Fig. 26w, x), Holotype
Original combination: *Stenostola basisuturale* Gressitt
Current name: same
Current placement: Cerambycidae: Lamiinae: Saperdini
Type locality: China: Sichuan Province, near Muping 3000–7400 ft.

***basivittatus* Breuning, 1940c**: 141 (Fig. 27a, b), Holotype
Original combination: *Prosoplus basivittatus* Breuning
Current name: same
Current placement: Cerambycidae: Lamiinae: Pterolophini
Type locality: Papua New Guinea
Remarks: Tavakilian and Chevillotte (2013) suggest the holotype is present in the Museum national d'Histoire naturelle, Paris, France, however the specimen and label match the published description.

***batesi* Chemsak, 1963**: 28 (Fig. 27c, d), Holotype
Original combination: *Tetraopes batesi* Chemsak
Current name: same
Current placement: Cerambycidae: Lamiinae: Tetraopini
Type locality: Mexico: Oaxaca, Tehuantepec

***batesi* Gressitt, 1937b**: 618 (Fig. 27e, f), Holotype
Original combination: *Nupserha batesi* Gressitt
Current name: *Nupserha testaceipes* Pic, 1926
Current placement: Cerambycidae: Lamiinae: Phytoeciini
Type locality: China: Jiangxi Province, Hong Shan

***bauhiniae* Fisher, 1934b**: 36 (Fig. 27g, h), Holotype
Original combination: *Exocentrus bauhiniae* Fisher
Current name: same
Current placement: Cerambycidae: Lamiinae: Acanthocinini
Type locality: Indonesia: Java, Kediri

***bauri* Linell, 1899**: 260 (Fig. 27i, j), Lectotype
Original combination: *Eburia bauri* Linell
Current name: *Eburia pilosa* (Erichson, 1834)
Current placement: Cerambycidae: Cerambycinae: Eburiini
Type locality: Ecuador: Galapagos Islands, Chatham Island
Remarks: This specimen is herein designated as the lectotype in order to stabilize the taxonomy and facilitate further identifications of this species.

***beardsleyi* Gressitt, 1956**: 120 (Fig. 27k, l), Holotype
Original combination: *Olenecamptus beardsleyi* Gressitt
Current name: same
Current placement: Cerambycidae: Lamiinae: Dorcaschematini
Type locality: Palau: Koror

***beckeri* Howden, 1960**: 175 (Fig. 27m, n), Holotype
Original combination: *Cyrtinus beckeri* Howden
Current name: same
Current placement: Cerambycidae: Lamiinae: Cyrtinini
Type locality: USA: Texas, Big Bend National Park, 5000 ft., Pine Canyon, Chisos Mts.

***beckeri* Knull, 1962**: 105 (Fig. 27o, p), Holotype
Original combination: *Obrium beckeri* Knull
Current name: same
Current placement: Cerambycidae: Cerambycinae: Obriini
Type locality: USA: Texas, Big Bend National Park, Tornillo Flat, 3200 ft.

***beebei* Fisher, 1944a**: 7 (Fig. 27q, r), Holotype
Original combination: *Ibidion beebei* Fisher
Current name: *Tapuruia beebei* (Fisher, 1944a)
Current placement: Cerambycidae: Cerambycinae: Hexoplini
Type locality: Venezuela: Monagas, Caripito

***bellina* Casey, 1913**: 265 (Fig. 27s, t), Holotype
Original combination: *Strophiona bellina* Casey
Current name: *Strophiona nitens* (Forster, 1771)
Current placement: Cerambycidae: Lepturinae: Lepturini
Type locality: USA: Colorado

***benguetana* Aurivillius, 1926**: 99 (Fig. 27u, v), Lectotype
Original combination: *Glenea benguetana* Aurivillius
Current name: same
Current placement: Cerambycidae: Cerambycinae: Saperdini
Type locality: Philippines: Luzon, Baguio, Benguet
Remarks: This specimen is herein designated as the lectotype in order to stabilize the taxonomy and facilitate further identifications of this species. Other paralectotype(s) are present in the Naturhistoriska Riksmuseet, Stockholm, Sweden.

***beyeri* Schaeffer, 1905b**: 163 (Fig. 27w, x), Lectotype
Original combination: *Stenosphenus beyeri* Schaeffer
Current name: same
Current placement: Cerambycidae: Cerambycinae: Elaphidiini
Type locality: USA: Arizona, Huachuca Mts, Palmerlee
Remarks: The lectotype was designated by Fisher (1946: 91).

***bicolor* Casey, 1912**: 239 (Fig. 28a, b), Holotype
Original combination: *Prionus (Prionus) bicolor* Casey
Current name: *Prionus pocularis* Dalman, 1817
Current placement: Cerambycidae: Prioninae: Prionini
Type locality: USA: Indiana

***bicolor* Fisher, 1925a**: 274 (Fig. 28c, d), Holotype
Original combination: *Languriomorpha bicolor* Fisher
Current name: same
Current placement: Cerambycidae: Lamiinae: Desmiphorini
Type locality: Malaysia: Borneo, Sandakan

***bicolor* Fisher, 1932**: 70 (Fig. 28e, f), Holotype
Original combination: *Trichrous bicolor* Fisher
Current name: *Trichrous fisheri* Monné & Giesbert, 1992
Current placement: Cerambycidae: Cerambycinae: Heteropsini
Type locality: Haiti: Port au Prince

***bicolor* Fisher, 1936d**: 346 (Fig. 28g, h), Holotype
Original combination: *Heterops bicolor* Fisher
Current name: same
Current placement: Cerambycidae: Cerambycinae: Heteropsini
Type locality: Cuba: Oriente, Pico Turquino

***bicolor* Fisher, 1946a**: 332 (Fig. 28i, j), Holotype
Original combination: *Cometes bicolor* Fisher
Current name: *America bicolor* (Fisher, 1946a)
Current placement: Disteniidae: Disteniinae: Disteniini
Type locality: Colombia

***bicolor* Kriesche, 1920**: 193 (Fig. 28k, l), Lectotype
Original combination: *Apriona bicolor* Kriesche
Current name: *Apriona sublaevis* Thomson, 1878
Current placement: Cerambycidae: Lamiinae: Batocerini
Type locality: Vietnam: Tonkin, Than-Moi
Remarks: This specimen is herein designated as the lectotype in order to stabilize the taxonomy and facilitate further identifications of this species. A paralectotype may also be present in Berlin, Germany. Tippmann labeled this specimen as having been obtained from the Kriesche Estate in 1955.

***bicolor* Melzer, 1934**: 75 (Fig. 28m, n), Holotype
Original combination: *Eucharassus bicolor* Melzer
Current name: same
Current placement: Cerambycidae: Cerambycinae: Necydalopsini
Type locality: Costa Rica: Cartago, 2200–3000 m, Vulcán Irazu

***bicolor* Melzer, 1935a**: 214 (Fig. 28o, p), Holotype
Original combination: *Merionoeda bicolor* Melzer
Current name: *Oxycoleus bicolor* (Melzer, 1935a)
Current placement: Cerambycidae: Cerambycinae: Oxycoleini
Type locality: Costa Rica: San José, 1000–1200 m.

***bicolor spissicornis* Casey, 1924**: 262 (Fig. 28q, r), Holotype
Original combination: *Ancylocera bicolor spissicornis* Casey
Current name: *Ancylocera bicolor* (Olivier, 1795)
Current placement: Cerambycidae: Cerambycinae: Trachyderini
Type locality: USA: Florida

***bicolor* var. *demaggii* Tippmann, 1956**: 484 (Fig. 28s, t), Holotype
Original combination: *Rhamnusium bicolor* var. *demaggii* Tippmann
Current name: *Rhamnusium bicolor demaggii* Tippmann, 1956
Current placement: Cerambycidae: Lepturinae: Lepturini
Type locality: Italy: Lazio, Roma Due Ponti
Remarks: The name following the binomen was not expressly given infrasubspecific rank and the work did not indicate unambiguously that the name was infrasubspecific, therefore, it is deemed a subspecies and available. Although published as "*de maggii*", we have removed the space following Article 32.5 (ICZN, 1999).

***bicolorella* Knull, 1934**: 336 (Fig. 28u, v), Holotype
Original combination: *Rhopalophora bicolorella* Knull
Current name: same
Current placement: Cerambycidae: Cerambycinae: Rhopalophorini
Type locality: USA: Arizona, Cochise Co., Palmerlee

***bicoloricornis* Galileo & Martins, 2009**: 8 (Fig. 28w, x), Holotype
Original combination: *Abanycha bicoloricornis* Galileo & Martins
Current name: same
Current placement: Cerambycidae: Lamiinae: Hemilophini
Type locality: Ecuador: Napo, Res. Ethnica Waorani, 1km S. Onkone Gare Camp, 220 m., 00°39′S 076°26′W

***bicordifera* Tippmann, 1953d**: 346 (Fig. 29a, b), Holotype
Original combination: *Calolamia bicordifera* Tippmann
Current name: same
Current placement: Cerambycidae: Lamiinae: Acanthocinini
Type locality: Brazil: Santa Catarina, Mafra, 800 m.

***bicristulata* Breuning, 1940d**: 422 (Fig. 29c, d), Holotype
Original combination: *Pterolophia bicristulata* Breuning
Current name: *Pterolophia (Ale) bicristulata* Breuning, 1940d
Current placement: Cerambycidae: Lamiinae: Pteropliini
Type locality: Vietnam: Chapa

bifasciata **Fisher, 1943**: 3 (Fig. 29e, f), Holotype
Original combination: *Niphonoclea bifasciata* Fisher
Current name: *Callimetopus irroratus* (Newman, 1842)
Current placement: Cerambycidae: Lamiinae: Pteropliini
Type locality: Philippines: Samar

bifasciatus **Dillon & Dillon, 1941**: 56 (Fig. 29g, h), Holotype
Original combination: *Ptychodes bifasciatus* Dillon & Dillon
Current name: same
Current placement: Cerambycidae: Lamiinae: Monochamini
Type locality: Mexico: Colima

bifasciatus **Martins & Galileo, 2009**: 66 (Fig. 29i, j), Holotype
Original combination: *Cyrtinus bifasciatus* Martins & Galileo
Current name: same
Current placement: Cerambycidae: Lamiinae: Cyrtinini
Type locality: Panama: Colon Pr., Fort Sherman, 9°17′N, 79°59′W

bifasciculata **Breuning, 1940d**: 422 (Fig. 29k, l), Holotype
Original combination: *Pterolophia bifasciculata* Breuning
Current name: *Pterolophia alboplagiata* Gahan, 1894
Current placement: Cerambycidae: Lamiinae: Pteropliini
Type locality: Laos

biformis **Casey, 1912**: 358 (Fig. 29m, n), Lectotype
Original combination: *Calloides biformis* Casey
Current name: *Calloides nobilis mormonus* Schaeffer, 1911
Current placement: Cerambycidae: Cerambycinae: Clytini
Type locality: USA: New Mexico
Remarks: This specimen is herein designated as the lectotype in order to stabilize the taxonomy and facilitate further identifications of this species.

bifossatum **Casey, 1912**: 287 (Fig. 29o, p), Holotype
Original combination: *Merium bifossatum* Casey
Current name: *Meriellum proteus* (Kirby, 1837)
Current placement: Cerambycidae: Cerambycinae: Callidiini
Type locality: USA: Wisconsin, Bayfield

bifurcatus **Dillon & Dillon, 1941**: 113 (Fig. 29q, r), Holotype
Original combination: *Deliathis bifurcatus* Dillon & Dillon
Current name: same
Current placement: Cerambycidae: Lamiinae: Monochamini
Type locality: Honduras: La Ceiba

biguttata **Aurivillius, 1925**: 522 (Fig. 29s, t), Syntype
Original combination: *Daphisia biguttata* Aurivillius
Current name: *Menesia bimaculata* Breuning, 1954
Current placement: Cerambycidae: Lamiinae: Saperdini
Type locality: Malaysia: Borneo, Sandakan
Remarks: Other syntypes are present in the Naturhistoriska Riksmuseet, Stockholm, Sweden.

biguttata **Aurivillius, 1927b**: 573 (Fig. 29u, v), Lectotype
Original combination: *Sybra biguttata* Aurivillius
Current name: same
Current placement: Cerambycidae: Lamiinae: Apomecynini
Type locality: Philippines: Mindanao, Surigao

biguttatus **Aurivillius, 1922**: 422 (Fig. 29w, x), Holotype
Original combination: *Demonax biguttatus* Aurivillius
Current name: same
Current placement: Cerambycidae: Cerambycinae: Clytini
Type locality: Philippines: Luzon, Mt. Banahao

bilobus gressitti **Dillon & Dillon, 1948**: 234 (Fig. 30a, b), Holotype
Original combination: *Olenecamptus bilobus gressitti* Dillon & Dillon
Current name: same
Current placement: Cerambycidae: Lamiinae: Dorcaschematini
Type locality: China: Sichuan Province, Suifu

bilobus tonkinus **Dillon & Dillon, 1948**: 230 (Fig. 30c, d), Holotype
Original combination: *Olenecamptus bilobus tonkinus* Dillon & Dillon
Current name: same
Current placement: Cerambycidae: Lamiinae: Dorcaschematini
Type locality: Vietnam: Tonkin, Cho Ganh

binocularis **Casey, 1913**: 328 (Fig. 30e, f), Holotype
Original combination: *Hyperplatys binocularis* Casey
Current name: *Hyperplatys maculata* Haldeman, 1847
Current placement: Cerambycidae: Lamiinae: Acanthocinini
Type locality: USA: New York

binoculatum **Linsley, 1935b**: 484 (Fig. 30g, h), Holotype
Original combination: *Ibidion binoculatum* Linsley
Current name: *Diasporidion duplicatum* (Gounelle, 1909)
Current placement: Cerambycidae: Cerambycinae: Neoibidionini
Type locality: El Salvador: San Salvador

***biocellata* Tippmann, 1960**: 177 (Fig. 30i, j), Holotype
Original combination: *Acanthoderes (Trichacanthoderes) biocellata* Tippmann
Current name: *Zikanita biocellata* (Tippmann, 1960)
Current placement: Cerambycidae: Lamiinae: Acanthoderini
Type locality: Peru: Junin, Satipo, Río Perené

***biocellatus* Breuning, 1940d**: 411 (Fig. 30k, l), Holotype
Original combination: *Dihammus biocellatus* Breuning
Current name: *Acalolepta biocellata* (Breuning, 1940d)
Current placement: Cerambycidae: Lamiinae: Monochamini
Type locality: Indonesia, Sumatra, Corinchi Lake

***biplagiata* Breuning, 1940**: 418 (Fig. 30m, n), Holotype
Original combination: *Anacasta biplagiata* Breuning
Current name: same
Current placement: Cerambycidae: Lamiinae: Rhodopinini
Type locality: Malaysia: Penang

***bisericans* Kriesche, 1936**: 68 (Fig. 30o, p), Holotype
Original combination: *Dihammus bisericans* Kriesche
Current name: *Acalolepta bisericans* (Kriesche, 1936)
Current placement: Cerambycidae: Lamiinae: Monochamini
Type locality: Papua New Guinea: Bougainville Island (Solomon Islands), Kieta
Remarks: Tippmann labeled this specimen as having been obtained from the Kriesche Estate in 1955.

***biustus* var. *cubanus* Fisher, 1926**: 22 (Fig. 30q, r), Holotype
Original combination: *Leptostylus biustus* var. *cubanus* Fisher
Current name: *Styloleptus biustus dorsalis* (Fisher, 1926)
Current placement: Cerambycidae: Lamiinae: Acanthocinini
Type locality: Cuba: Cayamas
Remarks: The name following the binomen was not expressly given infrasubspecific rank and the work did not indicate unambiguously that the name was infrasubspecific, therefore, it is deemed a subspecies and available.

***bivittata* Aurivillius, 1927b**: 561 (Fig. 30s, t), Syntype
Original combination: *Pterolophia bivittata* Aurivillius
Current name: *Pterolophia aurivillii* Breuning, 1961a
Current placement: Cerambycidae: Lamiinae: Pteropliini
Type locality: Philippines: Sibuyan, Romblon
Remarks: Tavakilian and Chevillotte (2013) state that syntypes are present in the Naturhistoriska Riksmuseet, Stockholm, Sweden as well as the USNM.

***bivittata* Melzer, 1933**: 373 (Fig. 30u, v), Holotype
Original combination: *Amphicnaeia bivittata* Melzer
Current name: *Amphicnaeia crustulata* Bates, 1872
Current placement: Cerambycidae: Lamiinae: Apomecynini
Type locality: Costa Rica: Cartago, Volcán de Irazú, 1500 m.

***blairi* Gressitt, 1937b**: 598 (Fig. 30w, x), Holotype
Original combination: *Rhodopis blairi* Gressitt
Current name: *Nanohammus sinicus* (Pic, 1926)
Current placement: Cerambycidae: Lamiinae: Lamiini
Type locality: China: Jiangxi Province, Hong Shan

***blandi* Casey, 1893**: 589 (Fig. 31a, b), Lectotype
Original combination: *Crossidius blandi* Casey
Current name: *Crossidius discoideus blandi* Casey, 1893
Current placement: Cerambycidae: Cerambycinae: Trachyderini
Type locality: USA: Utah
Remarks: This specimen is herein designated as the lectotype in order to stabilize the taxonomy and facilitate further identifications of this species.

***boharti* Gressitt, 1956**: 158 (Fig. 31c, d), Holotype
Original combination: *Sciadella (Micronesiella) boharti* Gressitt
Current name: *Sciades (Micronesiella) boharti* (Gressitt, 1956)
Current placement: Cerambycidae: Lamiinae: Acanthocinini
Type locality: Northern Mariana Islands: Agrihan Island

***boliviana* Breuning, 1940b**: 71 (Fig. 31e, f), Holotype
Original combination: *Estola boliviana* Breuning
Current name: same
Current placement: Cerambycidae: Lamiinae: Desmiphorini
Type locality: Bolivia: Argos

***boliviana* Tippmann, 1960**: 159 (Fig. 31g, h), Holotype
Original combination: *Oreodera boliviana* Tippmann
Current name: same
Current placement: Cerambycidae: Lamiinae: Acanthoderini
Type locality: Bolivia: Cochabamba, Reg. Chapare 400 m.

***boliviensis* Dillon & Dillon, 1941**: 33 (Fig. 31i, j), Holotype
Original combination: *Taeniotes boliviensis* Dillon & Dillon
Current name: same
Current placement: Cerambycidae: Lamiinae: Monochamini
Type locality: Bolivia: Tumupasa

***boreale* Casey, 1913**: 195 (Fig. 31k, l), Holotype
Original combination: *Rhagium (Hargium) boreale* Casey
Current name: *Rhagium inquisitor* (Linnaeus, 1758)
Current placement: Cerambycidae: Lepturinae: Rhagiini
Type locality: USA: Wisconsin, Bayfield

***boreale* Casey, 1924**: 228 (Fig. 31m, n), Holotype
Original combination: *Tetropium boreale* Casey
Current name: *Tetropium velutinum* LeConte, 1869
Current placement: Cerambycidae: Spondylidinae: Asemini
Type locality: USA: Idaho, Bonner Co., Priest Lake

***boricua* Micheli, 2003**: 200 (Fig. 31o, p), Holotype
Original combination: *Decarthria boricua* Micheli
Current name: same
Current placement: Cerambycidae: Lamiinae: Cyrtinini
Type locality: USA: Puerto Rico, Maricao Forest, Road 120, km 13.8, dead foliage

***borikensis* Micheli & Micheli, 2004**: 31 (Fig. 31q, r), Holotype
Original combination: *Urgleptes borikensis* Micheli & Micheli
Current name: same
Current placement: Cerambycidae: Lamiinae: Acanthocinini
Type locality: USA: Puerto Rico, Guilarte Forest

***borneensis* Fisher, 1925a**: 225 (Fig. 31s, t), Holotype
Original combination: *Ebaeides borneensis* Fisher
Current name: same
Current placement: Cerambycidae: Lamiinae: Acanthocinini
Type locality: Malaysia: Borneo, Sandakan

***borneensis* Fisher, 1935c**: 589 (Fig. 31u, v), Holotype
Original combination: *Chlorophorus borneensis* Fisher
Current name: same
Current placement: Cerambycidae: Cerambycinae: Clytini
Type locality: Malaysia: Borneo, Sabah, near Mt. Kinabalu, 4700–3000 (sic) ft., Tenompok-kian

***borneensis* Fisher, 1935c**: 591 (Fig. 31w, x), Holotype
Original combination: *Demonax borneensis* Fisher
Current name: same
Current placement: Cerambycidae: Cerambycinae: Clytini
Type locality: Malaysia: Borneo, Sabah, Mt. Kinabalu, 5500 ft., Lumu Lumu

***borneensis* Fisher, 1935c**: 611 (Fig. 32a, b), Holotype
Original combination: *Pterolophia borneensis* Fisher
Current name: *Pterolophia (Armatopraonetha) borneensis* Fisher, 1935
Current placement: Cerambycidae: Lamiinae: Pteropliini
Type locality: Malaysia: Borneo, Sabah, Mt. Kinabalu, 7000 ft., Kamborangah

***borneensis* Fisher, 1935c**: 621 (Fig. 32c, d), Holotype
Original combination: *Glenea borneensis* Fisher
Current name: same
Current placement: Cerambycidae: Lamiinae: Saperdini
Type locality: Malaysia: Borneo, Sabah, Mt. Kinabalu, Kenokok, 3300 ft.

***boucheri* Santos-Silva & Lezama, 2010** (Fig. 32e, f), Holotype
Original combination: *Birandra boucheri* Santos-Silva & Lezama
Current name: same
Current placement: Cerambycidae: Parandrinae: Parandrini
Type locality: Panama: Darién Province, Cana, Field Station

***bougainvillea* Gressitt, 1951a**: 3 (Fig. 32g, h), Holotype
Original combination: *Distenia bougainvillea* Gressitt
Current name: *Distenia bougainvilleana* Schwarzer, 1923
Current placement: Disteniidae: Disteniinae: Disteniini
Type locality: Papua New Guinea: Bougainville Island (Solomon Islands)

***boulderensis* Casey, 1913**: 252 (Fig. 32i, j), Holotype
Original combination: *Brachyleptura boulderensis* Casey
Current name: *Anastrangalia sanguinea* (LeConte, 1859)
Current placement: Cerambycidae: Lepturinae: Lepturini
Type locality: USA: Colorado, Boulder Co.

***boyacana* Galileo & Martins, 2008b**: 33 (Fig. 32k, l), Holotype
Original combination: *Nyctonympha boyacana* Galileo & Martins
Current name: same
Current placement: Cerambycidae: Lamiinae: Forsteriini
Type locality: Colombia: Boyacá SFF, Iguaque Qda. Los Francos, 5°25′N, 73°27′W

***brasiliensis* Fisher, 1938**: 148 (Fig. 32m, n), Holotype
Original combination: *Eupromerella brasiliensis* Fisher
Current name: *Eupromerella versicolor* (Melzer, 1935b)
Current placement: Cerambycidae: Lamiinae: Acanthoderini
Type locality: Brazil: Santa Catarina, Nova Teutonia

bredini **Chemsak, 1966**: 217 (Fig. 32o, p), Holotype
Original combination: *Leptostylus bredini* Chemsak
Current name: *Styloleptus inermis* (Fabricius, 1801)
Current placement: Cerambycidae: Lamiinae: Acanthocinini
Type locality: Antigua and Barbuda: Antigua, English Harbor

breiti **Tippmann, 1946a**: 126 (Fig. 32q, r), Holotype
Original combination: *Evodinus breiti* Tippmann
Current name: same
Current placement: Cerambycidae: Lepturinae: Rhagiini
Type locality: Russia: Siberia, Tunkun Sajan

breviceps **Casey, 1924**: 250 (Fig. 32s, t), Holotype
Original combination: *Haplidus breviceps* Casey
Current name: *Haplidus testaceus* LeConte, 1873
Current placement: Cerambycidae: Cerambycinae: Hesperophanini
Type locality: USA: New Mexico, Fort Wingate

brevicollis **Casey, 1912**: 344 (Fig. 32u, v), Holotype
Original combination: *Tylosis brevicollis* Casey
Current name: *Tylosis maculatus* LeConte, 1850
Current placement: Cerambycidae: Cerambycinae: Trachyderini
Type locality: USA: New Mexico

brevicollis **Casey, 1913**: 214 (Fig. 32w, x), Lectotype
Original combination: *Stenocorus (Stenocorus) brevicollis* Casey
Current name: *Stenocorus obtusus* (LeConte, 1873)
Current placement: Cerambycidae: Lepturinae: Rhagiini
Type locality: USA: Washington
Remarks: This specimen is herein designated as the lectotype in order to stabilize the taxonomy and facilitate further identifications of this species.

brevicollis **Casey, 1913**: 300 (Fig. 33a, b), Holotype
Original combination: *Synaphaeta brevicollis* Casey
Current name: *Synaphaeta guexi* (LeConte, 1852)
Current placement: Cerambycidae: Lamiinae: Mesosini
Type locality: USA: California

brevicollis **Casey, 1913**: 362 (Fig. 33c, d), Holotype
Original combination: *Mecas brevicollis* Casey
Current name: *Mecas cana saturnina* (LeConte, 1859)
Current placement: Cerambycidae: Lamiinae: Phytoeciini
Type locality: USA: Kansas

brevicorne **Casey, 1912**: 261 (Fig. 33e, f), Holotype
Original combination: *Asemum brevicorne* Casey
Current name: *Asemum striatum* (Linnaeus, 1758)
Current placement: Cerambycidae: Spondylidinae: Asemini
Type locality: Canada: Ontario

brevicornis **Casey, 1893**: 586 (Fig. 33g, h), Holotype
Original combination: *Ancylocera brevicornis* Casey
Current name: *Ancylocera bicolor* (Olivier, 1795)
Current placement: Cerambycidae: Cerambycinae: Trachyderini
Type locality: USA: Florida

brevicornis **Casey, 1912**: 301 (Fig. 33i, j), Holotype
Original combination: *Hypermallus brevicornis* Casey
Current name: *Parelaphidion incertum* (Newman, 1840)
Current placement: Cerambycidae: Cerambycinae: Elaphidiini
Type locality: USA: Indiana

brevicornis **Casey, 1913**: 226 (Fig. 33k, l), Holotype
Original combination: *Brachysomida brevicornis* Casey
Current name: *Brachysomida californica* (LeConte, 1851)
Current placement: Cerambycidae: Lepturinae: Rhagiini
Type locality: USA: California, Siskiyou Co.

brevicornis **Chemsak, 2000**: 47 (Fig. 33m, n), Holotype
Original combination: *Phaea brevicornis* Chemsak
Current name: same
Current placement: Cerambycidae: Lamiinae: Tetraopini
Type locality: Venezuela: Guarico, Mato Masaguaral (44 km S. Calabozo)

brevicornis **Fisher, 1944b**: 1 (Fig. 33o, p), Holotype
Original combination: *Anatinomma brevicornis* Fisher
Current name: same
Current placement: Cerambycidae: Cerambycinae: Hesperophanini
Type locality: Costa Rica

brevidens **Schaeffer, 1908**: 333 (Fig. 33q, r), Holotype
Original combination: *Elaphidion brevidens* Schaeffer
Current name: *Anelaphus brevidens* (Schaeffer, 1908)
Current placement: Cerambycidae: Cerambycinae: Elaphidiini
Type locality: USA: Arizona, Huachuca Mts.

***brevipennis* v. *flavofemorata* Fisher, 1937d**: 150 (Fig. 33s, t), Holotype
Original combination: *Ommata (Eclipta) brevipennis* v. *flavofemorata* Fisher
Current name: *Eclipta brevipennis* (Melzer, 1935a)
Current placement: Cerambycidae: Cerambycinae: Rhinotragini
Type locality: Brazil: Santa Catarina, Nova Teutonia
Remarks: The name following the binomen was not expressly given infrasubspecific rank and the work did not indicate unambiguously that the name was infrasubspecific, therefore, it is deemed a subspecies and available.

***brevipennis* v. *fulvipes* Fisher, 1947a**: 50 (Fig. 33u, v), Holotype
Original combination: *Ommata (Eclipta) brevipennis* v. *fulvipes* Fisher
Current name: *Eclipta brevipennis* (Melzer, 1935a)
Current placement: Cerambycidae: Cerambycinae: Rhinotragini
Type locality: Brazil: Santa Catarina, Nova Teutonia, 27°11´B, 52°23´L
Remarks: The name following the binomen was not expressly given infrasubspecific rank and the work did not indicate unambiguously that the name was infrasubspecific, therefore, it is deemed a subspecies and available.

***brevipennis* v. *sanguinicollis* Fisher, 1947a**: 50 (Fig. 33w, x), Holotype
Original combination: *Ommata (Eclipta) brevipennis* v. *sanguinicollis* Fisher
Current name: *Ommata (Eclipta) brevipennis* Melzer, 1934
Current placement: Cerambycidae: Cerambycinae: Rhinotragini
Type locality: Brazil: Santa Catarina, Nova Teutonia, 27°11´B, 52°23´L
Remarks: The name following the binomen was not expressly given infrasubspecific rank and the work did not indicate unambiguously that the name was infrasubspecific, therefore, it is deemed a subspecies and available.

***brevis* Casey, 1913**: 342 (Fig. 34a, b), Lectotype
Original combination: *Dectes brevis* Casey
Current name: *Dectes texanus* LeConte, 1862
Current placement: Cerambycidae: Lamiinae: Acanthocinini
Type locality: USA: New York, Long Island, Willets Point
Remarks: This specimen is herein designated as the lectotype in order to stabilize the taxonomy and facilitate further identifications of this species.

***brevis* Casey, 1924**: 267 (Fig. 34c, d), Holotype
Original combination: *Amannus brevis* Casey
Current name: *Ammanus vittiger* LeConte, 1858
Current placement: Cerambycidae: Cerambycinae: Trachyderini
Type locality: USA: Utah, St. George

***brevisetosus* Casey, 1913**: 343 (Fig. 34e, f), Lectotype
Original combination: *Dectes brevisetosus* Casey
Current name: *Dectes texanus* LeConte, 1862
Current placement: Cerambycidae: Lamiinae: Acanthocinini
Type locality: USA: Arizona
Remarks: This specimen is herein designated as the lectotype in order to stabilize the taxonomy and facilitate further identifications of this species.

***brevisetosus* Casey, 1913**: 381 (Fig. 34g, h), Lectotype
Original combination: *Tetraopes brevisetosus* Casey
Current name: *Tetraopes femoratus* LeConte, 1847
Current placement: Cerambycidae: Lamiinae: Tetraopini
Type locality: USA: Kansas
Remarks: This specimen is herein designated as the lectotype in order to stabilize the taxonomy and facilitate further identifications of this species.

***brevispinis* Gressitt, 1951a**: 8 (Fig. 34i, j), Holotype
Original combination: *Hoplocerambyx brevispinis* Gressitt
Current name: *Aeolesthes inhirsutus* Matsushita, 1932
Current placement: Cerambycidae: Cerambycinae: Cerambycini
Type locality: Palau: Pelelius Island

***brevitarsis* Casey, 1912**: 375 (Fig. 34k, l), Lectotype
Original combination: *Paraclytus brevitarsis* Casey
Current name: *Triodoclytus lanifer* (LeConte, 1873)
Current placement: Cerambycidae: Cerambycinae: Clytini
Type locality: USA: California
Remarks: This specimen is herein designated as the lectotype in order to stabilize the taxonomy and facilitate further identifications of this species.

***brevithorax* Galileo & Martins, 2007a**: 350 (Fig. 34m, n), Holotype
Original combination: *Hippopsis brevithorax* Galileo & Martins
Current name: same
Current placement: Cerambycidae: Lamiinae: Agapanthiini
Type locality: Colombia: Amazon River, Isla de los Monos near Leticia

***breviusculus* Casey, 1914**: 364 (Fig. 34o, p), Holotype
Original combination: *Hypermallus breviusculus* Casey
Current name: *Anelaphus villosus* (Fabricius, 1793)
Current placement: Cerambycidae: Cerambycinae: Elaphidiini
Type locality: USA: Massachusetts, Bedford

***bruneri* Fisher, 1926**: 23 (Fig. 34q, r), Holotype
Original combination: *Leptostylus bruneri* Fisher
Current name: *Fisherostylus bruneri* (Fisher, 1926)
Current placement: Cerambycidae: Lamiinae: Acanthocinini
Type locality: Cuba: Oriente, Nagua

***bruneri* Fisher, 1932**: 60 (Fig. 34s, t), Holotype
Original combination: *Tilloclytus bruneri* Fisher
Current name: same
Current placement: Cerambycidae: Cerambycinae: Anaglyptini
Type locality: Cuba: Sierra Maestra, Pico Turquino, 1100–1300 m.

***brunnea* Aurivillius, 1922**: 432 (Fig. 34u, v), Holotype
Original combination: *Onychoglenea brunnea* Aurivillius
Current name: same
Current placement: Cerambycidae: Lamiinae: Onychogleneini
Type locality: Philippines: Mindanao

***brunnea* Fisher, 1935c**: 601 (Fig. 34w, x), Holotype
Original combination: *Loxotropoides brunnea* Fisher
Current name: *Celosterna stolzi* Ritsema, 1911
Current placement: Cerambycidae: Lamiinae: Monochamini
Type locality: Malaysia: Borneo, Sabah, Mt. Kinabalu, Kiau, 3000 ft.

***brunneofasciatus* Fisher, 1935c**: 205 (Fig. 35a, b), Holotype
Original combination: *Leptostylus brunneofasciatus* Fisher
Current name: *Antilleptostylus brunneofasciatus* (Fisher, 1935c)
Current placement: Cerambycidae: Lamiinae: Acanthocinini
Type locality: Jamaica: Mandeville

***brunneovariegata* Breuning, 1940b**: 71 (Fig. 35c, d), Holotype
Original combination: *Estola brunneovariegata* Breuning
Current name: same
Current placement: Cerambycidae: Lamiinae: Desmiphorini
Type locality: Venezuela

***brunneum* Schaeffer, 1908**: 337 (Fig. 35e, f), Syntype
Original combination: *Malobidion brunneum* Schaeffer
Current name: same
Current placement: Cerambycidae: Cerambycinae: Hesperophanini
Type locality: USA: Arizona, Huachuca Mts.

***brunneus* Dillon & Dillon, 1941**: 92 (Fig. 35g, h), Holotype
Original combination: *Plagiohammus brunneus* Dillon & Dillon
Current name: same
Current placement: Cerambycidae: Lamiinae: Monochamini
Type locality: Mexico: Chiapas, Pacific Slope Cordillera Sur, 800–1000 m.

***brunneus* Fisher, 1935a**: 58 (Fig. 35i, j), Holotype
Original combination: *Eugamandus brunneus* Fisher
Current name: *Eugamandus oakleyi* Fisher, 1935a
Current placement: Cerambycidae: Lamiinae: Acanthocinini
Type locality: USA: Puerto Rico, Augustin Finca, Yauco, Vegetative Debris

***brunneus* Gressitt, 1938a**: 49 (Fig. 35k, l), Holotype
Original combination: *Echthistatodes brunneus* Gressitt
Current name: same
Current placement: Cerambycidae: Lamiinae: Morimopsini
Type locality: China: Sichuan Province, Shin Kai Si, Mt. Omei, 4400 ft.

***bulbifera* Breuning, 1940d**: 411 (Fig. 35m, n), Holotype
Original combination: *Centruroides bulbifera* Breuning
Current name: same
Current placement: Cerambycidae: Lamiinae: Morimopsini
Type locality: Myanmar: Birmah

***buqueti aureobrunneus* Gressitt, 1934**: 166 (Fig. 35o, p), Holotype
Original combination: *Xylotrechus buqueti aureobrunneus* Gressitt
Current name: *Xylotrechus formosanus* Schwarzer, 1925
Current placement: Cerambycidae: Cerambycinae: Clytini
Type locality: China: Taiwan, Kuraru

***buscki* Fisher, 1930a**: 4 (Fig. 35q, r), Holotype
Original combination: *Odontocera buscki* Fisher
Current name: same
Current placement: Cerambycidae: Cerambycinae: Rhinotragini
Type locality: Panama: Trinidad River

***buscki* Linsley, 1935b**: 482 (Fig. 35s, t), Holotype
Original combination: *Heterachthes buscki* Linsley
Current name: *Pygmodeon buscki* (Linsley, 1935b)
Current placement: Cerambycidae: Cerambycinae: Neoibidionini
Type locality: Panama: Bocas del Toro, Trinidad River

caerulea **Casey, 1913**: 227 (Fig. 35u, v), Lectotype
Original combination: *Brachysomida caerulea* Casey
Current name: *Brachysomida californica* (LeConte, 1851)
Current placement: Cerambycidae: Lepturinae: Rhagiini
Type locality: USA: California, Santa Cruz Mts.
Remarks: This specimen is herein designated as the lectotype in order to stabilize the taxonomy and facilitate further identifications of this species.

caerulea chalybea **Casey, 1913**: 227 (Fig. 35w, x), Holotype
Original combination: *Brachysomida caerulea chalybea* Casey
Current name: *Brachysomida californica* (LeConte, 1851)
Current placement: Cerambycidae: Lepturinae: Rhagiini
Type locality: USA: California, Santa Cruz Mts.

calca **Galileo & Martins, 2005a**: 278 (Fig. 36a, b), Holotype
Original combination: *Adesmus calca* Galileo & Martins
Current name: same
Current placement: Cerambycidae: Lamiinae: Hemilophini
Type locality: Peru: Cusco, 80 km north of Calca

californica **Casey, 1891**: 51 (Fig. 36c, d), Lectotype
Original combination: *Hyperplatys californica* Casey
Current name: same
Current placement: Cerambycidae: Lamiinae: Acanthocinini
Type locality: USA: California, Santa Cruz Mts.
Remarks: This specimen is herein designated as the lectotype in order to stabilize the taxonomy and facilitate further identifications of this species.

californica proxima **Casey, 1913**: 225 (Fig. 36e, f), Lectotype
Original combination: *Brachysomida californica proxima* Casey
Current name: *Brachysomida californica* (LeConte, 1851)
Current placement: Cerambycidae: Lepturinae: Rhagiini
Type locality: USA: California
Remarks: This specimen is herein designated as the lectotype in order to stabilize the taxonomy and facilitate further identifications of this species.

californicum **Casey, 1912**: 289 (Fig. 36g, h), Holotype
Original combination: *Callidium californicum* Casey
Current name: same
Current placement: Cerambycidae: Cerambycinae: Callidiini
Type locality: USA: California

californicum **Casey, 1913**: 195 (Fig. 36i, j), Lectotype
Original combination: *Rhagium (Hargium) californicum* Casey
Current name: *Rhagium inquisitor* (Linnaeus, 1758)
Current placement: Cerambycidae: Lepturinae: Rhagiini
Type locality: USA: California, Sacramento Co.
Remarks: This specimen is herein designated as the lectotype in order to stabilize the taxonomy and facilitate further identifications of this species.

californicus compar **Casey, 1924**: 217 (Fig. 36k, l), Lectotype
Original combination: *Prionus (Prionus) californicus compar* Casey
Current name: *Prionus (Prionus) californicus* Motschulsky, 1845
Current placement: Cerambycidae: Prioninae: Prionini
Type locality: USA: Utah, North Fork, Provo Canyon
Remarks: This specimen is herein designated as the lectotype in order to stabilize the taxonomy and facilitate further identifications of this species.

californicus ineptus **Casey, 1912**: 242 (Fig. 36m, n), Holotype
Original combination: *Prionus (Prionus) californicus ineptus* Casey
Current name: *Prionus californicus* Motschulsky, 1845
Current placement: Cerambycidae: Prioninae: Prionini
Type locality: USA: California

californicus ovipennis **Casey, 1924**: 217 (Fig. 36o, p), Holotype
Original combination: *Prionus (Prionus) californicus ovipennis* Casey
Current name: *Prionus californicus* Motschulsky, 1845
Current placement: Cerambycidae: Prioninae: Prionini
Type locality: USA: Utah, North Fork, Provo Canyon

californicus punctulatus **Casey, 1912**: 243 (Fig. 36q, r), Holotype
Original combination: *Prionus (Prionus) californicus punctulatus* Casey
Current name: *Prionus californicus* Motschulsky, 1845
Current placement: Cerambycidae: Prioninae: Prionini
Type locality: USA: Arizona

caligans **Casey, 1913**: 273 (Fig. 36s, t), Lectotype
Original combination: *Typocerus caligans* Casey
Current name: *Typocerus confluens* Casey, 1913
Current placement: Cerambycidae: Lepturinae: Lepturini
Type locality: USA: Kansas
Remarks: This specimen is herein designated as the lectotype in order to stabilize the taxonomy and facilitate further identifications of this species.

***caliginosum* Martins, 1970a**: 896 (Fig. 36u, v), Holotype
Original combination: *Corimbion caliginosum* Martins
Current name: same
Current placement: Cerambycidae: Cerambycinae: Neoibidionini
Type locality: Venezuela: Maracaibo, Zulia

***caliginosus* Lingafelter & Micheli, 2009**: 33 (Fig. 36w, x), Holotype
Original combination: *Leptostylopsis caliginosus* Lingafelter & Micheli
Current name: same
Current placement: Cerambycidae: Lamiinae: Acanthocinini
Type locality: Dominican Republic: La Vega Province, Pico Duarte Trail, 3300 ft., Los Tablones, 19°08.222′N, 70°27.736′W.

***calligramma* var. *consobrina* Melzer, 1934**: 109 (Fig. 37a, b), Holotype
Original combination: *Nyssodrys calligramma* var. *consobrina* Melzer
Current name: *Stenolis calligramma* (Bates, 1872)
Current placement: Cerambycidae: Lamiinae: Acanthocinini
Type locality: Costa Rica: Limón, Hamburg Farm, Reventazón, Ebene
Remarks: The name following the binomen was not expressly given infrasubspecific rank and the work did not indicate unambiguously that the name was infrasubspecific, therefore, it is deemed a subspecies and available.

***campanifera* Casey, 1913**: 256 (Fig. 37c, d), Holotype
Original combination: *Parallelina campanifera* Casey
Current name: *Grammoptera rhodopus* (LeConte, 1874)
Current placement: Cerambycidae: Lepturinae: Lepturini
Type locality: USA: California, Calaveras Co., Mokelumne Hill

***campanulifera* Aurivillius, 1922**: 412 (Fig. 37e, f), Holotype
Original combination: *Rhaphuma campanulifera* Aurivillius
Current name: same
Current placement: Cerambycidae: Cerambycinae: Clytini
Type locality: Philippines: Luzon, Mt. Banahao

***campbelli* Gressitt, 1937b**: 613 (Fig. 37g, h), Holotype
Original combination: *Acanthocinus campbelli* Gressitt
Current name: Leiopus (Carinopus) campbelli (Gressitt, 1937b)
Current placement: Cerambycidae: Lamiinae: Acanthocinini
Type locality: China: Jiangxi Province, Hong Shan

***canadensis* Casey, 1924**: 238 (Fig. 37i, j), Lectotype
Original combination: *Anocomis canadensis* Casey
Current name: *Semanotus litigiosus* (Casey, 1891)
Current placement: Cerambycidae: Cerambycinae: Callidiini
Type locality: Canada: Ontario, Ottawa
Remarks: This specimen is herein designated as the lectotype in order to stabilize the taxonomy and facilitate further identifications of this species.

***canadensis* Fisher, 1945**: 56 (Fig. 37k, l), Holotype
Original combination: *Oberea canadensis* Fisher
Current name: *Oberea caseyi* Plavilstshikov, 1926
Current placement: Cerambycidae: Lamiinae: Phytoeciini
Type locality: Canada: Ontario, Ojibway

***canescens* Dillon, 1955**: 184 (Fig. 37m, n), Holotype
Original combination: *Canidiopsis canescens* Dillon
Current name: *Canidia canescens* (Dillon, 1955)
Current placement: Cerambycidae: Lamiinae: Acanthocinini
Type locality: Mexico: Nayarit, 3 mi. south Tepic

***canescens fontinalis* Casey, 1913**: 378 (Fig. 37o, p), Lectotype
Original combination: *Tetraopes canescens fontinalis* Casey
Current name: *Tetraopes annulatus* LeConte, 1847
Current placement: Cerambycidae: Lamiinae: Tetraopini
Type locality: USA: New Mexico, Near hot springs, Las Vegas, 7000 ft.
Remarks: This specimen is herein designated as the lectotype in order to stabilize the taxonomy and facilitate further identifications of this species.

***canidia* Dillon & Dillon, 1946**: 365 (Fig. 37q, r), Holotype
Original combination: *Oncideres canidia* Dillon & Dillon
Current name: same
Current placement: Cerambycidae: Lamiinae: Onciderini
Type locality: Jamaica: Kingston, Stony Hill

***canotiae* Fisher, 1920**: 154 (Fig. 37s, t), Holotype
Original combination: *Anoplocurius canotiae* Fisher
Current name: same
Current placement: Cerambycidae: Cerambycinae: Elaphidiini
Type locality: USA: Arizona, Cotton City

***canus* Casey, 1913**: 317 (Fig. 37u, v), Holotype
Original combination: *Lepturges canus* Casey
Current name: *Lepturges angulatus* (LeConte, 1852)
Current placement: Cerambycidae: Lamiinae: Acanthocinini
Type locality: USA: Texas

capilosum **Martins & Galileo, 2008c**: 278 (Fig. 37w, x), Holotype
Original combination: *Acrepidopterum capilosum* Martins & Galileo
Current name: same
Current placement: Cerambycidae: Lamiinae: Apomecynini
Type locality: Honduras: Roatan Island (west)

captiocula **Giesbert, 1996**: 317 (Fig. 38a, b), Holotype
Original combination: *Amplitempora captiocula* Giesbert
Current name: same
Current placement: Cerambycidae: Lamiinae: Laticraniini
Type locality: Panama: Panama Province, 8–10 km north of El Llano

carenatum **Martins, 1969**: 767 (Fig. 38c, d), Holotype
Original combination: *Compsibidion carenatum* Martins
Current name: same
Current placement: Cerambycidae: Cerambycinae: Neoibidionini
Type locality: Brazil: Espirito Santo

carinatus **Aurivillius, 1913**: 236 (Fig. 38e, f), Holotype
Original combination: *Chlorophorus carinatus* Aurivillius
Current name: same
Current placement: Cerambycidae: Cerambycinae: Clytini
Type locality: Tanzania: Ufiome

carinatus **Aurivillius, 1922**: 419 (Fig. 38g, h), Holotype
Original combination: *Demonax? carinatus* Aurivillius
Current name: *Demonax polyzonus* Pascoe, 1869
Current placement: Cerambycidae: Cerambycinae: Clytini
Type locality: Malaysia: Borneo, Sandakan
Remarks: The genus was indicated as "*Demonax*?" in the original publication.

carinatus **Casey, 1913**: 346 (Fig. 38i, j), Lectotype
Original combination: *Pogonocherus carinatus* Casey
Current name: *Pogonocherus penicillatus* LeConte in Agassiz, 1850
Current placement: Cerambycidae: Lamiinae: Pogonocherini
Type locality: USA: Colorado
Remarks: This specimen is herein designated as the lectotype in order to stabilize the taxonomy and facilitate further identifications of this species.

carinifrons **Aurivillius, 1922**: 433 (Fig. 38k, l), Holotype
Original combination: *Menesida carinifrons* Aurivillius
Current name: same
Current placement: Cerambycidae: Lamiinae: Saperdini
Type locality: Malaysia: Borneo, Sandakan

cariniventre **Casey, 1913**: 196 (Fig. 38m, n), Lectotype
Original combination: *Rhagium (Hargium) cariniventre* Casey
Current name: *Rhagium inquisitor* (Linnaeus, 1758)
Current placement: Cerambycidae: Lepturinae: Rhagiini
Type locality: USA: Pennsylvania, State College
Remarks: This specimen is herein designated as the lectotype in order to stabilize the taxonomy and facilitate further identifications of this species.

carolinae **Casey, 1913**: 277 (Fig. 38o, p), Holotype
Original combination: *Ophistomis carolinae* Casey
Current name: *Strangalia luteicornis* (Fabricius, 1775)
Current placement: Cerambycidae: Lepturinae: Lepturini
Type locality: USA: South Carolina

carolinensis **Casey, 1912**: 372 (Fig. 38q, r), Holotype
Original combination: *Clytus carolinensis* Casey
Current name: *Clytus marginicollis* Castelnau & Gory, 1836
Current placement: Cerambycidae: Cerambycinae: Clytini
Type locality: USA: North Carolina, Southern Pines

carolinensis **Casey, 1924**: 275 (Fig. 38s, t), Lectotype
Original combination: *Evodinus carolinensis* Casey
Current name: *Evodinus monticola* (Randall, 1838)
Current placement: Cerambycidae: Lepturinae: Rhagiini
Type locality: USA: North Carolina, summit of Black Mts.
Remarks: This specimen is herein designated as the lectotype in order to stabilize the taxonomy and facilitate further identifications of this species.

carolinum **Casey, 1924**: 227 (Fig. 38u, v), Lectotype
Original combination: *Asemum carolinum* Casey
Current name: *Asemum striatum* (Linnaeus, 1758)
Current placement: Cerambycidae: Spondylidinae: Asemini
Type locality: USA: North Carolina, Southern Pines
Remarks: This specimen is herein designated as the lectotype in order to stabilize the taxonomy and facilitate further identifications of this species.

casilelium **Gressitt, 1956**: 79 (Fig. 38w, x), Holotype
Original combination: *Ceresium (Ceresium) casilelium* Gressitt
Current name: same
Current placement: Cerambycidae: Cerambycinae: Callidiopini
Type locality: Micronesia: Pohnpei (formerly Ponape Island), Colonia Agricultural Experiment Station.

***castanea* Santos-Silva & Hovore, 2008a**: 10 (Fig. 39a, b), Holotype
Original combination: *Elytrimitatrix (Grossifemora) castanea* Santos-Silva & Hovore
Current name: same
Current placement: Disteniidae: Disteniinae: Disteniini
Type locality: Panama: Canal Zone, Barro Colorado Island, 9°9′N 79°51′W

***castaneum* Casey, 1924**: 225 (Fig. 39c, d), Lectotype
Original combination: *Mallodon castaneum* Casey
Current name: *Nothopleurus castaneus* (Casey, 1924)
Current placement: Cerambycidae: Prioninae: Macrotomini
Type locality: Mexico: Baja California Norte, San Felipe
Remarks: This specimen is herein designated as the lectotype in order to stabilize the taxonomy and facilitate further identifications of this species.

***castaneus* Casey, 1924**: 253 (Fig. 39e, f), Holotype
Original combination: *Brothylus castaneus* Casey
Current name: *Brothylus gemmulatus* LeConte, 1859
Current placement: Cerambycidae: Cerambycinae: Hesperophanini
Type locality: USA: Utah, Provo

***castaneus* Casey, 1924**: 269 (Fig. 39g, h), Holotype
Original combination: *Stenosphenus castaneus* Casey
Current name: *Stenosphenus dolosus* Horn, 1885
Current placement: Cerambycidae: Cerambycinae: Elaphidiini
Type locality: USA: Texas

***catalinae* Casey, 1913**: 281 (Fig. 39i, j), Holotype
Original combination: *Ipochus catalinae* Casey
Current name: *Ipochus fasciatus* LeConte, 1852
Current placement: Cerambycidae: Lamiinae: Parmenini
Type locality: USA: California, Island of Santa Catalina

***catemaco* Martins & Galileo, 2005a**: 26 (Fig. 39k, l), Holotype
Original combination: *Adetus catemaco* Martins & Galileo
Current name: same
Current placement: Cerambycidae: Cerambycinae: Apomecynini
Type locality: Mexico: Veracruz, Coyame (Lago Catemaco)

***catulus* Casey, 1912**: 370 (Fig. 39m, n), Holotype
Original combination: *Clytanthus catulus* Casey
Current name: *Clytus ruricola* (Olivier, 1795)
Current placement: Cerambycidae: Cerambycinae: Clytini
Type locality: USA: Wisconsin, Bayfield

***caudalis* Schwarzer, 1929**: 359 (Fig. 39o, p), Holotype
Original combination: *Amphionthe caudalis* Schwarzer
Current name: same
Current placement: Cerambycidae: Cerambycinae: Trachyderini
Type locality: Costa Rica: Ramal Parismina, Santa Clara

***caudatus* Fisher, 1935c**: 595 (Fig. 39q, r), Holotype
Original combination: *Anxylotoles caudatus* Fisher
Current name: same
Current placement: Cerambycidae: Lamiinae: Apomecynini
Type locality: Malaysia: Borneo, Sabah, Mt. Kinabalu, Pakka, 10,000 ft.

***cavicollis* Casey, 1912**: 266 (Fig. 39s, t), Holotype
Original combination: *Criocephalus cavicollis* Casey
Current name: *Arhopalus rusticus montanus* (LeConte, 1873)
Current placement: Cerambycidae: Spondylidinae: Asemini
Type locality: USA: Colorado

***cavicollis* Casey, 1913**: 244 (Fig. 39u, v), Lectotype
Original combination: *Acmaeops cavicollis* Casey
Current name: *Acmaeops proteus* (Kirby, 1837)
Current placement: Cerambycidae: Lepturinae: Lepturini
Type locality: USA: Maine, Paris
Remarks: This specimen is herein designated as the lectotype in order to stabilize the taxonomy and facilitate further identifications of this species.

***cayamae* Fisher, 1926**: 35 (Fig. 39w, x), Holotype
Original combination: *Eugamandus cayamae* Fisher
Current name: same
Current placement: Cerambycidae: Lamiinae: Acanthocinini
Type locality: Cuba: Cayamas

***cayamae* Fisher, 1932**: 24 (Fig. 40a, b), Holotype
Original combination: *Elaphidion cayamae* Fisher
Current name: same
Current placement: Cerambycidae: Cerambycinae: Elaphidiini
Type locality: Cuba: Cayamas

***cayensis* Chemsak & Feller, 1988**: 182 (Fig. 40c, d), Holotype
Original combination: *Ataxia cayensis* Chemsak & Feller
Current name: same
Current placement: Cerambycidae: Lamiinae: Pteropliini
Type locality: Belize: Dangriga District, Twin Cays

ceara **Dillon & Dillon, 1946**: 280 (Fig. 40e, f), Holotype
Original combination: *Trestonia ceara* Dillon & Dillon
Current name: same
Current placement: Cerambycidae: Lamiinae: Onciderini
Type locality: Brazil: Ceara

celebiana **Tippmann, 1953a**: 103 (Fig. 40g, h), Holotype
Original combination: *Doesburgia celebiana* Tippmann
Current name: same
Current placement: Cerambycidae: Lamiinae: Batocerini
Type locality: Indonesia: Sulawesi (formerly Celebes), Todjamboe

celestina **Casey, 1913**: 222 (Fig. 40i, j), Lectotype
Original combination: *Brachysomida celestina* Casey
Current name: *Brachysomida californica* (LeConte, 1851)
Current placement: Cerambycidae: Lepturinae: Rhagiini
Type locality: USA: California
Remarks: This specimen is herein designated as the lectotype in order to stabilize the taxonomy and facilitate further identifications of this species.

celtis **Schaeffer, 1905b**: 168 (Fig. 40k, l), Holotype
Original combination: *Lepturges celtis* Schaeffer
Current name: *Urgleptes celtis* (Schaeffer, 1905b)
Current placement: Cerambycidae: Lamiinae: Acanthocinini
Type locality: USA: Texas, Brownsville, Esperanza Ranch

centrale **Martins, 1962c**: 158 (Fig. 40m, n), Holotype
Original combination: *Ibidion centrale* Martins
Current name: *Tropidion centrale* (Martins, 1962c)
Current placement: Cerambycidae: Cerambycinae: Neoibidionini
Type locality: Panama: La Chorrera

cereicola **Fisher, 1936b**: 7 (Fig. 40o, p), Holotype
Original combination: *Alcidion cereicola* Fisher
Current name: *Nealcidion cereicola* (Fisher, 1936b)
Current placement: Cerambycidae: Lamiinae: Acanthocinini
Type locality: Argentina: Salta, Guemes

cervicalis **Casey, 1912**: 256 (Fig. 40q, r), Lectotype
Original combination: *Pyrodes cervicalis* Casey
Current name: *Scatopyrodes tenuicornis marginatus* (White, 1853)
Current placement: Cerambycidae: Prioninae: Mallaspini
Type locality: Honduras: Unknown
Remarks: This specimen is herein designated as the lectotype in order to stabilize the taxonomy and facilitate further identifications of this species.

cervicornis melzeri **Tippmann, 1953c**: 188 (Fig. 40s, t), Holotype
Original combination: *Macrodontia cervicornis melzeri* Tippmann
Current name: *Macrodontia cervicornis* (Linnaeus, 1758)
Current placement: Cerambycidae: Prioninae: Macrodontiini
Type locality: Peru: Iquitos, Rio Marañon, ca. 150 m.

cervicornis zischkai **Tippmann, 1960**: 92 (Fig. 40u, v), Holotype
Original combination: *Macrodontia cervicornis zischkai* Tippmann
Current name: *Macrodontia zischkai* Tippmann, 1960
Current placement: Cerambycidae: Prioninae: Macrodontiini
Type locality: Bolivia: Cochabamba, Reg. Chapare 400 m.

cervinus **Casey, 1912**: 267 (Fig. 40w, x), Holotype
Original combination: *Criocephalus cervinus* Casey
Current name: *Arhopalus foveicollis* (Haldeman, 1847)
Current placement: Cerambycidae: Spondylidinae: Asemini
Type locality: USA: Indiana

chamaeropsis **Fisher, 1926**: 27 (Fig. 41a, b), Holotype
Original combination: *Lepturges chamaeropsis* Fisher
Current name: *Urgleptes chamaeropsis* (Fisher, 1926)
Current placement: Cerambycidae: Lamiinae: Acanthocinini
Type locality: Cuba: Cayamas

chamorro **Gressitt, 1956**: 142 (Fig. 41c, d), Holotype
Original combination: *Sybra chamorro* Gressitt
Current name: same
Current placement: Cerambycidae: Lamiinae: Apomecynini
Type locality: Northern Mariana Islands: Saipan Island, Mt. Tapotohau

champlaini **Casey, 1913**: 254 (Fig. 41e, f), Holotype
Original combination: *Brachyleptura champlaini* Casey
Current name: same
Current placement: Cerambycidae: Lepturinae: Lepturini
Type locality: USA: Pennsylvania, Carlisle

champlainianus **Casey, 1924**: 231 (Fig. 41g, h), Holotype
Original combination: *Criocephalus champlainianus* Casey
Current name: *Asemum australe* LeConte, 1850
Current placement: Cerambycidae: Spondylidinae: Asemini
Type locality: USA: New York, Lake George

chanleri **Linell, 1896**: 694 (Fig. 41i, j), Holotype
Original combination: *Alphitopola chanleri* Linell
Current name: *Prosopocera (Alphitopola) chanleri* (Linell, 1896)
Current placement: Cerambycidae: Lamiinae: Prosopocerini
Type locality: Kenya: Tana River

chaparensis **Tippmann, 1953c**: 216 (Fig. 41k, l), Holotype
Original combination: *Distenia chaparensis* Tippmann
Current name: *Distenia (Basisvallis) chaparensis* Tippmann, 1953c
Current placement: Disteniidae: Disteniinae: Disteniini
Type locality: Bolivia: Cochabamba, Reg. Chapare 400 m.

chaparensis **Tippmann, 1960**: 199 (Fig. 41m, n), Holotype
Original combination: *Ites chaparensis* Tippmann
Current name: same
Current placement: Cerambycidae: Lamiinae: Hemilophini
Type locality: Bolivia: Cochabamba, Reg. Chapare 400 m.

chara genitiva **Casey, 1912**: 355 (Fig. 41o, p), Holotype
Original combination: *Cyllene chara genitiva* Casey
Current name: *Megacyllene decora* (Olivier, 1795)
Current placement: Cerambycidae: Cerambycinae: Clytini
Type locality: USA: Texas

chara kansana **Casey, 1912**: 355 (Fig. 41q, r), Lectotype
Original combination: *Cyllene chara kansana* Casey
Current name: *Megacyllene decora* (Olivier, 1795)
Current placement: Cerambycidae: Cerambycinae: Clytini
Type locality: USA: Kansas
Remarks: This specimen is herein designated as the lectotype in order to stabilize the taxonomy and facilitate further identifications of this species.

chara solida **Casey, 1912**: 355 (Fig. 41s, t), Lectotype
Original combination: *Cyllene chara solida* Casey
Current name: *Megacyllene decora* (Olivier, 1795)
Current placement: Cerambycidae: Cerambycinae: Clytini
Type locality: USA: Iowa
Remarks: This specimen is herein designated as the lectotype in order to stabilize the taxonomy and facilitate further identifications of this species.

charynae **Lingafelter, 2008**: 361 (Fig. 41u, v), Holotype
Original combination: *Nesiosphaerion charynae* Lingafelter
Current name: same
Current placement: Cerambycidae: Cerambycinae: Elaphidiini
Type locality: Dominican Republic: Pedernales Province, 150 m. north of bridge on road Cabo Rojo-Aceitillar, 16m, 17°58.530′N 71°39.034′W

charynae **Micheli, 2003**: 192 (Fig. 41w, x), Holotype
Original combination: *Merostenus charynae* Micheli
Current name: *Caribbomerus charynae* (Micheli, 2003)
Current placement: Cerambycidae: Cerambycinae: Graciliini
Type locality: USA: Puerto Rico, Maricao Forest, Road 120, km 13.8

chemsaki **Nearns & Ray, 2006**: 51 (Fig. 42a, b), Holotype
Original combination: *Curius chemsaki* Nearns & Ray
Current name: same
Current placement: Cerambycidae: Cerambycinae: Curiini
Type locality: Venezuela: Aragua, Rancho Grande, collected at blacklight

chemsaki **Tyson, 1973**: 130 (Fig. 42c, d), Holotype
Original combination: *Spalacopsis chemsaki* Tyson
Current name: *Spalacopsis chemsaki* Tyson, 1973
Current placement: Cerambycidae: Lamiinae: Agapanthiini
Type locality: USA: Florida, Lake Ashby, Volusia County

chemsaki **Wappes & Lingafelter, 2005**: 4 (Fig. 42e, f), Holotype
Original combination: *Canidia chemsaki* Wappes & Lingafelter
Current name: same
Current placement: Cerambycidae: Lamiinae: Acanthocinini
Type locality: Mexico: Chiapas, San Jeronimo, Volcán Tacana
Remarks: on long term loan from UNAM (Mexico)

chicacaoensis **Galileo & Martins, 2011**:184 (Fig. 42g, h), Holotype
Original combination: *Esthlogena chicacaoensis* Galileo & Martins
Current name: same
Current placement: Cerambycidae: Lamiinae: Pteropliini
Type locality: Guatemala: 10 km northeast of Chicacao, Suchitepéquez, 5000 ft.

chinensis **Breuning, 1940b**: 48 (Fig. 42i, j), Holotype
Original combination: *Sophronica chinensis* Breuning
Current name: same
Current placement: Cerambycidae: Lamiinae: Desmiphorini
Type locality: China: Guizhou Province

chinensis **Breuning, 1942c**: 132 (Fig. 42k, l), Holotype
Original combination: *Parechthistatus chinensis* Breuning
Current name: same
Current placement: Cerambycidae: Lamiinae: Lamiini
Type locality: China: Shaanxi Province, Taipai Shan

chlorescens **Lingafelter & Micheli, 2009**: 37 (Fig. 42m, n), Holotype
Original combination: *Leptostylopsis chlorescens* Lingafelter & Micheli
Current name: same
Current placement: Cerambycidae: Lamiinae: Acanthocinini
Type locality: Dominican Republic: Pedernales Province, Parque Nacional Sierra de Baoruco, Las Abejas, 1150 m, 18°09.011′N, 71°37.342′W

chocolata **Gressitt, 1939a**: 88 (Fig. 42o, p), Holotype
Original combination: *Pothyne chocolata* Gressitt
Current name: same
Current placement: Cerambycidae: Lamiinae: Agapanthiini
Type locality: China: Guangdong Province, Tsin Leong Shan

ciliata **Fisher, 1938**: 152 (Fig. 42q, r), Holotype
Original combination: *Erana ciliata* Fisher
Current name: *Eranina ciliata* (Fisher, 1938)
Current placement: Cerambycidae: Lamiinae: Hemilophini
Type locality: Brazil: Santa Catarina, Nova Teutonia

ciliatissimus **Gressitt, 1956**: 156 (Fig. 42s, t), Holotype
Original combination: *Exocentrus ciliatissimus* Gressitt
Current name: *Exocentrus (Camptomyne) ciliatissimus* Gressitt, 1956
Current placement: Cerambycidae: Lamiinae: Acanthocinini
Type locality: Palau: Koror

ciliatus **Breuning, 1942a**: 141 (Fig. 42u, v), Holotype
Original combination: *Stenocoptoides ciliatus* Breuning
Current name: same
Current placement: Cerambycidae: Lamiinae: Apomecynini
Type locality: Kenya: Sagala

cincta **Martins & Galileo, 2007c**: 1149 (Fig. 42w, x), Holotype
Original combination: *Unaporanga cincta* Martins & Galileo
Current name: same
Current placement: Cerambycidae: Lamiinae: Hemilophini
Type locality: Panama: Darién Province, Cana ANCON Station, 500 m., 07°45.323′N 77°41.069′W, General day collecting

cinctus **var.** ***sonorensis*** **Schaeffer, 1908**: 332 (Fig. 43a, b), Lectotype
Original combination: *Chion cinctus* var. *sonorensis* Schaeffer
Current name: *Knulliana cincta sonorensis* (Schaeffer, 1908)
Current placement: Cerambycidae: Cerambycinae: Bothriospilini
Type locality: USA: Arizona, Huachuca Mts.
Remarks: The name following the binomen was not expressly given infrasubspecific rank and the work did not indicate unambiguously that the name was infrasubspecific, therefore, it is deemed a subspecies and available. This specimen is herein designated as the lectotype in order to stabilize the taxonomy and facilitate further identifications of this species.

cineracea **Casey, 1913**: 361 (Fig. 43c, d), Lectotype
Original combination: *Mecas cineracea* Casey
Current name: same
Current placement: Cerambycidae: Lamiinae: Phytoeciini
Type locality: USA: Texas, Harris Co.
Remarks: This specimen is herein designated as the lectotype in order to stabilize the taxonomy and facilitate further identifications of this species.

cingulatus pallescens **Casey, 1913**: 353 (Fig. 43e, f), Holotype
Original combination: *Oncideres cingulatus pallescens* Casey
Current name: *Oncideres c. cingulata* (Say, 1827)
Current placement: Cerambycidae: Lamiinae: Onciderini
Type locality: USA: New York

clara **Breuning, 1940d**: 414 (Fig. 43g, h), Holotype
Original combination: *Pharsalia clara* Breuning
Current name: *Pharsalia (Eopharsalia) clara* Breuning, 1940d
Current placement: Cerambycidae: Lamiinae: Lamiini
Type locality: Indonesia: Sumatra, Medan

clarkei **Chemsak, 1966**: 218 (Fig. 43i, j), Holotype
Original combination: *Urgleptes clarkei* Chemsak
Current name: same
Current placement: Cerambycidae: Lamiinae: Acanthocinini
Type locality: British Virgin Islands: Peter Island, Little Bay

clarkei **Gressitt, 1956**: 84 (Fig. 43k, l), Holotype
Original combination: *Ceresium clarkei* Gressitt
Current name: same
Current placement: Cerambycidae: Cerambycinae: Callidiopini
Type locality: Micronesia: Kusaie Hill 1010, 300 m.

clavata **Santos-Silva & Hovore, 2008a**: 5 (Fig. 43m, n), Holotype
Original combination: *Elytrimitatrix (Grossifemora) clavata* Santos-Silva & Hovore
Current name: same
Current placement: Disteniidae: Disteniinae: Disteniini
Type locality: Mexico: Veracruz, Los Tuxtlas, Biological Station

clavula **Martins & Galileo, 2005b**: 766 (Fig. 43o, p), Holotype
Original combination: *Ectenesseca clavula* Martins & Galileo
Current name: same
Current placement: Cerambycidae: Cerambycinae: Ectenessini
Type locality: Brazil: Rondônia, Ariquemes (62 km SW, Fazenda Rancho Grande)

coccineus **Casey, 1893**: 588 (Fig. 43q, r), Lectotype
Original combination: *Oxoplus coccineus* Casey
Current name: *Crossidius coralinus coccineus* (Casey, 1893)
Current placement: Cerambycidae: Cerambycinae: Trachyderini
Type locality: USA: Utah
Remarks: This specimen is herein designated as the lectotype in order to stabilize the taxonomy and facilitate further identifications of this species.

coccineus **Casey, 1913**: 382 (Fig. 43s, t), Lectotype
Original combination: *Tetraopes coccineus* Casey
Current name: *Tetraopes basalis* LeConte, 1852
Current placement: Cerambycidae: Lamiinae: Tetraopini
Type locality: USA: California
Remarks: This specimen is herein designated as the lectotype in order to stabilize the taxonomy and facilitate further identifications of this species.

cochabambae **Tippmann, 1960**: 146 (Fig. 43u, v), Holotype
Original combination: *Trachyderes cochabambae* Tippmann
Current name: *Trachyderes distinctus* Bosq, 1951
Current placement: Cerambycidae: Cerambycinae: Trachyderini
Type locality: Bolivia: Cochabamba, 2600 m.

cochisensis **Casey, 1912**: 296 (Fig. 43w, x), Holotype
Original combination: *Aneflus cochisensis* Casey
Current name: *Aneflus protensus* (LeConte, 1858)
Current placement: Cerambycidae: Cerambycinae: Elaphidiini
Type locality: USA: Arizona, Douglas, San Bernardino Ranch, 3750 ft.

coeruleoantennatus **Breuning, 1946**: 23 (Fig. 44a, b), Holotype
Original combination: *Melanauster coeruleoantennatus* Breuning
Current name: *Anoplophora coeruleoantennata* (Breuning, 1946)
Current placement: Cerambycidae: Lamiinae: Monochamini
Type locality: China: Sichuan Province, Tatsienlu Kiulung

collaris **Casey, 1912**: 218 (Fig. 44c, d), Holotype
Original combination: *Spondylis collaris* Casey
Current name: *Neospondylis upiformis* (Mannerheim, 1843)
Current placement: Cerambycidae: Spondylidinae: Spondylidini
Type locality: USA: Utah

collaris **Casey, 1924**: 240 (Fig. 44e, f), Holotype
Original combination: *Phymatodes (Phymatodina) collaris* Casey
Current name: *Phymatodes nitidus* LeConte, 1874
Current placement: Cerambycidae: Cerambycinae: Callidiini
Type locality: USA: California, Humboldt Co.

colombiana **Galileo & Martins, 2009**: 3 (Fig. 44g, h), Holotype
Original combination: *Tyrinthia colombiana* Galileo & Martins
Current name: same
Current placement: Cerambycidae: Lamiinae: Hemilophini
Type locality: Colombia: Río Putumayo

coloradensis **Casey, 1912**: 358 (Fig. 44i, j), Lectotype
Original combination: *Calloides coloradensis* Casey
Current name: *Calloides nobilis mormonus* Schaeffer, 1911
Current placement: Cerambycidae: Cerambycinae: Clytini
Type locality: USA: Colorado
Remarks: This specimen is herein designated as the lectotype in order to stabilize the taxonomy and facilitate further identifications of this species.

coloradensis **Casey, 1913**: 244 (Fig. 44k, l), Lectotype
Original combination: *Acmaeops coloradensis* Casey
Current name: *Acmaeops proteus* (Kirby, 1837)
Current placement: Cerambycidae: Lepturinae: Lepturini
Type locality: USA: Colorado, Fraser, 8000 ft.
Remarks: This specimen is herein designated as the lectotype in order to stabilize the taxonomy and facilitate further identifications of this species.

***coloradensis* Casey, 1924**: 231 (Fig. 44m, n), Holotype
Original combination: *Criocephalus coloradensis* Casey
Current name: *Arhopalus rusticus montanus* (LeConte, 1873)
Current placement: Cerambycidae: Spondylidinae: Asemini
Type locality: USA: Colorado, Florissant

***colossus* Casey, 1924**: 289 (Fig. 44o, p), Holotype
Original combination: *Moneilema (Collapteryx) colossus* Casey
Current name: *Moneilema gigas* LeConte, 1873
Current placement: Cerambycidae: Lamiinae: Moneilemini
Type locality: USA: Arizona, near Tuscson

***columbianum* Casey, 1912**: 268 (Fig. 44q, r), Holotype
Original combination: *Tetropium columbianum* Casey
Current name: *Tetropium velutinum* LeConte, 1869
Current placement: Cerambycidae: Spondylidinae: Asemini
Type locality: Canada: British Columbia, Inverness

***columbianus* Breuning, 1942a**: 161 (Fig. 44s, t), Holotype
Original combination: *Eupogonius columbianus* Breuning
Current name: same
Current placement: Cerambycidae: Lamiinae: Desmiphorini
Type locality: Colombia

***columbianus* Casey, 1912**: 361 (Fig. 44u, v), Holotype
Original combination: *Xylotrechus columbianus* Casey
Current name: *Xylotrechus longitarsis* Casey, 1912
Current placement: Cerambycidae: Cerambycinae: Clytini
Type locality: Canada: British Columbia, Aldermere

***columbica* Casey, 1913**: 261 (Fig. 44w, x), Holotype
Original combination: *Xestoleptura columbica* Casey
Current name: *Xestoleptura tibialis* (LeConte, 1850)
Current placement: Cerambycidae: Lepturinae: Lepturini
Type locality: Canada: British Columbia

***comanchei* Rice & Morris, 1992**: 200 (Fig. 45a, b), Holotype
Original combination: *Megacyllene comanchei* Rice & Morris
Current name: same
Current placement: Cerambycidae: Cerambycinae: Clytini
Type locality: USA: Texas, Crosby, Co., White River Lake

***compactus* Casey, 1914**: 364 (Fig. 45c, d), Holotype
Original combination: *Hypermallus compactus* Casey
Current name: *Parelaphidion aspersum* (Haldeman, 1847)
Current placement: Cerambycidae: Cerambycinae: Elaphidiini
Type locality: USA: Texas

***complicatus* Gressitt, 1948a**: 49 (Fig. 45e, f), Holotype
Original combination: *Stenhomalus complicatus* Gressitt
Current name: same
Current placement: Cerambycidae: Cerambycinae: Obriini
Type locality: China: Yunnan Province

***compressipenne* Fisher, 1932**: 35 (Fig. 45g, h), Holotype
Original combination: *Elaphidion compressipenne* Fisher
Current name: *Ceresium compressipenne* (Fisher, 1932)
Current placement: Cerambycidae: Cerambycinae: Elaphidiini
Type locality: Haiti: Mariani

***compressipes* Casey, 1912**: 269 (Fig. 45i, j), Holotype
Original combination: *Physocnemum compressipes* Casey
Current name: *Physocnemum brevilineum* (Say, 1824)
Current placement: Cerambycidae: Cerambycinae: Callidiini
Type locality: USA: Wisconsin, Bayfield

***compressipes* Casey, 1912**: 314 (Fig. 45k, l), Holotype
Original combination: *Callichroma compressipes* Casey
Current name: *Callichroma holochlorum* Bates, 1872
Current placement: Cerambycidae: Cerambycinae: Callichromatini
Type locality: Guatemala: Esquintla

***concinnus* Casey, 1912**: 278 (Fig. 45m, n), Holotype
Original combination: *Phymatodes (Poecilium) concinnus* Casey
Current name: *Phymatodes vulneratus* (LeConte, 1857)
Current placement: Cerambycidae: Cerambycinae: Callidiini
Type locality: USA: California, Santa Cruz Mts.

***confinus* Aurivillius, 1924**: 446 (Fig. 45o, p), Holotype
Original combination: *Demonax confinus* Aurivillius
Current name: *Demonax seriatopunctatus* Aurivillius, 1922
Current placement: Cerambycidae: Cerambycinae: Clytini
Type locality: Philippines: Luzon, Mt. Makiling

***confluens* Casey, 1913**: 272 (Fig. 45q, r), Lectotype
Original combination: *Typocerus confluens* Casey
Current name: same
Current placement: Cerambycidae: Lepturinae: Lepturini
Type locality: USA: Kansas, Hamilton, Co., 3350 ft.
Remarks: This specimen is herein designated as the lectotype in order to stabilize the taxonomy and facilitate further identifications of this species.

***confusa* Casey, 1924**: 293 (Fig. 45s, t), Holotype
Original combination: *Spalacopsis confusa* Casey
Current name: *Spalacopsis filum costulatum* Casey, 1913
Current placement: Cerambycidae: Lamiinae: Agapanthiini
Type locality: USA: Florida, Cape Sable

***confusum* Fisher, 1932**: 32 (Fig. 45u, v), Holotype
Original combination: *Elaphidion confusum* Fisher
Current name: *Elaphidion fullonium* Newman, 1841
Current placement: Cerambycidae: Cerambycinae: Elaphidiini
Type locality: Haiti: Hinche

***confusum* Fisher, 1944a**: 6 (Fig. 45w, x), Holotype
Original combination: *Gnomidolon confusum* Fisher
Current name: *Gnomidolon musivum* (Erichson, 1847)
Current placement: Cerambycidae: Cerambycinae: Hexoplini
Type locality: Venezuela: Monagas, Caripito

***congoana* Breuning, 1935a**: 172 (Fig. 46a, b), Holotype
Original combination: *Acmocera congoana* Breuning
Current name: *Acmocera inermis* Thomson, 1858
Current placement: Cerambycidae: Lamiinae: Acmocerini
Type locality: Gabon: Brazzaville (formerly French Congo), West Africa
Remarks: Tavakilian and Chevillotte (2013) indicate the holotype is in Basel, Switzerland, however, this specimen matches the description.

***congoensis* Breuning, 1942a**: 126 (Fig. 46c, d), Holotype
Original combination: *Docus congoensis* Breuning
Current name: *Planodema congoensis* (Breuning, 1942a)
Current placement: Cerambycidae: Lamiinae: Theocridini
Type locality: Congo: Leopoldville

***conicicollis* Tippmann, 1953c**: 199 (Fig. 46e, f), Holotype
Original combination: *Gigantotrichoderes conicicollis* Tippmann
Current name: same
Current placement: Cerambycidae: Cerambycinae: Torneutini
Type locality: Colombia: Cundinamarca, Fusagasuga

***conicola* Fisher, 1915**: 78 (Fig. 46g, h), Holotype
Original combination: *Paratimia conicola* Fisher
Current name: same
Current placement: Cerambycidae: Spondylidinae: Atimiini
Type locality: USA: California, Monumental Mines

***conjuncta* Casey, 1912**: 344 (Fig. 46i, j), Holotype
Original combination: *Tylosis conjuncta* Casey
Current name: *Tylosis maculatus* LeConte, 1850
Current placement: Cerambycidae: Cerambycinae: Trachyderini
Type locality: USA: New Mexico

***conjuncta parva* Casey, 1912**: 345 (Fig. 46k, l), Holotype
Original combination: *Tylosis conjuncta parva* Casey
Current name: *Tylosis maculatus* LeConte, 1850
Current placement: Cerambycidae: Cerambycinae: Trachyderini
Type locality: USA: New Mexico, Jemez Springs

***conjunctus occiduus* Casey, 1912**: 369 (Fig. 46m, n), Lectotype
Original combination: *Neoclytus conjunctus occiduus* Casey
Current name: *Neoclytus conjunctus* (LeConte, 1857)
Current placement: Cerambycidae: Cerambycinae: Clytini
Type locality: USA: California
Remarks: This specimen is herein designated as the lectotype in order to stabilize the taxonomy and facilitate further identifications of this species.

***consobrina* Gressitt, 1956**: 142 (Fig. 46o, p), Holotype
Original combination: *Sybra consobrina* Gressitt
Current name: same
Current placement: Cerambycidae: Lamiinae: Apomecynini
Type locality: Palau: Peleliu Island

***consors* Casey, 1912**: 240 (Fig. 46q, r), Holotype
Original combination: *Prionus (Prionus) consors* Casey
Current name: *Prionus californicus* Motschulsky, 1845
Current placement: Cerambycidae: Prioninae: Prionini
Type locality: USA: Colorado, Boulder, Co.

***consors* Casey, 1924**: 254 (Fig. 46s, t), Holotype
Original combination: *Brothylus consors* Casey
Current name: *Brothylus gemmulatus* LeConte, 1859
Current placement: Cerambycidae: Cerambycinae: Hesperophanini
Type locality: USA: Utah, Eureka

***consors acomanus* Casey, 1912**: 241 (Fig. 46u, v), Holotype
Original combination: *Prionus (Prionus) consors acomanus* Casey
Current name: *Prionus californicus* Motschulsky, 1845
Current placement: Cerambycidae: Prioninae: Prionini
Type locality: USA: New Mexico, Jemez Springs

consors proximans **Casey, 1912**: 241 (Fig. 46w, x), Holotype
Original combination: *Prionus (Prionus) consors proximans* Casey
Current name: *Prionus californicus* Motschulsky, 1845
Current placement: Cerambycidae: Prioninae: Prionini
Type locality: USA: New Mexico, Fort Wingate

conspersa **Aurivillius, 1927b**: 563 (Fig. 47a, b), Lectotype
Original combination: *Doliops conspersa* Aurivillius
Current name: same
Current placement: Cerambycidae: Lamiinae: Apomecynini
Type locality: Philippines: Sibuyan
Remarks: This specimen is herein designated as the lectotype in order to stabilize the taxonomy and facilitate further identifications of this species. Other paralectotype(s) are present in the Naturhistoriska Riksmuseet, Stockholm, Sweden.

conspersa **Aurivillius, 1927b**: 564 (Fig. 47c, d), Syntype
Original combination: *Plocia conspersa* Aurivillius
Current name: *Plociella conspersa* (Aurivillius, 1927b)
Current placement: Cerambycidae: Lamiinae: Apomecynini
Type locality: Philippines: Mindanao, Butuan
Remarks: Other syntypes are present in the USNM and the Naturhistoriska Riksmuseet, Stockholm, Sweden.

conspicua **Tippmann, 1960**: 89 (Fig. 47e, f), Holotype
Original combination: *Parandra (Archandra) conspicua* Tippmann
Current name: same
Current placement: Cerambycidae: Parandrinae: Parandrini
Type locality: Bolivia: Cochabamba, Yungas de Totora - Soluenca, 2100 m.

constricta **Casey, 1913**: 285 (Fig. 47g, h), Holotype
Original combination: *Moneilema (Collapteryx) constricta* Casey
Current name: *Moneilema gigas* LeConte, 1873
Current placement: Cerambycidae: Lamiinae: Moneilemini
Type locality: USA: Arizona

constricticolle **Schaeffer, 1908**: 338 (Fig. 47i, j), Lectotype
Original combination: *Obrium constricticolle* Schaeffer
Current name: same
Current placement: Cerambycidae: Cerambycinae: Obriini
Type locality: USA: Arizona, Cochise Co., Palmerlee
Remarks: This specimen is herein designated as the lectotype in order to stabilize the taxonomy and facilitate further identifications of this species.

constricticollis **Gressitt, 1951a**: 20 (Fig. 47k, l), Holotype
Original combination: *Iphrobrium constricticollis* Gressitt
Current name: same
Current placement: Cerambycidae: Cerambycinae: Obriini
Type locality: Indonesia: Papua, Irian Jaya, Jayapura (formerly Hollandia)

constricticollis **Schaeffer, 1908**: 351 (Fig. 47m, n), Holotype
Original combination: *Methia constricticollis* Schaeffer
Current name: *Methia necydalea* (Fabricius, 1798)
Current placement: Cerambycidae: Cerambycinae: Methiini
Type locality: USA: Texas, Brownsville

contestata **Lane, 1970**: 371 (Fig. 47o, p), Holotype
Original combination: *Cauca contestata* Lane
Current name: same
Current placement: Cerambycidae: Lamiinae: Apomecynini
Type locality: Colombia: Cauca

contortus **Lingafelter, 2013**: 63 (Fig. 47q, r), Holotype
Original combination: *Calocosmus contortus* Lingafelter
Current name: same
Current placement: Cerambycidae: Lamiinae: Hemilophini
Type locality: Dominican Republic: San Cristobal Province, Colonia Ramfis

contracticorne **Martins, 1960c**: 95 (Fig. 47s, t), Holotype
Original combination: *Ibidion contracticorne* Martins
Current name: *Rhysium bimaculatum* Pascoe, 1866
Current placement: Cerambycidae: Cerambycinae: Neoibidionini
Type locality: Bolivia: La Paz Dept., Yanachi, Yungas

convergens **Casey, 1912**: 269 (Fig. 47u, v), Holotype
Original combination: *Smodicum convergens* Casey
Current name: *Smodicum cucujiforme* (Say, 1827)
Current placement: Cerambycidae: Cerambycinae: Smodicini
Type locality: USA: Texas

convexa **Casey, 1924**: 289 (Fig. 47w, x), Holotype
Original combination: *Moneilema (Collapteryx) convexa* Casey
Current name: *Moneilema armatum* LeConte, 1853
Current placement: Cerambycidae: Lamiinae: Moneilemini
Type locality: USA: Texas, Bethage

convexa **Gressitt, 1956**: 142 (Fig. 48a, b), Holotype
Original combination: *Sybra convexa* Gressitt
Current name: same
Current placement: Cerambycidae: Lamiinae: Apomecynini
Type locality: Micronesia: Pohnpei (formerly Ponape Island), Mt. Tanatamansakir

convolvens **Casey, 1913**: 250 (Fig. 48c, d), Lectotype
Original combination: *Judolia convolvens* Casey
Current name: *Brachyleptura vexatrix* (Mannerheim, 1853)
Current placement: Cerambycidae: Lepturinae: Lepturini
Type locality: USA: California, Siskiyou Co.
Remarks: This specimen is herein designated as the lectotype in order to stabilize the taxonomy and facilitate further identifications of this species.

coquilletti **Linell, 1897a**: 398 (Fig. 48e, f), Holotype
Original combination: *Leptura coquilletti* Linell
Current name: *Stenostrophia coquilletti* (Linell, 1897a)
Current placement: Cerambycidae: Lepturinae: Lepturini
Type locality: USA: California, Los Angeles County

coquus filicorne **Casey, 1912**: 323 (Fig. 48g, h), Lectotype
Original combination: *Tragidion coquus filicorne* Casey
Current name: *Tragidion coquus* (Linnaeus, 1758)
Current placement: Cerambycidae: Cerambycinae: Trachyderini
Type locality: USA: New Mexico, Las Vegas
Remarks: This specimen is herein designated as the lectotype in order to stabilize the taxonomy and facilitate further identifications of this species.

coralinus obfuscipennis **Chemsak & Linsley, 1959**: 179 (Fig. 48i, j), Holotype
Original combination: *Crossidius coralinus obfuscipennis* Chemsak & Linsley
Current name: same
Current placement: Cerambycidae: Cerambycinae: Trachyderini
Type locality: USA: Idaho, Boise

coriacecollis **Aurivillius, 1922**: 417 (Fig. 48k, l), Holotype
Original combination: *Demonax coriacecollis* Aurivillius
Current name: same
Current placement: Cerambycidae: Cerambycinae: Clytini
Type locality: Philippines: Mindanao, Kolambugan

cornuticeps federalis **Dillon & Dillon, 1946**: 197 (Fig. 48m, n), Holotype
Original combination: *Lochmaeocles cornuticeps federalis* Dillon & Dillon
Current name: same
Current placement: Cerambycidae: Lamiinae: Onciderini
Type locality: Mexico: Cuernavaca

cornuticeps pacificus **Dillon & Dillon, 1946**: 198 (Fig. 48o, p), Holotype
Original combination: *Lochmaeocles cornuticeps pacificus* Dillon & Dillon
Current name: same
Current placement: Cerambycidae: Lamiinae: Onciderini
Type locality: Mexico: Sinaloa, Venodio

corpulenta **Casey, 1913**: 224 (Fig. 48q, r), Holotype
Original combination: *Brachysomida corpulenta* Casey
Current name: *Brachysomida californica* (LeConte, 1851)
Current placement: Cerambycidae: Lepturinae: Rhagiini
Type locality: USA: California

corrugans **Casey, 1913**: 284 (Fig. 48s, t), Holotype
Original combination: *Moneilema (Moneilema) corrugans* Casey
Current name: *Moneilema appressum* LeConte, 1852
Current placement: Cerambycidae: Lamiinae: Moneilemini
Type locality: USA: Arizona
Remarks: Although the original description suggests there may be more than one specimen in the original series, there is only one in the USNM collection.

corrugans ovipennis **Casey, 1913**: 284 (Fig. 48u, v), Holotype
Original combination: *Moneilema (Moneilema) corrugans ovipennis* Casey
Current name: *Moneilema appressum* LeConte, 1852
Current placement: Cerambycidae: Lamiinae: Moneilemini
Type locality: USA: Arizona

corticaria **Tippmann, 1960**: 171 (Fig. 48w, x), Holotype
Original combination: *Acanthoderes (Psapharochrus) corticaria* Tippmann
Current name: *Psapharochrus corticaria* (Tippmann, 1960)
Current placement: Cerambycidae: Lamiinae: Acanthoderini
Type locality: Bolivia: Province Chapare, Cochabamba

corusca **Casey, 1913**: 261 (Fig. 49a, b), Lectotype
Original combination: *Xestoleptura corusca* Casey
Current name: *Xestoleptura crassicornis* (LeConte, 1873)
Current placement: Cerambycidae: Lepturinae: Lepturini
Type locality: USA: California, Grant Forest
Remarks: This specimen is herein designated as the lectotype in order to stabilize the taxonomy and facilitate further identifications of this species.

***costaricense* Martins, 1962a**: 272 (Fig. 49c, d), Holotype
Original combination: *Gnomidolon costaricense* Martins
Current name: *Gnomidolon melanosomum* Bates, 1870
Current placement: Cerambycidae: Cerambycinae: Hexoplini
Type locality: Costa Rica: Limón, Hamburg Farm, Reventazón, Ebene

***costaricensis* Breuning, 1942a**: 140 (Fig. 49e, f), Holotype
Original combination: *Amphicnaeia costaricensis* Breuning
Current name: *Amphicnaeia lepida* Melzer, 1933
Current placement: Cerambycidae: Lamiinae: Apomecynini
Type locality: Costa Rica: Cartago, Turrialba

***costaricensis* Melzer, 1934**: 91 (Fig. 49g, h), Holotype
Original combination: *Aletretia costaricensis* Melzer
Current name: *Rosalba costaricensis* (Melzer, 1934)
Current placement: Cerambycidae: Lamiinae: Apomecynini
Type locality: Costa Rica: Limón, Hamburg Farm, Reventazón, Ebene

***costipennis* Breuning, 1940d**: 417 (Fig. 49i, j), Holotype
Original combination: *Tmesisternus costipennis* Breuning
Current name: same
Current placement: Cerambycidae: Lamiinae: Tmesisternini
Type locality: Papua New Guinea

***costipennis* Fisher, 1935c**: 583 (Fig. 49k, l), Holotype
Original combination: *Neosarmydus costipennis* Fisher
Current name: same
Current placement: Cerambycidae: Prioninae: Anacolini
Type locality: Malaysia: Borneo, Sabah, Mt. Kinabalu, 5500 ft., Lumu Lumu

***costipennis* Fisher, 1935c**: 629 (Fig. 49m, n), Holotype
Original combination: *Astathes costipennis* Fisher
Current name: *Astathes (Tetraophthalmus) costipennis* Fisher, 1935c
Current placement: Cerambycidae: Lamiinae: Astathini
Type locality: Malaysia: Borneo, Sabah, near Mt. Kinabalu, Kabayau, 600 ft.

***costulata* Casey, 1913**: 355 (Fig. 49o, p), Lectotype
Original combination: *Spalacopsis (Euthuorus) costulata* Casey
Current name: *Spalacopsis filum costulatum* Casey, 1913
Current placement: Cerambycidae: Lamiinae: Agapanthiini
Type locality: USA: Florida, Biscayne
Remarks: This specimen is herein designated as the lectotype in order to stabilize the taxonomy and facilitate further identifications of this species.

***costulatum* Casey, 1912**: 260 (Fig. 49q, r), Holotype
Original combination: *Asemum costulatum* Casey
Current name: *Asemum striatum* (Linnaeus, 1758)
Current placement: Cerambycidae: Spondylidinae: Asemini
Type locality: Canada: British Columbia, Aldermere

***crassipes* Casey, 1891**: 32 (Fig. 49s, t), Holotype
Original combination: *Crossidius crassipes* Casey
Current name: *Crossidius punctatus* LeConte, 1873
Current placement: Cerambycidae: Cerambycinae: Trachyderini
Type locality: USA: Washington

***crassipes* Casey, 1913**: 195 (Fig. 49u, v), Holotype
Original combination: *Rhagium (Hargium) crassipes* Casey
Current name: *Rhagium inquisitor* (Linnaeus, 1758)
Current placement: Cerambycidae: Lepturinae: Rhagiini
Type locality: USA: Massachusetts, Medford

***crassipes* Fisher, 1931c**: 197 (Fig. 49w, x), Holotype
Original combination: *Moneilema (Collapteryx) crassipes* Fisher
Current name: *Moneilema crassipes* Fisher, 1931c
Current placement: Cerambycidae: Lamiinae: Moneilemini
Type locality: Mexico: Tamaulipas, Palmillas

***crassipes shastana* Casey, 1913**: 263 (Fig. 50a, b), Holotype
Original combination: *Xestoleptura crassipes shastana* Casey
Current name: *Xestoleptura crassipes* (LeConte, 1857)
Current placement: Cerambycidae: Lepturinae: Lepturini
Type locality: USA: California, Shasta

***crescenticus* Casey, 1913**: 306 (Fig. 50c, d), Holotype
Original combination: *Leptostylus crescenticus* Casey
Current name: *Leptostylopsis planidorsus* (LeConte, 1873)
Current placement: Cerambycidae: Lamiinae: Acanthocinini
Type locality: USA: Florida, Crescent City

***cribrata* Casey, 1912**: 331 (Fig. 50e, f), Holotype
Original combination: *Stenobatyle cribrata* Casey
Current name: *Stenobatyle mineatocollis* (Chevrolat, 1862)
Current placement: Cerambycidae: Cerambycinae: Trachyderini
Type locality: Mexico: Guerrero

***cribripennis* Casey, 1913**: 198 (Fig. 50g, h), Holotype
Original combination: *Pyrotrichus cribripennis* Casey
Current name: *Pyrotrichus vitticollis* LeConte, 1862
Current placement: Cerambycidae: Lepturinae: Lepturini
Type locality: USA: California, Los Angeles Co.

***cribripennis divisa* Casey, 1924**: 281 (Fig. 50i, j), Lectotype
Original combination: *Brachyleptura (Stictoleptura) cribripennis divisa* Casey
Current name: *Stictoleptura canadensis cribripennis* (LeConte, 1859)
Current placement: Cerambycidae: Lepturinae: Lepturini
Type locality: USA: Idaho, Priest Lake, Bonner Co.
Remarks: This specimen is herein designated as the lectotype in order to stabilize the taxonomy and facilitate further identifications of this species.

***crinicorne* Casey, 1912**: 298 (Fig. 50k, l), Holotype
Original combination: *Romaleum crinicorne* Casey
Current name: *Enaphalodes rufulus* (Haldeman, 1847)
Current placement: Cerambycidae: Cerambycinae: Elaphidiini
Type locality: USA: Texas

***crinicornis* Casey, 1924**: 291 (Fig. 50m, n), Holotype
Original combination: *Leiopus crinicornis* Casey
Current name: *Eutrichillus biguttatus* (LeConte, 1852)
Current placement: Cerambycidae: Lamiinae: Acanthocinini
Type locality: USA: Virginia, Nelson Co.

***crispulum* Fisher, 1947c**: 30 (Fig. 50o, p), Holotype
Original combination: *Elaphidion crispulum* Fisher
Current name: *Anelaphus crispulum* (Fisher, 1947c)
Current placement: Cerambycidae: Cerambycinae: Elaphidiini
Type locality: USA: Puerto Rico, Guánica Forest

***cristata* Fisher, 1936a**: 183 (Fig. 50q, r), Holotype
Original combination: *Paracoptops cristata* Fisher
Current name: *Mutatocoptops alboapicalis* Pic, 1925
Current placement: Cerambycidae: Lamiinae: Mesosini
Type locality: Indonesia: Java, Mt. Tangkoeban Prahoe, Preanger 4000–5000 ft.

***cristatus* Fisher, 1935c**: 613 (Fig. 50s, t), Holotype
Original combination: *Ancornallis cristatus* Fisher
Current name: same
Current placement: Cerambycidae: Lamiinae: Apomecynini
Type locality: Malaysia: Borneo, Sabah, Mt. Kinabalu, 7000 ft.

***crucialis* Casey, 1912**: 282 (Fig. 50u, v), Holotype
Original combination: *Phymatodes (Phymatodina) crucialis* Casey
Current name: *Phymatodes nitidus* LeConte, 1874
Current placement: Cerambycidae: Cerambycinae: Callidiini
Type locality: USA: California, Santa Cruz Mts.

***crucialis* Casey, 1912**: 375 (Fig. 50w, x), Holotype
Original combination: *Paraclytus crucialis* Casey
Current name: *Triodoclytus lanifer* (LeConte, 1873)
Current placement: Cerambycidae: Cerambycinae: Clytini
Type locality: USA: California, Siskiyou Co.

***cryptica* Casey, 1913**: 327 (Fig. 51a, b), Lectotype
Original combination: *Hyperplatys cryptica* Casey
Current name: *Hyperplatys maculata* Haldeman, 1847
Current placement: Cerambycidae: Lamiinae: Acanthocinini
Type locality: USA: Indiana
Remarks: This specimen is herein designated as the lectotype in order to stabilize the taxonomy and facilitate further identifications of this species.

***cubae* Fisher, 1932**: 16 (Fig. 51c, d), Holotype
Original combination: *Eburia cubae* Fisher
Current name: same
Current placement: Cerambycidae: Cerambycinae: Eburiini
Type locality: Cuba: Baragua

***cubae* Fisher, 1932**: 27 (Fig. 51e, f), Holotype
Original combination: *Elaphidion cubae* Fisher
Current name: same
Current placement: Cerambycidae: Cerambycinae: Elaphidiini
Type locality: Cuba: Santiago

***cubae* Fisher, 1932**: 63 (Fig. 51g, h), Holotype
Original combination: *Tilloclytus cubae* Fisher
Current name: same
Current placement: Cerambycidae: Cerambycinae: Anaglyptini
Type locality: Cuba: Oriente, Pinares

***cubaecola* Fisher, 1947c**: 36 (Fig. 51i, j), Holotype
Original combination: *Heterops cubaecola* Fisher
Current name: same
Current placement: Cerambycidae: Cerambycinae: Heteropsini
Type locality: Cuba: Pinar del Río, Cayo de San Felipe

***cubanus* Fisher, 1926**: 26 (Fig. 51k, l), Holotype
Original combination: *Leiopus cubanus* Fisher
Current name: *Styloleptus cubanus* (Fisher, 1926)
Current placement: Cerambycidae: Lamiinae: Acanthocinini
Type locality: Cuba: Cayamas

***cubensis* Casey, 1912**: 224 (Fig. 51m, n), Lectotype
Original combination: *Stenodontes cubensis* Casey
Current name: *Stenodontes chevrolati* Gahan, 1890
Current placement: Cerambycidae: Prioninae: Mallodonini
Type locality: Cuba: Guantanamo
Remarks: This specimen is herein designated as the lectotype in order to stabilize the taxonomy and facilitate further identifications of this species.

cuernavacae **Dillon, 1955**: 177 (Fig. 51o, p), Holotype
Original combination: *Pseudocanidia cuernavacae* Dillon
Current name: *Canidia spinicornis* (Bates, 1881)
Current placement: Cerambycidae: Lamiinae: Acanthocinini
Type locality: Mexico: Morelos, Cuernavaca

cuneata **Casey, 1913**: 232 (Fig. 51q, r), Holotype
Original combination: *Leptacmaeops (Leptacmaeops) cuneata* Casey
Current name: *Cortodera cubitalis* (LeConte, 1861)
Current placement: Cerambycidae: Lepturinae: Rhagiini
Type locality: USA: California, Santa Barbara

cuneatula **Casey, 1913**: 253 (Fig. 51s, t), Lectotype
Original combination: *Brachyleptura cuneatula* Casey
Current name: *Brachyleptura vagans* (Olivier, 1795)
Current placement: Cerambycidae: Lepturinae: Lepturini
Type locality: USA: Wisconsin, Bayfield
Remarks: A lectotype label was placed on this specimen by Laplante in 1991, however, it was not formally designated until now. This specimen is herein designated as the lectotype in order to stabilize the taxonomy and facilitate further identifications of this species.

curticollis **Casey, 1912**: 247 (Fig. 51u, v), Holotype
Original combination: *Prionus (Riponus) curticollis* Casey
Current name: *Prionus (Neopolyarthron) curticollis* Casey, 1912
Current placement: Cerambycidae: Prioninae: Prionini
Type locality: Mexico: Chihuahua, Meadow Valley (6 mi. south Colonia García) (Sierra Madre Mtns)

curticollis **Casey, 1924**: 265 (Fig. 51w, x), Holotype
Original combination: *Crossidius curticollis* Casey
Current name: *Crossidius hirtipes wickhami* Casey, 1912
Current placement: Cerambycidae: Cerambycinae: Trachyderini
Type locality: USA: Utah, Parowan Canyon

curtipenne **Casey, 1912**: 258 (Fig. 52a, b), Holotype
Original combination: *Asemum curtipenne* Casey
Current name: *Asemum striatum* (Linnaeus, 1758)
Current placement: Cerambycidae: Spondylidinae: Asemini
Type locality: USA: Wisconsin, Bayfield

curu **Martins & Galileo, 2012**: 107 (Fig. 52c, d), Holotype
Original combination: *Etyma curu* Martins & Galileo
Current name: same
Current placement: Cerambycidae: Lamiinae: Desmiphorini
Type locality: Guatemala: Zacapa, 19 km north Estacion de la Virgen, 6000 ft.

curvatocostata **Aurivillius, 1927b**: 559 (Fig. 52e, f), Syntype
Original combination: *Pterolophia curvatocostata* Aurivillius
Current name: same
Current placement: Cerambycidae: Lamiinae: Pteropliini
Type locality: Philippines: Sibuyan, Romblon
Remarks: Tavakilian and Chevillotte (2013) state that syntypes are present in the Naturhistoriska Riksmuseet, Stockholm, Sweden as well as the USNM.

cylindrella **Casey, 1893**: 587 (Fig. 52g, h), Holotype
Original combination: *Batyle cylindrella* Casey
Current name: same
Current placement: Cerambycidae: Cerambycinae: Trachyderini
Type locality: USA: Texas

cylindrica **Gressitt, 1940**: 172 (Fig. 52i, j), Holotype
Original combination: *Atimura cylindrica* Gressitt
Current name: same
Current placement: Cerambycidae: Lamiinae: Apomecynini
Type locality: China: Hainan Province, Hainan Island, Ta Hau

cylindricollis **Casey, 1913**: 283 (Fig. 52k, l), Holotype
Original combination: *Moneilema cylindricollis* Casey
Current name: *Moneilema appressum* LeConte, 1852
Current placement: Cerambycidae: Lamiinae: Moneilemini
Type locality: USA: New Mexico, Luna

cylindricollis **Casey, 1924**: 267 (Fig. 52m, n), Holotype
Original combination: *Taranomis cylindricollis* Casey
Current name: *Sphaenothecus bilineatus* (Gory, 1831)
Current placement: Cerambycidae: Cerambycinae: Trachyderini
Type locality: USA: Arizona

cylindrus **Casey, 1912**: 364 (Fig. 52o, p), Holotype
Original combination: *Xylotrechus cylindrus* Casey
Current name: *Xylotrechus sagittatus* (Germar, 1821)
Current placement: Cerambycidae: Cerambycinae: Clytini
Type locality: USA: Wisconsin, Bayfield

cyrtomera **Aurivillius, 1922**: 428 (Fig. 52q, r), Lectotype
Original combination: *Driopea cyrtomera* Aurivillius
Current name: *Driopea (Inermodriopea) cyrtomera* Aurivillius, 1922
Current placement: Cerambycidae: Lamiinae: Acanthocinini
Type locality: Malaysia: Borneo, Sandakan
Remarks: This specimen is herein designated as the lectotype in order to stabilize the taxonomy and facilitate further identifications of this species. Other paralectotype(s) are present in the Naturhistoriska Riksmuseet, Stockholm, Sweden.

***dalmeidai* Dillon & Dillon, 1946**: 161 (Fig. 52s, t), Holotype
Original combination: *Eumimesis dalmeidai* Dillon & Dillon
Current name: *Harringtonia dalmeidai* (Dillon & Dillon, 1946)
Current placement: Cerambycidae: Lamiinae: Calliini
Type locality: Panama: Río Trinidad

***damarensis* Breuning, 1948b**: 37 (Fig. 52u, v), Holotype
Original combination: *Hyllisia damarensis* Breuning
Current name: same
Current placement: Cerambycidae: Lamiinae: Agapanthiini
Type locality: Namibia: Windhuk

***darlingtoni* Fisher, 1930a**: 9 (Fig. 52w, x), Holotype
Original combination: *Odontocera darlingtoni* Fisher
Current name: same
Current placement: Cerambycidae: Cerambycinae: Rhinotragini
Type locality: Colombia: Magdalena, Río Frio

***dasycerus* var. *texanus* Schaeffer, 1908**: 347 (Fig. 53a, b), Lectotype
Original combination: *Ecyrus dasycerus* var. *texanus* Schaeffer
Current name: *Ecyrus arcuatus* Gahan, 1892
Current placement: Cerambycidae: Lamiinae: Pogonocherini
Type locality: USA: Texas, Brownsville, Esperanza Ranch
Remarks: The name following the binomen was not expressly given infrasubspecific rank and the work did not indicate unambiguously that the name was infrasubspecific, therefore, it is deemed a subspecies and available. This specimen is herein designated as the lectotype in order to stabilize the taxonomy and facilitate further identifications of this species.

***debile* Casey, 1912**: 222 (Fig. 53c, d), Holotype
Original combination: *Mallodon debile* Casey
Current name: *Mallodon dasystomus* (Say, 1824)
Current placement: Cerambycidae: Prioninae: Mallodonini
Type locality: USA: Alabama

***debiliceps* Casey, 1912**: 255 (Fig. 53e, f), Holotype
Original combination: *Homaesthesis debiliceps* Casey
Current name: *Prionus (Homaesthesis) emarginatus* Say, 1824
Current placement: Cerambycidae: Prioninae: Prionini
Type locality: USA: Colorado

***debilicollis* Casey, 1912**: 316 (Fig. 53g, h), Lectotype
Original combination: *Dendrobias debilicollis* Casey
Current name: *Trachyderes (Dendrobias) mandibularis* (Dupont, 1834)
Current placement: Cerambycidae: Cerambycinae: Trachyderini
Type locality: USA: New Mexico
Remarks: This specimen is herein designated as the lectotype in order to stabilize the taxonomy and facilitate further identifications of this species.

***debilis* Casey, 1891**: 21 (Fig. 53i, j), Lectotype
Original combination: *Prionus debilis* Casey
Current name: *Prionus (Neopolyarthron) debilis* Casey, 1891
Current placement: Cerambycidae: Prioninae: Prionini
Type locality: USA: Indiana
Remarks: This specimen is herein designated as the lectotype in order to stabilize the taxonomy and facilitate further identifications of this species.

***debilis* Casey, 1912**: 339 (Fig. 53k, l), Holotype
Original combination: *Crossidius debilis* Casey
Current name: *Crossidius pulchellus* LeConte, 1861
Current placement: Cerambycidae: Cerambycinae: Trachyderini
Type locality: USA: Kansas

***decemmaculatus* Breuning, 1942a**: 122 (Fig. 53m, n), Holotype
Original combination: *Notomulciber decemmaculatus* Breuning
Current name: *Notomulciber (Micromulciber) decemmaculatus* Breuning, 1942a
Current placement: Cerambycidae: Lamiinae: Homonoeini
Type locality: India: Travancore

***decemmaculatus* Martins, 1960e**: 175 (Fig. 53o, p), Holotype
Original combination: *Heterachthes decemmaculatus* Martins
Current name: *Compsibidion decemmaculatum* (Martins, 1960e)
Current placement: Cerambycidae: Cerambycinae: Neoibidionini
Type locality: Ecuador: Manabi Province

***deceptiva* Casey, 1924**: 283 (Fig. 53q, r), Holotype
Original combination: *Judolia deceptiva* Casey
Current name: *Judolia cordifera* (Olivier, 1795)
Current placement: Cerambycidae: Lepturinae: Lepturini
Type locality: USA: Massachusetts, Framingham

deceptum **Swift & Ray, 2008**: 17 (Fig. 53s, t), Neotype
Original combination: *Tragidion deceptum* Swift & Ray
Current name: same
Current placement: Cerambycidae: Cerambycinae: Trachyderini
Type locality: USA: Arizona, Huachuca Mtns.
Remarks: Although the authors indicated that they returned the holotype to the Smithsonian Institution, it could not be located. The neotype is here designated.

decolorata lacustris **Casey, 1918**: 416 (Fig. 53u, v), Lectotype
Original combination: *Centrodera decolorata lacustris* Casey
Current name: *Centrodera decolorata* (Harris, 1838)
Current placement: Cerambycidae: Lepturinae: Rhagiini
Type locality: USA: Michigan, Marquette
Remarks: This specimen is herein designated as the lectotype in order to stabilize the taxonomy and facilitate further identifications of this species.

defectus **Casey, 1912**: 303 (Fig. 53w, x), Holotype
Original combination: *Hypermallus defectus* Casey
Current name: *Anelaphus parallelus* (Newman, 1840)
Current placement: Cerambycidae: Cerambycinae: Elaphidiini
Type locality: No locality data

degener **Casey, 1912**: 334 (Fig. 54a, b), Holotype
Original combination: *Plionoma degener* Casey
Current name: *Plionoma suturalis* (LeConte, 1858)
Current placement: Cerambycidae: Cerambycinae: Trachyderini
Type locality: USA: Arizona

degenerata **Gressitt, 1956**: 182 (Fig. 54c, d), Holotype
Original combination: *Boninella degenerata* Gressitt
Current name: same
Current placement: Cerambycidae: Lamiinae: Acanthocinini
Type locality: Japan: Bonin Islands (Ogasawara Islands), Chichi-jima

dehiscens **Casey, 1912**: 312 (Fig. 54e, f), Lectotype
Original combination: *Callimus dehiscens* Casey
Current name: *Lampropterus cyanipennis* (LeConte, 1873)
Current placement: Cerambycidae: Cerambycinae: Hyboderini
Type locality: USA: California, Los Angeles Co.
Remarks: This specimen is herein designated as the lectotype in order to stabilize the taxonomy and facilitate further identifications of this species.

dehiscens **Chemsak, 1997**: 15 (Fig. 54g, h), Holotype
Original combination: *Acyphoderes dehiscens* Chemsak
Current name: same
Current placement: Cerambycidae: Cerambycinae: Rhinotragini
Type locality: El Salvador: Volcán Conchagua

dejeani borgmeieri **Tippmann, 1953c**: 190 (Fig. 54i, j), Holotype
Original combination: *Macrodontia dejeani borgmeieri* Tippmann
Current name: *Macrodontia batesi* Lameere, 1912
Current placement: Cerambycidae: Prioninae: Macrodontiini
Type locality: Honduras: Tegucigalpa, Sierra de Dipilto, ca. 600 m.

delicata **Casey, 1913**: 327 (Fig. 54k, l), Holotype
Original combination: *Hyperplatys delicata* Casey
Current name: *Hyperplatys maculata* Haldeman, 1847
Current placement: Cerambycidae: Lamiinae: Acanthocinini
Type locality: USA: Rhode Island, Boston Neck

delicata **Casey, 1924**: 257 (Fig. 54m, n), Holotype
Original combination: *Methia delicata* Casey
Current name: *Methia mormona* Linell, 1897a
Current placement: Cerambycidae: Cerambycinae: Methiini
Type locality: USA: Utah, Parowan
Remarks: Although the original description suggests there may be more than one specimen in the original series, there is only one in the USNM collection.

delicata **Monné, 2008**: 246 (Fig. 54o, p), Holotype
Original combination: *Colobothea delicata* Monné
Current name: same
Current placement: Cerambycidae: Cerambycinae: Colobotheini
Type locality: Panama: Panama Province, Cerro Campana, 850 m.

delicatula **Casey, 1913**: 368 (Fig. 54q, r), Holotype
Original combination: *Oberea delicatula* Casey
Current name: *Oberea perspicillata* Haldeman, 1847
Current placement: Cerambycidae: Lamiinae: Phytoeciini
Type locality: USA: District of Columbia

delicatulum **Linsley, 1935b**: 486 (Fig. 54s, t), Holotype
Original combination: *Ibidion delicatulum* Linsley
Current name: *Compsa macra* (Thomson, 1867)
Current placement: Cerambycidae: Cerambycinae: Neoibidionini
Type locality: Panama: Canal Zone, Tabernilla

***densa* Casey, 1924**: 259 (Fig. 54u, v), Lectotype
Original combination: *Eburia densa* Casey
Current name: *Eburia ovicollis* LeConte, 1873
Current placement: Cerambycidae: Cerambycinae: Eburiini
Type locality: USA: Texas
Remarks: This specimen is herein designated as the lectotype in order to stabilize the taxonomy and facilitate further identifications of this species.

***densemaculata* Breuning, 1940d**: 414 (Fig. 54w, x), Holotype
Original combination: *Anamera densemaculata* Breuning
Current name: same
Current placement: Cerambycidae: Lamiinae: Lamiini
Type locality: Vietnam: Lakhon (Tonkin)

***denserugata* Breuning, 1936b**: 303 (Fig. 55a, b), Holotype
Original combination: *Rosenbergia denserugata* Breuning
Current name: same
Current placement: Cerambycidae: Lamiinae: Batocerini
Type locality: Papua New Guinea: Milne Bay, Fergusson Island
Remarks: Tavakilian and Chevillotte (2013) suggest the holotype is present in the British Museum of Natural History, however, this specimen and its labels match the original description.

***densesulcata* Breuning, 1940d**: 427 (Fig. 55c, d), Holotype
Original combination: *Stathmodera densesulcata* Breuning
Current name: same
Current placement: Cerambycidae: Lamiinae: Apomecynini
Type locality: Ivory Coast: Dimbokro

***densicolle* Casey, 1924**: 250 (Fig. 55e, f), Holotype
Original combination: *Paranoplium densicolle* Casey
Current name: *Paranoplium gracile* (LeConte, 1881)
Current placement: Cerambycidae: Cerambycinae: Oemini
Type locality: No locality label

***densicollis* Casey, 1914**: 366 (Fig. 55g, h), Holotype
Original combination: *Hapalosalia densicollis* Casey
Current name: *Pidonia densicollis* (Casey, 1914)
Current placement: Cerambycidae: Lepturinae: Rhagiini
Type locality: USA: North Carolina, Black Mts.

***densipennis* Casey, 1912**: 284 (Fig. 55i, j), Lectotype
Original combination: *Phymatodes (Microcallidium) densipennis* Casey
Current name: *Phymatodes hirtellus* (LeConte, 1873)
Current placement: Cerambycidae: Cerambycinae: Callidiini
Type locality: USA: New Mexico
Remarks: This specimen is herein designated as the lectotype in order to stabilize the taxonomy and facilitate further identifications of this species.

***densiventre* Casey, 1912**: 322 (Fig. 55k, l), Holotype
Original combination: *Tragidion densiventre* Casey
Current name: *Tragidion auripenne* Casey, 1893
Current placement: Cerambycidae: Cerambycinae: Trachyderini
Type locality: USA: Arizona

***densum* Casey, 1912**: 270 (Fig. 55m, n), Holotype
Original combination: *Physocnemum densum* Casey
Current name: *Physocnemum brevilineum* (Say, 1824)
Current placement: Cerambycidae: Cerambycinae: Callidiini
Type locality: USA: Kansas

***densus* Casey, 1924**: 211 (Fig. 55o, p), Holotype
Original combination: *Prionus (Prionellus) densus* Casey
Current name: *Prionus laticollis* (Drury, 1773)
Current placement: Cerambycidae: Prioninae: Prionini
Type locality: No locality label or published data.

***dentatipennis* Gressitt, 1954**: 319 (Fig. 55q, r), Holotype
Original combination: *Thranius dentatipennis* Gressitt
Current name: same
Current placement: Cerambycidae: Cerambycinae: Thraniini
Type locality: Indonesia: Papua, Irian Jaya, Jayapura (formerly Hollandia)

***dentatus* Casey, 1913**: 311 (Fig. 55s, t), Holotype
Original combination: *Leiopus dentatus* Casey
Current name: *Sternidius alpha* (Say, 1827)
Current placement: Cerambycidae: Lamiinae: Acanthocinini
Type locality: USA: Illinois

***denticollis* Fisher, 1925a**: 264 (Fig. 55u, v), Holotype
Original combination: *Xaenapta denticollis* Fisher
Current name: same
Current placement: Cerambycidae: Lamiinae: Acanthocinini
Type locality: Malaysia: Borneo, Sandakan

***dentifer* Aurivillius, 1927b**: 551 (Fig. 55w, x), Lectotype
Original combination: *Dihammus dentifer* Aurivillius
Current name: *Acalolepta dentifera* (Aurivillius, 1927b)
Current placement: Cerambycidae: Lamiinae: Monochamini
Type locality: Philippines: Mindanao, Kolambugan
Remarks: This specimen is herein designated as the lectotype in order to stabilize the taxonomy and facilitate further identifications of this species. Other paralectotype(s) are present in the Naturhistoriska Riksmuseet, Stockholm, Sweden.

***deserta* Casey, 1924**: 252 (Fig. 56a, b), Holotype
Original combination: *Oeme deserta* Casey
Current name: *Oeme rigida deserta* Casey, 1924
Current placement: Cerambycidae: Cerambycinae: Oemini
Type locality: USA: Arizona

***deslauriersi* Alten, Alten & Ramey, 2009**: 1 (Fig. 56c, d), Holotype
Original combination: *Liberedaxia deslauriersi* Alten, Alten and Ramey
Current name: same
Current placement: Cerambycidae: Cerambycinae: Oemini
Type locality: USA: Utah, Kane County, Handcock Rd, 2 mi west junction HWY 89.
Remarks: Author elected to deposit in USNM although the original description does not indicate so.

***desmaresti caripariensis* Tippmann, 1960**: 97 (Fig. 56e, f), Holotype
Original combination: *Calocomus desmaresti caripariensis* Tippmann
Current name: *Calocomus desmarestii* (Guerin-Meneville, 1831)
Current placement: Cerambycidae: Prioninae: Calocomini
Type locality: Bolivia: Tarija Dept., Caripari, 1200 m.

***detritus africaeseptentrionalis* Tippmann, 1952**: 143 (Fig. 56g, h), Holotype
Original combination: *Plagionotus detritus africaeseptentrionalis* Tippmann
Current name: same
Current placement: Cerambycidae: Cerambycinae: Clytini
Type locality: Egypt: Alexandria
Remarks: Following Article 32.5.2 (ICZN, 1999), the hyphen was removed from the specific epithet (originally published between *africae* and *septentrionalis*).

***detritus uralensis* Tippmann, 1952**: 144 (Fig. 56i, j), Holotype
Original combination: *Plagionotus detritus uralensis* Tippmann
Current name: same
Current placement: Cerambycidae: Cerambycinae: Clytini
Type locality: Russia: Ufa, Ural

***devittata* Aurivillius, 1927b**: 561 (Fig. 56k, l), Syntype
Original combination: *Pterolophia devittata* Aurivillius
Current name: *Pterolophia (Hylobrotus) devittata* Aurivillius, 1927b
Current placement: Cerambycidae: Lamiinae: Pteropliini
Type locality: Philippines: Sibuyan, Romblon
Remarks: Tavakilian and Chevillotte (2013) state that syntypes are present in the Naturhistoriska Riksmuseet, Stockholm, Sweden as well as the USNM.

***diadela* Martins & Galileo, 1996**: 21 (Fig. 56m, n), Holotype
Original combination: *Apagomerina diadela* Martins & Galileo
Current name: same
Current placement: Cerambycidae: Lamiinae: Hemilophini
Type locality: Venezuela: Miranda, Guatire Valley (El Marquez)

***dilatata* Galileo & Martins, 2005b**: 64 (Fig. 56o, p), Holotype
Original combination: *Acapiata dilatata* Galileo & Martins
Current name: same
Current placement: Cerambycidae: Lamiinae: Hemilophini
Type locality: Ecuador: Napo, Reserva Etnica Waorani, 1 km S. Onkone Gare Camp, Trans. Ent. 220 m, 00°39′10S, 076°26′W

***dilaticornis* Melzer, 1935b**: 177 (Fig. 56q, r), Holotype
Original combination: *Compsa dilaticornis* Melzer
Current name: *Neocompsa lineolata* (Bates, 1870)
Current placement: Cerambycidae: Cerambycinae: Neoibidionini
Type locality: Costa Rica: Limón, Hamburg Farm, Reventazón, Ebene

***dilloniorum* Tippmann, 1960**: 156 (Fig. 56s, t), Holotype
Original combination: *Neocherentes dilloniorum* Tippmann
Current name: same
Current placement: Cerambycidae: Lamiinae: Onciderini
Type locality: Peru: Cusco, Cosnipata, 1700 m.

***dimidiatipennis* Melzer, 1932**: 216 (Fig. 56u, v), Holotype
Original combination: *Apelocera dimidiatipennis* Melzer
Current name: *Euderces dimidiatipennis* (Melzer, 1932)
Current placement: Cerambycidae: Cerambycinae: Tillomorphini
Type locality: Costa Rica: San Jose, 1000–1200 m.

***dimidiatus* Aurivillius, 1922**: 409 (Fig. 56w, x), Holotype
Original combination: *Chlorophorus dimidiatus* Aurivillius
Current name: same
Current placement: Cerambycidae: Cerambycinae: Clytini
Type locality: Malaysia: Sabah, Sandakan

***dimorpha* Gressitt, 1937b**: 612 (Fig. 57a, b), Holotype
Original combination: *Hippocephala dimorpha* Gressitt
Current name: same
Current placement: Cerambycidae: Lamiinae: Pteropliini
Type locality: China: Jiangxi Province, Hong Shan

***dimorphus* Fisher, 1921**: 207 (Fig. 57c, d), Holotype
Original combination: *Desmocerus dimorphus* Fisher
Current name: *Desmocerus californicus dimorphus* Fisher, 1921
Current placement: Cerambycidae: Lepturinae: Desmocerini
Type locality: USA: California, Sacramento

***diruptus* Casey, 1912**: 367 (Fig. 57e, f), Holotype
Original combination: *Xylotrechus diruptus* Casey
Current name: *Xylotrechus insignis* LeConte, 1873
Current placement: Cerambycidae: Cerambycinae: Clytini
Type locality: USA: California, Los Angeles Co.

***discale* Gressitt, 1935c**: 569 (Fig. 57g, h), Holotype
Original combination: *Anoplodermorpha discale* Gressitt
Current name: *Pseudalosterna discalis* (Gressitt, 1935c)
Current placement: Cerambycidae: Lepturinae: Lepturini
Type locality: China: Sichuan Province, near Muping, 7000–13,000 ft.
Remarks: The holotype label has the incorrect genus name.

***discolor* Casey, 1913**: 343 (Fig. 57i, j), Holotype
Original combination: *Dectes discolor* Casey
Current name: *Dectes texanus* LeConte, 1862
Current placement: Cerambycidae: Lamiinae: Acanthocinini
Type locality: USA: Arizona

***disjunctus* Martins & Galileo, 2012**: 110 (Fig. 57k, l), Holotype
Original combination: *Tesapeus disjunctus* Martins & Galileo
Current name: same
Current placement: Cerambycidae: Lamiinae: Desmiphorini
Type locality: Costa Rica: Limón, Hamburg Farm, Reventazón, Ebene
Remarks: The specific epithet is misspelled on the label.

***divaricatus* Galileo & Martins, 2012c**: 299 (Fig. 57m, n), Holotype
Original combination: *Craterícus divaricatus* Galileo & Martins
Current name: same
Current placement: Cerambycidae: Cerambycinae: Elaphidiini
Type locality: Jamaica: Clarendon, Parish of Clarendon, Claredon Park (2 km south of Rocky Point, near Jackson Bay Cave)

***diversus* Casey, 1912**: 247 (Fig. 57o, p), Lectotype
Original combination: *Prionus (Riponus) diversus* Casey
Current name: *Prionus (Neopolyarthron) imbricornis* (Linnaeus, 1767)
Current placement: Cerambycidae: Prioninae: Prionini
Type locality: USA: Indiana
Remarks: This specimen is herein designated as the lectotype in order to stabilize the taxonomy and facilitate further identifications of this species.

***diversus cuneatus* Casey, 1912**: 247 (Fig. 57q, r), Lectotype
Original combination: *Prionus (Riponus) diversus cuneatus* Casey
Current name: *Prionus (Neopolyarthron) imbricornis* (Linnaeus, 1767)
Current placement: Cerambycidae: Prioninae: Prionini
Type locality: USA: North Carolina, Southern Pines
Remarks: This specimen is herein designated as the lectotype in order to stabilize the taxonomy and facilitate further identifications of this species.

***divisus* Casey, 1913**: 306 (Fig. 57s, t), Holotype
Original combination: *Leptostylus divisus* Casey
Current name: *Leptostylus transversus* (Gyllenhal in Schoenherr, 1817)
Current placement: Cerambycidae: Lamiinae: Acanthocinini
Type locality: USA: Texas

***dolichoptera* Lingafelter, 2010**: 266 (Fig. 57u, v), Holotype
Original combination: *Methia dolichoptera* Lingafelter
Current name: same
Current placement: Cerambycidae: Cerambycinae: Methiini
Type locality: Dominican Republic: La Vega Province, Nacimiento Río Yuna, Loma Hundida, Parque Juan B. Perez Rancier, 18°49.750′N 70°38.057′W, 2139 m.

***dolosa* Casey, 1913**: 371 (Fig. 57w, x), Holotype
Original combination: *Oberea dolosa* Casey
Current name: *Oberea perspicillata* Haldeman, 1847
Current placement: Cerambycidae: Lamiinae: Phytoeciini
Type locality: USA: North Carolina, Southern Pines

***domingoensis* Fisher, 1922**: 52 (Fig. 58a, b), Holotype
Original combination: *Callichroma domingoensis* Fisher
Current name: *Plinthocoelium domingoensis* (Fisher, 1922)
Current placement: Cerambycidae: Cerambycinae: Callichromatini
Type locality: Dominican Republic: San Sidro

***domingoensis* Fisher, 1932**: 1 (Fig. 58c, d), Holotype
Original combination: *Xixuthrus domingoensis* Fisher
Current name: same
Current placement: Cerambycidae: Prioninae: Xixuthrini
Type locality: Dominican Republic: Santiago

***dominicana* Micheli, 1983**: 262 (Fig. 58e, f), Holotype
Original combination: *Curiosa dominicana* Micheli
Current name: *Plectromerus dominicanus* (Micheli, 1983)
Current placement: Cerambycidae: Cerambycinae: Plectromerini
Type locality: Dominican Republic: La Vega, 20 km southeast of Constanza

***dominicensis* Villiers, 1980**: 96 (Fig. 58g, h), Holotype
Original combination: *Anisopodus dominicensis* Villiers
Current name: same
Current placement: Cerambycidae: Lamiinae: Acanthocinini
Type locality: Dominica: Trafalgar

***dominicus* Fisher, 1926**: 30 (Fig. 58i, j), Holotype
Original combination: *Probatius dominicus* Fisher
Current name: *Alcidion dominicum* (Fisher, 1926)
Current placement: Cerambycidae: Lamiinae: Acanthocinini
Type locality: Dominica

***dozieri* Fisher, 1932**: 38 (Fig. 58k, l), Holotype
Original combination: *Elaphidion dozieri* Fisher
Current name: *Trichophoroides dozieri* (Fisher, 1932)
Current placement: Cerambycidae: Cerambycinae: Elaphidiini
Type locality: Haiti: Port-au-Prince

***dozieri* Fisher, 1932**: 86 (Fig. 58m, n), Holotype
Original combination: *Leiopus dozieri* Fisher
Current name: *Styloleptus dozieri* (Fisher, 1932)
Current placement: Cerambycidae: Lamiinae: Acanthocinini
Type locality: Haiti: Port-au-Prince

***dozieri* Fisher, 1947c**: 33 (Fig. 58o, p), Holotype
Original combination: *Stizocera dozieri* Fisher
Current name: same
Current placement: Cerambycidae: Cerambycinae: Elaphidiini
Type locality: Haiti: Kenscoff

***drescheri* Fisher, 1936a**: 176 (Fig. 58q, r), Holotype
Original combination: *Coloborhombus drescheri* Fisher
Current name: *Scalenus drescheri* (Fisher, 1936a)
Current placement: Cerambycidae: Cerambycinae: Callichromatini
Type locality: Indonesia: Java, Mt. Tangkoeban Prahoe, Preanger 4000–5000 ft.

***drescheri* Fisher, 1936a**: 179 (Fig. 58s, t), Holotype
Original combination: *Demonax drescheri* Fisher
Current name: same
Current placement: Cerambycidae: Cerambycinae: Clytini
Type locality: Indonesia: Java, Batoerraden G. Slamat Java

***drescheri* Fisher, 1936a**: 190 (Fig. 58u, v), Holotype
Original combination: *Enispia drescheri* Fisher
Current name: *Athylia drescheri* (Fisher, 1936a)
Current placement: Cerambycidae: Lamiinae: Pteropliini
Type locality: Indonesia: Java, Mt. Tangkoeban Prahoe, Preanger 4000–5000 ft.

***drescheri* Fisher, 1936a**: 195 (Fig. 58w, x), Holotype
Original combination: *Sybra drescheri* Fisher
Current name: same
Current placement: Cerambycidae: Lamiinae: Apomecynini
Type locality: Indonesia: Java, Batoerraden, Mt. Slamat

***drescheri* Fisher, 1937c**: 201 (Fig. 59a, b), Holotype
Original combination: *Oberea drescheri* Fisher
Current name: *Oberea neptis* Pascoe, 1867
Current placement: Cerambycidae: Lamiinae: Phytoeciini
Type locality: Indonesia: Java, Batoerraden, Mt. Slamat

***drescheri* var. *flavipes* Fisher, 1936a**: 177 (Fig. 59c, d), Holotype
Original combination: *Coloborhombus drescheri* var. *flavipes* Fisher
Current name: *Scalenus drescheri* (Fisher, 1936a)
Current placement: Cerambycidae: Cerambycinae: Callichromatini
Type locality: Indonesia: Java, Mt. Tangkoeban Prahoe, Preanger 4000–5000 ft.
Remarks: The name following the binomen was not expressly given infrasubspecific rank and the work did not indicate unambiguously that the name was infrasubspecific, therefore, it is deemed a subspecies and available.

***dubius* Aurivillius, 1907**: 104 (Fig. 59e, f), Syntype
Original combination: *Clytus (Plagionotulus?) dubius* Aurivillius
Current name: same
Current placement: Cerambycidae: Cerambycinae: Clytini
Type locality: Tanzania: Lukuledi
Remarks: Tavakilian and Chevillotte (2013) state that syntypes are present in the Zoologische Staatssammlung des Bayerischen Staates, Munich, Germany and the Naturhistoriska Riksmuseet, Stockholm, Sweden, but this is clearly one of the original type series.

***dubius* Aurivillius, 1928**: 318 (Fig. 59g, h), Syntype
Original combination: *Demonax dubius* Aurivillius
Current name: same
Current placement: Cerambycidae: Cerambycinae: Clytini
Type locality: Philippines: Sibuyan, Romblon

***dunniana* Casey, 1891**: 27 (Fig. 59i, j), Holotype
Original combination: *Chrotoma dunniana* Casey
Current name: same
Current placement: Cerambycidae: Cerambycinae: Bothriospilini
Type locality: USA: Texas, El Paso

***durga* Kriesche, 1920**: 195 (Fig. 59k, l), Syntype
Original combination: *Apriona durga* Kriesche
Current name: *Apriona neglecta* Ritsema, 1911
Current placement: Cerambycidae: Lamiinae: Batocerini
Type locality: Indonesia: Sumatra, Singalang 2000 ft.
Remarks: This is one of three recorded syntypes. Tippmann labeled this specimen as having been obtained from the Kriesche Estate in 1955.

***duvali* Fisher, 1926**: 19 (Fig. 59m, n), Holotype
Original combination: *Leptostylus duvali* Fisher
Current name: *Leptostylopsis duvali* (Fisher, 1926)
Current placement: Cerambycidae: Lamiinae: Acanthocinini
Type locality: Cuba

***duvali* Fisher, 1947c**: 32 (Fig. 59o, p), Holotype
Original combination: *Heterops duvali* Fisher
Current name: same
Current placement: Cerambycidae: Cerambycinae: Heteropsini
Type locality: Cuba: Santiago de Cuba, Oriente, Jarahueca

***ebenina* Linsley, 1935a**: 143 (Fig. 59q, r), Holotype
Original combination: *Psyrassa ebenina* Linsley
Current name: same
Current placement: Cerambycidae: Cerambycinae: Elaphidiini
Type locality: El Salvador: San Salvador

***ebenum* Casey, 1912**: 258 (Fig. 59s, t), Holotype
Original combination: *Asemum ebenum* Casey
Current name: *Asemum striatum* (Linnaeus, 1758)
Current placement: Cerambycidae: Spondylidinae: Asemini
Type locality: USA: Lake Superior

***eburata* Martins, 1970a**: 914 (Fig. 59u, v), Holotype
Original combination: *Heterocompsa eburata* Martins
Current name: same
Current placement: Cerambycidae: Cerambycinae: Neoibidionini
Type locality: Bolivia: Santa Cruz

***echinoides* Breuning, 1940d**: 407 (Fig. 59w, x), Holotype
Original combination: *Phrynidius echinoides* Breuning
Current name: same
Current placement: Cerambycidae: Lamiinae: Apomecynini
Type locality: Mexico: Cerro Zunil

***ecuatoriana* Martins & Galileo, 2007b**: 131 (Fig. 60a, b), Holotype
Original combination: *Sternycha ecuatoriana* Martins & Galileo
Current name: same
Current placement: Cerambycidae: Lamiinae: Onciderini
Type locality: Ecuador: Pichincha Province, Tinalandia, E. Alluriquin, 700 m.

***elegans* Aurivillius, 1927b**: 583 (Fig. 60c, d), Holotype
Original combination: *Serixia elegans* Aurivillius
Current name: same
Current placement: Cerambycidae: Lamiinae: Saperdini
Type locality: Philippines: Samar

***elegans* Casey, 1912**: 282 (Fig. 60e, f), Holotype
Original combination: *Phymatodes (Phymatodina) elegans* Casey
Current name: *Phymatodes nitidus* LeConte, 1874
Current placement: Cerambycidae: Cerambycinae: Callidiini
Type locality: USA

***elegans* Fisher, 1925a**: 211 (Fig. 60g, h), Holotype
Original combination: *Callienispia elegans* Fisher
Current name: *Egesina (Callienispia) elegans* (Fisher, 1925a)
Current placement: Cerambycidae: Lamiinae: Pteropliini
Type locality: Philippines: Mindanao, Iligan

***elegans* Fisher, 1932**: 68 (Fig. 60i, j), Holotype
Original combination: *Lamproclytus elegans* Fisher
Current name: same
Current placement: Cerambycidae: Cerambycinae: Tillomorphini
Type locality: USA: Puerto Rico, Bayamon

***elegans* Fisher, 1936a**: 178 (Fig. 60k, l), Holotype
Original combination: *Ipothalia elegans* Fisher
Current name: *Gauresthes elegans* (Fisher, 1936a)
Current placement: Cerambycidae: Cerambycinae: Callichromatini
Type locality: Indonesia: Java, Mt. Tangkoeban Prahoe, Preanger 4000–5000 ft.

***elegantissimus* Tippmann, 1953d**: 347 (Fig. 60m, n), Holotype
Original combination: *Astyochus elegantissimus* Tippmann
Current name: *Granastyochus elegantissimus* (Tippmann, 1953d)
Current placement: Cerambycidae: Lamiinae: Acanthocinini
Type locality: Peru: Ayacucho, Rio Sivia, 520 m.

***elegantula* Casey, 1913**: 232 (Fig. 60o, p), Holotype
Original combination: *Leptacmaeops (Leptacmaeops) elegantula* Casey
Current name: *Cortodera cubitalis* (LeConte, 1861)
Current placement: Cerambycidae: Lepturinae: Rhagiini
Type locality: USA: California, Calaveras County

***elegantula* Melzer, 1933**: 381 (Fig. 60q, r), Holotype
Original combination: *Phaea elegantula* Melzer
Current name: same
Current placement: Cerambycidae: Lamiinae: Tetraopini
Type locality: Costa Rica: Limón, Hamburg Farm, Reventazón, Ebene

***elegantulus* Galileo & Martins, 2005a**: 275 (Fig. 60s, t), Holotype
Original combination: *Malacoscylus elegantulus* Galileo & Martins
Current name: same
Current placement: Cerambycidae: Lamiinae: Hemilophini
Type locality: Peru: Cusco, 35 km southwest of Pilcopata

***elongata* Breuning, 1943c**: 17 (Fig. 60u, v), Holotype
Original combination: *Amblymora elongata* Breuning
Current name: same
Current placement: Cerambycidae: Lamiinae: Rhodopini
Type locality: Indonesia: Sulawesi (formerly Celebes)

***elongatus* Casey, 1924**: 230 (Fig. 60w, x), Lectotype
Original combination: *Criocephalus elongatus* Casey
Current name: *Arhopalus productus* (LeConte, 1850)
Current placement: Cerambycidae: Spondylidinae: Asemini
Type locality: USA: Utah, Stockton
Remarks: This specimen is herein designated as the lectotype in order to stabilize the taxonomy and facilitate further identifications of this species

***elongatus* Fisher, 1932**: 54 (Fig. 61a, b), Holotype
Original combination: *Merostenus elongatus* Fisher
Current name: *Caribbomerus elongatus* (Fisher, 1932)
Current placement: Cerambycidae: Cerambycinae: Graciliini
Type locality: Cuba: Wajay, Havana

***elongatus* Fisher, 1935b**: 208 (Fig. 61c, d), Holotype
Original combination: *Sciocyrtinus elongatus* Fisher
Current name: same
Current placement: Cerambycidae: Lamiinae: Cyrtinini
Type locality: Jamaica: Mandeville

***emarginata* Fisher, 1946a**: 331 (Fig. 61e, f), Holotype
Original combination: *Cometes emarginata* Fisher
Current name: *Elytrimitatrix (Grossifemora) chrysostigma* (Bates, 1872)
Current placement: Disteniidae: Disteniinae: Disteniini
Type locality: Costa Rica: Limón, Hamburg Farm, Reventazón, Ebene

***emarginata* Gressitt, 1956**: 142 (Fig. 61g, h), Holotype
Original combination: *Sybra emarginata* Gressitt
Current name: same
Current placement: Cerambycidae: Lamiinae: Apomecynini
Type locality: USA: Guam

***emarginatus* Casey, 1913**: 347 (Fig. 61i, j), Holotype
Original combination: *Pogonocherus emarginatus* Casey
Current name: *Pogonocherus pictus* Fall, 1910
Current placement: Cerambycidae: Lamiinae: Pogonocherini
Type locality: USA: Colorado

***endyba* Galileo & Martins, 2004b**: 385 (Fig. 61k, l), Holotype
Original combination: *Hemiloapis endyba* Galileo & Martins
Current name: same
Current placement: Cerambycidae: Lamiinae: Hemilophini
Type locality: Ecuador: Napo, Reserva Étnica Waorani (1 km south Onkone Gare Camp, 00°39'10´S, 76°26´W, 220 m.)

***ensifera* Tippmann, 1960**: 206 (Fig. 61m, n), Holotype
Original combination: *Phoebemima ensifera* Tippmann
Current name: same
Current placement: Cerambycidae: Lamiinae: Hemilophini
Type locality: Bolivia: Cochabamba, Reg. Chapare 400 m.

***equatorianum* Martins, 1960b**: 89 (Fig. 61o, p), Holotype
Original combination: *Octoplon equatorianum* Martins
Current name: *Asynapteron equatorianum* (Martins, 1960b)
Current placement: Cerambycidae: Cerambycinae: Neoibidionini
Type locality: Ecuador: Guayas, Posorja

***erebea* Casey, 1912**: 220 (Fig. 61q, r), Lectotype
Original combination: *Spondylis erebea* Casey
Current name: *Neospondylis mexicanus* (Bates, 1879)
Current placement: Cerambycidae: Spondylidinae: Spondylidini
Type locality: Mexico: Chihuahua, Meadow Valley (6 mi. south Colonia García) (Sierra Madre Mtns)
Remarks: This specimen is herein designated as the lectotype in order to stabilize the taxonomy and facilitate further identifications of this species.

***erectepilosa* Tippmann, 1951**: 309 (Fig. 61s, t), Holotype
Original combination: *Xylorhiza erectepilosa* Tippmann
Current name: same
Current placement: Cerambycidae: Lamiinae: Xylorhizini
Type locality: China: Guangxi Province, Kwangsi Mts., Toyen-chan

***erichsonii* var. *flavipes* Fisher, 1944a**: 6 (Fig. 61u, v), Holotype
Original combination: *Pantonyssus erichsonii* var. *flavipes* Fisher
Current name: *Pantonyssus flavipes* Fisher, 1944a
Current placement: Cerambycidae: Cerambycinae: Elaphidiini
Type locality: Venezuela: Monagas, Caripito
Remarks: The name following the binomen was not expressly given infrasubspecific rank and the work did not indicate unambiguously that the name was infrasubspecific, therefore, it is deemed a subspecies and available.

***ertli* Aurivillius, 1913**: 248 (Fig. 61w, x), Lectotype
Original combination: *Peritragus ertli* Aurivillius
Current name: *Chariesthes (Peritragus) ertli* (Aurivillius, 1913)
Current placement: Cerambycidae: Lamiinae: Tragocephalini
Type locality: Angola: Bailungo
Remarks: This specimen is herein designated as the lectotype in order to stabilize the taxonomy and facilitate further identifications of this species. Other paralectotypes are present in the Naturhistoriska Riksmuseet, Stockholm, Sweden.

***erythrurum* Martins, 1962a**: 276 (Fig. 62a, b), Holotype
Original combination: *Octoplon erythrurum* Martins
Current name: *Tropidion erythrurum* (Martins, 1962a)
Current placement: Cerambycidae: Cerambycinae: Neoibidionini
Type locality: Bolivia: Cochabamba, Reg. Chapare 400 m.

***esakii* Gressitt, 1956**: 158 (Fig. 62c, d), Holotype
Original combination: *Sciadella (Micronesiella) esakii* Gressitt
Current name: *Sciades (Micronesiella) esakii* (Gressitt, 1956)
Current placement: Cerambycidae: Lamiinae: Acanthocinini
Type locality: Micronesia: Truk, Tol Island, Mt. Unibot, Alt. 390 m.

***eugeniae* Fisher, 1935a**: 60 (Fig. 62e, f), Holotype
Original combination: *Cyrtinus eugeniae* Fisher
Current name: same
Current placement: Cerambycidae: Lamiinae: Cyrtinini
Type locality: USA: Puerto Rico, Diaz Finca, Aibonito

***eurynota* Tippmann, 1960**: 191 (Fig. 62g, h), Holotype
Original combination: *Lepturges eurynota* Tippmann
Current name: same
Current placement: Cerambycidae: Lamiinae: Acanthocinini
Type locality: Bolivia: Santa Cruz, 500 m.

***evanescens* Casey, 1913**: 276 (Fig. 62i, j), Holotype
Original combination: *Ophistomis evanescens* Casey
Current name: *Strangalia sexnotata* Haldeman, 1847
Current placement: Cerambycidae: Lepturinae: Lepturini
Type locality: USA: Texas, Harris Co.

***eversa* Casey, 1913**: 277 (Fig. 62k, l), Lectotype
Original combination: *Ophistomis eversa* Casey
Current name: *Strangalia luteicornis* (Fabricius, 1775)
Current placement: Cerambycidae: Lepturinae: Lepturini
Type locality: USA: Illinois
Remarks: This specimen is herein designated as the lectotype in order to stabilize the taxonomy and facilitate further identifications of this species.

***excavata* Breuning, 1942a**: 133 (Fig. 62m, n), Holotype
Original combination: *Batrachorhina (Soridus) excavata* Breuning
Current name: same
Current placement: Cerambycidae: Lamiinae: Pteropliini
Type locality: Tanzania: Mufindi

***exigua* Fisher, 1947a**: 52 (Fig. 62o, p), Holotype
Original combination: *Ischasia exigua* Fisher
Current name: same
Current placement: Cerambycidae: Cerambycinae: Rhinotragini
Type locality: Venezuela: Miranda, Guatira Valley, El Marquiz

***exilis* Casey, 1893**: 591 (Fig. 62q, r), Holotype
Original combination: *Euderces exilis* Casey
Current name: *Euderces reichei* LeConte, 1873
Current placement: Cerambycidae: Cerambycinae: Tillomorphini
Type locality: USA: Texas

***exilis* Casey, 1912**: 283 (Fig. 62s, t), Holotype
Original combination: *Phymatodes (Phymatodina) exilis* Casey
Current name: *Phymatodes nitidus* LeConte, 1874
Current placement: Cerambycidae: Cerambycinae: Callidiini
Type locality: USA: California

***exilis* Casey, 1913**: 368 (Fig. 62u, v), Lectotype
Original combination: *Oberea exilis* Casey
Current name: *Oberea perspicillata* Haldeman, 1847
Current placement: Cerambycidae: Lamiinae: Phytoeciini
Type locality: USA: Pennsylvania, Harrisburg
Remarks: This specimen is herein designated as the lectotype in order to stabilize the taxonomy and facilitate further identifications of this species.

***exilis* Casey, 1924**: 294 (Fig. 62w, x), Holotype
Original combination: *Spalacopsis exilis* Casey
Current name: *Spalacopsis suffusa* Newman, 1842
Current placement: Cerambycidae: Lamiinae: Agapanthiini
Type locality: USA: Florida, Collier Co., Chokoloskee

***exilis* Chemsak & Linsley, 1979**: 70 (Fig. 63a, b), Holotype
Original combination: *Tomopterus exilis* Chemsak & Linsley
Current name: same
Current placement: Cerambycidae: Cerambycinae: Rhinotragini
Type locality: Mexico: Quintana Roo, Cancún
Remarks: The label incorrectly states *gracilis* as the specific epithet.

***exilis* Fisher, 1947a**: 51 (Fig. 63c, d), Holotype
Original combination: *Odontocera exilis* Fisher
Current name: same
Current placement: Cerambycidae: Cerambycinae: Rhinotragini
Type locality: Honduras: Tela

***exilissimus* Casey, 1912**: 307 (Fig. 63e, f), Lectotype
Original combination: *Heterachthes exilissimus* Casey
Current name: *Heterachthes ebenus* Newman, 1840
Current placement: Cerambycidae: Cerambycinae: Neoibidionini
Type locality: USA: North Carolina, Southern Pines
Remarks: This specimen is herein designated as the lectotype in order to stabilize the taxonomy and facilitate further identifications of this species.

***expurgata* Casey, 1913**: 386 (Fig. 63g, h), Holotype
Original combination: *Tetrops expurgata* Casey
Current name: *Phaea monostigma* (Haldeman, 1847)
Current placement: Cerambycidae: Lamiinae: Tetraopini
Type locality: USA: Indiana

***faldermanni insularis* Gressitt, 1936**: 90 (Fig. 63i, j), Holotype
Original combination: *Aromia faldermanni insularis* Gressitt
Current name: *Aphrodisium (Tomentaromia) faldermannii yugaii* Kano, 1933
Current placement: Cerambycidae: Cerambycinae: Callichromatini
Type locality: China: Taiwan, Bukai

***fallax* Dillon & Dillon, 1959**: 25 (Fig. 63k, l), Holotype
Original combination: *Meliochamus fallax* Dillon & Dillon
Current name: same
Current placement: Cerambycidae: Lamiinae: Monochamini
Type locality: Cameroon: Mukonje Farm near Mundame

***fasciata* Fisher, 1931b**: 23 (Fig. 63m, n), Holotype
Original combination: *Cosmotoma fasciata* Fisher
Current name: same
Current placement: Cerambycidae: Lamiinae: Acanthocinini
Type locality: Costa Rica: Limón, Hamburg Farm, Reventazón, Ebene

***fasciata* Fisher, 1932**: 49 (Fig. 63o, p), Holotype
Original combination: *Cylindera fasciata* Fisher
Current name: *Curtomerus fasciatus* (Fisher, 1932)
Current placement: Cerambycidae: Cerambycinae: Elaphidiini
Type locality: Cuba: Guantanamo, San Carlos Station

***fasciata* Fisher, 1936a**: 191 (Fig. 63q, r), Holotype
Original combination: *Enispia fasciata* Fisher
Current name: *Athylia fasciata* (Fisher, 1936a)
Current placement: Cerambycidae: Lamiinae: Pteropliini
Type locality: Indonesia: Java, Batoerraden G. Slamat

fasciata **Fisher, 1947b**: 176 (Fig. 63s, t), Holotype
Original combination: *Dorcasta fasciata* Fisher
Current name: *Bebelis fasciata* (Fisher, 1947b)
Current placement: Cerambycidae: Lamiinae: Apomecynini
Type locality: Panama: Canal Zone, Paraiso

fasciata **Galileo & Martins, 2005a**: 276 (Fig. 63u, v), Holotype
Original combination: *Abanycha fasciata* Galileo & Martins
Current name: same
Current placement: Cerambycidae: Lamiinae: Hemilophini
Type locality: Venezuela: Aragua, Rancho Grande, 1100 m.

fasciata reducta **Casey, 1913**: 333 (Fig. 63w, x), Lectotype
Original combination: *Urographis fasciata reducta* Casey
Current name: *Graphisurus fasciatus* (Degeer, 1775)
Current placement: Cerambycidae: Lamiinae: Acanthocinini
Type locality: USA: New York, West Point
Remarks: This specimen is herein designated as the lectotype in order to stabilize the taxonomy and facilitate further identifications of this species.

fasciatum **Fisher, 1932**: 28 (Fig. 64a, b), Holotype
Original combination: *Elaphidion fasciatum* Fisher
Current name: *Anelaphus fasciatum* Fisher, 1932
Current placement: Cerambycidae: Cerambycinae: Elaphidiini
Type locality: Cuba: Central Constancia

fasciatus **Fisher, 1925a**: 222 (Fig. 64c, d), Holotype
Original combination: *Mimogyaritus fasciatus* Fisher
Current name: same
Current placement: Cerambycidae: Lamiinae: Rhodopini
Type locality: Malaysia: Borneo, Sandakan

fasciatus **Fisher, 1925a**: 224 (Fig. 64e, f), Holotype
Original combination: *Zeargyrodes fasciatus* Fisher
Current name: same
Current placement: Cerambycidae: Lamiinae: Apomecynini
Type locality: Malaysia: Borneo, Sandakan

fasciatus **Fisher, 1944a**: 9 (Fig. 64g, h), Holotype
Original combination: *Lissonotypus fasciatus* Fisher
Current name: *Lissonoschema fasciatum* (Fisher, 1944a)
Current placement: Cerambycidae: Cerambycinae: Trachyderini
Type locality: Venezuela: Monagas, Caripito

fasciatus **Hamilton, 1896**: 137 (Fig. 64i, j), Lectotype
Original combination: *Ecyrus fasciatus* Hamilton
Current name: *Ecyrus penicillatus* Bates, 1880
Current placement: Cerambycidae: Lamiinae: Pogonocherini
Type locality: USA: Texas, Brownsville

fasciatus **Martins & Galileo, 2010**: 89 (Fig. 64k, l), Holotype
Original combination: *Mimestoloides fasciatus* Martins & Galileo
Current name: same
Current placement: Cerambycidae: Lamiinae: Desmiphorini
Type locality: Mexico: Guerrero, Hwy 134, 67 km NE Jct 200

fasciatus **Tippmann, 1953c**: 203 (Fig. 64m, n), Holotype
Original combination: *Callistoprionus fasciatus* Tippmann
Current name: same
Current placement: Cerambycidae: Prioninae: Prionini
Type locality: Costa Rica: Cariblanco

fasciculata **Breuning, 1940d**: 420 (Fig. 64o, p), Holotype
Original combination: *Mimoniphona fasciculata* Breuning
Current name: same
Current placement: Cerambycidae: Lamiinae: Pteropliini
Type locality: India: Kodai Kamel

fasciculosa **Breuning, 1942b**: 17 (Fig. 64q, r), Holotype
Original combination: *Falsobiobessa fasciculosa* Breuning
Current name: same
Current placement: Cerambycidae: Lamiinae: Crossotini
Type locality: Tanzania: Tanganyika

fasciolata **Aurivillius, 1922**: 431 (Fig. 64s, t), Lectotype
Original combination: *Daphisia fasciolata* Aurivillius
Current name: *Menesia fasciolata* (Aurivillius, 1922)
Current placement: Cerambycidae: Lamiinae: Saperdini
Type locality: Malaysia: Borneo, Sandakan
Remarks: This specimen is herein designated as the lectotype in order to stabilize the taxonomy and facilitate further identifications of this species. Three paralectotypes are present in the USNM and others are in the Naturhistoriska Riksmuseet, Stockholm, Sweden.

fascipennis **Schiefer, 2001**: 533 (Fig. 64u, v), Holotype
Original combination: *Astylopsis fascipennis* Schiefer
Current name: same
Current placement: Cerambycidae: Lamiinae: Acanthocinini
Type locality: USA: Mississippi, Oktibbeha Co., 3 miles west of Adaton

***fascosternalis* Breuning, 1942a**: 148 (Fig. 64w, x), Holotype
Original combination: *Sybra fascosternalis* Breuning
Current name: *Sybra fuscosternalis* Breuning, 1942a
Current placement: Cerambycidae: Lamiinae: Apomecynini
Type locality: Malaysia: Penang, Pulau Pinang
Remarks: The specific epithet is misspelled in the original publication (but is correct on the label).

***fastigiatus* Casey, 1913**: 348 (Fig. 65a, b), Holotype
Original combination: *Pogonocherus fastigiatus* Casey
Current name: *Pogonocherus pictus* Fall, 1910
Current placement: Cerambycidae: Lamiinae: Pogonocherini
Type locality: USA: Montana

***favorabilis* Tippmann, 1960**: 195 (Fig. 65c, d), Holotype
Original combination: *Sympagus favorabilis* Tippmann
Current name: same
Current placement: Cerambycidae: Lamiinae: Acanthocinini
Type locality: Peru: Junin, Satipo, 700 m.

***fellerae* Chemsak, 1983**: 714 (Fig. 65e, f), Holotype
Original combination: *Derancistrus fellerae* Chemsak
Current name: *Elateropsis fellerae* (Chemsak, 1983)
Current placement: Cerambycidae: Prioninae: Solenopterini
Type locality: Belize: Twin Cays (15 miles from Mainland town of Dangriga)

***femoralis* Aurivillius, 1927b**: 553 (Fig. 65g, h), Syntype
Original combination: *Dihammus femoralis* Aurivillius
Current name: *Sternohammus femoralis* (Aurivillius, 1927b)
Current placement: Cerambycidae: Lamiinae: Monochamini
Type locality: Philippines: Mindanao, Butuan
Remarks: Given the condition of this specimen, it is best to designate a lectotype from one of the other syntypes that are present in the Naturhistoriska Riksmuseet, Stockholm, Sweden.

***femoralis* Breuning, 1940d**: 429 (Fig. 65i, j), Holotype
Original combination: *Sybra femoralis* Breuning
Current name: same
Current placement: Cerambycidae: Lamiinae: Apomecynini
Type locality: Indonesia: Java

***femoralis* Casey, 1913**: 288 (Fig. 65k, l), Holotype
Original combination: *Moneilema (Collapteryx) femoralis* Casey
Current name: *Moneilema armatum* LeConte, 1853
Current placement: Cerambycidae: Lamiinae: Moneilemini
Type locality: USA: Texas
Remarks: Although the original description suggests there may be more than one specimen in the original series, there is only one in the USNM collection.

***femoratum* Aurivillius, 1927a**: 529 (Fig. 65m, n), Lectotype
Original combination: *Ceresium femoratum* Aurivillius
Current name: same
Current placement: Cerambycidae: Cerambycinae: Callidiopini
Type locality: Philippines: Sibuyan
Remarks: This specimen is herein designated as the lectotype in order to stabilize the taxonomy and facilitate further identifications of this species.

***femoratus* Gressitt, 1938c**: 165 (Fig. 65o, p), Holotype
Original combination: *Rondibilis femoratus* Gressitt
Current name: same
Current placement: Cerambycidae: Lamiinae: Acanthocinini
Type locality: China: Taiwan, Rokki

***femoratus* Schaeffer, 1909**: 384 (Fig. 65q, r), Syntype
Original combination: *Metaleptus femoratus* Schaeffer
Current name: *Rhodoleptus femoratus* (Schaeffer, 1909)
Current placement: Cerambycidae: Cerambycinae: Trachyderini
Type locality: USA: Arizona, Baboquivaria Mts.

***femoratus amnicola* Casey, 1913**: 383 (Fig. 65s, t), Lectotype
Original combination: *Tetraopes femoratus amnicola* Casey
Current name: *Tetraopes femoratus* LeConte, 1847
Current placement: Cerambycidae: Lamiinae: Tetraopini
Type locality: USA: Iowa, Keokuk and vicinity
Remarks: This specimen is herein designated as the lectotype in order to stabilize the taxonomy and facilitate further identifications of this species.

***femoratus monticola* Casey, 1913**: 384 (Fig. 65u, v), Holotype
Original combination: *Tetraopes femoratus monticola* Casey
Current name: *Tetraopes femoratus* LeConte, 1847
Current placement: Cerambycidae: Lamiinae: Tetraopini
Type locality: USA: Utah

***ferruginea* Aurivillius, 1927b**: 588 (Fig. 65w, x), Syntype
Original combination: *Chreonoma ferruginea* Aurivillius
Current name: *Hispastathes ferruginea* (Aurivillius, 1927b)
Current placement: Cerambycidae: Lamiinae: Tetraopini
Type locality: Philippines: Mindanao, Butuan
Remarks: The specific epithet label is misspelled. There are also syntypes present in the Naturhistoriska Riksmuseet, Stockholm, Sweden.

ferruginea **Casey, 1913**: 366 (Fig. 66a, b), Lectotype
Original combination: *Oberea ferruginea* Casey
Current name: *Oberea caseyi* Plavilstshikov, 1926
Current placement: Cerambycidae: Lamiinae: Phytoeciini
Type locality: USA: Kansas
Remarks: Lectotype designated by Yanega (1996: 149)

filicorne **Casey, 1924**: 256 (Fig. 66c, d), Holotype
Original combination: *Chion filicorne* Casey
Current name: *Knulliana cincta* (Drury, 1773)
Current placement: Cerambycidae: Cerambycinae: Bothriospilini
Type locality: USA: North Carolina, Southern Pines

filicornis **Casey, 1912**: 352 (Fig. 66e, f), Holotype
Original combination: *Cyllene filicornis* Casey
Current name: *Placosternus difficilis* (Chevrolat, 1862)
Current placement: Cerambycidae: Cerambycinae: Clytini
Type locality: USA: Texas

filicornis **Casey, 1913**: 255 (Fig. 66g, h), Holotype
Original combination: *Parallelina filicornis* Casey
Current name: *Grammoptera subargentata* (Kirby, 1837)
Current placement: Cerambycidae: Lepturinae: Lepturini
Type locality: USA: Washington

filicornis **Gressitt, 1935d**: 380 (Fig. 66i, j), Holotype
Original combination: *Monochammus filicornis* Gressitt
Current name: *Xenohammus bimaculatus* Schwarzer, 1931
Current placement: Cerambycidae: Lamiinae: Monochamini
Type locality: China: Taiwan, Hori
Remarks: The genus *Monochamus* was misspelled as *Monochammus* in Gressitt (1935c).

filitarsis **Casey, 1912**: 266 (Fig. 66k, l), Holotype
Original combination: *Criocephalus filitarsis* Casey
Current name: *Arhopalus asperatus* (LeConte, 1859)
Current placement: Cerambycidae: Spondylidinae: Asemini
Type locality: USA: Texas, El Paso

filum **Casey, 1913**: 369 (Fig. 66m, n), Holotype
Original combination: *Oberea filum* Casey
Current name: *Oberea perspicillata* Haldeman, 1847
Current placement: Cerambycidae: Lamiinae: Phytoeciini
Type locality: USA: Pennsylvania, Harrisburg

filum brevialatum **Tyson, 1973**: 125 (Fig. 66o, p), Holotype
Original combination: *Spalacopsis (Euthuorus) filum brevialatum* Tyson
Current name: *Spalacopsis filum brevialatum* Tyson, 1973
Current placement: Cerambycidae: Lamiinae: Agapanthiini
Type locality: Dominican Republic: Samana, Samana Bay, Bosque del Infierno

fisheri **Dillon & Dillon, 1941**: 122 (Fig. 66q, r), Holotype
Original combination: *Goes fisheri* Dillon & Dillon
Current name: same
Current placement: Cerambycidae: Lamiinae: Monochamini
Type locality: USA: Texas, Uvalde

fisheri **Dillon & Dillon, 1946**: 395 (Fig. 66s, t), Holotype
Original combination: *Oncideres fisheri* Dillon & Dillon
Current name: same
Current placement: Cerambycidae: Lamiinae: Onciderini
Type locality: Guatemala: Cayuga

fisheri **Lane, 1972**: 356 (Fig. 66u, v), Holotype
Original combination: *Sepaicutea fisheri* Lane
Current name: same
Current placement: Cerambycidae: Cerambycinae: Oemini
Type locality: Guatemala: Sapacuite
Remarks: The genus name is misspelled as *Sepacuitea* on the label.

fisheri **Linsley, 1935b**: 485 (Fig. 66w, x), Holotype
Original combination: *Ibidion fisheri* Linsley
Current name: *Heterachthes x-notatus* (Linsley, 1935b)
Current placement: Cerambycidae: Cerambycinae: Neoibidionini
Type locality: Guatemala: Izabal, Livingston

fisheri **Russo, 1930**: 144 (Fig. 67a, b), Holotype
Original combination: *Eburia fisheri* Russo
Current name: same
Current placement: Cerambycidae: Cerambycinae: Eburiini
Type locality: Dominican Republic

fisheri formosae **Gressitt, 1938c**: 147 (Fig. 67c, d), Holotype
Original combination: *Macrotoma fisheri formosae* Gressitt
Current name: *Bandar pascoei formosae* (Gressitt, 1938c)
Current placement: Cerambycidae: Prioninae: Macrotomini
Type locality: China: Taiwan, Hori

fisheriana **Gressitt, 1938a**: 45 (Fig. 67e, f), Holotype
Original combination: *Leptura fisheriana* Gressitt
Current name: *Leptura rufoannulata* (Pic, 1933)
Current placement: Cerambycidae: Lepturinae: Lepturini
Type locality: China: Sichuan Province, southwest of Tatsienlu, 9,000–13,000 ft.

fissicornis parviceps **Casey, 1912**: 250 (Fig. 67g, h), Holotype
Original combination: *Prionus (Antennalia) fissicornis parviceps* Casey
Current name: *Prionus (Antennalia) fissicornis* Haldeman, 1848
Current placement: Cerambycidae: Prioninae: Prionini
Type locality: USA: Texas

fissicornis transversus **Casey, 1912**: 251 (Fig. 67i, j), Holotype
Original combination: *Prionus (Antennalia) fissicornis transversus* Casey
Current name: *Prionus (Antennalia) fissicornis* Haldeman, 1848
Current placement: Cerambycidae: Prioninae: Prionini
Type locality: USA: Texas

fissifrons **Casey, 1912**: 243 (Fig. 67k, l), Holotype
Original combination: *Prionus (Prionus) fissifrons* Casey
Current name: *Prionus californicus* Motschulsky, 1845
Current placement: Cerambycidae: Prioninae: Prionini
Type locality: USA: Arizona

flaccidus **Casey, 1913**: 212 (Fig. 67m, n), Holotype
Original combination: *Stenocorus (Stenocorus) flaccidus* Casey
Current name: *Stenocorus vestitus* (Haldeman, 1847)
Current placement: Cerambycidae: Lepturinae: Rhagiini
Type locality: USA: California
Remarks: Although the original description suggests there may be more than one specimen in the original series, there is only one in the USNM collection.

flava **Martins & Galileo, 2006a**: 212 (Fig. 67o, p), Holotype
Original combination: *Asemolea flava* Martins & Galileo
Current name: same
Current placement: Cerambycidae: Lamiinae: Calliini
Type locality: Costa Rica: Heredia, Estación Biológica La Selva

flavicollis **Fisher, 1938**: 135 (Fig. 67q, r), Holotype
Original combination: *Callideriphus flavicollis* Fisher
Current name: *Eryphus flavicollis* (Fisher, 1938)
Current placement: Cerambycidae: Cerambycinae: Heteropsini
Type locality: Brazil: Santa Catarina, Nova Teutonia

flavicorne **Aurivillius, 1927a**: 529 (Fig. 67s, t), Lectotype
Original combination: *Ceresium flavicorne* Aurivillius
Current name: same
Current placement: Cerambycidae: Cerambycinae: Callidiopini
Type locality: Malaysia: Borneo, Sandakan
Remarks: This specimen is herein designated as the lectotype in order to stabilize the taxonomy and facilitate further identifications of this species. Other paralectotype(s) are present in the Naturhistoriska Riksmuseet, Stockholm, Sweden.

flavicornis **Casey, 1924**: 257 (Fig. 67u, v), Holotype
Original combination: *Methia flavicornis* Casey
Current name: same
Current placement: Cerambycidae: Cerambycinae: Methiini
Type locality: USA: Arizona, Tucson

flavidosignatus **Aurivillius, 1927b**: 554 (Fig. 67w, x), Syntype
Original combination: *Orsidis (?) flavidosignatus* Aurivillius
Current name: *Acalolepta flavidosignata* (Aurivillius, 1927b)
Current placement: Cerambycidae: Lamiinae: Monochamini
Type locality: Philippines: Mindanao, Butuan
Remarks: Tavakilian and Chevillotte (2013) suggest syntypes are divided between the USNM and the Naturhistoriska Riksmuseet, Stockholm.

flavipennis **Lane, 1966a**: 271 (Fig. 68a, b), Holotype
Original combination: *Eulachnesia flavipennis* Lane
Current name: *Fredlanea flavipennis* (Lane, 1966a)
Current placement: Cerambycidae: Lamiinae: Hemilophini
Type locality: Ecuador: Pastaza

flavipes **Fisher, 1935a**: 59 (Fig. 68c, d), Holotype
Original combination: *Eugamandus flavipes* Fisher
Current name: same
Current placement: Cerambycidae: Lamiinae: Acanthocinini
Type locality: USA: Puerto Rico, Villalba, Insular Government Finca

flaviventris **Galileo & Martins, 2005b**: 66 (Fig. 68e, f), Holotype
Original combination: *Erana flaviventris* Galileo & Martins
Current name: *Eranina flaviventris* (Galileo & Martins, 2005b)
Current placement: Cerambycidae: Lamiinae: Hemilophini
Type locality: Honduras: Yoro, Parque Nacional Pico Pijol

flaviventris **Schaeffer, 1908**: 342 (Fig. 68g, h), Holotype
Original combination: *Leptura flaviventris* Schaeffer
Current name: *Judolia instabilis* (Haldeman, 1847)
Current placement: Cerambycidae: Lepturinae: Lepturini
Type locality: USA: Arizona, Huachuca Mts., near Carr Peak

flavocinctus **Tippmann, 1960**: 197 (Fig. 68i, j), Holotype
Original combination: *Acanthosphenopsilus flavocinctus* Tippmann
Current name: *Cobelura lorigera* Erichson, 1847
Current placement: Cerambycidae: Lamiinae: Acanthocinini
Type locality: Peru: Junin, Satipo, Rio Perené

flavofasciatus **Fisher, 1947a**: 56 (Fig. 68k, l), Holotype
Original combination: *Tomopterus flavofasciatus* Fisher
Current name: same
Current placement: Cerambycidae: Cerambycinae: Rhinotragini
Type locality: Brazil: Santa Catarina, Nova Teutonia, 27°11′B, 52°23′L

flavofasciculata **Breuning, 1948b**: 30 (Fig. 68m, n), Holotype
Original combination: *Aerenea flavofasciculata* Breuning
Current name: same
Current placement: Cerambycidae: Lamiinae: Compsosomatini
Type locality: Peru: Junin, Satipo

flavoguttata **Breuning, 1957a**: 15 (Fig. 68o, p), Holotype
Original combination: *Clodia flavoguttata* Breuning
Current name: same
Current placement: Cerambycidae: Lamiinae: Acanthocinini
Type locality: Philippines: Luzon, Vivac

flavoguttatus **Fisher, 1938**: 145 (Fig. 68q, r), Holotype
Original combination: *Lepturges flavoguttatus* Fisher
Current name: *Probatiomimus schwarzeri* Melzer, 1926
Current placement: Cerambycidae: Lamiinae: Acanthocinini
Type locality: Brazil: Santa Catarina, Nova Teutonia

flavomaculatus **Fisher, 1925a**: 266 (Fig. 68s, t), Holotype
Original combination: *Aegocidnus flavomaculatus* Fisher
Current name: *Sciades (Aegocidnus) flavomaculata* (Fisher, 1925a)
Current placement: Cerambycidae: Lamiinae: Acanthocinini
Type locality: Philippines: Luzon, Mt. Makiling

flavorubra **Gressitt, 1940**: 205 (Fig. 68u, v), Holotype
Original combination: *Glenea flavorubra* Gressitt
Current name: *Glenea (Aridoglenea) flavorubra* Gressitt, 1940
Current placement: Cerambycidae: Lamiinae: Saperdini
Type locality: China: Hainan Province, Hainan Island, Liamui

flavovittata **Breuning, 1940c**: 127 (Fig. 68w, x), Holotype
Original combination: *Ocularia flavovittata* Breuning
Current name: same
Current placement: Cerambycidae: Lamiinae: Oculariini
Type locality: Gabon: Ogavue

flavozonatum **Martins, 1960d**: 109 (Fig. 69a, b), Holotype
Original combination: *Prothoracibidion flavozonatum* Martins
Current name: same
Current placement: Cerambycidae: Cerambycinae: Neoibidionini
Type locality: Peru: Huadquina, 5000 ft.

flavus **Tippmann, 1953c**: 202 (Fig. 69c, d), Holotype
Original combination: *Cycloprionus flavus* Tippmann
Current name: same
Current placement: Cerambycidae: Prioninae: Anacolini
Type locality: French Guiana: Cayenne

flocculatus paucisetosus **Gressitt, 1938b**: 154 (Fig. 69e, f), Holotype
Original combination: *Dihammus flocculatus paucisetosus* Gressitt
Current name: *Acalolepta flocculata paucisetosa* (Gressitt, 1938b)
Current placement: Cerambycidae: Lamiinae: Monochamini
Type locality: China: Sichuan Province, SiGiPin 6,000–7,000 ft.

floridanus **Casey, 1912**: 307 (Fig. 69g, h), Lectotype
Original combination: *Heterachthes floridanus* Casey
Current name: *Heterachthes ebenus* Newman, 1840
Current placement: Cerambycidae: Cerambycinae: Neoibidionini
Type locality: USA: Florida, Biscayne
Remarks: This specimen is herein designated as the lectotype in order to stabilize the taxonomy and facilitate further identifications of this species.

floridanus **Casey, 1912**: 376 (Fig. 69i, j), Holotype
Original combination: *Agallissus floridanus* Casey
Current name: *Zagymnus clerinus* LeConte, 1873
Current placement: Cerambycidae: Cerambycinae: Agallisini
Type locality: USA: Florida, Palm Beach Florida

floridanus **Casey, 1913**: 336 (Fig. 69k, l), Lectotype
Original combination: *Graphisurus (Graphisurus) floridanus* Casey
Current name: *Acanthocinus obsoletus* (Olivier, 1795)
Current placement: Cerambycidae: Lamiinae: Acanthocinini
Type locality: USA: Florida
Remarks: This specimen is herein designated as the lectotype in order to stabilize the taxonomy and facilitate further identifications of this species.

flumineus **Casey, 1912**: 300 (Fig. 69m, n), Holotype
Original combination: *Hypermallus flumineus* Casey
Current name: *Parelaphidion aspersum* (Haldeman, 1847)
Current placement: Cerambycidae: Cerambycinae: Elaphidiini
Type locality: USA: Missouri

fluviatilis **Casey, 1912**: 308 (Fig. 69o, p), Holotype
Original combination: *Heterachthes fluviatilis* Casey
Current name: *Heterachthes quadrimaculatus* Haldeman, 1847
Current placement: Cerambycidae: Cerambycinae: Neoibidionini
Type locality: USA: Missouri

fontinalis **Casey, 1912**: 330 (Fig. 69q, r), Holotype
Original combination: *Batyleoma fontinalis* Casey
Current name: *Batyle suturalis* (Say, 1824)
Current placement: Cerambycidae: Cerambycinae: Trachyderini
Type locality: USA: New Mexico, Las Vegas

fontinalis **Casey, 1924**: 221 (Fig. 69s, t), Holotype
Original combination: *Prionus (Prionus) fontinalis* Casey
Current name: *Prionus heroicus* Semenov, 1908
Current placement: Cerambycidae: Prioninae: Prionini
Type locality: USA: New Mexico, Jemez Springs

formicaria **Lingafelter, 2011**: 60 (Fig. 69u, v), Holotype
Original combination: *Licracantha formicaria* Lingafelter
Current name: same
Current placement: Cerambycidae: Cerambycinae: Anaglyptini
Type locality: Dominican Republic: La Vega Province, 4.7 km. SE Jarabacoa
Remarks: On long-term loan from FSCA.

formosana burkwalli **Gressitt, 1940**: 54 (Fig. 69w, x), Holotype
Original combination: *Merionoeda (Ocytasia) formosana burkwalli* Gressitt
Current name: same
Current placement: Cerambycidae: Cerambycinae: Molorchini
Type locality: China: Hainan Province, Hainan Island, Nokyu Chun

forsteri **Tippmann, 1960**: 152 (Fig. 70a, b), Holotype
Original combination: *Ptericoptus forsteri* Tippmann
Current name: *Bisaltes buquetii* Thomson, 1868
Current placement: Cerambycidae: Lamiinae: Apomecynini
Type locality: Bolivia: Cochabamba, Reg. Chapare 400 m.

forsteri **Tippmann, 1960**: 158 (Fig. 70c, d), Holotype
Original combination: *Oreodera forsteri* Tippmann
Current name: same
Current placement: Cerambycidae: Lamiinae: Acanthoderini
Type locality: Bolivia: Yungas del Palmar, 1000 m.

forsteri **Tippmann, 1960**: 166 (Fig. 70e, f), Holotype
Original combination: *Acanthoderes forsteri* Tippmann
Current name: *Irundisaua forsteri* (Tippmann, 1960)
Current placement: Cerambycidae: Lamiinae: Acanthoderini
Type locality: Peru: Satipo, Junin, Rio Tercue, 600 m.

forticornis **Tippmann, 1960**: 102 (Fig. 70g, h), Holotype
Original combination: *Neoeme forticornis* Tippmann
Current name: *Temnopis forticornis* (Tippmann, 1960)
Current placement: Cerambycidae: Cerambycinae: Oemini
Type locality: Brazil: Goias, Silvania

fortis **Casey, 1913**: 384 (Fig. 70i, j), Holotype
Original combination: *Tetraopes fortis* Casey
Current name: *Tetraopes femoratus* LeConte, 1847
Current placement: Cerambycidae: Lamiinae: Tetraopini
Type locality: Mexico: Durango, Ciudád de Durango

fortunei **Lingafelter, 2013**: 58 (Fig. 70k, l), Holotype
Original combination: *Adesmus fortunei* Lingafelter
Current name: same
Current placement: Cerambycidae: Lamiinae: Hemilophini
Type locality: Dominican Republic: Pedernales Province, N. of Cabo Rojo, Parque Nacional Sierra de Bahoruco, km 26 on Carretera Alcoa, transitional forest (between dry and pine forests), elev. 691 m., 18.113500° –71.621100°

fossulatum **Breuning, 1936b**: 280 (Fig. 70m, n), Holotype
Original combination: *Parahepomidion fossulatum* Breuning
Current name: same
Current placement: Cerambycidae: Lamiinae: Dorcadionini
Type locality: Tanzania: Tanganyika, Kilimanjaro Sjostedt, Kibonoto kulturz.
Remarks: Tavakilian and Chevillotte (2013) suggest the holotype is present in the Naturhistoriska Riksmuseet, Stockholm, Sweden, however, this specimen and its labels match the original description.

foveatocollis **Hamilton in Leng & Hamilton, 1896**: 125 (Fig. 70o, p), Lectotype
Original combination: *Liopus foveatocollis* Hamilton in Leng & Hamilton
Current name: *Urgleptes foveatocollis* (Hamilton, 1896)
Current placement: Cerambycidae: Lamiinae: Acanthocinini
Type locality: USA: Florida, Biscayne
Remarks: This specimen is herein designated as the lectotype in order to stabilize the taxonomy and facilitate further identifications of this species.

foveolatus **Dillon, 1957**: 152 (Fig. 70q, r), Holotype
Original combination: *Lagocheirus foveolatus* Dillon
Current name: same
Current placement: Cerambycidae: Lamiinae: Acanthocinini
Type locality: Panama: Chiriquí, Chiriquicito

francisca **Casey, 1913**: 225 (Fig. 70s, t), Holotype
Original combination: *Brachysomida francisca* Casey
Current name: *Brachysomida californica* (LeConte, 1851)
Current placement: Cerambycidae: Lepturinae: Rhagiini
Type locality: USA: California, San Francisco Co.

frater **Aurivillius, 1924**: 446 (Fig. 70u, v), Holotype
Original combination: *Demonax frater* Aurivillius
Current name: same
Current placement: Cerambycidae: Cerambycinae: Clytini
Type locality: Philippines: Bukidnon, Tangcolan

frater salomonum **Kreische, 1936**: 68 (Fig. 70w, x), Holotype
Original combination: *Periaptodes frater salomonum* Kreische
Current name: same
Current placement: Cerambycidae: Lamiinae: Lamiini
Type locality: Papua New Guinea: Bougainville, Kieta

fraterculus **Martins & Galileo, 2008a**: 273 (Fig. 71a, b), Holotype
Original combination: *Neoclytus fraterculus* Martins & Galileo
Current name: same
Current placement: Cerambycidae: Cerambycinae: Clytini
Type locality: Venezuela: Guarico, Hato Masaguaral, 45 km south of Calabozo

frigida **Casey, 1913**: 328 (Fig. 71c, d), Lectotype
Original combination: *Hyperplatys frigida* Casey
Current name: *Hyperplatys maculata* Haldeman, 1847
Current placement: Cerambycidae: Lamiinae: Acanthocinini
Type locality: USA: Maine, Monmouth
Remarks: This specimen is herein designated as the lectotype in order to stabilize the taxonomy and facilitate further identifications of this species.

frigidum **Casey, 1912**: 289 (Fig. 71e, f), Holotype
Original combination: *Callidium frigidum* Casey
Current name: same
Current placement: Cerambycidae: Cerambycinae: Callidiini
Type locality: Canada

frosti **Casey, 1912**: 375 (Fig. 71g, h), Holotype
Original combination: *Microclytus frosti* Casey
Current name: *Microclytus compressicollis* (Castelnau & Gory, 1841)
Current placement: Cerambycidae: Cerambycinae: Anaglyptini
Type locality: USA: Maine, Monmouth

frosti **Casey, 1924**: 210 (Fig. 71i, j), Lectotype
Original combination: *Prionus (Prionellus) frosti* Casey
Current name: *Prionus laticollis* (Drury, 1773)
Current placement: Cerambycidae: Prioninae: Prionini
Type locality: USA: Massachusetts, Framingham
Remarks: This specimen is herein designated as the lectotype in order to stabilize the taxonomy and facilitate further identifications of this species.

frosti **Casey, 1924**: 240 (Fig. 71k, l), Lectotype
Original combination: *Phymatodes frosti* Casey
Current name: *Phymatodes dimidiatus* (Kirby, 1837)
Current placement: Cerambycidae: Cerambycinae: Callidiini
Type locality: USA: Maine, Monmouth
Remarks: This specimen is herein designated as the lectotype in order to stabilize the taxonomy and facilitate further identifications of this species.

fuchsii **Wickham, 1905**: 170 (Fig. 71m, n), Holotype
Original combination: *Elaphidion fuchsii* Wickham
Current name: *Neaneflus fuchsii* (Wickham, 1905)
Current placement: Cerambycidae: Cerambycinae: Elaphidiini
Type locality: USA: California, Independence

fulgens **Fisher, 1932**: 48 (Fig. 71o, p), Holotype
Original combination: *Heterachtes fulgens* Fisher
Current name: *Neocompsa fulgens* (Fisher, 1932)
Current placement: Cerambycidae: Cerambycinae: Neoibidionini
Type locality: Dominica
Remarks: The genus name was spelled incorrectly as *Heterachtes* and should be emended to *Heterachthes*.

fulguratus **Casey, 1912**: 367 (Fig. 71q, r), Holotype
Original combination: *Neoclytus fulguratus* Casey
Current name: *Neoclytus horridus* LeConte, 1862
Current placement: Cerambycidae: Cerambycinae: Clytini
Type locality: USA: Texas

fullonium **Newman, 1841**: 110 (Fig. 71s, t), Neotype
Original combination: *Elaphidion fullonium* Newman
Current name: same
Current placement: Cerambycidae: Cerambycinae: Elaphidiini
Type locality: Haiti: Hinche
Remarks: This specimen was designed as the neotype by Lingafelter (2008: 357).

fulmeki **Fisher, 1934a**: 5 (Fig. 71u, v), Holotype
Original combination: *Glenea fulmeki* Fisher
Current name: same
Current placement: Cerambycidae: Lamiinae: Saperdini
Type locality: Indonesia: Sumatra, Dolok Baros

fulva **Fisher, 1925a**: 226 (Fig. 71w, x), Holotype
Original combination: *Ebaeides fulva* Fisher
Current name: same
Current placement: Cerambycidae: Lamiinae: Acanthocinini
Type locality: Malaysia: Borneo, Sandakan

fulvicornis **Aurivillius, 1903**: 322 (Fig. 72a, b), Lectotype
Original combination: *Cyrtocris fulvicornis* Aurivillius
Current name: same
Current placement: Cerambycidae: Lamiinae: Theocridini
Type locality: Tanzania: Lukuledi
Remarks: This specimen is herein designated as the lectotype in order to stabilize the taxonomy and facilitate further identifications of this species. Other paralectotype(s) may be present in the Naturhistoriska Riksmuseet, Stockholm, Sweden.

fulvipenne **Casey, 1912**: 260 (Fig. 72c, d), Holotype
Original combination: *Asemum fulvipenne* Casey
Current name: *Asemum striatum* (Linnaeus, 1758)
Current placement: Cerambycidae: Spondylidinae: Asemini
Type locality: USA: Wyoming, Laramie

fulvipenne levipes **Casey, 1912**: 323 (Fig. 72e, f), Holotype
Original combination: *Tragidion fulvipenne levipes* Casey
Current name: *Tragidion coquus* (Linnaeus, 1758)
Current placement: Cerambycidae: Cerambycinae: Trachyderini
Type locality: USA: New Mexico, Las Vegas

fulvipenne nubifer **Casey, 1912**: 323 (Fig. 72g, h), Holotype
Original combination: *Tragidion fulvipenne nubifer* Casey
Current name: *Tragidion coquus* (Linnaeus, 1758)
Current placement: Cerambycidae: Cerambycinae: Trachyderini
Type locality: USA: Texas

fulvipennis **Gressitt, 1935g**: 267 (Fig. 72i, j), Holotype
Original combination: *Distenia fulvipennis* Gressitt
Current name: same
Current placement: Disteniidae: Disteniinae: Disteniini
Type locality: Thailand: Chiang Mai

fulvonotatus **Fisher, 1925a**: 251 (Fig. 72k, l), Holotype
Original combination: *Myromeus fulvonotatus* Fisher
Current name: same
Current placement: Cerambycidae: Lamiinae: Acanthocinini
Type locality: Malaysia: Borneo, Sandakan

fulvovestitus **Schaeffer, 1905a**: 134 (Fig. 72m, n), Syntype
Original combination: *Eupogonius fulvovestitus* Schaeffer
Current name: same
Current placement: Cerambycidae: Lamiinae: Desmiphorini
Type locality: USA: Texas, Brownsville, Las Boragos

fulvus **Gressitt, 1938a**: 53 (Fig. 72o, p), Holotype
Original combination: *Anaespogonius fulvus* Gressitt
Current name: same
Current placement: Cerambycidae: Lamiinae: Desmiphorini
Type locality: China: Sichuan Province, near Muping, 3000–7400 ft.

fumosus **Fisher, 1925a**: 249 (Fig. 72q, r), Holotype
Original combination: *Myromeus fumosus* Fisher
Current name: *Myromeus immaculicollis* Heller, 1924c
Current placement: Cerambycidae: Lamiinae: Acanthocinini
Type locality: Philippines: Mindanao, Dapitan

funebris acomana **Casey, 1912**: 313 (Fig. 72s, t), Holotype
Original combination: *Rosalia funebris acomana* Casey
Current name: *Rosalia funebris* Motschulsky, 1845
Current placement: Cerambycidae: Cerambycinae: Compsocerini
Type locality: USA: New Mexico, Jemez Springs

funebris mandibularis **Casey, 1912**: 313 (Fig. 72u, v), Lectotype
Original combination: *Rosalia funebris mandibularis* Casey
Current name: *Rosalia funebris* Motschulsky, 1845
Current placement: Cerambycidae: Cerambycinae: Compsocerini
Type locality: USA: California
Remarks: This specimen is herein designated as the lectotype in order to stabilize the taxonomy and facilitate further identifications of this species.

***funeraria* Lane, 1958**: 13 (Fig. 72w, x), Holotype
Original combination: *Orteguaza funeraria* Lane
Current name: *Clavidesmus funerarius* (Lane, 1958)
Current placement: Cerambycidae: Lamiinae: Apomecynini
Type locality: Bolivia: La Paz, Tumupasa

***fusca* Fisher, 1925a**: 209 (Fig. 73a, b), Holotype
Original combination: *Enispia fusca* Fisher
Current name: *Athylia fusca* (Fisher, 1925a)
Current placement: Cerambycidae: Lamiinae: Pteropliini
Type locality: Philippines: Negros Oriental, Cuernos Mountains

***fusca* Fisher, 1925a**: 219 (Fig. 73c, d), Holotype
Original combination: *Neoegesina fusca* Fisher
Current name: *Egesina fusca* (Fisher, 1925a)
Current placement: Cerambycidae: Lamiinae: Pteropliini
Type locality: Malaysia: Borneo, Sandakan

***fuscoapicalis* Breuning, 1942a**: 139 (Fig. 73e, f), Holotype
Original combination: *Adetus fuscoapicalis* Breuning
Current name: same
Current placement: Cerambycidae: Lamiinae: Apomecynini
Type locality: Brazil: Santa Catarina, Nova Teutonia

***fuscomarmorata* Breuning, 1940b**: 71 (Fig. 73g, h), Holotype
Original combination: *Estola fuscomarmorata* Breuning
Current name: same
Current placement: Cerambycidae: Lamiinae: Desmiphorini
Type locality: Bolivia

***fuscoplagiata* Breuning, 1942a**: 132 (Fig. 73i, j), Holotype
Original combination: *Batrachorhina fuscoplagiata* Breuning
Current name: *Batrachorhina (Coedomea) madagascariensis* (Thomson, 1868)
Current placement: Cerambycidae: Lamiinae: Pteropliini
Type locality: Madagascar: Ambovombe
Remarks: According to Tavakilian and Chevillotte (2013), the holotype is in the Institut Royal des Sciences Naturelles de Belgique, however this specimen matches the original data.

***fuscopunctatus* Aurivillius, 1927b**: 552 (Fig. 73k, l), Lectotype
Original combination: *Dihammus fuscopunctatus* Aurivillius
Current name: *Acalolepta fuscopunctata* (Aurivillius, 1927b)
Current placement: Cerambycidae: Lamiinae: Monochamini
Type locality: Philippines: Samar
Remarks: This specimen is herein designated as the lectotype in order to stabilize the taxonomy and facilitate further identifications of this species. Two paralectotypes are present in the USNM and others in the Naturhistoriska Riksmuseet, Stockholm, Sweden.

***fuscotriangularis* Breuning, 1939c**: 268 (Fig. 73m, n), Holotype
Original combination: *Sybra fuscotriangularis* Breuning
Current name: same
Current placement: Cerambycidae: Lamiinae: Apomecynini
Type locality: Indonesia: Java
Remarks: It is not certain if this is the primary type. The genus name (*Ropica*) is incorrect on the label. The original description states the type is in The Natural History Museum, London.

***fusella* Casey, 1913**: 254 (Fig. 73o, p), Lectotype
Original combination: *Brachyleptura fusella* Casey
Current name: *Brachyleptura vagans* (Olivier, 1795)
Current placement: Cerambycidae: Lepturinae: Lepturini
Type locality: There is no locality data associated with this specimen.
Remarks: A lectotype label was placed on this specimen by Laplante in 1991, however, it was not formally designated until now. This specimen is herein designated as the lectotype in order to stabilize the taxonomy and facilitate further identifications of this species.

***galapagoensis* Linell, 1899**: 259 (Fig. 73q, r), Lectotype
Original combination: *Achryson galapagoensis* Linell
Current name: *Achryson lineolatum* Erichson, 1847
Current placement: Cerambycidae: Cerambycinae: Achrysonini
Type locality: Ecuador: Galapagos Islands, Chatham Island
Remarks: This specimen is herein designated as the lectotype in order to stabilize the taxonomy and facilitate further identifications of this species.

***galapagoensis* Linell, 1899**: 261 (Fig. 73s, t), Lectotype
Original combination: *Acanthoderes galapagoensis* Linell
Current name: *Psapharochrus galapagoensis* (Linell, 1899)
Current placement: Cerambycidae: Lamiinae: Acanthoderini
Type locality: Ecuador: Galápagos Islands, Chatham Island
Remarks: There are two specimens with the same USNM type No. 1319 (male and female). Linell (1898) did not single out one as the holotype. This specimen figured here is designated as the lectotype in order to stabilize the taxonomy and facilitate further identifications of this species.

***gaucha* Galileo & Martins, 2008b**: 32 (Fig. 73u, v), Holotype
Original combination: *Drycothaea gaucha* Galileo & Martins
Current name: same
Current placement: Cerambycidae: Lamiinae: Calliini
Type locality: Brazil: Rio Grande Do Sul, Pelotas

gaurotoides **Casey, 1893**: 592 (Fig. 73w, x), Holotype
Original combination: *Leptura gaurotoides* Casey
Current name: *Judolia gaurotoides* (Casey, 1893)
Current placement: Cerambycidae: Lepturinae: Lepturini
Type locality: USA: Utah

gemellus **Casey, 1893**: 590 (Fig. 74a, b), Lectotype
Original combination: *Xylotrechus gemellus* Casey
Current name: same
Current placement: Cerambycidae: Cerambycinae: Clytini
Type locality: USA: Indiana
Remarks: Lectotype designated by Heffern (1998).

geminatum **Martins, 1969**: 840 (Fig. 74c, d), Holotype
Original combination: *Tetraopidion geminatum* Martins
Current name: same
Current placement: Cerambycidae: Cerambycinae: Neoibidionini
Type locality: Bolivia: Santa Cruz

geometrica **Tippmann, 1960**: 198 (Fig. 74e, f), Holotype
Original combination: *Nyssodrys geometrica* Tippmann
Current name: *Eucharitolus geometricus* (Tippmann, 1960)
Current placement: Cerambycidae: Lamiinae: Acanthocinini
Type locality: Bolivia: Santa Cruz, 500 m.

geometricus **Breuning, 1940d**: 416 (Fig. 74g, h), Holotype
Original combination: *Metamulciber geometricus* Breuning
Current name: *Metamulciber lineatus* (Aurivillius, 1920)
Current placement: Cerambycidae: Lamiinae: Homonoeini
Type locality: Papua New Guinea

giesberti **Galileo & Martins, 2007b**: 632 (Fig. 74i, j), Holotype
Original combination: *Nesozineus giesberti* Galileo & Martins
Current name: same
Current placement: Cerambycidae: Lamiinae: Acanthoderini
Type locality: Panama: Cocle Province, El Valle 2400 ft.

giesberti **Lingafelter & Nearns, 2007**: 180 (Fig. 74k, l), Holotype
Original combination: *Hormathus giesberti* Lingafelter & Nearns
Current name: same
Current placement: Cerambycidae: Cerambycinae: Neoibidionini
Type locality: Dominican Republic: La Altagracia, Punta Cana near Eco. Reserve, 0–5 m.

giesberti **Monné, 2011**: 41 (Fig. 74m, n), Holotype
Original combination: *Stenolis giesberti* Monné
Current name: same
Current placement: Cerambycidae: Cerambycinae: Acanthocinini
Type locality: Panama: Panama Province, Cerro Azul, Altos De Pacoras

giesberti **Monné & Monné, 2007**: 1039 (Fig. 74o, p), Holotype
Original combination: *Baryssinus giesberti* Monné & Monné
Current name: same
Current placement: Cerambycidae: Lamiinae: Acanthocinini
Type locality: Panama: Panama Province, km 8–11, El Llano-Cartí Road

giesberti **Wappes & Lingafelter, 2005**: 7 (Fig. 74q, r), Holotype
Original combination: *Canidia giesberti* Wappes & Lingafelter
Current name: same
Current placement: Cerambycidae: Lamiinae: Acanthocinini
Type locality: Mexico: Chiapas, Oaxaca border on Pan-Am Hwy., 800 m.
Remarks: On long term term loan from UNAM (Mexico)

gigas **Galileo & Martins, 2012a**: 63 (Fig. 74s, t), Holotype
Original combination: *Psapharochrus gigas* Galileo & Martins
Current name: same
Current placement: Cerambycidae: Lamiinae: Acanthoderini
Type locality: Peru: Sinchona

gigas **Martins & Galileo, 2007c**: 1146 (Fig. 74u, v), Holotype
Original combination: *Apagomerina gigas* Martins & Galileo
Current name: same
Current placement: Cerambycidae: Lamiinae: Hemilophini
Type locality: Brazil: Santa Catarina, Joinville

gilana **Casey, 1924**: 243 (Fig. 74w, x), Holotype
Original combination: *Aneflomorpha gilana* Casey
Current name: same
Current placement: Cerambycidae: Cerambycinae: Elaphidiini
Type locality: USA: Arizona, Phoenix

gilloglyi **Santos-Silva & Lezama, 2010**: 400 (Fig. 75a, b), Holotype
Original combination: *Parandra gilloglyi* Santos-Silva & Lezama
Current name: same
Current placement: Cerambycidae: Parandrinae: Parandrini
Type locality: Panama: Chiriquí Province, Reserva Fortuna, 4020 ft.

gilvolineata **Monné, 2011**: 43 (Fig. 75c, d), Holotype
Original combination: *Stenolis gilvolineata* Monné
Current name: same
Current placement: Cerambycidae: Lamiinae: Acanthocinini
Type locality: Panama: Panama Province, Cerro Jefe (9°12′N, 79°21′W)

glabra **Fisher, 1932**: 51 (Fig. 75e, f), Holotype
Original combination: *Cylindera glabra* Fisher
Current name: *Curtomerus glabrus* (Fisher, 1932)
Current placement: Cerambycidae: Cerambycinae: Elaphidiini
Type locality: Cuba: Cayamas

glabropleura **Gressitt, 1935g**: 275 (Fig. 75g, h), Holotype
Original combination: *Acantholeptura glabropleura* Gressitt
Current name: *Elacomia femorata* (Pascoe, 1869)
Current placement: Cerambycidae: Lepturinae: Lepturini
Type locality: Malaysia: Sabah, Sandakan, Borneo

glaphyra **Martins, 1970a**: 1100 (Fig. 75i, j), Holotype
Original combination: *Neocompsa glaphyra* Martins
Current name: same
Current placement: Cerambycidae: Cerambycinae: Neoibidionini
Type locality: Panama: Panama Province, La Campana

glauca **var.** ***jamaicensis*** **Fisher, 1942a**: 26 (Fig. 75k, l), Holotype
Original combination: *Oreodera glauca* var. *jamaicensis* Fisher
Current name: same
Current placement: Cerambycidae: Lamiinae: Acanthoderini
Type locality: Jamaica: Stewart Town
Remarks: The name following the binomen was not expressly given infrasubspecific rank and the work did not indicate unambiguously that the name was infrasubspecific, therefore, it is deemed a subspecies and available.

globicollis **Casey, 1913**: 280 (Fig. 75m, n), Holotype
Original combination: *Ipochus globicollis* Casey
Current name: *Ipochus fasciatus* LeConte, 1852
Current placement: Cerambycidae: Lamiinae: Parmenini
Type locality: USA: California, Los Angeles Co.

gloriosa **Schultze, 1922**: 571 (Fig. 75o, p), Lectotype
Original combination: *Euclea gloriosa* Schultze
Current name: *Callimetopus gloriosus* (Schultze, 1922)
Current placement: Cerambycidae: Lamiinae: Pteropliini
Type locality: Philippines: Mindanao, Surigao
Remarks: This specimen is herein designated as the lectotype in order to stabilize the taxonomy and facilitate further identifications of this species. Other paralectotype(s) are present in the Staatliches Museum für Tierkunde, Dresden, Germany.

gracilicorne **Casey, 1912**: 258 (Fig. 75q, r), Holotype
Original combination: *Asemum gracilicorne* Casey
Current name: *Asemum striatum* (Linnaeus, 1758)
Current placement: Cerambycidae: Spondylidinae: Asemini
Type locality: USA: Wisconsin, Bayfield

gracilicornis **Casey, 1912**: 371 (Fig. 75s, t), Lectotype
Original combination: *Clytoleptus gracilicornis* Casey
Current name: *Clytoleptus albofasciatus* (Castelnau & Gory, 1836)
Current placement: Cerambycidae: Cerambycinae: Clytini
Type locality: USA: Indiana
Remarks: This specimen is herein designated as the lectotype in order to stabilize the taxonomy and facilitate further identifications of this species.

gracilipes **Casey, 1912**: 265 (Fig. 75u, v), Holotype
Original combination: *Nothorhina gracilipes* Casey
Current name: *Arhopalus rusticus montanus* (LeConte, 1873)
Current placement: Cerambycidae: Spondylidinae: Asemini
Type locality: USA: New Mexico, Las Vegas

gracilis **Breuning, 1940d**: 423 (Fig. 75w, x), Holotype
Original combination: *Niphovelleda gracilis* Breuning
Current name: same
Current placement: Cerambycidae: Lamiinae: Pteropliini
Type locality: Ethiopia (formerly Abyssinia)

gracilis **Fisher, 1932**: 43 (Fig. 76a, b), Holotype
Original combination: *Elaphidion gracilis* Fisher
Current name: *Xeranoplium gracilis* (Fisher, 1932)
Current placement: Cerambycidae: Cerambycinae: Hesperophanini
Type locality: Cuba: Camaguey

***gracilis* Fisher, 1932**: 82 (Fig. 76c, d), Holotype
Original combination: *Dorcasta gracilis* Fisher
Current name: same
Current placement: Cerambycidae: Lamiinae: Apomecynini
Type locality: Haiti: Fourgu

***grahami* Gressitt, 1937a**: 89 (Fig. 76e, f), Holotype
Original combination: *Allotraeus grahami* Gressitt
Current name: same
Current placement: Cerambycidae: Cerambycinae: Phoracanthini
Type locality: China: Sichuan Province, Suifu

***grahamiana* Gressitt, 1938a**: 46 (Fig. 76g, h), Holotype
Original combination: *Leptura grahamiana* Gressitt
Current name: same
Current placement: Cerambycidae: Lepturinae: Lepturini
Type locality: China: Sichuan Province, Washan, 11,000 ft.

***grammicus marshalli* Dillon & Dillon, 1948**: 274 (Fig. 76i, j), Holotype
Original combination: *Cylindrepomus grammicus marshalli* Dillon & Dillon
Current name: *Cylindrepomus grammicus* Pascoe, 1860
Current placement: Cerambycidae: Lamiinae: Gnomiini
Type locality: Papua New Guinea: Manus Province (Admiralty Islands)

***grandaevum* Wickham, 1917**: 469 (Fig. 76k, l), Holotype
Original combination: *Callidium grandaevum* Wickham
Current name: *Phymatodes grandaevus* (Wickham, 1917)
Current placement: Cerambycidae: Cerambycinae: Callidiini
Type locality: USA: Colorado, Florissant
Remarks: This is one of four fossil specimens at the USNM and is housed in the Paleobiology (Paleoentomology) section.

***grandis* Casey, 1912**: 277 (Fig. 76m, n), Holotype
Original combination: *Phymatodes (Poecilium) grandis* Casey
Current name: *Phymatodes grandis* Casey, 1912
Current placement: Cerambycidae: Cerambycinae: Callidiini
Type locality: USA: California, San Diego

***granulicolle* Gressitt, 1938c**: 166 (Fig. 76o, p), Holotype
Original combination: *Miaenia granulicolle* Gressitt
Current name: *Sciades (Granulimiaenia) granulicollis* (Gressitt, 1938c)
Current placement: Cerambycidae: Lamiinae: Acanthocinini
Type locality: China: Taiwan, Kuraru
Remarks: Specific epithet is misspelled as "*granulicolle*" on label.

***granulipennis* Breuning, 1954**: 21 (Fig. 76q, r), Holotype
Original combination: *Neosybra granulipennis* Breuning
Current name: same
Current placement: Cerambycidae: Lamiinae: Apomecynini
Type locality: Papua New Guinea: East New Britain, Mope, Neupommern
Remarks: Genus name is incorrect on the labels.

***granulosa* Breuning, 1938b**: 232 (Fig. 76s, t), Holotype
Original combination: *Eczemotes granulosa* Breuning
Current name: same
Current placement: Cerambycidae: Lamiinae: Niphonini
Type locality: Indonesia: Yos Sudarso Bay (formerly known as Humboldt Bay)

***gravidula* Casey, 1912**: 216 (Fig. 76u, v), Lectotype
Original combination: *Parandra gravidula* Casey
Current name: *Neandra brunnea* (Fabricius, 1798)
Current placement: Cerambycidae: Parandrinae: Parandrini
Type locality: USA: Indiana
Remarks: This specimen is herein designated as the lectotype in order to stabilize the taxonomy and facilitate further identifications of this species.

***gregalis* Fisher, 1927a**: 80 (Fig. 76w, x), Holotype
Original combination: *Pterolophia gregalis* Fisher
Current name: *Pterolophia (Ale) gregalis* Fisher, 1927a
Current placement: Cerambycidae: Lamiinae: Pteropliini
Type locality: Indonesia: Sumatra, Bekalla

***gressitti* Tippmann, 1955a**: 124 (Fig. 77a, b), Holotype
Original combination: *Kamikiria gressitti* Tippmann
Current name: *Notomulciber (Micromulciber) gressitti* (Tippmann, 1955a)
Current placement: Cerambycidae: Lamiinae: Apomecynini
Type locality: China: Taiwan, Musha

***grisea* Dillon & Dillon, 1946**: 268 (Fig. 77c, d), Holotype
Original combination: *Cacostola grisea* Dillon & Dillon
Current name: same
Current placement: Cerambycidae: Lamiinae: Onciderini
Type locality: Guyana: Demerara River bank, 1 mi from Georgetown

***griseiventris* Breuning, 1942a**: 131 (Fig. 77e, f), Holotype
Original combination: *Batrachorhina griseiventris* Breuning
Current name: *Batrachorhina (Coedomea) griseiventris* Breuning, 1942a
Current placement: Cerambycidae: Lamiinae: Pteropliini
Type locality: Mauritius

***griseomaculatus* Fisher, 1926**: 28 (Fig. 77g, h), Holotype
Original combination: *Hyperplatys griseomaculatus* Fisher
Current name: same
Current placement: Cerambycidae: Lamiinae: Acanthocinini
Type locality: Trinidad and Tobago: Montserrat Ward

***griseomarmorata* Breuning, 1939b**: 517 (Fig. 77i, j), Holotype
Original combination: *Mesosa (Aphelocnemia) griseomarmorata* Breuning
Current name: same
Current placement: Cerambycidae: Lamiinae: Mesosini
Type locality: China: Guangdong Province, Kuangtung, Nanning

***griseonotata* Breuning, 1957a**: 11 (Fig. 77k, l), Holotype
Original combination: *Driopea (Inermodriopea) griseonotata* Breuning
Current name: same
Current placement: Cerambycidae: Lamiinae: Acanthocinini
Type locality: Philippines: Luzon, Mt. Banahao

***grisescens* Dillon & Dillon, 1951**: 21 (Fig. 77m, n), Holotype
Original combination: *Imantocera grisescens* Dillon & Dillon
Current name: same
Current placement: Cerambycidae: Lamiinae: Gnomini
Type locality: Indonesia: East Java, Nusa Tenggara Barat, Senggoro Mts. of Res. Passoeroean

***griseum* Schaeffer, 1908**: 344 (Fig. 77o, p), Holotype
Original combination: *Parmenosoma griseum* Schaeffer
Current name: same
Current placement: Cerambycidae: Lamiinae: Parmenini
Type locality: USA: Texas, Edingburg (Hidalgo)

***griseus* Fisher, 1926**: 11 (Fig. 77q, r), Holotype
Original combination: *Eupogonius griseus* Fisher
Current name: same
Current placement: Cerambycidae: Lamiinae: Desmiphorini
Type locality: Cuba: Cayamas

***griseus novaki* Tippmann, 1952**: 153 (Fig. 77s, t), Holotype
Original combination: *Acanthocinus griseus novaki* Tippmann
Current name: same
Current placement: Cerambycidae: Lamiinae: Acanthocinini
Type locality: Croatia: Split, Dalmatia (formerly Yugoslavia)

***grossa* Tippmann, 1960**: 157 (Fig. 77u, v), Holotype
Original combination: *Oreodera grossa* Tippmann
Current name: *Oreodera charisoma* Lane, 1955
Current placement: Cerambycidae: Lamiinae: Acanthoderini
Type locality: Brazil: São Paulo, Peruibe

***grossepunctata* Breuning, 1942a**: 175 (Fig. 77w, x), Holotype
Original combination: *Epectasis grossepunctata* Breuning
Current name: *Epectasis juncea* (Newman, 1840)
Current placement: Cerambycidae: Lamiinae: Pteropliini
Type locality: Peru: Junin, Satipo

***grossepunctatus* Breuning, 1938b**: 360 (Fig. 78a, b), Holotype
Original combination: *Prosoplus grossepunctatus* Breuning
Current name: same
Current placement: Cerambycidae: Lamiinae: Pterolophini
Type locality: Indonesia: Maluku
Remarks: Tavakilian and Chevillotte (2013) suggest the holotype is present in the the Staatliches Museum für Tierkunde, Dresden, Germany, however this specimen and its labels match the original description.

***grossepunctatus* Breuning, 1939c**: 180 (Fig. 78c, d), Holotype
Original combination: *Temnosternus grossepunctatus* Breuning
Current name: same
Current placement: Cerambycidae: Lamiinae: Tmesisternini
Type locality: Australia: Queensland
Remarks: Although the original description states the holotype is in the Staatliches Museum für Tierkunde, Dresden, Germany, the specimen and label match the description.

***grossescapus* Breuning, 1942a**: 117 (Fig. 78e, f), Holotype
Original combination: *Dihammus grossescapus* Breuning
Current name: *Acalolepta grossescapus* (Breuning, 1942a)
Current placement: Cerambycidae: Lamiinae: Monochamini
Type locality: China: Taiwan, Kosyum (=Kaohsiung?)

***guamum guamum* Gressitt, 1956**: 85 (Fig. 78g, h), Holotype
Original combination: *Ceresium (Ceresium) guamum guamum* Gressitt
Current name: *Ceresium guamum* Gressitt, 1956
Current placement: Cerambycidae: Cerambycinae: Callidiopini
Type locality: USA: Guam, Pt. Ritidian, Marianas

guanica **Micheli & Micheli, 2004**: 22 (Fig. 78i, j), Holotype
Original combination: *Leptostylopsis guanica* Micheli & Micheli
Current name: same
Current placement: Cerambycidae: Lamiinae: Acanthocinini
Type locality: USA: Puerto Rico, Guánica

guarani **Monné & Monné, 2011**: 1859 (Fig. 78k, l), Holotype
Original combination: *Atrocolus guarani* Monné & Monné
Current name: same
Current placement: Cerambycidae: Prioninae: Anacolini
Type locality: Paraguay: Cordillera, San Bernardino

guatemalensis **Casey, 1912**: 227 (Fig. 78m, n), Holotype
Original combination: *Aplagiognathus guatemalensis* Casey
Current name: *Physopleurus villardi* (Lameere, 1902)
Current placement: Cerambycidae: Prioninae: Macrotomini
Type locality: Guatemala: Escuintla

guatemalensis **Casey, 1913**: 303 (Fig. 78o, p), Holotype
Original combination: *Psapharochrus guatemalensis* Casey
Current name: same
Current placement: Cerambycidae: Lamiinae: Acanthoderini
Type locality: Guatemala: Villa Nueva

guilartensis **Micheli & Micheli, 2004**: 20 (Fig. 78q, r), Holotype
Original combination: *Antilleptostylus guilartensis* Micheli & Micheli
Current name: *Styloleptus guilartensis* (Micheli & Micheli, 2004)
Current placement: Cerambycidae: Lamiinae: Acanthocinini
Type locality: USA: Puerto Rico, Guilarte Forest

guttata **Aurivillius, 1927b**: 578 (Fig. 78s, t), Syntype
Original combination: *Clodia guttata* Aurivillius
Current name: *Pseudoclodia guttata* (Aurivillius, 1927b)
Current placement: Cerambycidae: Lamiinae: Acanthocinini
Type locality: Philippines: Sibuyan, Romblon
Remarks: Other syntype(s) are present in the Naturhistoriska Riksmuseet, Stockholm, Sweden.

guttata **Fisher, 1935c**: 585 (Fig. 78u, v), Holotype
Original combination: *Eburiomorpha guttata* Fisher
Current name: same
Current placement: Cerambycidae: Cerambycinae: Oemini
Type locality: Malaysia: Borneo, Sabah, Mt. Kinabalu, 5000 ft., Marei Parei

guttata **Galileo & Martins, 2006b**: 26 (Fig. 78w, x), Holotype
Original combination: *Xenofrea guttata* Galileo & Martins
Current name: same
Current placement: Cerambycidae: Lamiinae: Xenofreini
Type locality: Mexico: Chiapas, Parque Aguacero

guttaticolle yapense **Gressitt, 1956**: 86 (Fig. 79a, b), Holotype
Original combination: *Ceresium (Ceresium) guttaticolle yapense* Gressitt
Current name: same
Current placement: Cerambycidae: Cerambycinae: Callidiopini
Type locality: Micronesia: Yap Islands, Dugoi

guttatus **Galileo & Martins, 2005a**: 277 (Fig. 79c, d), Holotype
Original combination: *Adesmus guttatus* Galileo & Martins
Current name: same
Current placement: Cerambycidae: Lamiinae: Hemilophini
Type locality: Peru: Cusco, Valle de Lares, 75 km northwest of Calca, el. 2060 m.

guttulatus **Dillon & Dillon, 1941**: 57 (Fig. 79e, f), Holotype
Original combination: *Ptychodes guttulatus* Dillon & Dillon
Current name: same
Current placement: Cerambycidae: Lamiinae: Monochamini
Type locality: Mexico: Guerrero

habra **Martins, 1970a**: 1065 (Fig. 79g, h), Holotype
Original combination: *Neocompsa habra* Martins
Current name: same
Current placement: Cerambycidae: Cerambycinae: Neoibidionini
Type locality: Venezuela

hainana **Gressitt, 1940**: 18 (Fig. 79i, j), Holotype
Original combination: *Macrotoma (Zooblax) hainana* Gressitt
Current name: *Anomophysis hainana* (Gressitt, 1940)
Current placement: Cerambycidae: Prioninae: Macrotomini
Type locality: China: Hainan Province (formerly Guangdong), Dwa-Bi (Tai-pin-tsuen), near Loi Mother Mountain, central Hainan Island

hainanicus **Gressitt, 1940**: 75 (Fig. 79k, l), Holotype
Original combination: *Chlorophorus hainanicus* Gressitt
Current name: same
Current placement: Cerambycidae: Cerambycinae: Clytini
Type locality: China: Hainan Province, Hainan Island, Dwa Bi (Tai-pin-tsuen)

haitiensis **Fisher, 1932**: 73 (Fig. 79m, n), Holotype
Original combination: *Ataxia haitiensis* Fisher
Current name: same
Current placement: Cerambycidae: Lamiinae: Pteropliini
Type locality: Haiti: near Hinche

haitiensis **Fisher, 1935b**: 197 (Fig. 79o, p), Holotype
Original combination: *Eupogonius haitiensis* Fisher
Current name: same
Current placement: Cerambycidae: Lamiinae: Desmiphorini
Type locality: Haiti: Plaisance

hakka **Gressitt, 1937b**: 597 (Fig. 79q, r), Holotype
Original combination: *Falsomesosella hakka* Gressitt
Current name: *Falsomesosella nigronotata hakka* Gresitt, 1937
Current placement: Cerambycidae: Lamiinae: Mesosini
Type locality: China: Guangdong, Yim Na Shan

haldemani **Casey, 1891**: 42 (Fig. 79s, t), Lectotype
Original combination: *Leptura haldemani* Casey
Current name: *Anastrangalia haldemani* (Casey, 1891)
Current placement: Cerambycidae: Lepturinae: Lepturini
Type locality: USA: New Mexico
Remarks: This specimen is herein designated as the lectotype in order to stabilize the taxonomy and facilitate further identifications of this species.

haliki **Martins, 1962a**: 301 (Fig. 79u, v), Holotype
Original combination: *Ibidion haliki* Martins
Current name: *Tropidion sipolisi* (Gounelle, 1909)
Current placement: Cerambycidae: Cerambycinae: Neoibidionini
Type locality: Brazil: São Paulo
Remarks: The holotype for this species is also recorded for the Museu de Zoologia, Universidade de São Paulo, where the Jaroslav Halik collection was deposited.

hamatum **Linsley, 1935b**: 479 (Fig. 79w, x), Holotype
Original combination: *Gnomidolon hamatum* Linsley
Current name: same
Current placement: Cerambycidae: Cerambycinae: Hexoplini
Type locality: Panama: Coclé, Penonome

harfordi **Casey, 1912**: 279 (Fig. 80a, b), Lectotype
Original combination: *Phymatodes (Poecilium) harfordi* Casey
Current name: *Phymatodes obliquus* Casey, 1891
Current placement: Cerambycidae: Cerambycinae: Callidiini
Type locality: USA: California, Santa Clara Co.
Remarks: This specimen is herein designated as the lectotype in order to stabilize the taxonomy and facilitate further identifications of this species.

harringtoni **Fisher, 1947b**: 181 (Fig. 80c, d), Holotype
Original combination: *Dorcasta harringtoni* Fisher
Current name: *Bebelis pseudolignosa* Breuning, 1942a
Current placement: Cerambycidae: Lamiinae: Apomecynini
Type locality: Argentina: Salta, General Ballivián

hebes **Casey, 1913**: 333 (Fig. 80e, f), Lectotype
Original combination: *Urographis hebes* Casey
Current name: *Graphisurus despectus* (LeConte in Agassiz, 1850)
Current placement: Cerambycidae: Lamiinae: Acanthocinini
Type locality: USA: Iowa
Remarks: This specimen is herein designated as the lectotype in order to stabilize the taxonomy and facilitate further identifications of this species.

hebes **Dillon, 1955**: 185 (Fig. 80g, h), Holotype
Original combination: *Canidiopsis hebes* Dillon
Current name: *Canidia mexicana* Thomson, 1860
Current placement: Cerambycidae: Lamiinae: Acanthocinini
Type locality: Mexico: Morelos, Cuernavaca

hecabe **Dillon & Dillon, 1945**: 141 (Fig. 80i, j), Holotype
Original combination: *Midamus hecabe* Dillon & Dillon
Current name: *Midamiella hecabe* (Dillon & Dillon, 1945)
Current placement: Cerambycidae: Lamiinae: Onciderini
Type locality: Paraguay: San Bernardino

hefferni **Morati & Huet, 2004**: 102 (Fig. 80k, l), Holotype
Original combination: *Leptosiella hefferni* Morati & Huet
Current name: same
Current placement: Cerambycidae: Cerambycinae: Callichromatini
Type locality: Indonesia: West Sumatra, Muara-Labuh

herteli **Tippmann, 1953d**: 330 (Fig. 80m, n), Holotype
Original combination: *Trachyderes herteli* Tippmann
Current name: *Chydarteres octolineatus* (Thunberg, 1822)
Current placement: Cerambycidae: Cerambycinae: Trachyderini
Type locality: Brazil: Santa Catarina, Corupa

herteli **Tippmann, 1960**: 131 (Fig. 80o, p), Holotype
Original combination: *Closteropus herteli* Tippmann
Current name: same
Current placement: Cerambycidae: Cerambycinae: Rhopalophorini
Type locality: Peru: Junin, Satipo

herteli flavitarsis **Tippmann, 1960**: 132 (Fig. 80q, r), Holotype
Original combination: *Closteropus herteli flavitarsis* Tippmann
Current name: *Closteropus herteli* Tippmann, 1960
Current placement: Cerambycidae: Cerambycinae: Rhopalophorini
Type locality: Bolivia: Cochabamba, Reg. Chapare 400 m.

hesperus **Casey, 1913**: 210 (Fig. 80s, t), Holotype
Original combination: *Stenocorus (Stenocorus) hesperus* Casey
Current name: *Stenocorus nubifer* (LeConte, 1860)
Current placement: Cerambycidae: Lepturinae: Rhagiini
Type locality: USA: California, Lake Co.

hexagonum **Casey, 1924**: 228 (Fig. 80u, v), Holotype
Original combination: *Tetropium hexagonum* Casey
Current name: *Tetropium velutinum* LeConte, 1869
Current placement: Cerambycidae: Spondylidinae: Asemini
Type locality: USA: California, Humboldt Co.

heyrovskyi **Tippmann, 1946b**: 1 (Fig. 80w, x), Holotype
Original combination: *Xixuthrus heyrovskyi* Tippmann
Current name: *Xixuthrus terribilis* Thomson, 1877
Current placement: Cerambycidae: Prioninae: Xixuthrini
Type locality: Fiji: Viti Levu

hibisci **Gressitt, 1956**: 135 (Fig. 81a, b), Holotype
Original combination: *Prosoplus hibisci* Gressitt
Current name: same
Current placement: Cerambycidae: Lamiinae: Pterolophini
Type locality: Marshall Islands: Majuroa

hirayamai **Ohbayashi, 1948**: 13 (Fig. 81c, d), Holotype
Original combination: *Necydalis hirayamai* Ohbayashi
Current name: *Necydalis hirayamai* Ohbayashi, 1948
Current placement: Cerambycidae: Necydalinae: Necydalini
Type locality: China: Taiwan, Inouye

hirsuta **Breuning, 1943c**: 25 (Fig. 81e, f), Holotype
Original combination: *Pterolophia hirsuta* Breuning
Current name: *Pterolophia (Ale) hirsuta* Breuning, 1943c
Current placement: Cerambycidae: Lamiinae: Pteropliini
Type locality: Laos

hirsuta **Casey, 1913**: 228 (Fig. 81g, h), Lectotype
Original combination: *Brachysomida hirsuta* Casey
Current name: *Brachysomida californica* (LeConte, 1851)
Current placement: Cerambycidae: Lepturinae: Rhagiini
Type locality: USA: California
Remarks: This specimen is herein designated as the lectotype in order to stabilize the taxonomy and facilitate further identifications of this species.

hirsuta **Fisher, 1925a**: 230 (Fig. 81i, j), Holotype
Original combination: *Ebaeides hirsuta* Fisher
Current name: same
Current placement: Cerambycidae: Lamiinae: Acanthocinini
Type locality: Philippines: Mindanao, Iligan

hirsuta **Fisher, 1926**: 32 (Fig. 81k, l), Holotype
Original combination: *Eneodes hirsuta* Fisher
Current name: same
Current placement: Cerambycidae: Lamiinae: Acanthocinini
Type locality: Cuba: Cayamas

hirsuticolle **Casey, 1912**: 321 (Fig. 81m, n), Holotype
Original combination: *Tragidion hirsuticolle* Casey
Current name: *Tragidion auripenne* Casey, 1893
Current placement: Cerambycidae: Cerambycinae: Trachyderini
Type locality: USA: New Mexico, Fort Wingate

hirsuticollis **Casey, 1924**: 266 (Fig. 81o, p), Holotype
Original combination: *Crossidius hirsuticollis* Casey
Current name: *Crossidius punctatus* LeConte, 1873
Current placement: Cerambycidae: Cerambycinae: Trachyderini
Type locality: USA: Utah, Stockton

hirsutipes **Aurivillius, 1922**: 434 (Fig. 81q, r), Holotype
Original combination: *Ossonis hirsutipes* Aurivillius
Current name: same
Current placement: Cerambycidae: Lamiinae: Saperdini
Type locality: Malaysia: Borneo, Sandakan

hirticollis **Tippmann, 1953c**: 183 (Fig. 81s, t), Holotype
Original combination: *Strongylaspis hirticollis* Tippmann
Current name: same
Current placement: Cerambycidae: Prioninae: Macrotomini
Type locality: Colombia: Bogotá

hirticornis **Gressitt, 1936**: 102 (Fig. 81u, v), Holotype
Original combination: *Saimia hirticornis* Gressitt
Current name: *Agelasta (Parasaimia) hirticornis* (Gressitt, 1936)
Current placement: Cerambycidae: Lamiinae: Mesosini
Type locality: China: Taiwan, Hassenzan

hirtipes **Tippmann, 1960**: 144 (Fig. 81w, x), Holotype
Original combination: *Phaedinus hirtipes* Tippmann
Current name: same
Current placement: Cerambycidae: Cerambycinae: Trachyderini
Type locality: Bolivia: Cochabamba, Reg. Chapare 400 m.

***hirtus* Fisher, 1925a**: 245 (Fig. 82a, b), Holotype
Original combination: *Exocentrus hirtus* Fisher
Current name: *Exocentrus (Camptomyne) hirtus* Fisher, 1925a
Current placement: Cerambycidae: Lamiinae: Acanthocinini
Type locality: Malaysia: Borneo, Sandakan

***hispaniolae* Fisher, 1932**: 30 (Fig. 82c, d), Holotype
Original combination: *Elaphidion hispaniolae* Fisher
Current name: *Anelaphus hispaniolae* (Fisher, 1932)
Current placement: Cerambycidae: Cerambycinae: Elaphidiini
Type locality: Haiti: Port-au-Prince

***hispaniolae* Fisher, 1932**: 72 (Fig. 82e, f), Holotype
Original combination: *Heterops hispaniolae* Fisher
Current name: same
Current placement: Cerambycidae: Cerambycinae: Heteropsini
Type locality: Dominican Republic: Santo Domingo

***hispaniolae* Lingafelter, 2010**: 266 (Fig. 82g, h), Holotype
Original combination: *Tessaropa hispaniolae* Lingafelter
Current name: same
Current placement: Cerambycidae: Cerambycinae: Methiini
Type locality: Dominican Republic: Pedernales, 25.5 km. north Cabo Rojo

***hispidus* Casey, 1913**: 279 (Fig. 82i, j), Holotype
Original combination: *Ipochus hispidus* Casey
Current name: *Ipochus fasciatus* LeConte, 1852
Current placement: Cerambycidae: Lamiinae: Parmenini
Type locality: USA: California, San Diego

***histrio* Casey, 1913**: 302 (Fig. 82k, l), Holotype
Original combination: *Psapharochrus histrio* Casey
Current name: same
Current placement: Cerambycidae: Lamiinae: Acanthoderini
Type locality: Honduras

***hoehneli* Linell, 1896**: 693 (Fig. 82m, n), Holotype
Original combination: *Prosopocera hoehneli* Linell
Current name: *Prosopocera (Dalterus) cornifrons* Gahan, 1890
Current placement: Cerambycidae: Lamiinae: Prosopocerini
Type locality: Kenya: Tana River

***hoferi* Fisher, 1919**: 39 (Fig. 82o, p), Holotype
Original combination: *Elatotrypes hoferi* Fisher
Current name: same
Current placement: Cerambycidae: Cerambycinae: Callidiini
Type locality: USA: Colorado, Ute Pass

***hoffmanni* Fisher, 1932**: 78 (Fig. 82q, r), Holotype
Original combination: *Ecyrus hoffmanni* Fisher
Current name: *Ecyrus hirtipes* Gahan, 1895
Current placement: Cerambycidae: Lamiinae: Pogonocherini
Type locality: Haiti: Hinche

***hololeucus* Linell, 1896**: 693 (Fig. 82s, t), Holotype
Original combination: *Paraphosphorus hololeucus* Linell
Current name: same
Current placement: Cerambycidae: Lamiinae: Tragocephalini
Type locality: Kenya: Tana River, East Africa

***honduranus* Casey, 1912**: 265 (Fig. 82u, v), Holotype
Original combination: *Criocephalus honduranus* Casey
Current name: *Arhopalus asperatus* (LeConte, 1859)
Current placement: Cerambycidae: Spondylidinae: Asemini
Type locality: Honduras

***hondurenha* Santos-Silva & Hovore, 2008a**: 7 (Fig. 82w, x), Holotype
Original combination: *Elytrimitatrix (Grossifemora) hondurenha* Santos-Silva & Hovore
Current name: same
Current placement: Disteniidae: Disteniinae: Disteniini
Type locality: Honduras, Olancho, La Muralla

***horiensis hongshanus* Gressitt, 1937b**: 614 (Fig. 83a, b), Holotype
Original combination: *Rondibilis horiensis hongshanus* Gressitt
Current name: same
Current placement: Cerambycidae: Lamiinae: Acanthocinini
Type locality: China: Jiangxi Province, Hong Shan
Remarks: Subspecific epithet misspelled on holotype label.

***horioni* Tippmann, 1960**: 133 (Fig. 83c, d), Holotype
Original combination: *Megacyllene (Sierracyllene) horioni* Tippmann
Current name: *Megacyllene horioni* Tippmann, 1960
Current placement: Cerambycidae: Cerambycinae: Clytini
Type locality: Bolivia: Cochabamba, 2600 m.

***hornii* Joutel, 1902**: 33 (Fig. 83e, f), Holotype
Original combination: *Saperda hornii* Joutel
Current name: same
Current placement: Cerambycidae: Lamiinae: Saperdini
Type locality: USA: California, Los Angeles Co.

houstoni **Casey, 1913**: 315 (Fig. 83g, h), Lectotype
Original combination: *Leiopus houstoni* Casey
Current name: *Sternidius mimeticus* (Casey, 1891)
Current placement: Cerambycidae: Lamiinae: Acanthocinini
Type locality: USA: Texas, Brownsville
Remarks: This specimen is herein designated as the lectotype in order to stabilize the taxonomy and facilitate further identifications of this species.

hovorei **Lingafelter & Nearns, 2007**: 184 (Fig. 83i, j), Holotype
Original combination: *Ataxia hovorei* Lingafelter & Nearns
Current name: same
Current placement: Cerambycidae: Lamiinae: Pteropliini
Type locality: Dominican Republic: La Altagracia Province, Punta Cana near Ecological Reserve, 0–5 m, 18°30.477′N, 68°22.499′W

hovorei **Lingafelter & Woodley, 2007**: 167 (Fig. 83k, l), Holotype
Original combination: *Derancistrus hovorei* Lingafelter & Woodley
Current name: same
Current placement: Cerambycidae: Prioninae: Solenopterini
Type locality: Dominican Republic: Pedernales Province, 25 km north of Cabo Rojo, 679 m., 18°06.769′N, 71°37.245′W

hovorei **Santos-Silva, 2007**: 589 (Fig. 83m, n), Holotype
Original combination: *Novantinoe hovorei* Santos-Silva
Current name: same
Current placement: Disteniidae: Disteniinae: Disteniini
Type locality: Costa Rica: Puntarenas Province, Cerro Chai nr. summit, 8°58′N, 82°49′W 2000 m.

hovorei **Swift, 2007**: 287 (Fig. 83o, p), Holotype
Original combination: *Xylotrechus hovorei* Swift
Current name: same
Current placement: Cerambycidae: Cerambycinae: Clytini
Type locality: USA: California, San Bernardino Co., Mt. Baldy Road

huachucae **Casey, 1924**: 245 (Fig. 83q, r), Holotype
Original combination: *Anoplium huachucae* Casey
Current name: *Eustromula validum* (LeConte, 1858)
Current placement: Cerambycidae: Cerambycinae: Elaphidiini
Type locality: USA: Arizona, Huachuca Mts.

huachucae **Schaeffer, 1905b**: 164 (Fig. 83s, t), Syntype
Original combination: *Euryptera huachucae* Schaeffer
Current name: *Orthochoriolaus chihuahuae* (Bates, 1885)
Current placement: Cerambycidae: Lepturinae: Lepturini
Type locality: USA: Arizona, Huachuca Mts.

hubbardi **Fisher, 1924**: 253 (Fig. 83u, v), Holotype
Original combination: *Ataxia hubbardi* Fisher
Current name: same
Current placement: Cerambycidae: Lamiinae: Pteropliini
Type locality: USA: Arizona, Tucson

hubbardi **Fisher, 1926**: 38 (Fig. 83w, x), Holotype
Original combination: *Cyrtinus hubbardi* Fisher
Current name: same
Current placement: Cerambycidae: Lamiinae: Cyrtinini
Type locality: Montserrat

humeralis **Aurivillius, 1926**: 95 (Fig. 84a, b), Holotype
Original combination: *Glenea (Glenea) humeralis* Aurivillius
Current name: *Glenea humeralis* Aurivillius, 1926
Current placement: Cerambycidae: Lamiinae: Saperdini
Type locality: Philippines: Polillo

humeralis **Aurivillius, 1927b**: 572 (Fig. 84c, d), Syntype
Original combination: *Sybra humeralis* Aurivillius
Current name: same
Current placement: Cerambycidae: Lamiinae: Apomecynini
Type locality: Philippines: Mindanao, Kolambugan
Remarks: Tavakilian and Chevillotte (2013) states that syntypes are also present in the Naturhistoriska Riksmuseet, Stockholm, Sweden.

humeralis **Aurivillius, 1928**: 308 (Fig. 84e, f), Syntype
Original combination: *Xylotrechus humeralis* Aurivillius
Current name: same
Current placement: Cerambycidae: Cerambycinae: Clytini
Type locality: Philippines: Samar
Remarks: Tavakilian and Chevillotte (2013) state that syntypes are present in the Naturhistoriska Riksmuseet, Stockholm, Sweden as well as the USNM.

humeralis **Casey, 1913**: 299 (Fig. 84g, h), Holotype
Original combination: *Synaphaeta humeralis* Casey
Current name: *Synaphaeta guexi* (LeConte, 1852)
Current placement: Cerambycidae: Lamiinae: Mesosini
Type locality: USA: California

humeralis **Casey, 1913**: 385 (Fig. 84i, j), Holotype
Original combination: *Tetraopes humeralis* Casey
Current name: *Tetraopes tetrophthalmus* (Forster, 1771)
Current placement: Cerambycidae: Lamiinae: Tetraopini
Type locality: USA: South or North Dakota

humeralis **Casey, 1924**: 216 (Fig. 84k, l), Holotype
Original combination: *Prionus (Prionus) humeralis* Casey
Current name: *Prionus californicus* Motschulsky, 1845
Current placement: Cerambycidae: Prioninae: Prionini
Type locality: USA: California

humeridens **Breuning, 1940d**: 410 (Fig. 84m, n), Holotype
Original combination: *Mimovelleda humeridens* Breuning
Current name: same
Current placement: Cerambycidae: Lamiinae: Morimopsini
Type locality: Tanzania: Usambara, Lutindi

humerofulvus **Lingafelter & Micheli, 2009**: 39 (Fig. 84o, p), Holotype
Original combination: *Leptostylopsis humerofulvus* Lingafelter & Micheli
Current name: same
Current placement: Cerambycidae: Lamiinae: Acanthocinini
Type locality: Dominican Republic: Pedernales Province, Parque Nacional Sierra de Baoruco, Las Abejas, 1150 m, 18°09.011′N, 71°37.342′W

humerointerrupta **Breuning, 1956**: 798 (Fig. 84q, r), Holotype
Original combination: *Glenea (Glenea) humerointerrupta* Breuning
Current name: *Glenea humerointerrupta* Breuning, 1956
Current placement: Cerambycidae: Lamiinae: Saperdini
Type locality: Indonesia: Java, Mts. Tengger

iani **Santos-Silva & Tavakilian, 2009**: 23 (Fig. 84s, t), Holotype
Original combination: *Abauba iani* Santos-Silva & Tavakilian
Current name: same
Current placement: Disteniidae: Disteniinae: Disteniini
Type locality: Ecuador: Napo, Res. Ethnica Waorani, 1km S. Onkone Gare Camp, 220 m., 00°39′10S 076°26′W

icima **Martins & Galileo, 2012**: 108 (Fig. 84u, v), Holotype
Original combination: *Etyma icima* Martins & Galileo
Current name: same
Current placement: Cerambycidae: Lamiinae: Desmiphorini
Type locality: Mexico: Quintana Roo, Cancun Airport

idahoensis **Casey, 1913**: 259 (Fig. 84w, x), Lectotype
Original combination: *Strangalia idahoensis* Casey
Current name: *Leptura obliterata* (Haldeman, 1847)
Current placement: Cerambycidae: Lepturinae: Lepturini
Type locality: USA: Idaho, Bonner Co., Priest Lake
Remarks: This specimen is herein designated as the lectotype in order to stabilize the taxonomy and facilitate further identifications of this species.

idoneus **Fisher, 1927a**: 82 (Fig. 85a, b), Holotype
Original combination: *Pterolophia idoneus* Fisher
Current name: *Pterolophia (Ale) idioneus* Fisher, 1927a
Current placement: Cerambycidae: Lamiinae: Pteropliini
Type locality: Indonesia: Sumatra, Medan
Remarks: The genus name is misspelled as "*Pterlophia*" in the original publication.

ignita **Schaeffer, 1908**: 341 (Fig. 85c, d), Holotype
Original combination: *Leptura ignita* Schaeffer
Current name: *Megachoriolaus ignitus* (Schaeffer, 1908)
Current placement: Cerambycidae: Lepturinae: Lepturini
Type locality: USA: Arizona, Huachuca Mts.

ilaire **Dillon & Dillon, 1946**: 319 (Fig. 85e, f), Holotype
Original combination: *Oncideres ilaire* Dillon & Dillon
Current name: same
Current placement: Cerambycidae: Lamiinae: Onciderini
Type locality: Colombia: Buena Vista

iligana **Aurivillius, 1926**: 94 (Fig. 85g, h), Holotype
Original combination: *Glenea (Glenea) iligana* Aurivillius
Current name: *Glenea iligana* Aurivillius, 1926
Current placement: Cerambycidae: Lamiinae: Saperini
Type locality: Philippines: Mindanao, Lanao, Iligan

imbellis **Casey, 1914**: 363 (Fig. 85i, j), Holotype
Original combination: *Aneflomorpha imbellis* Casey
Current name: *Micraneflus imbellis* (Casey, 1914)
Current placement: Cerambycidae: Cerambycinae: Elaphidiini
Type locality: USA: California, San Diego

imbricornis brunneus **Casey, 1912**: 248 (Fig. 85k, l), Holotype
Original combination: *Prionus (Riponus) imbricornis brunneus* Casey
Current name: *Prionus (Neopolyarthron) imbricornis* Linnaeus, 1767
Current placement: Cerambycidae: Prioninae: Prionini
Type locality: USA: North Carolina, Southern Pines

imbricornis mimus **Casey, 1912**: 248 (Fig. 85m, n), Holotype
Original combination: *Prionus (Riponus) imbricornis mimus* Casey
Current name: *Prionus (Neopolyarthron) imbricornis* Linnaeus, 1767
Current placement: Cerambycidae: Prioninae: Prionini
Type locality: No locality label or published data.

immaculipenne **var.** ***humeromaculatum*** **Tippmann, 1960**: 107 (Fig. 85o, p), Holotype
Original combination: *Achryson immaculipenne* var. *humeromaculatum* Tippmann
Current name: *Achryson immaculipenne* Gounelle, 1909
Current placement: Cerambycidae: Cerambycinae: Achrysonini
Type locality: Brazil
Remarks: The name following the binomen was not expressly given infrasubspecific rank and the work did not indicate unambiguously that the name was infrasubspecific, therefore, it is deemed a subspecies and available.

***impressicollis* Chemsak, 1966**: 211 (Fig. 85q, r), Holotype
Original combination: *Methia impressicollis* Chemsak
Current name: *Methia necydalea* (Fabricius, 1798)
Current placement: Cerambycidae: Cerambycinae: Methiini
Type locality: British Virgin Islands: Jost van Dyke, Little Harbor

***inaequalis* Casey, 1924**: 233 (Fig. 85s, t), Lectotype
Original combination: *Hylotrupes inaequalis* Casey
Current name: *Hylotrupes bajulus* (Linnaeus, 1758)
Current placement: Cerambycidae: Cerambycinae: Hylotrupini
Type locality: USA: District of Columbia
Remarks: This specimen is herein designated as the lectotype in order to stabilize the taxonomy and facilitate further identifications of this species.

***inarmata* Gressitt, 1940**: 92 (Fig. 85u, v), Holotype
Original combination: *Psacothea inarmata* Gressitt
Current name: *Macrochenus tonkinensis inarmata* (Gressitt, 1940)
Current placement: Cerambycidae: Lamiinae: Monochamini
Type locality: China: Hainan Province, Hainan Island, Dwa Bi

***inarmatum* Martins, 1960c**: 94 (Fig. 85w, x), Holotype
Original combination: *Ibidion inarmatum* Martins
Current name: *Rhysium bivulneratum* (Thomson, 1867)
Current placement: Cerambycidae: Cerambycinae: Neoibidionini
Type locality: Colombia: Antioquia, Angelopolis

***inca* Martins & Galileo, 2005a**: 26 (Fig. 86a, b), Holotype
Original combination: *Adetus inca* Martins & Galileo
Current name: same
Current placement: Cerambycidae: Lamiinae: Apomecynini
Type locality: Peru: Cusco: Pilcopata

***incaniformis* Breuning, 1954**: 22 (Fig. 86c, d), Holotype
Original combination: *Sybra incaniformis* Breuning
Current name: same
Current placement: Cerambycidae: Lamiinae: Apomecynini
Type locality: Papua New Guinea: East New Britain, Mope

***incanoides* Breuning, 1942a**: 150 (Fig. 86e, f), Holotype
Original combination: *Sybra incanoides* Breuning
Current name: *Sybra arator* Pascoe, 1865
Current placement: Cerambycidae: Lamiinae: Apomecynini
Type locality: India: Darjeeling

***incertula* Casey, 1924**: 252 (Fig. 86g, h), Holotype
Original combination: *Oeme incertula* Casey
Current name: *Oeme rigida deserta* Casey, 1924
Current placement: Cerambycidae: Cerambycinae: Oemini
Type locality: USA: Arizona

***incertus* Casey, 1924**: 234 (Fig. 86i, j), Holotype
Original combination: *Hylotrupes incertus* Casey
Current name: *Hylotrupes bajulus* (Linnaeus, 1758)
Current placement: Cerambycidae: Cerambycinae: Hylotrupini
Type locality: USA: New Jersey

***incertus externus* Casey, 1912**: 300 (Fig. 86k, l), Holotype
Original combination: *Hypermallus incertus externus* Casey
Current name: *Parelaphidion aspersum* (Haldeman, 1847)
Current placement: Cerambycidae: Cerambycinae: Elaphidiini
Type locality: USA: Wisconsin

***includens* Aurivillius, 1928**: 322 (Fig. 86m, n), Lectotype
Original combination: *Demonax includens* Aurivillius
Current name: same
Current placement: Cerambycidae: Cerambycinae: Clytini
Type locality: Philippines: Sibuyan
Remarks: This specimen is herein designated as the lectotype in order to stabilize the taxonomy and facilitate further identifications of this species.

***incompta* Gressitt, 1935e**: 181 (Fig. 86o, p), Holotype
Original combination: *Noemia incompta* Gressitt
Current name: same
Current placement: Disteniidae: Disteniinae: Disteniini
Type locality: China: Taiwan, Hassenzan

***incongruens* Casey, 1912**: 366 (Fig. 86q, r), Lectotype
Original combination: *Xylotrechus incongruens* Casey
Current name: *Xylotrechus insignis* LeConte, 1873
Current placement: Cerambycidae: Cerambycinae: Clytini
Type locality: USA: California
Remarks: This specimen is herein designated as the lectotype in order to stabilize the taxonomy and facilitate further identifications of this species.

***indica* Breuning, 1938b**: 387 (Fig. 86s, t), Holotype
Original combination: *Egesina indica* Breuning
Current name: *Egesina (Niijimaia) indica* Breuning, 1938b
Current placement: Cerambycidae: Lamiinae: Pteropliini
Type locality: India: Tamil Nadu, Madurai
Remarks: Although Tavakilian and Chevillotte (2013) indicate that this holotype is in Basel, Switzerland, the Smithsonian Institution specimen matches the original description and data.

***indistincta* Lingafelter & Nearns, 2007**: 188 (Fig. 86u, v), Holotype
Original combination: *Drycothaea indistincta* Lingafelter & Nearns
Current name: same
Current placement: Cerambycidae: Lamiinae: Calliini
Type locality: Dominican Republic: La Altagracia Province, Punta Cana near Ecological Reserve, 0–5 m, 18°30.477′N, 68°22.499′W

***ineptus ambiguus* Casey, 1924**: 219 (Fig. 86w, x), Lectotype
Original combination: *Prionus (Prionus) ineptus ambiguus* Casey
Current name: *Prionus californicus* Motschulsky, 1845
Current placement: Cerambycidae: Prioninae: Prionini
Type locality: USA: Utah, Eureka
Remarks: This specimen is herein designated as the lectotype in order to stabilize the taxonomy and facilitate further identifications of this species.

***ineptus uintanus* Casey, 1924**: 220 (Fig. 87a, b), Lectotype
Original combination: *Prionus (Prionus) ineptus uintanus* Casey
Current name: *Prionus californicus* Motschulsky, 1845
Current placement: Cerambycidae: Prioninae: Prionini
Type locality: USA: Utah, North Fork, Provo Canyon
Remarks: This specimen is herein designated as the lectotype in order to stabilize the taxonomy and facilitate further identifications of this species.

***inerme* Martins, 1965**: 124 (Fig. 87c, d), Holotype
Original combination: *Ophtalmoplon inerme* Martins
Current name: same
Current placement: Cerambycidae: Cerambycinae: Neoibidionini
Type locality: French Guiana: St. Jean, St. Laurent du Maroni
Remarks: Although the genus was misspelled as *Optalmoplon*, it was emended by Martins to *Ophtalmoplon*, now widely used, and considered a justified emendation.

***infans* Casey, 1912**: 368 (Fig. 87e, f), Holotype
Original combination: *Neoclytus infans* Casey
Current name: *Neoclytus muricatulus infans* Casey, 1912
Current placement: Cerambycidae: Cerambycinae: Clytini
Type locality: USA: California, Siskiyou Co.

***inflaticollis* Casey, 1912**: 360 (Fig. 87g, h), Holotype
Original combination: *Xylotrechus inflaticollis* Casey
Current name: *Xylotrechus longitarsis* Casey, 1912
Current placement: Cerambycidae: Cerambycinae: Clytini
Type locality: USA: Colorado, Boulder Co.

***inflaticollis* Chemsak, 1966**: 215 (Fig. 87i, j), Holotype
Original combination: *Leptostylus inflaticollis* Chemsak
Current name: *Styloleptoides inflaticollis* (Chemsak, 1966)
Current placement: Cerambycidae: Lamiinae: Acanthocinini
Type locality: British Virgin Islands: Peter Island, Little Bay

***inflaticollis* Chemsak & Linsley, 1976a**: 174 (Fig. 87k, l), Holotype
Original combination: *Pseudotypocerus inflaticollis* Chemsak & Linsley
Current name: same
Current placement: Cerambycidae: Lepturinae: Lepturini
Type locality: Panama: Arraiján

***infuscatus* Fisher, 1932**: 84 (Fig. 87m, n), Holotype
Original combination: *Leiopus infuscatus* Fisher
Current name: *Styloleptus infuscatus* (Fisher, 1932)
Current placement: Cerambycidae: Lamiinae: Acanthocinini
Type locality: Dominican Republic: Samana

***inornatus* Casey, 1891**: 50 (Fig. 87o, p), Holotype
Original combination: *Valenus inornatus* Casey
Current name: same
Current placement: Cerambycidae: Lamiinae: Acanthocinini
Type locality: USA: Texas, El Paso

***insignis* Casey, 1913**: 370 (Fig. 87q, r), Lectotype
Original combination: *Oberea insignis* Casey
Current name: *Oberea perspicillata* Haldeman, 1847
Current placement: Cerambycidae: Lamiinae: Phytoeciini
Type locality: USA: North Carolina, Southern Pines
Remarks: This specimen is herein designated as the lectotype in order to stabilize the taxonomy and facilitate further identifications of this species.

***insignis* Casey, 1924**: 262 (Fig. 87s, t), Holotype
Original combination: *Stenaspis insignis* Casey
Current name: *Stenaspis verticalis insignis* Casey, 1924
Current placement: Cerambycidae: Cerambycinae: Trachyderini
Type locality: USA: Texas, Comal Co.

***insinuans* Casey, 1893**: 591 (Fig. 87u, v), Holotype
Original combination: *Cyrtophorus insinuans* Casey
Current name: *Microclytus compressicollis* (Castelnau & Gory, 1841)
Current placement: Cerambycidae: Cerambycinae: Anaglyptini
Type locality: Canada: Ontario

***insulana* Fisher, 1943**: 12 (Fig. 87w, x), Holotype
Original combination: *Ceragenia insulana* Fisher
Current name: same
Current placement: Cerambycidae: Cerambycinae: Trachyderini
Type locality: Panama: Canal Zone, Barro Colorado Island

***insularis* Chemsak & Linsley, 1964a**: 217 (Fig. 88a, b), Holotype
Original combination: *Anatinomma insularis* Chemsak & Linsley
Current name: same
Current placement: Cerambycidae: Cerambycinae: Hesperophanini
Type locality: Mexico: Quintana Roo, Celerain Point, Cozumel Island

***insularis* Fisher, 1925a**: 273 (Fig. 88c, d), Holotype
Original combination: *Camptomyne insularis* Fisher
Current name: *Exocentrus (Camptomyne) insularis* (Fisher, 1925a)
Current placement: Cerambycidae: Lamiinae: Acanthocinini
Type locality: Philippines: Sibuyan, Romblon

***insularis* Fisher, 1925b**: 8 (Fig. 88e, f), Holotype
Original combination: *Trypanidius insularis* Fisher
Current name: same
Current placement: Cerambycidae: Lamiinae: Acanthocinini
Type locality: Cuba: Santiago de Cuba, Oriente, Nagua

***insularis* Fisher, 1930b**: 9 (Fig. 88g, h), Holotype
Original combination: *Parmenonta insularis* Fisher
Current name: same
Current placement: Cerambycidae: Lamiinae: Apomecynini
Type locality: Cuba: Santiago de Las Vegas, Habana

***insularis* Fisher, 1932**: 56 (Fig. 88i, j), Holotype
Original combination: *Ophistomis insularis* Fisher
Current name: *Strangalia insularis* (Fisher, 1932)
Current placement: Cerambycidae: Lepturinae: Lepturini
Type locality: Dominica

***insularis* Fisher, 1932**: 76 (Fig. 88k, l), Holotype
Original combination: *Ecyrus insularis* Fisher
Current name: *Ecyrus hirtipes* Gahan, 1895
Current placement: Cerambycidae: Lamiinae: Pogonocherini
Type locality: Cuba: Baraguá

***insularis* Fisher, 1934b**: 36 (Fig. 88m, n), Holotype
Original combination: *Exocentrus insularis* Fisher
Current name: *Exocentrus insulicola* Breuning, 1958
Current placement: Cerambycidae: Lamiinae: Acanthocinini
Type locality: Indonesia: Mount Mangkol, east of Sumatra, 200 m.

***insularis* Fisher, 1934a**: 4 (Fig. 88o, p), Holotype
Original combination: *Aegocidnus insularis* Fisher
Current name: *Sciades (Aegocidnus) insularis* (Fisher, 1934a)
Current placement: Cerambycidae: Lamiinae: Acanthocinini
Type locality: Indonesia: Sumatra, Dolok Baros

***insularis* Fisher, 1935b**: 195 (Fig. 88q, r), Holotype
Original combination: *Haruspex insularis* Fisher
Current name: *Haruspex inscriptus* Gahan, 1895
Current placement: Cerambycidae: Cerambycinae: Piezocerini
Type locality: Barbados

***insularis* Fisher, 1935b**: 199 (Fig. 88s, t), Holotype
Original combination: *Hypsioma insularis* Fisher
Current name: *Hypsioma grisea* (Fleutiaux & Sallé, 1889)
Current placement: Cerambycidae: Lamiinae: Onciderini
Type locality: Barbados: Chelston

***insularis* Fisher, 1937c**: 198 (Fig. 88u, v), Holotype
Original combination: *Oberea insularis* Fisher
Current name: *Oberea rubetra* Pascoe, 1858
Current placement: Cerambycidae: Lamiinae: Phytoeciini
Type locality: Indonesia: Jawa Tengah, Noesa Kambangan Island

***insularum* Chemsak, 1966**: 210 (Fig. 88w, x), Holotype
Original combination: *Methia insularum* Chemsak
Current name: *Methia necydalea* (Fabricius, 1798)
Current placement: Cerambycidae: Cerambycinae: Methiini
Type locality: British Virgin Islands: Virgin Gorda, Prickley Pear Island, Vixen Point

***insulicola* Fisher, 1936a**: 173 (Fig. 89a, b), Holotype
Original combination: *Necydalis insulicola* Fisher
Current name: same
Current placement: Cerambycidae: Necydalinae: Necydalini
Type locality: Indonesia: Java, Mt. Tangkoeban Prahoe, Preanger 4000–5000 ft.

insulorum **Monné & Tavakilian, 2011**: 461 (Fig. 89c, d), Holotype
Original combination: *Nyssodrysternum insulorum* Monné & Tavakilian
Current name: same
Current placement: Cerambycidae: Lamiinae: Acanthocinini
Type locality: Panama: Panama Province

insulsus **Martins & Galileo, 2005b**: 766 (Fig. 89e, f), Holotype
Original combination: *Haruspex insulsus* Martins & Galileo
Current name: same
Current placement: Cerambycidae: Cerambycinae: Piezocerini
Type locality: Panama: Panama Province, El Llano-Carti road, km 8, 9°16′N, 78°57′W

integra **Breuning, 1938c**: 469 (Fig. 89g, h), Holotype
Original combination: *Pseudostixis integra* Breuning
Current name: same
Current placement: Cerambycidae: Lamiinae: Phrissomini
Type locality: Congo (as per published description)
Remarks: It is unknown if this is the true type since Tavakilian and Chevillotte (2013) indicates it is present in Tervuren.

intermedia **Casey, 1913**: 236 (Fig. 89i, j), Holotype
Original combination: *Leptacmaeops (Leptacmaeops) intermedia* Casey
Current name: *Cortodera subpilosa* (LeConte, 1850)
Current placement: Cerambycidae: Lepturinae: Rhagiini
Type locality: No locality data

intricata **Casey, 1913**: 350 (Fig. 89k, l), Holotype
Original combination: *Desmiphora intricata* Casey
Current name: *Desmiphora hirticollis* (Olivier, 1795)
Current placement: Cerambycidae: Lamiinae: Desmiphorini
Type locality: USA: Texas, Brownsville, Esperanza Ranch

invicta **Galileo & Martins, 1996**: 244 (Fig. 89m, n), Holotype
Original combination: *Isomerida invicta* Galileo & Martins
Current name: same
Current placement: Cerambycidae: Lamiinae: Hemilophini
Type locality: Peru: Junin, Piches & Perene Valleys, 2000–3000 ft.

iowensis **Casey, 1913**: 370 (Fig. 89o, p), Lectotype
Original combination: *Oberea iowensis* Casey
Current name: *Oberea perspicillata* Haldeman, 1847
Current placement: Cerambycidae: Lamiinae: Phytoeciini
Type locality: USA: Iowa, Keokuk
Remarks: This specimen is herein designated as the lectotype in order to stabilize the taxonomy and facilitate further identifications of this species.

iridescens **Dillon & Dillon, 1941**: 26 (Fig. 89q, r), Holotype
Original combination: *Taeniotes iridescens* Dillon & Dillon
Current name: same
Current placement: Cerambycidae: Lamiinae: Monochamini
Type locality: Colombia: Medellin

irma **Kriesche, 1920**: 194 (Fig. 89s, t), Holotype
Original combination: *Apriona irma* Kriesche
Current name: same
Current placement: Cerambycidae: Lamiinae: Batocerini
Type locality: Indonesia: Sumatra, Singalang, 2000 ft.
Remarks: Tippmann labeled this specimen as having been obtained from the Kriesche Estate in 1955.

itzingeri **Tippmann, 1951**: 326 (Fig. 89u, v), Holotype
Original combination: *Guitelia itzingeri* Tippmann
Current name: same
Current placement: Cerambycidae: Cerambycinae: Callichromatini
Type locality: Tanzania: Ukerewe Island, Lake Victoria

itzingeri **Tippmann, 1953d**: 328 (Fig. 89w, x), Holotype
Original combination: *Trachyderes itzingeri* Tippmann
Current name: *Deretrachys montanus* (Tippmann, 1953d)
Current placement: Cerambycidae: Cerambycinae: Trachyderini
Type locality: Peru: Huacaña a Querobamba, 3720 m.

iuasanga **Martins & Galileo, 2007a**: 484 (Fig. 90a, b), Holotype
Original combination: *Hippopsis iuasanga* Martins & Galileo
Current name: same
Current placement: Cerambycidae: Cerambycinae: Agapanthiini
Type locality: Ecuador: Pichincha, 15 km east of Santo Domingo, Tinalandia

iviei **Lingafelter, 2008**: 367 (Fig. 90c, d), Holotype
Original combination: *Elaphidion iviei* Lingafelter
Current name: same
Current placement: Cerambycidae: Cerambycinae: Elaphidiini
Type locality: Dominican Republic: Barahona Province, Vic. Filipinas, 1700 ft.
Remarks: On long term loan from FSCA, pending revision of *Elaphidion* by Lingafelter, in prep.

jamaicae **Fisher, 1942a**: 6 (Fig. 90e, f), Holotype
Original combination: *Eburia jamaicae* Fisher
Current name: same
Current placement: Cerambycidae: Cerambycinae: Eburiini
Type locality: Jamaica: Brumelia, near Mandeville

jamaicensis **Fisher, 1942a**: 18 (Fig. 90g, h), Holotype
Original combination: *Acrepidopterum jamaicensis* Fisher
Current name: same
Current placement: Cerambycidae: Lamiinae: Apomecynini
Type locality: Jamaica: Ocho Rios

jamaicensis **Toledo & Hovore, 2005**: 30 (Fig. 90i, j), Holotype
Original combination: *Lagocheirus jamaicensis* Toledo & Hovore
Current name: same
Current placement: Cerambycidae: Lamiinae: Acanthocinini
Type locality: Jamaica: Green Hills

japonica **Breuning, 1940d**: 414 (Fig. 90k, l), Holotype
Original combination: *Rhodopis japonica* Breuning
Current name: *Rhodopina lewisii* (Bates, 1873)
Current placement: Cerambycidae: Lamiinae: Rhodopini
Type locality: Japan: Honshu, Nagato

jaumei **Fisher, 1935b**: 193 (Fig. 90m, n), Holotype
Original combination: *Anepsyra jaumei* Fisher
Current name: *Psyrassa jaumei* (Fisher, 1935b)
Current placement: Cerambycidae: Cerambycinae: Elaphidiini
Type locality: Cuba: Corralillo, Santa Clara

javana **Fisher, 1936a**: 193 (Fig. 90o, p), Holotype
Original combination: *Neoegesina javana* Fisher
Current name: *Egesina javanica* Breuning, 1961b
Current placement: Cerambycidae: Lamiinae: Pteropliini
Type locality: Indonesia: Java, Mt. Tangkoeban Prahoe, Preanger 4000–5000 ft.

javanica **Breuning, 1943a**: 146 (Fig. 90q, r), Holotype
Original combination: *Amechana javanica* Breuning
Current name: same
Current placement: Cerambycidae: Lamiinae: Lamiini
Type locality: Indonesia: Java

javanicus **Breuning, 1940d**: 424 (Fig. 90s, t), Holotype
Original combination: *Sthenias javanicus* Breuning
Current name: same
Current placement: Cerambycidae: Lamiinae: Niphonini
Type locality: Indonesia: Java, Soekaboemi

javanicus **Fisher, 1936a**: 170 (Fig. 90u, v), Holotype
Original combination: *Zatrephus javanicus* Fisher
Current name: same
Current placement: Cerambycidae: Cerambycinae: Cerambycini
Type locality: Indonesia: Java, Batoerraden, Mt. Slamat

javanicus **Fisher, 1936a**: 180 (Fig. 90w, x), Holotype
Original combination: *Demonax javanicus* Fisher
Current name: same
Current placement: Cerambycidae: Cerambycinae: Clytini
Type locality: Indonesia: Java, Batoerraden G. Slamat

javicola **Fisher, 1936a**: 187 (Fig. 91a, b), Holotype
Original combination: *Pterolophia javicola* Fisher
Current name: same
Current placement: Cerambycidae: Lamiinae: Pteropliini
Type locality: Indonesia: Java, Batoerraden, Mt. Slamat

jolyi **Santos-Silva & Hovore, 2007b**: 587 (Fig. 91c, d), Holotype
Original combination: *Novantinoe jolyi* Santos-Silva & Hovore
Current name: same
Current placement: Disteniidae: Disteniinae: Disteniini
Type locality: Venezuela: Aragua, Rancho Grande, 1200–1400 m.

jubapennis **Fisher, 1938**: 142 (Fig. 91e, f), Holotype
Original combination: *Ozineus jubapennis* Fisher
Current name: same
Current placement: Cerambycidae: Lamiinae: Acanthocinini
Type locality: Brazil: Santa Catarina, Nova Teutonia

jugalis samar **Dillon & Dillon, 1951**: 47 (Fig. 91g, h), Holotype
Original combination: *Gnoma jugalis samar* Dillon & Dillon
Current name: *Gnoma jugalis* Newman, 1842
Current placement: Cerambycidae: Lamiinae: Gnomini
Type locality: Philippines: Samar

julio **Lingafelter & Micheli, 2004**: 40 (Fig. 91i, j), Holotype
Original combination: *Elateropsis julio* Lingafelter & Micheli
Current name: same
Current placement: Cerambycidae: Prioninae: Solenopterini
Type locality: USA: Puerto Rico, Municipio de Ponce, Tuque, elevation 70–80 m, 17°58′42″N, 66°40′21″W

junctus **Casey, 1913**: 380 (Fig. 91k, l), Holotype
Original combination: *Tetraopes junctus* Casey
Current name: *Tetraopes basalis* LeConte, 1852
Current placement: Cerambycidae: Lamiinae: Tetraopini
Type locality: USA: Colorado (?)

juniperi **Fisher, 1915**: 77 (Fig. 91m, n), Holotype
Original combination: *Hylotrupes juniperi* Fisher
Current name: *Semanotus juniperi* (Fisher, 1915)
Current placement: Cerambycidae: Cerambycinae: Callidiini
Type locality: USA: Arizona, Santa Catalina Mts.

***juniperi* Fisher, 1920**: 157 (Fig. 91o, p), Holotype
Original combination: *Callidium juniperi* Fisher
Current name: same
Current placement: Cerambycidae: Cerambycinae: Callidiini
Type locality: USA: New Mexico, Maxwell

***juno* Fisher, 1938**: 139 (Fig. 91q, r), Holotype
Original combination: *Acanthoderes juno* Fisher
Current name: *Psapharochrus juno* (Fisher, 1938)
Current placement: Cerambycidae: Lamiinae: Acanthoderini
Type locality: Brazil: Santa Catarina, Nova Teutonia

***kalshoveni* Fisher, 1934b**: 36 (Fig. 91s, t), Holotype
Original combination: *Exocentrus kalshoveni* Fisher
Current name: *Exocentrus (Camptomyne) kalshoveni* Fisher, 1934b
Current placement: Cerambycidae: Lamiinae: Acanthocinini
Type locality: Indonesia: Java, Rembang

***kankauensis chungkonensis* Gressitt, 1940**: 73 (Fig. 91u, v), Holotype
Original combination: *Perissus kankauensis chungkonensis* Gressitt
Current name: *Perissus laetus* Lameere, 1893
Current placement: Cerambycidae: Cerambycinae: Clytini
Type locality: China: Hainan Province, Hainan Island, Chung Kon

***kartaboensis* Fisher, 1944b**: 1 (Fig. 91w, x), Holotype
Original combination: *Stromatium kartaboensis* Fisher
Current name: *Achryson quadrimaculatum* (Fabricius, 1793)
Current placement: Cerambycidae: Cerambycinae: Achrysonini
Type locality: Guyana: Bartica District, Kartabo

***keeni* Casey, 1913**: 257 (Fig. 92a, b), Holotype
Original combination: *Strangalepta keeni* Casey
Current name: *Neoalosterna rubida* (LeConte, 1873)
Current placement: Cerambycidae: Lepturinae: Lepturini
Type locality: Canada: British Columbia, Inverness

***kempi* Casey, 1912**: 233 (Fig. 92c, d), Holotype
Original combination: *Prionus (Prionus) kempi* Casey
Current name: *Prionus laticollis* (Drury, 1773)
Current placement: Cerambycidae: Prioninae: Prionini
Type locality: USA: New York, Adirondack

***kempiana* Casey, 1924**: 281 (Fig. 92e, f), Lectotype
Original combination: *Cercolia kempiana* Casey
Current name: *Trigonarthris proxima* (Say, 1824)
Current placement: Cerambycidae: Lepturinae: Lepturini
Type locality: USA: New York, Lake George
Remarks: Lectotype labeled by Laplante in 1991.

***kenyensis* Breuning, 1940d**: 409 (Fig. 92g, h), Holotype
Original combination: *Monoxenus (Dityloderus) kenyensis* Breuning
Current name: same
Current placement: Cerambycidae: Lamiinae: Parmenini
Type locality: Kenya: Mt. Kenya

***kenyensis* Breuning, 1948b**: 36 (Fig. 92i, j), Holotype
Original combination: *Hyllisia kenyensis* Breuning
Current name: same
Current placement: Cerambycidae: Lamiinae: Agapanthiini
Type locality: Kenya: Sagola

***kiangsina* Gressitt, 1937b**: 601 (Fig. 92k, l), Holotype
Original combination: *Pterolophia kiangsina* Gressitt
Current name: same
Current placement: Cerambycidae: Lamiinae: Pteropliini
Type locality: China: Jiangxi Province, Hong Shan

***kinabaluensis* Fisher, 1935c**: 622 (Fig. 92m, n), Holotype
Original combination: *Glenea kinabaluensis* Fisher
Current name: same
Current placement: Cerambycidae: Lamiinae: Saperdini
Type locality: Malaysia: Borneo, Sabah, Mt. Kinabalu, 5000 ft., Marei Parei

***klapperichi* Breuning, 1957a**: 12 (Fig. 92o, p), Holotype
Original combination: *Exocentrus klapperichi* Breuning
Current name: *Exocentrus klapperichi* Breuning, 1957a
Current placement: Cerambycidae: Lamiinae: Acanthocinini
Type locality: Afghanistan: Nuristan, Kutiau, 1500 m.

***knabi* Fisher, 1925c**: 15 (Fig. 92q, r), Holotype
Original combination: *Eupogonius knabi* Fisher
Current name: same
Current placement: Cerambycidae: Lamiinae: Desmiphorini
Type locality: Mexico: Veracruz - Llave

***knulli* Fisher, 1925d**: 103 (Fig. 92s, t), Holotype
Original combination: *Leptostylus knulli* Fisher
Current name: *Astylopsis arcuata* (LeConte, 1878)
Current placement: Cerambycidae: Lamiinae: Acanthocinini
Type locality: USA: Maryland, Dorchester County, near Lloyds

***koreana* Gressitt, 1951b**: 508 (Fig. 92u, v), Holotype
Original combination: *Sophronica koreana* Gressitt
Current name: same
Current placement: Cerambycidae: Lamiinae: Desmiphorini
Type locality: South Korea: Suigen

***lactiflua coerulescens* Kriesche, 1940**: 116 (Fig. 92w, x), Holotype
Original combination: *Rosenbergia lactiflua coerulescens* Kriesche
Current name: same
Current placement: Cerambycidae: Lamiinae: Batocerini
Type locality: Papua New Guinea: Bougainville Island (Solomon Islands), Kieta
Remarks: Tavakilian and Chevillotte (2013) suggest the holotype is present in Berlin, however, this specimen and its labels match the original description.

***lacustre* Casey, 1912**: 289 (Fig. 93a, b), Holotype
Original combination: *Callidium lacustre* Casey
Current name: *Callidium frigidum* Casey, 1912
Current placement: Cerambycidae: Cerambycinae: Callidiini
Type locality: USA: Wisconsin, Bayfield

***lacustrina* Casey, 1913**: 242 (Fig. 93c, d), Holotype
Original combination: *Acmaeops lacustrina* Casey
Current name: *Acmaeops proteus* (Kirby, 1837)
Current placement: Cerambycidae: Lepturinae: Rhagiini
Type locality: USA: Lake Superior

***lacustrinus* Casey, 1912**: 267 (Fig. 93e, f), Holotype
Original combination: *Criocephalus lacustrinus* Casey
Current name: *Arhopalus foveicollis* (Haldeman, 1847)
Current placement: Cerambycidae: Spondylidinae: Asemini
Type locality: USA: Wisconsin, Bayfield

***lacustris* Casey, 1891**: 43 (Fig. 93g, h), Lectotype
Original combination: *Leptura lacustris* Casey
Current name: *Pygoleptura nigrella* (Say, 1827)
Current placement: Cerambycidae: Lepturinae: Lepturini
Type locality: USA: Michigan, Marquette
Remarks: This specimen is herein designated as the lectotype in order to stabilize the taxonomy and facilitate further identifications of this species.

***lacustris* Casey, 1912**: 302 (Fig. 93i, j), Holotype
Original combination: *Hypermallus lacustris* Casey
Current name: *Anelaphus villosus* (Fabricius, 1793)
Current placement: Cerambycidae: Cerambycinae: Elaphidiini
Type locality: USA: Wisconsin

***lacustris* Martins, 1977**: 110 (Fig. 93k, l), Holotype
Original combination: *Oemida lacustris* Martins
Current name: same
Current placement: Cerambycidae: Cerambycinae: Oemini
Type locality: Tanzania: Ukerewe Island, Lake Victoria

***laeviceps* Casey, 1913**: 201 (Fig. 93m, n), Holotype
Original combination: *Hapalosalia laeviceps* Casey
Current name: *Pidonia ruficollis* (Say, 1824)
Current placement: Cerambycidae: Lepturinae: Rhagiini
Type locality: USA: Wisconsin, Bayfield

***laevicollis* var. *rubrosignata* Tippmann, 1953c**: 215 (Fig. 93o, p), Holotype
Original combination: *Isthmiade laevicollis* var. *rubrosignata* Tippmann
Current name: *Isthmiade laevicollis* Tippmann, 1953c
Current placement: Cerambycidae: Cerambycinae: Rhinotragini
Type locality: Peru: Pucallpa
Remarks: The name following the binomen was not expressly given infrasubspecific rank and the work did not indicate unambiguously that the name was infrasubspecific, therefore, it is deemed a subspecies and available.

***lama* Breuning, 1943c**: 20 (Fig. 93q, r), Holotype
Original combination: *Pterolophia lama* Breuning
Current name: same
Current placement: Cerambycidae: Lamiinae: Pteropliini
Type locality: China: Tibet

***lanaticollis* Casey, 1912**: 339 (Fig. 93s, t), Holotype
Original combination: *Crossidius lanaticollis* Casey
Current name: *Crossidius pulchellus* LeConte, 1861
Current placement: Cerambycidae: Cerambycinae: Trachyderini
Type locality: No published locality data or label associated with the specimen.

***lanatula* Casey, 1913**: 224 (Fig. 93u, v), Holotype
Original combination: *Brachysomida lanatula* Casey
Current name: *Brachysomida californica* (LeConte, 1851)
Current placement: Cerambycidae: Lepturinae: Rhagiini
Type locality: USA: Colorado

***lanceolata* Martins & Galileo, 2007c**: 1148 (Fig. 93w, x), Holotype
Original combination: *Unaporanga lanceolata* Martins & Galileo
Current name: same
Current placement: Cerambycidae: Lamiinae: Hemilophini
Type locality: Ecuador: Los Rios, Rio Palenque Biol. Station 40 km north of Quevedo

lanceolatum **Monné & Tavakilian, 2011**: 462 (Fig. 94a, b), Holotype
Original combination: *Nyssodrysternum lanceolatum* Monné & Tavakilian
Current name: same
Current placement: Cerambycidae: Lamiinae: Acanthocinini
Type locality: Panama: Panama Province, km 8–13, El Llano-Cartí Road

lanigera **Linell, 1899**: 259 (Fig. 94c, d), Holotype
Original combination: *Eburia lanigera* Linell
Current name: same
Current placement: Cerambycidae: Cerambycinae: Eburiini
Type locality: Ecuador, Galapagos Islands, Charles Island

laosensis **Breuning, 1942a**: 172 (Fig. 94e, f), Holotype
Original combination: *Paranandra laosensis* Breuning
Current name: same
Current placement: Cerambycidae: Lamiinae: Agapanthiini
Type locality: Laos

lara **Galileo & Martins, 2012c**: 297 (Fig. 94g, h), Holotype
Original combination: *Eligmoderma lara* Galileo & Martins
Current name: same
Current placement: Cerambycidae: Cerambycinae: Eligmodermini
Type locality: Venezuela: Lara, Yacambu Natural Park (13 km SE Sanare)

laterale **Casey, 1914**: 365 (Fig. 94i, j), Lectotype
Original combination: *Anoplium laterale* Casey
Current name: *Anelaphus moestus* (LeConte, 1854)
Current placement: Cerambycidae: Cerambycinae: Elaphidiini
Type locality: USA: Texas
Remarks: This specimen is herein designated as the lectotype in order to stabilize the taxonomy and facilitate further identifications of this species.

lateralis **Aurivillius, 1927b**: 578 (Fig. 94k, l), Syntype
Original combination: *Clodia lateralis* Aurivillius
Current name: *Pseudoclodia lateralis* (Aurivillius, 1927b)
Current placement: Cerambycidae: Lamiinae: Acanthocinini
Type locality: Philippines: Sibuyan, Romblon
Remarks: Other syntype(s) are present in the Naturhistoriska Riksmuseet, Stockholm, Sweden.

lateralis **Casey, 1891**: 37 (Fig. 94m, n), Holotype
Original combination: *Toxotus lateralis* Casey
Current name: *Stenocorus nubifer* (LeConte, 1859)
Current placement: Cerambycidae: Lepturinae: Lepturini
Type locality: USA: California, Mt. Diablo, Contra Costa Co.

lateralis **Casey, 1924**: 287 (Fig. 94o, p), Holotype
Original combination: *Moneilema (Collapteryx) lateralis* Casey
Current name: *Moneilema subrugosum* Bland, 1862
Current placement: Cerambycidae: Lamiinae: Moneilemini
Type locality: Mexico: Baja California Sur, El Taste

lateralis **Fisher, 1946a**: 329 (Fig. 94q, r), Holotype
Original combination: *Distenia lateralis* Fisher
Current name: *Distenia suturalis* Bates, 1870
Current placement: Disteniidae: Disteniinae: Disteniini
Type locality: Bolivia: El Beni, Rurrenabaque

laterialba **Gressitt, 1937b**: 611 (Fig. 94s, t), Holotype
Original combination: *Pothyne (Laterialba) laterialba* Gressitt
Current name: same
Current placement: Cerambycidae: Lamiinae: Agapanthiini
Type locality: China: Jiangxi Province, Tai Au Hong

laterialba **Martins & Galileo, 2001**: 103 (Fig. 94u, v), Holotype
Original combination: *Acakyra laterialba* Martins & Galileo
Current name: same
Current placement: Cerambycidae: Lamiinae: Acanthoderini
Type locality: Ecuador: Pastaza

laterimaculata **Breuning, 1943c**: 48 (Fig. 94w, x), Holotype
Original combination: *Diadelia (Trichodiadelia) laterimaculata* Breuning
Current name: *Diadelia (Setodiadelia) laterimaculata* Breuning, 1943c
Current placement: Cerambycidae: Lamiinae: Rhodopini
Type locality: Madagascar: Midongy

lateriplagiatus **Breuning, 1942a**: 116 (Fig. 95a, b), Holotype
Original combination: *Monochamus lateriplagiatus* Breuning
Current name: *Monochamus (Laertochamus) pictor* Bates, 1884
Current placement: Cerambycidae: Lamiinae: Monochamini
Type locality: Exact type locality unknown. The original description states, "wahrscheinlich aus sudasien" (probably South Asian)

laticinctus **Dillon & Dillon, 1946**: 200 (Fig. 95c, d), Holotype
Original combination: *Lochmaeocles laticinctus* Dillon & Dillon
Current name: same
Current placement: Cerambycidae: Lamiinae: Onciderini
Type locality: Guatemala: Cayaga

***laticollis* Casey, 1913**: 341 (Fig. 95e, f), Holotype
Original combination: *Graphisurus (Tylocerina) laticollis* Casey
Current name: *Acanthocinus nodosus* (Fabricius, 1775)
Current placement: Cerambycidae: Lamiinae: Acanthocinini
Type locality: USA: North Carolina, Southern Pines

***laticollis* Tippmann, 1953d**: 314 (Fig. 95g, h), Holotype
Original combination: *Andraegoidus laticollis* Tippmann
Current name: same
Current placement: Cerambycidae: Cerambycinae: Trachyderini
Type locality: Colombia: La Garita

***laticollis oblongus* Casey, 1912**: 234 (Fig. 95i, j), Holotype
Original combination: *Prionus (Prionus) laticollis oblongus* Casey
Current name: *Prionus laticollis* (Drury, 1773)
Current placement: Cerambycidae: Prioninae: Prionini
Type locality: USA: Indiana

***laticornis* Tippmann, 1960**: 148 (Fig. 95k, l), Holotype
Original combination: *Palaeotrachyderes laticornis* Tippmann
Current name: same
Current placement: Cerambycidae: Cerambycinae: Trachyderini
Type locality: Panama: Chiriquí

***latifrons* Breuning, 1940c**: 121 (Fig. 95m, n), Holotype
Original combination: *Mecynome (Niphoparmena) latifrons* Breuning
Current name: *Niphoparmena latifrons* (Breuning, 1940c)
Current placement: Cerambycidae: Lamiinae: Parmenini
Type locality: Ethiopia
Remarks: Although Tavakilian and Chevillotte (2013) suggest the holotype is present in the BMNH, their catalog does not indicate so, and this specimen and its labels match the original description.

***latifrons* Chemsak, 1977**: 271 (Fig. 95o, p), Holotype
Original combination: *Phaea latifrons* Chemsak
Current name: *Phaea hogei* Bates, 1881
Current placement: Cerambycidae: Lamiinae: Tetraopini
Type locality: Mexico: Oaxaca, Puerto Angel

***latior* Casey, 1924**: 296 (Fig. 95q, r), Lectotype
Original combination: *Tetraopes latior* Casey
Current name: *Tetraopes basalis* LeConte, 1852
Current placement: Cerambycidae: Lamiinae: Tetraopini
Type locality: USA: California, Summit Mt. Wilson
Remarks: This specimen is herein designated as the lectotype in order to stabilize the taxonomy and facilitate further identifications of this species.

***latipennis* Casey, 1912**: 280 (Fig. 95s, t), Lectotype
Original combination: *Phymatodes (Poecilium) latipennis* Casey
Current name: *Phymatodes varius* (Fabricius, 1776)
Current placement: Cerambycidae: Cerambycinae: Callidiini
Type locality: USA: Massachusetts, Cohasset
Remarks: This specimen is herein designated as the lectotype in order to stabilize the taxonomy and facilitate further identifications of this species.

***latispina* Gressitt, 1956**: 157 (Fig. 95u, v), Holotype
Original combination: *Sciadella (Acanthosciades) latispina* Gressitt
Current name: *Sciades (Acanthosciades) latispina* (Gressitt, 1956)
Current placement: Cerambycidae: Lamiinae: Acanthocinini
Type locality: Micronesia: Ponape Island, Nett District, Nanpil

***latitarsis* Casey, 1913**: 342 (Fig. 95w, x), Holotype
Original combination: *Dectes latitarsis* Casey
Current name: *Dectes texanus* LeConte, 1862
Current placement: Cerambycidae: Lamiinae: Acanthocinini
Type locality: USA: Texas, Brownsville

***latithorax* Martins & Galileo, 2008a**: 276 (Fig. 96a, b), Holotype
Original combination: *Mecometopus latithorax* Martins & Galileo
Current name: *Pirangoclytus latithorax* (Martins & Galileo, 2008)
Current placement: Cerambycidae: Cerambycinae: Clytini
Type locality: Panama: Canal Zone, Barro Colorado Island

***latiuscula* Aurivillius, 1927b**: 571 (Fig. 96c, d), Lectotype
Original combination: *Sybra latiuscula* Aurivillius
Current name: *Sybra alternans* (Wiedemann, 1823)
Current placement: Cerambycidae: Lamiinae: Apomecynini
Type locality: Philippines: Sibuyan, Romblon

***latus* Casey, 1924**: 233 (Fig. 96e, f), Holotype
Original combination: *Hylotrupes latus* Casey
Current name: *Hylotrupes bajulus* (Linnaeus, 1758)
Current placement: Cerambycidae: Cerambycinae: Hylotrupini
Type locality: USA: Rhode Island

***latus* Chemsak & Feller, 1988**: 182 (Fig. 96g, h), Holotype
Original combination: *Leptostylopsis latus* Chemsak & Feller
Current name: same
Current placement: Cerambycidae: Lamiinae: Acanthocinini
Type locality: Belize: Twin Cays

***laurentica* Casey, 1913**: 271 (Fig. 96i, j), Holotype
Original combination: *Leptura (Dorcasina) laurentica* Casey
Current name: *Lepturopsis biforis* (Newman, 1841)
Current placement: Cerambycidae: Lepturinae: Lepturini
Type locality: USA: New York, Bluff Point, Lake Champlain, 3 miles below Plattsburg

***laurenticus* Casey, 1913**: 297 (Fig. 96k, l), Holotype
Original combination: *Goes laurenticus* Casey
Current name: *Goes pulverulentus* (Haldeman, 1847)
Current placement: Cerambycidae: Lamiinae: Monochamini
Type locality: USA: New York, Bluff Point, Lake Champlain

***lecontei* Casey, 1912**: 325 (Fig. 96m, n), Holotype
Original combination: *Purpuricenus lecontei* Casey
Current name: *Metaleptus lecontei* (Casey, 1912)
Current placement: Cerambycidae: Cerambycinae: Trachyderini
Type locality: Honduras

***lecontei* Casey, 1913**: 219 (Fig. 96o, p), Holotype
Original combination: *Gaurotes lecontei* Casey
Current name: *Pseudogaurotina cressoni lecontei* (Casey, 1913)
Current placement: Cerambycidae: Lepturinae: Rhagiini
Type locality: USA: Oregon

***lecontei* Casey, 1913**: 305 (Fig. 96q, r), Holotype
Original combination: *Leptostylus lecontei* Casey
Current name: *Leptostylopsis planidorsus* (LeConte, 1873)
Current placement: Cerambycidae: Lamiinae: Acanthocinini
Type locality: USA: Florida, Lake Worth

***leguminicola* Linell, 1896**: 43 (Fig. 96s, t), Holotype
Original combination: *Baryssinus leguminicola* Linell
Current name: *Lophopoeum timbouvae* Lameere, 1884
Current placement: Cerambycidae: Lamiinae: Acanthocinini
Type locality: Paraguay

***leiopinus* Casey, 1913**: 308 (Fig. 96u, v), Holotype
Original combination: *Astylidius leiopinus* Casey
Current name: *Astylidius parvus* (LeConte, 1873)
Current placement: Cerambycidae: Lamiinae: Acanthocinini
Type locality: USA: Texas, Columbus

***lengi* Joutel, 1911**: 200 (Fig. 96w, x), Syntype
Original combination: *Phymatodes lengi* Joutel
Current name: same
Current placement: Cerambycidae: Cerambycinae: Callidiini
Type locality: USA: New Jersey, Fort Lee

***lenticula* Galileo & Martins, 2006a**: 13 (Fig. 97a, b), Holotype
Original combination: *Parmenonta lenticula* Galileo & Martins
Current name: same
Current placement: Cerambycidae: Lamiinae: Apomecynini
Type locality: Paraguay: Paraguari, 5 km southeast of Ybycuí

***lentiginosa* Casey, 1913**: 325 (Fig. 97c, d), Holotype
Original combination: *Hyperplatys lentiginosa* Casey
Current name: *Hyperplatys maculata* Haldeman, 1847
Current placement: Cerambycidae: Lamiinae: Acanthocinini
Type locality: USA: Rhode Island, Watch Hill

***leopoldi* Fisher, 1934c**: 4 (Fig. 97e, f), Holotype
Original combination: *Daphysia leopoldi* Fisher
Current name: *Menesia transversonotata* Heller, 1924c
Current placement: Cerambycidae: Lamiinae: Saperdini
Type locality: Philippines: Samar

***lepesmei* Tippmann, 1953c**: 196 (Fig. 97g, h), Holotype
Original combination: *Derobrachus lepesmei* Tippmann
Current name: *Derobrachus megacles* Bates, 1884
Current placement: Cerambycidae: Prioninae: Macrotomini
Type locality: Mexico: Durango

***lepida* Melzer, 1933**: 375 (Fig. 97i, j), Holotype
Original combination: *Amphicnaeia lepida* Melzer
Current name: same
Current placement: Cerambycidae: Lamiinae: Apomecynini
Type locality: Costa Rica: Cartago, Turrialba, 800 m.

***lepidula* Casey, 1913**: 229 (Fig. 97k, l), Lectotype
Original combination: *Brachysomida lepidula* Casey
Current name: *Brachysomida californica* (LeConte, 1851)
Current placement: Cerambycidae: Lepturinae: Rhagiini
Type locality: USA: California
Remarks: This specimen is herein designated as the lectotype in order to stabilize the taxonomy and facilitate further identifications of this species.

***leptissimus* Gressitt, 1936**: 103 (Fig. 97m, n), Holotype
Original combination: *Pseudocalamobius leptissimus* Gressitt
Current name: same
Current placement: Cerambycidae: Lamiinae: Agapanthiini
Type locality: China: Taiwan, Arisan

***lessoni sericeiceps* Kriesche, 1936**: 68 (Fig. 97o, p), Holotype
Original combination: *Dihammus lessoni sericeiceps* Kriesche
Current name: *Acalolepta sericeiceps* (Kriesche, 1936)
Current placement: Cerambycidae: Lamiinae: Monochamini
Type locality: Papua New Guinea: Bougainville Island (Solomon Islands)
Remarks: Tippmann labeled this specimen as having been obtained from the Kriesche Estate in 1955.

***leucostictica* Breuning, 1942a**: 126 (Fig. 97q, r), Holotype
Original combination: *Ichthyodes (Ichthyodes) leucostictica* Breuning
Current name: *Ichthyodes leucostictica* Breuning, 1942a
Current placement: Cerambycidae: Lamiinae: Apomecynini
Type locality: Papua New Guinea: Rossel Island

***leucostictica rustica* Gressitt, 1940**: 124 (Fig. 97s, t), Holotype
Original combination: *Coptops leucostictica rustica* Gressitt
Current name: same
Current placement: Cerambycidae: Lamiinae: Mesosini
Type locality: China: Hainan Island, Dwa-Bi (Tai-pin-tsuen)

***leucostictum* Martins, 1959**: 343 (Fig. 97u, v), Holotype
Original combination: *Hexoplon leucostictum* Martins
Current name: same
Current placement: Cerambycidae: Cerambycinae: Hexoplini
Type locality: Colombia: Hacienda Pehlke

***levettei* Casey, 1891**: 29 (Fig. 97w, x), Lectotype
Original combination: *Elaphidion levettei* Casey
Current name: *Aneflus levettei* (Casey, 1891)
Current placement: Cerambycidae: Cerambycinae: Elaphidiini
Type locality: USA: Arizona
Remarks: This specimen is herein designated as the lectotype in order to stabilize the taxonomy and facilitate further identifications of this species.

***lignea parvicollis* Casey, 1912**: 272 (Fig. 98a, b), Holotype
Original combination: *Anocomis lignea parvicollis* Casey
Current name: *Semanotus ligneus parvicollis* (Casey, 1912)
Current placement: Cerambycidae: Cerambycinae: Callidiini
Type locality: USA: Colorado

***linearis* Breuning, 1943c**: 20 (Fig. 98c, d), Holotype
Original combination: *Falsozorilispe linearis* Breuning
Current name: same
Current placement: Cerambycidae: Lamiinae: Pteropliini
Type locality: New Caledonia

***lineata* Fisher, 1926**: 14 (Fig. 98e, f), Holotype
Original combination: *Jamesia lineata* Fisher
Current name: same
Current placement: Cerambycidae: Lamiinae: Onciderini
Type locality: Saint Lucia

***lineata* Hamilton, 1896**: 142 (Fig. 98g, h), Lectotype
Original combination: *Aporataxia lineata* Hamilton
Current name: *Cacostola lineata* (Hamilton, 1896)
Current placement: Cerambycidae: Lamiinae: Onciderini
Type locality: USA: Texas, Brownsville

***lineaticollis* Chemsak & Linsley, 1974**: 36 (Fig. 98i, j), Holotype
Original combination: *Megachoriolaus lineaticollis* Chemsak & Linsley
Current name: same
Current placement: Cerambycidae: Lepturinae: Lepturini
Type locality: Panama: Chiriquí, Volcán de Chiriqui, 1000 m.

***lineatipennis* Breuning, 1950a**: 306 (Fig. 98k, l), Holotype
Original combination: *Emphreus lineatipennis* Breuning
Current name: same
Current placement: Cerambycidae: Lamiinae: Stenobiini
Type locality: Kenya: Sagala

***lineellus* Fisher, 1938**: 143 (Fig. 98m, n), Holotype
Original combination: *Ozineus lineellus* Fisher
Current name: *Sternacutus zikani* (Melzer, 1935b)
Current placement: Cerambycidae: Lamiinae: Acanthocinini
Type locality: Brazil: Santa Catarina, Nova Teutonia

***lineicornis* Casey, 1913**: 201 (Fig. 98o, p), Holotype
Original combination: *Hapalosalia lineicornis* Casey
Current name: *Pidonia ruficollis* (Say, 1824)
Current placement: Cerambycidae: Lepturinae: Rhagiini
Type locality: USA: Wisconsin, Bayfield

***linelli* Casey, 1924**: 246 (Fig. 98q, r), Lectotype
Original combination: *Anoplium linelli* Casey
Current name: *Anelaphus albofasciatus* (Linell, 1897a)
Current placement: Cerambycidae: Cerambycinae: Elaphidiini
Type locality: USA: Arizona, Tucson
Remarks: This specimen is herein designated as the lectotype in order to stabilize the taxonomy and facilitate further identifications of this species.

***lineolatum* var. *andinum* Tippmann, 1960**: 108 (Fig. 98s, t), Holotype
Original combination: *Achryson lineolatum* var. *andinum* Tippmann
Current name: *Achryson lineolatum* Erichson, 1847
Current placement: Cerambycidae: Cerambycinae: Achrysonini
Type locality: Bolivia: Cochabamba, 260 m.
Remarks: The name following the binomen was not expressly given infrasubspecific rank and the work did not indicate unambiguously that the name was infrasubspecific, therefore, it is deemed a subspecies and available.

***lineolatum* var. *nigricans* Tippmann, 1960**: 109 (Fig. 98u, v), Holotype
Original combination: *Achryson lineolatum* var. *nigricans* Tippmann
Current name: *Achryson lutarium* Burmeister, 1865
Current placement: Cerambycidae: Cerambycinae: Achrysonini
Type locality: Argentina: Cordoba, Coronel Olmedo
Remarks: The name following the binomen was not expressly given infrasubspecific rank and the work did not indicate unambiguously that the name was infrasubspecific, therefore, it is deemed a subspecies and available.

***lingafelteri* Hovore & Santos-Silva, 2007**: 79 (Fig. 98w, x), Holotype
Original combination: *Cometes lingafelteri* Hovore & Santos-Silva
Current name: same
Current placement: Disteniidae: Disteniinae: Disteniini
Type locality: Panama: Bocas del Toro, Almirate

***lingafelteri* Micheli & Nearns, 2005**: 25 (Fig. 99a, b), Holotype
Original combination: *Plectromerus lingafelteri* Micheli & Nearns
Current name: same
Current placement: Cerambycidae: Cerambycinae: Plectromerini
Type locality: Dominican Republic: La Vega Province, Pico Duarte Trail, 3300 ft., Los Tablones, beating, 19°08.222′N, 70°27.736′W

***lingafelteri* Santos-Silva, 2007**: 88 (Fig. 99c, d), Holotype
Original combination: *Derobrachus lingafelteri* Santos-Silva
Current name: same
Current placement: Cerambycidae: Prioninae: Macrotomini
Type locality: Mexico: Veracruz, 14–16 km west of Sntecompn

***lingafelteri* Santos-Silva & Hovore, 2007b**: 583 (Fig. 99e, f), Holotype
Original combination: *Novantinoe lingafelteri* Santos-Silva & Hovore
Current name: same
Current placement: Disteniidae: Disteniinae: Disteniini
Type locality: Panama: Cocle Province, El Valle

***linsleyi* Fisher, 1947a**: 47 (Fig. 99g, h), Holotype
Original combination: *Cantharoxylymna linsleyi* Fisher
Current name: *Stultutragus linsleyi* (Fisher, 1947a)
Current placement: Cerambycidae: Cerambycinae: Rhinotragini
Type locality: Costa Rica: Limón, Hamburg Farm, Reventazón, Ebene

***linsleyi* Santos-Silva & Hovore, 2008a**: 6 (Fig. 99i, j), Holotype
Original combination: *Elytrimitatrix (Grossifemora) linsleyi* Santos-Silva & Hovore
Current name: same
Current placement: Disteniidae: Disteniinae: Disteniini
Type locality: Mexico: Guerrero, hwy 134, 67.4 km NE jct. hwy 200

***lirissa* Dillon & Dillon, 1945**: 153 (Fig. 99k, l), Holotype
Original combination: *Hesychotypa lirissa* Dillon & Dillon
Current name: same
Current placement: Cerambycidae: Lamiinae: Onciderini
Type locality: Brazil: Mato Grosso, Chapada

***litigiosus* Casey, 1891**: 25 (Fig. 99m, n), Lectotype
Original combination: *Hylotrupes litigiosus* Casey
Current name: *Semanotus litigiosus* (Casey, 1891)
Current placement: Cerambycidae: Cerambycinae: Callidiini
Type locality: USA: California
Remarks: This specimen is herein designated as the lectotype in order to stabilize the taxonomy and facilitate further identifications of this species.

***litturatum* Martins, 1960b**: 91 (Fig. 99o, p), Holotype
Original combination: *Octoplon litturatum* Martins
Current name: *Compsibidion litturatum* (Martins, 1960b)
Current placement: Cerambycidae: Cerambycinae: Neoibidionini
Type locality: Venezuela: Aragua, Maracay

***litura* Dillon & Dillon, 1945**: 62 (Fig. 99q, r), Holotype
Original combination: *Charoides litura* Dillon & Dillon
Current name: *Tulcus liturus* (Dillon & Dillon, 1945)
Current placement: Cerambycidae: Lamiinae: Onciderini
Type locality: Bolivia: Beni or Pando Dept., Cachuela

liui **Gressitt, 1948a**: 51 (Fig. 99s, t), Holotype
Original combination: *Molorchus liui* Gressitt
Current name: same
Current placement: Cerambycidae: Cerambycinae: Molorchini
Type locality: China: Yunnan Province

lobata **Aurivillius, 1926**: 103 (Fig. 99u, v), Lectotype
Original combination: *Glenea lobata* Aurivillius
Current name: *Glenea fissicauda* Aurivillius, 1926
Current placement: Cerambycidae: Lamiinae: Saperdini
Type locality: Philippines: Negros, Cuernos Mts.
Remarks: This specimen is herein designated as the lectotype in order to stabilize the taxonomy and facilitate further identifications of this species. Other paralectotype(s) are present in the Naturhistoriska Riksmuseet, Stockholm, Sweden.

lobata **Breuning, 1942a**: 143 (Fig. 99w, x), Holotype
Original combination: *Ogmodera lobata* Breuning
Current name: same
Current placement: Cerambycidae: Lamiinae: Apomecynini
Type locality: Tanzania: Ukerewe Island, Lake Victoria

longeantennatus **Breuning, 1938b**: 369 (Fig. 100a, b), Holotype
Original combination: *Sthenias longeantennatus* Breuning
Current name: same
Current placement: Cerambycidae: Lamiinae: Niphonini
Type locality: India: Assam, Naga Hills

longicollis **Breuning, 1942a**: 128 (Fig. 100c, d), Holotype
Original combination: *Niphotragulus longicollis* Breuning
Current name: same
Current placement: Cerambycidae: Lamiinae: Pteropliini
Type locality: Tanzania: Ukami

longicollis **Casey, 1891**: 34 (Fig. 100e, f), Holotype
Original combination: *Stenosphenus longicollis* Casey
Current name: *Stenosphenus dolosus* Horn, 1885
Current placement: Cerambycidae: Cerambycinae: Elaphidiini
Type locality: USA: Texas

longicollis **Casey, 1912**: 310 (Fig. 100g, h), Holotype
Original combination: *Callimus longicollis* Casey
Current name: *Lampropterus ruficollis* (LeConte, 1873)
Current placement: Cerambycidae: Cerambycinae: Hyboderini
Type locality: USA: California, Kaweah

longicollis **Casey, 1924**: 254 (Fig. 100i, j), Holotype
Original combination: *Brothylus longicollis* Casey
Current name: *Brothylus gemmulatus* LeConte, 1859
Current placement: Cerambycidae: Cerambycinae: Hesperophanini
Type locality: USA: Utah, Provo

longicollis **Gressitt, 1935b**: 287 (Fig. 100k, l), Holotype
Original combination: *Neoserixia longicollis* Gressitt
Current name: same
Current placement: Cerambycidae: Lamiinae: Saperdini
Type locality: China: Taiwan, Chirifu

longicollis **Heller, 1915**: 35 (Fig. 100m, n), Lectotype
Original combination: *Epania longicollis* Heller
Current name: *Leptepania longicollis* (Heller, 1915)
Current placement: Cerambycidae: Cerambycinae: Molorchini
Type locality: Philippines: Luzon, Mt. Makiling

longicollis **Heller, 1916**: 302 (Fig. 100o, p), Holotype
Original combination: *Demonax longicollis* Heller
Current name: same
Current placement: Cerambycidae: Cerambycinae: Clytini
Type locality: Philippines: Mindanao, Davao

longicollis infasciatus **Gressitt, 1935b**: 289 (Fig. 100q, r), Holotype
Original combination: *Neoserixia longicollis infasciatus* Gressitt
Current name: same
Current placement: Cerambycidae: Lamiinae: Saperdini
Type locality: China: Taiwan, Hassenzan

longicorne **Gressitt, 1935d**: 382 (Fig. 100s, t), Holotype
Original combination: *Strangalia (Strangalina) longicorne* Gressitt
Current name: *Mimostrangalia longicornis* (Gressitt, 1935d)
Current placement: Cerambycidae: Lepturinae: Lepturini
Type locality: Japan: Okinawa, Amami-Oshima, Loochoo Island
Remarks: The specific epithet is misspelled on label.

longicornis **Fisher, 1925a**: 275 (Fig. 100u, v), Holotype
Original combination: *Emeopedus longicornis* Fisher
Current name: *Emeopedus (Longicornemeopedus) longicornis* Fisher, 1925a
Current placement: Cerambycidae: Lamiinae: Acanthocinini
Type locality: Malaysia: Borneo, Sandakan

***longicornis* Fisher, 1926**: 15 (Fig. 100w, x), Holotype
Original combination: *Leptostylus longicornis* Fisher
Current name: *Leptostylopsis longicornis* (Fisher, 1926)
Current placement: Cerambycidae: Lamiinae: Acanthocinini
Type locality: USA: Puerto Rico, Río Piedras

***longicornis* Fisher, 1932**: 9 (Fig. 101a, b), Holotype
Original combination: *Eburia longicornis* Fisher
Current name: same
Current placement: Cerambycidae: Cerambycinae: Eburiini
Type locality: Cuba: Camaguey Province, Central Florida

***longior* Casey, 1924**: 284 (Fig. 101c, d), Holotype
Original combination: *Judolia longior* Casey
Current name: *Judolia cordiera* (Olivier, 1795)
Current placement: Cerambycidae: Lepturinae: Lepturini
Type locality: USA: Pennsylvania, Carlisle Junction

***longipennis* Casey, 1891**: 31 (Fig. 101e, f), Lectotype
Original combination: *Crossidius longipennis* Casey
Current name: *Crossidius suturalis* LeConte, 1858
Current placement: Cerambycidae: Cerambycinae: Trachyderini
Type locality: USA: New Mexico
Remarks: This specimen is herein designated as the lectotype in order to stabilize the taxonomy and facilitate further identifications of this species.

***longipennis* Casey, 1912**: 294 (Fig. 101g, h), Holotype
Original combination: *Aneflomorpha longipennis* Casey
Current name: *Aneflomorpha lineare* (LeConte, 1859)
Current placement: Cerambycidae: Cerambycinae: Elaphidiini
Type locality: USA: California, Siskiyou Co.

***longipilis* Fisher, 1926**: 7 (Fig. 101i, j), Holotype
Original combination: *Ischnolea longipilis* Fisher
Current name: *Desmiphora longipilis* (Fisher, 1926)
Current placement: Cerambycidae: Lamiinae: Desmiphorini
Type locality: Cuba: Cayamas

***longiscapus* Breuning, 1943c**: 50 (Fig. 101k, l), Holotype
Original combination: *Sophronica longiscapus* Breuning
Current name: same
Current placement: Cerambycidae: Lamiinae: Desmiphorini
Type locality: Tanzania: Tanganyika, Mufindi, 2000 m.

***longissima* Tippmann, 1960**: 124 (Fig. 101m, n), Holotype
Original combination: *Cylindrommata longissima* Tippmann
Current name: same
Current placement: Cerambycidae: Cerambycinae: Rhinotragini
Type locality: Brazil: Santa Catarina, Mafra

***longissimum* Tippmann, 1953c**: 206 (Fig. 101o, p), Holotype
Original combination: *Elaterosoma longissimum* Tippmann
Current name: *Metopocoilus longissimum* (Tippmann, 1953c)
Current placement: Cerambycidae: Cerambycinae: Trachyderini
Type locality: Venezuela: Upper Orinoco

***longitarse* Casey, 1924**: 239 (Fig. 101q, r), Holotype
Original combination: *Physocnemum longitarse* Casey
Current name: *Physocnemum brevilineum* (Say, 1824)
Current placement: Cerambycidae: Cerambycinae: Callidiini
Type locality: USA: Nebraska

***longitarsis* Casey, 1912**: 362 (Fig. 101s, t), Lectotype
Original combination: *Xylotrechus longitarsis* Casey
Current name: same
Current placement: Cerambycidae: Cerambycinae: Clytini
Type locality: USA: California
Remarks: This specimen is herein designated as the lectotype in order to stabilize the taxonomy and facilitate further identifications of this species.

***longitarsus* Gressitt, 1935b**: 273 (Fig. 101u, v), Holotype
Original combination: *Stenoleptura longitarsus* Gressitt
Current name: same
Current placement: Cerambycidae: Lepturinae: Lepturini
Type locality: Malaysia: Borneo, Sandakan

***loochooanum taiwanum* Gressitt, 1936**: 91 (Fig. 101w, x), Holotype
Original combination: *Chloridolum loochooanum taiwanum* Gressitt
Current name: *Chloridolum taiwanum* Gressitt, 1936
Current placement: Cerambycidae: Cerambycinae: Callichromatini
Type locality: China: Taiwan, Bukai

***lucanus* Casey, 1912**: 317 (Fig. 102a, b), Holotype
Original combination: *Lissonotus lucanus* Casey
Current name: *Lissonotus flavocinctus puncticollis* Bates, 1885
Current placement: Cerambycidae: Cerambycinae: Lissonotini
Type locality: Mexico: Baja California Sur, San Jose

***lucanus* Casey, 1912**: 346 (Fig. 102c, d), Holotype
Original combination: *Stenosphenus lucanus* Casey
Current name: *Stenosphenus cribripennis novatus* Horn, 1885
Current placement: Cerambycidae: Cerambycinae: Elaphidiini
Type locality: Mexico: Baja California Sur

luchti **Fisher, 1937c**: 200 (Fig. 102e, f), Holotype
Original combination: *Oberea luchti* Fisher
Current name: *Obereopsis luchti* (Fisher, 1937c)
Current placement: Cerambycidae: Lamiinae: Phytoeciini
Type locality: Indonesia: Java, Mt. Raoeng (Idjen-Kendang Mountains)

lugubris **Casey, 1912**: 318 (Fig. 102g, h), Lectotype
Original combination: *Stenaspis lugubris* Casey
Current name: *Stenaspis solitaria* (Say, 1824)
Current placement: Cerambycidae: Cerambycinae: Trachyderini
Type locality: USA: Arizona, Santa Rita Mts.
Remarks: This specimen is herein designated as the lectotype in order to stabilize the taxonomy and facilitate further identifications of this species.

lumulumuensis **Fisher, 1935c**: 623 (Fig. 102i, j), Holotype
Original combination: *Glenea lumulumuensis* Fisher
Current name: *Glenea despecta* Pascoe, 1858
Current placement: Cerambycidae: Lamiinae: Saperdini
Type locality: Malaysia: Borneo, Sabah, Mt. Kinabalu, 5,500 ft., Lumu Lumu

lunatus fractus **Hoffman, 1958**: 220 (Fig. 102k, l), Holotype
Original combination: *Typocerus lunatus fractus* Hoffman
Current name: *Typocerus lunulatus* (Swederus, 1787)
Current placement: Cerambycidae: Lepturinae: Lepturini
Type locality: USA: Virginia, Blacksburg

luscus strictulus **Casey, 1912**: 369 (Fig. 102m, n), Holotype
Original combination: *Neoclytus luscus strictulus* Casey
Current name: *Neoclytus mucronatus* (Fabricius, 1775)
Current placement: Cerambycidae: Cerambycinae: Clytini
Type locality: No published locality data or label associated with the specimen.

luteicolle **Gressitt, 1935e**: 190 (Fig. 102o, p), Holotype
Original combination: *Glenea luteicolle* Gressitt
Current name: same
Current placement: Cerambycidae: Lamiinae: Saperdini
Type locality: China: Taiwan, Hassenzan

luteofasciatus **Gressitt, 1941**: 337 (Fig. 102q, r), Holotype
Original combination: *Perihammus luteofasciatus* Gressitt
Current name: *Blepephaeus luteofasciatus* (Gressitt, 1941)
Current placement: Cerambycidae: Lamiinae: Monochamini
Type locality: Thailand: Khun Tan Mts., 4000 ft.

luteonotata **Monné & Monné, 2009**: 36 (Fig. 102s, t), Holotype
Original combination: *Sangaris luteonotata* Monné & Monné
Current name: same
Current placement: Cerambycidae: Lamiinae: Colobotheini
Type locality: Ecuador: Napo, Yasuni Research Station, 76°36′W, 0°38′S, 250 m.

luzonica **Breuning, 1957a**: 10 (Fig. 102u, v), Holotype
Original combination: *Mimosybra luzonica* Breuning
Current name: same
Current placement: Cerambycidae: Lamiinae: Apomecynini
Type locality: Philippines: Luzon, Heightsplan

luzonica **Santos-Silva, Heffern & Matsuda, 2010**: 9 (Fig. 102w, x), Holotype
Original combination: *Komiyandra luzonica* Santos-Silva, Heffern, Matsuda
Current name: same
Current placement: Cerambycidae: Parandrinae: Parandrini
Type locality: Philippines: Luzon, Baguio

luzonica bilaras **Dillon & Dillon, 1951**: 50 (Fig. 103a, b), Holotype
Original combination: *Gnoma luzonica bilaras* Dillon & Dillon
Current name: *Gnoma luzonicum* Erichson, 1834
Current placement: Cerambycidae: Lamiinae: Gnomini
Type locality: Philippines: Bohol Province, Bilar

luzonica transiens **Kriesche, 1936**: 69 (Fig. 103c, d), Holotype
Original combination: *Gnoma luzonica transiens* Kriesche
Current name: *Gnoma luzonicum* Erichson, 1834
Current placement: Cerambycidae: Lamiinae: Gnomini
Type locality: Philippines: Leyte, Santa Cruz
Remarks: Tippmann labeled this specimen as having been obtained from the Kriesche Estate in 1955.

luzonicus **Fisher, 1925a**: 252 (Fig. 103e, f), Holotype
Original combination: *Myromeus luzonicus* Fisher
Current name: same
Current placement: Cerambycidae: Lamiinae: Acanthocinini
Type locality: Philippines: Luzon, Mt. Makiling

luzonicus **Fisher, 1925a**: 254 (Fig. 103g, h), Holotype
Original combination: *Enes luzonicus* Fisher
Current name: *Enes (Sibuyenes) luzonicus* Fisher, 1925a
Current placement: Cerambycidae: Lamiinae: Acanthocinini
Type locality: Philippines: Luzon, Mt. Makiling

***lycimnia* Dillon & Dillon, 1945**: 72 (Fig. 103i, j), Holotype
Original combination: *Charoides lycimnia* Dillon & Dillon
Current name: *Tulcus lycimnius* (Dillon & Dillon, 1945)
Current placement: Cerambycidae: Lamiinae: Onciderini
Type locality: Costa Rica: Limón, Hamburg Farm, Reventazón, Ebene

***lycoides* Chemsak & Linsley, 1971**: 120 (Fig. 103k, l), Holotype
Original combination: *Methia lycoides* Chemsak & Linsley
Current name: same
Current placement: Cerambycidae: Cerambycinae: Methiini
Type locality: Mexico: Tamaulipas, Tampico

***macoris* Lingafelter, 2011**: 63 (Fig. 103m, n), Holotype
Original combination: *Calliclytus macoris* Lingafelter
Current name: same
Current placement: Cerambycidae: Cerambycinae: Tillomorphini
Type locality: Dominican Republic: San Pedro de Macoris, 12 km. west of San Pedro de Macoris
Remarks: On long-term loan from FSCA.

***macrocera* Casey, 1913**: 270 (Fig. 103o, p), Lectotype
Original combination: *Leptura (Dorcasina) macrocera* Casey
Current name: *Dorcasina matthewsii* (LeConte, 1869)
Current placement: Cerambycidae: Lepturinae: Lepturini
Type locality: USA: California, Mendocino County, Gualala
Remarks: This specimen is herein designated as the lectotype in order to stabilize the taxonomy and facilitate further identifications of this species

***macrophthalmus* Breuning, 1954**: 11 (Fig. 103q, r), Holotype
Original combination: *Sulenus macrophthalmus* Breuning
Current name: same
Current placement: Cerambycidae: Lamiinae: Rhodopini
Type locality: Madagascar: Bezanozano

***macrotomoides* Tippmann, 1953c**: 186 (Fig. 103s, t), Holotype
Original combination: *Strongylaspis macrotomoides* Tippmann
Current name: same
Current placement: Cerambycidae: Prioninae: Macrotomini
Type locality: Colombia: Hacienda Pehlke

***maculicollis* Casey, 1912**: 337 (Fig. 103u, v), Lectotype
Original combination: *Crossidius maculicollis* Casey
Current name: *Crossidius testaceus maculicollis* Casey, 1912
Current placement: Cerambycidae: Cerambycinae: Trachyderini
Type locality: USA: California, Los Angeles Co.
Remarks: This specimen is herein designated as the lectotype in order to stabilize the taxonomy and facilitate further identifications of this species.

***maculosa* Fisher, 1935c**: 614 (Fig. 103w, x), Holotype
Original combination: *Ectatosia maculosa* Fisher
Current name: same
Current placement: Cerambycidae: Lamiinae: Desmiphorini
Type locality: Malaysia: Borneo, Sabah, Mt. Kinabalu, Kenokok, 3300 ft.

***maculosa* Martins, 1976**: 353 (Fig. 104a, b), Holotype
Original combination: *Gorybia maculosa* Martins
Current name: same
Current placement: Cerambycidae: Cerambycinae: Piezocerini
Type locality: Paraguay: Central, Asuncion

***madericus* Skiles, 1978**: 414 (Fig. 104c, d), Holotype
Original combination: *Stenodontes (Mallodon) madericus* Skiles
Current name: *Nothopleurus madericus* (Skiles, 1978)
Current placement: Cerambycidae: Prioninae: Mallodonini
Type locality: USA: Arizona, Santa Cruz Co., Madera Canyon

***magnificus* Fisher, 1932**: 89 (Fig. 104e, f), Holotype
Original combination: *Calocosmus magnificus* Fisher
Current name: same
Current placement: Cerambycidae: Lamiinae: Hemilophini
Type locality: Haiti: Valley Rivere Froide

***magnificus* Tippmann, 1953d**: 351 (Fig. 104g, h), Holotype
Original combination: *Neotoronaeus magnificus* Tippmann
Current name: *Toronaeus magnificus* (Tippmann, 1953d)
Current placement: Cerambycidae: Lamiinae: Acanthocinini
Type locality: Peru: Junin, Satipo

***magnus* Schaeffer, 1904**: 224 (Fig. 104i, j), Holotype
Original combination: *Neoclytus magnus* Schaeffer
Current name: same
Current placement: Cerambycidae: Cerambycinae: Clytini
Type locality: Mexico: Baja California Sur, Ensenada

major **Gressitt, 1956**: 136 (Fig. 104k, l), Holotype
Original combination: *Prosoplus major* Gressitt
Current name: same
Current placement: Cerambycidae: Lamiinae: Pterolophini
Type locality: Marshall Islands: Arno Atoll, Ine Island

malkini **Fisher, 1942c**: 51 (Fig. 104m, n), Holotype
Original combination: *Phymatodes malkini* Fisher
Current name: *Eumichthus oedipus* LeConte, 1873
Current placement: Cerambycidae: Cerambycinae: Eumichthini
Type locality: USA: Oregon, Eugene

mancum **Casey, 1912**: 298 (Fig. 104o, p), Holotype
Original combination: *Romaleum mancum* Casey
Current name: *Enaphalodes rufulus* (Haldeman, 1847)
Current placement: Cerambycidae: Cerambycinae: Elaphidiini
Type locality: USA: Rhode Island

mandarinus **Gressitt, 1939b**: 209 (Fig. 104q, r), Holotype
Original combination: *Neocerambyx mandarinus* Gressitt
Current name: *Sinopachys mandarinus* (Gressitt, 1939b)
Current placement: Cerambycidae: Cerambycinae: Cerambycini
Type locality: China: Shaanxi Province, Chin-ling Mts.

mandibularis **Chemsak, 1963**: 47 (Fig. 104s, t), Holotype
Original combination: *Tetraopes mandibularis* Chemsak
Current name: same
Current placement: Cerambycidae: Lamiinae: Tetraopini
Type locality: USA: Texas, Childress

maneei **Casey, 1914**: 367 (Fig. 104u, v), Holotype
Original combination: *Strangalia maneei* Casey
Current name: *Pseudostrangalia cruentata* (Haldeman, 1847)
Current placement: Cerambycidae: Lepturinae: Lepturini
Type locality: USA: North Carolina, Southern Pines

manni **Fisher, 1930a**: 15 (Fig. 104w, x), Holotype
Original combination: *Phygopoda manni* Fisher
Current name: *Epimelitta manni* (Fisher, 1930a)
Current placement: Cerambycidae: Cerambycinae: Rhinotragini
Type locality: Bolivia: El Beni, Rurrenabaque

marginalis **Fisher, 1947c**: 29 (Fig. 105a, b), Holotype
Original combination: *Eburia marginalis* Fisher
Current name: same
Current placement: Cerambycidae: Cerambycinae: Eburiini
Type locality: Cuba: Matanzas, central Mercedes

marginalis **Ohbayashi, 1948**: 14 (Fig. 105c, d), Holotype
Original combination: *Necydalis marginalis* Ohbayashi
Current name: *Necydalis gigantea* Kano, 1933
Current placement: Cerambycidae: Necydalinae: Necydalini
Type locality: Japan: Yamagata Prefecture, Hanazawa, Yonezawa

marginellus **Casey, 1924**: 274 (Fig. 105e, f), Holotype
Original combination: *Stenocorus marginellus* Casey
Current name: *Stenocorus nubifer* (LeConte, 1859)
Current placement: Cerambycidae: Lepturinae: Lepturini
Type locality: USA: California, Mt. Wilson

marginipennis **Gressitt, 1948a**: 47 (Fig. 105g, h), Holotype
Original combination: *Necydalis marginipennis* Gressitt
Current name: same
Current placement: Cerambycidae: Necydalinae: Necydalini
Type locality: China: Sichuan Province, south of Suifu

marianae **Martins & Monné, 1974**: 330 (Fig. 105i, j), Holotype
Original combination: *Baryssinus marianae* Martins & Monné
Current name: *Neobaryssinus marianae* (Martins & Monné, 1974)
Current placement: Cerambycidae: Lamiinae: Acanthocinini
Type locality: French Guiana: St-Laurent-du-Maroni, St. Jean

marmoratus **Casey, 1913**: 351 (Fig. 105k, l), Lectotype
Original combination: *Lochmaeocles marmoratus* Casey
Current name: same
Current placement: Cerambycidae: Lamiinae: Onciderini
Type locality: USA: Arizona, Cochise Co., Saint Bernardino Ranch, 2750 ft.
Remarks: This specimen is herein designated as the lectotype in order to stabilize the taxonomy and facilitate further identifications of this species.

marmoratus **Fisher, 1925c**: 16 (Fig. 105m, n), Holotype
Original combination: *Eupogonius marmoratus* Fisher
Current name: *Eupogonius infimus* (Thomson, 1868)
Current placement: Cerambycidae: Lamiinae: Desmiphorini
Type locality: Mexico: Veracruz, Cordoba

marmoratus **Fisher, 1925a**: 255 (Fig. 105o, p), Holotype
Original combination: *Enes marmoratus* Fisher
Current name: *Enes (Sibuyenes) marmoratus* Fisher, 1925a
Current placement: Cerambycidae: Lamiinae: Acanthocinini
Type locality: Philippines: Mindanao, Dapitan

marmoratus **Fisher, 1935c**: 602 (Fig. 105q, r), Holotype
Original combination: *Dihammus marmoratus* Fisher
Current name: *Acalolepta marmorata* (Fisher, 1935c)
Current placement: Cerambycidae: Lamiinae: Monochamini
Type locality: Malaysia: Borneo, Sabah, Mt. Kinabalu, 5500 ft., Lumu Lumu

marmoratus **Gressitt, 1940**: 158 (Fig. 105s, t), Holotype
Original combination: *Presates marmoratus* Gressitt
Current name: *Pseudoterinaea bicoloripes* (Pic, 1926)
Current placement: Cerambycidae: Lamiinae: Pteropliini
Type locality: China: Hainan Province, Hainan Island, Dwa Bi
Remarks: The labels indicate "*Phesates marmoratus*" but it was published as "*Presates marmoratus*".

marmorea **Breuning, 1942a**: 129 (Fig. 105u, v), Holotype
Original combination: *Rhytiphora (Rhytiphora) marmorea* Breuning
Current name: same
Current placement: Cerambycidae: Lamiinae: Pteropliini
Type locality: Australia: Western Australia

marshallensis **Gressitt, 1956**: 151 (Fig. 105w, x), Holotype
Original combination: *Oopsis marshallensis* Gressitt
Current name: same
Current placement: Cerambycidae: Lamiinae: Apomecynini
Type locality: Marshall Islands: Medyado Is., Jaluit Atoll

masbatensis **Breuning, 1957a**: 14 (Fig. 106a, b), Holotype
Original combination: *Aegocidnus (Miaena) masbatensis* Breuning
Current name: *Sciades (Miaenia) masbatensis* (Breuning, 1957a)
Current placement: Cerambycidae: Lamiinae: Acanthocinini
Type locality: Philippines: Masbate Island, Aroroy

matsudai **Gressitt, 1938c**: 163 (Fig. 106c, d), Holotype
Original combination: *Euseboides matsudai* Gressitt
Current name: same
Current placement: Cerambycidae: Lamiinae: Rhodopini
Type locality: China: Taiwan, Horisha

matsushitai **Gressitt, 1936**: 97 (Fig. 106e, f), Holotype
Original combination: *Demonax matsushitai* Gressitt
Current name: *Demonax sauteri* Matsushita, 1933
Current placement: Cerambycidae: Cerambycinae: Clytini
Type locality: China: Taiwan, Hassenzan

maximus **Tippmann, 1953d**: 339 (Fig. 106g, h), Holotype
Original combination: *Pseudaethomerus maximus* Tippmann
Current name: same
Current placement: Cerambycidae: Lamiinae: Acanthoderini
Type locality: Bolivia: Cochabamba, Reg. Chapare 400 m.

mayesae **Ivie, 2007**: 58 (Fig. 106i, j), Holotype
Original combination: *Elaphidion mayesae* Ivie
Current name: same
Current placement: Cerambycidae: Cerambycinae: Elaphidiini
Type locality: USA: Virgin Island, St. Thomas, Est. St. Peter, ca 1450 ft., 3-H-4 North Star

mecasoides **Casey, 1913**: 359 (Fig. 106k, l), Holotype
Original combination: *Saperda mecasoides* Casey
Current name: *Saperda inornata* Say, 1824
Current placement: Cerambycidae: Lamiinae: Saperdini
Type locality: USA: New York, New York City and vicinity

medenbachi **var.** ***sumatrensis*** **Fisher, 1934a**: 3 (Fig. 106m, n), Holotype
Original combination: *Melanauster medenbachi* var. *sumatrensis* Fisher
Current name: *Anoplophora medenbachii* (Ritsema, 1881)
Current placement: Cerambycidae: Lamiinae: Monochamini
Type locality: Indonesia: Sumatra, Dolok Baros
Remarks: The name following the binomen was not expressly given infrasubspecific rank and the work did not indicate unambiguously that the name was infrasubspecific, therefore, it is deemed a subspecies and available.

medialis **Casey, 1912**: 302 (Fig. 106o, p), Holotype
Original combination: *Hypermallus medialis* Casey
Current name: *Anelaphus parallelus* (Newman, 1840)
Current placement: Cerambycidae: Cerambycinae: Elaphidiini
Type locality: USA: North Carolina

medialis **Casey, 1912**: 344 (Fig. 106q, r), Lectotype
Original combination: *Tylosis medialis* Casey
Current name: *Tylosis maculatus* LeConte, 1850
Current placement: Cerambycidae: Cerambycinae: Trachyderini
Type locality: USA: New Mexico, Jemez Springs
Remarks: This specimen is herein designated as the lectotype in order to stabilize the taxonomy and facilitate further identifications of this species.

medialis bifasciata **Casey, 1912**: 344 (Fig. 106s, t), Holotype
Original combination: *Tylosis medialis bifasciata* Casey
Current name: *Tylosis maculatus* LeConte, 1850
Current placement: Cerambycidae: Cerambycinae: Trachyderini
Type locality: USA: New Mexico, Jemez Springs

medialis canadensis **Casey, 1912**: 303 (Fig. 106u, v), Holotype
Original combination: *Hypermallus medialis canadensis* Casey
Current name: *Anelaphus parallelus* (Newman, 1840)
Current placement: Cerambycidae: Cerambycinae: Elaphidiini
Type locality: Canada: Ontario

medialis densicollis **Casey, 1912**: 303 (Fig. 106w, x), Holotype
Original combination: *Hypermallus medialis densicollis* Casey
Current name: *Anelaphus parallelus* (Newman, 1840)
Current placement: Cerambycidae: Cerambycinae: Elaphidiini
Type locality: USA: Florida

mediofasciata **Breuning, 1942d**: 142 (Fig. 107a, b), Holotype
Original combination: *Mesosa mediofasciata* Breuning
Current name: *Mesosa mediofasciata* Breuning, 1942d
Current placement: Cerambycidae: Lamiinae: Mesosini
Type locality: Japan: Mt. Takao

meeskei **Casey, 1891**: 30 (Fig. 107c, d), Lectotype
Original combination: *Rhopalophora meeskei* Casey
Current name: same
Current placement: Cerambycidae: Cerambycinae: Rhopalophorini
Type locality: USA: New Mexico, Las Vegas
Remarks: This specimen is herein designated as the lectotype in order to stabilize the taxonomy and facilitate further identifications of this species.

megacephala **Breuning, 1940b**: 83 (Fig. 107e, f), Holotype
Original combination: *Falsohippopsis megacephala* Breuning
Current name: *Aerenicopsis megacephala* (Breuning, 1940b)
Current placement: Cerambycidae: Lamiinae: Aerenicini
Type locality: Brazil: Santa Catarina

megalophthalmus **Tippmann, 1953c**: 194 (Fig. 107g, h), Holotype
Original combination: *Derobrachus megalophthalmus* Tippmann
Current name: *Derobrachus inaequalis* Bates, 1872
Current placement: Cerambycidae: Prioninae: Prionini
Type locality: Colombia: Antioquia, Medellin

megalops **Hamilton, 1896**: 127 (Fig. 107i, j), Lectotype
Original combination: *Lepturges megalops* Hamilton
Current name: same
Current placement: Cerambycidae: Lamiinae: Acanthocinini
Type locality: USA: Florida, Biscayne
Remarks: This specimen is herein designated as the lectotype in order to stabilize the taxonomy and facilitate further identifications of this species.

megalopus **Casey, 1924**: 271 (Fig. 107k, l), Holotype
Original combination: *Euryscelis megalopus* Casey
Current name: *Euryscelis suturalis* (Olivier, 1795)
Current placement: Cerambycidae: Cerambycinae: Clytini
Type locality: USA: Florida

melanoscelis **Aurivillius, 1922**: 434 (Fig. 107m, n), Holotype
Original combination: *Nupserha melanoscelis* Aurivillius
Current name: same
Current placement: Cerambycidae: Lamiinae: Phytoeciini
Type locality: Philippines: Mindanao, Kolambugan

mendosa **Martins & Galileo, 1998**: 97 (Fig. 107o, p), Holotype
Original combination: *Aerenicopsis mendosa* Martins & Galileo
Current name: same
Current placement: Cerambycidae: Lamiinae: Aerenicini
Type locality: Mexico: Veracruz, Boca del Río, Veracruz

meridianum **Fisher, 1938**: 141 (Fig. 107q, r), Holotype
Original combination: *Lophopoeum meridianum* Fisher
Current name: same
Current placement: Cerambycidae: Lamiinae: Acanthocinini
Type locality: Brazil: Santa Catarina, Nova Teutonia

meridionalis **Fisher, 1930a**: 14 (Fig. 107s, t), Holotype
Original combination: *Bromiades meridionalis* Fisher
Current name: *Bromiades brachyptera* (Chevrolat, 1838)
Current placement: Cerambycidae: Cerambycinae: Rhinotragini
Type locality: Panama: Cabima

metallica **Galileo & Martins, 2008b**: 33 (Fig. 107u, v), Holotype
Original combination: *Callia metallica* Galileo & Martins
Current name: same
Current placement: Cerambycidae: Lamiinae: Calliini
Type locality: Ecuador: Sucumbios, San Rafael Falls, 1100 m.

***metalliceps* Breuning, 1940d**: 417 (Fig. 107w, x), Holotype
Original combination: *Tmesisternus metalliceps* Breuning
Current name: same
Current placement: Cerambycidae: Lamiinae: Tmesisternini
Type locality: Indonesia: Papua, Irian Jaya, Mt. Goliath, 5000–7000 ft.

***mexicanum* Casey, 1913**: 197 (Fig. 108a, b), Lectotype
Original combination: *Rhagium (Hargium) mexicanum* Casey
Current name: *Rhagium inquisitor* (Linnaeus, 1758)
Current placement: Cerambycidae: Lepturinae: Rhagiini
Type locality: Mexico: Guerrero
Remarks: This specimen is herein designated as the lectotype in order to stabilize the taxonomy and facilitate further identifications of this species.

***mexicanum* Fisher, 1927b**: 215 (Fig. 108c, d), Holotype
Original combination: *Moneilema (Collapteryx) mexicanum* Fisher
Current name: *Moneilema mexicanum* Fisher, 1927b
Current placement: Cerambycidae: Lamiinae: Moneilemini
Type locality: Mexico: Pachuca

***mexicanum* Galileo & Martins, 2007c**: 706 (Fig. 108e, f), Holotype
Original combination: *Sphallambyx mexicanum* Galileo & Martins
Current name: same
Current placement: Cerambycidae: Cerambycinae: Cerambycini
Type locality: Mexico: Veracruz, 14–16 km west of Santecomapan

***mexicanus* Dillon & Dillon, 1946**: 240 (Fig. 108g, h), Holotype
Original combination: *Trachysomus mexicanus* Dillon & Dillon
Current name: same
Current placement: Cerambycidae: Lamiinae: Onciderini
Type locality: Mexico: Colima, Ciudad de Colima

***micans* Fisher, 1937d**: 146 (Fig. 108i, j), Holotype
Original combination: *Gnomidolon micans* Fisher
Current name: *Gnomidolon varians* Gounelle, 1909
Current placement: Cerambycidae: Cerambycinae: Hexoplini
Type locality: Brazil: Santa Catarina, Nova Teutonia

***micans* Santos-Silva & Tavakilian, 2009**: 27 (Fig. 108k, l), Holotype
Original combination: *Paracometes micans* Santos-Silva & Tavakilian
Current name: same
Current placement: Disteniidae: Disteniinae: Disteniini
Type locality: Panama: Panama Province, Canal Zone, Gatun Spillway

***micans* v. *brunneicaudatum* Fisher, 1937d**: 147 (Fig. 108m, n), Holotype
Original combination: *Gnomidolon micans* v. *brunneicaudatum* Fisher
Current name: *Gnomidolon varians* Gounelle, 1909
Current placement: Cerambycidae: Cerambycinae: Hexoplini
Type locality: Brazil: Santa Catarina, Nova Teutonia
Remarks: The name following the binomen was not expressly given infrasubspecific rank and the work did not indicate unambiguously that the name was infrasubspecific, therefore, it is deemed a subspecies and available.

***michelii* Chemsak, 1979**: 125 (Fig. 108o, p), Holotype
Original combination: *Derancistrus michelii* Chemsak
Current name: *Solenoptera michelii* (Chemsak, 1979)
Current placement: Cerambycidae: Prioninae: Solenopterini
Type locality: USA: Puerto Rico, Rd. 10, Km. 21, 1500 ft.

***michelii* Ivie, 2007**: 56 (Fig. 108q, r), Holotype
Original combination: *Elaphidion michelii* Ivie
Current name: same
Current placement: Cerambycidae: Cerambycinae: Elaphidiini
Type locality: USA: Puerto Rico, Toa Baja

***microphthalma* Breuning, 1942a**: 163 (Fig. 108s, t), Holotype
Original combination: *Estola microphthalma* Breuning
Current name: same
Current placement: Cerambycidae: Lamiinae: Desmiphorini
Type locality: Brazil: Espirito Santo, Tijuco Preto

***microsticta* Casey, 1913**: 283 (Fig. 108u, v), Holotype
Original combination: *Moneilema (Moneilema) microsticta* Casey
Current name: *Moneilema appressum* LeConte, 1852
Current placement: Cerambycidae: Lamiinae: Moneilemini
Type locality: USA: Arizona, Douglas, San Bernardino Ranch, 3750 ft.

***militaris* Casey, 1914**: 364 (Fig. 108w, x), Holotype
Original combination: *Hypermallus militaris* Casey
Current name: *Anelaphus villosus* (Fabricius, 1793)
Current placement: Cerambycidae: Cerambycinae: Elaphidiini
Type locality: USA: New York, West Point

***milleri* Fisher, 1932**: 83 (Fig. 109a, b), Holotype
Original combination: *Leptostylus milleri* Fisher
Current name: *Leptostylopsis milleri* (Fisher, 1932)
Current placement: Cerambycidae: Lamiinae: Acanthocinini
Type locality: Dominican Republic: Río San Juan

***mimetica* Lane, 1970**: 387 (Fig. 109c, d), Holotype
Original combination: *Oreodera mimetica* Lane
Current name: same
Current placement: Cerambycidae: Lamiinae: Acanthoderini
Type locality: Costa Rica: Limón, Hamburg Farm, Reventazón, Ebene

***mimeticum* Schaeffer, 1905a**: 132 (Fig. 109e, f), Lectotype
Original combination: *Elaphidion mimeticum* Schaeffer
Current name: same
Current placement: Cerambycidae: Cerambycinae: Elaphidiini
Type locality: USA: Texas, Brownsville, Esperanza Ranch
Remarks: This specimen is herein designated as the lectotype in order to stabilize the taxonomy and facilitate further identifications of this species.

***mimeticus* Casey, 1891**: 48 (Fig. 109g, h), Holotype
Original combination: *Leiopus mimeticus* Casey
Current name: *Sternidius mimeticus* (Casey, 1891)
Current placement: Cerambycidae: Lamiinae: Acanthocinini
Type locality: USA: Texas, Columbus
Remarks: Although the original description suggests there may be more than one specimen in the original series, there is only one in the USNM collection.

***mimica* Tippmann, 1960**: 162 (Fig. 109i, j), Holotype
Original combination: *Oreodera mimica* Tippmann
Current name: *Psapharochrus vetustus* (Bates, 1880)
Current placement: Cerambycidae: Lamiinae: Acanthoderini
Type locality: Brazil: Santa Catarina, Corupá (formerly Hansa-Humboldt)

***mindanaoensis* Fisher, 1925a**: 239 (Fig. 109k, l), Holotype
Original combination: *Exocentrus mindanaoensis* Fisher
Current name: *Exocentrus (Centenexocentrus) mindanaoensis* Fisher, 1925a
Current placement: Cerambycidae: Lamiinae: Acanthocinini
Type locality: Philippines: Mindanao, Davao

***mindanaoensis* Gressitt, 1935g**: 268 (Fig. 109m, n), Holotype
Original combination: *Noemia mindanaoensis* Gressitt
Current name: same
Current placement: Disteniidae: Disteniinae: Disteniini
Type locality: Philippines: Mindanao, Zamboanga del Sur

***mindanaonis* Aurivillius, 1928**: 309 (Fig. 109o, p), Syntype
Original combination: *Xylotrechus mindanaonis* Aurivillius
Current name: same
Current placement: Cerambycidae: Cerambycinae: Clytini
Type locality: Philippines: Mindanao, Surigao
Remarks: Tavakilian and Chevillotte (2013) state that syntypes are present in the Naturhistoriska Riksmuseet, Stockholm, Sweden as well as the USNM.

***mindanaonis* Breuning, 1957a**: 14 (Fig. 109q, r), Holotype
Original combination: *Aegocidnus (Miaena) mindanaonis* Breuning
Current name: *Sciades (Miaenia) mindanaonis* (Breuning, 1957)
Current placement: Cerambycidae: Lamiinae: Acanthocinini
Type locality: Philippines: Mindanao, Kolambugan

***mindoroanus* Breuning, 1957a**: 13 (Fig. 109s, t), Holotype
Original combination: *Exocentrus (Pseudocentrus) mindoroanus* Breuning
Current name: same
Current placement: Cerambycidae: Lamiinae: Acanthocinini
Type locality: Philippines: Mindoro, Calapan

***minima* Casey, 1913**: 234 (Fig. 109u, v), Holotype
Original combination: *Leptacmaeops minima* Casey
Current name: *Cortodera cubitalis* (LeConte, 1861)
Current placement: Cerambycidae: Lepturinae: Rhagiini
Type locality: USA: California

***minimus* Galileo & Martins, 2012a**: 67 (Fig. 109w, x), Holotype
Original combination: *Miriochrus minimus* Galileo & Martins
Current name: same
Current placement: Cerambycidae: Lamiinae: Acanthoderini
Type locality: Paraguay: Dept. Concepcion, Ciudad Concepcion

***minnesotana* Casey, 1913**: 269 (Fig. 110a, b), Lectotype
Original combination: *Leptura (Cercolia) minnesotana* Casey
Current name: *Trigonarthris minnesotana* (Casey, 1913)
Current placement: Cerambycidae: Lepturinae: Lepturini
Type locality: USA: Wisconsin, Bayfield
Remarks: Although Laplante placed a lectotype label on this specimen in 1991, no formal lectotype designation was published. Therefore, this specimen is herein designated as the lectotype in order to stabilize the taxonomy and facilitate further identifications of this species.

***minuens* Hamilton in Leng & Hamilton, 1896**: 123 (Fig. 110c, d), Lectotype
Original combination: *Liopus minuens* Hamilton in Leng & Hamilton
Current name: *Styloleptus biustus* (LeConte, 1852)
Current placement: Cerambycidae: Lamiinae: Acanthocinini
Type locality: USA: Florida, Biscayne
Remarks: This specimen is herein designated as the lectotype in order to stabilize the taxonomy and facilitate further identifications of this species.

***minuscula* Casey, 1913**: 260 (Fig. 110e, f), Lectotype
Original combination: *Strangalia minuscula* Casey
Current name: *Leptura propinqua* Bland, 1865
Current placement: Cerambycidae: Lepturinae: Lepturini
Type locality: USA: Colorado
Remarks: This specimen is herein designated as the lectotype in order to stabilize the taxonomy and facilitate further identifications of this species.

***minuta* Casey, 1924**: 284 (Fig. 110g, h), Holotype
Original combination: *Judolia minuta* Casey
Current name: *Judolia sexspilota* (LeConte, 1859)
Current placement: Cerambycidae: Lepturinae: Lepturini
Type locality: USA: Texas

***minuta* Fisher, 1925a**: 212 (Fig. 110i, j), Holotype
Original combination: *Callienispia minuta* Fisher
Current name: *Egesina (Callienispia) minuta* (Fisher, 1925a)
Current placement: Cerambycidae: Lamiinae: Pteropliini
Type locality: Singapore

***minuta* Fisher, 1936a**: 197 (Fig. 110k, l), Holotype
Original combination: *Miaenia minuta* Fisher
Current name: *Sciades (Miaenia) minutus* (Fisher, 1936a)
Current placement: Cerambycidae: Lamiinae: Acanthocinini
Type locality: Indonesia: Java, Mt. Tangkoeban Prahoe, Preanger 4000–5000 ft.

***minutivestis* Casey, 1924**: 251 (Fig. 110m, n), Holotype
Original combination: *Oeme minutivestis* Casey
Current name: *Oeme costata* LeConte, 1873
Current placement: Cerambycidae: Cerambycinae: Oemini
Type locality: USA: Nevada, Las Vegas

***minutum* Casey, 1924**: 261 (Fig. 110o, p), Holotype
Original combination: *Poecilobrium minutum* Casey
Current name: *Poecilobrium chalybeum* (LeConte, 1873)
Current placement: Cerambycidae: Cerambycinae: Eumichthini
Type locality: USA: California, Eldorado Co.

***minutum* Fisher, 1926**: 6 (Fig. 110q, r), Holotype
Original combination: *Acrepidopterum minutum* Fisher
Current name: same
Current placement: Cerambycidae: Lamiinae: Apomecynini
Type locality: Cuba: Cayamas

***minutum* var. *apicalis* Fisher, 1935b**: 197 (Fig. 110s, t), Holotype
Original combination: *Acrepidopterum minutum* var. *apicalis* Fisher
Current name: *Acrepidopterum minutum* Fisher, 1926
Current placement: Cerambycidae: Lamiinae: Apomecynini
Type locality: Jamaica: Mandeville
Remarks: The name following the binomen was not expressly given infrasubspecific rank and the work did not indicate unambiguously that the name was infrasubspecific, therefore, it is deemed a subspecies and available.

***minutus* Casey, 1924**: 235 (Fig. 110u, v), Holotype
Original combination: *Hylotrupes minutus* Casey
Current name: *Hylotrupes bajulus* (Linnaeus, 1758)
Current placement: Cerambycidae: Cerambycinae: Hylotrupini
Type locality: Central Europe (country unknown)

***miranda* Tippmann, 1960**: 128 (Fig. 110w, x), Holotype
Original combination: *Epimelitta miranda* Tippmann
Current name: *Epimelitta meliponica* Bates, 1870
Current placement: Cerambycidae: Cerambycinae: Rhinotragini
Type locality: Peru: Loreto, Pucallpa

***miwai* Gressitt, 1936**: 100 (Fig. 111a, b), Holotype
Original combination: *Chlorophorus miwai* Gressitt
Current name: same
Current placement: Cerambycidae: Cerambycinae: Clytini
Type locality: China: Taiwan, Rarasan

mm-ornata **Tippmann, 1960**: 146 (Fig. 111c, d), Holotype
Original combination: *Trachyderes mm-ornata* Tippmann
Current name: *Trachyderes succinctus duponti* Aurivillius, 1912
Current placement: Cerambycidae: Cerambycinae: Trachyderini
Type locality: Brazil: Espirito Santo, Viana
Remarks: Following ICZN Article 32.5.2.4.3, MM refers to the pattern of the elytra that look like two Ms. Therefore, the name is corrected to all lower case letters, however, the hyphen is retained.

moderator **Casey, 1913**: 314 (Fig. 111e, f), Holotype
Original combination: *Leiopus moderator* Casey
Current name: *Sternidius mimeticus* (Casey, 1891)
Current placement: Cerambycidae: Lamiinae: Acanthocinini
Type locality: USA: District of Columbia

mokelumne **Casey, 1912**: 262 (Fig. 111g, h), Lectotype
Original combination: *Liasemum mokelumne* Casey
Current name: *Asemum nitidum* LeConte, 1873
Current placement: Cerambycidae: Spondylidinae: Asemini
Type locality: USA: California, Mokelumne Hill
Remarks: This specimen is herein designated as the lectotype in order to stabilize the taxonomy and facilitate further identifications of this species.

molliculus **Casey, 1912**: 304 (Fig. 111i, j), Holotype
Original combination: *Hypermallus molliculus* Casey
Current name: *Anelaphus parallelus* (Newman, 1840)
Current placement: Cerambycidae: Cerambycinae: Elaphidiini
Type locality: USA: Texas

monin **Micheli & Micheli, 2004**: 27 (Fig. 111k, l), Holotype
Original combination: *Leptostylopsis monin* Micheli & Micheli
Current name: same
Current placement: Cerambycidae: Lamiinae: Acanthocinini
Type locality: USA: Puerto Rico, Guilarte Forest

montana **Casey, 1891**: 40 (Fig. 111m, n), Holotype
Original combination: *Strangalia montana* Casey
Current name: *Strangalia sexnotata* Haldeman, 1847
Current placement: Cerambycidae: Lepturinae: Lepturini
Type locality: USA: New Mexico

montana **Casey, 1913**: 325 (Fig. 111o, p), Holotype
Original combination: *Hyperplatys montana* Casey
Current name: same
Current placement: Cerambycidae: Lamiinae: Acanthocinini
Type locality: USA: Colorado, Boulder Co.

montana **Casey, 1913**: 369 (Fig. 111q, r), Lectotype
Original combination: *Oberea montana* Casey
Current name: *Oberea perspicillata* Haldeman, 1847
Current placement: Cerambycidae: Lamiinae: Phytoeciini
Type locality: USA: Utah, Nephi
Remarks: This specimen is herein designated as the lectotype in order to stabilize the taxonomy and facilitate further identifications of this species.

montana **Fisher, 1925a**: 231 (Fig. 111s, t), Holotype
Original combination: *Ebaeides montana* Fisher
Current name: same
Current placement: Cerambycidae: Lamiinae: Acanthocinini
Type locality: Philippines: Negros Oriental, Cuernos Mountains

montanicus **Casey, 1912**: 361 (Fig. 111u, v), Holotype
Original combination: *Xylotrechus montanicus* Casey
Current name: *Xylotrechus longitarsis* Casey, 1912
Current placement: Cerambycidae: Cerambycinae: Clytini
Type locality: USA: Idaho, Coeur d'Alene

montanum **Casey, 1913**: 197 (Fig. 111w, x), Lectotype
Original combination: *Rhagium (Hargium) montanum* Casey
Current name: *Rhagium inquisitor* (Linnaeus, 1758)
Current placement: Cerambycidae: Lepturinae: Rhagiini
Type locality: USA: Colorado, Boulder Co.
Remarks: This specimen is herein designated as the lectotype in order to stabilize the taxonomy and facilitate further identifications of this species.

montanus **Tippmann, 1953d**: 327 (Fig. 112a, b), Holotype
Original combination: *Trachyderes montanus* Tippmann
Current name: *Deretrachys montanus* (Tippmann, 1953d)
Current placement: Cerambycidae: Cerambycinae: Trachyderini
Type locality: Peru: Ocaña, 2600 m.

monticola **Casey, 1913**: 293 (Fig. 112c, d), Lectotype
Original combination: *Monochamus monticola* Casey
Current name: *Monochamus scutellatus oregonensis* LeConte, 1873
Current placement: Cerambycidae: Lamiinae: Monochamini
Type locality: USA: Colorado, Rocky Mountains
Remarks: This specimen is herein designated as the lectotype in order to stabilize the taxonomy and facilitate further identifications of this species.

monticola **Fisher, 1935b**: 191 (Fig. 112e, f), Holotype
Original combination: *Elaphidion monticola* Fisher
Current name: *Anopliomorpha antillarum* (Fisher, 1932)
Current placement: Cerambycidae: Cerambycinae: Elaphidiini
Type locality: Cuba: Pan de Matanzas Mt.

***monticola* Fisher, 1935b**: 201 (Fig. 112g, h), Holotype
Original combination: *Leptostylus monticola* Fisher
Current name: *Leptostylopsis monticola* (Fisher, 1935b)
Current placement: Cerambycidae: Lamiinae: Acanthocinini
Type locality: Cuba: Matanzas, Pan de Matanzas Mt.

***monticola* Fisher, 1935c**: 610 (Fig. 112i, j), Holotype
Original combination: *Pterolophia monticola* Fisher
Current name: *Pterolophia illicita* (Pascoe, 1865)
Current placement: Cerambycidae: Lamiinae: Pteropliini
Type locality: Malaysia: Borneo, Sabah, Mt. Kinabalu, 3000 ft., Kiau

***monticola* Fisher, 1935c**: 627 (Fig. 112k, l), Holotype
Original combination: *Oberea monticola* Fisher
Current name: same
Current placement: Cerambycidae: Lamiinae: Phytoeciini
Type locality: Malaysia: Borneo, Sabah, Mt. Kinabalu, 5500 ft., Lumu Lumu

***monticola* Fisher, 1936a**: 192 (Fig. 112m, n), Holotype
Original combination: *Callienispia monticola* Fisher
Current name: *Egesina (Callienispia) monticola* (Fisher, 1936a)
Current placement: Cerambycidae: Lamiinae: Pteropliini
Type locality: Indonesia: Java, Dreanger, G. Tangkoeban Prahoe, 4000–5000 ft.

***monticola* Tippmann, 1960**: 153 (Fig. 112o, p), Holotype
Original combination: *Bisaltes monticola* Tippmann
Current name: same
Current placement: Cerambycidae: Lamiinae: Apomecynini
Type locality: Bolivia: Cochabamba, Tiraque, 3200 m.

***montivagans* Fisher, 1937c**: 202 (Fig. 112q, r), Holotype
Original combination: *Oberea montivagans* Fisher
Current name: *Oberea bivittata montivagans* Fisher, 1937c
Current placement: Cerambycidae: Lamiinae: Phytoeciini
Type locality: Indonesia: Java, Mt. Tangkoeban Prahoe, Preanger, 4000–5000 ft.

***morata* Casey, 1913**: 222 (Fig. 112s, t), Holotype
Original combination: *Brachysomida morata* Casey
Current name: *Brachysomida atra* (LeConte, 1850)
Current placement: Cerambycidae: Lepturinae: Rhagiini
Type locality: USA: Colorado, Top of Mt. Arapahoe, 13000 ft.

***morio* Casey, 1924**: 274 (Fig. 112u, v), Lectotype
Original combination: *Stenocorus morio* Casey
Current name: *Stenocorus nubifer* (LeConte, 1859)
Current placement: Cerambycidae: Lepturinae: Lepturini
Type locality: USA: Utah, Provo
Remarks: This specimen is herein designated as the lectotype in order to stabilize the taxonomy and facilitate further identifications of this species.

***morrisi* Nearns & Branham, 2008**: 43 (Fig. 112w, x), Holotype
Original combination: *Plectromerus morrisi* Nearns & Branham
Current name: same
Current placement: Cerambycidae: Cerambycinae: Plectromerini
Type locality: Panama: Panama Province, 12 km north of El Llano

***morrisi* Santos-Silva & Hovore, 2007b**: 586 (Fig. 113a, b), Holotype
Original combination: *Novantinoe morrisi* Santos-Silva & Hovore
Current name: same
Current placement: Disteniidae: Disteniinae: Disteniini
Type locality: Mexico: Hidalgo, 2.4 miles N. Tlanchinol

***mozinnae* Linell, 1897a**: 395 (Fig. 113c, d), Lectotype
Original combination: *Obrium mozinnae* Linell
Current name: same
Current placement: Cerambycidae: Cerambycinae: Obriini
Type locality: USA: Texas, Brownsville
Remarks: This specimen is herein designated as the lectotype in order to stabilize the taxonomy and facilitate further identifications of this species.

***muliebris* Casey, 1912**: 358 (Fig. 113e, f), Holotype
Original combination: *Calloides muliebris* Casey
Current name: *Calloides nobilis mormonus* Schaeffer, 1911
Current placement: Cerambycidae: Cerambycinae: Clytini
Type locality: USA: New Mexico

***muliebris* Casey, 1913**: 263 (Fig. 113g, h), Holotype
Original combination: *Xestoleptura muliebris* Casey
Current name: *Xestoleptura crassipes* (LeConte, 1857)
Current placement: Cerambycidae: Lepturinae: Lepturini
Type locality: USA: California

***multifasciculata* Breuning, 1943c**: 33 (Fig. 113i, j), Holotype
Original combination: *Dystasia multifasciculata* Breuning
Current name: same
Current placement: Cerambycidae: Lamiinae: Pteropliini
Type locality: Vietnam: Annam

***multifasciculatus* Breuning, 1939a**: 65 (Fig. 113k, l), Holotype
Original combination: *Anipocregyes multifasciculatus* Breuning
Current name: same
Current placement: Cerambycidae: Lamiinae: Mesosini
Type locality: Indonesia: Sumatra, Palemlang

***multimaculata* Gressitt, 1935g**: 279 (Fig. 113m, n), Holotype
Original combination: *Paraphemone multimaculata* Gressitt
Current name: same
Current placement: Cerambycidae: Lamiinae: Pteropliini
Type locality: Thailand: Trong

***multimaculatus* Breuning, 1943c**: 31 (Fig. 113o, p), Holotype
Original combination: *Prosoplus multimaculatus* Breuning
Current name: same
Current placement: Cerambycidae: Lamiinae: Pterolophini
Type locality: Papua New Guinea

***mundelli* Fisher, 1931c**: 200 (Fig. 113q, r), Holotype
Original combination: *Moneilema (Collapteryx) mundelli* Fisher
Current name: *Moneilema blapsides ulkei* Horn, 1885
Current placement: Cerambycidae: Lamiinae: Moneilemini
Type locality: Mexico: Tamaulipas, Gonzalez

***mupinensis* Gressitt, 1935c**: 570 (Fig. 113s, t), Holotype
Original combination: *Anoploderamorpha mupinensis* Gressitt
Current name: *Pseudalosterna mupinensis* (Gressitt, 1935c)
Current placement: Cerambycidae: Lepturinae: Lepturini
Type locality: China: Sichuan Province, near Muping, 2,000–8,000 ft.

***mutabile* Melzer, 1935b**: 179 (Fig. 113u, v), Holotype
Original combination: *Heterachthes mutabile* Melzer
Current name: *Pygmodeon mutabile* (Melzer, 1935b)
Current placement: Cerambycidae: Cerambycinae: Neoibidionini
Type locality: Costa Rica: Guayabillos, southwest Abhang Irazu, 2200 m.

***mutilus* Casey, 1913**: 307 (Fig. 113w, x), Holotype
Original combination: *Leptostylus mutilus* Casey
Current name: *Leptostylopsis terraecolor* (Horn, 1880)
Current placement: Cerambycidae: Lamiinae: Acanthocinini
Type locality: USA: Florida, Key Largo

***myia* Dillon & Dillon, 1946**: 163 (Fig. 114a, b), Holotype
Original combination: *Eumimesis myia* Dillon & Dillon
Current name: *Harringtonia myia* (Dillon & Dillon, 1946)
Current placement: Cerambycidae: Lamiinae: Calliini
Type locality: Bolivia: Dept. La Paz, San Jose

***mysticus* Tippmann, 1960**: 119 (Fig. 114c, d), Holotype
Original combination: *Necydalosaurus mysticus* Tippmann
Current name: same
Current placement: Cerambycidae: Cerambycinae: Oemini
Type locality: Peru: Loreto, Pucallpa, 200 m.

***nanulum* Casey, 1924**: 247 (Fig. 114e, f), Lectotype
Original combination: *Anoplium nanulum* Casey
Current name: *Anelaphus tuckeri* (Casey, 1924)
Current placement: Cerambycidae: Cerambycinae: Elaphidiini
Type locality: USA: Arizona, Tucson
Remarks: This specimen is herein designated as the lectotype in order to stabilize the taxonomy and facilitate further identifications of this species.

***nanulus* Casey, 1913**: 376 (Fig. 114g, h), Holotype
Original combination: *Tetraopes nanulus* Casey
Current name: *Tetraopes discoideus* LeConte, 1858
Current placement: Cerambycidae: Lamiinae: Tetraopini
Type locality: USA: Arizona
Remarks: Although the original description suggests there may be more than one specimen in the original series, there is only one in the USNM collection.

***nanus* Casey, 1924**: 222 (Fig. 114i, j), Holotype
Original combination: *Prionus (Prionus) nanus* Casey
Current name: *Prionus californicus* Motschulsky, 1845
Current placement: Cerambycidae: Prioninae: Prionini
Type locality: USA: Utah, Eureka

***napoensis* Santos-Silva & Tavakilian, 2009**: 21 (Fig. 114k, l), Holotype
Original combination: *Abauba napoensis* Santos-Silva & Tavakilian
Current name: same
Current placement: Disteniidae: Disteniinae: Disteniini
Type locality: Ecuador: Napo, Res. Ethnica Waorani, 1km S. Onkone Gare Camp, 220 m., 00°39′10S 076°26′W

***navajasi* Martins, 1959**: 344 (Fig. 114m, n), Holotype
Original combination: *Hexoplon navajasi* Martins
Current name: same
Current placement: Cerambycidae: Cerambycinae: Hexoplini
Type locality: Panama: La Chorrera

navassae **Nearns & Steiner, 2006**: 63 (Fig. 114o, p), Holotype
Original combination: *Plectromerus navassae* Nearns & Steiner
Current name: same
Current placement: Cerambycidae: Cerambycinae: Plectromerini
Type locality: Navassa Island: near lighthouse, 80 m., 18°23.82´N, 75°00.74´W, at black light in open weedy scrub near mixed forest (*Ficus*, *Metopium*, *Thrinax*) on limestone and red oolitic soil

nayani **Santos-Silva, Heffern & Matsuda, 2010**: 18 (Fig. 114q, r), Holotype
Original combination: *Komiyandra nayani* Santos-Silva, Heffern, Matsuda
Current name: same
Current placement: Cerambycidae: Parandrinae: Parandrini
Type locality: Malaysia: Borneo, Sabah, Mt. Trus-Madi

neali **Danilevsky, 2004**: 2 (Fig. 114s, t), Holotype
Original combination: *Cortodera neali* Danilevsky
Current name: same
Current placement: Cerambycidae: Lepturinae: Rhagiini
Type locality: Iran: Kordistan Province, 66 km. north-northwest of Sanandadj

nearnsi **Lingafelter, 2008**: 370 (Fig. 114u, v), Holotype
Original combination: *Elaphidion nearnsi* Lingafelter
Current name: same
Current placement: Cerambycidae: Cerambycinae: Elaphidiini
Type locality: Dominican Republic: La Altagracia Province, Punta Cana Near Ecological Reserve, 0–5 meters, 18°30.477´N, 68°22.499´W

nearnsi **Santos-Silva & Hovore, 2008a**: 23 (Fig. 114w, x), Holotype
Original combination: *Elytrimitatrix (Grossifemora) nearnsi* Santos-Silva & Hovore
Current name: same
Current placement: Disteniidae: Disteniinae: Disteniini
Type locality: Mexico: Veracruz, Los Tuxtlas, Biological Station

nebulosa silvestrii **Tippmann, 1956**: 492 (Fig. 115a, b), Holotype
Original combination: *Mesosa nebulosa silvestrii* Tippmann
Current name: same
Current placement: Cerambycidae: Lamiinae: Mesosini
Type locality: Croatia: Križevci

negundo **Schaeffer, 1905b**: 164 (Fig. 115c, d), Syntype
Original combination: *Pogonocherus negundo* Schaeffer
Current name: *Poliaenus negundo* (Schaeffer, 1905b)
Current placement: Cerambycidae: Lamiinae: Pogonocherini
Type locality: USA: Arizona, Huachuca Mts.

neiba **Lingafelter, 2011**: 69 (Fig. 115e, f), Holotype
Original combination: *Tilloclytus neiba* Lingafelter
Current name: same
Current placement: Cerambycidae: Cerambycinae: Anaglyptini
Type locality: Dominican Republic: San Juan Province, Sierra de Neiba, trail to Sabana del Silencio, 10 km SSW of El Cercado, 1650–1700 m.

nelsonicus **Casey, 1924**: 291 (Fig. 115g, h), Holotype
Original combination: *Leiopus nelsonicus* Casey
Current name: *Sternidius alpha* (Say, 1827)
Current placement: Cerambycidae: Lamiinae: Acanthocinini
Type locality: USA: Virginia, Nelson Co.

nematocera **Casey, 1924**: 268 (Fig. 115i, j), Lectotype
Original combination: *Taranomis nematocera* Casey
Current name: *Sphaenothecus bilineatus* (Gory, 1831)
Current placement: Cerambycidae: Cerambycinae: Trachyderini
Type locality: USA: Arizona, near Tucson

nematocerus **Casey, 1924**: 234 (Fig. 115k, l), Holotype
Original combination: *Hylotrupes nematocerus* Casey
Current name: *Hylotrupes bajulus* (Linnaeus, 1758)
Current placement: Cerambycidae: Cerambycinae: Hylotrupini
Type locality: USA

neomexicanus **Casey, 1890**: 491 (Fig. 115m, n), Lectotype
Original combination: *Ergates (Trichocnemis) neomexicanus* Casey
Current name: *Trichocnemis spiculatus neomexicanus* (Casey, 1890)
Current placement: Cerambycidae: Prioninae: Callipogonini
Type locality: USA: New Mexico, Las Vegas
Remarks: This specimen is herein designated as the lectotype in order to stabilize the taxonomy and facilitate further identifications of this species

neopomerianus **Breuning, 1957a**: 12 (Fig. 115o, p), Holotype
Original combination: *Exocentrus (Camptomyme) neopomerianus* Breuning
Current name: *Exocentrus (Centenexocentrus) neopomerianus* Breuning, 1957a
Current placement: Cerambycidae: Lamiinae: Acanthocinini
Type locality: Papua New Guinea: East New Britain, Mope

***neopomerianus* Breuning, 1957a**: 14 (Fig. 115q, r), Holotype
Original combination: *Aegocidnus (Pomeranaegocidnus) neopomerianus* Breuning
Current name: *Sciades (Pomeranaegocidnus) neopomerianus* (Breuning, 1957a)
Current placement: Cerambycidae: Lamiinae: Acanthocinini
Type locality: Papua New Guinea: East New Britain, Mope

***neotropica* Tippmann, 1960**: 138 (Fig. 115s, t), Holotype
Original combination: *Corallancyla neotropica* Tippmann
Current name: same
Current placement: Cerambycidae: Cerambycinae: Rhinotragini
Type locality: Brazil: Santa Catarina, Mafra

***nesiope* Dillon & Dillon, 1945**: 33 (Fig. 115u, v), Holotype
Original combination: *Hypsioma nesiope* Dillon & Dillon
Current name: same
Current placement: Cerambycidae: Lamiinae: Onciderini
Type locality: Panama: Cabima

***nevermanni* Fisher, 1930a**: 8 (Fig. 115w, x), Holotype
Original combination: *Odontocera nevermanni* Fisher
Current name: same
Current placement: Cerambycidae: Cerambycinae: Rhinotragini
Type locality: Costa Rica: Limón, Hamburg Farm, Reventazón, Ebene

***nevermanni* Fisher, 1947a**: 53 (Fig. 116a, b), Holotype
Original combination: *Ischasia nevermanni* Fisher
Current name: same
Current placement: Cerambycidae: Cerambycinae: Rhinotragini
Type locality: Costa Rica: Limón, Hamburg Farm, Reventazón, Ebene

***nevermanni* Martins & Galileo, 2008a**: 272 (Fig. 116c, d), Holotype
Original combination: *Megacyllene (Megacyllene) nevermanni* Martins & Galileo
Current name: *Megacyllene nevermanni* Martins & Galileo, 2008a
Current placement: Cerambycidae: Cerambycinae: Clytini
Type locality: Costa Rica: Limón, Hamburg Farm, Reventazón, Ebene

***nevermanni* Melzer, 1932**: 236 (Fig. 116e, f), Holotype
Original combination: *Phaea nevermanni* Melzer
Current name: *Phaea rubella* Bates, 1881
Current placement: Cerambycidae: Lamiinae: Tetraopini
Type locality: Costa Rica: Cartago, Turrialba

***nevermanni* Melzer, 1934**: 74 (Fig. 116g, h), Holotype
Original combination: *Eucharassus nevermanni* Melzer
Current name: same
Current placement: Cerambycidae: Cerambycinae: Necydalopsini
Type locality: Costa Rica: Cartago, 2000–3000 m, Vulkán Irazu

***nevermanni* Schwarzer, 1929**: 353 (Fig. 116i, j), Holotype
Original combination: *Bothrocerambyx nevermanni* Schwarzer
Current name: same
Current placement: Cerambycidae: Cerambycinae: Cerambycini
Type locality: Costa Rica: Pozo Azul

***ngauchiliae* Gressitt, 1940**: 163 (Fig. 116k, l), Holotype
Original combination: *Ropica ngauchiliae* Gressitt
Current name: same
Current placement: Cerambycidae: Lamiinae: Apomecynini
Type locality: China: Hainan Province, Hainan Island, Ta Hian

***niger* Breuning, 1942a**: 126 (Fig. 116m, n), Holotype
Original combination: *Docus niger* Breuning
Current name: *Planodema niger* (Breuning, 1942a)
Current placement: Cerambycidae: Lamiinae: Theocridini
Type locality: Congo: Dugo

***nigerrima flavipubescens* Fisher, 1947a**: 56 (Fig. 116o, p), Holotype
Original combination: *Epimelitta nigerrima flavipubescens* Fisher
Current name: same
Current placement: Cerambycidae: Cerambycinae: Rhinotragini
Type locality: Costa Rica: La Caja bei, San Jose

***nigra* Aurivillius, 1922**: 430 (Fig. 116q, r), Lectotype
Original combination: *Daphisia nigra* Aurivillius
Current name: *Menesia nigra* (Aurivillius, 1922)
Current placement: Cerambycidae: Lamiinae: Saperdini
Type locality: Malaysia: Borneo, Sandakan
Remarks: This specimen is herein designated as the lectotype in order to stabilize the taxonomy and facilitate further identifications of this species. One paralectotype is present in the USNM and others are in the Naturhistoriska Riksmuseet, Stockholm, Sweden.

***nigrescans* Casey, 1924**: 211 (Fig. 116s, t), Lectotype
Original combination: *Prionus (Prionellus) nigrescans* Casey
Current name: *Prionus laticollis* (Drury, 1773)
Current placement: Cerambycidae: Prioninae: Prionini
Type locality: USA: New York, Lake George
Remarks: This specimen is herein designated as the lectotype in order to stabilize the taxonomy and facilitate further identifications of this species.

***nigrescens* Breuning, 1943c**: 57 (Fig. 116u, v), Holotype
Original combination: *Estola nigrescens* Breuning
Current name: same
Current placement: Cerambycidae: Lamiinae: Desmiphorini
Type locality: Venezuela: Aragua, Maracay

***nigricans* Fisher, 1935a**: 55 (Fig. 116w, x), Holotype
Original combination: *Leptostylus nigricans* Fisher
Current name: *Styloleptus nigricans* (Fisher, 1935a)
Current placement: Cerambycidae: Lamiinae: Acanthocinini
Type locality: USA: Puerto Rico, Villalba, Insular Government Finca

***nigricollis* Casey, 1913**: 376 (Fig. 117a, b), Lectotype
Original combination: *Tetraopes nigricollis* Casey
Current name: *Tetraopes discoideus* LeConte, 1858
Current placement: Cerambycidae: Lamiinae: Tetraopini
Type locality: USA: New Mexico, Fort Wingate
Remarks: This specimen is herein designated as the lectotype in order to stabilize the taxonomy and facilitate further identifications of this species.

***nigricornis* Fisher, 1946b**: 90 (Fig. 117c, d), Holotype
Original combination: *Stenosphenus nigricornis* Fisher
Current name: *Stenosphenus debilis* Horn, 1885
Current placement: Cerambycidae: Cerambycinae: Elaphidiini
Type locality: USA: California, Yuma

***nigripennis* Aurivillius, 1915**: 321 (Fig. 117e, f), Holotype
Original combination: *Allogaster nigripennis* Aurivillius
Current name: same
Current placement: Cerambycidae: Cerambycinae: Achrysonini
Type locality: Sudan: Wau (east Sudan)
Remarks: Tavakilian and Chevillotte (2013) indicate the holotype is in the Zoologische Staatssammlung des Bayerischen Staates, Munich, Germany, however, this specimen and its labels agree with the original publication.

***nigripes* Casey, 1913**: 377 (Fig. 117g, h), Holotype
Original combination: *Tetraopes nigripes* Casey
Current name: *Tetraopes femoratus* LeConte, 1847
Current placement: Cerambycidae: Lamiinae: Tetraopini
Type locality: Mexico: Chihuahua, Meadow Valley (6 mi. south Colonia García) (Sierra Madre Mtns., 7300 ft.)

***nigritarsis* Fisher, 1926**: 10 (Fig. 117i, j), Holotype
Original combination: *Eupogonius nigritarsis* Fisher
Current name: same
Current placement: Cerambycidae: Lamiinae: Desmiphorini
Type locality: Cuba: Cayamas

***nigriventris* Fisher, 1927b**: 215 (Fig. 117k, l), Holotype
Original combination: *Moneilema (Moneilema) nigriventris* Fisher
Current name: *Moneilema appressum* LeConte, 1852
Current placement: Cerambycidae: Lamiinae: Moneilemini
Type locality: USA: Texas, Panhandle

***nigrofasciata* Aurivillius, 1927b**: 566 (Fig. 117m, n), Syntype
Original combination: *Sybra nigrofasciata* Aurivillius
Current name: *Sybra binotata* Gahan, 1907
Current placement: Cerambycidae: Lamiinae: Apomecynini
Type locality: Philippines: Bukidnon, Tangcolan
Remarks: Tavakilian and Chevillotte (2013) state that syntypes are present in the Naturhistoriska Riksmuseet, Stockholm, Sweden as well as the USNM.

***nigrolineatus* Martins & Galileo, 2008b**: 548 (Fig. 117o, p), Holotype
Original combination: *Adesmus nigrolineatus* Martins & Galileo
Current name: same
Current placement: Cerambycidae: Lamiinae: Hemilophini
Type locality: Mexico: Oaxaca, 33 km. north Tepanetapec

***nigromaculata* var. *trimaculata* Aurivillius, 1915**: 330 (Fig. 117q, r), Holotype
Original combination: *Eunidia nigromaculata* var. *trimaculata* Aurivillius
Current name: *Eunidia batesii* Olliff, 1889
Current placement: Cerambycidae: Lamiinae: Apomecynini
Type locality: Tanzania: Lukule
Remarks: Although Tavakilian and Chevillotte (2013) indicate the holotype is in the Zoologische Staatssammlung des Bayerischen Staates, Munich, Germany, the Smithsonian specimen matches the description and data. The name following the binomen was not expressly given infrasubspecific rank and the work did not indicate unambiguously that the name was infrasubspecific, therefore, it is deemed a subspecies and available.

***nigroocellata* Tippmann, 1960**: 170 (Fig. 117s, t), Holotype
Original combination: *Acanthoderes (Psapharochrus) nigroocellata* Tippmann
Current name: *Psapharochrus nigroocellatus* (Tippmann, 1960)
Current placement: Cerambycidae: Lamiinae: Acanthoderini
Type locality: Peru: Junin, Satipo

nigropunctata **Tippmann, 1960**: 169 (Fig. 117u, v), Holotype
Original combination: *Acanthoderes (Psapharochrus) nigropunctata* Tippmann
Current name: *Psapharochrus nigropunctatus* (Tippmann, 1960)
Current placement: Cerambycidae: Lamiinae: Acanthoderini
Type locality: Bolivia: Santa Cruz
Remarks: The specific epithet is misspelled on the label.

nigrosulphureus **Gressitt, 1940**: 67 (Fig. 117w, x), Holotype
Original combination: *Xylotrechus nigrosulphureus* Gressitt
Current name: same
Current placement: Cerambycidae: Cerambycinae: Clytini
Type locality: China: Hainan Island, Dwa-Bi (Tai-pin-tsuen)
Remarks: Label incorrectly states *Clytobius*, described in *Xylotrechus*

nigrosuturalis **Galileo & Martins, 2011**:183 (Fig. 118a, b), Holotype
Original combination: *Esthlogena nigrosuturalis* Galileo & Martins
Current name: same
Current placement: Cerambycidae: Lamiinae: Pteropliini
Type locality: Panama: Chiriquí Province, Cerro Punta Area

nigrotarsis **Fisher, 1937d**: 149 (Fig. 118c, d), Holotype
Original combination: *Ommata (Rhopalessa) nigrotarsis* Fisher
Current name: *Rhopalessa nigrotarsis* (Fisher, 1937d)
Current placement: Cerambycidae: Cerambycinae: Rhinotragini
Type locality: Brazil: Santa Catarina, Nova Teutonia

nigrovittata **Breuning, 1935a**: 160 (Fig. 118e, f), Holotype
Original combination: *Frea nigrovittata* Breuning
Current name: *Frea nigrovittata* Breuning, 1935a
Current placement: Cerambycidae: Lamiinae: Crossotini
Type locality: Congo: Kondue
Remarks: Although Tavakilian and Chevillotte (2013) indicates the holotype is in Belgium, the Smithsonian specimen matches the original description and data.

nigrovittata **Fisher, 1937d**: 151 (Fig. 118g, h), Holotype
Original combination: *Dihammaphora nigrovittata* Fisher
Current name: same
Current placement: Cerambycidae: Cerambycinae: Rhopalophorini
Type locality: Brazil: Santa Catarina, Nova Teutonia

nilseni **Micheli, 2003**: 198 (Fig. 118i, j), Holotype
Original combination: *Boricyrtinus nilseni* Micheli
Current name: same
Current placement: Cerambycidae: Lamiinae: Cyrtinini
Type locality: USA: Puerto Rico, Maricao Forest, Road 120, km 13.8

nimbata **Casey, 1912**: 373 (Fig. 118k, l), Lectotype
Original combination: *Clytopsis nimbata* Casey
Current name: *Clytopsis dimidiaticornis* (Chevrolat, 1860)
Current placement: Cerambycidae: Cerambycinae: Clytini
Type locality: Mexico: Guerrero
Remarks: This specimen is herein designated as the lectotype in order to stabilize the taxonomy and facilitate further identifications of this species.

nitidicollis **Casey, 1891**: 33 (Fig. 118m, n), Lectotype
Original combination: *Crossidius nitidicollis* Casey
Current name: *Crossidius pulchellus* LeConte, 1861
Current placement: Cerambycidae: Cerambycinae: Trachyderini
Type locality: USA: Arizona
Remarks: This specimen is herein designated as the lectotype in order to stabilize the taxonomy and facilitate further identifications of this species.

nitidicollis **Chemsak & Linsley, 1976a**: 172 (Fig. 118o, p), Holotype
Original combination: *Pseudotypocerus nitidicollis* Chemsak & Linsley
Current name: same
Current placement: Cerambycidae: Lepturinae: Lepturini
Type locality: El Salvador: La Unión, Volcán Conchagua

nitidipennis **Casey, 1913**: 236 (Fig. 118q, r), Lectotype
Original combination: *Leptacmaeops (Leptacmaeops) nitidipennis* Casey
Current name: *Cortodera nitidipennis* (Casey, 1913)
Current placement: Cerambycidae: Lepturinae: Rhagiini
Type locality: USA: Washington, Seattle
Remarks: This specimen is herein designated as the lectotype in order to stabilize the taxonomy and facilitate further identifications of this species.

nivea **Martins, 1962c**: 143 (Fig. 118s, t), Holotype
Original combination: *Compsa nivea* Martins
Current name: *Compsibidion niveum* (Martins, 1962c)
Current placement: Cerambycidae: Cerambycinae: Neoibidionini
Type locality: Brazil: Santa Catarina, Mafra

***niveicollis* Linell, 1897a**: 396 (Fig. 118u, v), Holotype
Original combination: *Tetranodus niveicollis* Linell
Current name: same
Current placement: Cerambycidae: Cerambycinae: Tillomorphini
Type locality: USA: Texas, Brownsville

***niveisparsis adspersus* Dillon & Dillon, 1941**: 58 (Fig. 118w, x), Holotype
Original combination: *Ptychodes niveisparsis adspersus* Dillon & Dillon
Current name: *Ptychodes taeniotoides* Thomson, 1865
Current placement: Cerambycidae: Lamiinae: Monochamini
Type locality: Colombia: Bogotá

***niveitectum* Schaeffer, 1905b**: 160 (Fig. 119a, b), Syntype
Original combination: *Romaleum niveitectum* Schaeffer
Current name: *Enaphalodes niveitectus* (Schaeffer, 1905b)
Current placement: Cerambycidae: Cerambycinae: Elaphidiini
Type locality: USA: Arizona, Cochise Co., Palmerlee

***niveivestitum* Schaeffer, 1905a**: 132 (Fig. 119c, d), Lectotype
Original combination: *Elaphidion niveivestitum* Schaeffer
Current name: *Anelaphus niveivestitus* (Schaeffer, 1905a)
Current placement: Cerambycidae: Cerambycinae: Elaphidiini
Type locality: USA: Texas, Brownsville, Esperanza Ranch
Remarks: This specimen is herein designated as the lectotype in order to stabilize the taxonomy and facilitate further identifications of this species.

***niveocincta* Lane, 1959c**: 222 (Fig. 119e, f), Holotype
Original combination: *Ischnolea niveocincta* Lane
Current name: *Desmiphora niveocincta* (Lane, 1959c)
Current placement: Cerambycidae: Lamiinae: Desmiphorini
Type locality: Panama: Canal Zone, Barro Colorado Island

***niveoguttata* Aurivillius, 1925**: 522 (Fig. 119g, h), Lectotype
Original combination: *Daphisia niveoguttata* Aurivillius
Current name: *Menesia niveoguttata* (Aurivillius, 1925)
Current placement: Cerambycidae: Lamiinae: Saperdini
Type locality: Malaysia: Borneo, Sandakan (according to original description); Island of Penang according to label.
Remarks: This specimen is herein designated as the lectotype in order to stabilize the taxonomy and facilitate further identifications of this species. Other paralectotype(s) are present in the Naturhistoriska Riksmuseet, Stockholm, Sweden.

***niveus* Linsley, 1935a**: 140 (Fig. 119i, j), Holotype
Original combination: *Trichophoroides niveus* Linsley
Current name: same
Current placement: Cerambycidae: Cerambycinae: Elaphidiini
Type locality: Mexico: Morelos, Cuernavaca

***nobilis* Tippmann, 1960**: 142 (Fig. 119k, l), Holotype
Original combination: *Weyrauchia nobilis* Tippmann
Current name: same
Current placement: Cerambycidae: Cerambycinae: Trachyderini
Type locality: Peru: Cajamarca, Jaen, Bellavista Marañon, 700 m.

***nobilis* var. *mormonus* Schaeffer, 1911**: 124 (Fig. 119m, n), Holotype
Original combination: *Calloides nobilis* var. *mormonus* Schaeffer
Current name: *Calloides nobilis mormonus* Schaeffer, 1911
Current placement: Cerambycidae: Cerambycinae: Clytini
Type locality: USA: Utah, Beaver Canyon
Remarks: The name following the binomen was not expressly given infrasubspecific rank and the work did not indicate unambiguously that the name was infrasubspecific, therefore, it is deemed a subspecies and available.

***nocturnus* Fisher, 1942b**: 38 (Fig. 119o, p), Holotype
Original combination: *Trypanidius nocturnus* Fisher
Current name: same
Current placement: Cerambycidae: Lamiinae: Acanthocinini
Type locality: USA: Puerto Rico, Villalba
Remarks: Interestingly, this specimen (originally described in 1942, type number 58124) also represents the holotype of the species that was redescribed under the same name in 1947 (Type No. 55677)!

***notabiloides* Gressitt, 1936**: 96 (Fig. 119q, r), Holotype
Original combination: *Raphuma notabiloides* Gressitt
Current name: *Rhaphuma virens* Matsushita, 1931
Current placement: Cerambycidae: Cerambycinae: Clytini
Type locality: China: Taiwan, Sakahen
Remarks: The generic name is misspelled on the label.

***notatus* Fisher, 1936a**: 186 (Fig. 119s, t), Holotype
Original combination: *Micromulciber notatus* Fisher
Current name: *Notomulciber (Micromulciber) notatus* (Fisher, 1936a)
Current placement: Cerambycidae: Lamiinae: Homonoeini
Type locality: Indonesia: Jawa Tengah, Noesa Kambangan Island

***novateutoniae* Fisher, 1938**: 137 (Fig. 119u, v), Holotype
Original combination: *Plaumanniella novateutoniae* Fisher
Current name: same
Current placement: Cerambycidae: Lamiinae: Parmenini
Type locality: Brazil: Santa Catarina, Nova Teutonia

***nubecula* Casey, 1913**: 282 (Fig. 119w, x), Lectotype
Original combination: *Moneilema (Moneilema) nubecula* Casey
Current name: *Moneilema annulatum* Say, 1824
Current placement: Cerambycidae: Lamiinae: Moneilemini
Type locality: USA: Kansas, Hamilton Co., 3350 ft.
Remarks: This specimen is herein designated as the lectotype in order to stabilize the taxonomy and facilitate further identifications of this species.

***nubecula demissa* Casey, 1913**: 282 (Fig. 120a, b), Holotype
Original combination: *Moneilema (Moneilema) nubecula demissa* Casey
Current name: *Moneilema annulatum* Say, 1824
Current placement: Cerambycidae: Lamiinae: Moneilemini
Type locality: USA: Kansas

***oakleyi* Fisher, 1935a**: 52 (Fig. 120c, d), Holotype
Original combination: *Lamproclytus oakleyi* Fisher
Current name: *Lamproclytus elegans* Fisher, 1932
Current placement: Cerambycidae: Cerambycinae: Tillomorphini
Type locality: USA: Puerto Rico, Ponce, Guánica Central Finca, Ponce, on Tabebuia sp.

***oakleyi* Fisher, 1935a**: 54 (Fig. 120e, f), Holotype
Original combination: *Leptostylus oakleyi* Fisher
Current name: *Leptostylopsis gundlachi* (Fisher, 1925b)
Current placement: Cerambycidae: Lamiinae: Acanthocinini
Type locality: USA: Puerto Rico, Bayamor

***oakleyi* Fisher, 1935a**: 57 (Fig. 120g, h), Holotype
Original combination: *Eugamandus oakleyi* Fisher
Current name: same
Current placement: Cerambycidae: Lamiinae: Acanthocinini
Type locality: USA: Puerto Rico, Insular Government Finca, Matrullas

***oakleyi* Fisher, 1935a**: 62 (Fig. 120i, j), Holotype
Original combination: *Cyrtinus oakleyi* Fisher
Current name: same
Current placement: Cerambycidae: Lamiinae: Cyrtinini
Type locality: USA: Puerto Rico, Augustin Finca, Yauco

***oaxaca* Martins & Galileo, 2006b**: 176 (Fig. 120k, l), Holotype
Original combination: *Coscinedes oaxaca* Martins & Galileo
Current name: same
Current placement: Cerambycidae: Cerambycinae: Neocorini
Type locality: Mexico: Oaxaca, 7 km north-northwest of Díaz Ordaz (2200 m.)

***obesa* Casey, 1924**: 287 (Fig. 120m, n), Holotype
Original combination: *Moneilema (Collapteryx) obesa* Casey
Current name: *Moneilema armatum* LeConte, 1853
Current placement: Cerambycidae: Lamiinae: Moneilemini
Type locality: USA: Texas

***obliquus* Casey, 1891**: 26 (Fig. 120o, p), Holotype
Original combination: *Phymatodes obliquus* Casey
Current name: same
Current placement: Cerambycidae: Cerambycinae: Callidiini
Type locality: USA: California, Santa Clara Co.
Remarks: Although the original description suggests there may be more than one specimen in the original series, there is only one in the USNM collection.

***obliquus chihuahuae* Casey, 1913**: 337 (Fig. 120q, r), Holotype
Original combination: *Graphisurus (Graphisurus) obliquus chihuahuae* Casey
Current name: *Acanthocinus obliquus* (LeConte, 1862)
Current placement: Cerambycidae: Lamiinae: Acanthocinini
Type locality: Mexico: Chihuahua

***obliquus sedulus* Casey, 1913**: 337 (Fig. 120s, t), Lectotype
Original combination: *Graphisurus (Graphisurus) obliquus sedulus* Casey
Current name: *Acanthocinus obliquus* (LeConte, 1862)
Current placement: Cerambycidae: Lamiinae: Acanthocinini
Type locality: USA: New Mexico, Fort Wingate
Remarks: This specimen is herein designated as the lectotype in order to stabilize the taxonomy and facilitate further identifications of this species.

***oblonga* Casey, 1912**: 329 (Fig. 120u, v), Holotype
Original combination: *Batyle oblonga* Casey
Current name: *Batyle ignicollis oblonga* Casey, 1912
Current placement: Cerambycidae: Cerambycinae: Trachyderini
Type locality: USA: Utah

***oblongus* Casey, 1913**: 218 (Fig. 120w, x), Holotype
Original combination: *Gaurotes oblongus* Casey
Current name: *Brachysomida bivittata* (Say, 1824)
Current placement: Cerambycidae: Lepturinae: Rhagiini
Type locality: USA: Nebraska

***obscura* Swaine & Hopping, 1928**: 56 (Fig. 121a, b), Holotype
Original combination: *Anoplodera obscura* Swaine & Hopping
Current name: *Ortholeptura obscura* (Swaine & Hopping, 1928)
Current placement: Cerambycidae: Lepturinae: Lepturini
Type locality: USA: Washington, Blue Mountains

***obscurellus* Casey, 1913**: 313 (Fig. 121c, d), Holotype
Original combination: *Leiopus obscurellus* Casey
Current name: *Sternidius alpha* (Say, 1827)
Current placement: Cerambycidae: Lamiinae: Acanthocinini
Type locality: USA: New York, Bluff Point, Lake Champlain

***obscurellus* Casey, 1924**: 255 (Fig. 121e, f), Holotype
Original combination: *Osmidus obscurellus* Casey
Current name: *Osmidus guttatus* LeConte, 1873
Current placement: Cerambycidae: Cerambycinae: Hesperophanini
Type locality: USA: Arizona, Baboquivari Mts.

***obscuroides* Breuning, 1942a**: 163 (Fig. 121g, h), Holotype
Original combination: *Estola obscuroides* Breuning
Current name: same
Current placement: Cerambycidae: Lamiinae: Desmiphorini
Type locality: Paraguay: Alto Parana

***obsolescens* Casey, 1912**: 217 (Fig. 121i, j), Holotype
Original combination: *Parandra obsolescens* Casey
Current name: *Parandra glabra* (DeGeer, 1774)
Current placement: Cerambycidae: Parandrinae: Parandrini
Type locality: Panama: Chiriquí (erroneously published as Costa Rica)

***obsolescens* Casey, 1924**: 244 (Fig. 121k, l), Holotype
Original combination: *Hypermallus obsolescens* Casey
Current name: *Parelaphidion incertum* (Newman, 1840)
Current placement: Cerambycidae: Cerambycinae: Elaphidiini
Type locality: USA: Pennsylvania, Harrisburg

***obsoleta* Casey, 1913**: 243 (Fig. 121m, n), Holotype
Original combination: *Acmaeops obsoleta* Casey
Current name: *Acmaeops proteus* (Kirby, 1837)
Current placement: Cerambycidae: Lepturinae: Rhagiini
Type locality: USA: Colorado
Remarks: Antennae, hind leg missing

***obsoletus* Casey, 1913**: 382 (Fig. 121o, p), Lectotype
Original combination: *Tetraopes obsoletus* Casey
Current name: *Tetraopes basalis* LeConte, 1852
Current placement: Cerambycidae: Lamiinae: Tetraopini
Type locality: USA: California, Siskiyou Co.
Remarks: This specimen is herein designated as the lectotype in order to stabilize the taxonomy and facilitate further identifications of this species.

***obtusus* Casey, 1891**: 47 (Fig. 121q, r), Lectotype
Original combination: *Monochamus obtusus* Casey
Current name: *Monochamus obtusus* Casey, 1891
Current placement: Cerambycidae: Lamiinae: Monochamini
Type locality: USA: California, Siskiyou Co.
Remarks: This specimen is herein designated as the lectotype in order to stabilize the taxonomy and facilitate further identifications of this species.

***obtusus* Casey, 1924**: 292 (Fig. 121s, t), Holotype
Original combination: *Graphisurus obtusus* Casey
Current name: *Acanthocinus obliquus* (LeConte, 1862)
Current placement: Cerambycidae: Lamiinae: Acanthocinini
Type locality: USA: Idaho

***obyuna* Martins & Galileo, 2005c**: 245 (Fig. 121u, v), Holotype
Original combination: *Malthonea obyuna* Martins & Galileo
Current name: same
Current placement: Cerambycidae: Lamiinae: Desmiphorini
Type locality: Peru: La Libertad, Cumpang (above Uctabamba, on trail to Ongón, ca. 2615 m.)

***occipitalis* Gressitt, 1935f**: 255 (Fig. 121w, x), Holotype
Original combination: *Pseudopidonia occipitalis* Gressitt
Current name: *Pidonia occipitalis* (Gressitt, 1935f)
Current placement: Cerambycidae: Lepturinae: Rhagiini
Type locality: China: Taiwan, Taiheizan

***ocellata plagiata* Casey, 1913**: 371 (Fig. 122a, b), Lectotype
Original combination: *Oberea ocellata plagiata* Casey
Current name: *Oberea ocellata* Haldeman, 1847
Current placement: Cerambycidae: Lamiinae: Phytoeciini
Type locality: USA: North Carolina, Southern Pines
Remarks: This specimen is herein designated as the lectotype in order to stabilize the taxonomy and facilitate further identifications of this species.

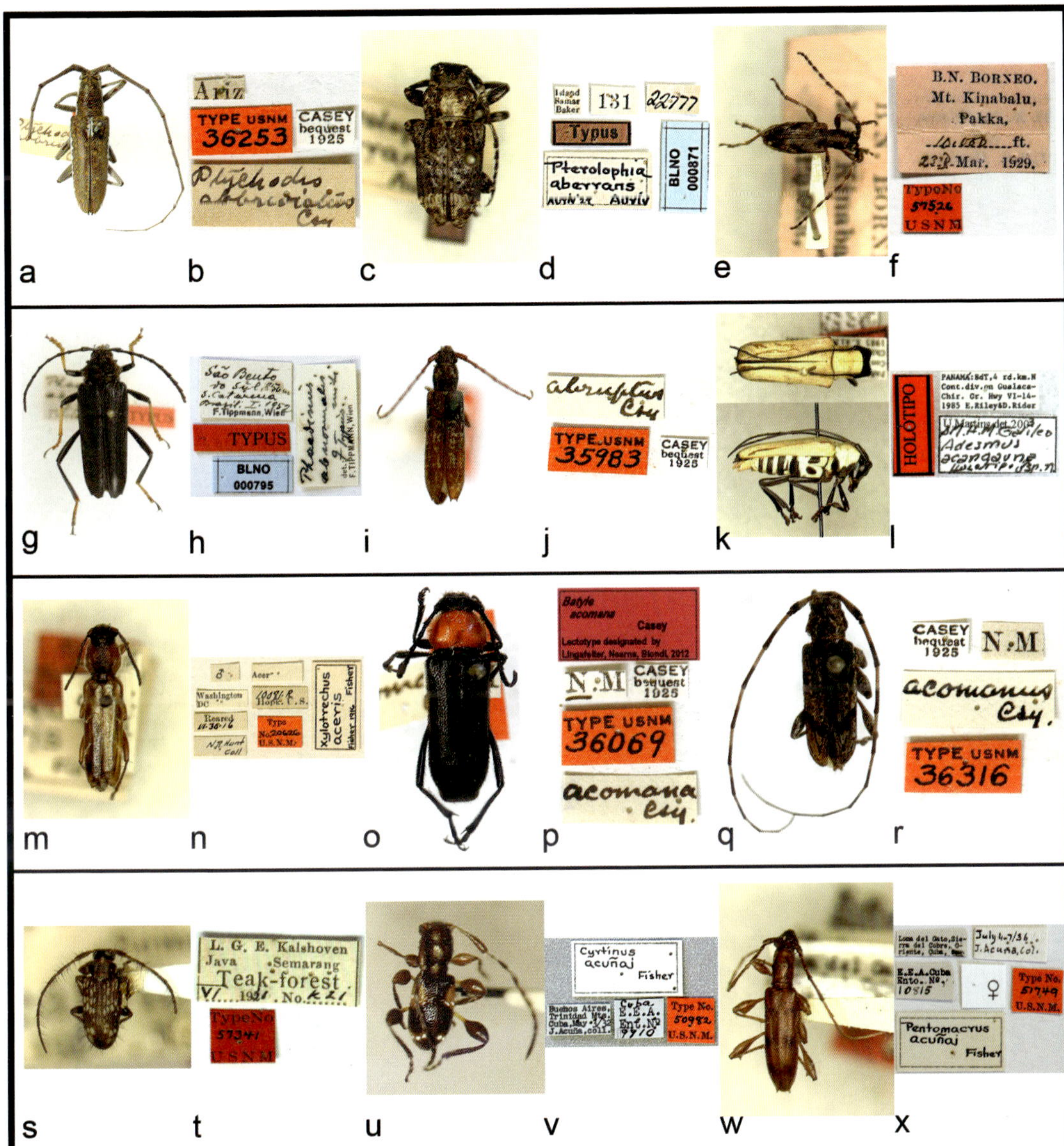

Figure 8. Primary types of Cerambycidae of the Smithsonian Institution (for each pair of images, dorsal habitus left, labels right): **(a, b)** *Ptychodes abbreviatus* Casey, 1893, Holotype; **(c, d)** *Pterolophia aberrans* Aurivillius, 1927, Holotype; **(e, f)** *Heteropalpoides aberrans* Fisher, 1935, Holotype; **(g, h)** *Phaedinus abnormalis* Tippmann, 1953, Holotype; **(i, j)** *Hypermallus abruptus* Casey, 1912, Holotype; **(k, l)** *Adesmus acangauna* Martins & Galileo, 2004, Holotype; **(m, n)** *Xylotrechus aceris* Fisher, 1917, Holotype; **(o, p)** *Batyle acomana* Casey, 1912, Lectotype; **(q, r)** *Graphisurus (Graphisurus) acomanus* Casey, 1913, Holotype; **(s, t)** *Exocentrus actinophorae* Fisher, 1934, Holotype; **(u, v)** *Cyrtinus acunai* Fisher, 1935, Holotype; **(w, x)** *Pentomacrus acunai* Fisher, 1936, Holotype.

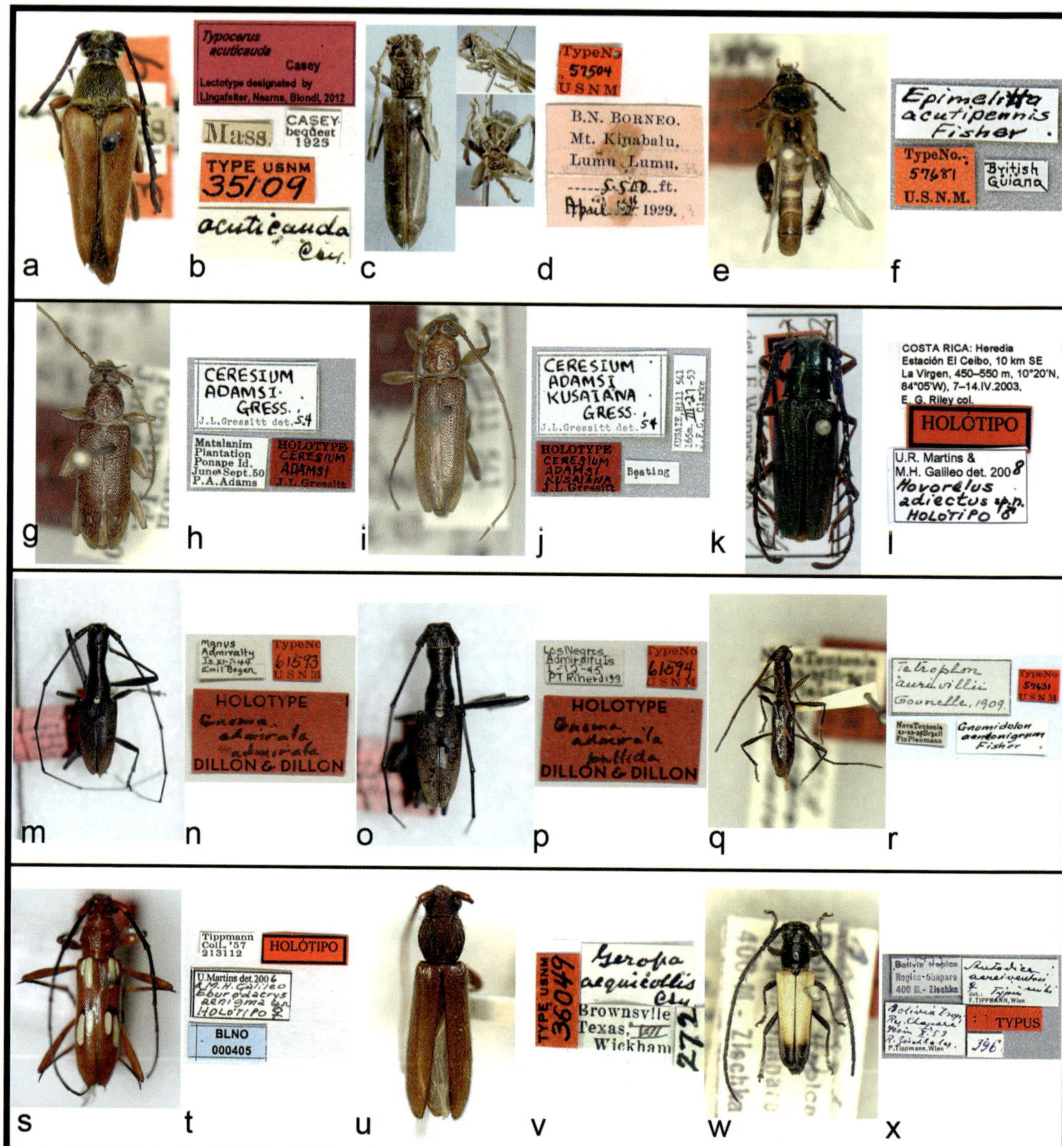

Figure 9. Primary types of Cerambycidae of the Smithsonian Institution (for each pair of images, dorsal habitus left, labels right): **(a, b)** *Typocerus acuticauda* Casey, 1913, Lectotype; **(c, d)** *Dymasius acutipennis* Fisher, 1935, Holotype; **(e, f)** *Epimelitta acutipennis* Fisher, 1947, Holotype; **(g, h)** *Ceresium (Ceresium) adamsi adamsi* Gressitt, 1956, Holotype; **(i, j)** *Ceresium (Ceresium) adamsi kusiana* Gressitt, 1956, Holotype; **(k, l)** *Hovorelus adiectus* Galileo & Martins, 2010, Holotype; **(m, n)** *Gnoma admirala admirala* Dillon & Dillon, 1951, Holotype; **(o, p)** *Gnoma admirala pallida* Dillon & Dillon, 1951, Holotype; **(q, r)** *Gnomidolon aeneonigrum* Fisher, 1937, Holotype; **(s, t)** *Eburodacrys aenigma* Galileo & Martins, 2006, Holotype; **(u, v)** *Geropa aequicollis* Casey, 1912, Holotype; **(w, x)** *Antodice aereiventris* Tippmann, 1960, Holotype.

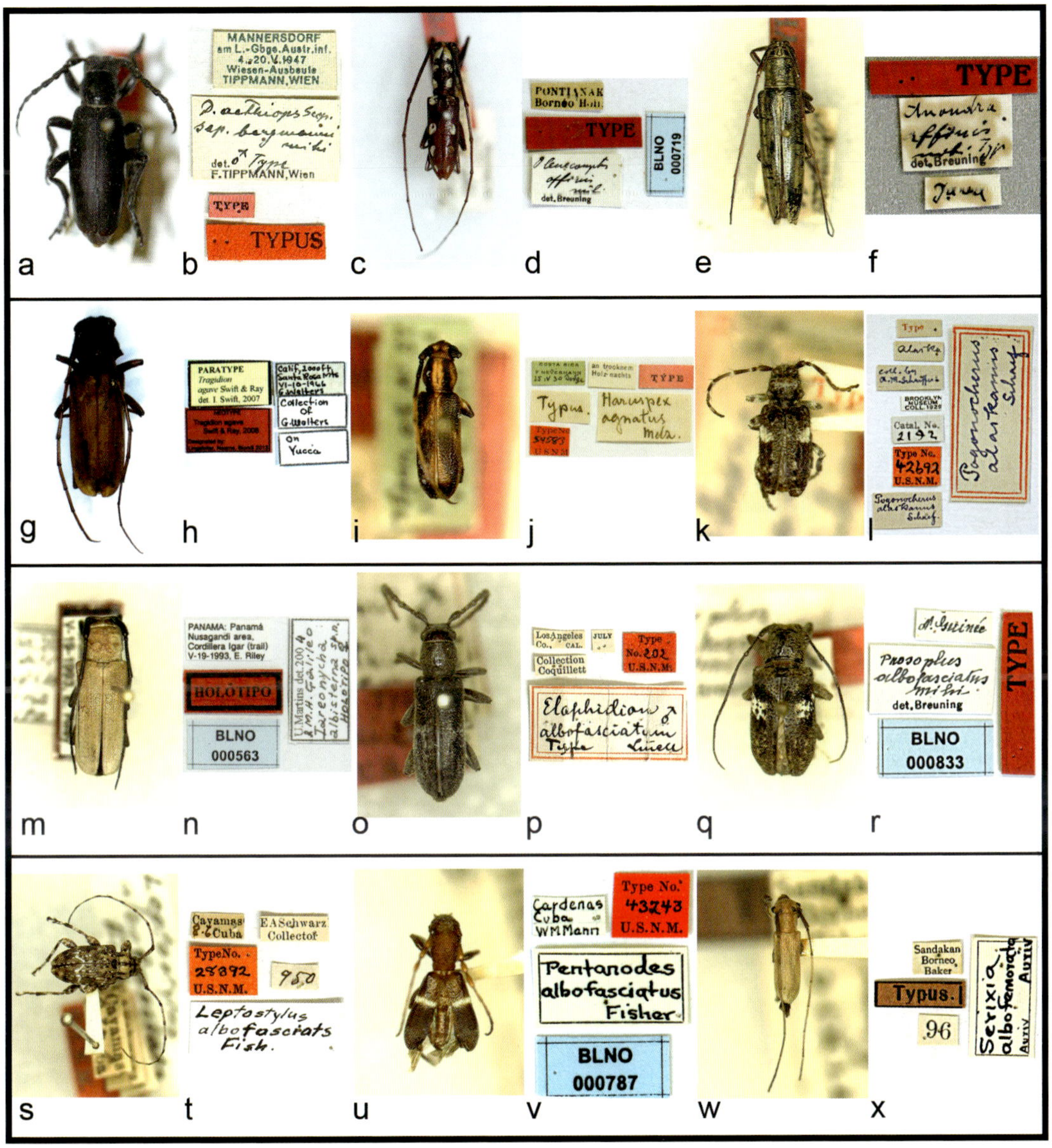

Figure 10. Primary types of Cerambycidae of the Smithsonian Institution (for each pair of images, dorsal habitus left, labels right): **(a, b)** *Dorcadion (Carinatorcadion) aethiops bergmanni* Tippmann, 1958, Holotype; **(c, d)** *Olenecamptus affinis* Breuning, 1936, Holotype; **(e, f)** *Anandra affinis* Breuning, 1940, Holotype; **(g, h)** *Tragidion agave* Swift & Ray, 2008, Neotype; **(i, j)** *Haruspex agnatus* Melzer, 1934, Holotype; **(k, l)** *Pogonocherus alaskanus* Schaeffer, 1909, Holotype; **(m, n)** *Iareonycha albisterna* Martins & Galileo, 2004, Holotype; **(o, p)** *Elaphidion albofasciatum* Linell, 1897, Holotype; **(q, r)** *Prosoplus albofasciatus* Breuning, 1938, Holotype; **(s, t)** *Leptostylus albofasciatus* Fisher, 1926, Holotype; **(u, v)** *Pentanodes albofasciatus* Fisher, 1932, Holotype; **(w, x)** *Serixia albofemorata* Aurivillius, 1927, Holotype.

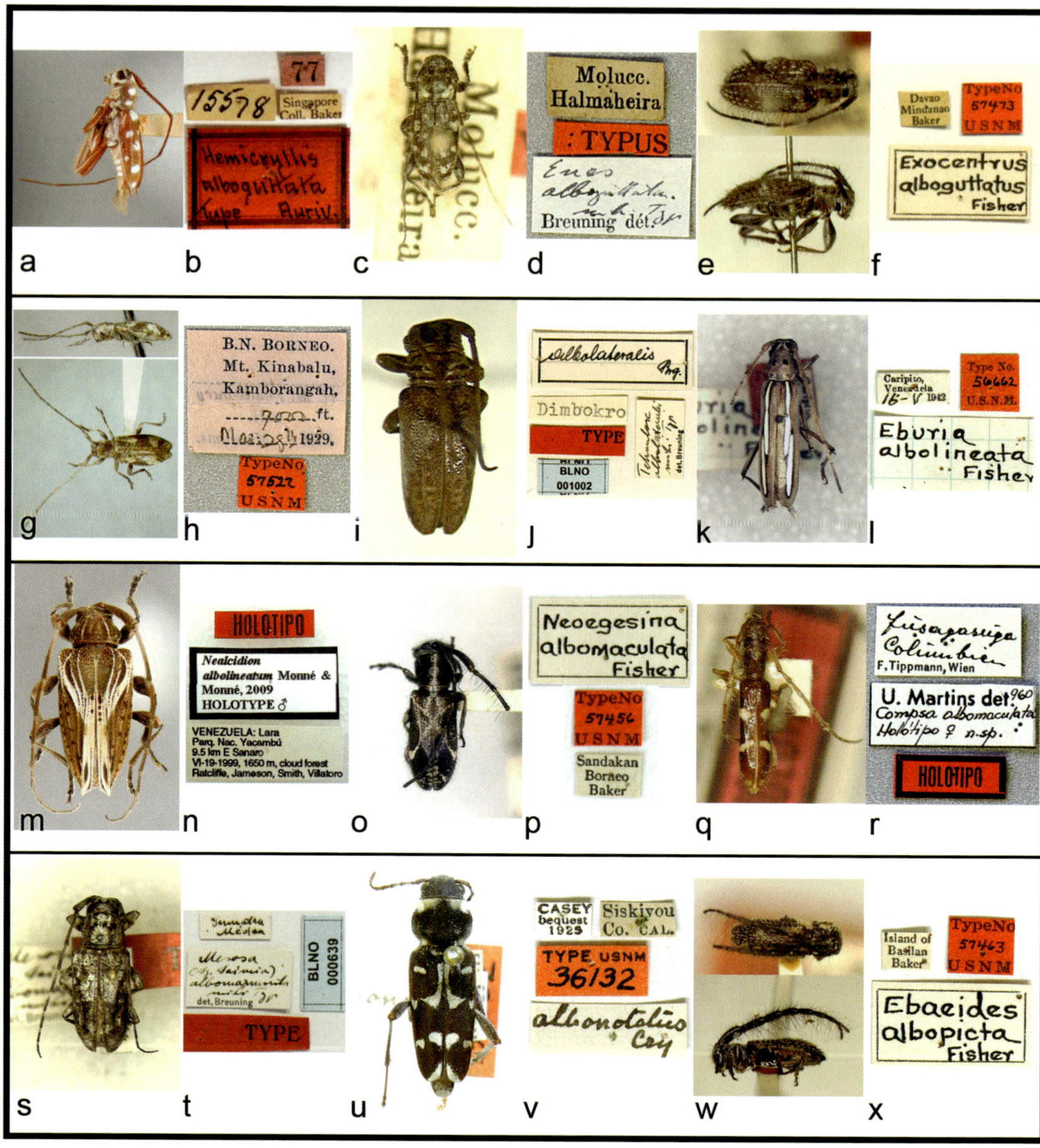

Figure 11. Primary types of Cerambycidae of the Smithsonian Institution (for each pair of images, dorsal habitus left, labels right): **(a, b)** *Hemicryllis alboguttata* Aurivillius, 1922, Holotype; **(c, d)** *Enes alboguttatus* Breuning, 1957, Holotype; **(e, f)** *Exocentrus alboguttatus* Fisher, 1925, Holotype; **(g, h)** *Emeopedus alboguttatus* Fisher, 1935, Holotype; **(i, j)** *Tetraulax albolateralis* Breuning, 1940, Holotype; **(k, l)** *Eburia albolineata* Fisher, 1944, Holotype; **(m, n)** *Nealcidion albolineatum* Monné & Monné, 2009, Holotype; **(o, p)** *Neoegesina albomaculata* Fisher, 1925, Holotype; **(q, r)** *Compsa albomaculata* Martins, 1962, Holotype; **(s, t)** *Mesosa (Saimia) albomarmorata* Breuning, 1939, Holotype; **(u, v)** *Xylotrechus albonotatus* Casey, 1912, Holotype; **(w, x)** *Ebaeides albopicta* Fisher, 1925, Holotype.

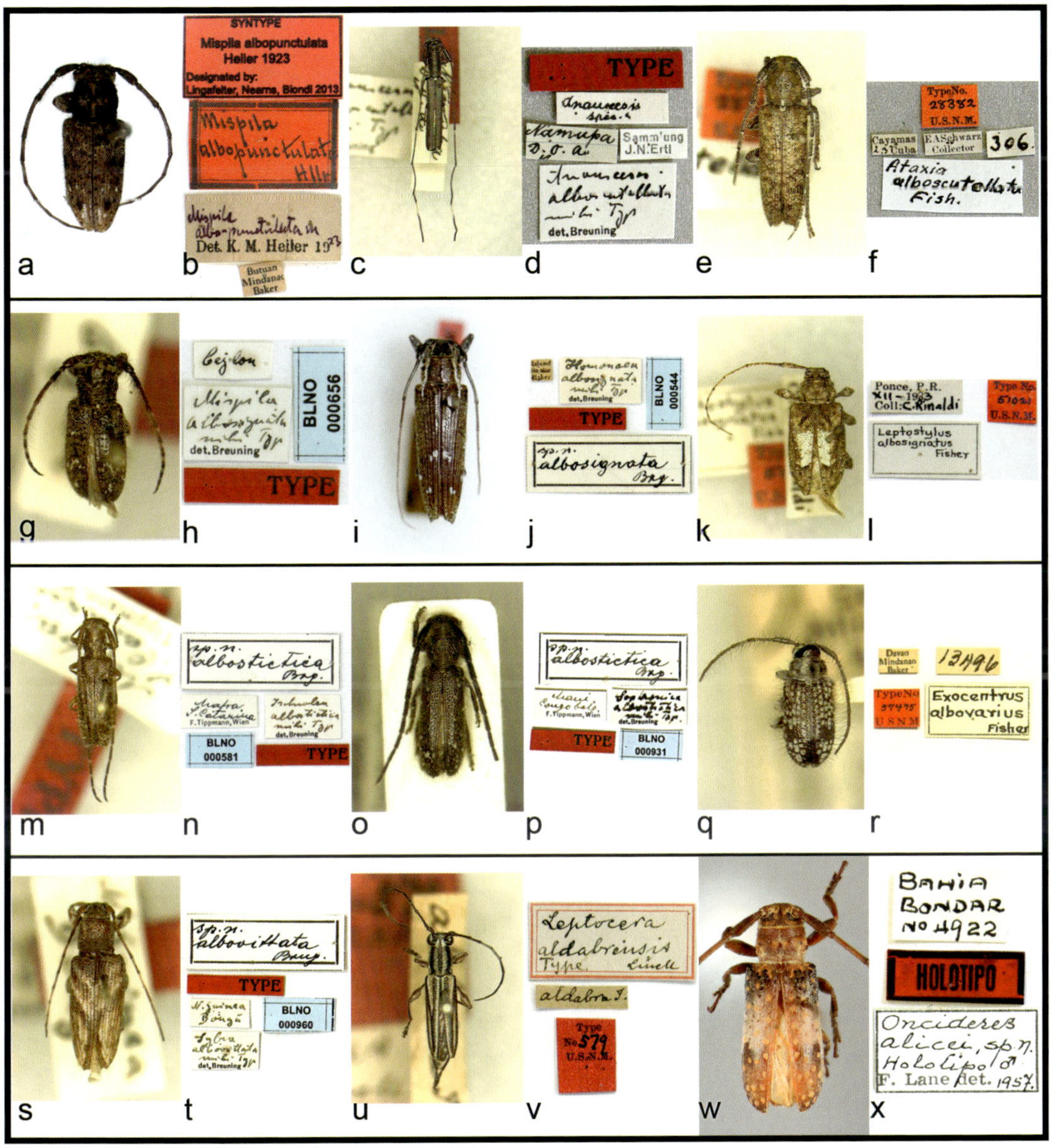

Figure 12. Primary types of Cerambycidae of the Smithsonian Institution (for each pair of images, dorsal habitus left, labels right): **(a, b)** *Mispila albopunctulata* Heller, 1923, Syntype; **(c, d)** *Anauxesis alboscutellaris* Breuning, 1938, Holotype; **(e, f)** *Ataxia alboscutellata* Fisher, 1926, Holotype; **(g, h)** *Mispila albosignata* Breuning, 1940, Holotype; **(i, j)** *Homonoea (Homonoea) albosignata* Breuning, 1950, Holotype; **(k, l)** *Leptostylus albosignatus* Fisher, 1935, Holotype; **(m, n)** *Ischnolea albostictica* Breuning, 1942, Holotype; **(o, p)** *Sophronica albostictica* Breuning, 1942, Holotype; **(q, r)** *Exocentrus albovarius* Fisher, 1925, Holotype; **(s, t)** *Sybra albovittata* Breuning, 1943, Holotype; **(u, v)** *Leptocera aldabrensis* Linell, 1897, Holotype; **(w, x)** *Oncideres alicei* Lane, 1977, Holotype.

Figure 13. Primary types of Cerambycidae of the Smithsonian Institution (for each pair of images, dorsal habitus left, labels right): **(a, b)** *Liopus alpha* var. *floridanus* Hamilton, 1896, Lectotype; **(c, d)** *Pterolophia alternata* Gressitt, 1938, Holotype; **(e, f)** *Leptacmaeops (Leptacmaeops) alticola* Casey, 1913, Holotype; **(g, h)** *Dectes alticola* Casey, 1913, Lectotype; **(i, j)** *Leptura alticola* Gressitt, 1948, Holotype; **(k, l)** *Prionus (Prionus) alutaceus* Casey, 1912, Holotype; **(m, n)** *Udeteromorphus amabilis* Tippmann, 1960, Holotype; **(o, p)** *Esthlogena amaliae* Galileo & Martins, 2011, Holotype; **(q, r)** *Eburia (Eburostola) amazonica* Tippmann, 1960, Holotype; **(s, t)** *Pithomictus amboinicus* Breuning, 1957, Holotype; **(u, v)** *Hyperplatys amnicola* Casey, 1913, Holotype; **(w, x)** *Compsa amoena* Fisher, 1937, Holotype.

Figure 14. Primary types of Cerambycidae of the Smithsonian Institution (for each pair of images, dorsal habitus left, labels right): **(a, b)** *Amphionycha amoena* Hamilton, 1896, Holotype; **(c, d)** *Adesmus amoenoides* Fisher, 1938, Holotype; **(e, f)** *Phymatodes (Microcallidium) amoenus hudsonicus* Casey, 1912, Lectotype; **(g, h)** *Anocomis ampla* Casey, 1912, Lectotype; **(i, j)** *Anocomis ampla angusta* Casey, 1912, Holotype; **(k, l)** *Anocomis ampla arizonae* Casey, 1924, Holotype; **(m, n)** *Anocomis ampla basalis* Casey, 1924, Holotype; **(o, p)** *Anocomis ampla conformis* Casey, 1924, Lectotype; **(q, r)** *Anocomis ampla obliviosa* Casey, 1924, Holotype; **(s, t)** *Anocomis ampla terracensis* Casey, 1924, Lectotype; **(u, v)** *Orthosoma amplians* Casey, 1912, Holotype; **(w, x)** *Parandra ampliceps* Casey, 1912, Lectotype.

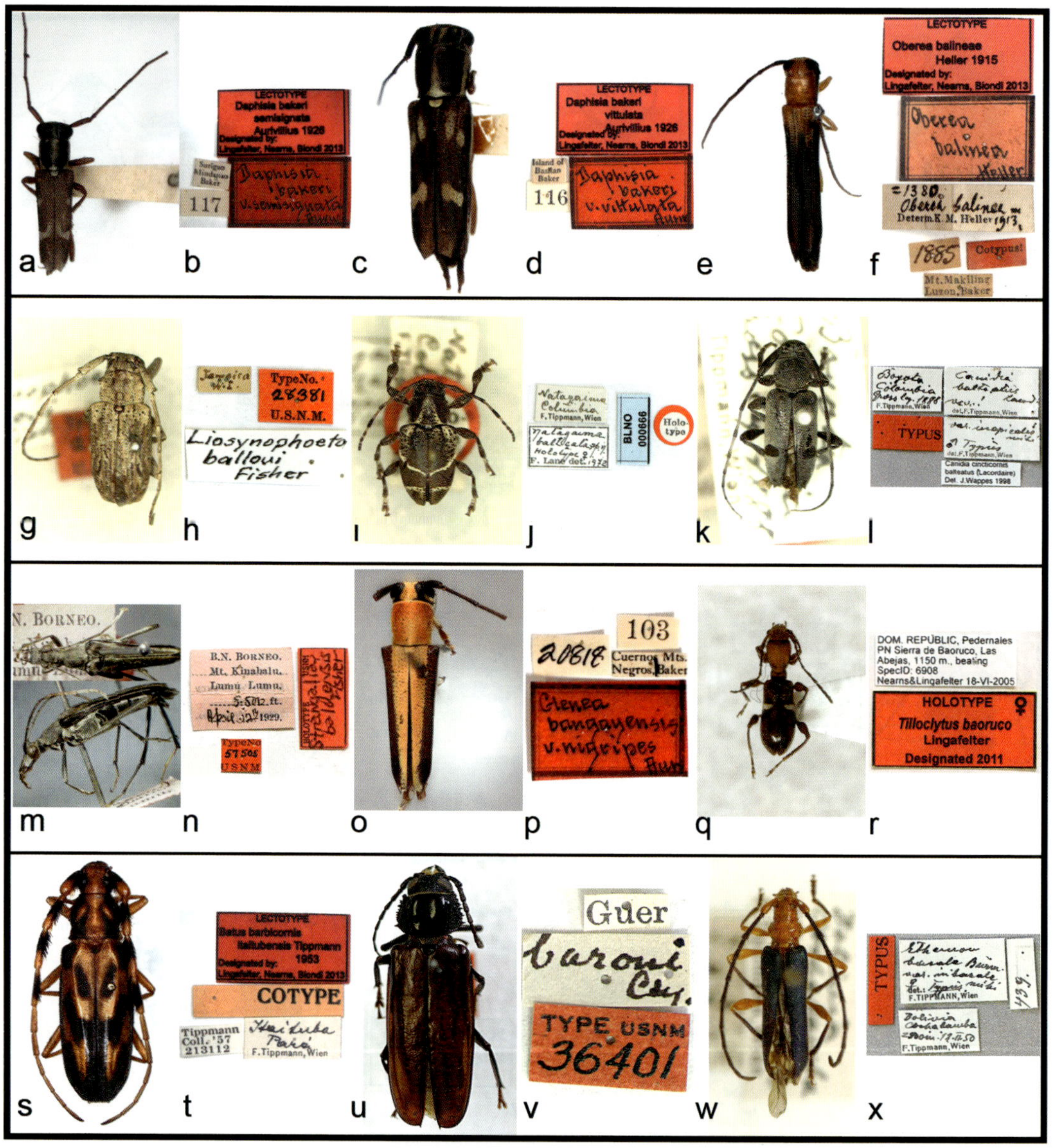

Figure 25. Primary types of Cerambycidae of the Smithsonian Institution (for each pair of images, dorsal habitus left, labels right): **(a, b)** *Daphisia bakeri* var. *semisignata* Aurivillius, 1926, Lectotype; **(c, d)** *Daphisia bakeri* var. *vittulata* Aurivillius, 1926, Lectotype; **(e, f)** *Oberea balineae* Heller, 1915, Lectotype; **(g, h)** *Liosynaphaeta balloui* Fisher, 1926, Holotype; **(i, j)** *Natagaima balteata* Lane, 1972, Holotype; **(k, l)** *Dectes (Canidia) balteata* var. *inapicalis* Tippmann, 1960, Holotype; **(m, n)** *Strangalia baluensis* Fisher, 1935, Holotype; **(o, p)** *Glenea bangueyensis nigripes* Aurivillius, 1926, Holotype; **(q, r)** *Tilloclytus baoruco* Lingafelter, 2011, Holotype; **(s, t)** *Batus barbicornis itaitubensis* Tippmann, 1953, Lectotype; **(u, v)** *Mallodon baroni* Casey, 1912, Holotype; **(w, x)** *Ethemon basale* var. *inbasale* Tippmann, 1960, Holotype.

Figure 14. Primary types of Cerambycidae of the Smithsonian Institution (for each pair of images, dorsal habitus left, labels right): **(a, b)** *Amphionycha amoena* Hamilton, 1896, Holotype; **(c, d)** *Adesmus amoenoides* Fisher, 1938, Holotype; **(e, f)** *Phymatodes (Microcallidium) amoenus hudsonicus* Casey, 1912, Lectotype; **(g, h)** *Anocomis ampla* Casey, 1912, Lectotype; **(i, j)** *Anocomis ampla angusta* Casey, 1912, Holotype; **(k, l)** *Anocomis ampla arizonae* Casey, 1924, Holotype; **(m, n)** *Anocomis ampla basalis* Casey, 1924, Holotype; **(o, p)** *Anocomis ampla conformis* Casey, 1924, Lectotype; **(q, r)** *Anocomis ampla obliviosa* Casey, 1924, Holotype; **(s, t)** *Anocomis ampla terracensis* Casey, 1924, Lectotype; **(u, v)** *Orthosoma amplians* Casey, 1912, Holotype; **(w, x)** *Parandra ampliceps* Casey, 1912, Lectotype.

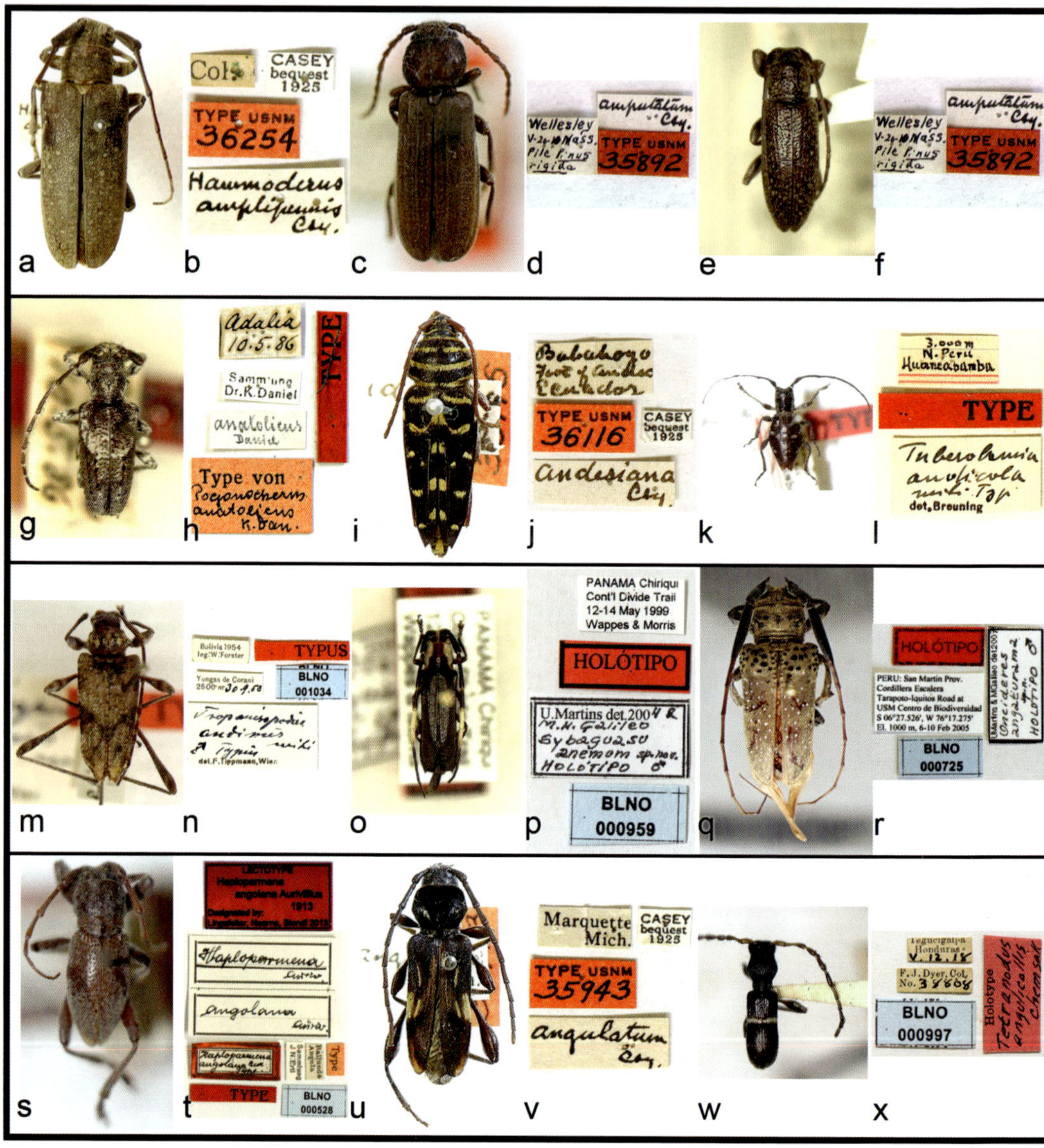

Figure 15. Primary types of Cerambycidae of the Smithsonian Institution (for each pair of images, dorsal habitus left, labels right): **(a, b)** *Hammoderus amplipennis* Casey, 1913, Holotype; **(c, d)** *Asemum amputatum* Casey, 1912, Holotype; **(e, f)** *Sybra anatahana* Gressitt, 1956, Holotype; **(g, h)** *Pogonochaerus anatolicus* Daniel & Daniel, 1898, Holotype; **(i, j)** *Cyllene andesiana* Casey, 1912, Holotype; **(k, l)** *Tuberolamia andicola* Breuning, 1940, Holotype; **(m, n)** *Tropanisopodus andinus* Tippmann, 1960, Holotype; **(o, p)** *Sybaguasu anemum* Martins & Galileo, 2004, Holotype; **(q, r)** *Oncideres angaturama* Galileo & Martins, 2008, Holotype; **(s, t)** *Haploparmena angolana* Aurivillius, 1913, Lectotype; **(u, v)** *Physocnemum angulatum* Casey, 1924, Holotype; **(w, x)** *Tetranodus angulicollis* Chemsak, 1969, Holotype.

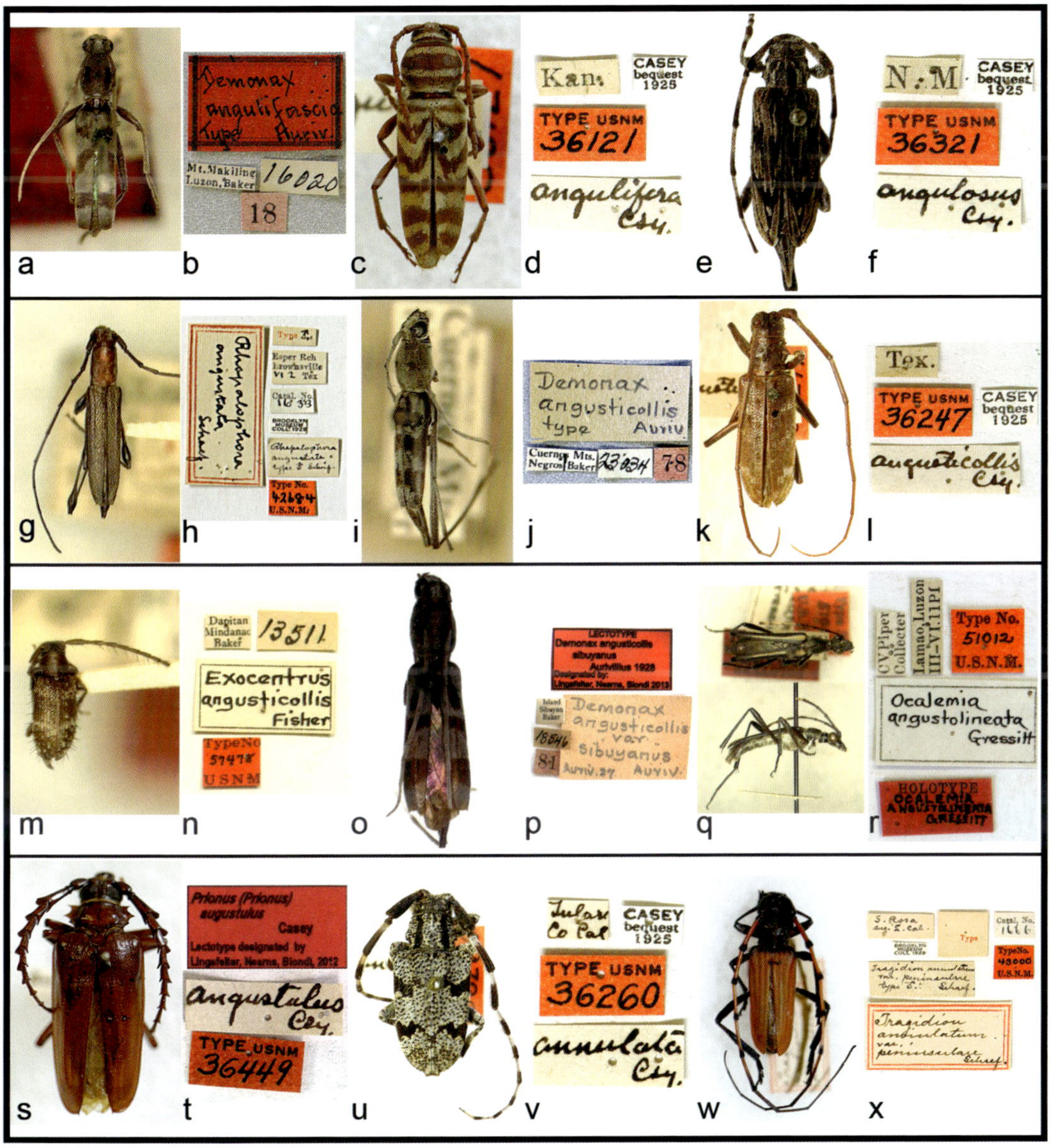

Figure 16. Primary types of Cerambycidae of the Smithsonian Institution (for each pair of images, dorsal habitus left, labels right): **(a, b)** *Demonax angulifascia* Aurivillius, 1922, Holotype; **(c, d)** *Cyllene angulifera* Casey, 1912, Holotype; **(e, f)** *Graphisurus (Graphisurus) angulosus* Casey, 1913, Holotype; **(g, h)** *Rhopalophora angustata* Schaeffer, 1905, Syntype; **(i, j)** *Demonax angusticollis* Aurivillius, 1928, Holotype; **(k, l)** *Monochamus angusticollis* Casey, 1913, Holotype; **(m, n)** *Exocentrus angusticollis* Fisher, 1925, Holotype; **(o, p)** *Demonax angusticollis* var. *sibuyanus* Aurivillius, 1928, Lectotype; **(q, r)** *Ocalemia angustolineata* Gressitt, 1935, Holotype; **(s, t)** *Prionus (Prionus) angustulus* Casey, 1912, Lectotype; **(u, v)** *Synaphaeta annulata* Casey, 1913, Holotype; **(w, x)** *Tragidion annulatum* var. *peninsulare* Schaeffer, 1908, Holotype.

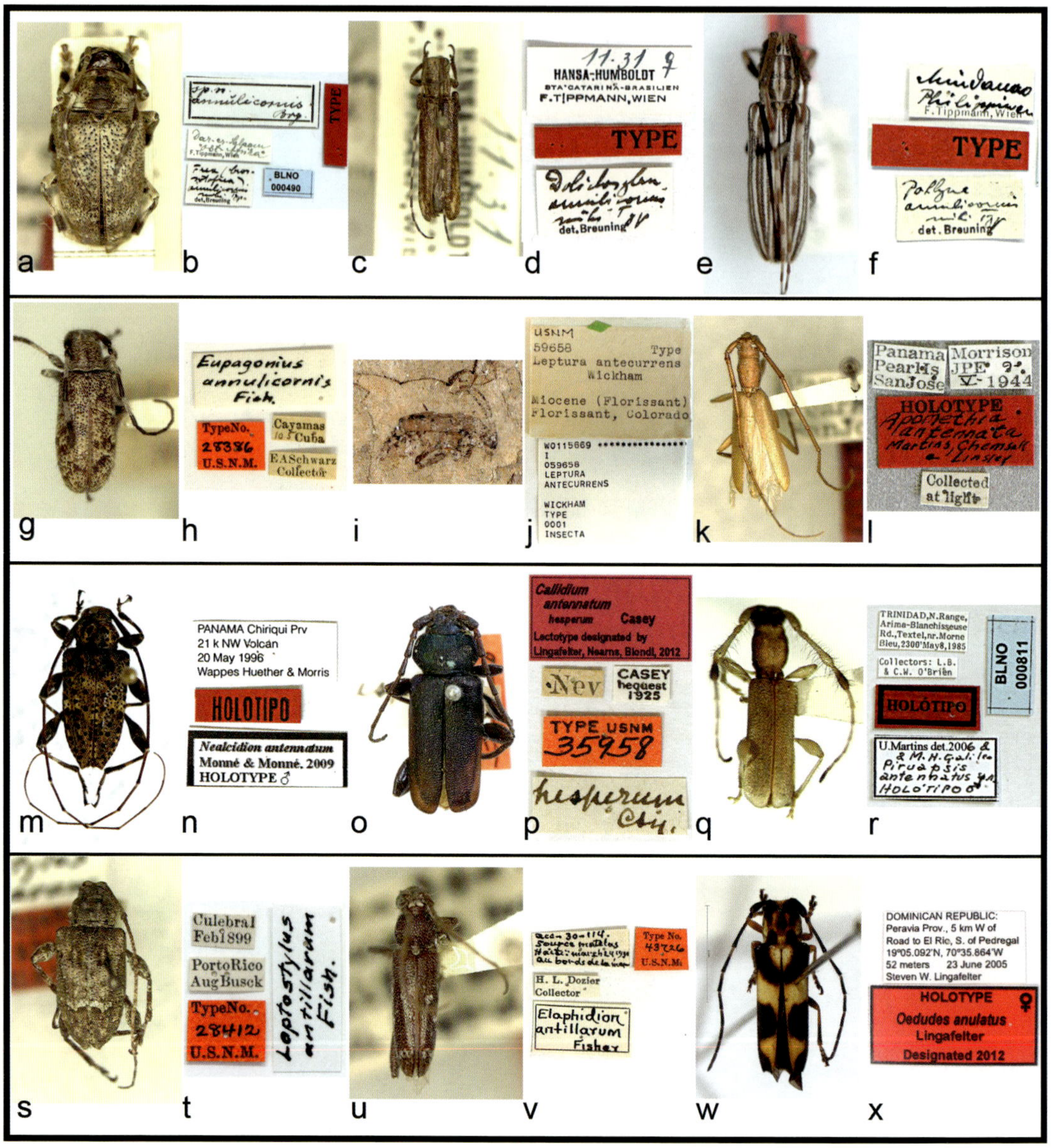

Figure 17. Primary types of Cerambycidae of the Smithsonian Institution (for each pair of images, dorsal habitus left, labels right): **(a, b)** *Frea (Crossotofrea) annulicornis* Breuning, 1942, Holotype; **(c, d)** *Dolichosybra annulicornis* Breuning, 1942, Holotype; **(e, f)** *Pothyne annulicornis* Breuning, 1948, Holotype; **(g, h)** *Eupogonius annulicornis* Fisher, 1926, Holotype; **(i, j)** *Leptura antecurrens* Wickham, 1913, Holotype; **(k, l)** *Apomethia antennata* Martins, Chemsak & Linsley, 1966, Holotype; **(m, n)** *Nealcidion antennatum* Monné & Monné, 2009, Holotype; **(o, p)** *Callidium antennatum hesperum* Casey, 1912, Lectotype; **(q, r)** *Piruapsis antennatus* Galileo & Martins, 2007, Holotype; **(s, t)** *Leptostylus antillarum* Fisher, 1925, Holotype; **(u, v)** *Elaphidion antillarum* Fisher, 1932, Holotype; **(w, x)** *Oedudes anulatus* Lingafelter, 2013, Holotype.

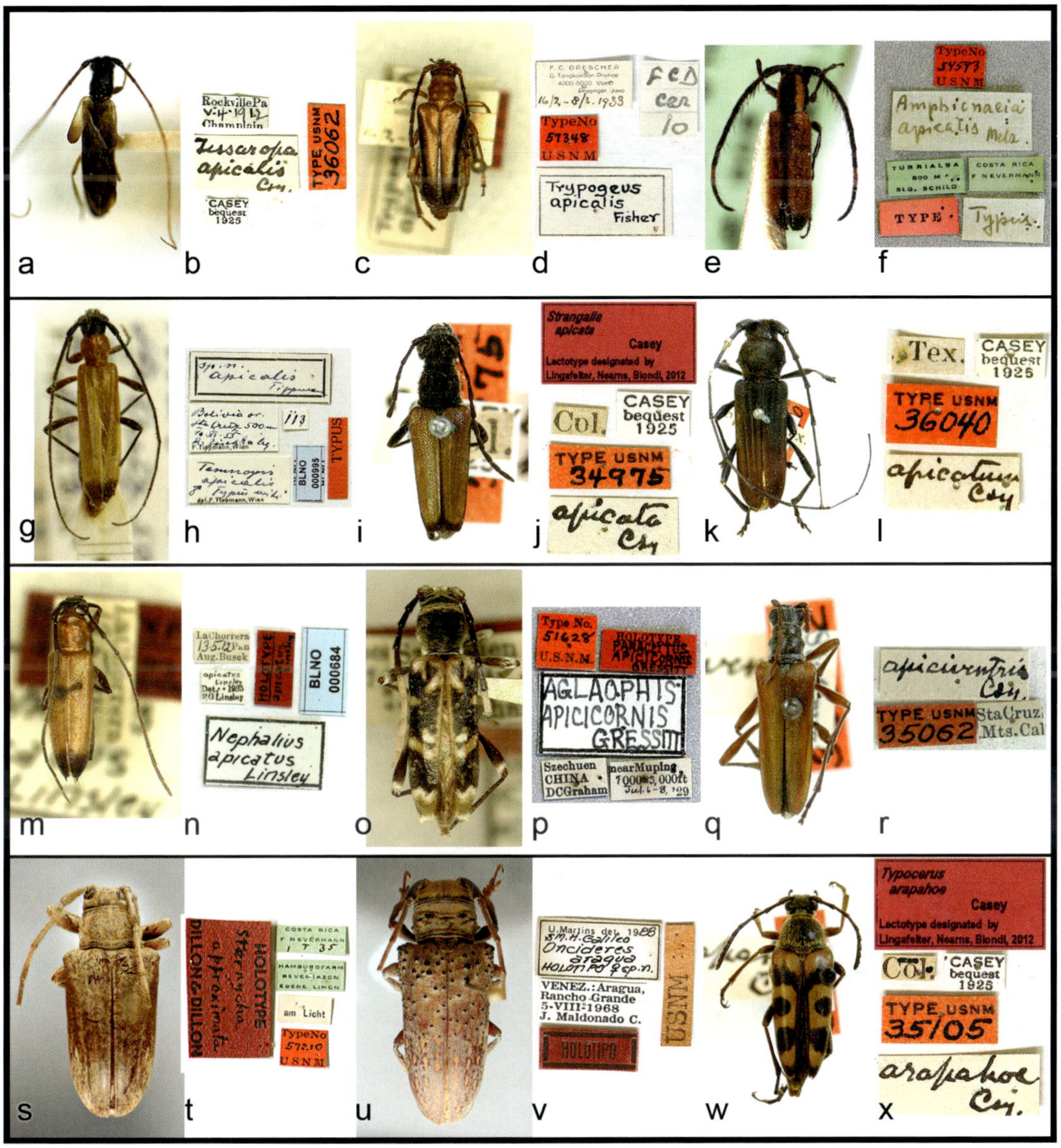

Figure 18. Primary types of Cerambycidae of the Smithsonian Institution (for each pair of images, dorsal habitus left, labels right): **(a, b)** *Tessaropa apicalis* Casey, 1914, Holotype; **(c, d)** *Trypogeus apicalis* Fisher, 1936, Holotype; **(e, f)** *Amphicnaeia apicalis* Melzer, 1933, Holotype; **(g, h)** *Temnopis apicalis* Tippmann, 1960, Holotype; **(i, j)** *Strangalia (Anastrangalia) apicata* Casey, 1924, Lectotype; **(k, l)** *Tragidion apicatum* Casey, 1912, Holotype; **(m, n)** *Nephalius apicatus* Linsley, 1935, Holotype; **(o, p)** *Aglaophis apicicornis* Gressitt, 1937, Holotype; **(q, r)** *Stenocorus (Stenocorus) apiciventris* Casey, 1913, Holotype; **(s, t)** *Sternycha approximata* Dillon & Dillon, 1945, Holotype; **(u, v)** *Oncideres aragua* Martins & Galileo, 1990, Holotype; **(w, x)** *Typocerus arapahoe* Casey, 1924, Lectotype.

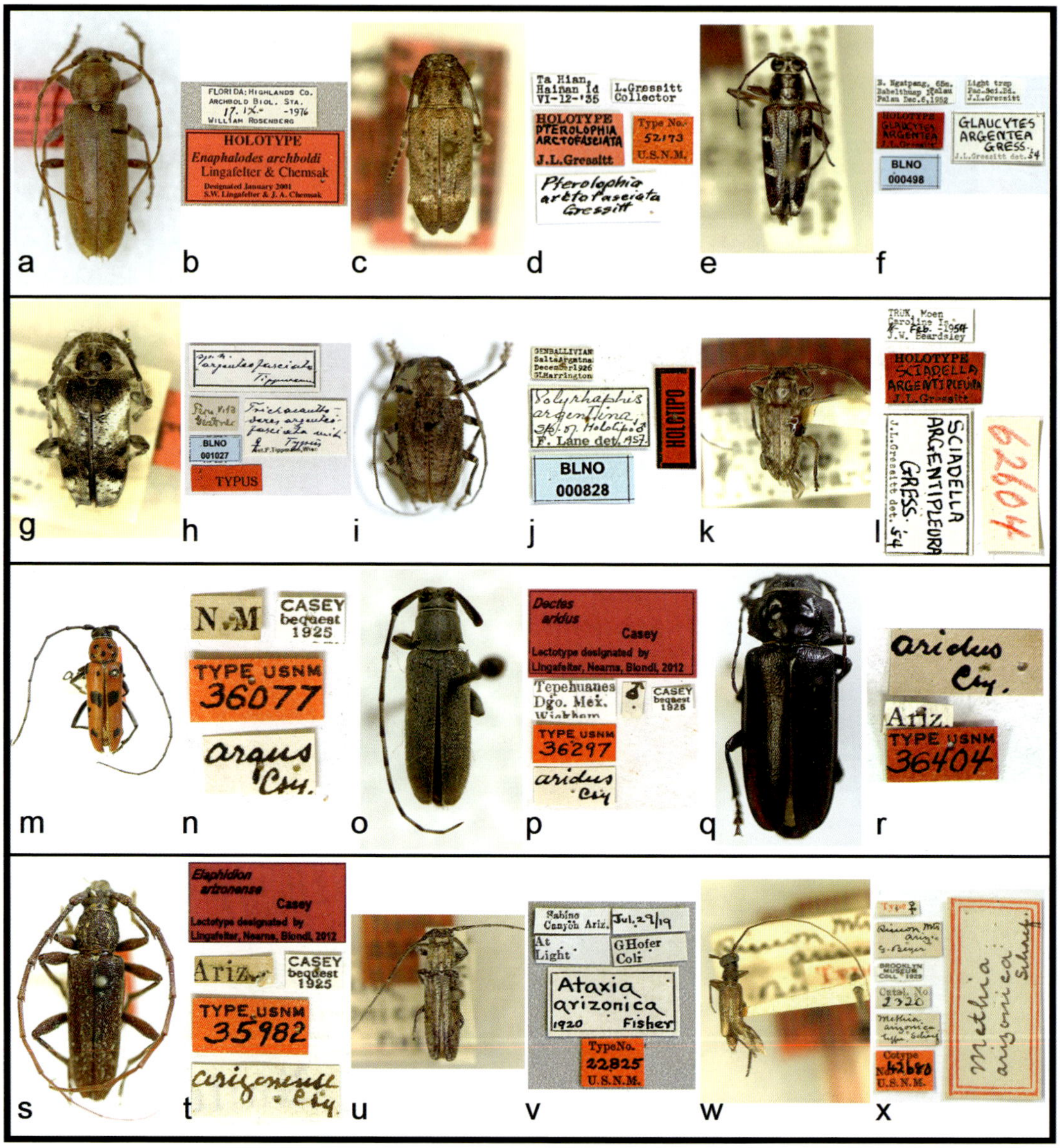

Figure 19. Primary types of Cerambycidae of the Smithsonian Institution (for each pair of images, dorsal habitus left, labels right): **(a, b)** *Enaphalodes archboldi* Lingafelter & Chemsak, 2002, Holotype; **(c, d)** *Pterolophia arctofasciata* Gressitt, 1940, Holotype; **(e, f)** *Glaucytes argentea* Gressitt, 1956, Holotype; **(g, h)** *Acanthoderes (Trichacanthoderes) argenteofasciata* Tippmann, 1960, Holotype; **(i, j)** *Polyrhaphis argentina* Lane, 1978, Holotype; **(k, l)** *Sciadella (Sciadella) argentipleura* Gressitt, 1956, Holotype; **(m, n)** *Tylosis argus* Casey, 1912, Holotype; **(o, p)** *Dectes aridus* Casey, 1913, Lectotype; **(q, r)** *Archodontes aridus* Casey, 1924, Holotype; **(s, t)** *Elaphidion arizonense* Casey, 1891, Lectotype; **(u, v)** *Ataxia arizonica* Fisher, 1920, Holotype; **(w, x)** *Methia arizonica* Schaeffer, 1908, Syntype.

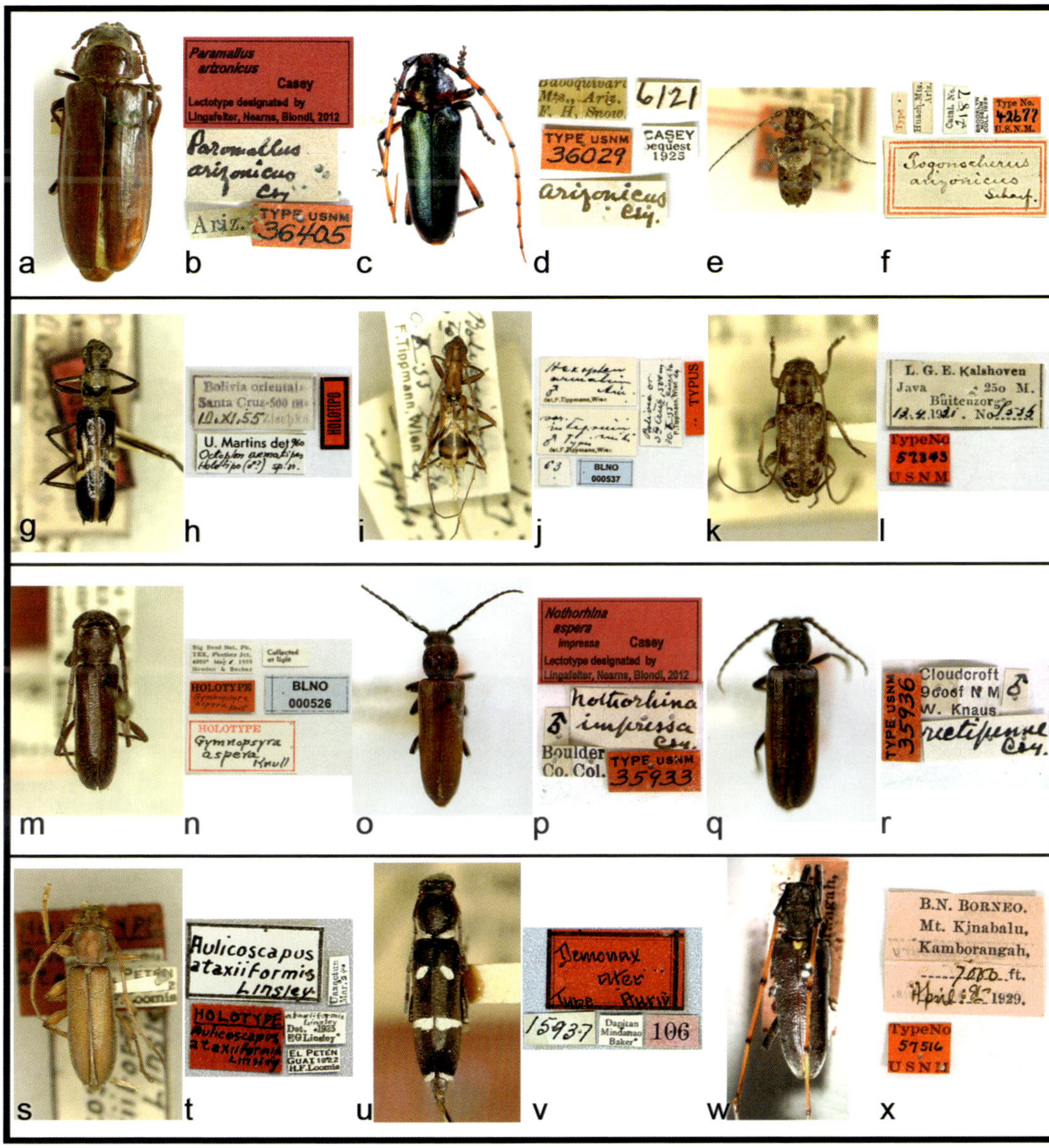

Figure 20. Primary types of Cerambycidae of the Smithsonian Institution (for each pair of images, dorsal habitus left, labels right): **(a, b)** *Paramallus arizonicus* Casey, 1912, Lectotype; **(c, d)** *Stenaspis arizonicus* Casey, 1912, Holotype; **(e, f)** *Pogonocherus arizonicus* Schaeffer, 1908, Holotype; **(g, h)** *Octoplon armatipes* Martins, 1962, Holotype; **(i, j)** *Hexoplon armatum* var. *integrum* Tippmann, 1960, Holotype; **(k, l)** *Exocentrus artocarpi* Fisher, 1934, Holotype; **(m, n)** *Gymnopsyra aspera* Knull, 1962, Holotype; **(o, p)** *Nothorhina aspera impressa* Casey, 1912, Lectotype; **(q, r)** *Nothorhina aspera rectipennis* Casey, 1912, Holotype; **(s, t)** *Aulacoscapus ataxiiformis* Linsley, 1935, Holotype; **(u, v)** *Demonax ater* Aurivillius, 1922, Holotype; **(w, x)** *Dihammus ater* Fisher, 1935, Holotype.

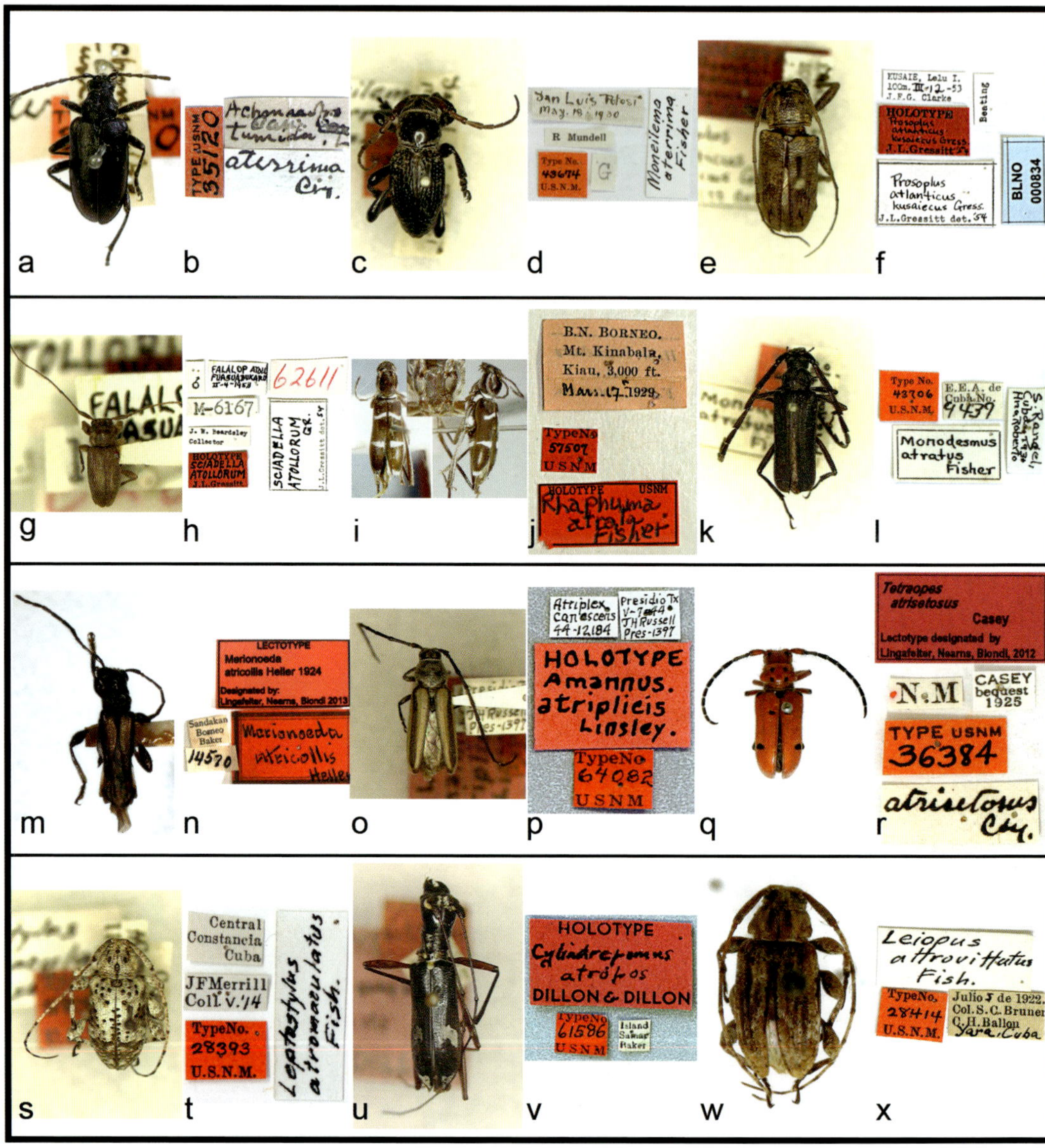

Figure 21. Primary types of Cerambycidae of the Smithsonian Institution (for each pair of images, dorsal habitus left, labels right): **(a, b)** *Brachysomida aterrima* Casey, 1924, Holotype; **(c, d)** *Moneilema (Collapteryx) aterrima* Fisher, 1931, Holotype; **(e, f)** *Prosoplus atlanticus kusaiecus* Gressitt, 1956, Holotype; **(g, h)** *Sciadella (Micronesiella) atollorum* Gressitt, 1956, Holotype; **(i, j)** *Rhaphuma atrata* Fisher, 1935, Holotype; **(k, l)** *Monodesmus atratus* Fisher, 1932, Holotype; **(m, n)** *Merionoeda atricollis* Heller, 1924, Lectotype; **(o, p)** *Amannus atriplicis* Linsley, 1957, Holotype; **(q, r)** *Tetraopes atrisetosus* Casey, 1913, Lectotype; **(s, t)** *Leptostylus atromaculatus* Fisher, 1926, Holotype; **(u, v)** *Cylindrepomus atropos* Dillon & Dillon, 1948, Holotype; **(w, x)** *Leiopus atrovittatus* Fisher, 1925, Holotype.

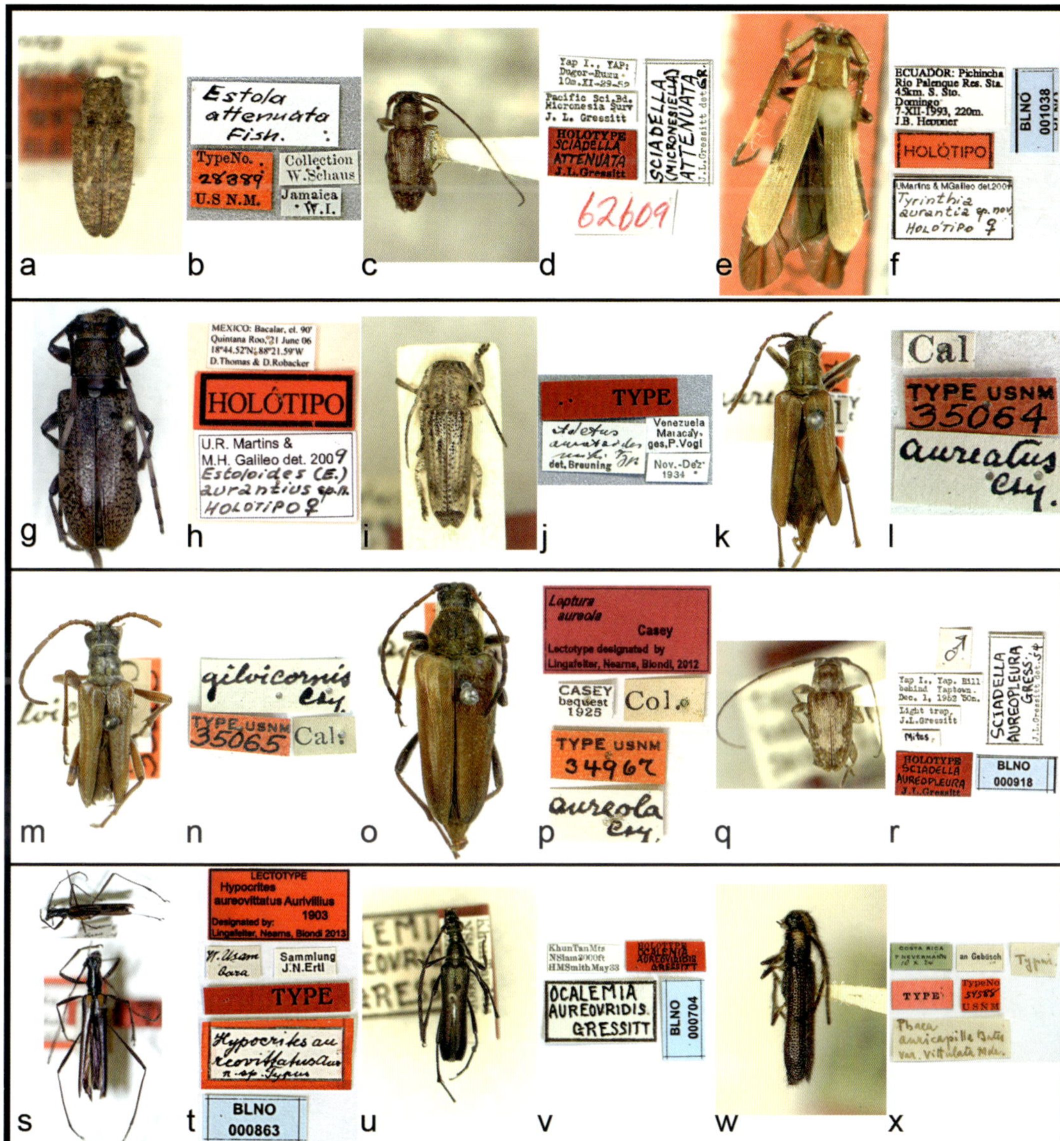

Figure 22. Primary types of Cerambycidae of the Smithsonian Institution (for each pair of images, dorsal habitus left, labels right): **(a, b)** *Estola attenuata* Fisher, 1926, Holotype; **(c, d)** *Sciadella (Micronesiella) attenuata* Gressitt, 1956, Holotype; **(e, f)** *Tyrinthia aurantia* Martins & Galileo, 2007, Holotype; **(g, h)** *Estoloides (Estoloides) aurantius* Martins & Galileo, 2010, Holotype; **(i, j)** *Adetus auratoides* Breuning, 1943, Holotype; **(k, l)** *Stenocorus (Stenocorus) aureatus* Casey, 1913, Holotype; **(m, n)** *Stenocorus (Stenocorus) aureatus gilvicornis* Casey, 1913, Holotype; **(o, p)** *Leptura (Cosmosalia) aureola* Casey, 1913, Lectotype; **(q, r)** *Sciadella (Sciadella) aureopleura* Gressitt, 1956, Holotype; **(s, t)** *Hypocrites aureovittatus* Aurivillius, 1903, Lectotype; **(u, v)** *Ocalemia aureovridis* Gressitt, 1941, Holotype; **(w, x)** *Phaea auricapilla* var. *vittulata* Melzer, 1935, Holotype.

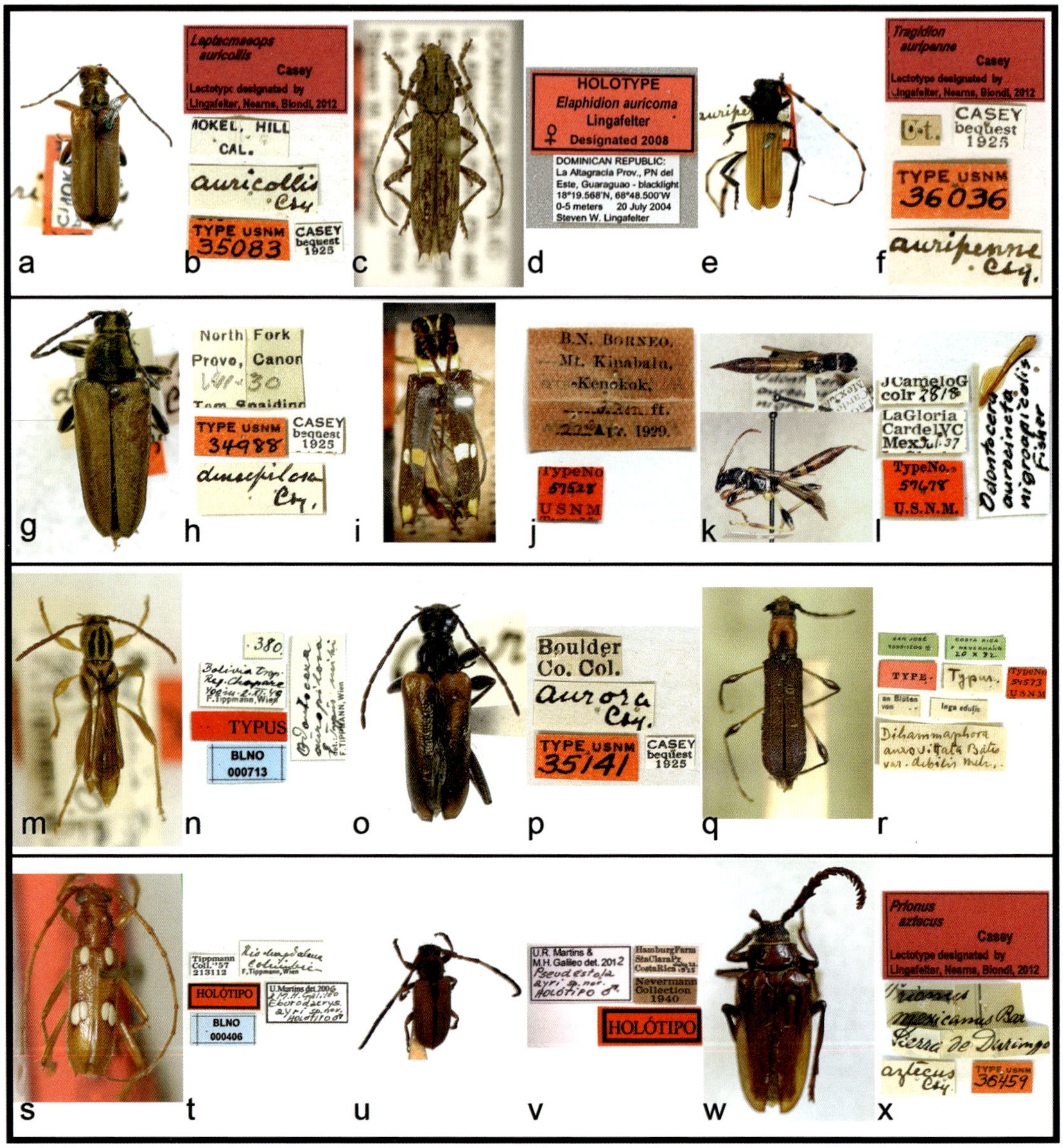

Figure 23. Primary types of Cerambycidae of the Smithsonian Institution (for each pair of images, dorsal habitus left, labels right): **(a, b)** *Leptacmaeops (Leptacmaeops) auricollis* Casey, 1913, Lectotype; **(c, d)** *Elaphidion auricoma* Lingafelter, 2008, Holotype; **(e, f)** *Tragidion auripenne* Casey, 1893, Lectotype; **(g, h)** *Cosmosalia auripilis densepilosa* Casey, 1924, Holotype; **(i, j)** *Glenea aurivillii* Fisher, 1935, Holotype; **(k, l)** *Odontocera aurocincta nigroapicalis* Fisher, 1947, Holotype; **(m, n)** *Odontocera auropilosa* Tippmann, 1953, Holotype; **(o, p)** *Acmaeops aurora* Casey, 1913, Holotype; **(q, r)** *Dihammaphora aurovittata* var. *debilis* Melzer, 1935, Holotype; **(s, t)** *Eburodacrys ayri* Galileo & Martins, 2006, Holotype; **(u, v)** *Pseudestola ayri* Martins & Galileo, 2012, Holotype; **(w, x)** *Prionus (Riponus) aztecus* Casey, 1912, Lectotype.

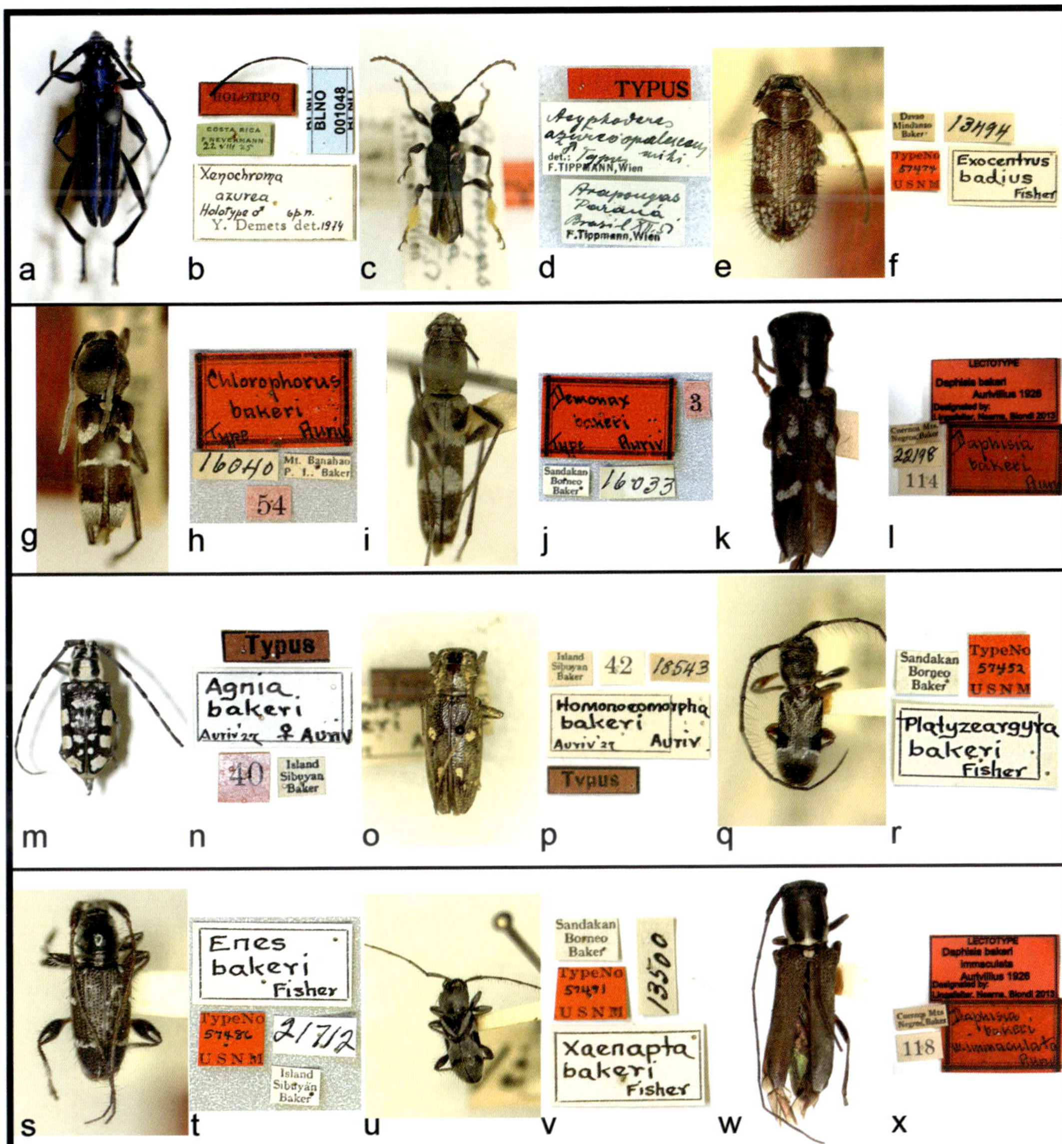

Figure 24. Primary types of Cerambycidae of the Smithsonian Institution (for each pair of images, dorsal habitus left, labels right): **(a, b)** *Xenochroma azurea* Demets, 1976, Holotype; **(c, d)** *Acyphoderes azureoopalescens* Tippmann, 1953, Holotype; **(e, f)** *Exocentrus badius* Fisher, 1925, Holotype; **(g, h)** *Chlorophorus bakeri* Aurivillius, 1922, Holotype; **(i, j)** *Demonax bakeri* Aurivillius, 1922, Holotype; **(k, l)** *Daphisia bakeri* Aurivillius, 1926, Lectotype; **(m, n)** *Agnia bakeri* Aurivillius, 1927, Holotype; **(o, p)** *Homonaeomorpha bakeri* Aurivillius, 1927, Holotype; **(q, r)** *Platyzeargyra bakeri* Fisher, 1925, Holotype; **(s, t)** *Enes bakeri* Fisher, 1925, Holotype; **(u, v)** *Xaenapta bakeri* Fisher, 1925, Holotype; **(w, x)** *Daphisia bakeri* var. *immaculata* Aurivillius, 1926, Lectotype.

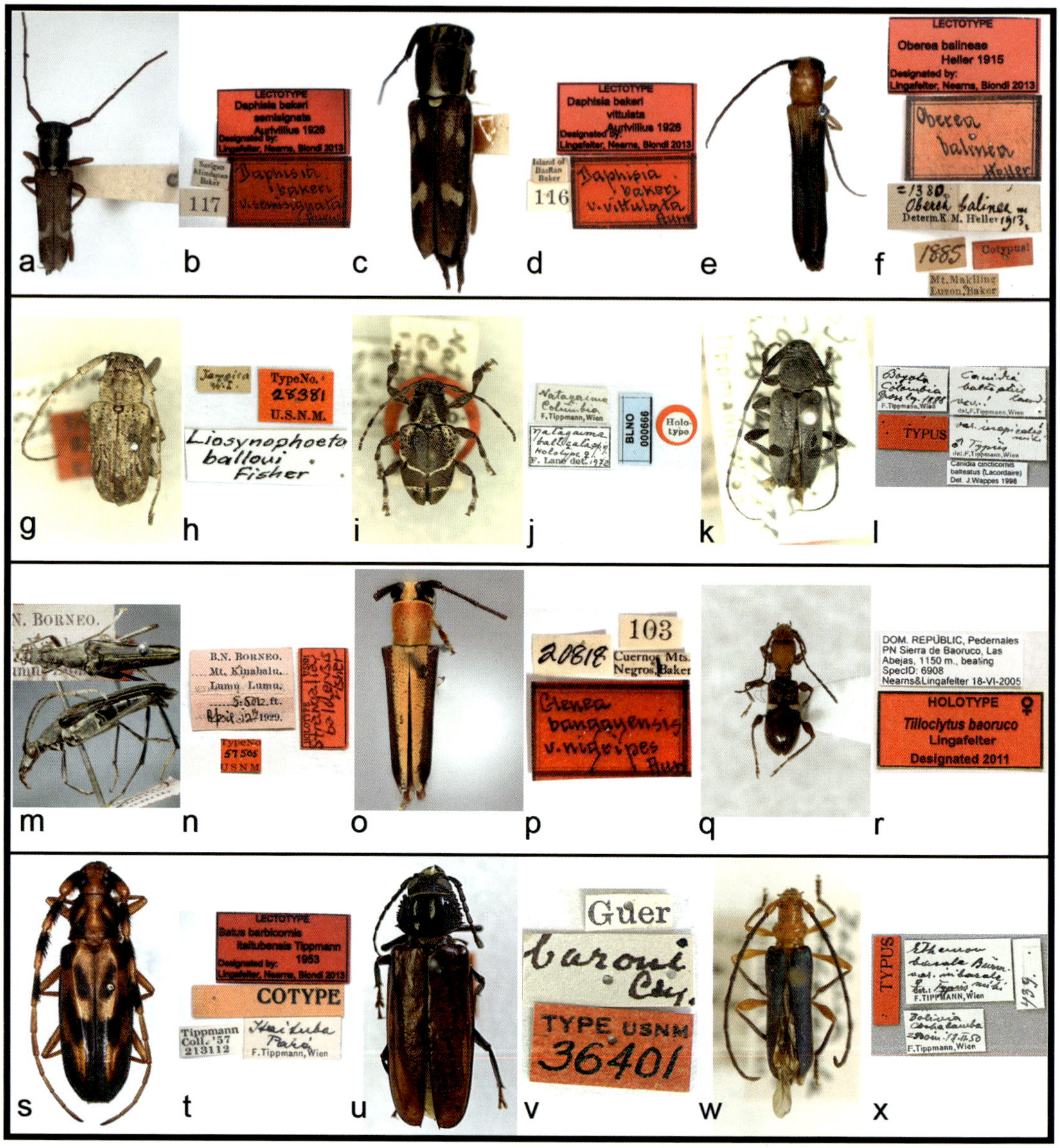

Figure 25. Primary types of Cerambycidae of the Smithsonian Institution (for each pair of images, dorsal habitus left, labels right): **(a, b)** *Daphisia bakeri* var. *semisignata* Aurivillius, 1926, Lectotype; **(c, d)** *Daphisia bakeri* var. *vittulata* Aurivillius, 1926, Lectotype; **(e, f)** *Oberea balineae* Heller, 1915, Lectotype; **(g, h)** *Liosynaphaeta balloui* Fisher, 1926, Holotype; **(i, j)** *Natagaima balteata* Lane, 1972, Holotype; **(k, l)** *Dectes (Canidia) balteata* var. *inapicalis* Tippmann, 1960, Holotype; **(m, n)** *Strangalia baluensis* Fisher, 1935, Holotype; **(o, p)** *Glenea bangueyensis nigripes* Aurivillius, 1926, Holotype; **(q, r)** *Tilloclytus baoruco* Lingafelter, 2011, Holotype; **(s, t)** *Batus barbicornis itaitubensis* Tippmann, 1953, Lectotype; **(u, v)** *Mallodon baroni* Casey, 1912, Holotype; **(w, x)** *Ethemon basale* var. *inbasale* Tippmann, 1960, Holotype.

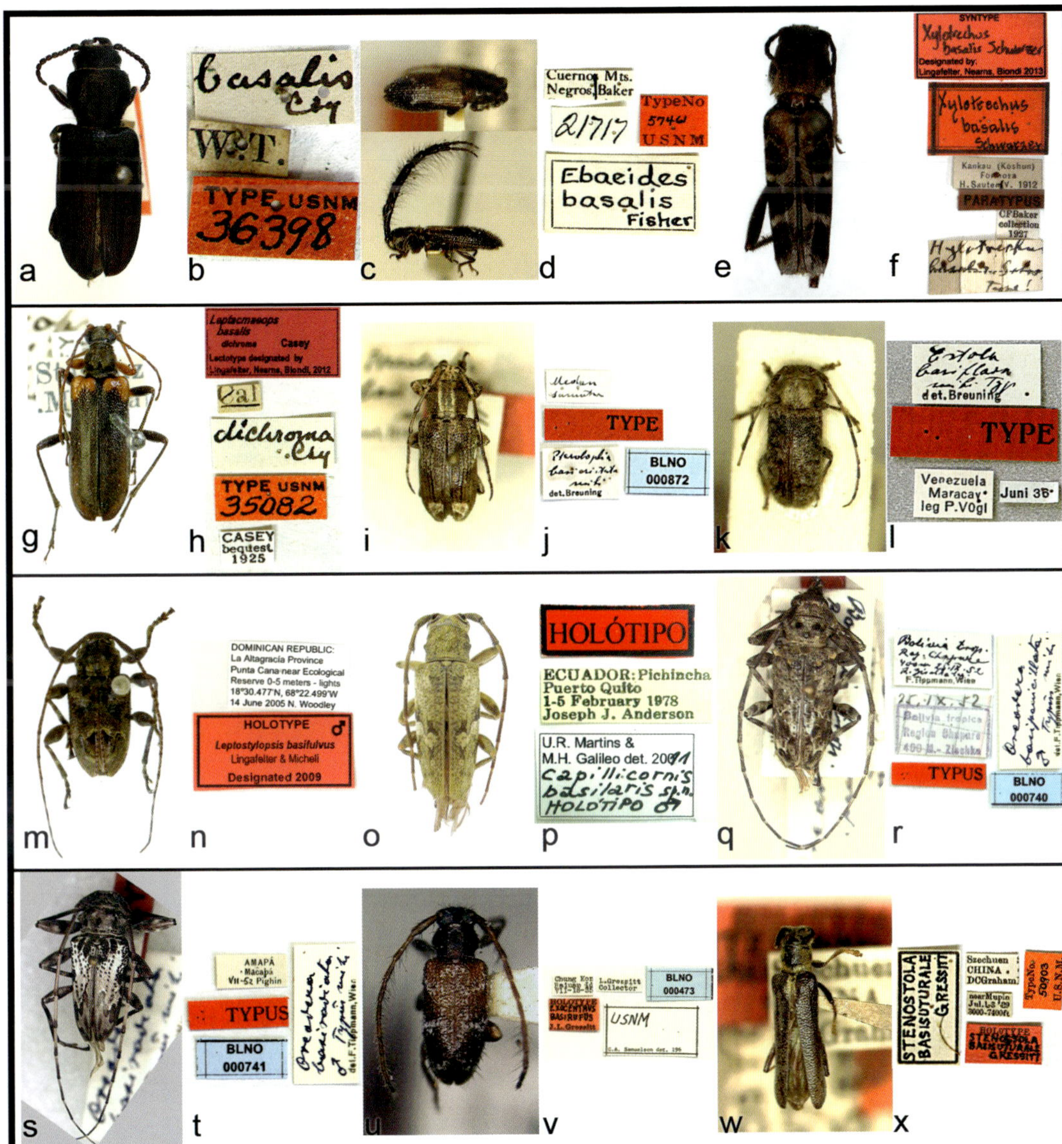

Figure 26. Primary types of Cerambycidae of the Smithsonian Institution (for each pair of images, dorsal habitus left, labels right): **(a, b)** *Spondylis basalis* Casey, 1912, Holotype; **(c, d)** *Ebaeides basalis* Fisher, 1925, Holotype; **(e, f)** *Xylotrechus basalis* Schwarzer, 1925, Syntype; **(g, h)** *Leptacmaeops (Leptacmaeops) basalis dichroma* Casey, 1913, Lectotype; **(i, j)** *Pterolophia basicristata* Breuning, 1938, Holotype; **(k, l)** *Estola basiflava* Breuning, 1943, Holotype; **(m, n)** *Leptostylopsis basifulvus* Lingafelter & Micheli, 2009, Holotype; **(o, p)** *Capillicornis basilaris* Galileo & Martins, 2012, Holotype; **(q, r)** *Oreodera basipenicillata* Tippmann, 1960, Holotype; **(s, t)** *Oreodera basiradiata* Tippmann, 1960, Holotype; **(u, v)** *Exocentrus basirufus* Gressitt, 1940, Holotype; **(w, x)** *Stenostola basisuturale* Gressitt, 1935, Holotype.

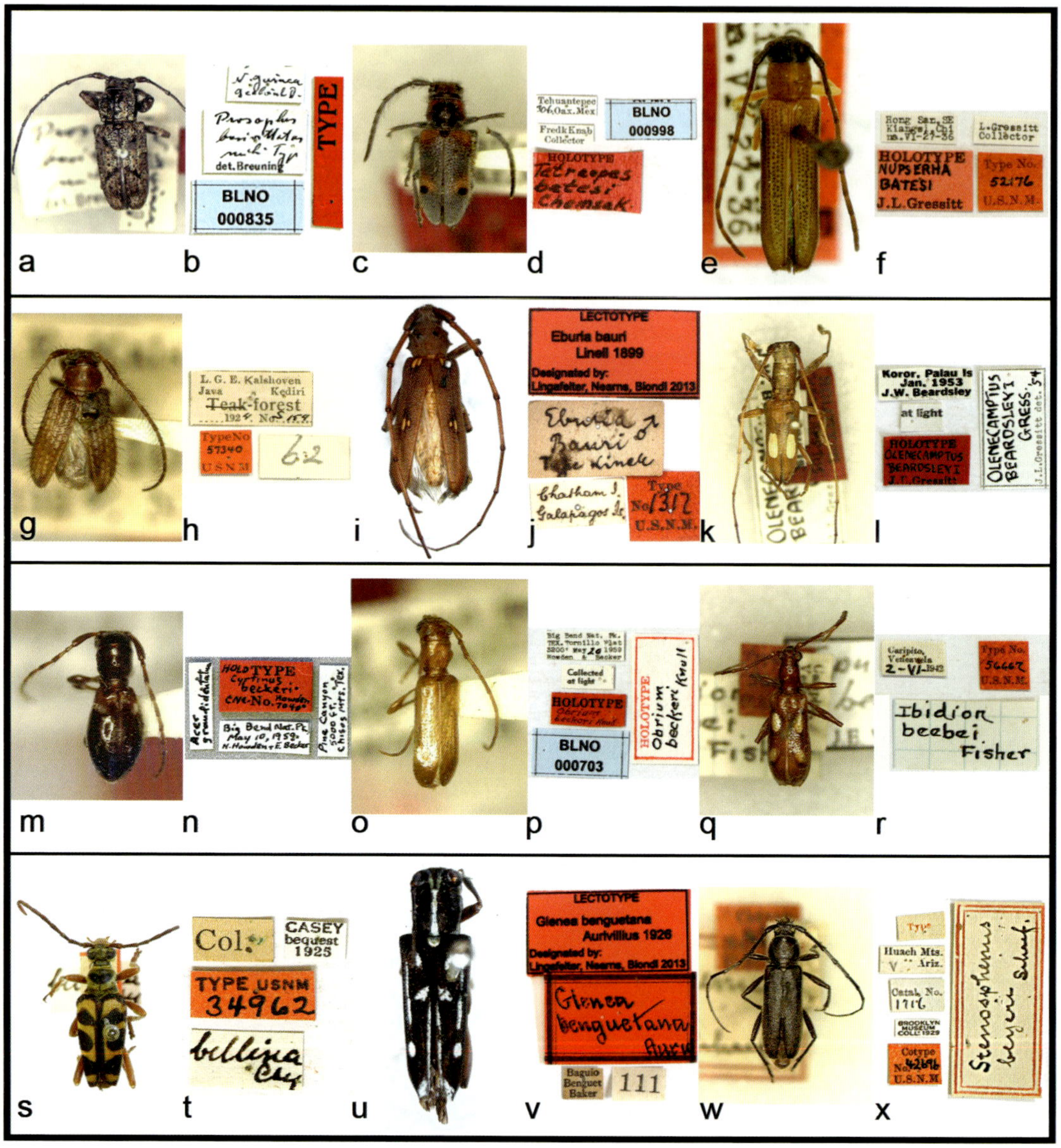

Figure 27. Primary types of Cerambycidae of the Smithsonian Institution (for each pair of images, dorsal habitus left, labels right): **(a, b)** *Prosoplus basivittatus* Breuning, 1940, Holotype; **(c, d)** *Tetraopes batesi* Chemsak, 1963, Holotype; **(e, f)** *Nupserha batesi* Gressitt, 1937, Holotype; **(g, h)** *Exocentrus bauhiniae* Fisher, 1934, Holotype; **(i, j)** *Eburia bauri* Linell, 1899, Lectotype; **(k, l)** *Olenecamptus beardsleyi* Gressitt, 1956, Holotype; **(m, n)** *Cyrtinus beckeri* Howden, 1960, Holotype; **(o, p)** *Obrium beckeri* Knull, 1962, Holotype; **(q, r)** *Ibidion beebei* Fisher, 1944, Holotype; **(s, t)** *Strophiona bellina* Casey, 1913, Holotype; **(u, v)** *Glenea benguetana* Aurivillius, 1926, Lectotype; **(w, x)** *Stenosphenus beyeri* Schaeffer, 1905, Lectotype.

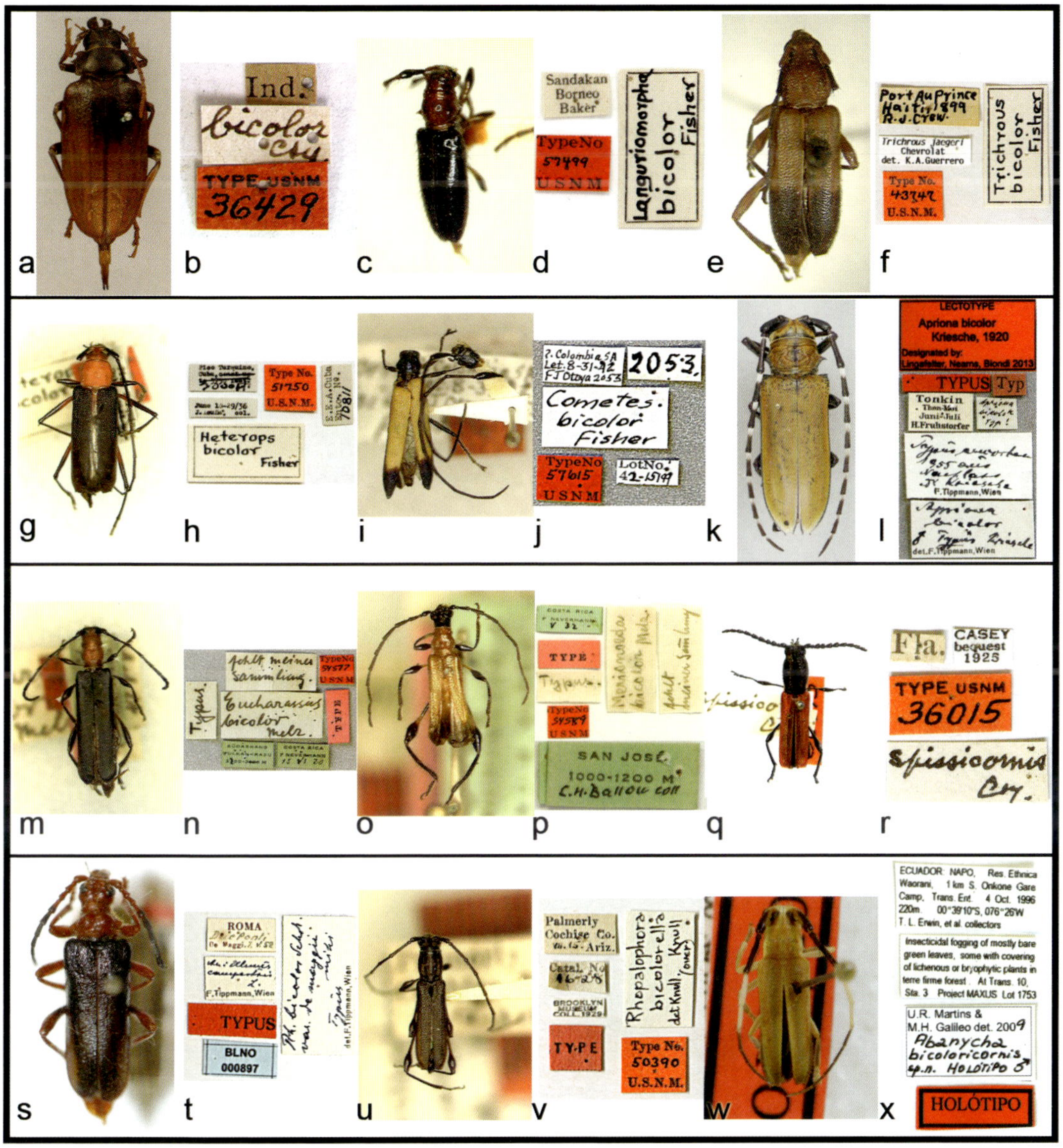

Figure 28. Primary types of Cerambycidae of the Smithsonian Institution (for each pair of images, dorsal habitus left, labels right): **(a, b)** *Prionus (Prionus) bicolor* Casey, 1912, Holotype; **(c, d)** *Languriomorpha bicolor* Fisher, 1925, Holotype; **(e, f)** *Trichrous bicolor* Fisher, 1932, Holotype; **(g, h)** *Heterops bicolor* Fisher, 1936, Holotype; **(i, j)** *Cometes bicolor* Fisher, 1946, Holotype; **(k, l)** *Apriona bicolor* Kriesche, 1920, Lectotype; **(m, n)** *Eucharassus bicolor* Melzer, 1934, Holotype; **(o, p)** *Merionoeda bicolor* Melzer, 1935, Holotype; **(q, r)** *Ancylocera bicolor spissicornis* Casey, 1924, Holotype; **(s, t)** *Rhamnusium bicolor* var. *demaggii* Tippmann, 1956, Holotype; **(u, v)** *Rhopalophora bicolorella* Knull, 1934, Holotype; **(w, x)** *Abanycha bicoloricornis* Galileo & Martins, 2009, Holotype.

Figure 29. Primary types of Cerambycidae of the Smithsonian Institution (for each pair of images, dorsal habitus left, labels right): **(a, b)** *Calolamia bicordifera* Tippmann, 1953, Holotype; **(c, d)** *Pterolophia bicristulata* Breuning, 1940, Holotype; **(e, f)** *Niphonoclea bifasciata* Fisher, 1943, Holotype; **(g, h)** *Ptychodes bifasciatus* Dillon & Dillon, 1941, Holotype; **(i, j)** *Cyrtinus bifasciatus* Martins & Galileo, 2009, Holotype; **(k, l)** *Pterolophia bifasciculata* Breuning, 1940, Holotype; **(m, n)** *Calloides biformis* Casey, 1912, Lectotype; **(o, p)** *Merium bifossatum* Casey, 1912, Holotype; **(q, r)** *Deliathis bifurcatus* Dillon & Dillon, 1941, Holotype; **(s, t)** *Daphisia biguttata* Aurivillius, 1925, Syntype; **(u, v)** *Sybra biguttata* Aurivillius, 1927, Lectotype; **(w, x)** *Demonax biguttatus* Aurivillius, 1922, Holotype.

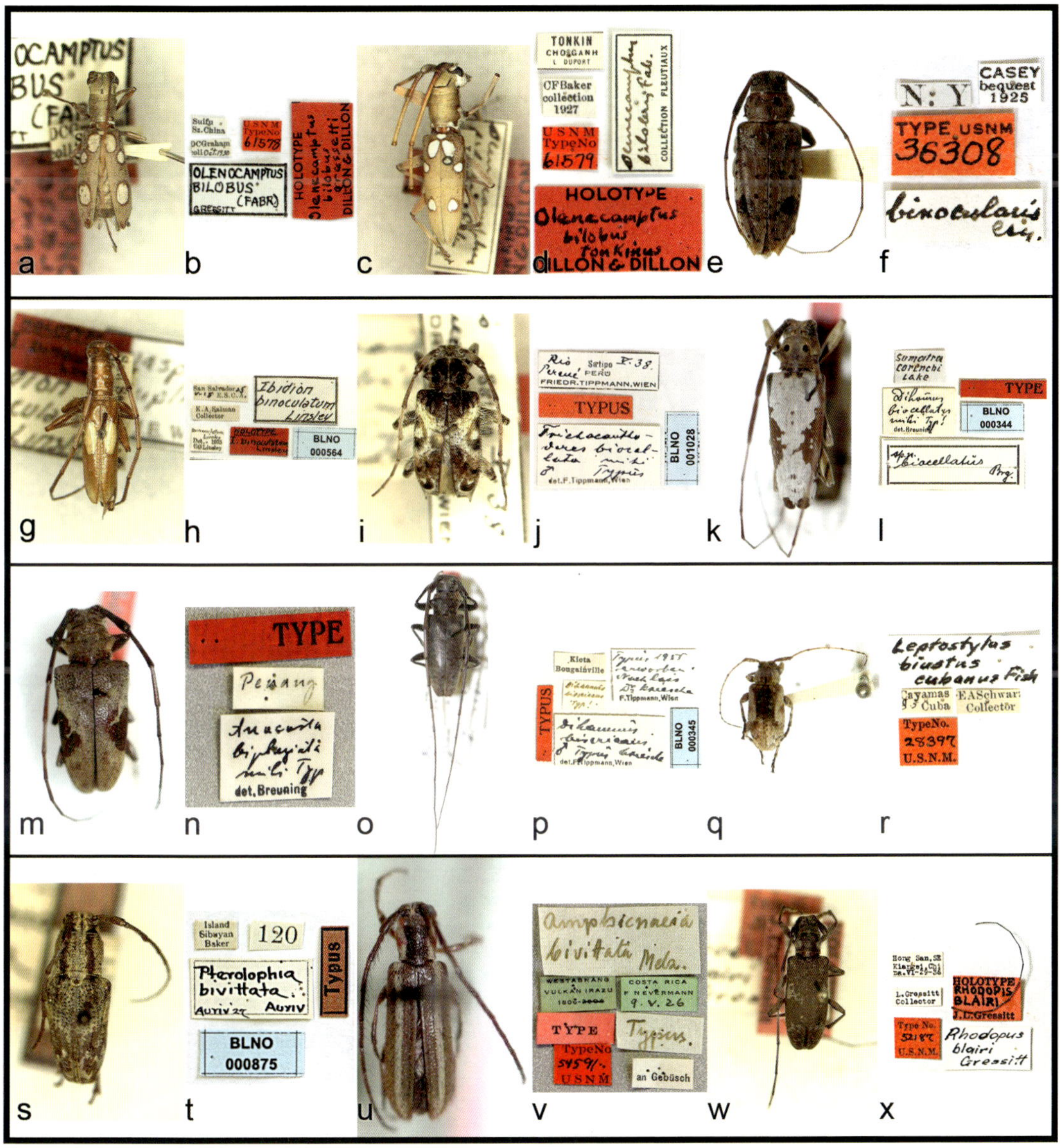

Figure 30. Primary types of Cerambycidae of the Smithsonian Institution (for each pair of images, dorsal habitus left, labels right): **(a, b)** *Olenecamptus bilobus gressitti* Dillon & Dillon, 1948, Holotype; **(c, d)** *Olenecamptus bilobus tonkinus* Dillon & Dillon, 1948, Holotype; **(e, f)** *Hyperplatys binocularis* Casey, 1913, Holotype; **(g, h)** *Ibidion binoculatum* Linsley, 1935, Holotype; **(i, j)** *Acanthoderes (Trichacanthoderes) biocellata* Tippmann, 1960, Holotype; **(k, l)** *Dihammus biocellatus* Breuning, 1940, Holotype; **(m, n)** *Anacasta biplagiata* Breuning, 1940, Holotype; **(o, p)** *Dihammus bisericans* Kriesche, 1936, Holotype; **(q, r)** *Leptostylus biustus* var. *cubanus* Fisher, 1926, Holotype; **(s, t)** *Pterolophia bivittata* Aurivillius, 1927, Syntype; **(u, v)** *Amphicnaeia bivittata* Melzer, 1933, Holotype; **(w, x)** *Rhodopis blairi* Gressitt, 1937, Holotype.

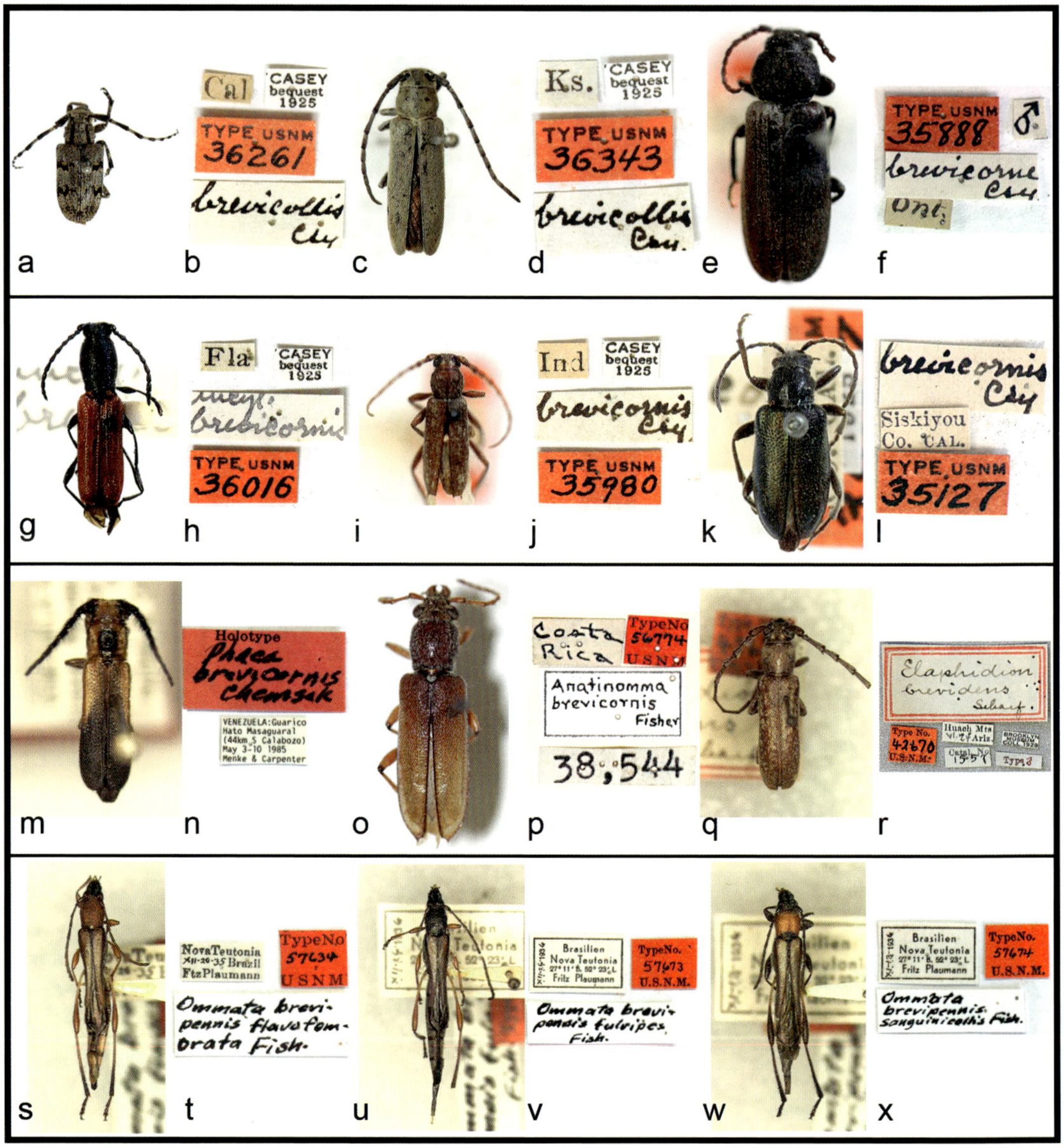

Figure 33. Primary types of Cerambycidae of the Smithsonian Institution (for each pair of images, dorsal habitus left, labels right): **(a, b)** *Synaphaeta brevicollis* Casey, 1913, Holotype; **(c, d)** *Mecas brevicollis* Casey, 1913, Holotype; **(e, f)** *Asemum brevicorne* Casey, 1912, Holotype; **(g, h)** *Ancylocera brevicornis* Casey, 1893, Holotype; **(i, j)** *Hypermallus brevicornis* Casey, 1912, Holotype; **(k, l)** *Brachysomida brevicornis* Casey, 1913, Holotype; **(m, n)** *Phaea brevicornis* Chemsak, 2000, Holotype; **(o, p)** *Anatinomma brevicornis* Fisher, 1944, Holotype; **(q, r)** *Elaphidion brevidens* Schaeffer, 1908, Holotype; **(s, t)** *Ommata (Eclipta) brevipennis* v. *flavofemorata* Fisher, 1937, Holotype; **(u, v)** *Ommata (Eclipta) brevipennis* v. *fulvipes* Fisher, 1947, Holotype; **(w, x)** *Ommata (Eclipta) brevipennis* v. *sanguinicollis* Fisher, 1947, Holotype.

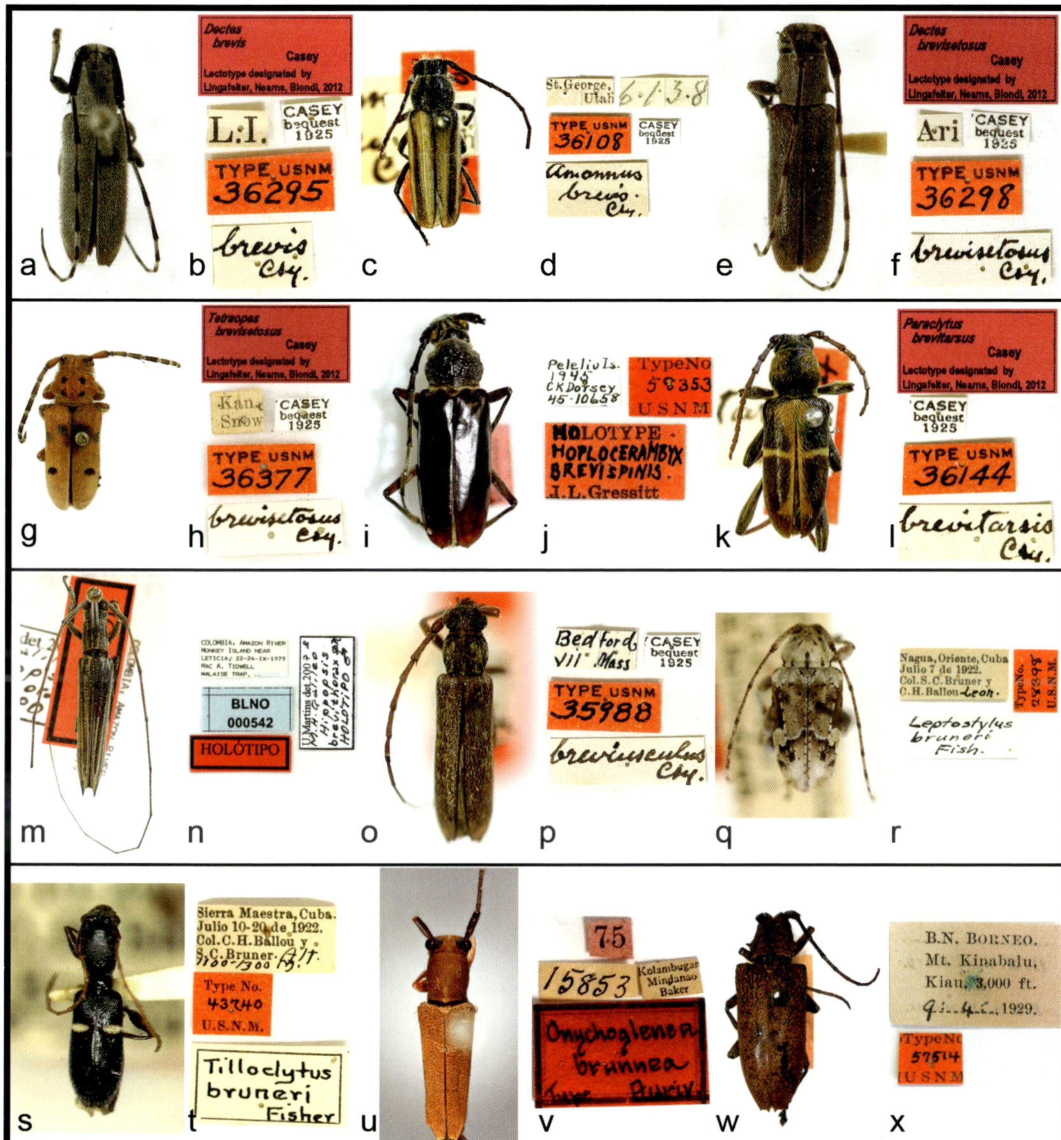

Figure 34. Primary types of Cerambycidae of the Smithsonian Institution (for each pair of images, dorsal habitus left, labels right): **(a, b)** *Dectes brevis* Casey, 1913, Lectotype; **(c, d)** *Amannus brevis* Casey, 1924, Holotype; **(e, f)** *Dectes brevisetosus* Casey, 1913, Lectotype; **(g, h)** *Tetraopes brevisetosus* Casey, 1913, Lectotype; **(i, j)** *Hoplocerambyx brevispinis* Gressitt, 1951, Holotype; **(k, l)** *Paraclytus brevitarsis* Casey, 1912, Lectotype; **(m, n)** *Hippopsis brevithorax* Galileo & Martins, 2007, Holotype; **(o, p)** *Hypermallus breviusculus* Casey, 1914, Holotype; **(q, r)** *Leptostylus bruneri* Fisher, 1926, Holotype; **(s, t)** *Tilloclytus bruneri* Fisher, 1932, Holotype; **(u, v)** *Onychoglenea brunnea* Aurivillius, 1922, Holotype; **(w, x)** *Loxotropoides brunnea* Fisher, 1935, Holotype.

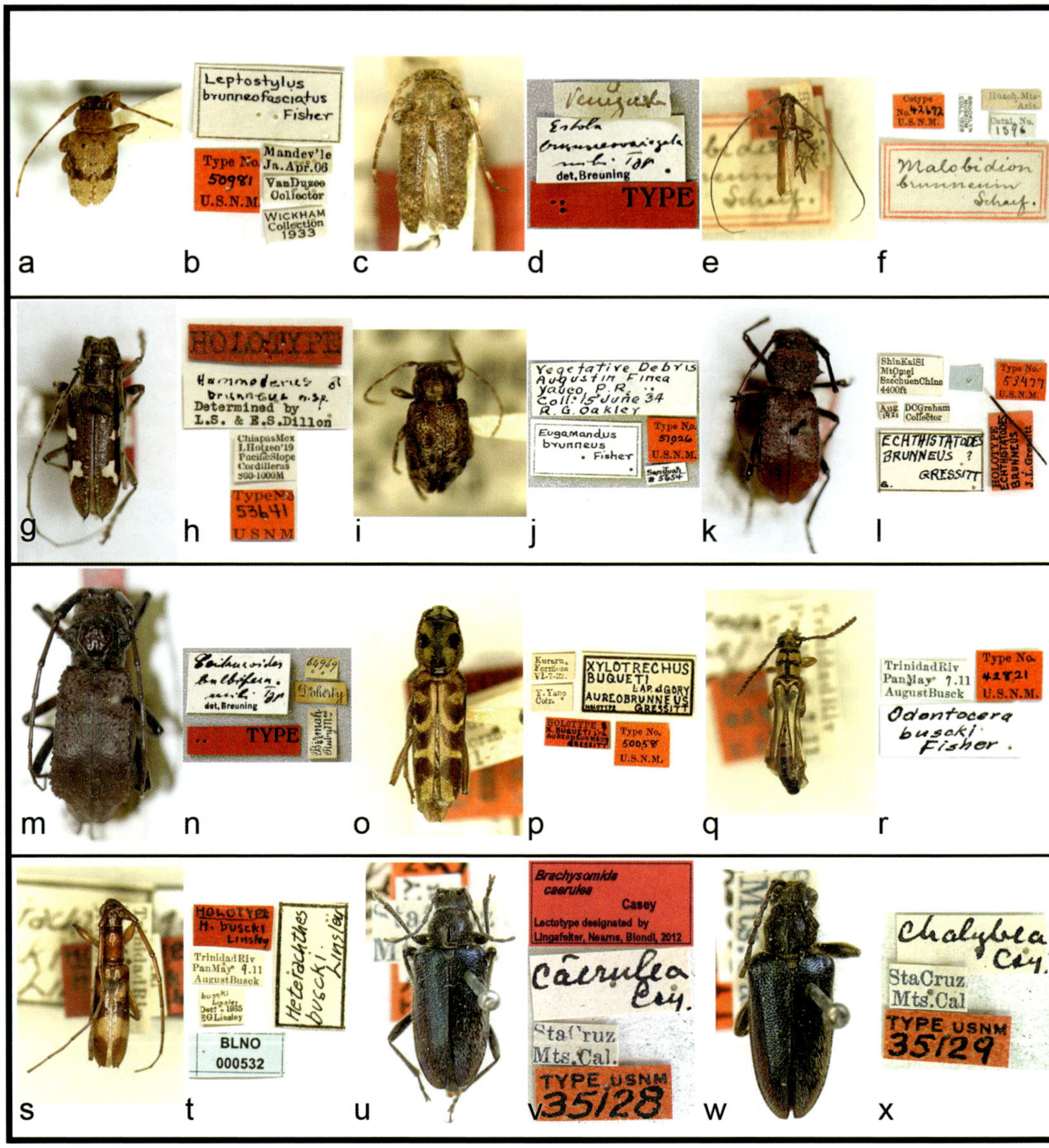

Figure 35. Primary types of Cerambycidae of the Smithsonian Institution (for each pair of images, dorsal habitus left, labels right): **(a, b)** *Leptostylus brunneofasciatus* Fisher, 1935, Holotype; **(c, d)** *Estola brunneovariegata* Breuning, 1940, Holotype; **(e, f)** *Malobidion brunneum* Schaeffer, 1908, Syntype; **(g, h)** *Plagiohammus brunneus* Dillon & Dillon, 1941, Holotype; **(i, j)** *Eugamandus brunneus* Fisher, 1935, Holotype; **(k, l)** *Echthistatodes brunneus* Gressitt, 1938, Holotype; **(m, n)** *Centruroides bulbifera* Breuning, 1940, Holotype; **(o, p)** *Xylotrechus buqueti aureobrunneus* Gressitt, 1934, Holotype; **(q, r)** *Odontocera buscki* Fisher, 1930, Holotype; **(s, t)** *Heterachthes buscki* Linsley, 1935, Holotype; **(u, v)** *Brachysomida caerulea* Casey, 1913, Lectotype; **(w, x)** *Brachysomida caerulea chalybea* Casey, 1913, Holotype.

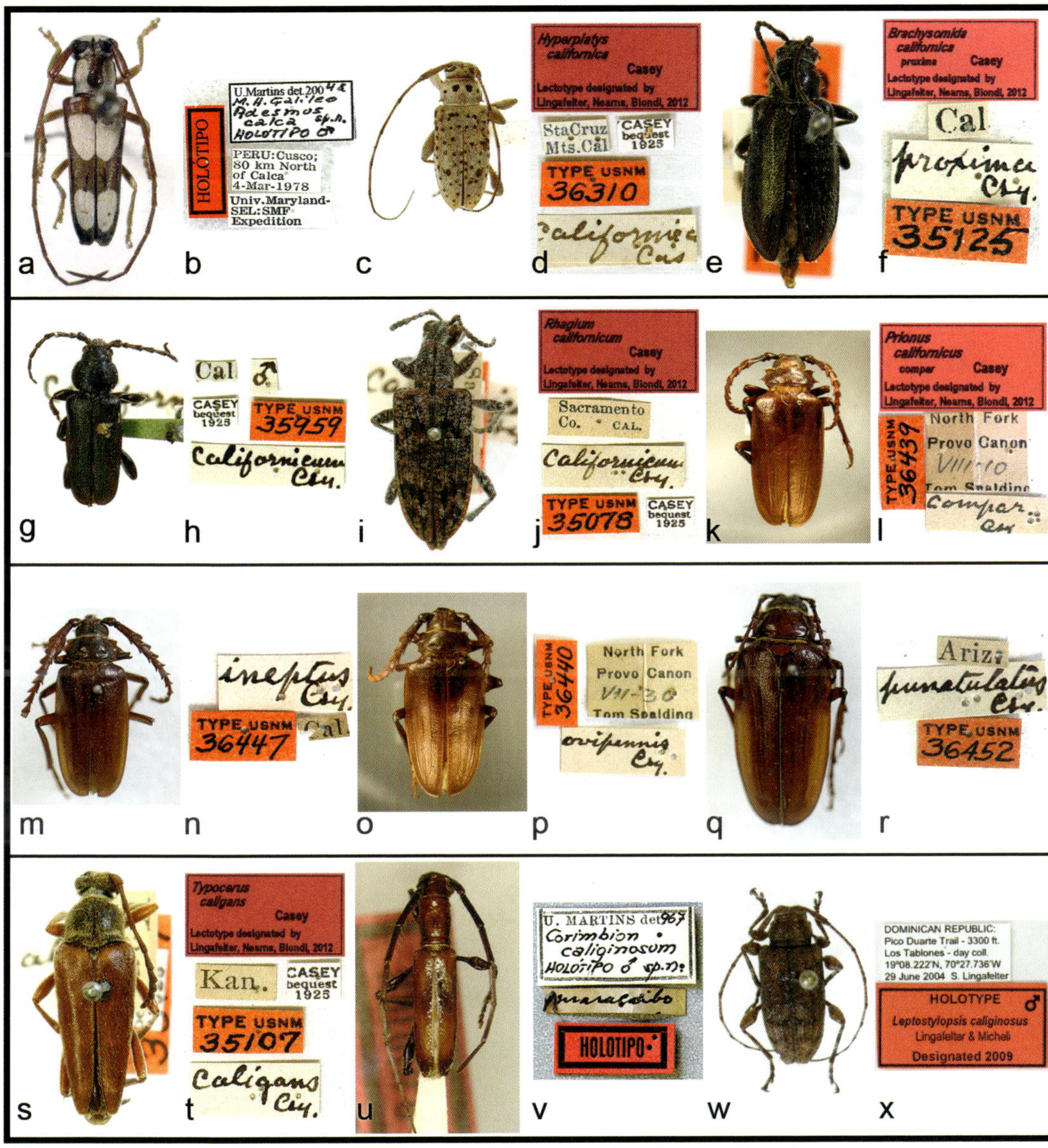

Figure 36. Primary types of Cerambycidae of the Smithsonian Institution (for each pair of images, dorsal habitus left, labels right): **(a, b)** *Adesmus calca* Galileo & Martins, 2005, Holotype; **(c, d)** *Hyperplatys californica* Casey, 1891, Lectotype; **(e, f)** *Brachysomida californica proxima* Casey, 1913, Lectotype; **(g, h)** *Callidium californicum* Casey, 1912, Holotype; **(i, j)** *Rhagium (Hargium) californicum* Casey, 1913, Lectotype; **(k, l)** *Prionus (Prionus) californicus compar* Casey, 1924, Lectotype; **(m, n)** *Prionus (Prionus) californicus ineptus* Casey, 1912, Holotype; **(o, p)** *Prionus (Prionus) californicus ovipennis* Casey, 1924, Holotype; **(q, r)** *Prionus (Prionus) californicus punctulatus* Casey, 1912, Holotype; **(s, t)** *Typocerus caligans* Casey, 1913, Lectotype; **(u, v)** *Corimbion caliginosum* Martins, 1970, Holotype; **(w, x)** *Leptostylopsis caliginosus* Lingafelter & Micheli, 2009, Holotype.

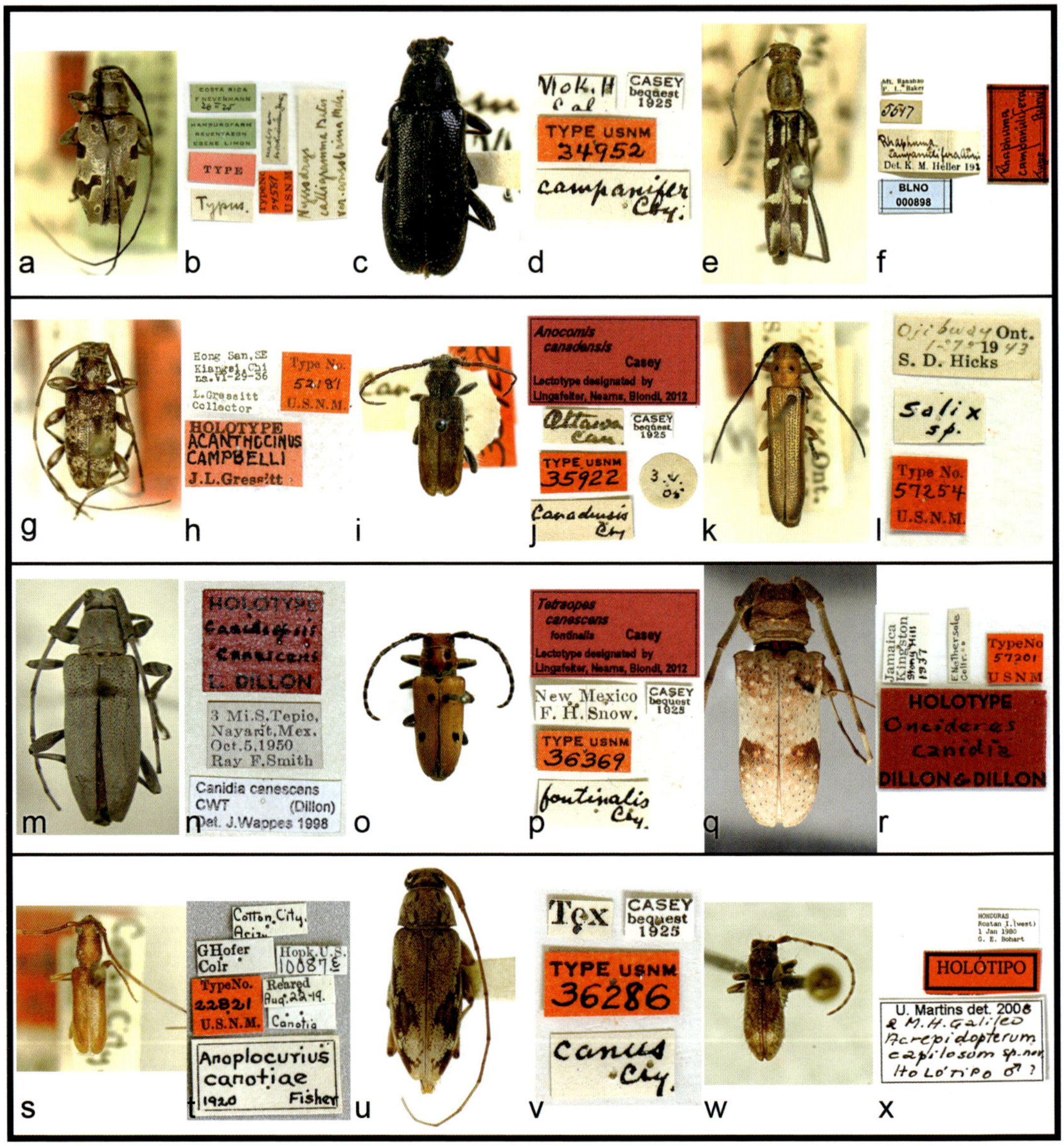

Figure 37. Primary types of Cerambycidae of the Smithsonian Institution (for each pair of images, dorsal habitus left, labels right): **(a, b)** *Nyssodrys calligramma* var. *consobrina* Melzer, 1934, Holotype; **(c, d)** *Parallelina campanifera* Casey, 1913, Holotype; **(e, f)** *Rhaphuma campanulifera* Aurivillius, 1922, Holotype; **(g, h)** *Acanthocinus campbelli* Gressitt, 1937, Holotype; **(i, j)** *Anocomis canadensis* Casey, 1924, Lectotype; **(k, l)** *Oberea canadensis* Fisher, 1945, Holotype; **(m, n)** *Canidiopsis canescens* Dillon, 1955, Holotype; **(o, p)** *Tetraopes canescens fontinalis* Casey, 1913, Lectotype; **(q, r)** *Oncideres canidia* Dillon & Dillon, 1946, Holotype; **(s, t)** *Anoplocurius canotiae* Fisher, 1920, Holotype; **(u, v)** *Lepturges canus* Casey, 1913, Holotype; **(w, x)** *Acrepidopterum capilosum* Martins & Galileo, 2008, Holotype.

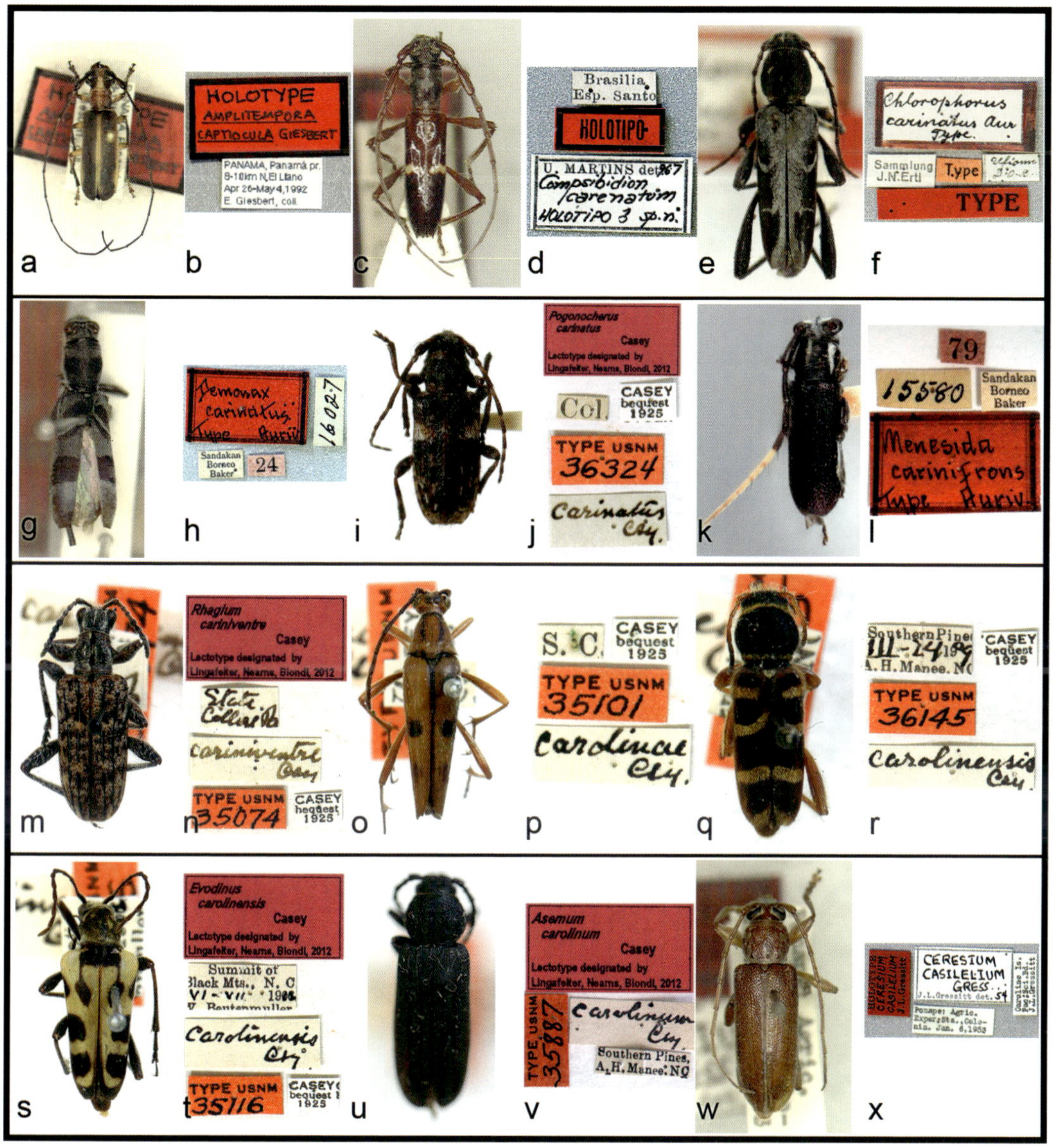

Figure 38. Primary types of Cerambycidae of the Smithsonian Institution (for each pair of images, dorsal habitus left, labels right): **(a, b)** *Amplitempora captiocula* Giesbert, 1996, Holotype; **(c, d)** *Compsibidion carenatum* Martins, 1969, Holotype; **(e, f)** *Chlorophorus carinatus* Aurivillius, 1913, Holotype; **(g, h)** *Demonax? carinatus* Aurivillius, 1922, Holotype; **(i, j)** *Pogonocherus carinatus* Casey, 1913, Lectotype; **(k, l)** *Menesida carinifrons* Aurivillius, 1922, Holotype; **(m, n)** *Rhagium (Hargium) cariniventre* Casey, 1913, Lectotype; **(o, p)** *Ophistomis carolinae* Casey, 1913, Holotype; **(q, r)** *Clytus carolinensis* Casey, 1912, Holotype; **(s, t)** *Evodinus carolinensis* Casey, 1924, Lectotype; **(u, v)** *Asemum carolinum* Casey, 1924, Lectotype; **(w, x)** *Ceresium (Ceresium) casilelium* Gressitt, 1956, Holotype.

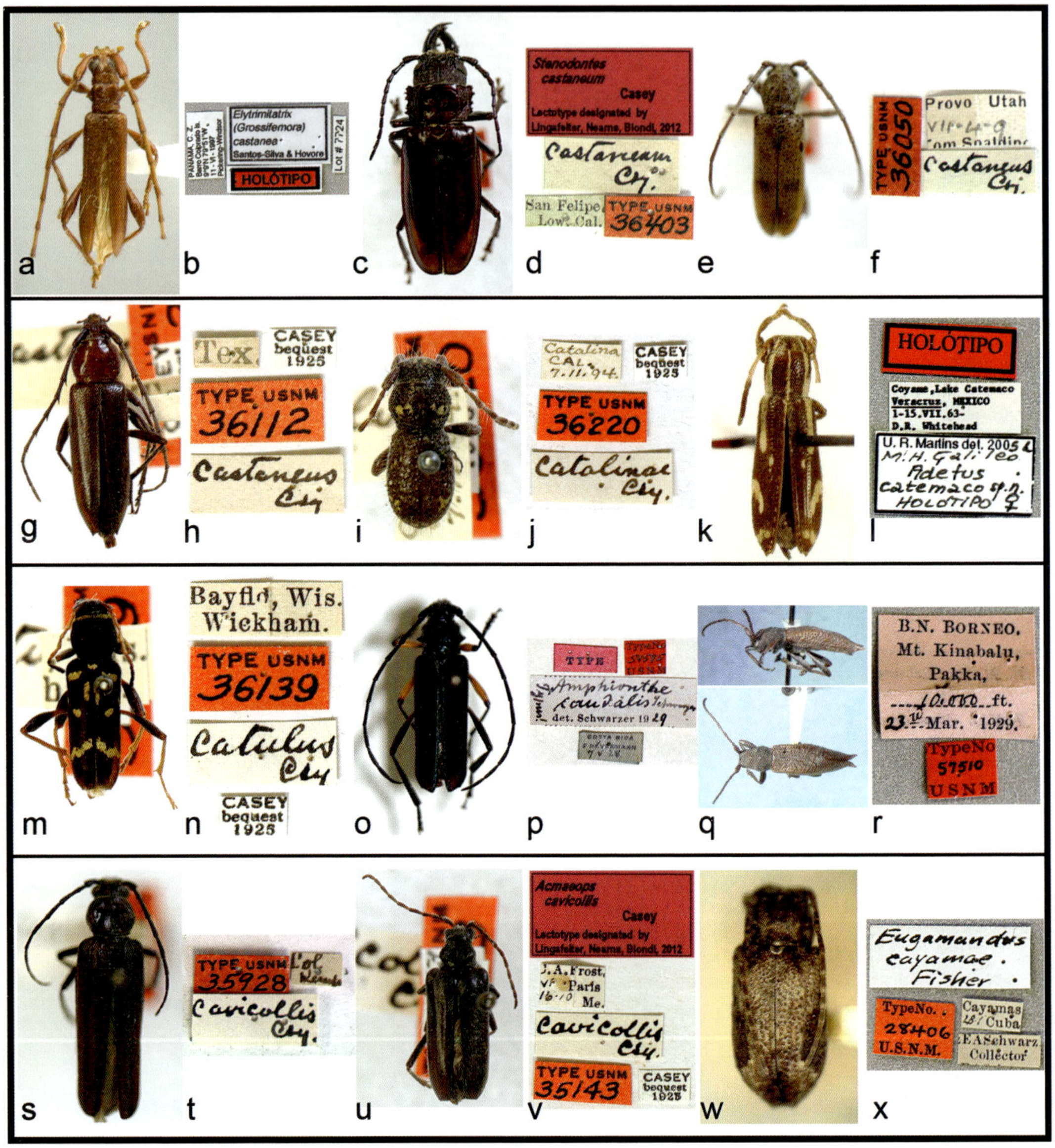

Figure 39. Primary types of Cerambycidae of the Smithsonian Institution (for each pair of images, dorsal habitus left, labels right): **(a, b)** *Elytrimitatrix (Grossifemora) castanea* Santos-Silva & Hovore, 2008, Holotype; **(c, d)** *Mallodon castaneum* Casey, 1924, Lectotype; **(e, f)** *Brothylus castaneus* Casey, 1924, Holotype; **(g, h)** *Stenosphenus castaneus* Casey, 1924, Holotype; **(i, j)** *Ipochus catalinae* Casey, 1913, Holotype; **(k, l)** *Adetus catemaco* Martins & Galileo, 2005, Holotype; **(m, n)** *Clytanthus catulus* Casey, 1912, Holotype; **(o, p)** *Amphionthe caudalis* Schwarzer, 1929, Holotype; **(q, r)** *Anxylotoles caudatus* Fisher, 1935, Holotype; **(s, t)** *Criocephalus cavicollis* Casey, 1912, Holotype; **(u, v)** *Acmaeops cavicollis* Casey, 1913, Lectotype; **(w, x)** *Eugamandus cayamae* Fisher, 1926, Holotype.

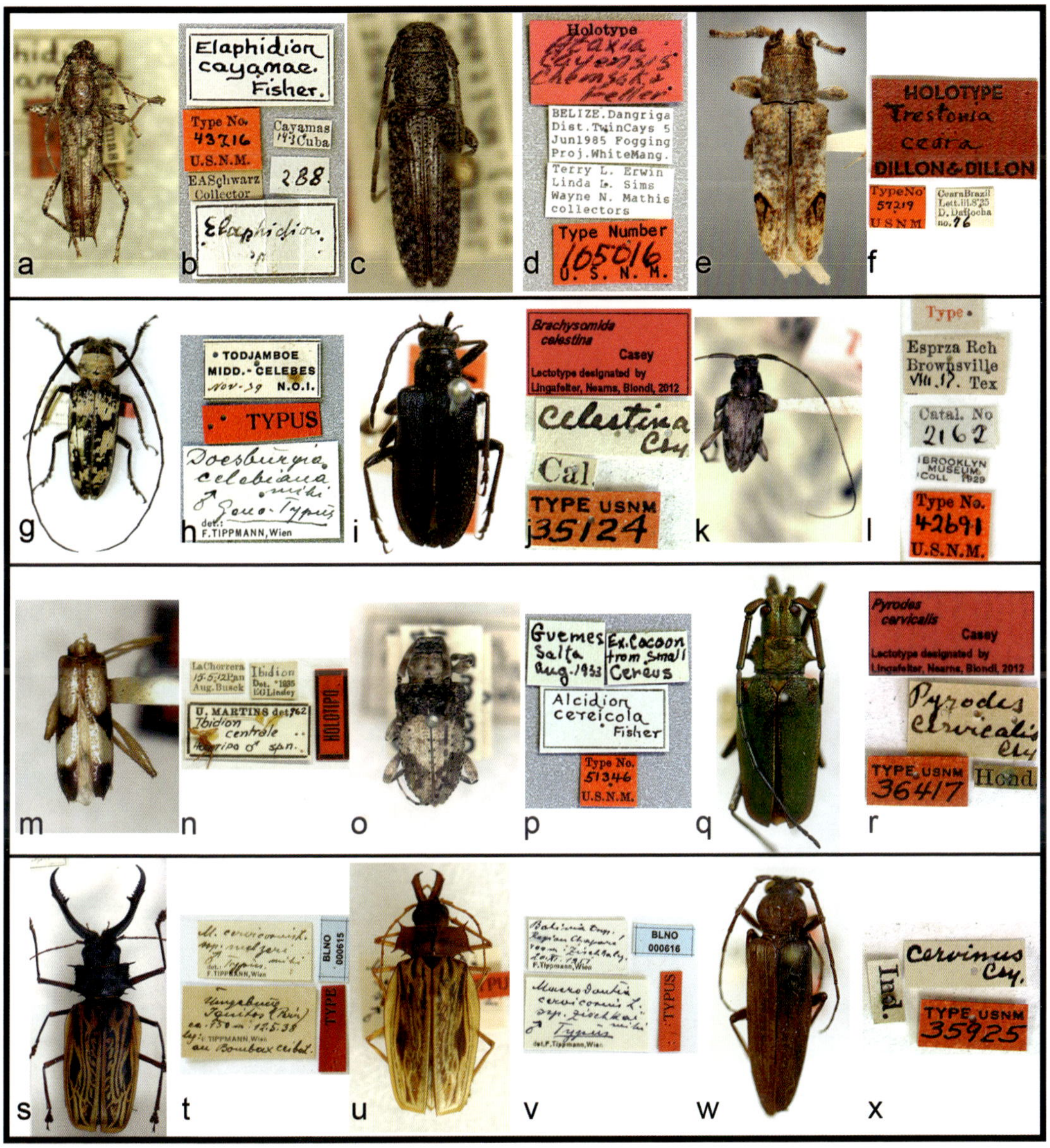

Figure 40. Primary types of Cerambycidae of the Smithsonian Institution (for each pair of images, dorsal habitus left, labels right): **(a, b)** *Elaphidion cayamae* Fisher, 1932, Holotype; **(c, d)** *Ataxia cayensis* Chemsak & Feller, 1988, Holotype; **(e, f)** *Trestonia ceara* Dillon & Dillon, 1946, Holotype; **(g, h)** *Doesburgia celebiana* Tippmann, 1953, Holotype; **(i, j)** *Brachysomida celestina* Casey, 1913, Lectotype; **(k, l)** *Lepturges celtis* Schaeffer, 1905, Holotype; **(m, n)** *Ibidion centrale* Martins, 1962, Holotype; **(o, p)** *Alcidion cereicola* Fisher, 1936, Holotype; **(q, r)** *Pyrodes cervicalis* Casey, 1912, Lectotype; **(s, t)** *Macrodontia cervicornis melzeri* Tippmann, 1953, Holotype; **(u, v)** *Macrodontia cervicornis zischkai* Tippmann, 1960, Holotype; **(w, x)** *Criocephalus cervinus* Casey, 1912, Holotype.

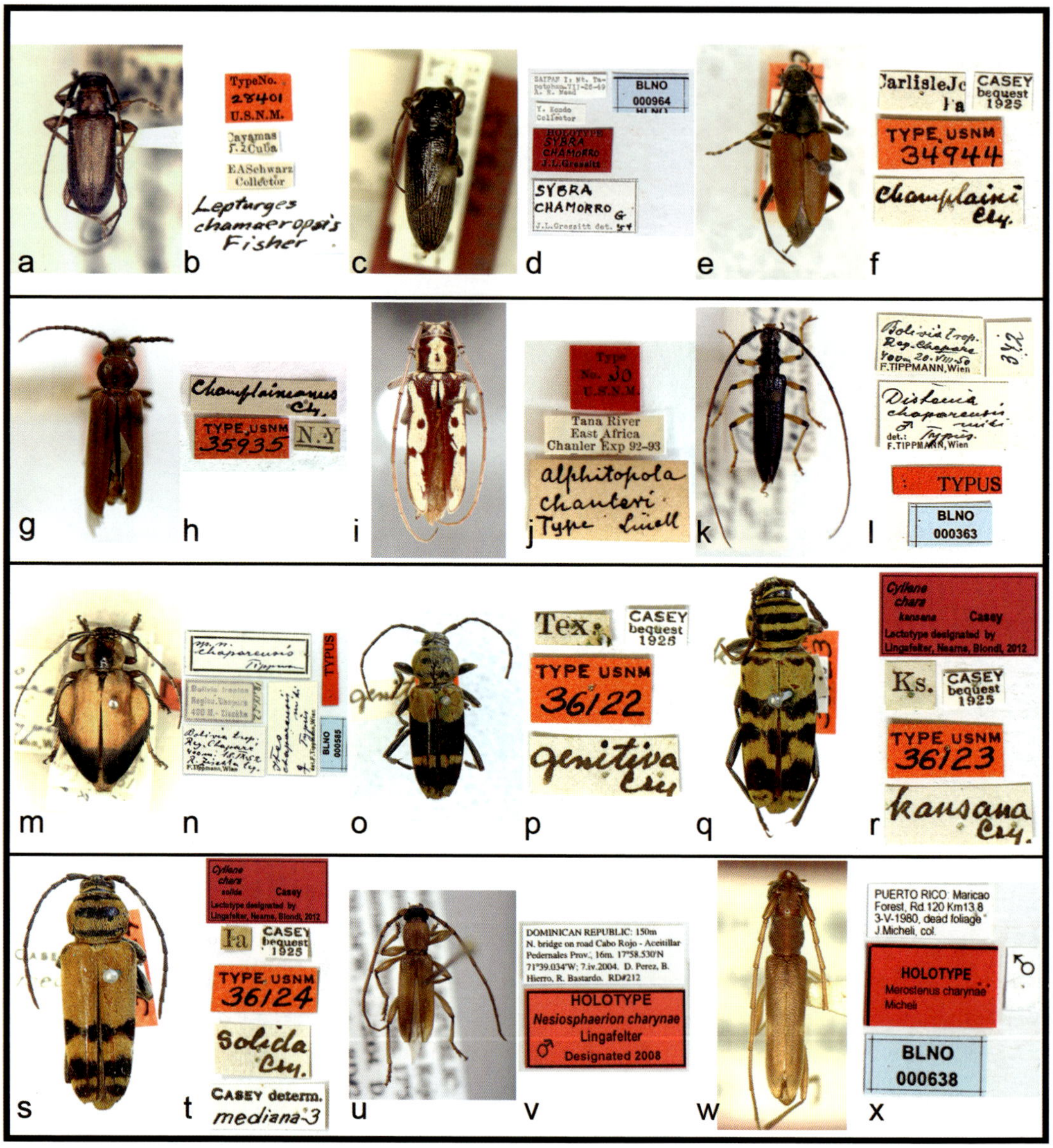

Figure 41. Primary types of Cerambycidae of the Smithsonian Institution (for each pair of images, dorsal habitus left, labels right): **(a, b)** *Lepturges chamaeropsis* Fisher, 1926, Holotype; **(c, d)** *Sybra chamorro* Gressitt, 1956, Holotype; **(e, f)** *Brachyleptura champlaini* Casey, 1913, Holotype; **(g, h)** *Criocephalus champlainianus* Casey, 1924, Holotype; **(i, j)** *Alphitopola chanleri* Linell, 1896, Holotype; **(k, l)** *Distenia chaparensis* Tippmann, 1953, Holotype; **(m, n)** *Ites chaparensis* Tippmann, 1960, Holotype; **(o, p)** *Cyllene chara genitiva* Casey, 1912, Holotype; **(q, r)** *Cyllene chara kansana* Casey, 1912, Lectotype; **(s, t)** *Cyllene chara solida* Casey, 1912, Lectotype; **(u, v)** *Nesiosphaerion charynae* Lingafelter, 2008, Holotype; **(w, x)** *Merostenus charynae* Micheli, 2003, Holotype.

Figure 42. Primary types of Cerambycidae of the Smithsonian Institution (for each pair of images, dorsal habitus left, labels right): **(a, b)** *Curius chemsaki* Nearns & Ray, 2006, Holotype; **(c, d)** *Spalacopsis chemsaki* Tyson, 1973, Holotype; **(e, f)** *Canidia chemsaki* Wappes & Lingafelter, 2005, Holotype; **(g, h)** *Esthlogena chicacaoensis* Galileo & Martins, 2011, Holotype; **(i, j)** *Sophronica chinensis* Breuning, 1940, Holotype; **(k, l)** *Parechthistatus chinensis* Breuning, 1942, Holotype; **(m, n)** *Leptostylopsis chlorescens* Lingafelter & Micheli, 2009, Holotype; **(o, p)** *Pothyne chocolata* Gressitt, 1939, Holotype; **(q, r)** *Erana ciliata* Fisher, 1938, Holotype; **(s, t)** *Exocentrus ciliatissimus* Gressitt, 1956, Holotype; **(u, v)** *Stenocoptoides ciliatus* Breuning, 1942, Holotype; **(w, x)** *Unaporanga cincta* Martins & Galileo, 2007, Holotype.

Figure 43. Primary types of Cerambycidae of the Smithsonian Institution (for each pair of images, dorsal habitus left, labels right): **(a, b)** *Chion cinctus* var. *sonorensis* Schaeffer, 1908, Lectotype; **(c, d)** *Mecas cineracea* Casey, 1913, Lectotype; **(e, f)** *Oncideres cingulatus pallescens* Casey, 1913, Holotype; **(g, h)** *Pharsalia clara* Breuning, 1940, Holotype; **(i, j)** *Urgleptes clarkei* Chemsak, 1966, Holotype; **(k, l)** *Ceresium clarkei* Gressitt, 1956, Holotype; **(m, n)** *Elytrimitatrix (Grossifemora) clavata* Santos-Silva & Hovore, 2008, Holotype; **(o, p)** *Ectenesseca clavula* Martins & Galileo, 2005, Holotype; **(q, r)** *Oxoplus coccineus* Casey, 1893, Lectotype; **(s, t)** *Tetraopes coccineus* Casey, 1913, Lectotype; **(u, v)** *Trachyderes cochabambae* Tippmann, 1960, Holotype; **(w, x)** *Aneflus cochisensis* Casey, 1912, Holotype.

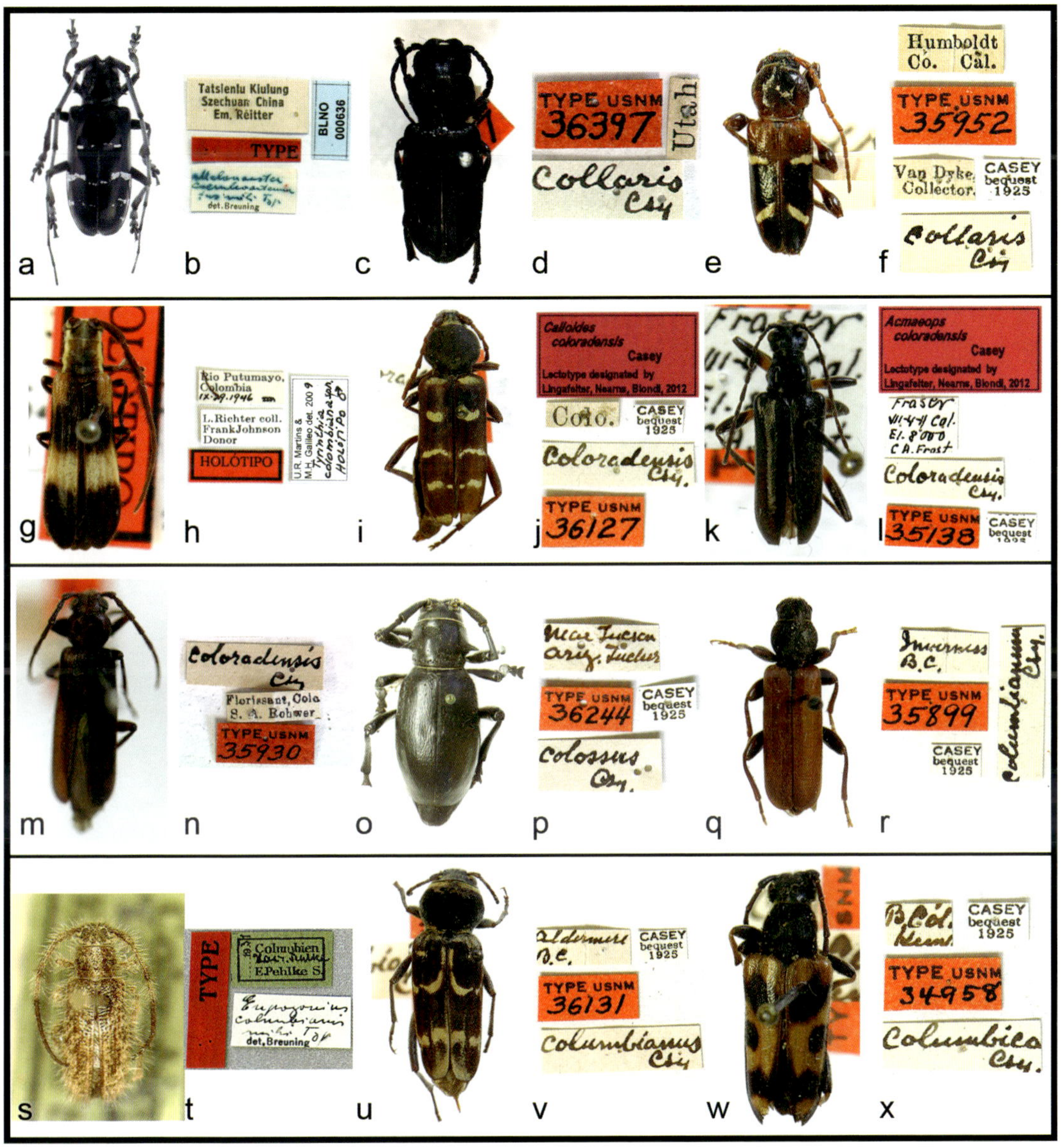

Figure 44. Primary types of Cerambycidae of the Smithsonian Institution (for each pair of images, dorsal habitus left, labels right): **(a, b)** *Melanauster coeruleoantennatus* Breuning, 1946, Holotype; **(c, d)** *Spondylis collaris* Casey, 1912, Holotype; **(e, f)** *Phymatodes (Phymatodina) collaris* Casey, 1924, Holotype; **(g, h)** *Tyrinthia colombiana* Galileo & Martins, 2009, Holotype; **(i, j)** *Calloides coloradensis* Casey, 1912, Lectotype; **(k, l)** *Acmaeops coloradensis* Casey, 1913, Lectotype; **(m, n)** *Criocephalus coloradensis* Casey, 1924, Holotype; **(o, p)** *Moneilema (Collapteryx) colossus* Casey, 1924, Holotype; **(q, r)** *Tetropium columbianum* Casey, 1912, Holotype; **(s, t)** *Eupogonius columbianus* Breuning, 1942, Holotype; **(u, v)** *Xylotrechus columbianus* Casey, 1912, Holotype; **(w, x)** *Xestoleptura columbica* Casey, 1913, Holotype.

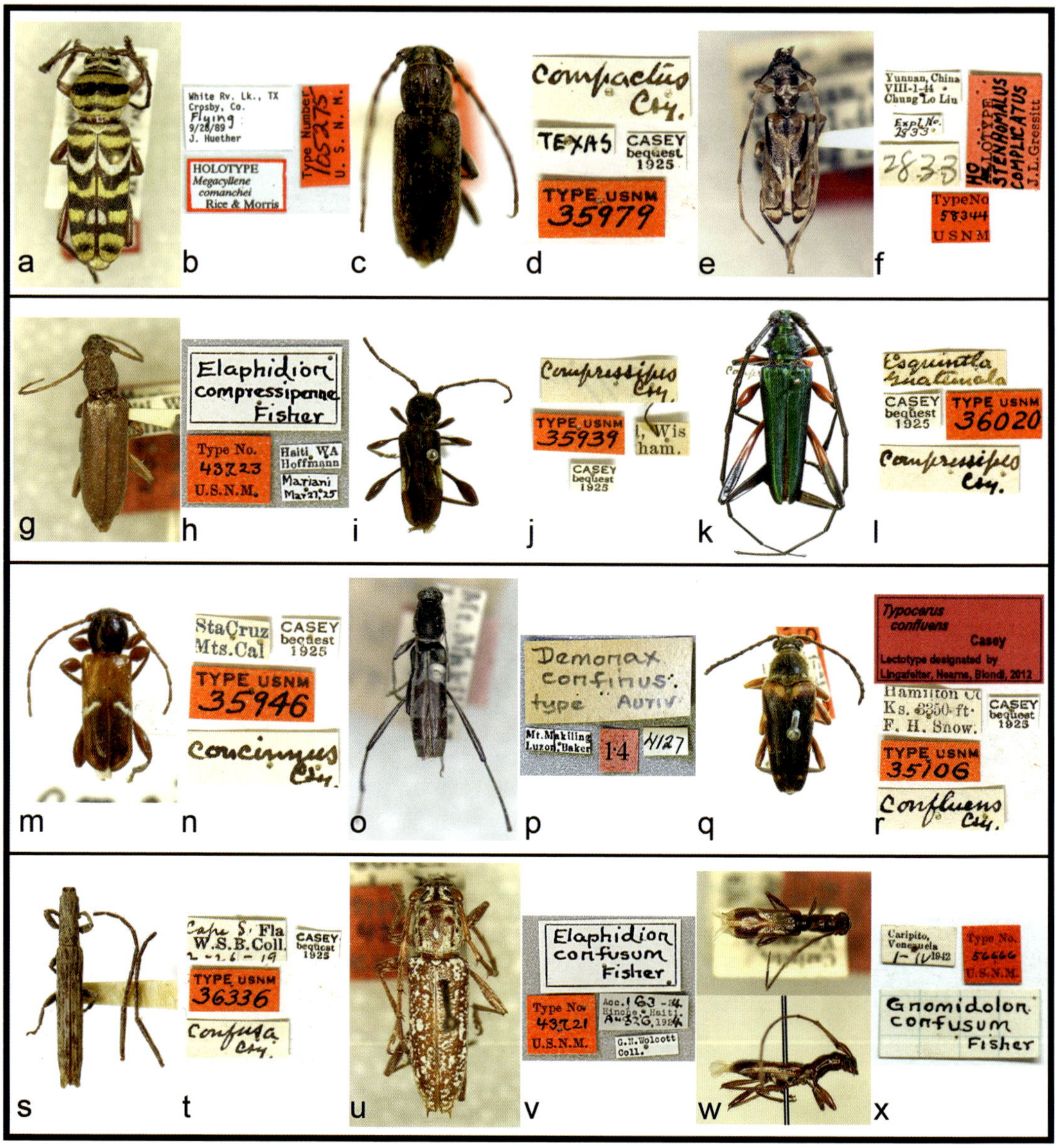

Figure 45. Primary types of Cerambycidae of the Smithsonian Institution (for each pair of images, dorsal habitus left, labels right): **(a, b)** *Megacyllene comanchei* Rice & Morris, 1992, Holotype; **(c, d)** *Hypermallus compactus* Casey, 1914, Holotype; **(e, f)** *Stenhomalus complicatus* Gressitt, 1948, Holotype; **(g, h)** *Elaphidion compressipenne* Fisher, 1932, Holotype; **(i, j)** *Physocnemum compressipes* Casey, 1912, Holotype; **(k, l)** *Callichroma compressipes* Casey, 1912, Holotype; **(m, n)** *Phymatodes (Poecilium) concinnus* Casey, 1912, Holotype; **(o, p)** *Demonax confinus* Aurivillius, 1924, Holotype; **(q, r)** *Typocerus confluens* Casey, 1913, Lectotype; **(s, t)** *Spalacopsis confusa* Casey, 1924, Holotype; **(u, v)** *Elaphidion confusum* Fisher, 1932, Holotype; **(w, x)** *Gnomidolon confusum* Fisher, 1944, Holotype.

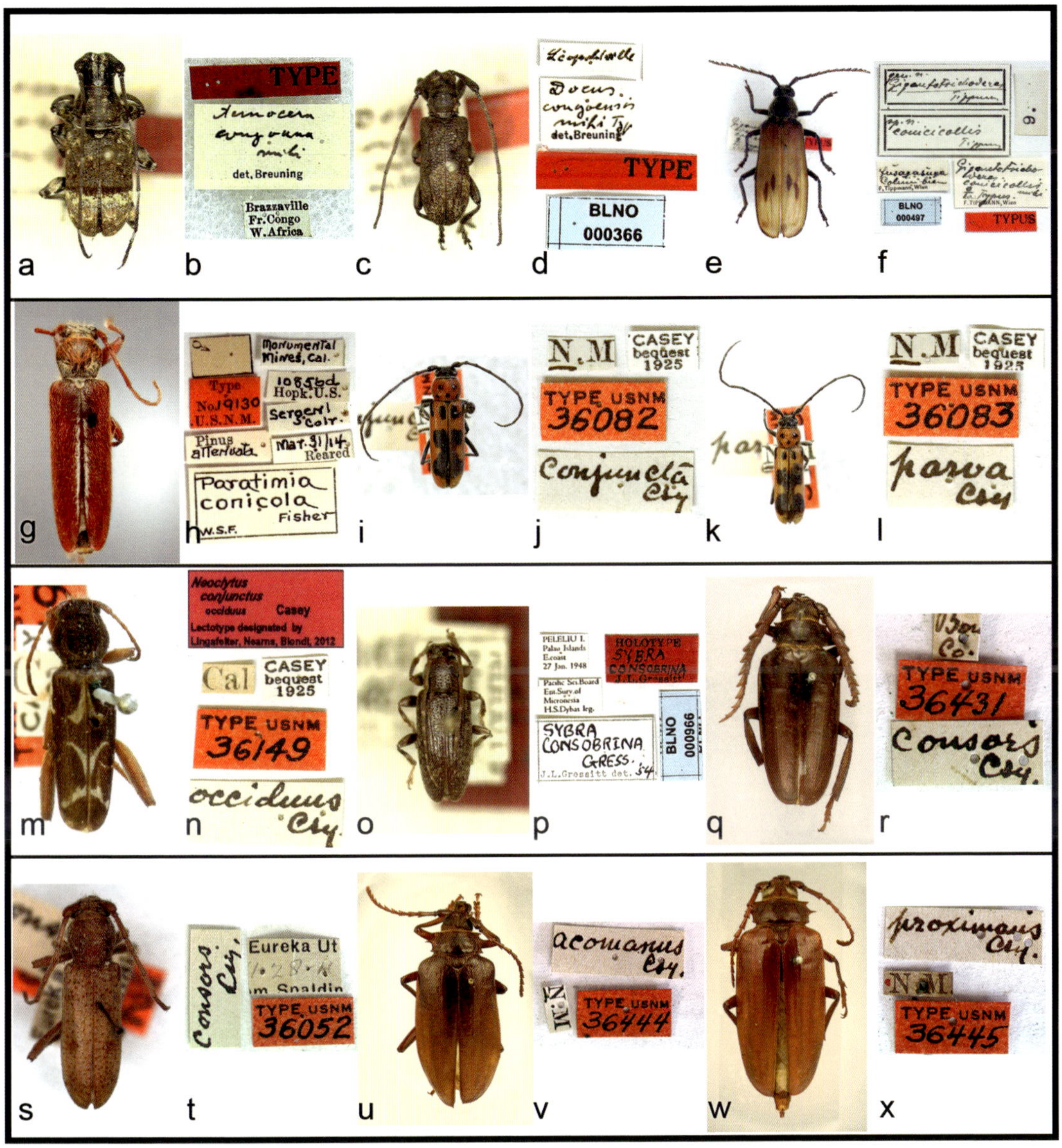

Figure 46. Primary types of Cerambycidae of the Smithsonian Institution (for each pair of images, dorsal habitus left, labels right): **(a, b)** *Acmocera congoana* Breuning, 1935, Holotype; **(c, d)** *Docus congoensis* Breuning, 1942, Holotype; **(e, f)** *Gigantotrichoderes conicicollis* Tippmann, 1953, Holotype; **(g, h)** *Paratimia conicola* Fisher, 1915, Holotype; **(i, j)** *Tylosis conjuncta* Casey, 1912, Holotype; **(k, l)** *Tylosis conjuncta parva* Casey, 1912, Holotype; **(m, n)** *Neoclytus conjunctus occiduus* Casey, 1912, Lectotype; **(o, p)** *Sybra consobrina* Gressitt, 1956, Holotype; **(q, r)** *Prionus (Prionus) consors* Casey, 1912, Holotype; **(s, t)** *Brothylus consors* Casey, 1924, Holotype; **(u, v)** *Prionus (Prionus) consors acomanus* Casey, 1912, Holotype; **(w, x)** *Prionus (Prionus) consors proximans* Casey, 1912, Holotype.

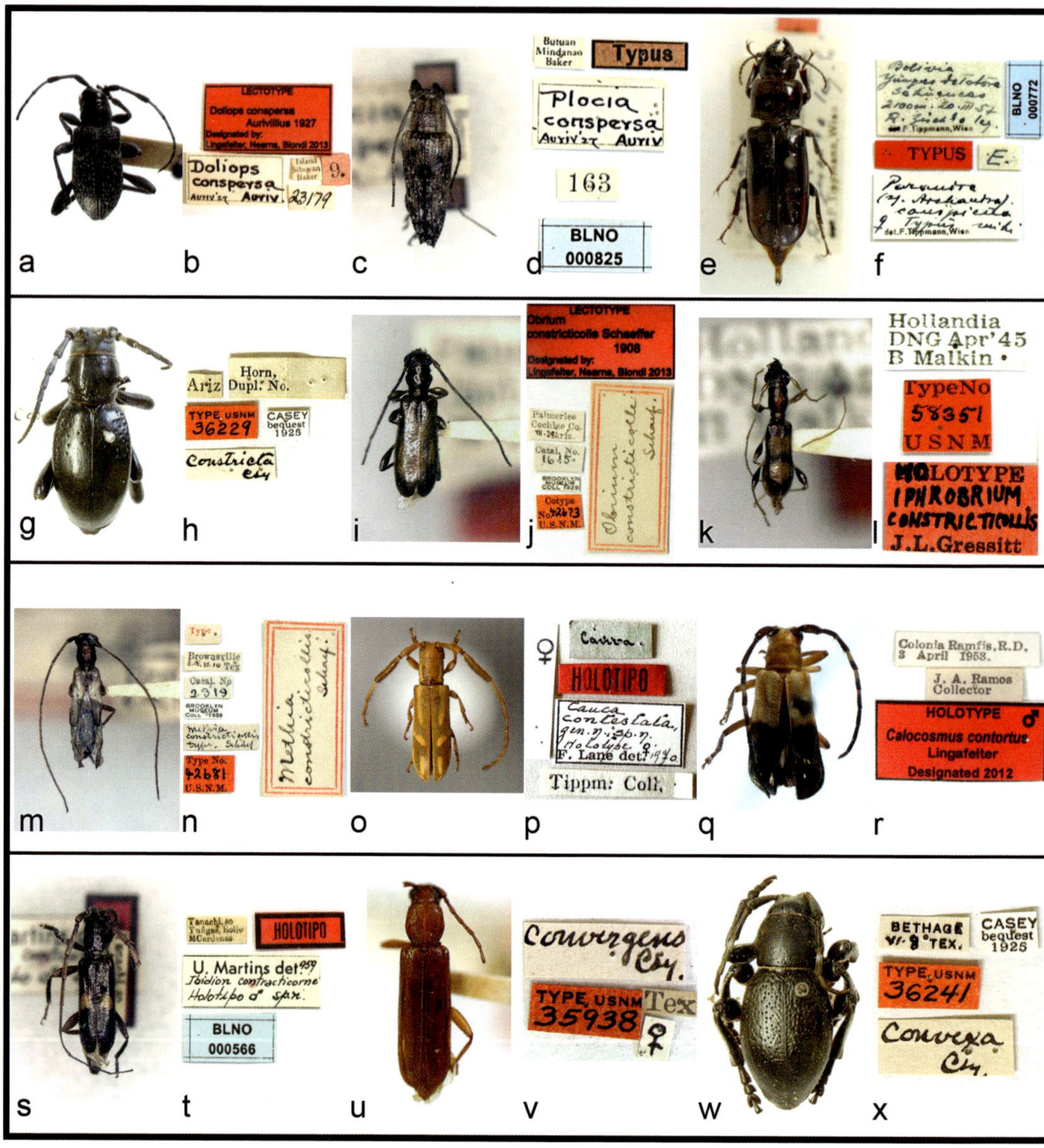

Figure 47. Primary types of Cerambycidae of the Smithsonian Institution (for each pair of images, dorsal habitus left, labels right): **(a, b)** *Doliops conspersa* Aurivillius, 1927, Lectotype; **(c, d)** *Plocia conspersa* Aurivillius, 1927, Syntype; **(e, f)** *Parandra (Archandra) conspicua* Tippmann, 1960, Holotype; **(g, h)** *Moneilema (Collapteryx) constricta* Casey, 1913, Holotype; **(i, j)** *Obrium constricticolle* Schaeffer, 1908, Lectotype; **(k, l)** *Iphrobrium constricticollis* Gressitt, 1951, Holotype; **(m, n)** *Methia constricticollis* Schaeffer, 1908, Holotype; **(o, p)** *Cauca contestata* Lane, 1970, Holotype; **(q, r)** *Calocosmus contortus* Lingafelter, 2013, Holotype; **(s, t)** *Ibidion contracticorne* Martins, 1960, Holotype; **(u, v)** *Smodicum convergens* Casey, 1912, Holotype; **(w, x)** *Moneilema (Collapteryx) convexa* Casey, 1924, Holotype.

Figure 48. Primary types of Cerambycidae of the Smithsonian Institution (for each pair of images, dorsal habitus left, labels right): **(a, b)** *Sybra convexa* Gressitt, 1956, Holotype; **(c, d)** *Judolia convolvens* Casey, 1913, Lectotype; **(e, f)** *Leptura coquilletti* Linell, 1897, Holotype; **(g, h)** *Tragidion coquus filicorne* Casey, 1912, Lectotype; **(i, j)** *Crossidius coralinus obfuscipennis* Chemsak & Linsley, 1959, Holotype; **(k, l)** *Demonax coriacecollis* Aurivillius, 1922, Holotype; **(m, n)** *Lochmaeocles cornuticeps federalis* Dillon & Dillon, 1946, Holotype; **(o, p)** *Lochmaeocles cornuticeps pacificus* Dillon & Dillon, 1946, Holotype; **(q, r)** *Brachysomida corpulenta* Casey, 1913, Holotype; **(s, t)** *Moneilema (Moneilema) corrugans* Casey, 1913, Holotype; **(u, v)** *Moneilema (Moneilema) corrugans ovipennis* Casey, 1913, Holotype; **(w, x)** *Acanthoderes (Psapharochrus) corticaria* Tippmann, 1960, Holotype.

Figure 49. Primary types of Cerambycidae of the Smithsonian Institution (for each pair of images, dorsal habitus left, labels right): **(a, b)** *Xestoleptura corusca* Casey, 1913, Lectotype; **(c, d)** *Gnomidolon costaricense* Martins, 1962, Holotype; **(e, f)** *Amphicnaeia costaricensis* Breuning, 1942, Holotype; **(g, h)** *Aletretia costaricensis* Melzer, 1934, Holotype; **(i, j)** *Tmesisternus costipennis* Breuning, 1940, Holotype; **(k, l)** *Neosarmydus costipennis* Fisher, 1935, Holotype; **(m, n)** *Astathes costipennis* Fisher, 1935, Holotype; **(o, p)** *Spalacopsis (Euthuorus) costulata* Casey, 1913, Lectotype; **(q, r)** *Asemum costulatum* Casey, 1912, Holotype; **(s, t)** *Crossidius crassipes* Casey, 1891, Holotype; **(u, v)** *Rhagium (Hargium) crassipes* Casey, 1913, Holotype; **(w, x)** *Moneilema (Collapteryx) crassipes* Fisher, 1931, Holotype.

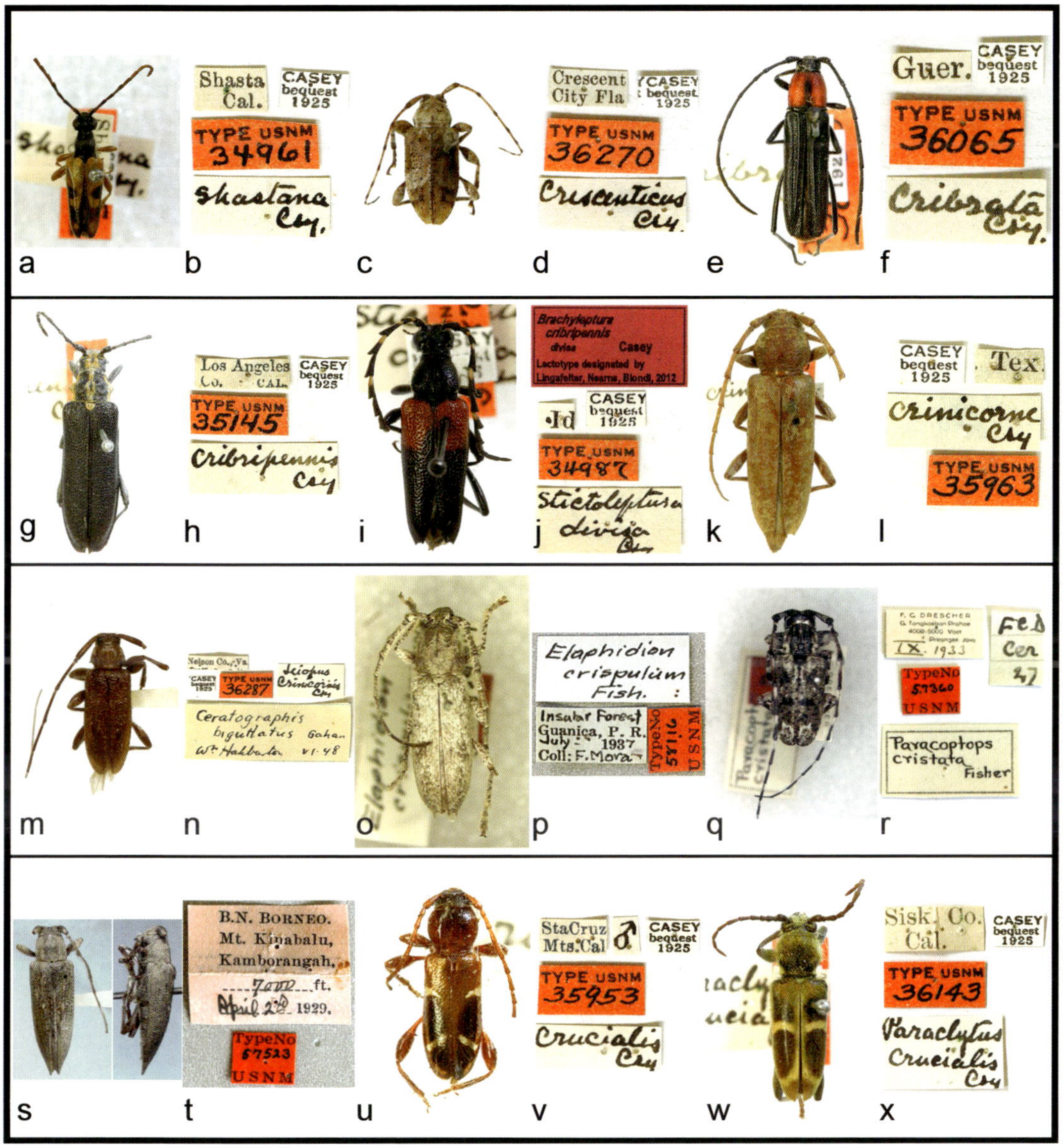

Figure 50. Primary types of Cerambycidae of the Smithsonian Institution (for each pair of images, dorsal habitus left, labels right): **(a, b)** *Xestoleptura crassipes shastana* Casey, 1913, Holotype; **(c, d)** *Leptostylus crescenticus* Casey, 1913, Holotype; **(e, f)** *Stenobatyle cribrata* Casey, 1912, Holotype; **(g, h)** *Pyrotrichus cribripennis* Casey, 1913, Holotype; **(i, j)** *Brachyleptura (Stictoleptura) cribripennis divisa* Casey, 1924, Lectotype; **(k, l)** *Romaleum crinicorne* Casey, 1912, Holotype; **(m, n)** *Leiopus crinicornis* Casey, 1924, Holotype; **(o, p)** *Elaphidion crispulum* Fisher, 1947, Holotype; **(q, r)** *Paracoptops cristata* Fisher, 1936, Holotype; **(s, t)** *Ancornallis cristatus* Fisher, 1935, Holotype; **(u, v)** *Phymatodes (Phymatodina) crucialis* Casey, 1912, Holotype; **(w, x)** *Paraclytus crucialis* Casey, 1912, Holotype.

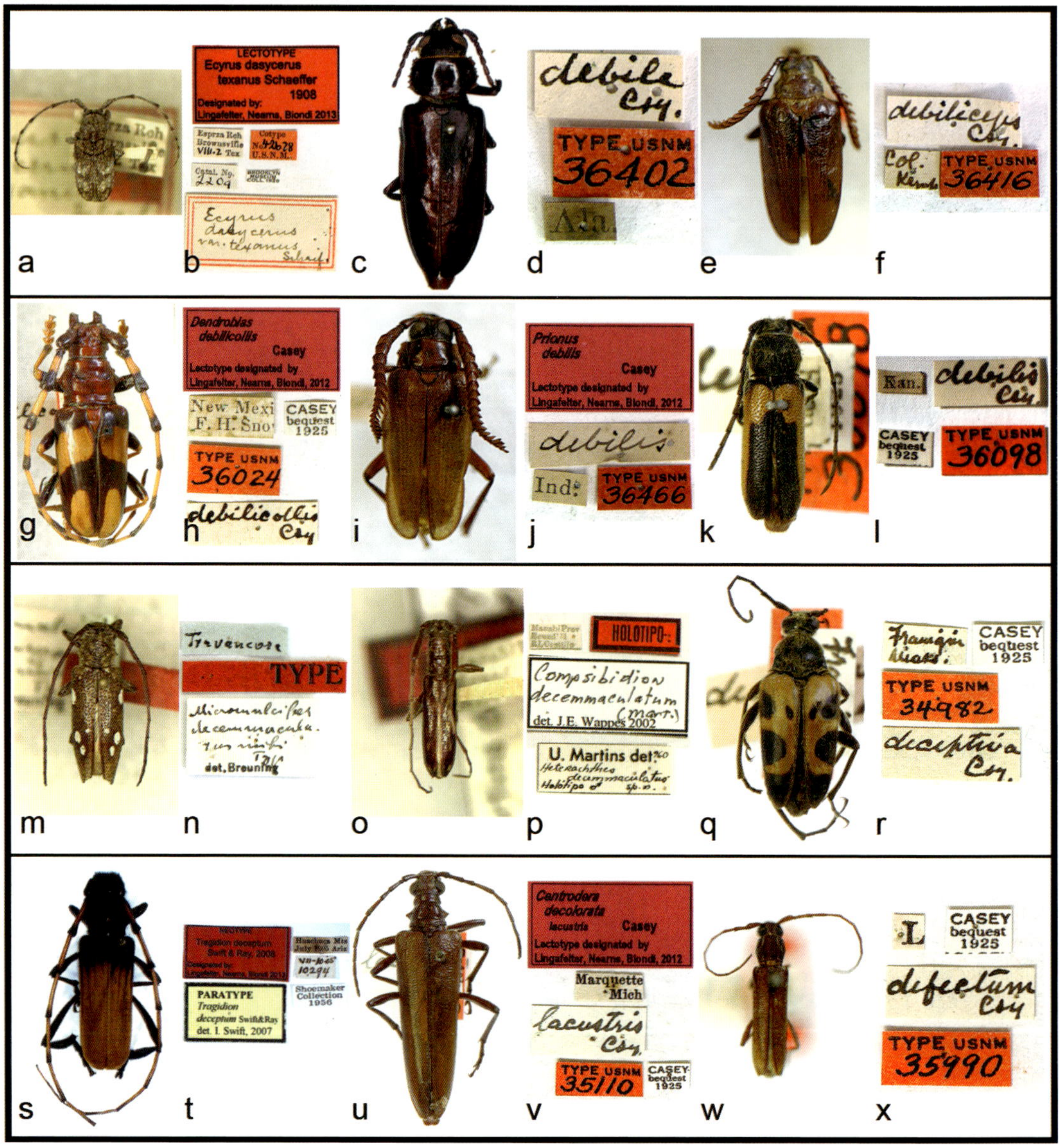

Figure 53. Primary types of Cerambycidae of the Smithsonian Institution (for each pair of images, dorsal habitus left, labels right): **(a, b)** *Ecyrus dasycerus* var. *texanus* Schaeffer, 1908, Lectotype; **(c, d)** *Mallodon debile* Casey, 1912, Holotype; **(e, f)** *Homaesthesis debiliceps* Casey, 1912, Holotype; **(g, h)** *Dendrobias debilicollis* Casey, 1912, Lectotype; **(i, j)** *Prionus debilis* Casey, 1891, Lectotype; **(k, l)** *Crossidius debilis* Casey, 1912, Holotype; **(m, n)** *Notomulciber decemmaculatus* Breuning, 1942, Holotype; **(o, p)** *Heterachthes decemmaculatus* Martins, 1960, Holotype; **(q, r)** *Judolia deceptiva* Casey, 1924, Holotype; **(s, t)** *Tragidion deceptum* Swift & Ray, 2008, Neotype; **(u, v)** *Centrodera decolorata lacustris* Casey, 1918, Lectotype; **(w, x)** *Hypermallus defectus* Casey, 1912, Holotype.

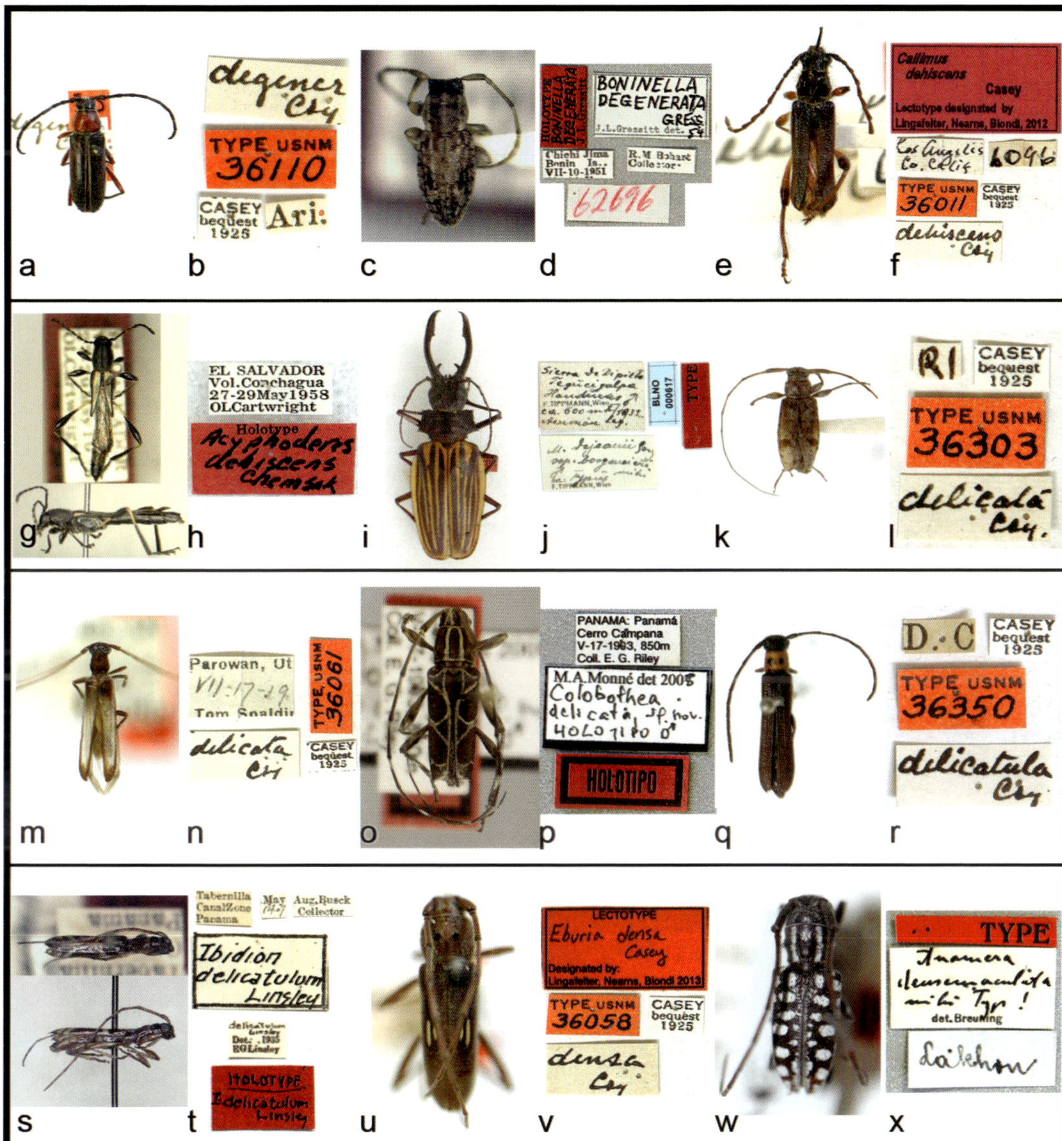

Figure 54. Primary types of Cerambycidae of the Smithsonian Institution (for each pair of images, dorsal habitus left, labels right): **(a, b)** *Plionoma degener* Casey, 1912, Holotype; **(c, d)** *Boninella degenerata* Gressitt, 1956, Holotype; **(e, f)** *Callimus dehiscens* Casey, 1912, Lectotype; **(g, h)** *Acyphoderes dehiscens* Chemsak, 1997, Holotype; **(i, j)** *Macrodontia dejeani borgmeieri* Tippmann, 1953, Holotype; **(k, l)** *Hyperplatys delicata* Casey, 1913, Holotype; **(m, n)** *Methia delicata* Casey, 1924, Holotype; **(o, p)** *Colobothea delicata* Monné, 2008, Holotype; **(q, r)** *Oberea delicatula* Casey, 1913, Holotype; **(s, t)** *Ibidion delicatulum* Linsley, 1935, Holotype; **(u, v)** *Eburia densa* Casey, 1924, Lectotype; **(w, x)** *Anamera densemaculata* Breuning, 1940, Holotype.

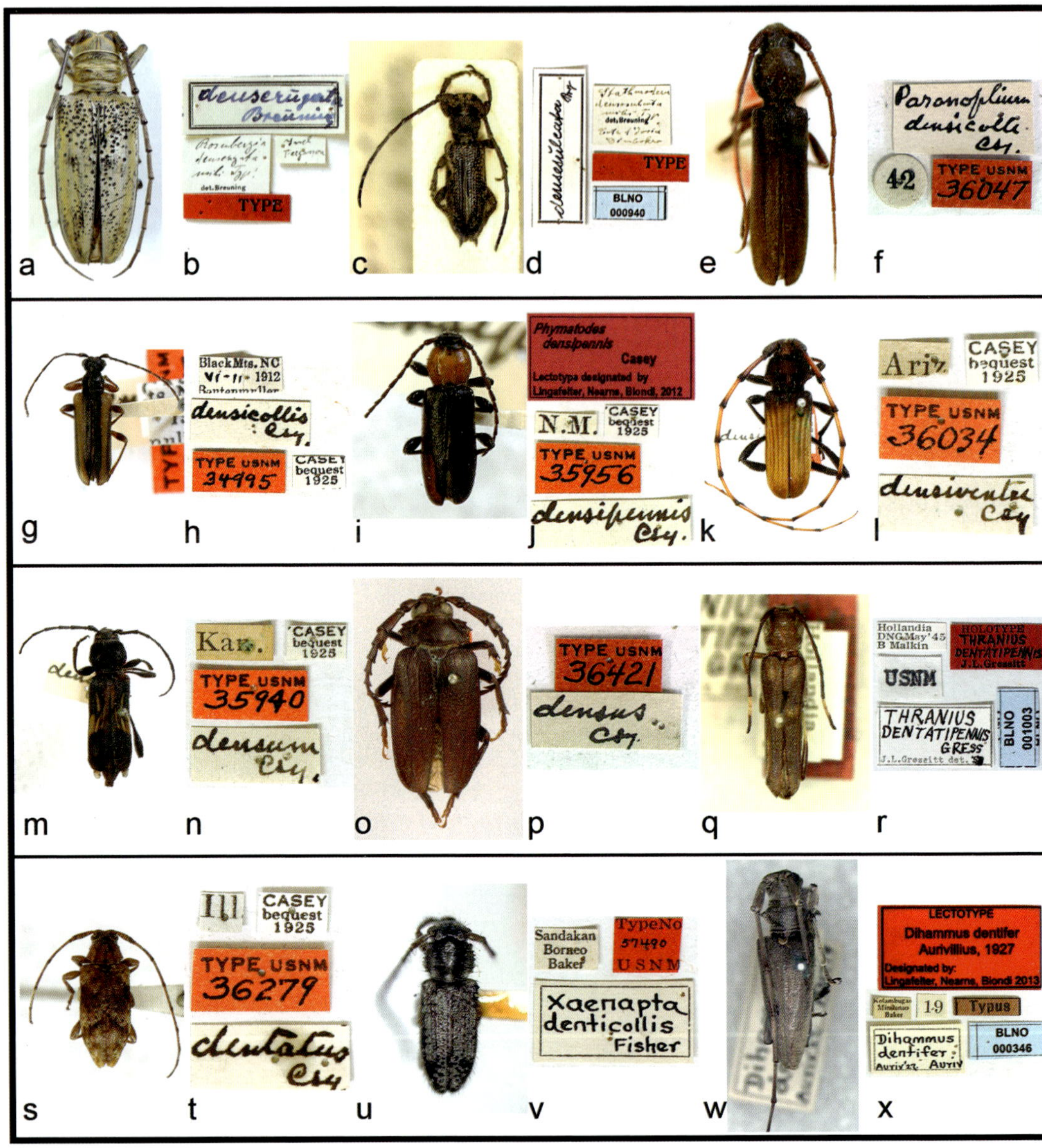

Figure 55. Primary types of Cerambycidae of the Smithsonian Institution (for each pair of images, dorsal habitus left, labels right): **(a, b)** *Rosenbergia denserugata* Breuning, 1936, Holotype; **(c, d)** *Stathmodera densesulcata* Breuning, 1940, Holotype; **(e, f)** *Paranoplium densicolle* Casey, 1924, Holotype; **(g, h)** *Hapalosalia densicollis* Casey, 1914, Holotype; **(i, j)** *Phymatodes (Microcallidium) densipennis* Casey, 1912, Lectotype; **(k, l)** *Tragidion densiventre* Casey, 1912, Holotype; **(m, n)** *Physocnemum densum* Casey, 1912, Holotype; **(o, p)** *Prionus (Prionellus) densus* Casey, 1924, Holotype; **(q, r)** *Thranius dentatipennis* Gressitt, 1954, Holotype; **(s, t)** *Leiopus dentatus* Casey, 1913, Holotype; **(u, v)** *Xaenapta denticollis* Fisher, 1925, Holotype; **(w, x)** *Dihammus dentifer* Aurivillius, 1927, Lectotype.

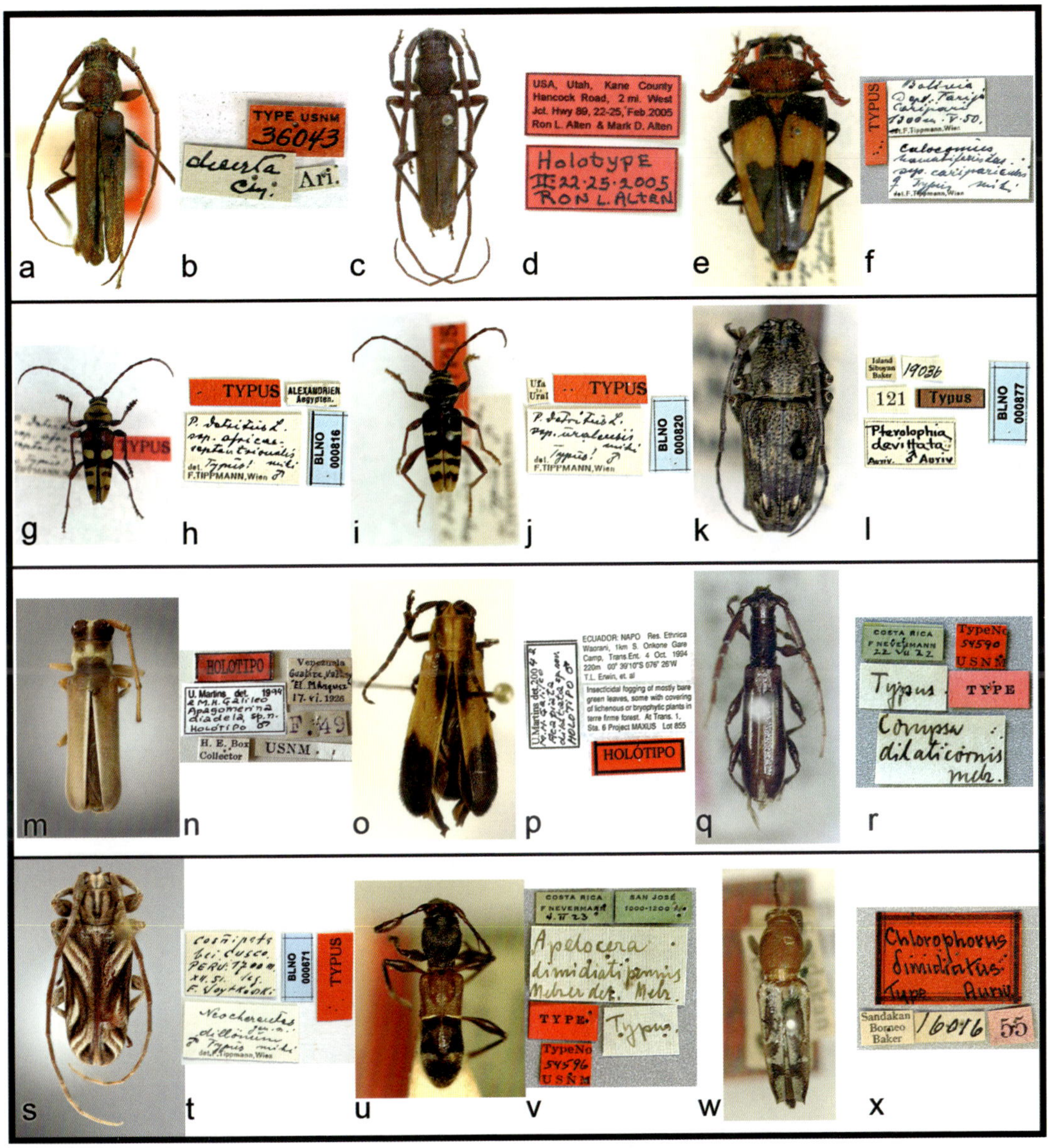

Figure 56. Primary types of Cerambycidae of the Smithsonian Institution (for each pair of images, dorsal habitus left, labels right): **(a, b)** *Oeme deserta* Casey, 1924, Holotype; **(c, d)** *Liberedaxia deslauriersi* Alten, Alten & Ramey, 2009, Holotype; **(e, f)** *Calocomus desmaresti caripariensis* Tippmann, 1960, Holotype; **(g, h)** *Plagionotus detritus africaeseptentrionalis* Tippmann, 1952, Holotype; **(i, j)** *Plagionotus detritus uralensis* Tippmann, 1952, Holotype; **(k, l)** *Pterolophia devittata* Aurivillius, 1927, Syntype; **(m, n)** *Apagomerina diadela* Martins & Galileo, 1996, Holotype; **(o, p)** *Acapiata dilatata* Galileo & Martins, 2005, Holotype; **(q, r)** *Compsa dilaticornis* Melzer, 1935, Holotype; **(s, t)** *Neocherentes dilloniorum* Tippmann, 1960, Holotype; **(u, v)** *Apelocera dimidiatipennis* Melzer, 1932, Holotype; **(w, x)** *Chlorophorus dimidiatus* Aurivillius, 1922, Holotype.

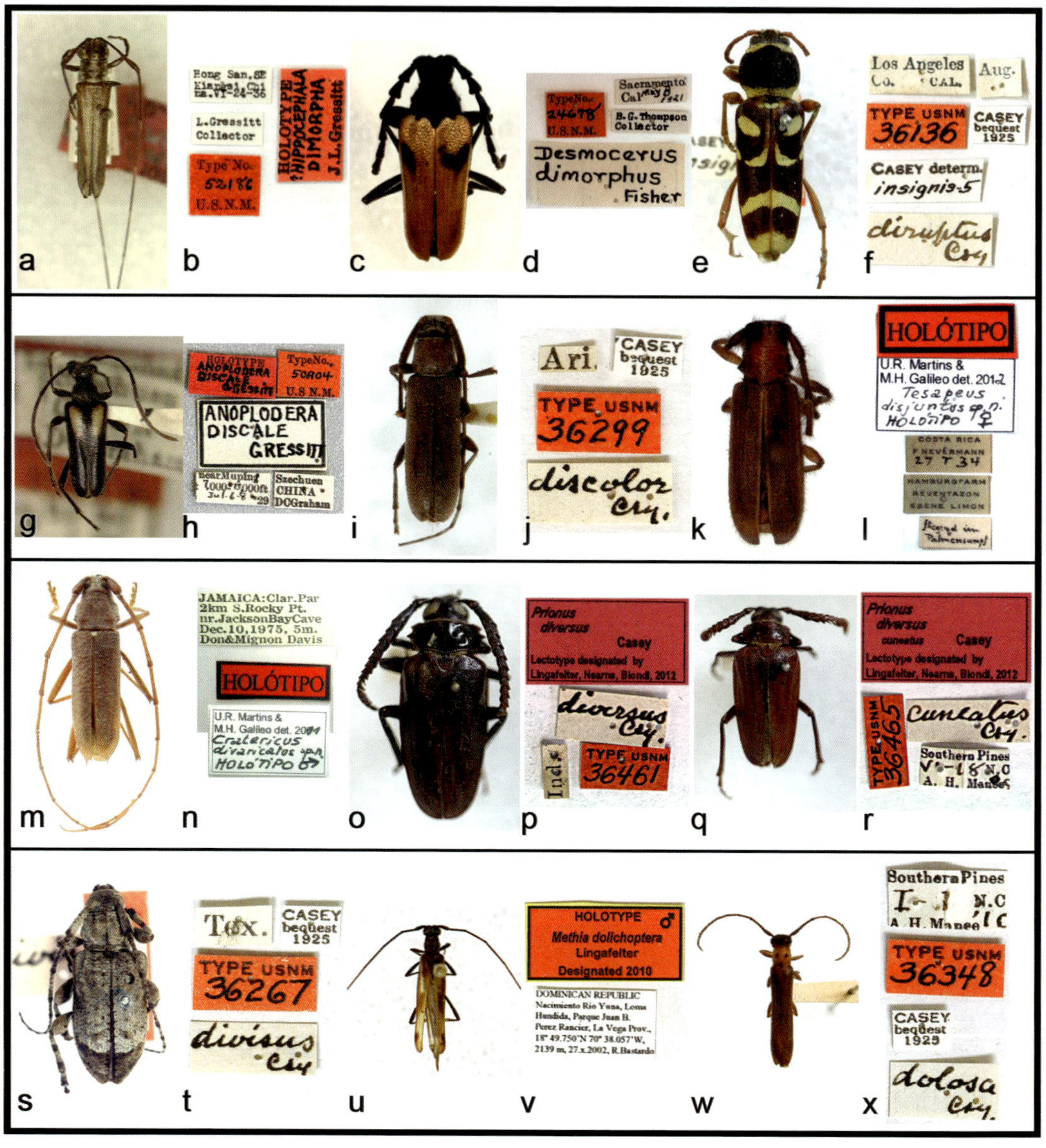

Figure 57. Primary types of Cerambycidae of the Smithsonian Institution (for each pair of images, dorsal habitus left, labels right): **(a, b)** *Hippocephala dimorpha* Gressitt, 1937, Holotype; **(c, d)** *Desmocerus dimorphus* Fisher, 1921, Holotype; **(e, f)** *Xylotrechus diruptus* Casey, 1912, Holotype; **(g, h)** *Anoplodermorpha discale* Gressitt, 1935, Holotype; **(i, j)** *Dectes discolor* Casey, 1913, Holotype; **(k, l)** *Tesapeus disjunctus* Martins & Galileo, 2012, Holotype; **(m, n)** *Cratericus divaricatus* Galileo & Martins, 2012, Holotype; **(o, p)** *Prionus (Riponus) diversus* Casey, 1912, Lectotype; **(q, r)** *Prionus (Riponus) diversus cuneatus* Casey, 1912, Lectotype; **(s, t)** *Leptostylus divisus* Casey, 1913, Holotype; **(u, v)** *Methia dolichoptera* Lingafelter, 2010, Holotype; **(w, x)** *Oberea dolosa* Casey, 1913, Holotype.

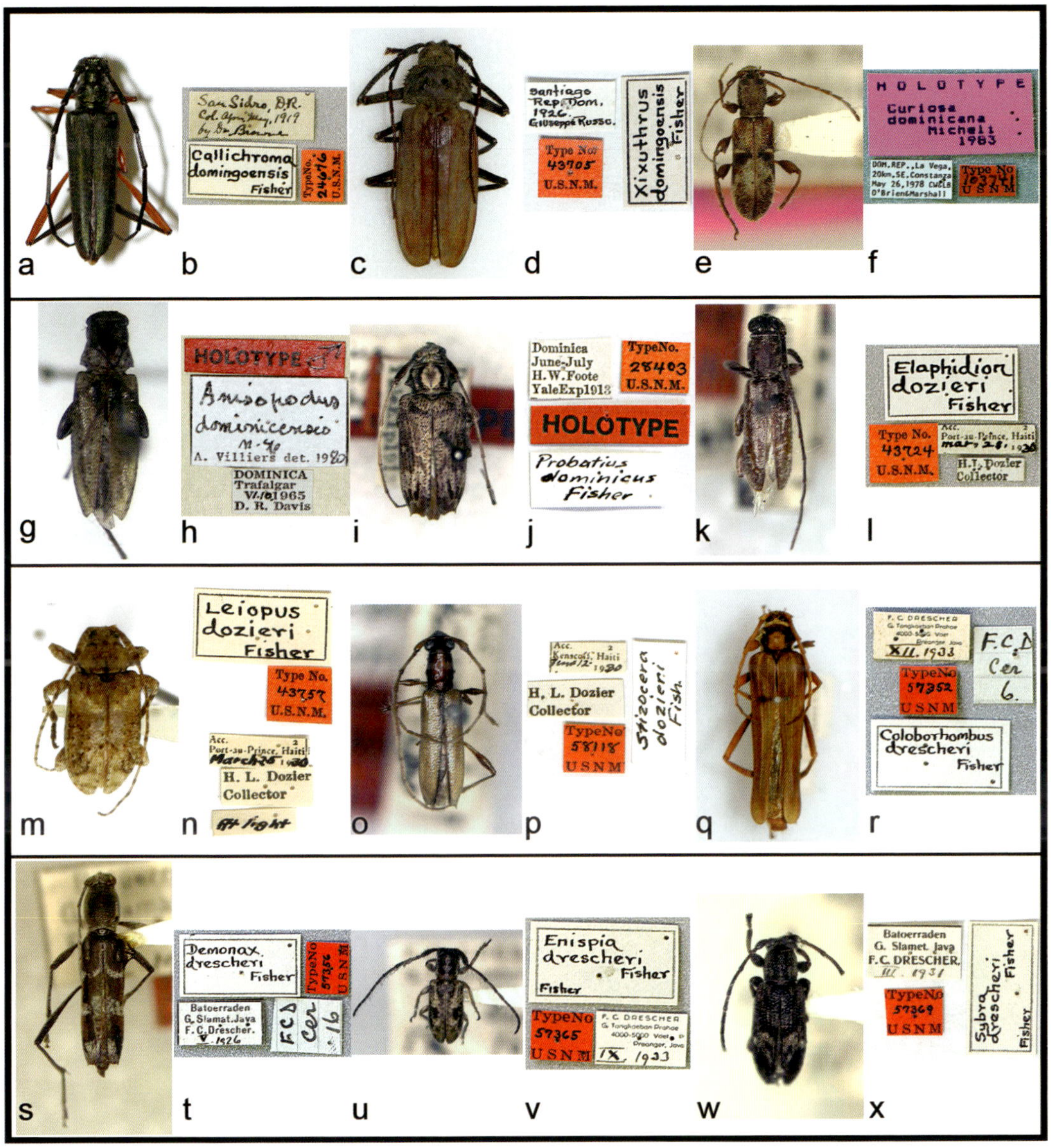

Figure 58. Primary types of Cerambycidae of the Smithsonian Institution (for each pair of images, dorsal habitus left, labels right): **(a, b)** *Callichroma domingoensis* Fisher, 1922, Holotype; **(c, d)** *Xixuthrus domingoensis* Fisher, 1932, Holotype; **(e, f)** *Curiosa dominicana* Micheli, 1983, Holotype; **(g, h)** *Anisopodus dominicensis* Villiers, 1980, Holotype; **(i, j)** *Probatius dominicus* Fisher, 1926, Holotype; **(k, l)** *Elaphidion dozieri* Fisher, 1932, Holotype; **(m, n)** *Leiopus dozieri* Fisher, 1932, Holotype; **(o, p)** *Stizocera dozieri* Fisher, 1947, Holotype; **(q, r)** *Coloborhombus drescheri* Fisher, 1936, Holotype; **(s, t)** *Demonax drescheri* Fisher, 1936, Holotype; **(u, v)** *Enispia drescheri* Fisher, 1936, Holotype; **(w, x)** *Sybra drescheri* Fisher, 1936, Holotype.

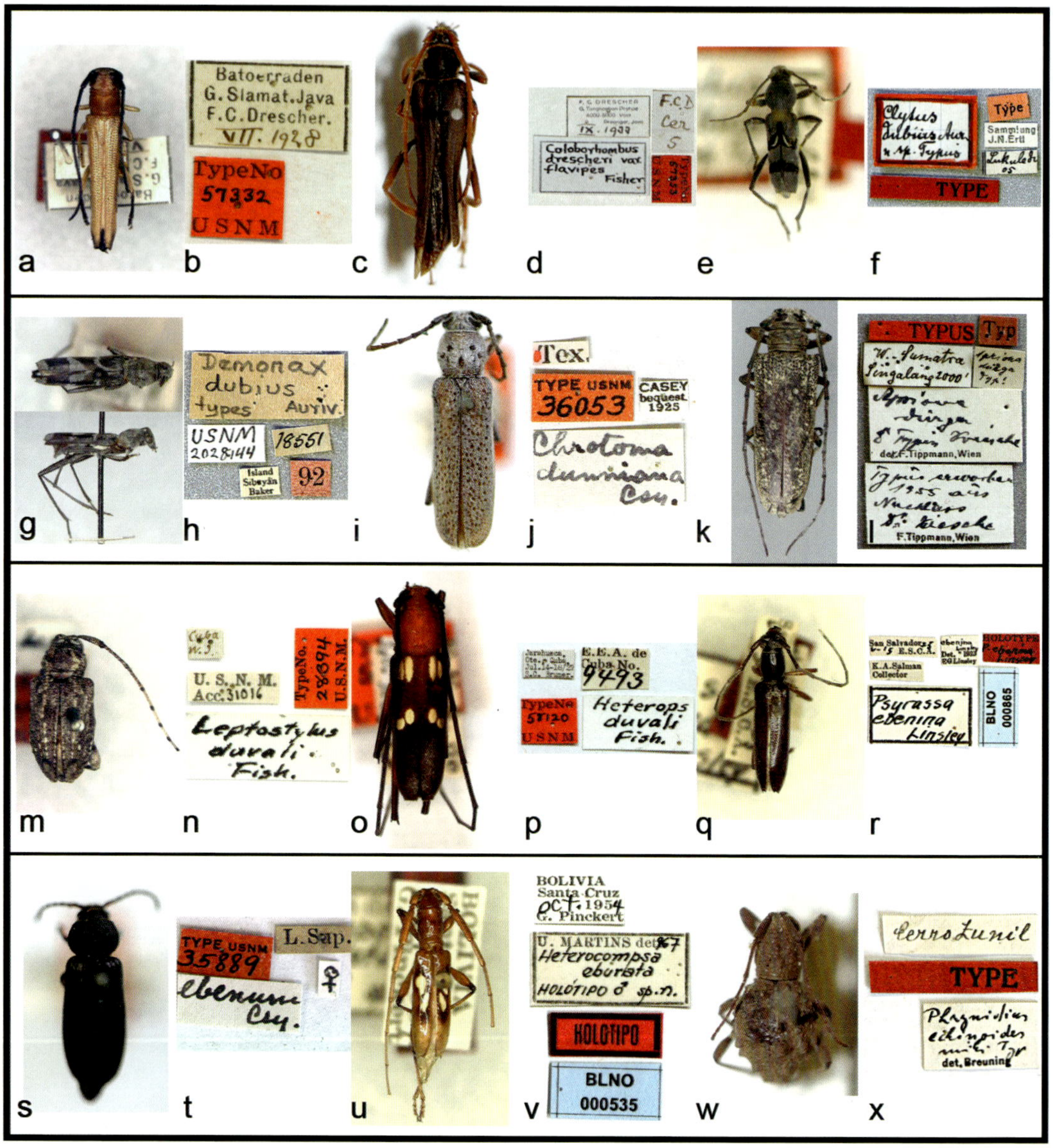

Figure 59. Primary types of Cerambycidae of the Smithsonian Institution (for each pair of images, dorsal habitus left, labels right): **(a, b)** *Oberea drescheri* Fisher, 1937, Holotype; **(c, d)** *Coloborhombus drescheri* var. *flavipes* Fisher, 1936, Holotype; **(e, f)** *Clytus (Plagionotulus?) dubius* Aurivillius, 1907, Syntype; **(g, h)** *Demonax dubius* Aurivillius, 1928, Syntype; **(i, j)** *Chrotoma dunniana* Casey, 1891, Holotype; **(k, l)** *Apriona durga* Kriesche, 1920, Syntype; **(m, n)** *Leptostylus duvali* Fisher, 1926, Holotype; **(o, p)** *Heterops duvali* Fisher, 1947, Holotype; **(q, r)** *Psyrassa ebenina* Linsley, 1935, Holotype; **(s, t)** *Asemum ebenum* Casey, 1912, Holotype; **(u, v)** *Heterocompsa eburata* Martins, 1970, Holotype; **(w, x)** *Phrynidius echinoides* Breuning, 1940, Holotype.

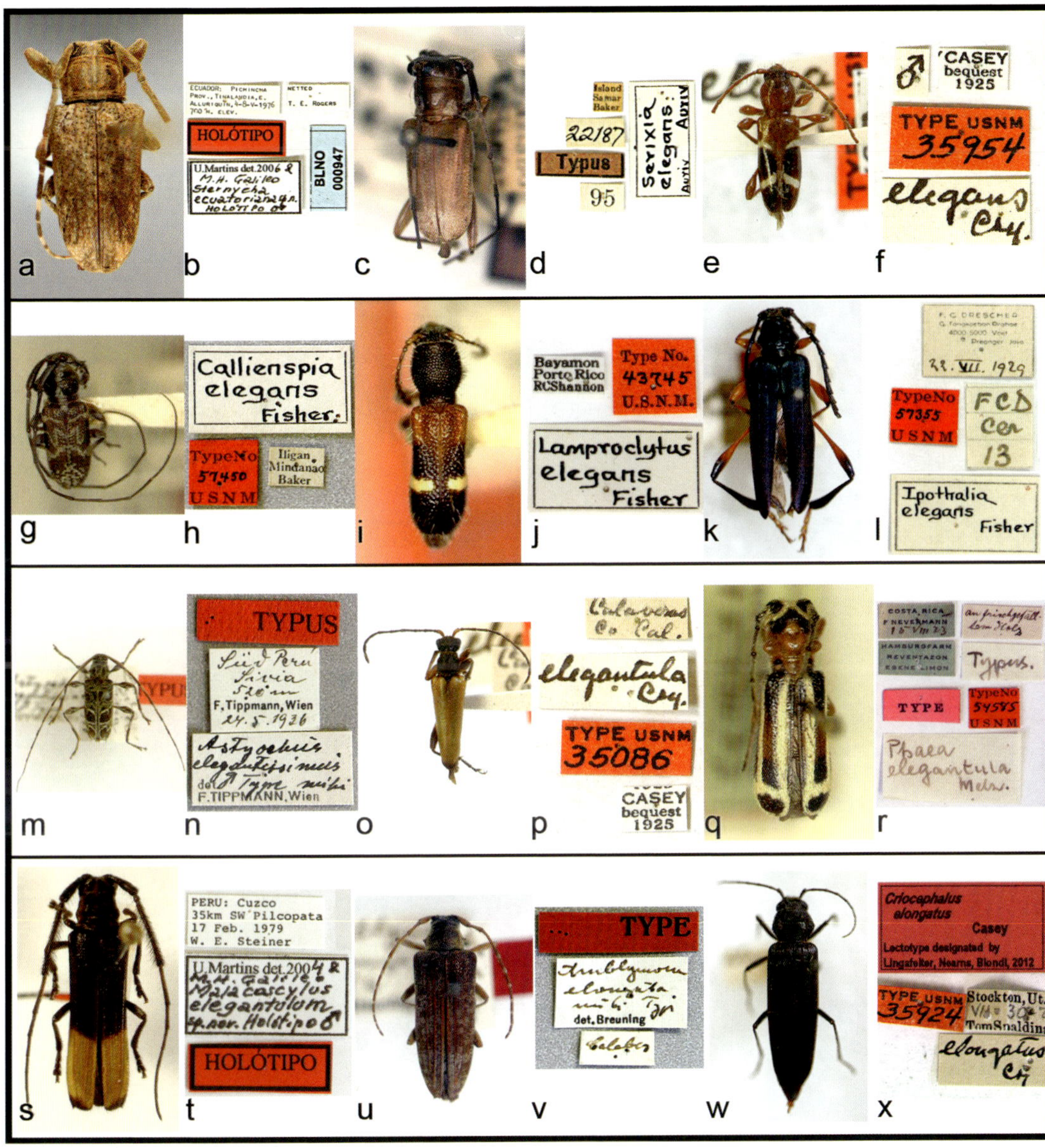

Figure 60. Primary types of Cerambycidae of the Smithsonian Institution (for each pair of images, dorsal habitus left, labels right): **(a, b)** *Sternycha ecuatoriana* Martins & Galileo, 2007, Holotype; **(c, d)** *Serixia elegans* Aurivillius, 1927, Holotype; **(e, f)** *Phymatodes (Phymatodina) elegans* Casey, 1912, Holotype; **(g, h)** *Callienispia elegans* Fisher, 1925, Holotype; **(i, j)** *Lamproclytus elegans* Fisher, 1932, Holotype; **(k, l)** *Ipothalia elegans* Fisher, 1936, Holotype; **(m, n)** *Astyochus elegantissimus* Tippmann, 1953, Holotype; **(o, p)** *Leptacmaeops (Leptacmaeops) elegantula* Casey, 1913, Holotype; **(q, r)** *Phaea elegantula* Melzer, 1933, Holotype; **(s, t)** *Malacoscylus elegantulus* Galileo & Martins, 2005, Holotype; **(u, v)** *Amblymora elongata* Breuning, 1943, Holotype; **(w, x)** *Criocephalus elongatus* Casey, 1924, Lectotype.

Figure 61. Primary types of Cerambycidae of the Smithsonian Institution (for each pair of images, dorsal habitus left, labels right): **(a, b)** *Merostenus elongatus* Fisher, 1932, Holotype; **(c, d)** *Sciocyrtinus elongatus* Fisher, 1935, Holotype; **(e, f)** *Cometes emarginata* Fisher, 1946, Holotype; **(g, h)** *Sybra emarginata* Gressitt, 1956, Holotype; **(i, j)** *Pogonocherus emarginatus* Casey, 1913, Holotype; **(k, l)** *Hemiloapis endyba* Galileo & Martins, 2004, Holotype; **(m, n)** *Phoebemima ensifera* Tippmann, 1960, Holotype; **(o, p)** *Octoplon equatorianum* Martins, 1960, Holotype; **(q, r)** *Spondylis erebea* Casey, 1912, Lectotype; **(s, t)** *Xylorhiza erectepilosa* Tippmann, 1951, Holotype; **(u, v)** *Pantonyssus erichsonii* var. *flavipes* Fisher, 1944, Holotype; **(w, x)** *Peritragus ertli* Aurivillius, 1913, Lectotype.

Figure 62. Primary types of Cerambycidae of the Smithsonian Institution (for each pair of images, dorsal habitus left, labels right): **(a, b)** *Octoplon erythrurum* Martins, 1962, Holotype; **(c, d)** *Sciadella (Micronesiella) esakii* Gressitt, 1956, Holotype; **(e, f)** *Cyrtinus eugeniae* Fisher, 1935, Holotype; **(g, h)** *Lepturges eurynota* Tippmann, 1960, Holotype; **(i, j)** *Ophistomis evanescens* Casey, 1913, Holotype; **(k, l)** *Ophistomis eversa* Casey, 1913, Lectotype; **(m, n)** *Batrachorhina (Soridus) excavata* Breuning, 1942, Holotype; **(o, p)** *Ischasia exigua* Fisher, 1947, Holotype; **(q, r)** *Euderces exilis* Casey, 1893, Holotype; **(s, t)** *Phymatodes (Phymatodina) exilis* Casey, 1912, Holotype; **(u, v)** *Oberea exilis* Casey, 1913, Lectotype; **(w, x)** *Spalacopsis exilis* Casey, 1924, Holotype.

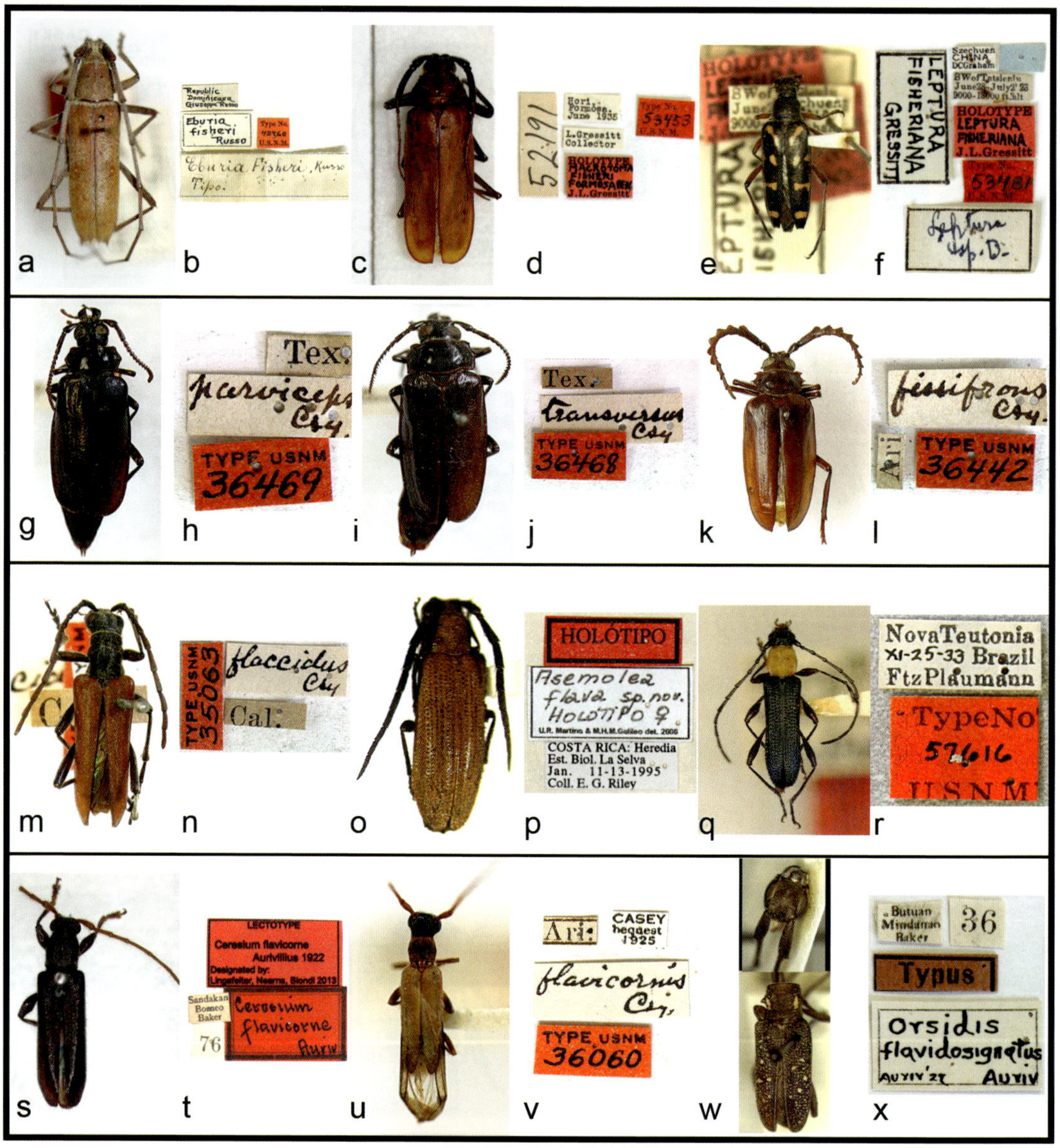

Figure 67. Primary types of Cerambycidae of the Smithsonian Institution (for each pair of images, dorsal habitus left, labels right): **(a, b)** *Eburia fisheri* Russo, 1930, Holotype; **(c, d)** *Macrotoma fisheri formosae* Gressitt, 1938, Holotype; **(e, f)** *Leptura fisheriana* Gressitt, 1938, Holotype; **(g, h)** *Prionus (Antennalia) fissicornis parviceps* Casey, 1912, Holotype; **(i, j)** *Prionus (Antennalia) fissicornis transversus* Casey, 1912, Holotype; **(k, l)** *Prionus (Prionus) fissifrons* Casey, 1912, Holotype; **(m, n)** *Stenocorus (Stenocorus) flaccidus* Casey, 1913, Holotype; **(o, p)** *Asemolea flava* Martins & Galileo, 2006, Holotype; **(q, r)** *Callideriphus flavicollis* Fisher, 1938, Holotype; **(s, t)** *Ceresium flavicorne* Aurivillius, 1927, Lectotype; **(u, v)** *Methia flavicornis* Casey, 1924, Holotype; **(w, x)** *Orsidis (?) flavidosignatus* Aurivillius, 1927, Syntype.

Figure 62. Primary types of Cerambycidae of the Smithsonian Institution (for each pair of images, dorsal habitus left, labels right): **(a, b)** *Octoplon erythrurum* Martins, 1962, Holotype; **(c, d)** *Sciadella (Micronesiella) esakii* Gressitt, 1956, Holotype; **(e, f)** *Cyrtinus eugeniae* Fisher, 1935, Holotype; **(g, h)** *Lepturges eurynota* Tippmann, 1960, Holotype; **(i, j)** *Ophistomis evanescens* Casey, 1913, Holotype; **(k, l)** *Ophistomis eversa* Casey, 1913, Lectotype; **(m, n)** *Batrachorhina (Soridus) excavata* Breuning, 1942, Holotype; **(o, p)** *Ischasia exigua* Fisher, 1947, Holotype; **(q, r)** *Euderces exilis* Casey, 1893, Holotype; **(s, t)** *Phymatodes (Phymatodina) exilis* Casey, 1912, Holotype; **(u, v)** *Oberea exilis* Casey, 1913, Lectotype; **(w, x)** *Spalacopsis exilis* Casey, 1924, Holotype.

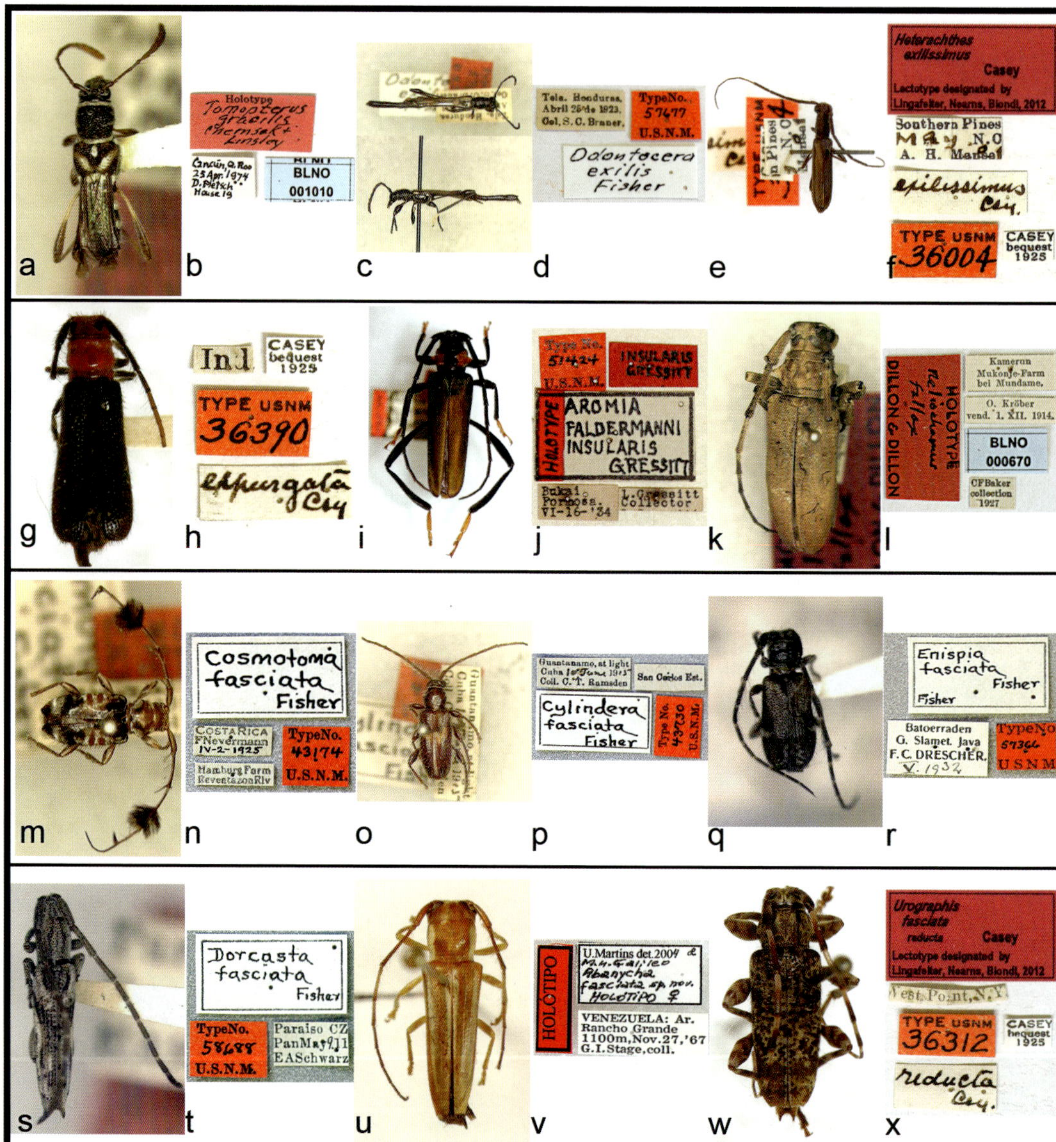

Figure 63. Primary types of Cerambycidae of the Smithsonian Institution (for each pair of images, dorsal habitus left, labels right): **(a, b)** *Tomopterus exilis* Chemsak & Linsley, 1979, Holotype; **(c, d)** *Odontocera exilis* Fisher, 1947, Holotype; **(e, f)** *Heterachthes exilissimus* Casey, 1912, Lectotype; **(g, h)** *Tetrops expurgata* Casey, 1913, Holotype; **(i, j)** *Aromia faldermanni insularis* Gressitt, 1936, Holotype; **(k, l)** *Meliochamus fallax* Dillon & Dillon, 1959, Holotype; **(m, n)** *Cosmotoma fasciata* Fisher, 1931, Holotype; **(o, p)** *Cylindera fasciata* Fisher, 1932, Holotype; **(q, r)** *Enispia fasciata* Fisher, 1936, Holotype; **(s, t)** *Dorcasta fasciata* Fisher, 1947, Holotype; **(u, v)** *Abanycha fasciata* Galileo & Martins, 2005, Holotype; **(w, x)** *Urographis fasciata reducta* Casey, 1913, Lectotype.

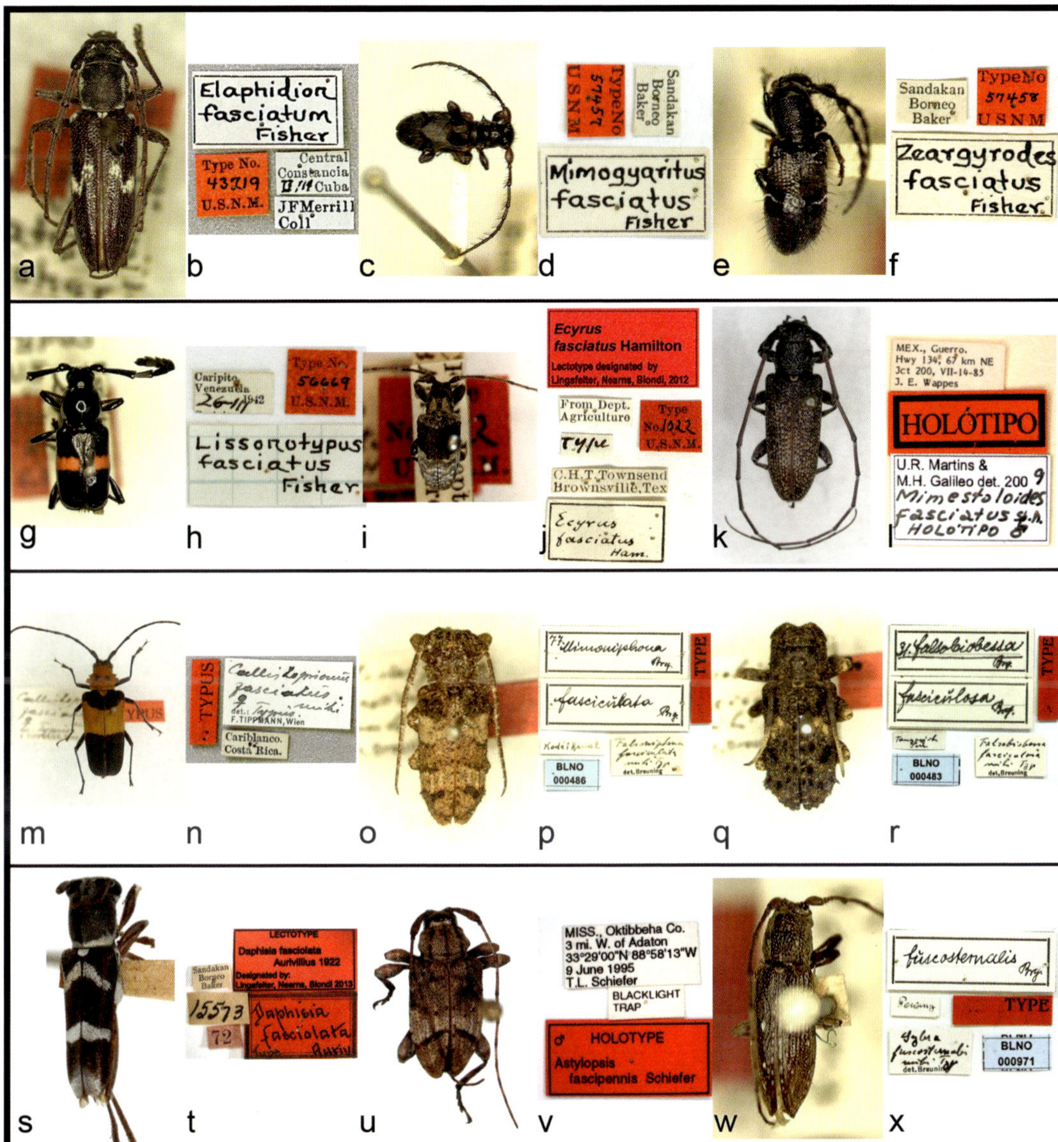

Figure 64. Primary types of Cerambycidae of the Smithsonian Institution (for each pair of images, dorsal habitus left, labels right): **(a, b)** *Elaphidion fasciatum* Fisher, 1932, Holotype; **(c, d)** *Mimogyaritus fasciatus* Fisher, 1925, Holotype; **(e, f)** *Zeargyrodes fasciatus* Fisher, 1925, Holotype; **(g, h)** *Lissonotypus fasciatus* Fisher, 1944, Holotype; **(i, j)** *Ecyrus fasciatus* Hamilton, 1896, Lectotype; **(k, l)** *Mimestoloides fasciatus* Martins & Galileo, 2010, Holotype; **(m, n)** *Callistoprionus fasciatus* Tippmann, 1953, Holotype; **(o, p)** *Mimoniphona fasciculata* Breuning, 1940, Holotype; **(q, r)** *Falsobiobessa fasciculosa* Breuning, 1942, Holotype; **(s, t)** *Daphisia fasciolata* Aurivillius, 1922, Lectotype; **(u, v)** *Astylopsis fascipennis* Schiefer, 2001, Holotype; **(w, x)** *Sybra fascosternalis* Breuning, 1942, Holotype.

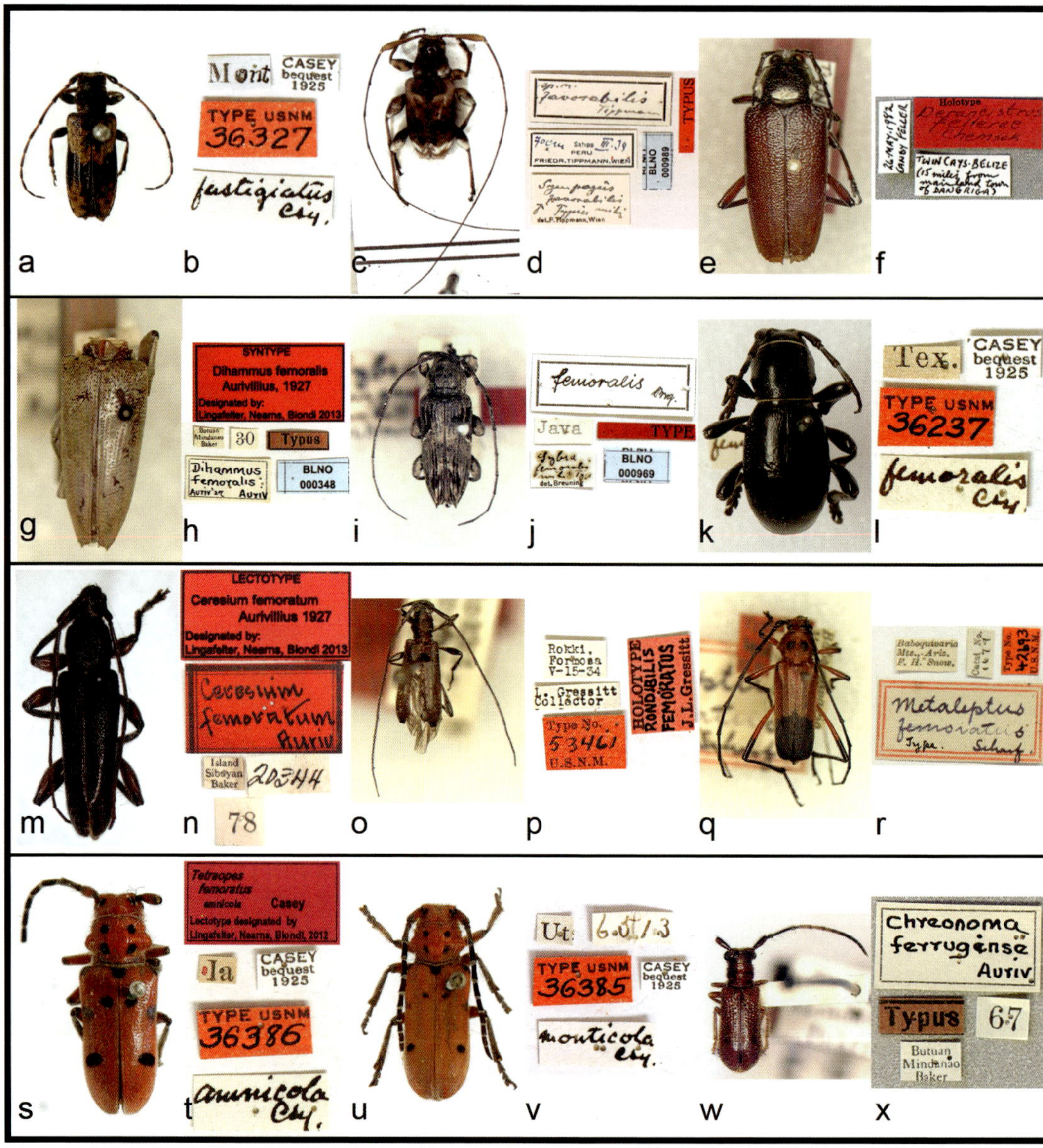

Figure 65. Primary types of Cerambycidae of the Smithsonian Institution (for each pair of images, dorsal habitus left, labels right): **(a, b)** *Pogonocherus fastigiatus* Casey, 1913, Holotype; **(c, d)** *Sympagus favorabilis* Tippmann, 1960, Holotype; **(e, f)** *Derancistrus fellerae* Chemsak, 1983, Holotype; **(g, h)** *Dihammus femoralis* Aurivillius, 1927, Syntype; **(i, j)** *Sybra femoralis* Breuning, 1940, Holotype; **(k, l)** *Moneilema (Collapteryx) femoralis* Casey, 1913, Holotype; **(m, n)** *Ceresium femoratum* Aurivillius, 1927, Lectotype; **(o, p)** *Rondibilis femoratus* Gressitt, 1938, Holotype; **(q, r)** *Metaleptus femoratus* Schaeffer, 1909, Syntype; **(s, t)** *Tetraopes femoratus amnicola* Casey, 1913, Lectotype; **(u, v)** *Tetraopes femoratus monticola* Casey, 1913, Holotype; **(w, x)** *Chreonoma ferruginea* Aurivillius, 1927, Syntype.

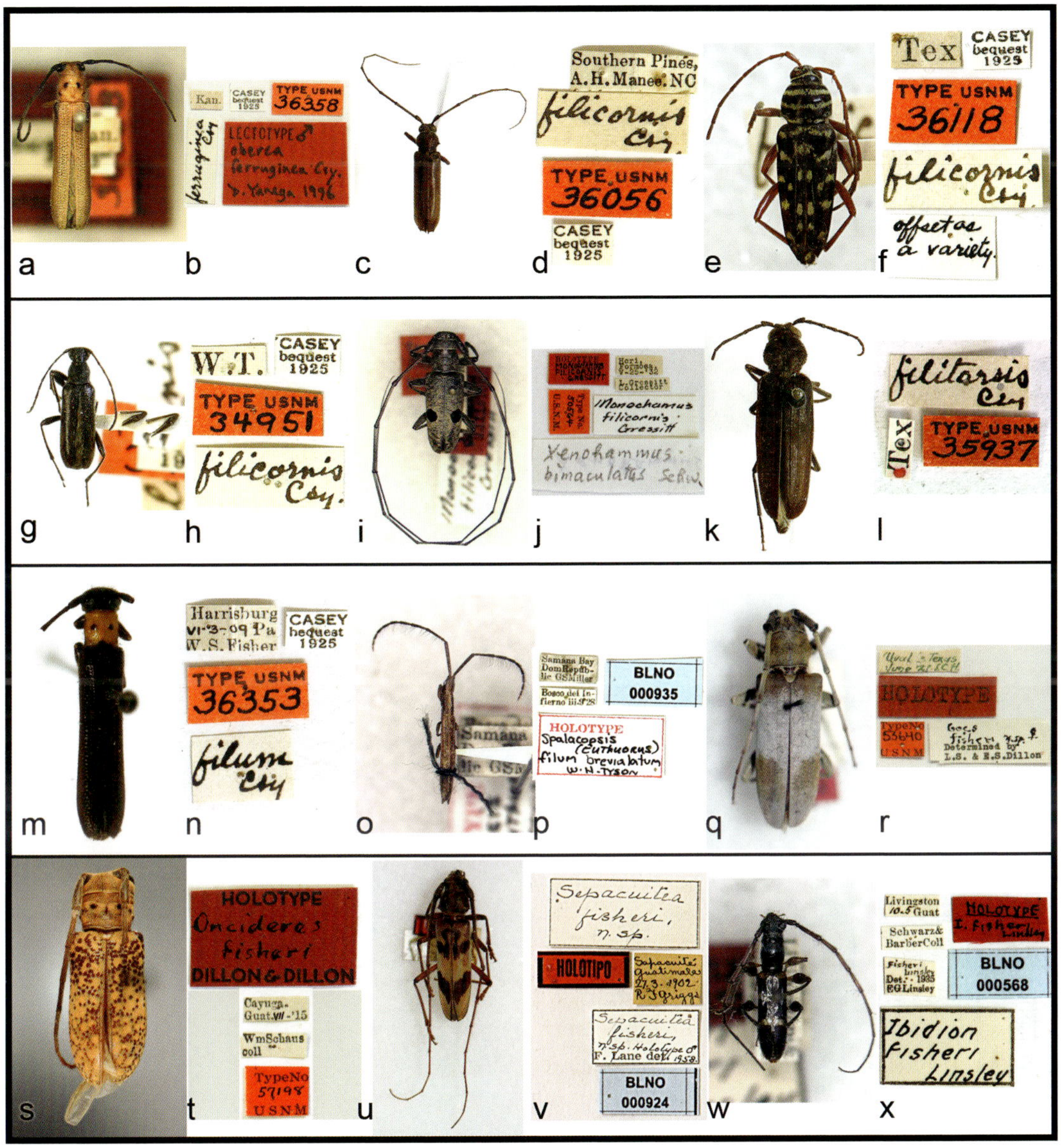

Figure 66. Primary types of Cerambycidae of the Smithsonian Institution (for each pair of images, dorsal habitus left, labels right): **(a, b)** *Oberea ferruginea* Casey, 1913, Lectotype; **(c, d)** *Chion filicorne* Casey, 1924, Holotype; **(e, f)** *Cyllene filicornis* Casey, 1912, Holotype; **(g, h)** *Parallelina filicornis* Casey, 1913, Holotype; **(i, j)** *Monochammus filicornis* Gressitt, 1935, Holotype; **(k, l)** *Criocephalus filitarsis* Casey, 1912, Holotype; **(m, n)** *Oberea filum* Casey, 1913, Holotype; **(o, p)** *Spalacopsis (Euthuorus) filum brevialatum* Tyson, 1973, Holotype; **(q, r)** *Goes fisheri* Dillon & Dillon, 1941, Holotype; **(s, t)** *Oncideres fisheri* Dillon & Dillon, 1946, Holotype; **(u, v)** *Sepaicutea fisheri* Lane, 1972, Holotype; **(w, x)** *Ibidion fisheri* Linsley, 1935, Holotype.

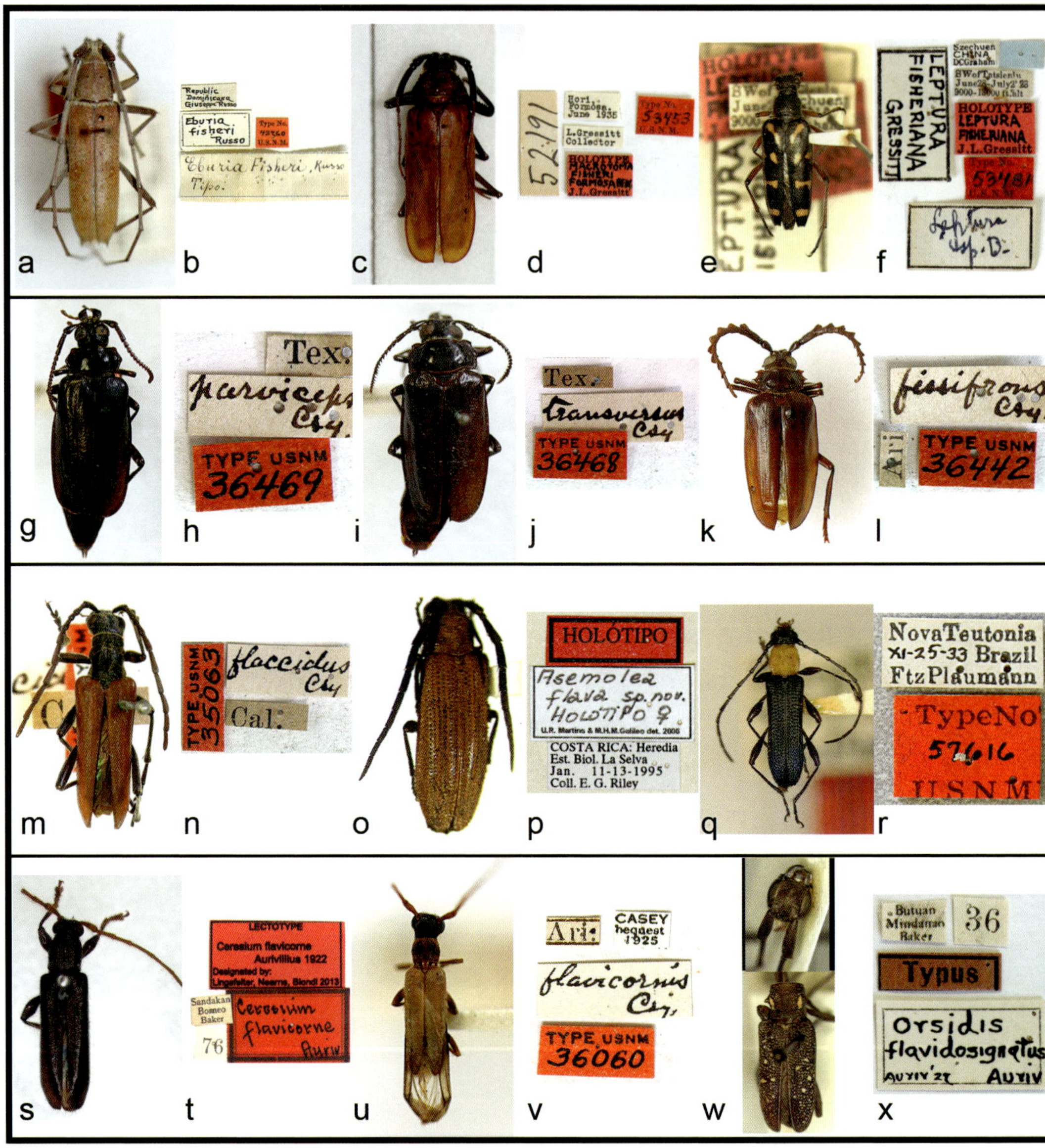

Figure 67. Primary types of Cerambycidae of the Smithsonian Institution (for each pair of images, dorsal habitus left, labels right): **(a, b)** *Eburia fisheri* Russo, 1930, Holotype; **(c, d)** *Macrotoma fisheri formosae* Gressitt, 1938, Holotype; **(e, f)** *Leptura fisheriana* Gressitt, 1938, Holotype; **(g, h)** *Prionus (Antennalia) fissicornis parviceps* Casey, 1912, Holotype; **(i, j)** *Prionus (Antennalia) fissicornis transversus* Casey, 1912, Holotype; **(k, l)** *Prionus (Prionus) fissifrons* Casey, 1912, Holotype; **(m, n)** *Stenocorus (Stenocorus) flaccidus* Casey, 1913, Holotype; **(o, p)** *Asemolea flava* Martins & Galileo, 2006, Holotype; **(q, r)** *Callideriphus flavicollis* Fisher, 1938, Holotype; **(s, t)** *Ceresium flavicorne* Aurivillius, 1927, Lectotype; **(u, v)** *Methia flavicornis* Casey, 1924, Holotype; **(w, x)** *Orsidis (?) flavidosignatus* Aurivillius, 1927, Syntype.

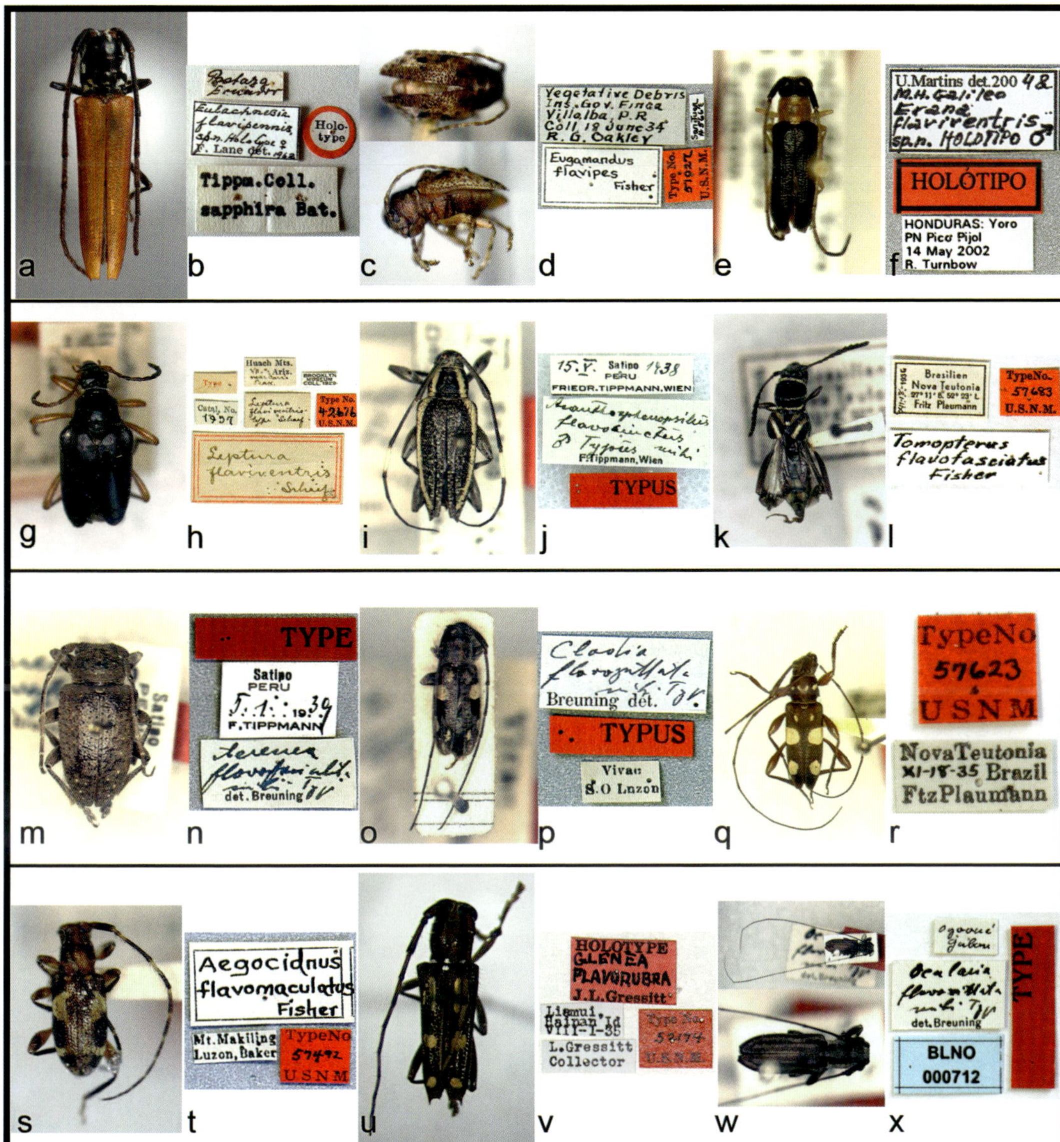

Figure 68. Primary types of Cerambycidae of the Smithsonian Institution (for each pair of images, dorsal habitus left, labels right): **(a, b)** *Eulachnesia flavipennis* Lane, 1966, Holotype; **(c, d)** *Eugamandus flavipes* Fisher, 1935, Holotype; **(e, f)** *Erana flaviventris* Galileo & Martins, 2005, Holotype; **(g, h)** *Leptura flaviventris* Schaeffer, 1908, Holotype; **(i, j)** *Acanthosphenopsilus flavocinctus* Tippmann, 1960, Holotype; **(k, l)** *Tomopterus flavofasciatus* Fisher, 1947, Holotype; **(m, n)** *Aerenea flavofasciculata* Breuning, 1948, Holotype; **(o, p)** *Clodia flavoguttata* Breuning, 1957, Holotype; **(q, r)** *Lepturges flavoguttatus* Fisher, 1938, Holotype; **(s, t)** *Aegocidnus flavomaculatus* Fisher, 1925, Holotype; **(u, v)** *Glenea flavorubra* Gressitt, 1940, Holotype; **(w, x)** *Ocularia flavovittata* Breuning, 1940, Holotype.

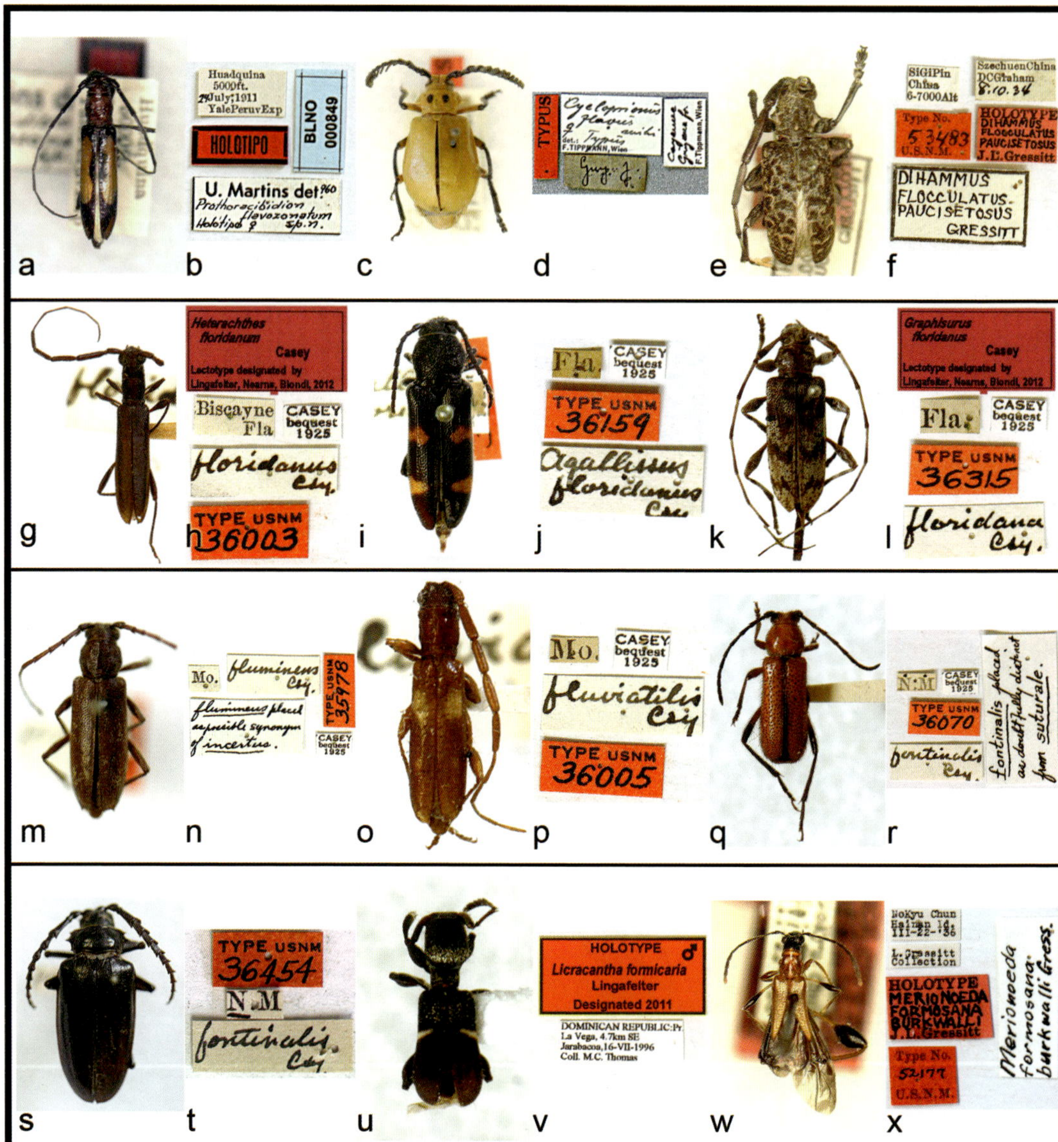

Figure 69. Primary types of Cerambycidae of the Smithsonian Institution (for each pair of images, dorsal habitus left, labels right): **(a, b)** *Prothoracibidion flavozonatum* Martins, 1960, Holotype; **(c, d)** *Cycloprionus flavus* Tippmann, 1953, Holotype; **(e, f)** *Dihammus flocculatus paucisetosus* Gressitt, 1938, Holotype; **(g, h)** *Heterachthes floridanus* Casey, 1912, Lectotype; **(i, j)** *Agallissus floridanus* Casey, 1912, Holotype; **(k, l)** *Graphisurus (Graphisurus) floridanus* Casey, 1913, Lectotype; **(m, n)** *Hypermallus flumineus* Casey, 1912, Holotype; **(o, p)** *Heterachthes fluviatilis* Casey, 1912, Holotype; **(q, r)** *Batyleoma fontinalis* Casey, 1912, Holotype; **(s, t)** *Prionus (Prionus) fontinalis* Casey, 1924, Holotype; **(u, v)** *Licracantha formicaria* Lingafelter, 2011, Holotype; **(w, x)** *Merionoeda (Ocytasia) formosana burkwalli* Gressitt, 1940, Holotype.

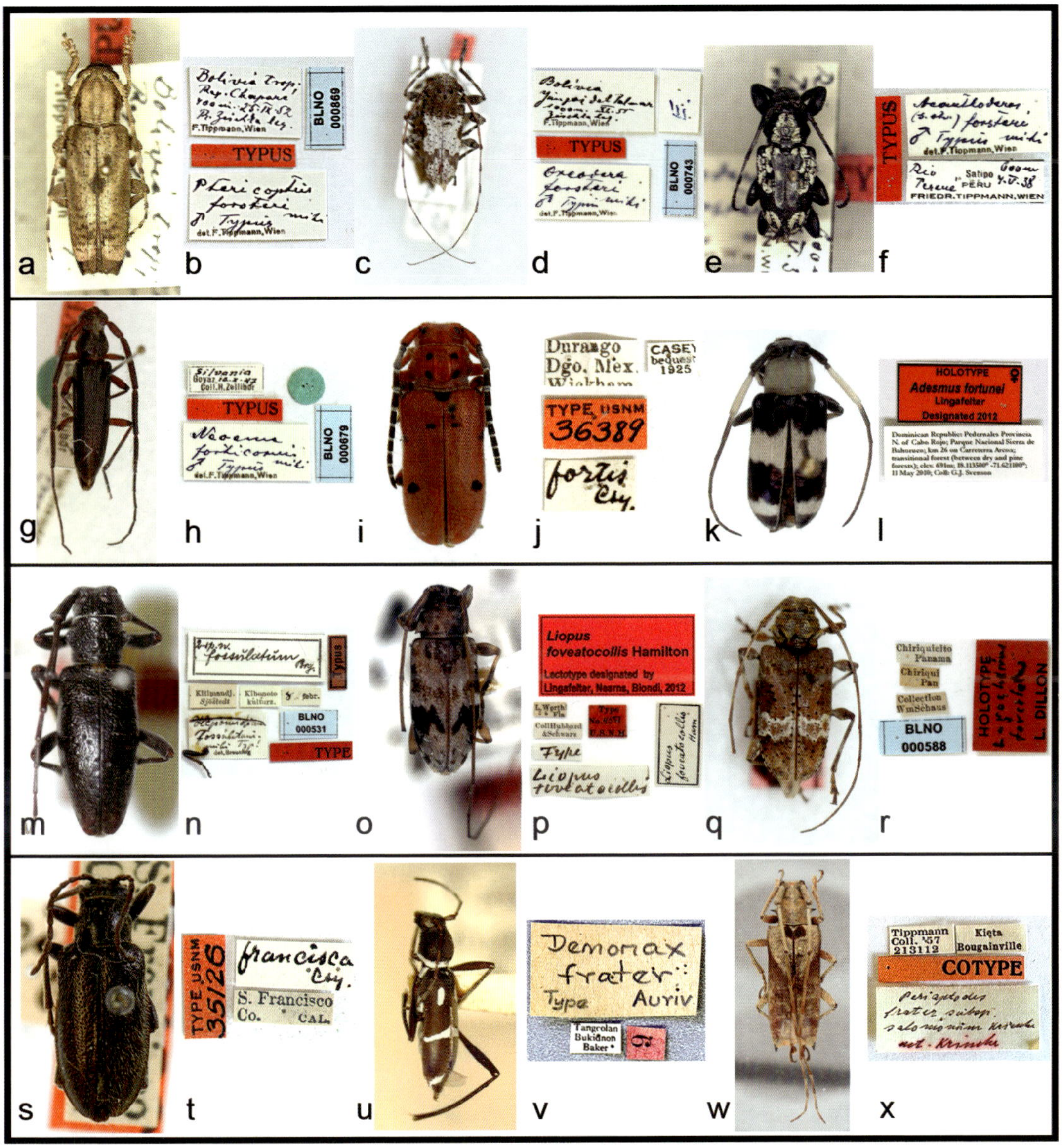

Figure 70. Primary types of Cerambycidae of the Smithsonian Institution (for each pair of images, dorsal habitus left, labels right): **(a, b)** *Ptericoptus forsteri* Tippmann, 1960, Holotype; **(c, d)** *Oreodera forsteri* Tippmann, 1960, Holotype; **(e, f)** *Acanthoderes forsteri* Tippmann, 1960, Holotype; **(g, h)** *Neoeme forticornis* Tippmann, 1960, Holotype; **(i, j)** *Tetraopes fortis* Casey, 1913, Holotype; **(k, l)** *Adesmus fortunei* Lingafelter, 2013, Holotype; **(m, n)** *Parahepomidion fossulatum* Breuning, 1936, Holotype; **(o, p)** *Liopus foveatocollis* Hamilton in Leng & Hamilton, 1896, Lectotype; **(q, r)** *Lagocheirus foveolatus* Dillon, 1957, Holotype; **(s, t)** *Brachysomida francisca* Casey, 1913, Holotype; **(u, v)** *Demonax frater* Aurivillius, 1924, Holotype; **(w, x)** *Periaptodes frater salomonum* Kreische, 1936, Holotype.

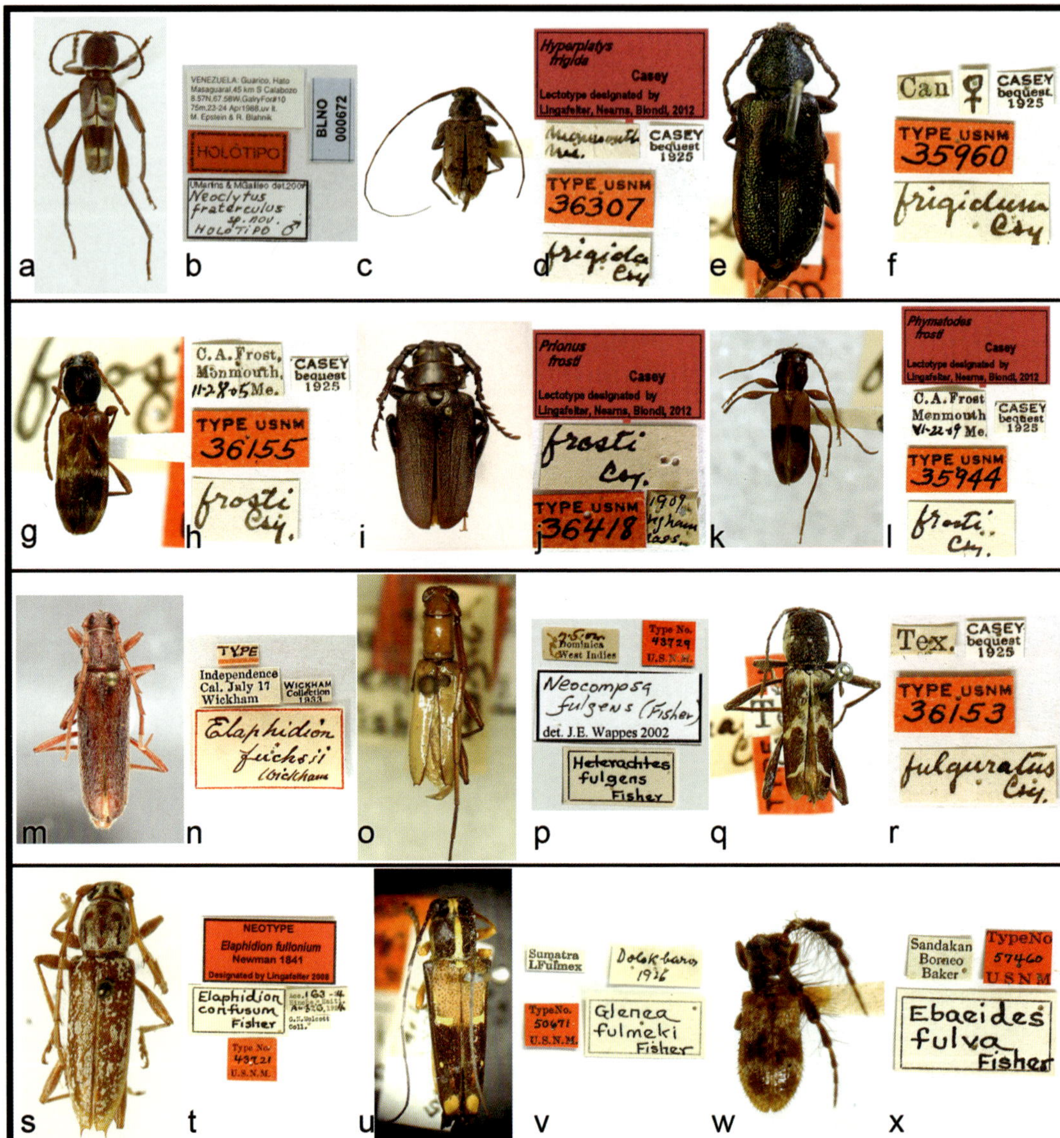

Figure 71. Primary types of Cerambycidae of the Smithsonian Institution (for each pair of images, dorsal habitus left, labels right): **(a, b)** *Neoclytus fraterculus* Martins & Galileo, 2008, Holotype; **(c, d)** *Hyperplatys frigida* Casey, 1913, Lectotype; **(e, f)** *Callidium frigidum* Casey, 1912, Holotype; **(g, h)** *Microclytus frosti* Casey, 1912, Holotype; **(i, j)** *Prionus (Prionellus) frosti* Casey, 1924, Lectotype; **(k, l)** *Phymatodes frosti* Casey, 1924, Lectotype; **(m, n)** *Elaphidion fuchsii* Wickham, 1905, Holotype; **(o, p)** *Heterachtes fulgens* Fisher, 1932, Holotype; **(q, r)** *Neoclytus fulguratus* Casey, 1912, Holotype; **(s, t)** *Elaphidion fullonium* Newman, 1841, Neotype; **(u, v)** *Glenea fulmeki* Fisher, 1934, Holotype; **(w, x)** *Ebaeides fulva* Fisher, 1925, Holotype.

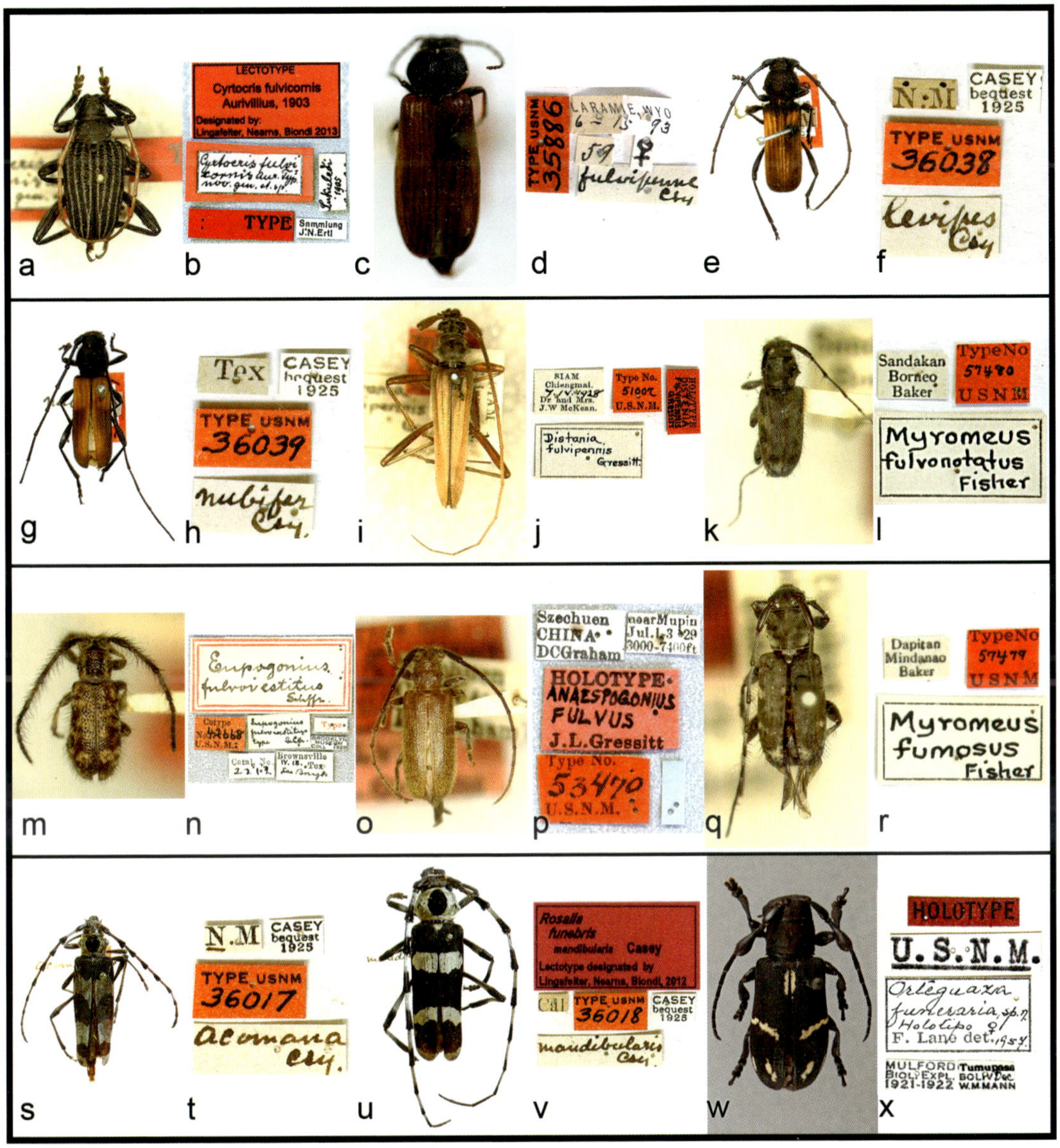

Figure 72. Primary types of Cerambycidae of the Smithsonian Institution (for each pair of images, dorsal habitus left, labels right): **(a, b)** *Cyrtocris fulvicornis* Aurivillius, 1903, Lectotype; **(c, d)** *Asemum fulvipenne* Casey, 1912, Holotype; **(e, f)** *Tragidion fulvipenne levipes* Casey, 1912, Holotype; **(g, h)** *Tragidion fulvipenne nubifer* Casey, 1912, Holotype; **(i, j)** *Distenia fulvipennis* Gressitt, 1935, Holotype; **(k, l)** *Myromeus fulvonotatus* Fisher, 1925, Holotype; **(m, n)** *Eupogonius fulvovestitus* Schaeffer, 1905, Syntype; **(o, p)** *Anaespogonius fulvus* Gressitt, 1938, Holotype; **(q, r)** *Myromeus fumosus* Fisher, 1925, Holotype; **(s, t)** *Rosalia funebris acomana* Casey, 1912, Holotype; **(u, v)** *Rosalia funebris mandibularis* Casey, 1912, Lectotype; **(w, x)** *Orteguaza funeraria* Lane, 1958, Holotype.

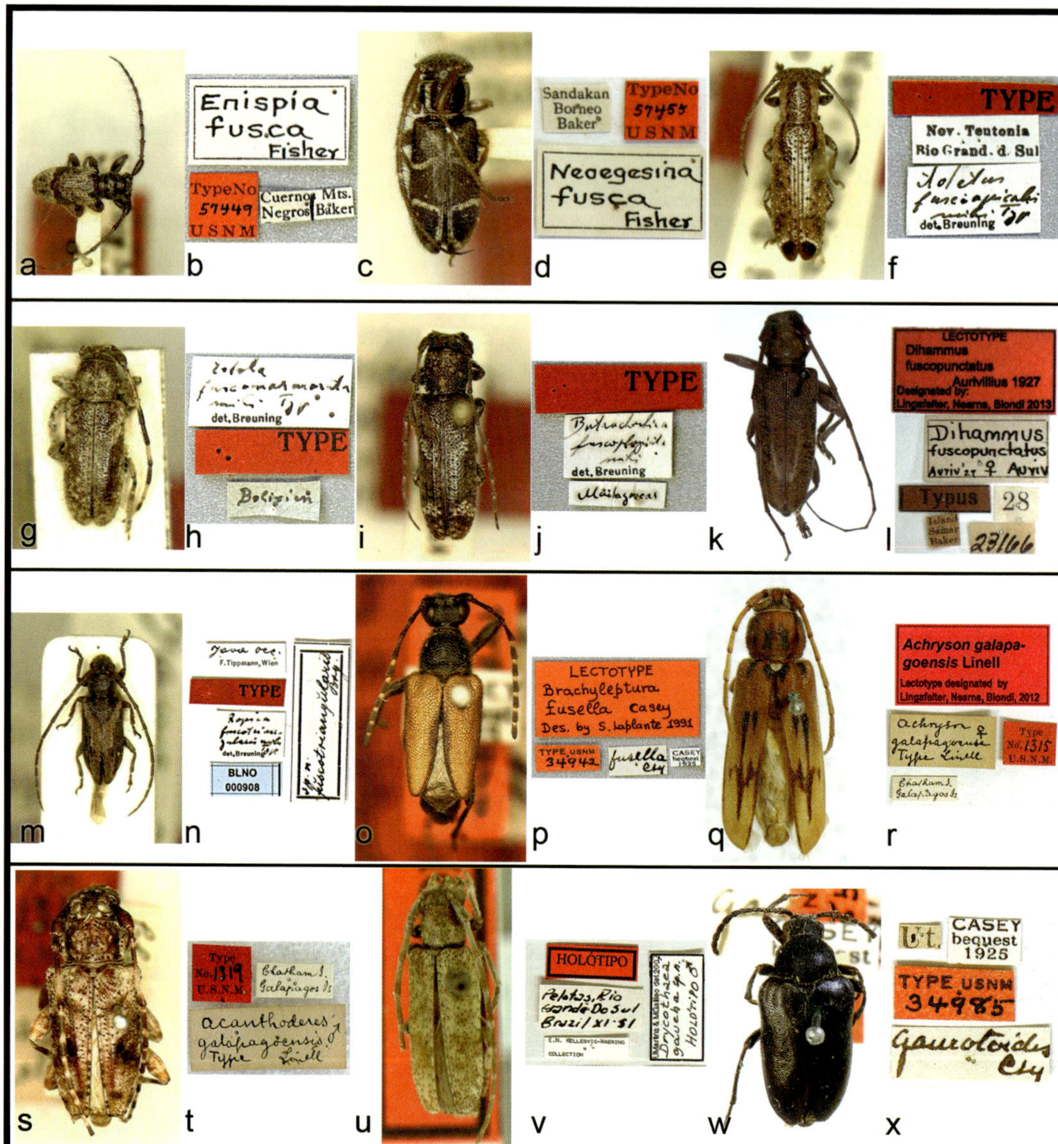

Figure 73. Primary types of Cerambycidae of the Smithsonian Institution (for each pair of images, dorsal habitus left, labels right): **(a, b)** *Enispia fusca* Fisher, 1925, Holotype; **(c, d)** *Neoegesina fusca* Fisher, 1925, Holotype; **(e, f)** *Adetus fuscoapicalis* Breuning, 1942, Holotype; **(g, h)** *Estola fuscomarmorata* Breuning, 1940, Holotype; **(i, j)** *Batrachorhina fuscoplagiata* Breuning, 1942, Holotype; **(k, l)** *Dihammus fuscopunctatus* Aurivillius, 1927, Lectotype; **(m, n)** *Sybra fuscotriangularis* Breuning, 1939, Holotype; **(o, p)** *Brachyleptura fusella* Casey, 1913, Lectotype; **(q, r)** *Achryson galapagoensis* Linell, 1899, Lectotype; **(s, t)** *Acanthoderes galapagoensis* Linell, 1899, Lectotype; **(u, v)** *Drycothaea gaucha* Galileo & Martins, 2008, Holotype; **(w, x)** *Leptura gaurotoides* Casey, 1893, Holotype.

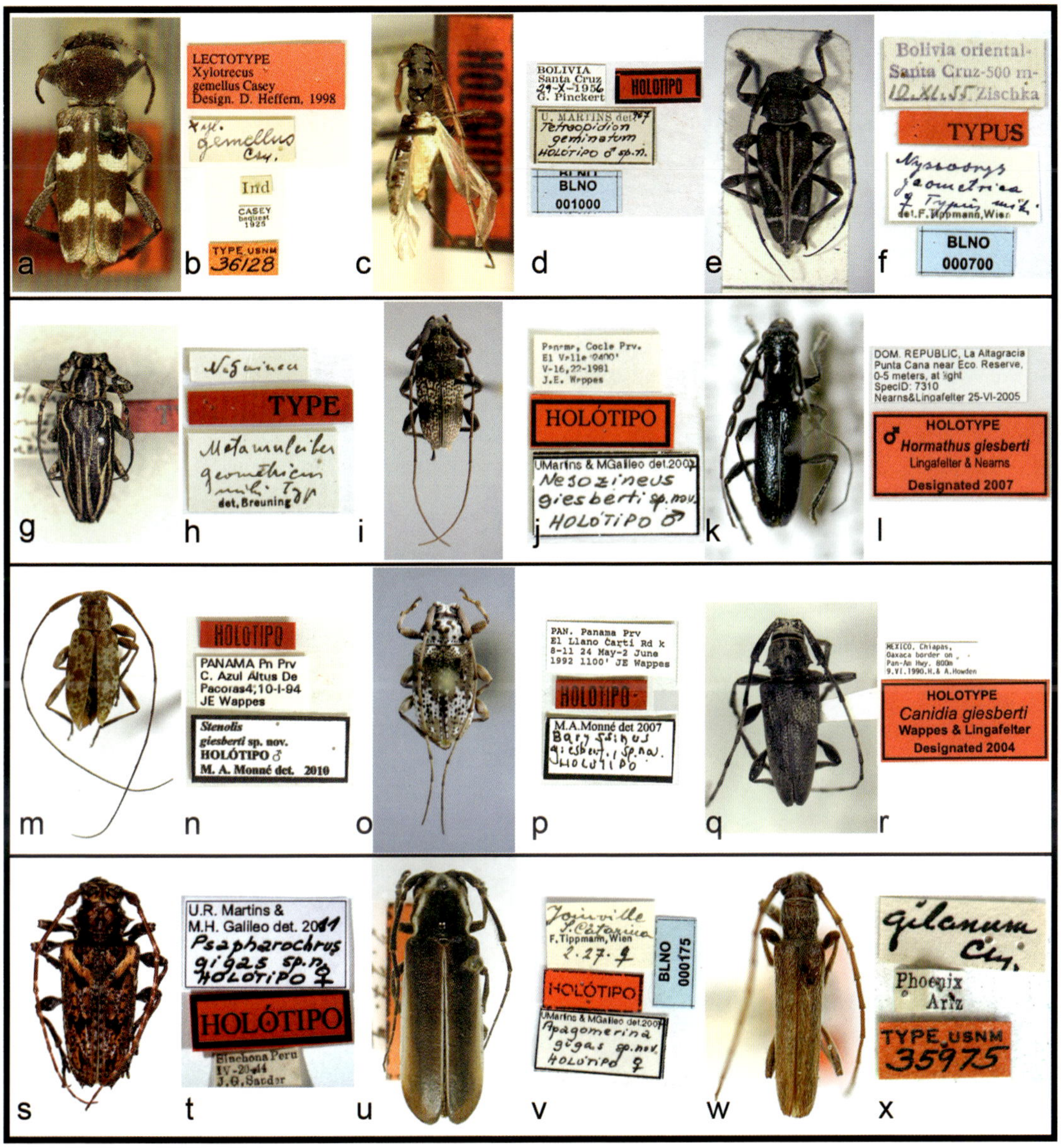

Figure 74. Primary types of Cerambycidae of the Smithsonian Institution (for each pair of images, dorsal habitus left, labels right): **(a, b)** *Xylotrechus gemellus* Casey, 1893, Lectotype; **(c, d)** *Tetraopidion geminatum* Martins, 1969, Holotype; **(e, f)** *Nyssodrys geometrica* Tippmann, 1960, Holotype; **(g, h)** *Metamulciber geometricus* Breuning, 1940, Holotype; **(i, j)** *Nesozineus giesberti* Galileo & Martins, 2007, Holotype; **(k, l)** *Hormathus giesberti* Lingafelter & Nearns, 2007, Holotype; **(m, n)** *Stenolis giesberti* Monné, 2011, Holotype; **(o, p)** *Baryssinus giesberti* Monné & Monné, 2007, Holotype; **(q, r)** *Canidia giesberti* Wappes & Lingafelter, 2005, Holotype; **(s, t)** *Psapharochrus gigas* Galileo & Martins, 2012, Holotype; **(u, v)** *Apagomerina gigas* Martins & Galileo, 2007, Holotype; **(w, x)** *Aneflomorpha gilana* Casey, 1924, Holotype.

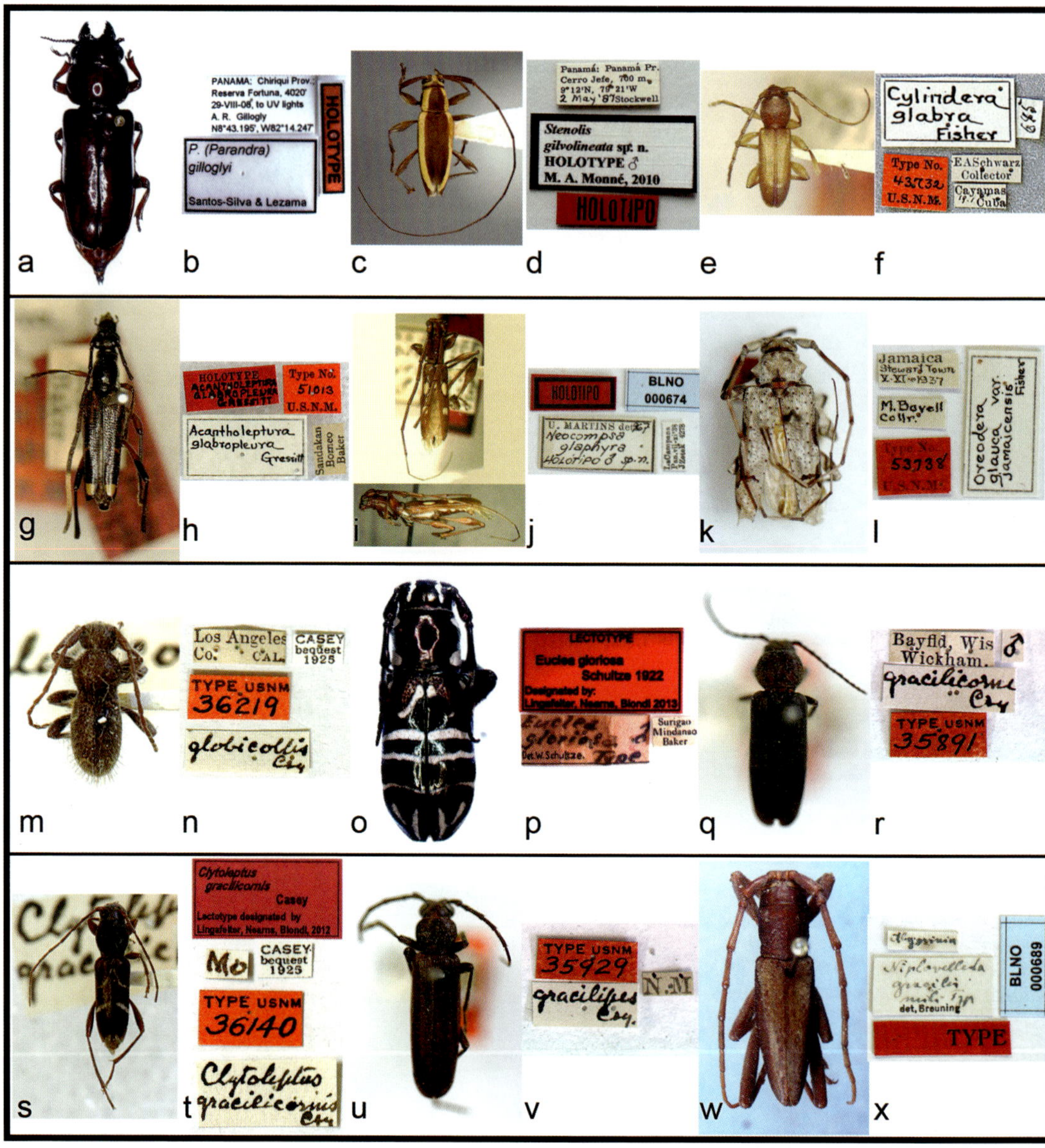

Figure 75. Primary types of Cerambycidae of the Smithsonian Institution (for each pair of images, dorsal habitus left, labels right): **(a, b)** *Parandra gilloglyi* Santos-Silva & Lezama, 2010, Holotype; **(c, d)** *Stenolis gilvolineata* Monné, 2011, Holotype; **(e, f)** *Cylindera glabra* Fisher, 1932, Holotype; **(g, h)** *Acantholeptura glabropleura* Gressitt, 1935, Holotype; **(i, j)** *Neocompsa glaphyra* Martins, 1970, Holotype; **(k, l)** *Oreodera glauca* var. *jamaicensis* Fisher, 1942, Holotype; **(m, n)** *Ipochus globicollis* Casey, 1913, Holotype; **(o, p)** *Euclea gloriosa* Schultze, 1922, Lectotype; **(q, r)** *Asemum gracilicorne* Casey, 1912, Holotype; **(s, t)** *Clytoleptus gracilicornis* Casey, 1912, Lectotype; **(u, v)** *Nothorhina gracilipes* Casey, 1912, Holotype; **(w, x)** *Niphovelleda gracilis* Breuning, 1940, Holotype.

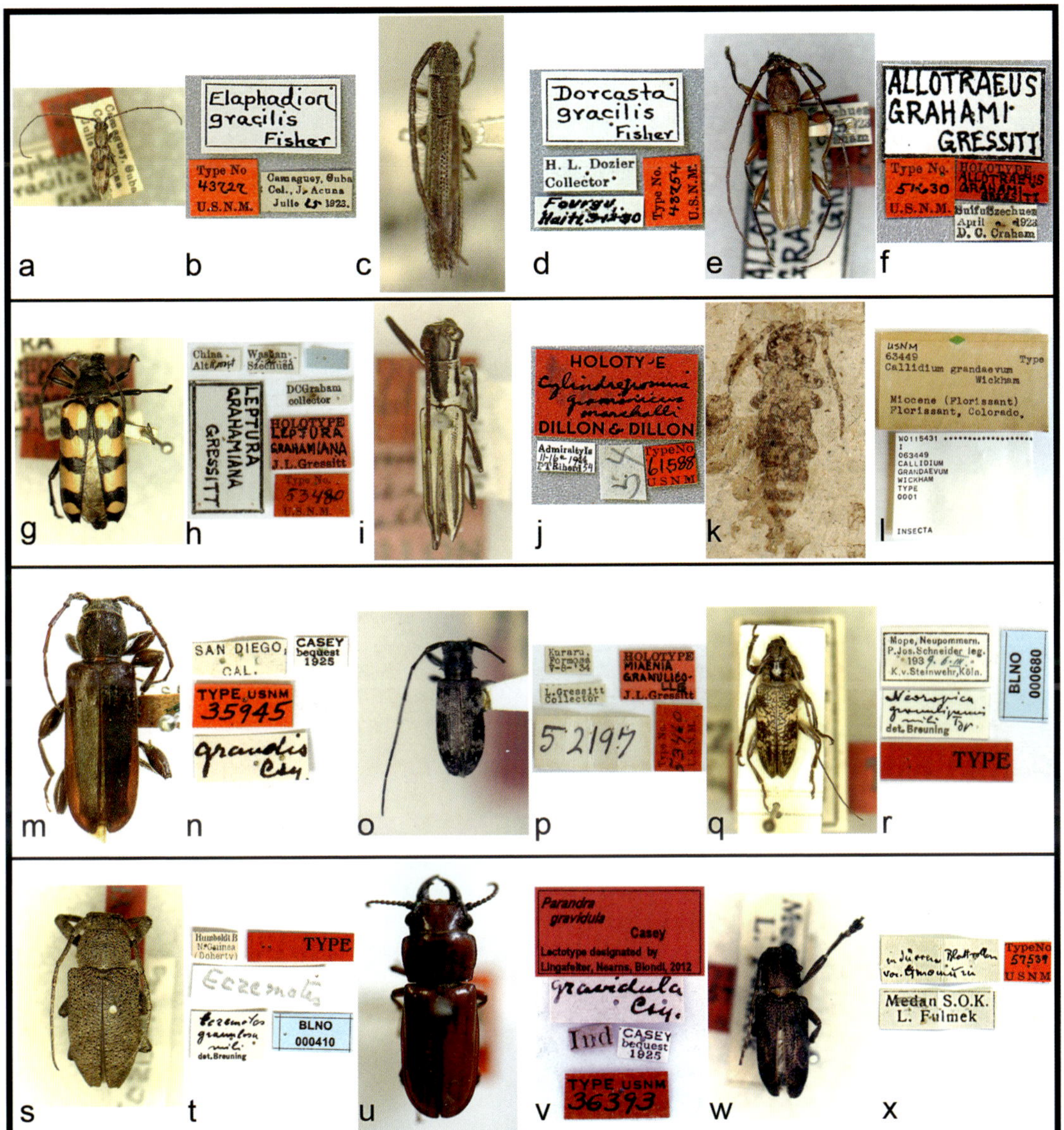

Figure 76. Primary types of Cerambycidae of the Smithsonian Institution (for each pair of images, dorsal habitus left, labels right): **(a, b)** *Elaphidion gracilis* Fisher, 1932, Holotype; **(c, d)** *Dorcasta gracilis* Fisher, 1932, Holotype; **(e, f)** *Allotraeus grahami* Gressitt, 1937, Holotype; **(g, h)** *Leptura grahamiana* Gressitt, 1938, Holotype; **(i, j)** *Cylindrepomus grammicus marshalli* Dillon & Dillon, 1948, Holotype; **(k, l)** *Callidium grandaevum* Wickham, 1917, Holotype; **(m, n)** *Phymatodes (Poecilium) grandis* Casey, 1912, Holotype; **(o, p)** *Miaenia granulicolle* Gressitt, 1938, Holotype; **(q, r)** *Neosybra granulipennis* Breuning, 1954, Holotype; **(s, t)** *Eczemotes granulosa* Breuning, 1938, Holotype; **(u, v)** *Parandra gravidula* Casey, 1912, Lectotype; **(w, x)** *Pterolophia gregalis* Fisher, 1927, Holotype.

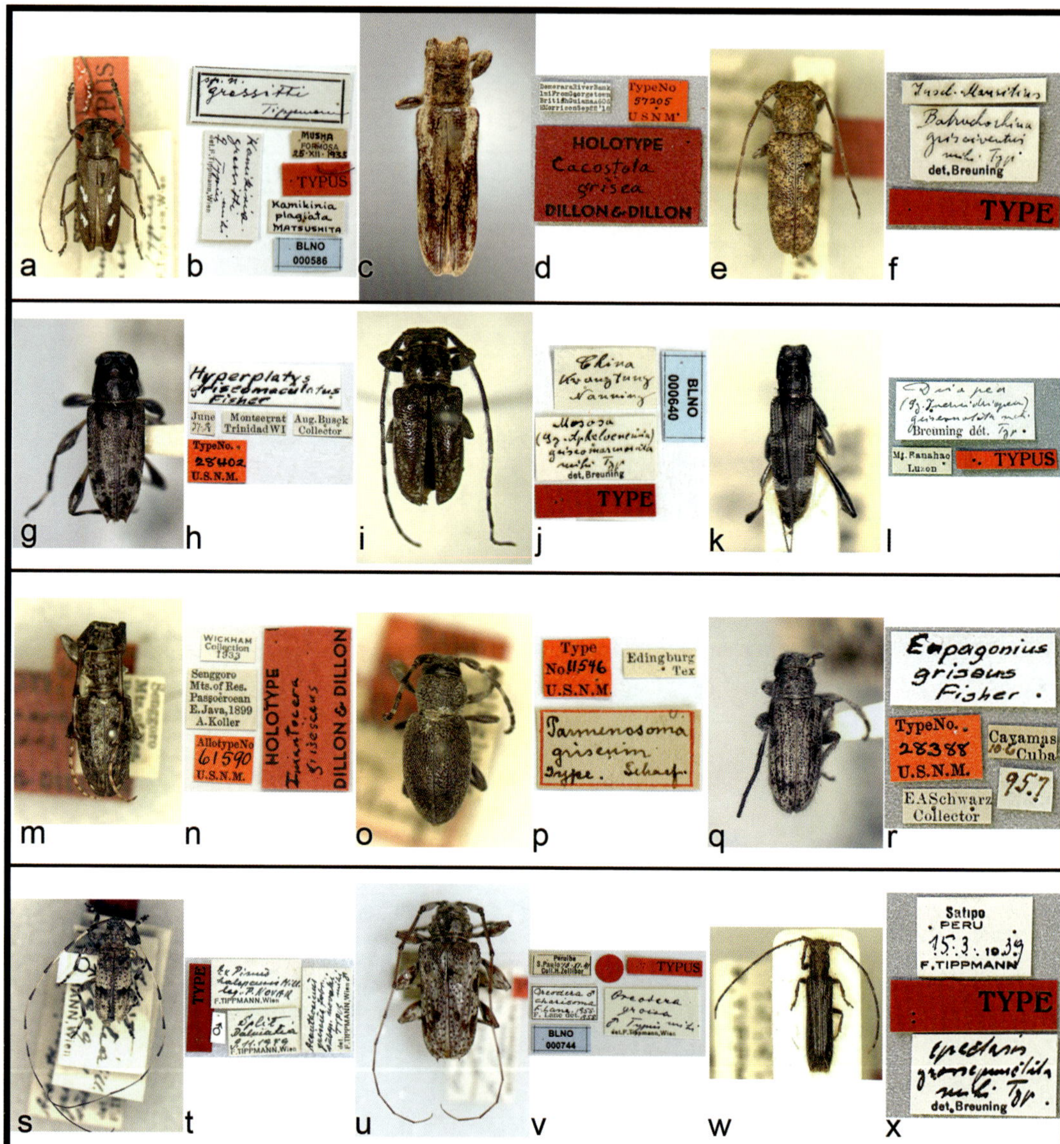

Figure 77. Primary types of Cerambycidae of the Smithsonian Institution (for each pair of images, dorsal habitus left, labels right): **(a, b)** *Kamikiria gressitti* Tippmann, 1955, Holotype; **(c, d)** *Cacostola grisea* Dillon & Dillon, 1946, Holotype; **(e, f)** *Batrachorhina griseiventris* Breuning, 1942, Holotype; **(g, h)** *Hyperplatys griseomaculatus* Fisher, 1926, Holotype; **(i, j)** *Mesosa (Aphelocnemia) griseomarmorata* Breuning, 1939, Holotype; **(k, l)** *Driopea (Inermodriopea) griseonotata* Breuning, 1957, Holotype; **(m, n)** *Imantocera grisescens* Dillon & Dillon, 1951, Holotype; **(o, p)** *Parmenosoma griseum* Schaeffer, 1908, Holotype; **(q, r)** *Eupogonius griseus* Fisher, 1926, Holotype; **(s, t)** *Acanthocinus griseus novaki* Tippmann, 1952, Holotype; **(u, v)** *Oreodera grossa* Tippmann, 1960, Holotype; **(w, x)** *Epectasis grossepunctata* Breuning, 1942, Holotype.

Figure 78. Primary types of Cerambycidae of the Smithsonian Institution (for each pair of images, dorsal habitus left, labels right): **(a, b)** *Prosoplus grossepunctatus* Breuning, 1938, Holotype; **(c, d)** *Temnosternus grossepunctatus* Breuning, 1939, Holotype; **(e, f)** *Dihammus grossescapus* Breuning, 1942, Holotype; **(g, h)** *Ceresium (Ceresium) guamum guamum* Gressitt, 1956, Holotype; **(i, j)** *Leptostylopsis guanica* Micheli & Micheli, 2004, Holotype; **(k, l)** *Atrocolus guarani* Monné & Monné, 2011, Holotype; **(m, n)** *Aplagiognathus guatemalensis* Casey, 1912, Holotype; **(o, p)** *Psapharochrus guatemalensis* Casey, 1913, Holotype; **(q, r)** *Antilleptostylus guilartensis* Micheli & Micheli, 2004, Holotype; **(s, t)** *Clodia guttata* Aurivillius, 1927, Syntype; **(u, v)** *Eburiomorpha guttata* Fisher, 1935, Holotype; **(w, x)** *Xenofrea guttata* Galileo & Martins, 2006, Holotype.

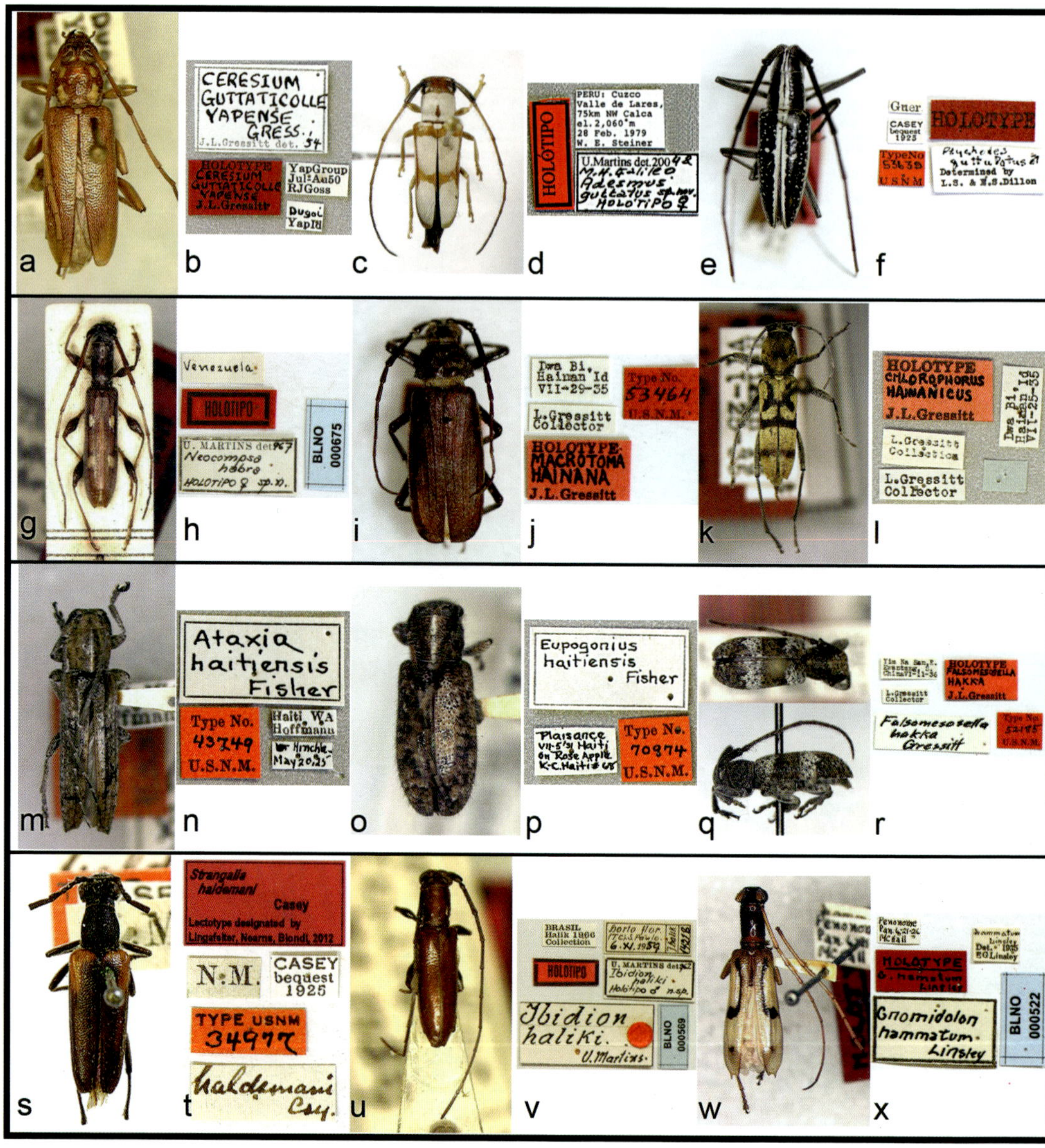

Figure 79. Primary types of Cerambycidae of the Smithsonian Institution (for each pair of images, dorsal habitus left, labels right): **(a, b)** *Ceresium (Ceresium) guttaticolle yapense* Gressitt, 1956, Holotype; **(c, d)** *Adesmus guttatus* Galileo & Martins, 2005, Holotype; **(e, f)** *Ptychodes guttulatus* Dillon & Dillon, 1941, Holotype; **(g, h)** *Neocompsa habra* Martins, 1970, Holotype; **(i, j)** *Macrotoma (Zooblax) hainana* Gressitt, 1940, Holotype; **(k, l)** *Chlorophorus hainanicus* Gressitt, 1940, Holotype; **(m, n)** *Ataxia haitiensis* Fisher, 1932, Holotype; **(o, p)** *Eupogonius haitiensis* Fisher, 1935, Holotype; **(q, r)** *Falsomesosella hakka* Gressitt, 1937, Holotype; **(s, t)** *Leptura haldemani* Casey, 1891, Lectotype; **(u, v)** *Ibidion haliki* Martins, 1962, Holotype; **(w, x)** *Gnomidolon hamatum* Linsley, 1935, Holotype.

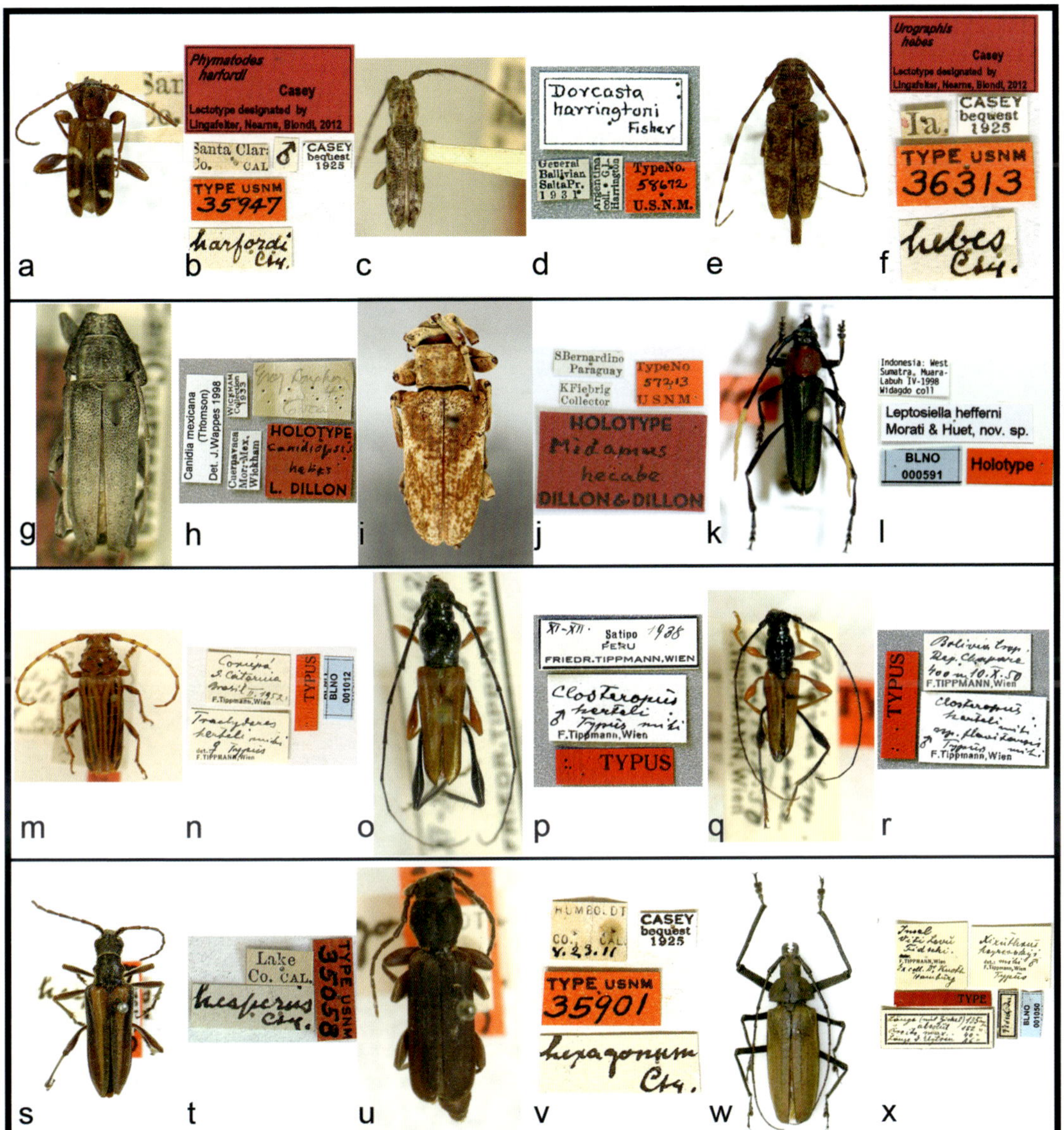

Figure 80. Primary types of Cerambycidae of the Smithsonian Institution (for each pair of images, dorsal habitus left, labels right): **(a, b)** *Phymatodes (Poecilium) harfordi* Casey, 1912, Lectotype; **(c, d)** *Dorcasta harringtoni* Fisher, 1947, Holotype; **(e, f)** *Urographis hebes* Casey, 1913, Lectotype; **(g, h)** *Canidiopsis hebes* Dillon, 1955, Holotype; **(i, j)** *Midamus hecabe* Dillon & Dillon, 1945, Holotype; **(k, l)** *Leptosiella hefferni* Morati & Huet, 2004, Holotype; **(m, n)** *Trachyderes herteli* Tippmann, 1953, Holotype; **(o, p)** *Closteropus herteli* Tippmann, 1960, Holotype; **(q, r)** *Closteropus herteli flavitarsis* Tippmann, 1960, Holotype; **(s, t)** *Stenocorus (Stenocorus) hesperus* Casey, 1913, Holotype; **(u, v)** *Tetropium hexagonum* Casey, 1924, Holotype; **(w, x)** *Xixuthrus heyrovskyi* Tippmann, 1946, Holotype.

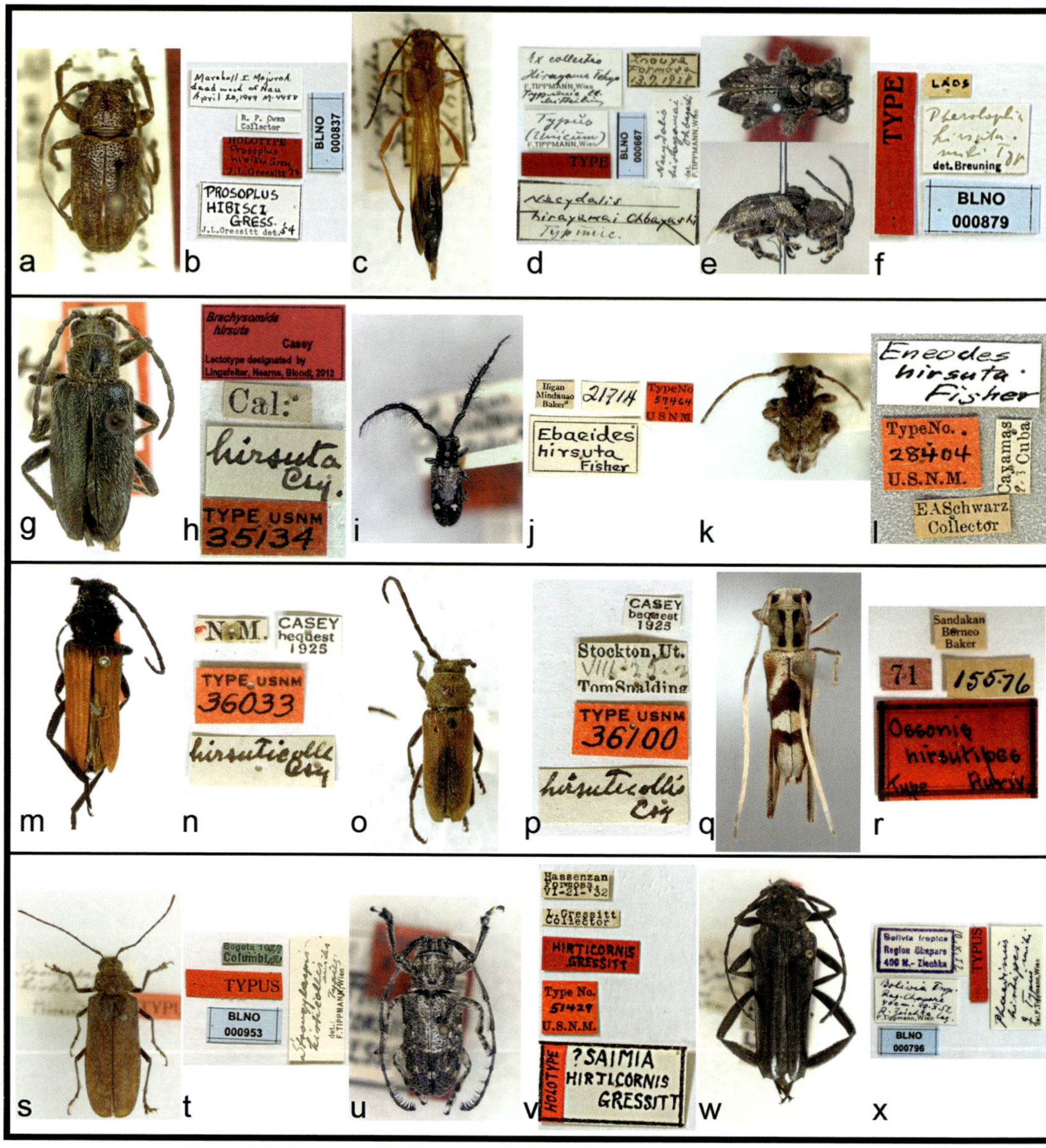

Figure 81. Primary types of Cerambycidae of the Smithsonian Institution (for each pair of images, dorsal habitus left, labels right): **(a, b)** *Prosoplus hibisci* Gressitt, 1956, Holotype; **(c, d)** *Necydalis hirayamai* Ohbayashi, 1948, Holotype; **(e, f)** *Pterolophia hirsuta* Breuning, 1943, Holotype; **(g, h)** *Brachysomida hirsuta* Casey, 1913, Lectotype; **(i, j)** *Ebaeides hirsuta* Fisher, 1925, Holotype; **(k, l)** *Eneodes hirsuta* Fisher, 1926, Holotype; **(m, n)** *Tragidion hirsuticolle* Casey, 1912, Holotype; **(o, p)** *Crossidius hirsuticollis* Casey, 1924, Holotype; **(q, r)** *Ossonis hirsutipes* Aurivillius, 1922, Holotype; **(s, t)** *Strongylaspis hirticollis* Tippmann, 1953, Holotype; **(u, v)** *Saimia hirticornis* Gressitt, 1936, Holotype; **(w, x)** *Phaedinus hirtipes* Tippmann, 1960, Holotype.

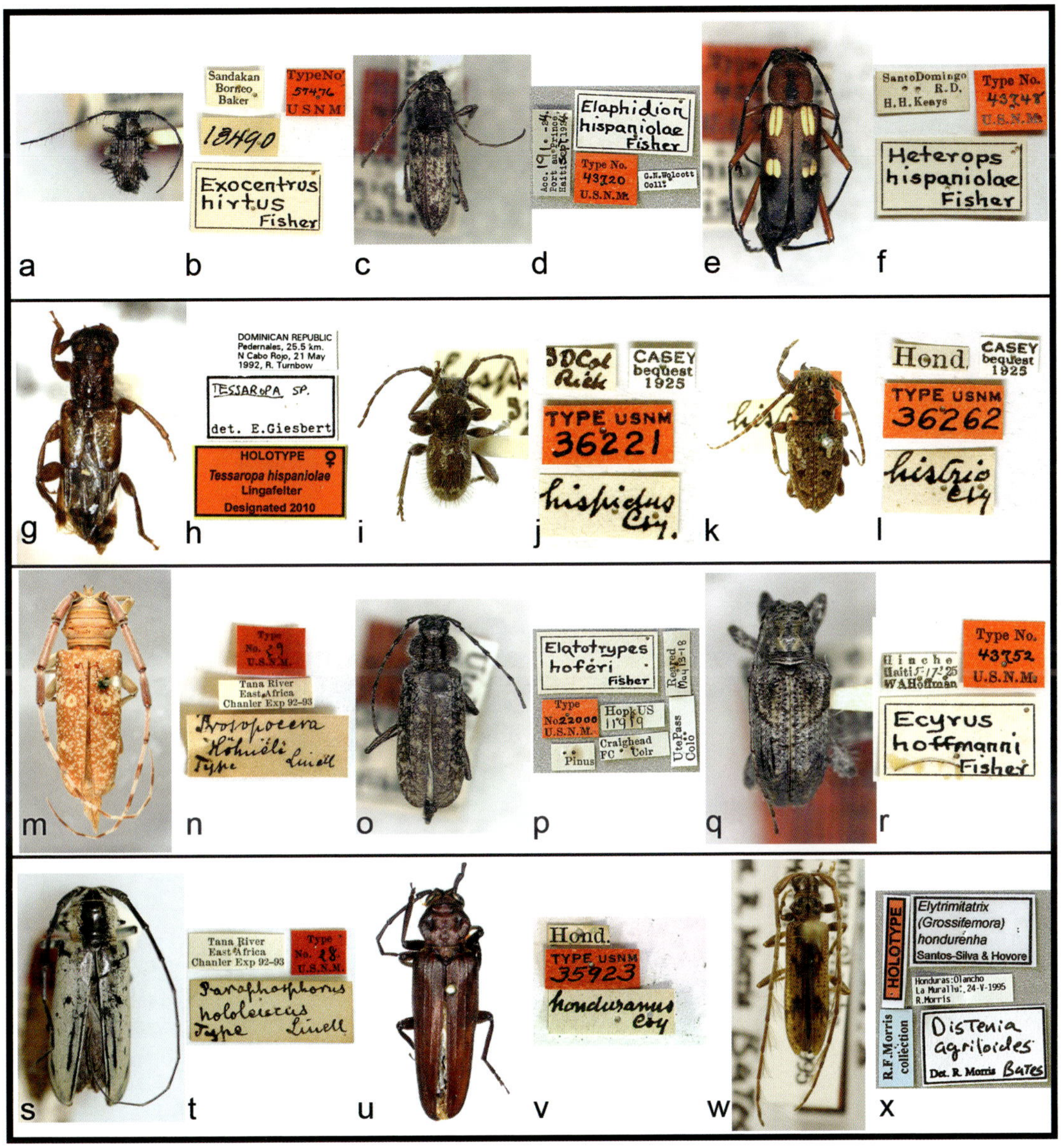

Figure 82. Primary types of Cerambycidae of the Smithsonian Institution (for each pair of images, dorsal habitus left, labels right): **(a, b)** *Exocentrus hirtus* Fisher, 1925, Holotype; **(c, d)** *Elaphidion hispaniolae* Fisher, 1932, Holotype; **(e, f)** *Heterops hispaniolae* Fisher, 1932, Holotype; **(g, h)** *Tessaropa hispaniolae* Lingafelter, 2010, Holotype; **(i, j)** *Ipochus hispidus* Casey, 1913, Holotype; **(k, l)** *Psapharochrus histrio* Casey, 1913, Holotype; **(m, n)** *Prosopocera hoehneli* Linell, 1896, Holotype; **(o, p)** *Elatotrypes hoferi* Fisher, 1919, Holotype; **(q, r)** *Ecyrus hoffmanni* Fisher, 1932, Holotype; **(s, t)** *Paraphosphorus hololeucus* Linell, 1896, Holotype; **(u, v)** *Criocephalus honduranus* Casey, 1912, Holotype; **(w, x)** *Elytrimitatrix (Grossifemora) hondurenha* Santos-Silva & Hovore, 2008, Holotype.

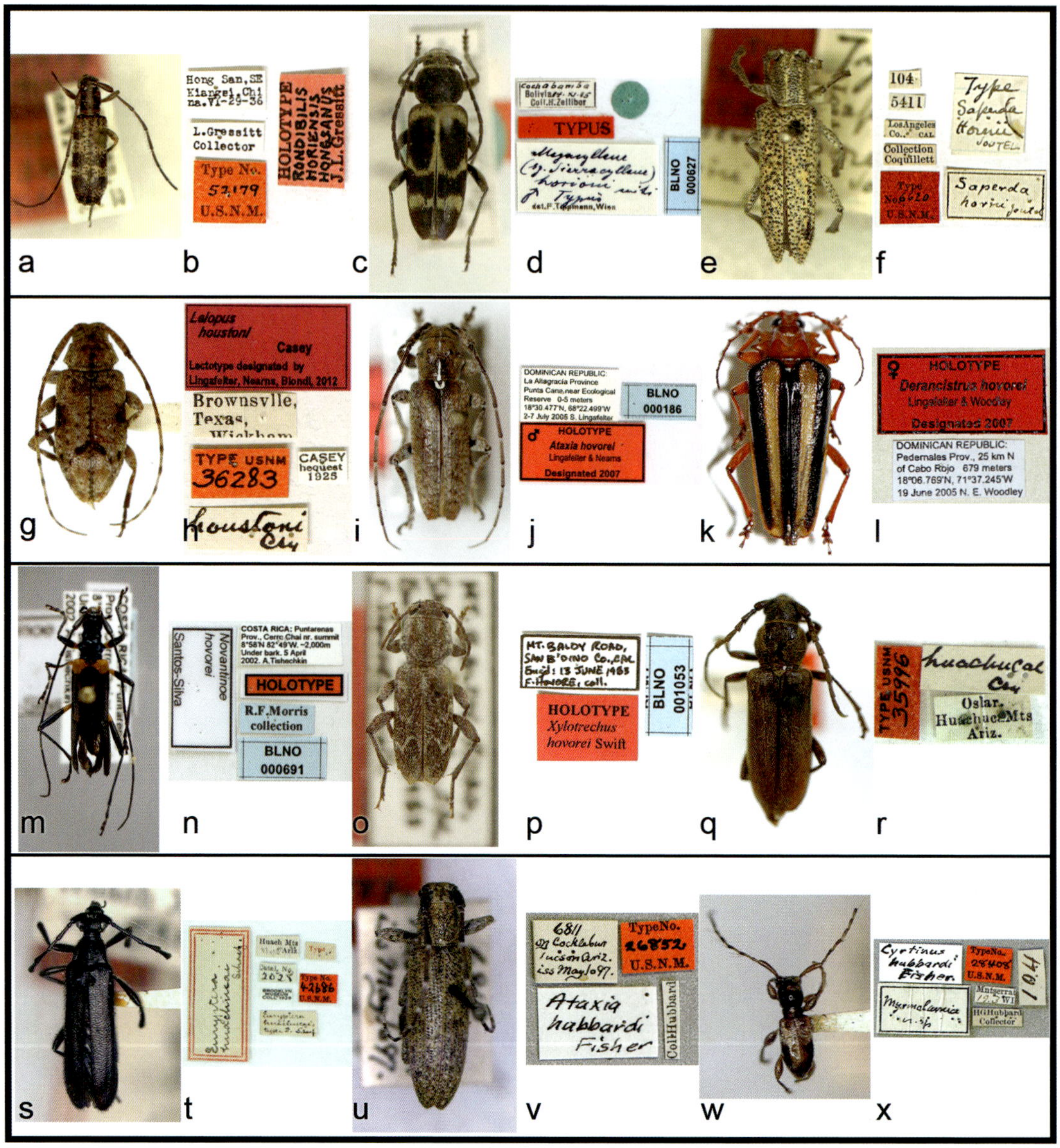

Figure 83. Primary types of Cerambycidae of the Smithsonian Institution (for each pair of images, dorsal habitus left, labels right): **(a, b)** *Rondibilis horiensis hongshanus* Gressitt, 1937, Holotype; **(c, d)** *Megacyllene (Sierracyllene) horioni* Tippmann, 1960, Holotype; **(e, f)** *Saperda hornii* Joutel, 1902, Holotype; **(g, h)** *Leiopus houstoni* Casey, 1913, Lectotype; **(i, j)** *Ataxia hovorei* Lingafelter & Nearns, 2007, Holotype; **(k, l)** *Derancistrus hovorei* Lingafelter & Woodley, 2007, Holotype; **(m, n)** *Novantinoe hovorei* Santos-Silva, 2007, Holotype; **(o, p)** *Xylotrechus hovorei* Swift, 2007, Holotype; **(q, r)** *Anoplium huachucae* Casey, 1924, Holotype; **(s, t)** *Euryptera huachucae* Schaeffer, 1905, Syntype; **(u, v)** *Ataxia hubbardi* Fisher, 1924, Holotype; **(w, x)** *Cyrtinus hubbardi* Fisher, 1926, Holotype.

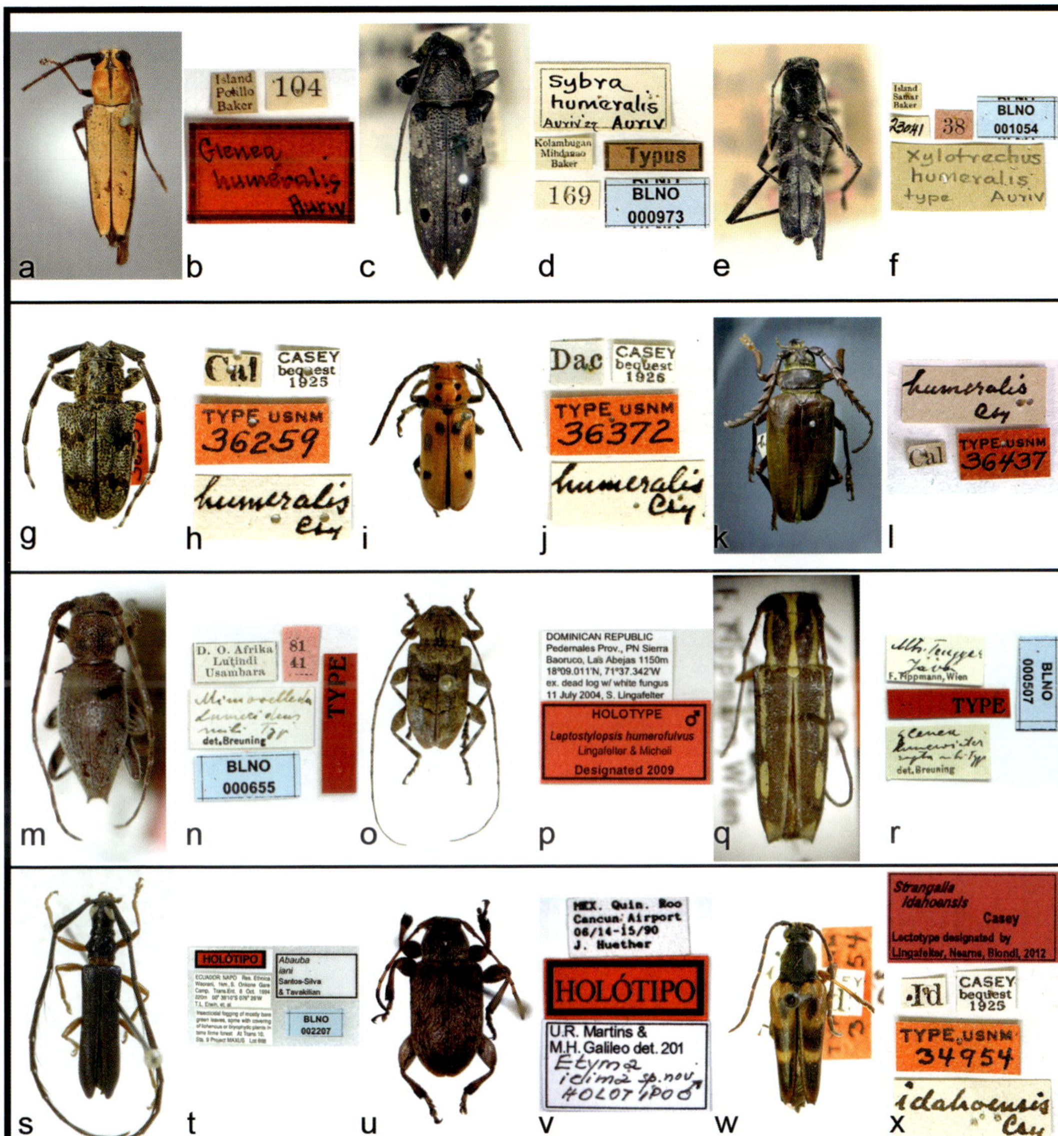

Figure 84. Primary types of Cerambycidae of the Smithsonian Institution (for each pair of images, dorsal habitus left, labels right): **(a, b)** *Glenea (Glenea) humeralis* Aurivillius, 1926, Holotype; **(c, d)** *Sybra humeralis* Aurivillius, 1927, Syntype; **(e, f)** *Xylotrechus humeralis* Aurivillius, 1928, Syntype; **(g, h)** *Synaphaeta humeralis* Casey, 1913, Holotype; **(i, j)** *Tetraopes humeralis* Casey, 1913, Holotype; **(k, l)** *Prionus (Prionus) humeralis* Casey, 1924, Holotype; **(m, n)** *Mimovelleda humeridens* Breuning, 1940, Holotype; **(o, p)** *Leptostylopsis humerofulvus* Lingafelter & Micheli, 2009, Holotype; **(q, r)** *Glenea (Glenea) humerointerrupta* Breuning, 1956, Holotype; **(s, t)** *Abauba iani* Santos-Silva & Tavakilian, 2009, Holotype; **(u, v)** *Etyma icima* Martins & Galileo, 2012, Holotype; **(w, x)** *Strangalia idahoensis* Casey, 1913, Lectotype.

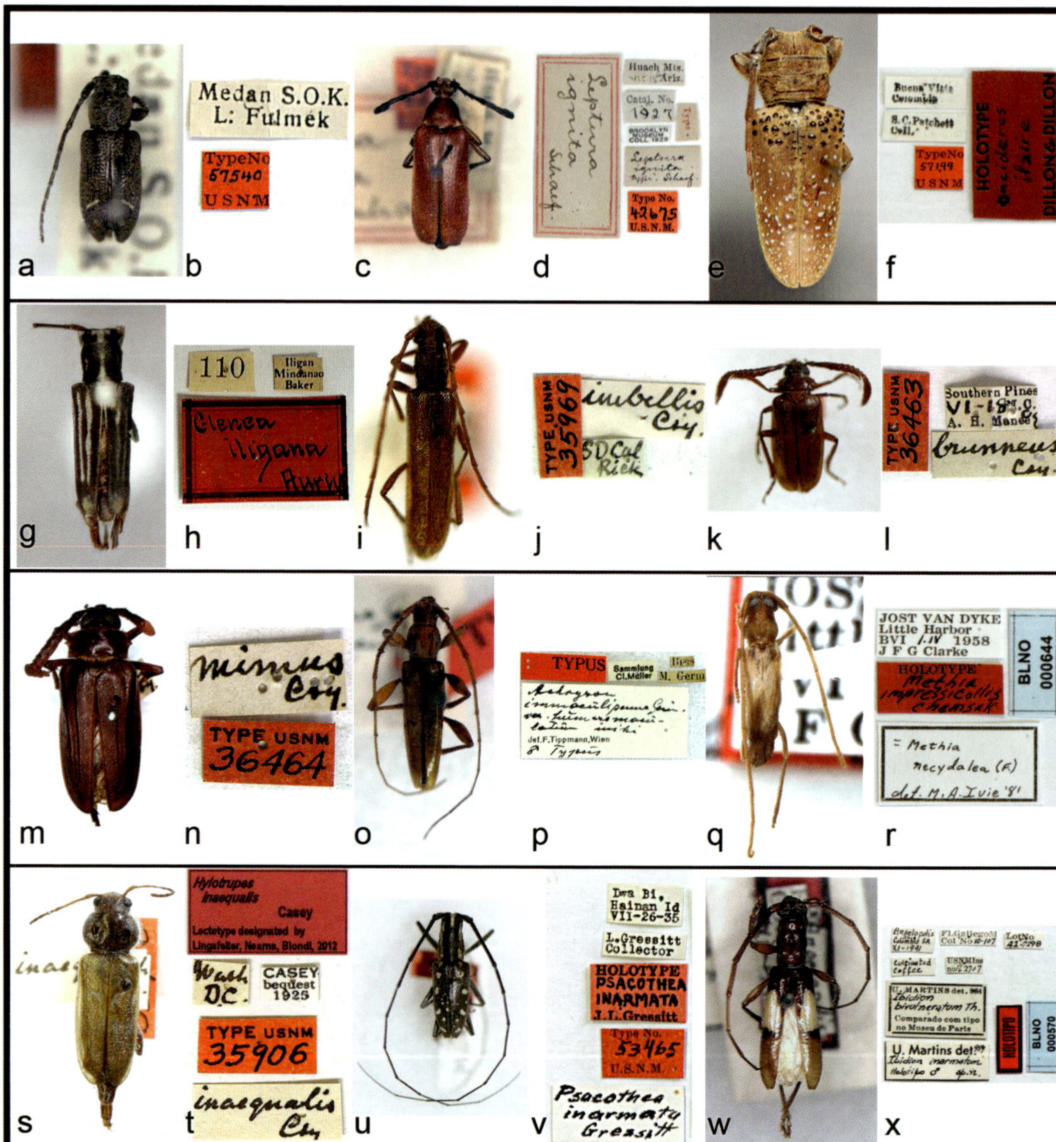

Figure 85. Primary types of Cerambycidae of the Smithsonian Institution (for each pair of images, dorsal habitus left, labels right): **(a, b)** *Pterolophia idoneus* Fisher, 1927, Holotype; **(c, d)** *Leptura ignita* Schaeffer, 1908, Holotype; **(e, f)** *Oncideres ilaire* Dillon & Dillon, 1946, Holotype; **(g, h)** *Glenea (Glenea) iligana* Aurivillius, 1926, Holotype; **(i, j)** *Aneflomorpha imbellis* Casey, 1914, Holotype; **(k, l)** *Prionus (Riponus) imbricornis brunneus* Casey, 1912, Holotype; **(m, n)** *Prionus (Riponus) imbricornis mimus* Casey, 1912, Holotype; **(o, p)** *Achryson immaculipenne* var. *humeromaculatum* Tippmann, 1960, Holotype; **(q, r)** *Methia impressicollis* Chemsak, 1966, Holotype; **(s, t)** *Hylotrupes inaequalis* Casey, 1924, Lectotype; **(u, v)** *Psacothea inarmata* Gressitt, 1940, Holotype; **(w, x)** *Ibidion inarmatum* Martins, 1960, Holotype.

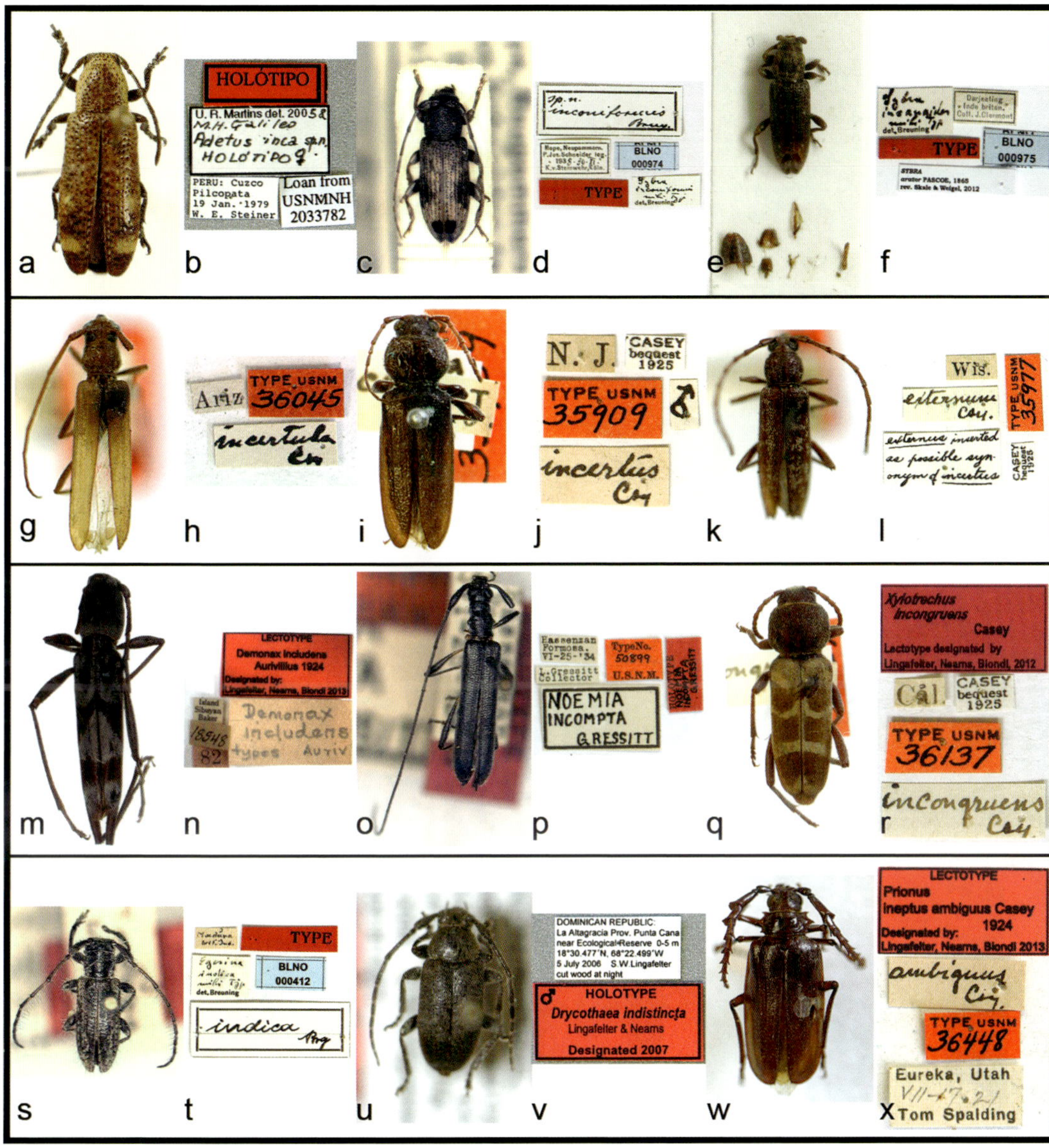

Figure 86. Primary types of Cerambycidae of the Smithsonian Institution (for each pair of images, dorsal habitus left, labels right): **(a, b)** *Adetus inca* Martins & Galileo, 2005, Holotype; **(c, d)** *Sybra incaniformis* Breuning, 1954, Holotype; **(e, f)** *Sybra incanoides* Breuning, 1942, Holotype; **(g, h)** *Oeme incertula* Casey, 1924, Holotype; **(i, j)** *Hylotrupes incertus* Casey, 1924, Holotype; **(k, l)** *Hypermallus incertus externus* Casey, 1912, Holotype; **(m, n)** *Demonax includens* Aurivillius, 1928, Lectotype; **(o, p)** *Noemia incompta* Gressitt, 1935, Holotype; **(q, r)** *Xylotrechus incongruens* Casey, 1912, Lectotype; **(s, t)** *Egesina indica* Breuning, 1938, Holotype; **(u, v)** *Drycothaea indistincta* Lingafelter & Nearns, 2007, Holotype; **(w, x)** *Prionus (Prionus) ineptus ambiguus* Casey, 1924, Lectotype.

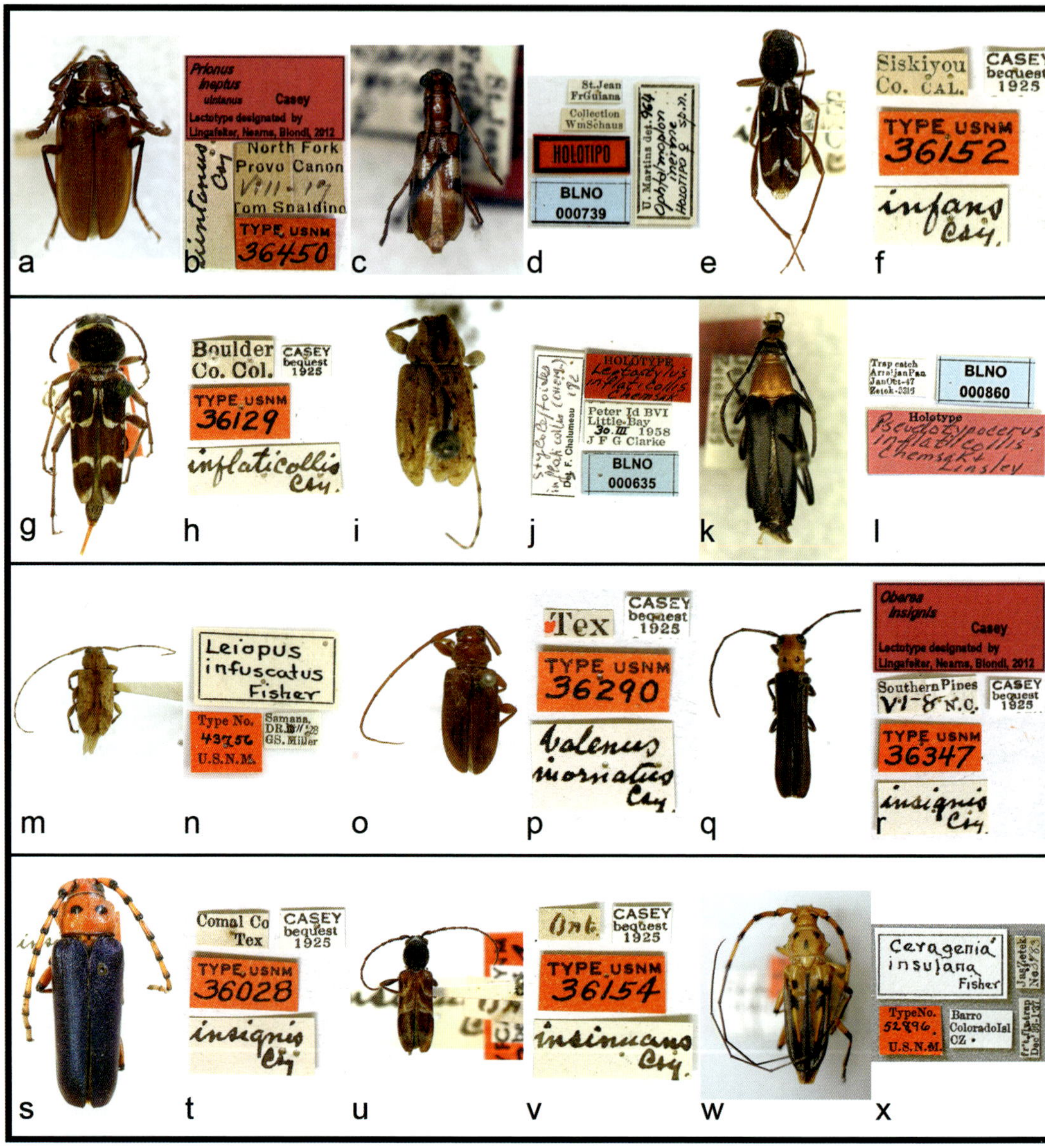

Figure 87. Primary types of Cerambycidae of the Smithsonian Institution (for each pair of images, dorsal habitus left, labels right): **(a, b)** *Prionus (Prionus) ineptus uintanus* Casey, 1924, Lectotype; **(c, d)** *Ophtalmoplon inerme* Martins, 1965, Holotype; **(e, f)** *Neoclytus infans* Casey, 1912, Holotype; **(g, h)** *Xylotrechus inflaticollis* Casey, 1912, Holotype; **(i, j)** *Leptostylus inflaticollis* Chemsak, 1966, Holotype; **(k, l)** *Pseudotypocerus inflaticollis* Chemsak & Linsley, 1976, Holotype; **(m, n)** *Leiopus infuscatus* Fisher, 1932, Holotype; **(o, p)** *Valenus inornatus* Casey, 1891, Holotype; **(q, r)** *Oberea insignis* Casey, 1913, Lectotype; **(s, t)** *Stenaspis insignis* Casey, 1924, Holotype; **(u, v)** *Cyrtophorus insinuans* Casey, 1893, Holotype; **(w, x)** *Ceragenia insulana* Fisher, 1943, Holotype.

Figure 88. Primary types of Cerambycidae of the Smithsonian Institution (for each pair of images, dorsal habitus left, labels right): **(a, b)** *Anatinomma insularis* Chemsak & Linsley, 1964, Holotype; **(c, d)** *Camptomyne insularis* Fisher, 1925, Holotype; **(e, f)** *Trypanidius insularis* Fisher, 1925, Holotype; **(g, h)** *Parmenonta insularis* Fisher, 1930, Holotype; **(i, j)** *Ophistomis insularis* Fisher, 1932, Holotype; **(k, l)** *Ecyrus insularis* Fisher, 1932, Holotype; **(m, n)** *Exocentrus insularis* Fisher, 1934, Holotype; **(o, p)** *Aegocidnus insularis* Fisher, 1934, Holotype; **(q, r)** *Haruspex insularis* Fisher, 1935, Holotype; **(s, t)** *Hypsioma insularis* Fisher, 1935, Holotype; **(u, v)** *Oberea insularis* Fisher, 1937, Holotype; **(w, x)** *Methia insularum* Chemsak, 1966, Holotype.

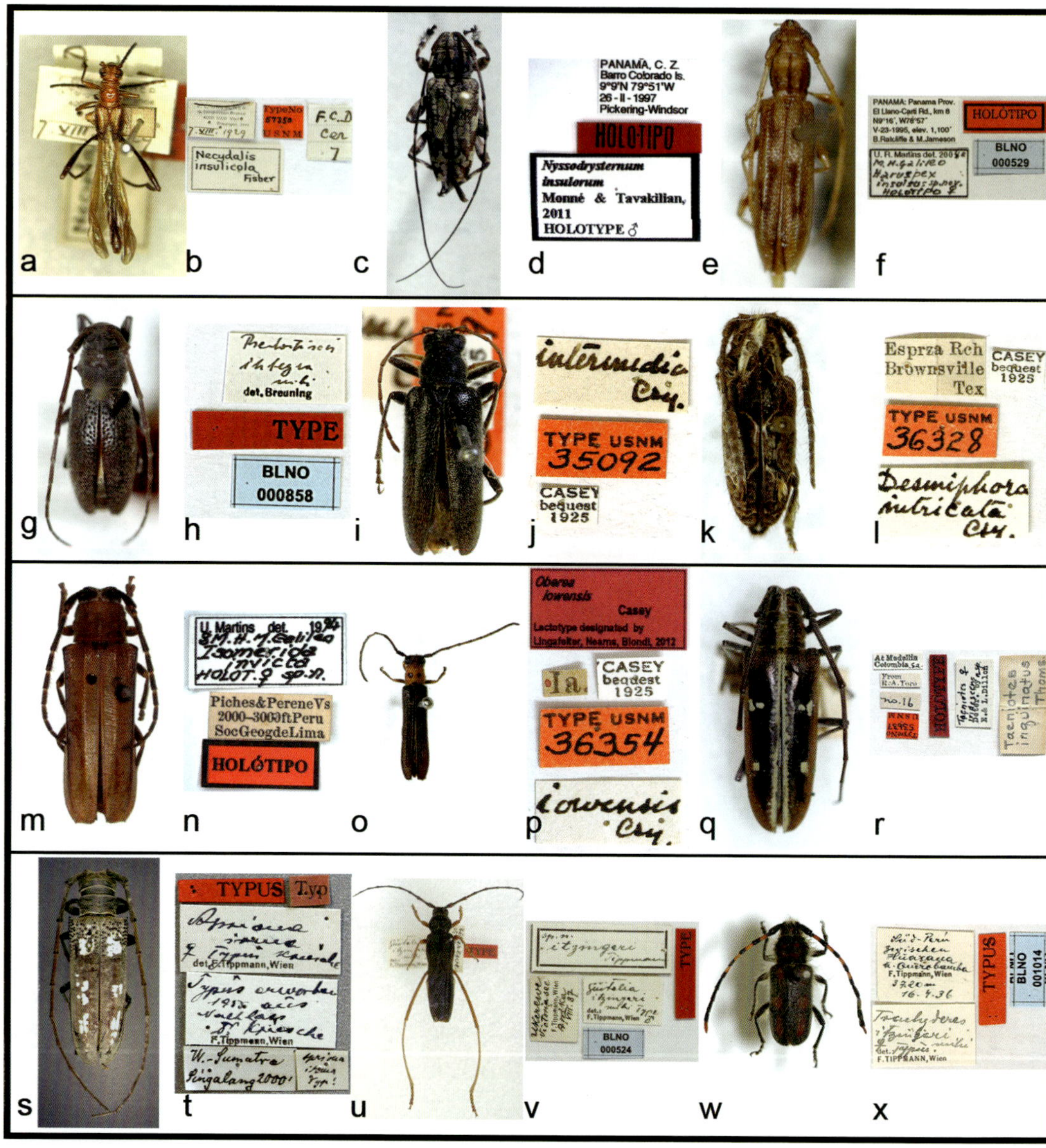

Figure 89. Primary types of Cerambycidae of the Smithsonian Institution (for each pair of images, dorsal habitus left, labels right): **(a, b)** *Necydalis insulicola* Fisher, 1936, Holotype; **(c, d)** *Nyssodrysternum insulorum* Monné & Tavakilian, 2011, Holotype; **(e, f)** *Haruspex insulsus* Martins & Galileo, 2005, Holotype; **(g, h)** *Pseudostixis integra* Breuning, 1938, Holotype; **(i, j)** *Leptacmaeops (Leptacmaeops) intermedia* Casey, 1913, Holotype; **(k, l)** *Desmiphora intricata* Casey, 1913, Holotype; **(m, n)** *Isomerida invicta* Galileo & Martins, 1996, Holotype; **(o, p)** *Oberea iowensis* Casey, 1913, Lectotype; **(q, r)** *Taeniotes iridescens* Dillon & Dillon, 1941, Holotype; **(s, t)** *Apriona irma* Kriesche, 1920, Holotype; **(u, v)** *Guitelia itzingeri* Tippmann, 1951, Holotype; **(w, x)** *Trachyderes itzingeri* Tippmann, 1953, Holotype.

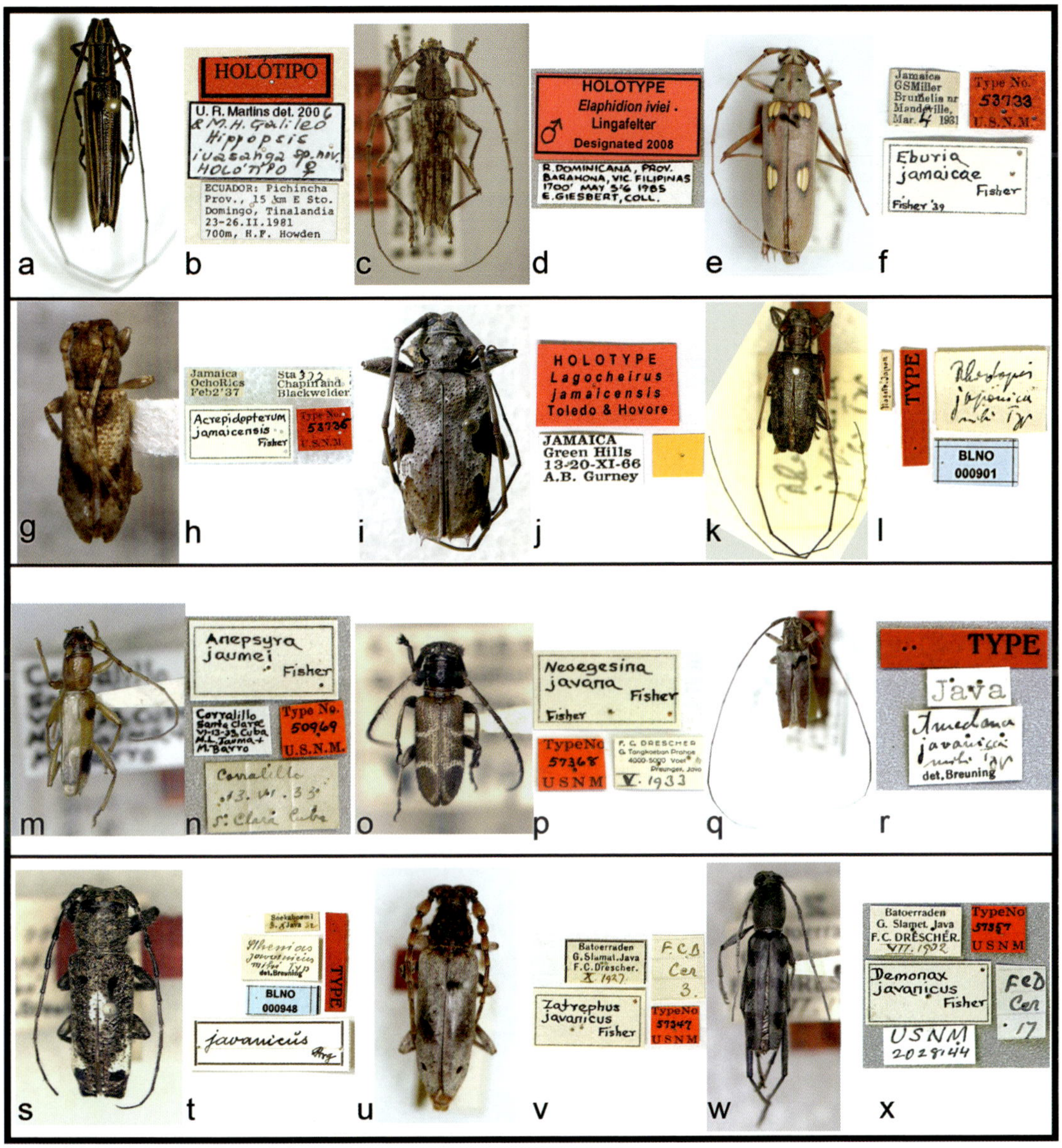

Figure 90. Primary types of Cerambycidae of the Smithsonian Institution (for each pair of images, dorsal habitus left, labels right): **(a, b)** *Hippopsis iuasanga* Martins & Galileo, 2007, Holotype; **(c, d)** *Elaphidion iviei* Lingafelter, 2008, Holotype; **(e, f)** *Eburia jamaicae* Fisher, 1942, Holotype; **(g, h)** *Acrepidopterum jamaicensis* Fisher, 1942, Holotype; **(i, j)** *Lagocheirus jamaicensis* Toledo & Hovore, 2005, Holotype; **(k, l)** *Rhodopis japonica* Breuning, 1940, Holotype; **(m, n)** *Anepsyra jaumei* Fisher, 1935, Holotype; **(o, p)** *Neoegesina javana* Fisher, 1936, Holotype; **(q, r)** *Amechana javanica* Breuning, 1943, Holotype; **(s, t)** *Sthenias javanicus* Breuning, 1940, Holotype; **(u, v)** *Zatrephus javanicus* Fisher, 1936, Holotype; **(w, x)** *Demonax javanicus* Fisher, 1936, Holotype.

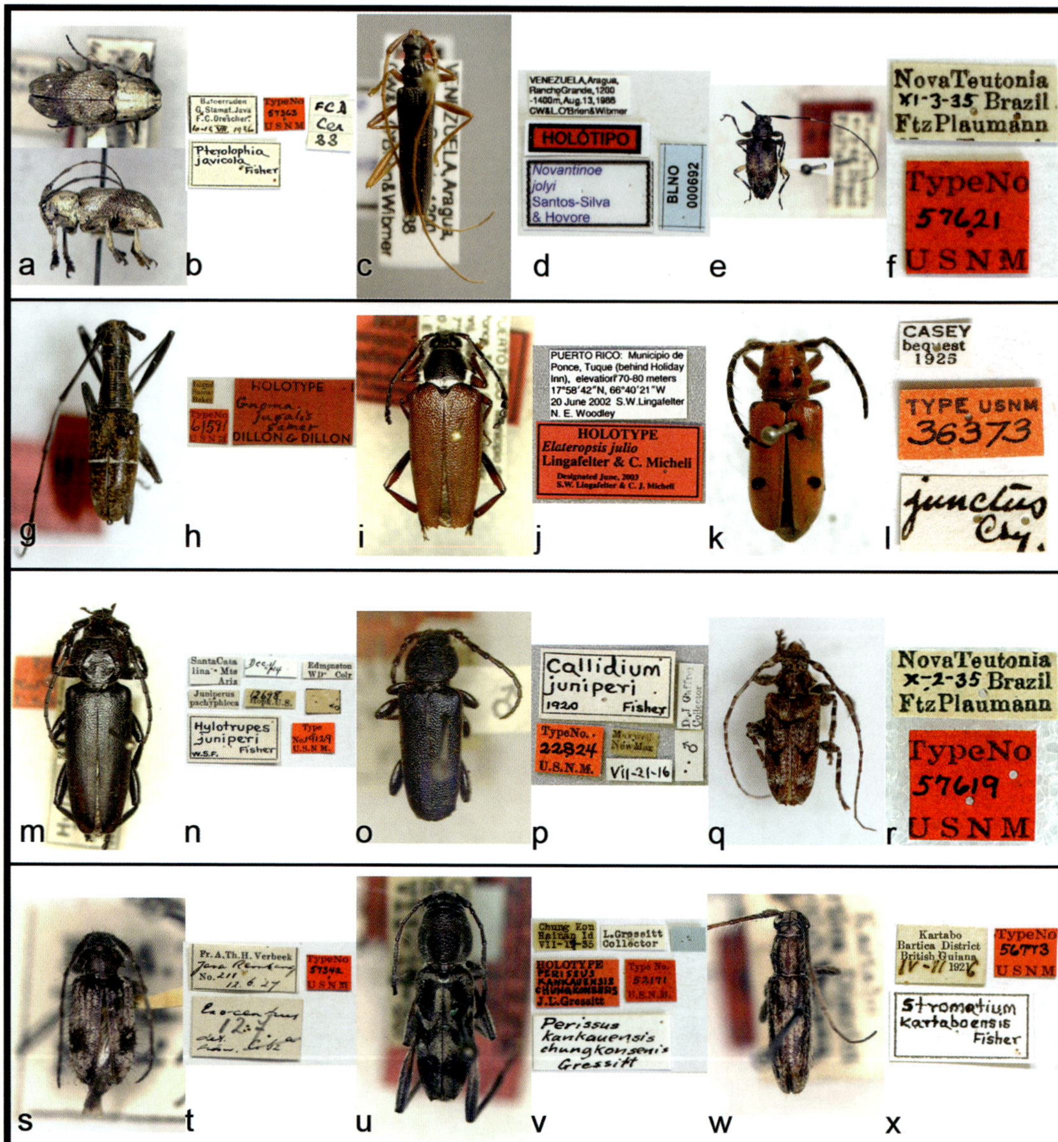

Figure 91. Primary types of Cerambycidae of the Smithsonian Institution (for each pair of images, dorsal habitus left, labels right): **(a, b)** *Pterolophia javicola* Fisher, 1936, Holotype; **(c, d)** *Novantinoe jolyi* Santos-Silva & Hovore, 2007, Holotype; **(e, f)** *Ozineus jubapennis* Fisher, 1938, Holotype; **(g, h)** *Gnoma jugalis samar* Dillon & Dillon, 1951, Holotype; **(i, j)** *Elateropsis julio* Lingafelter & Micheli, 2004, Holotype; **(k, l)** *Tetraopes junctus* Casey, 1913, Holotype; **(m, n)** *Hylotrupes juniperi* Fisher, 1915, Holotype; **(o, p)** *Callidium juniperi* Fisher, 1920, Holotype; **(q, r)** *Acanthoderes juno* Fisher, 1938, Holotype; **(s, t)** *Exocentrus kalshoveni* Fisher, 1934, Holotype; **(u, v)** *Perissus kankauensis chungkonensis* Gressitt, 1940, Holotype; **(w, x)** *Stromatium kartaboensis* Fisher, 1944, Holotype.

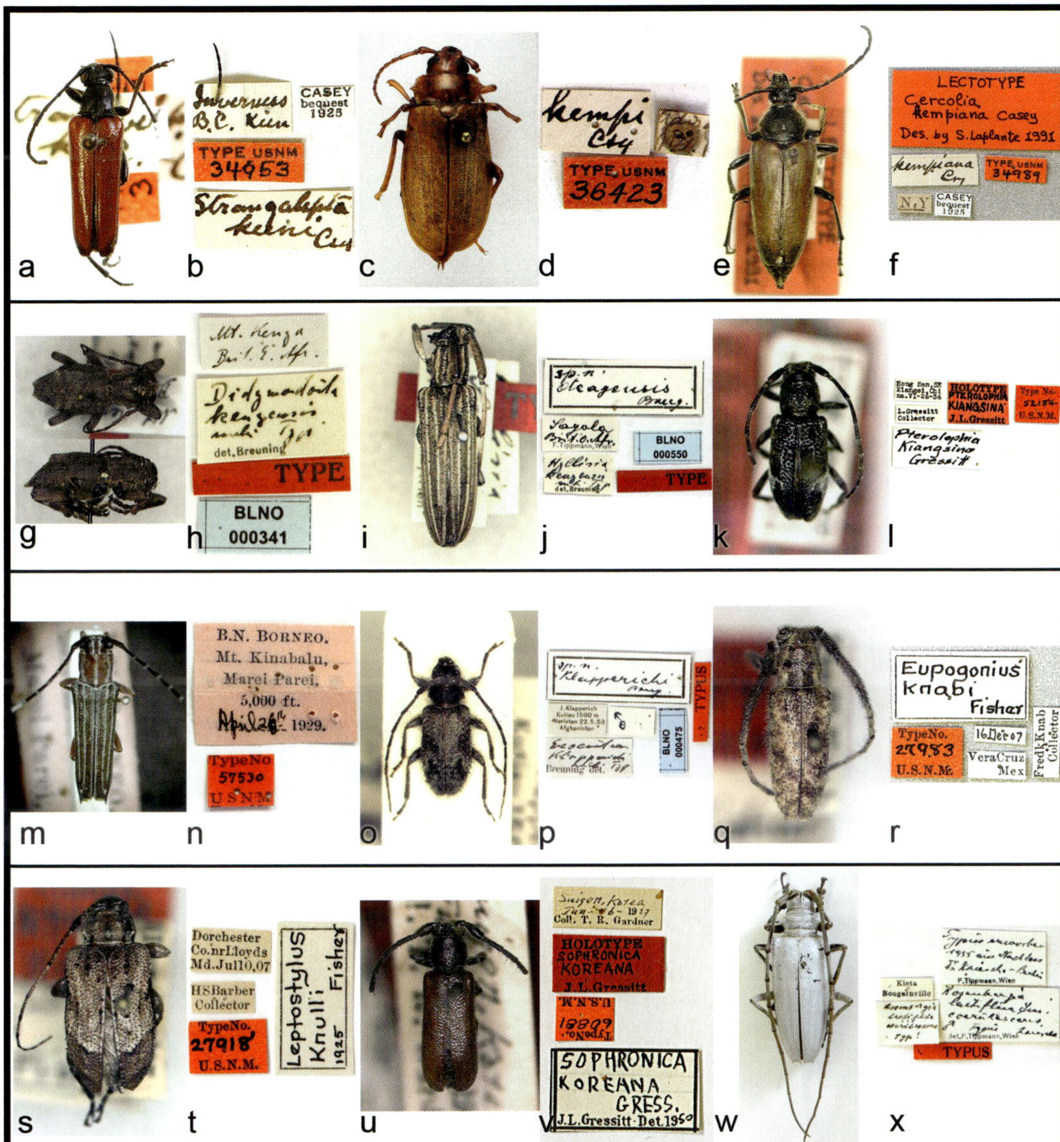

Figure 92. Primary types of Cerambycidae of the Smithsonian Institution (for each pair of images, dorsal habitus left, labels right): **(a, b)** *Strangalepta keeni* Casey, 1913, Holotype; **(c, d)** *Prionus (Prionus) kempi* Casey, 1912, Holotype; **(e, f)** *Cercolia kempiana* Casey, 1924, Lectotype; **(g, h)** *Monoxenus (Dityloderus) kenyensis* Breuning, 1940, Holotype; **(i, j)** *Hyllisia kenyensis* Breuning, 1948, Holotype; **(k, l)** *Pterolophia kiangsina* Gressitt, 1937, Holotype; **(m, n)** *Glenea kinabaluensis* Fisher, 1935, Holotype; **(o, p)** *Exocentrus klapperichi* Breuning, 1957, Holotype; **(q, r)** *Eupogonius knabi* Fisher, 1925, Holotype; **(s, t)** *Leptostylus knulli* Fisher, 1925, Holotype; **(u, v)** *Sophronica koreana* Gressitt, 1951, Holotype; **(w, x)** *Rosenbergia lactiflua coerulescens* Kriesche, 1940, Holotype.

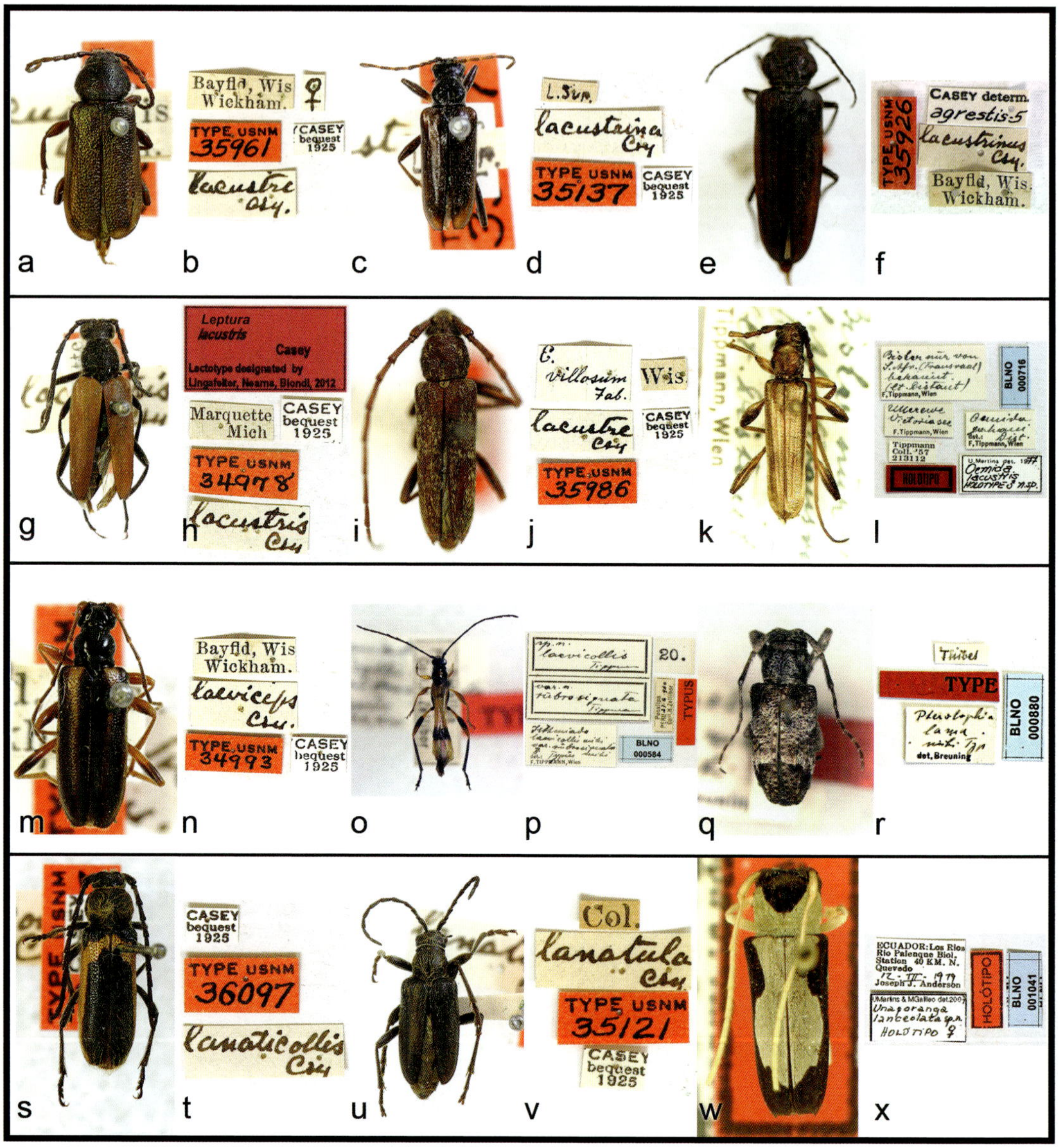

Figure 93. Primary types of Cerambycidae of the Smithsonian Institution (for each pair of images, dorsal habitus left, labels right): **(a, b)** *Callidium lacustre* Casey, 1912, Holotype; **(c, d)** *Acmaeops lacustrina* Casey, 1913, Holotype; **(e, f)** *Criocephalus lacustrinus* Casey, 1912, Holotype; **(g, h)** *Leptura lacustris* Casey, 1891, Lectotype; **(i, j)** *Hypermallus lacustris* Casey, 1912, Holotype; **(k, l)** *Oemida lacustris* Martins, 1977, Holotype; **(m, n)** *Hapalosalia laeviceps* Casey, 1913, Holotype; **(o, p)** *Isthmiade laevicollis* var. *rubrosignata* Tippmann, 1953, Holotype; **(q, r)** *Pterolophia lama* Breuning, 1943, Holotype; **(s, t)** *Crossidius lanaticollis* Casey, 1912, Holotype; **(u, v)** *Brachysomida lanatula* Casey, 1913, Holotype; **(w, x)** *Unaporanga lanceolata* Martins & Galileo, 2007, Holotype.

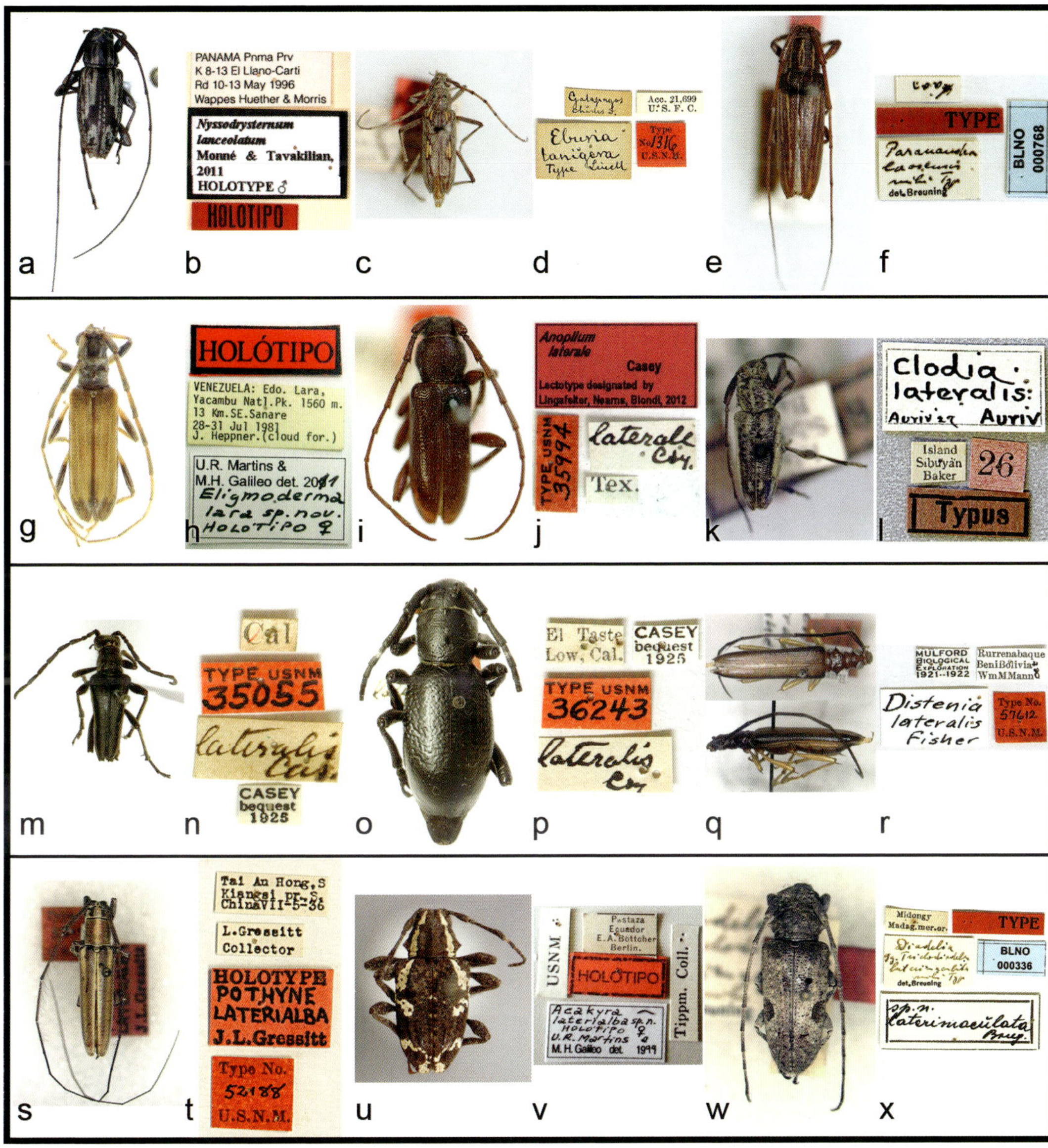

Figure 94. Primary types of Cerambycidae of the Smithsonian Institution (for each pair of images, dorsal habitus left, labels right): **(a, b)** *Nyssodrysternum lanceolatum* Monné & Tavakilian, 2011, Holotype; **(c, d)** *Eburia lanigera* Linell, 1899, Holotype; **(e, f)** *Paranandra laosensis* Breuning, 1942, Holotype; **(g, h)** *Eligmoderma lara* Galileo & Martins, 2012, Holotype; **(i, j)** *Anoplium laterale* Casey, 1914, Lectotype; **(k, l)** *Clodia lateralis* Aurivillius, 1927, Syntype; **(m, n)** *Toxotus lateralis* Casey, 1891, Holotype; **(o, p)** *Moneilema (Collapteryx) lateralis* Casey, 1924, Holotype; **(q, r)** *Distenia lateralis* Fisher, 1946, Holotype; **(s, t)** *Pothyne (Laterialba) laterialba* Gressitt, 1937, Holotype; **(u, v)** *Acakyra laterialba* Martins & Galileo, 2001, Holotype; **(w, x)** *Diadelia (Trichodiadelia) laterimaculata* Breuning, 1943, Holotype.

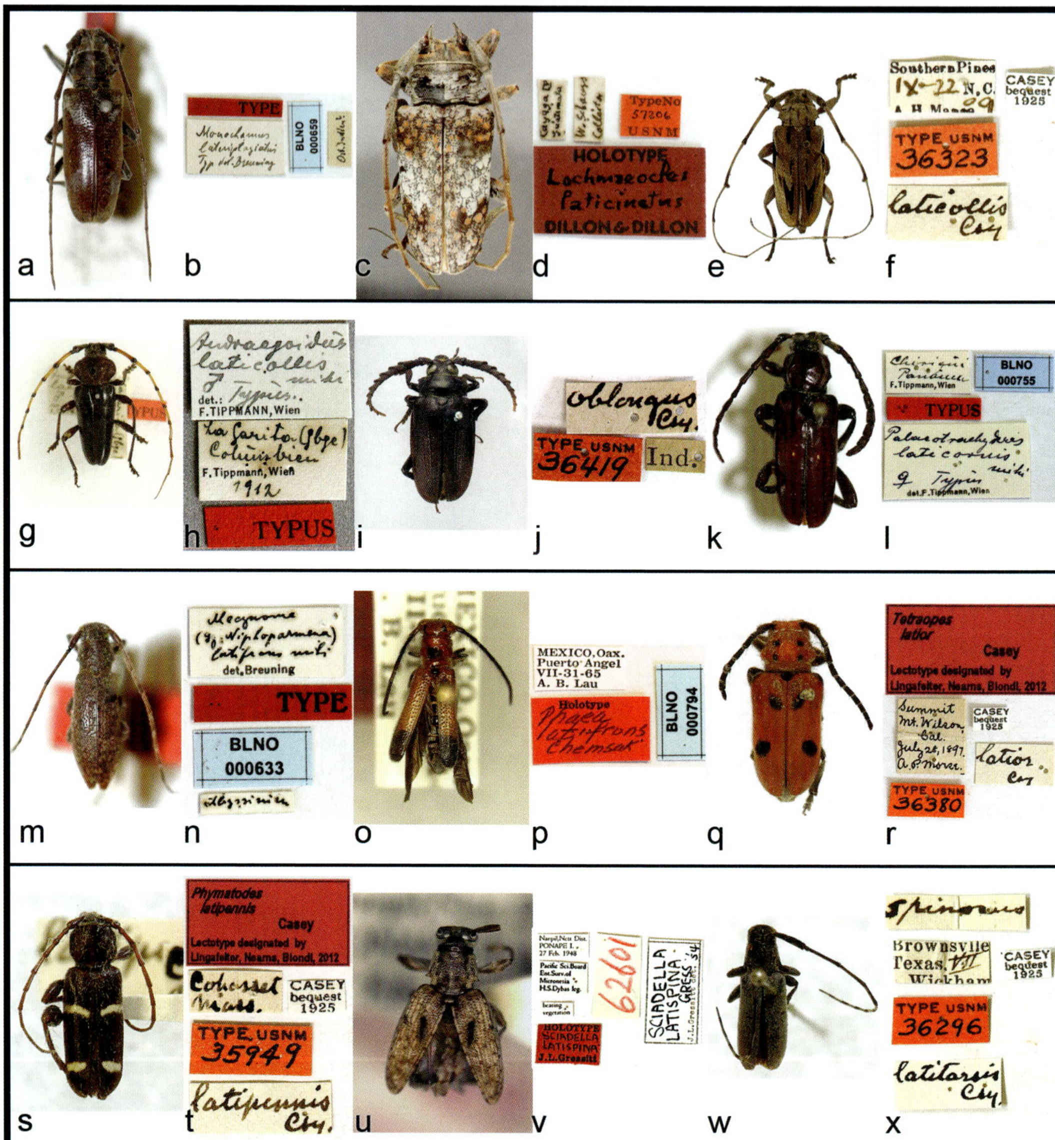

Figure 95. Primary types of Cerambycidae of the Smithsonian Institution (for each pair of images, dorsal habitus left, labels right): **(a, b)** *Monochamus lateriplagiatus* Breuning, 1942, Holotype; **(c, d)** *Lochmaeocles laticinctus* Dillon & Dillon, 1946, Holotype; **(e, f)** *Graphisurus (Tylocerina) laticollis* Casey, 1913, Holotype; **(g, h)** *Andraegoidus laticollis* Tippmann, 1953, Holotype; **(i, j)** *Prionus (Prionus) laticollis oblongus* Casey, 1912, Holotype; **(k, l)** *Palaeotrachyderes laticornis* Tippmann, 1960, Holotype; **(m, n)** *Mecynome (Niphoparmena) latifrons* Breuning, 1940, Holotype; **(o, p)** *Phaea latifrons* Chemsak, 1977, Holotype; **(q, r)** *Tetraopes latior* Casey, 1924, Lectotype; **(s, t)** *Phymatodes (Poecilium) latipennis* Casey, 1912, Lectotype; **(u, v)** *Sciadella (Acanthosciades) latispina* Gressitt, 1956, Holotype; **(w, x)** *Dectes latitarsis* Casey, 1913, Holotype.

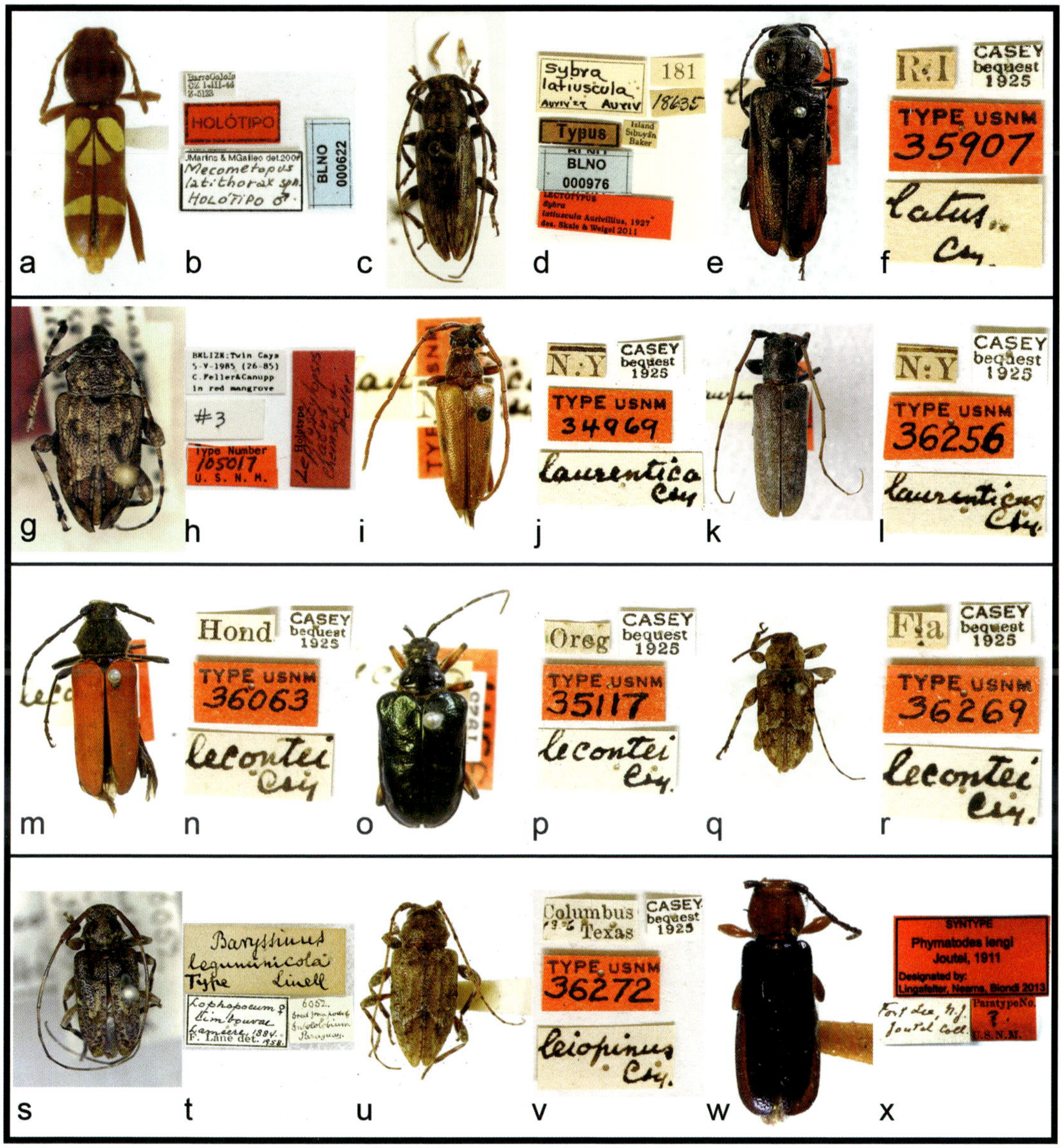

Figure 96. Primary types of Cerambycidae of the Smithsonian Institution (for each pair of images, dorsal habitus left, labels right): **(a, b)** *Mecometopus latithorax* Martins & Galileo, 2008, Holotype; **(c, d)** *Sybra latiuscula* Aurivillius, 1927, Lectotype; **(e, f)** *Hylotrupes latus* Casey, 1924, Holotype; **(g, h)** *Leptostylopsis latus* Chemsak & Feller, 1988, Holotype; **(i, j)** *Leptura (Dorcasina) laurentica* Casey, 1913, Holotype; **(k, l)** *Goes laurenticus* Casey, 1913, Holotype; **(m, n)** *Purpuricenus lecontei* Casey, 1912, Holotype; **(o, p)** *Gaurotes lecontei* Casey, 1913, Holotype; **(q, r)** *Leptostylus lecontei* Casey, 1913, Holotype; **(s, t)** *Baryssinus leguminicola* Linell, 1896, Holotype; **(u, v)** *Astylidius leiopinus* Casey, 1913, Holotype; **(w, x)** *Phymatodes lengi* Joutel, 1911, Syntype.

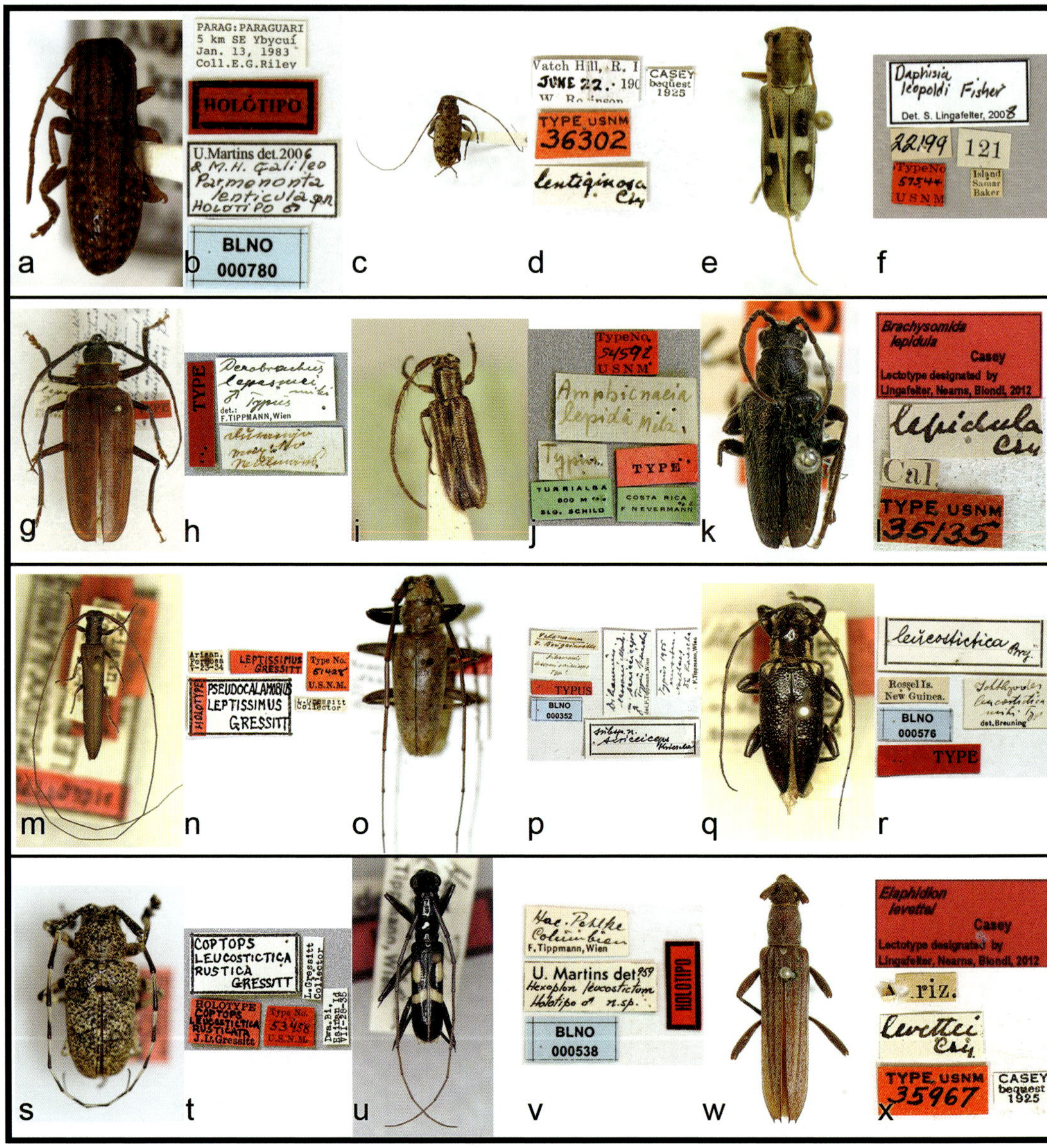

Figure 97. Primary types of Cerambycidae of the Smithsonian Institution (for each pair of images, dorsal habitus left, labels right): **(a, b)** *Parmenonta lenticula* Galileo & Martins, 2006, Holotype; **(c, d)** *Hyperplatys lentiginosa* Casey, 1913, Holotype; **(e, f)** *Daphysia leopoldi* Fisher, 1934, Holotype; **(g, h)** *Derobrachus lepesmei* Tippmann, 1953, Holotype; **(i, j)** *Amphicnaeia lepida* Melzer, 1933, Holotype; **(k, l)** *Brachysomida lepidula* Casey, 1913, Lectotype; **(m, n)** *Pseudocalamobius leptissimus* Gressitt, 1936, Holotype; **(o, p)** *Dihammus lessoni sericeiceps* Kriesche, 1936, Holotype; **(q, r)** *Ichthyodes (Ichthyodes) leucostictica* Breuning, 1942, Holotype; **(s, t)** *Coptops leucostictica rustica* Gressitt, 1940, Holotype; **(u, v)** *Hexoplon leucostictum* Martins, 1959, Holotype; **(w, x)** *Elaphidion levettei* Casey, 1891, Lectotype.

Figure 98. Primary types of Cerambycidae of the Smithsonian Institution (for each pair of images, dorsal habitus left, labels right): **(a, b)** *Anocomis lignea parvicollis* Casey, 1912, Holotype; **(c, d)** *Falsozorilispe linearis* Breuning, 1943, Holotype; **(e, f)** *Jamesia lineata* Fisher, 1926, Holotype; **(g, h)** *Aporataxia lineata* Hamilton, 1896, Lectotype; **(i, j)** *Megachoriolaus lineaticollis* Chemsak & Linsley, 1974, Holotype; **(k, l)** *Emphreus lineatipennis* Breuning, 1950, Holotype; **(m, n)** *Ozineus lineellus* Fisher, 1938, Holotype; **(o, p)** *Hapalosalia lineicornis* Casey, 1913, Holotype; **(q, r)** *Anoplium linelli* Casey, 1924, Lectotype; **(s, t)** *Achryson lineolatum* var. *andinum* Tippmann, 1960, Holotype; **(u, v)** *Achryson lineolatum* var. *nigricans* Tippmann, 1960, Holotype; **(w, x)** *Cometes lingafelteri* Hovore & Santos-Silva, 2007, Holotype.

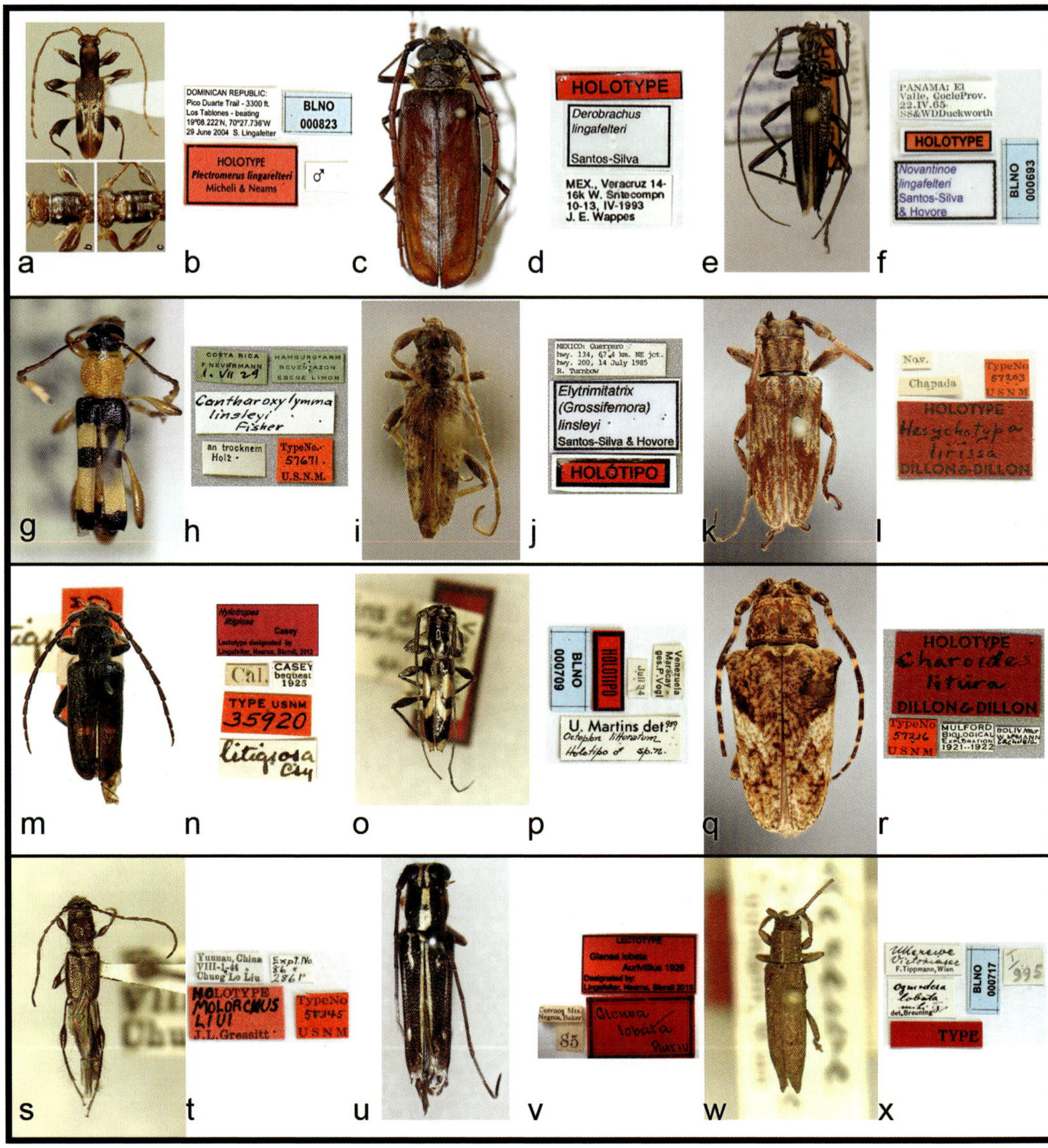

Figure 99. Primary types of Cerambycidae of the Smithsonian Institution (for each pair of images, dorsal habitus left, labels right): **(a, b)** *Plectromerus lingafelteri* Micheli & Nearns, 2005, Holotype; **(c, d)** *Derobrachus lingafelteri* Santos-Silva, 2007, Holotype; **(e, f)** *Novantinoe lingafelteri* Santos-Silva & Hovore, 2007, Holotype; **(g, h)** *Cantharoxylymna linsleyi* Fisher, 1947, Holotype; **(i, j)** *Elytrimitatrix (Grossifemora) linsleyi* Santos-Silva & Hovore, 2008, Holotype; **(k, l)** *Hesychotypa lirissa* Dillon & Dillon, 1945, Holotype; **(m, n)** *Hylotrupes litigiosus* Casey, 1891, Lectotype; **(o, p)** *Octoplon litturatum* Martins, 1960, Holotype; **(q, r)** *Charoides litura* Dillon & Dillon, 1945, Holotype; **(s, t)** *Molorchus liui* Gressitt, 1948, Holotype; **(u, v)** *Glenea lobata* Aurivillius, 1926, Lectotype; **(w, x)** *Ogmodera lobata* Breuning, 1942, Holotype.

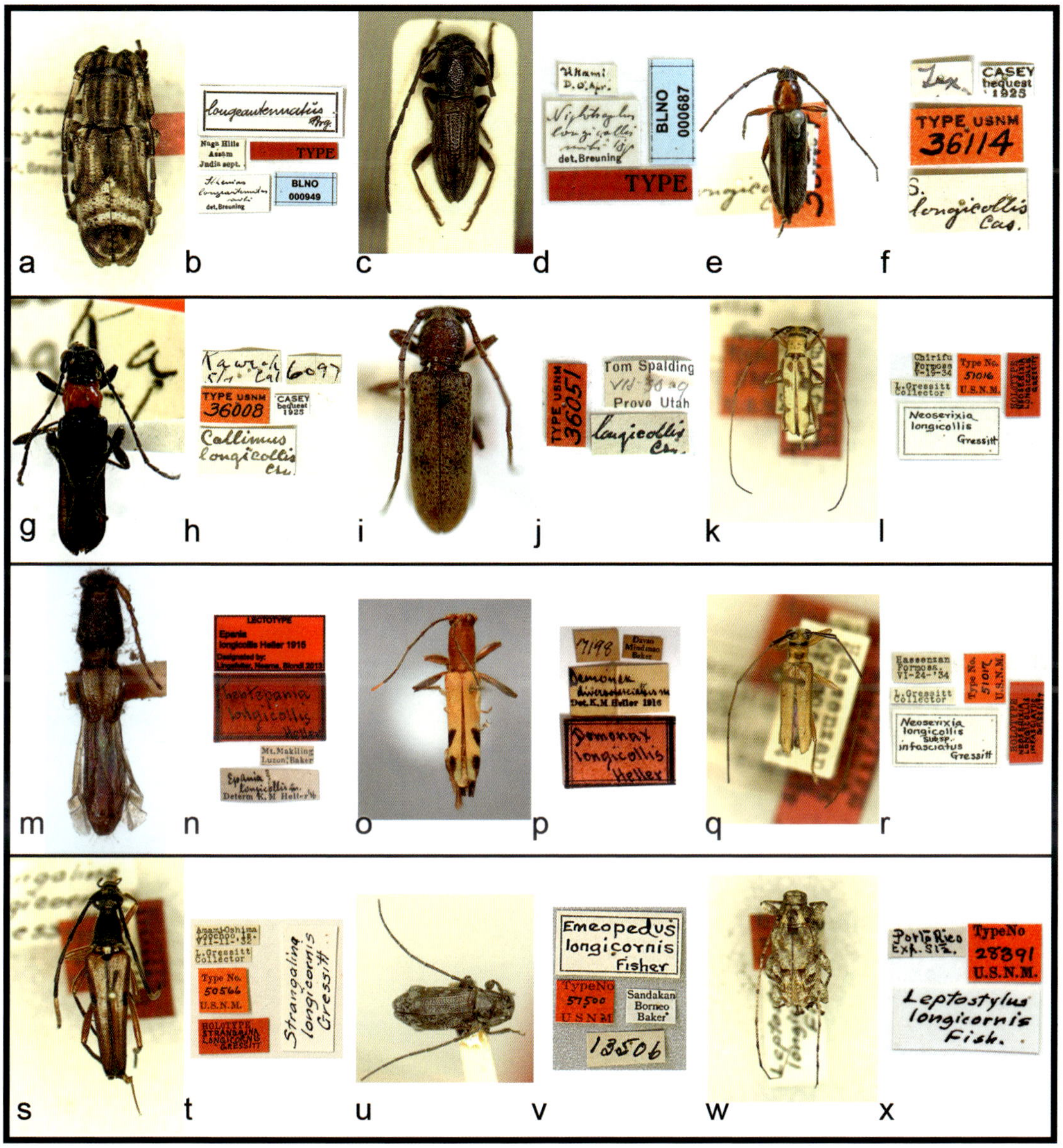

Figure 100. Primary types of Cerambycidae of the Smithsonian Institution (for each pair of images, dorsal habitus left, labels right): **(a, b)** *Sthenias longeantennatus* Breuning, 1938, Holotype; **(c, d)** *Niphotragulus longicollis* Breuning, 1942, Holotype; **(e, f)** *Stenosphenus longicollis* Casey, 1891, Holotype; **(g, h)** *Callimus longicollis* Casey, 1912, Holotype; **(i, j)** *Brothylus longicollis* Casey, 1924, Holotype; **(k, l)** *Neoserixia longicollis* Gressitt, 1935, Holotype; **(m, n)** *Epania longicollis* Heller, 1915, Lectotype; **(o, p)** *Demonax longicollis* Heller, 1916, Holotype; **(q, r)** *Neoserixia longicollis infasciatus* Gressitt, 1935, Holotype; **(s, t)** *Strangalia (Strangalina) longicorne* Gressitt, 1935, Holotype; **(u, v)** *Emeopedus longicornis* Fisher, 1925, Holotype; **(w, x)** *Leptostylus longicornis* Fisher, 1926, Holotype.

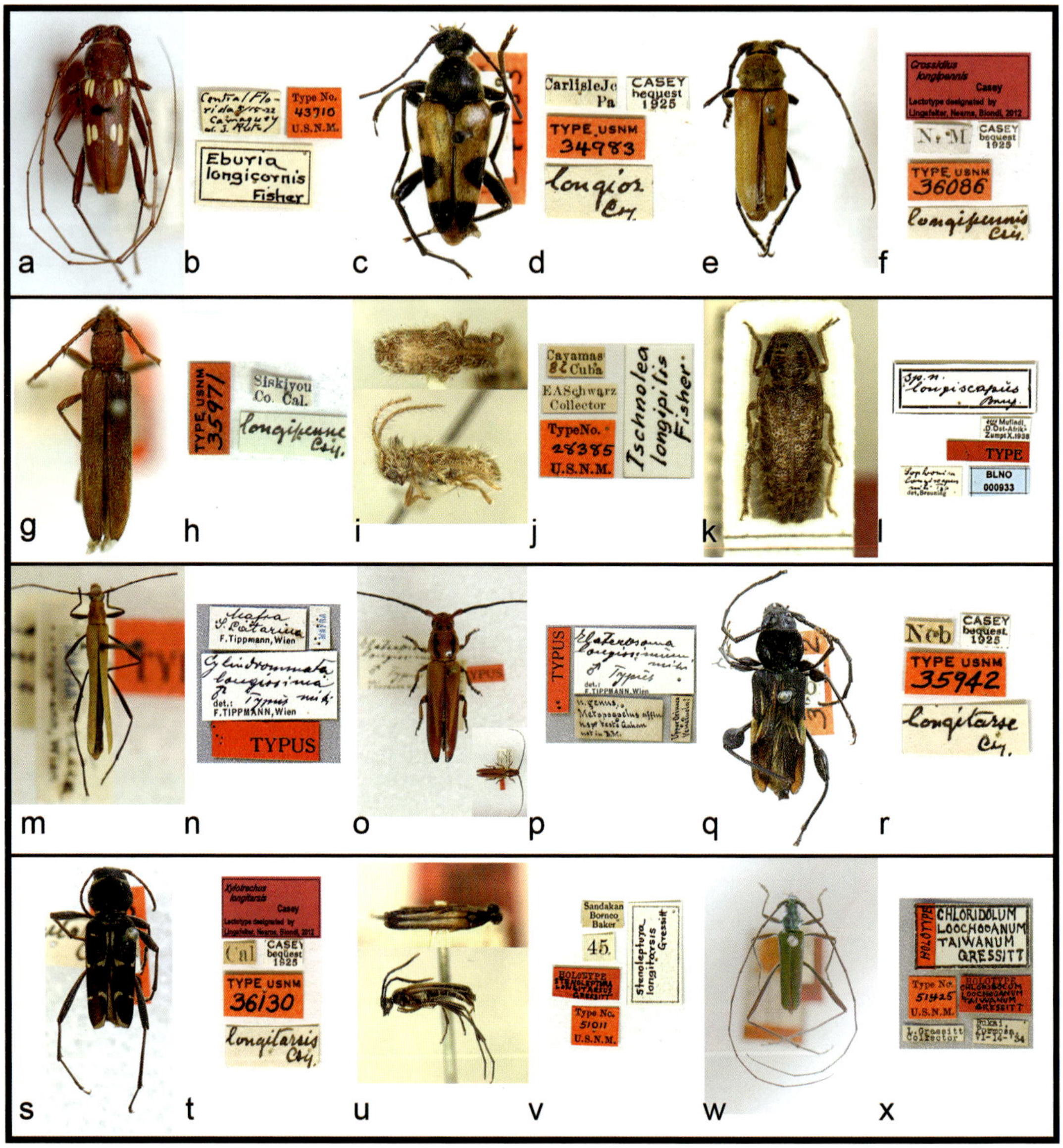

Figure 101. Primary types of Cerambycidae of the Smithsonian Institution (for each pair of images, dorsal habitus left, labels right): **(a, b)** *Eburia longicornis* Fisher, 1932, Holotype; **(c, d)** *Judolia longior* Casey, 1924, Holotype; **(e, f)** *Crossidius longipennis* Casey, 1891, Lectotype; **(g, h)** *Aneflomorpha longipennis* Casey, 1912, Holotype; **(i, j)** *Ischnolea longipilis* Fisher, 1926, Holotype; **(k, l)** *Sophronica longiscapus* Breuning, 1943, Holotype; **(m, n)** *Cylindrommata longissima* Tippmann, 1960, Holotype; **(o, p)** *Elaterosoma longissimum* Tippmann, 1953, Holotype; **(q, r)** *Physocnemum longitarse* Casey, 1924, Holotype; **(s, t)** *Xylotrechus longitarsis* Casey, 1912, Lectotype; **(u, v)** *Stenoleptura longitarsus* Gressitt, 1935, Holotype; **(w, x)** *Chloridolum loochooanum taiwanum* Gressitt, 1936, Holotype.

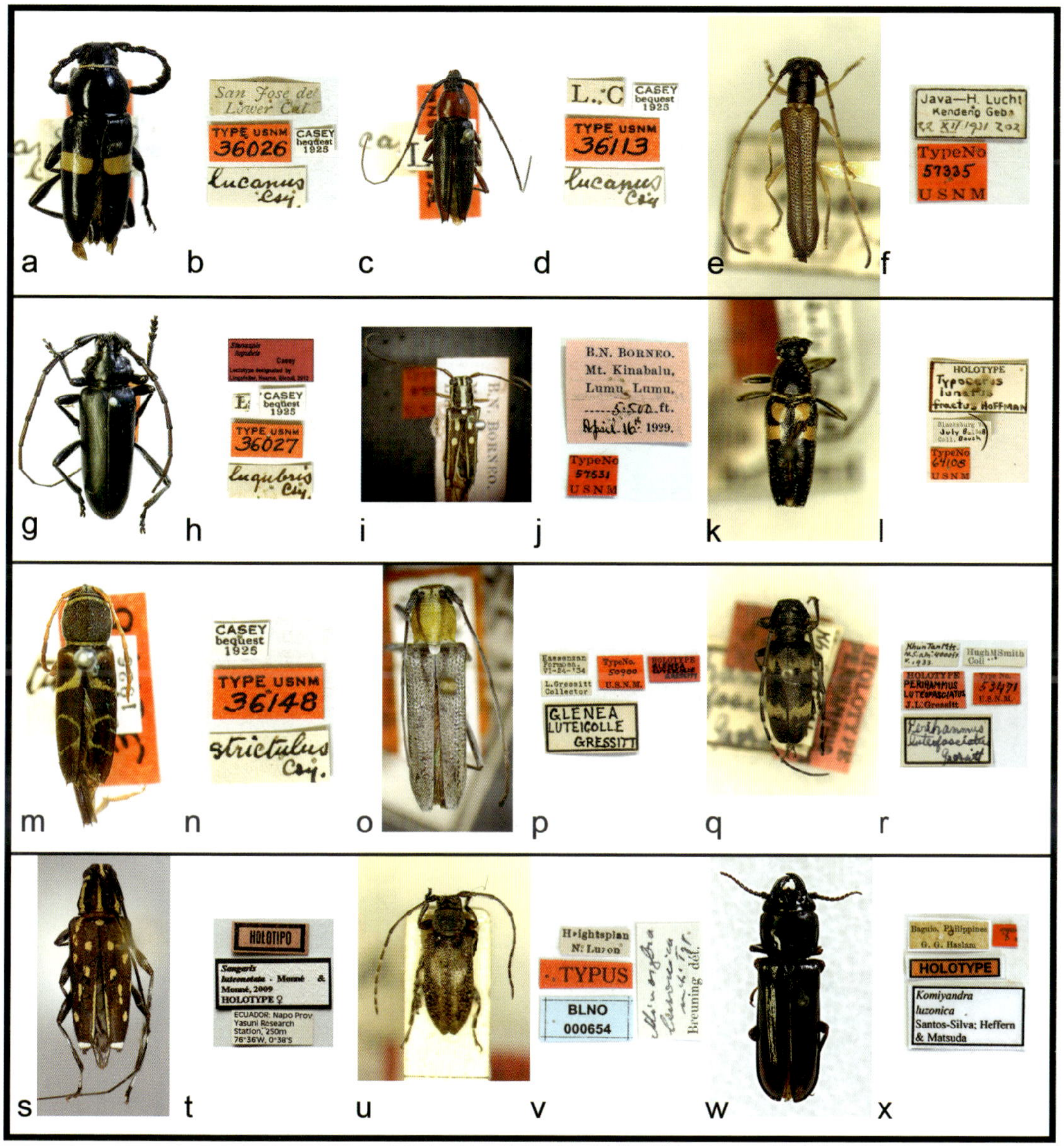

Figure 102. Primary types of Cerambycidae of the Smithsonian Institution (for each pair of images, dorsal habitus left, labels right): **(a, b)** *Lissonotus lucanus* Casey, 1912, Holotype; **(c, d)** *Stenosphenus lucanus* Casey, 1912, Holotype; **(e, f)** *Oberea luchti* Fisher, 1937, Holotype; **(g, h)** *Stenaspis lugubris* Casey, 1912, Lectotype; **(i, j)** *Glenea lumulumuensis* Fisher, 1935, Holotype; **(k, l)** *Typocerus lunatus fractus* Hoffman, 1958, Holotype; **(m, n)** *Neoclytus luscus strictulus* Casey, 1912, Holotype; **(o, p)** *Glenea luteicolle* Gressitt, 1935, Holotype; **(q, r)** *Perihammus luteofasciatus* Gressitt, 1941, Holotype; **(s, t)** *Sangaris luteonotata* Monné & Monné, 2009, Holotype; **(u, v)** *Mimosybra luzonica* Breuning, 1957, Holotype; **(w, x)** *Komiyandra luzonica* Santos-Silva, Heffern & Matsuda, 2010, Holotype.

Figure 103. Primary types of Cerambycidae of the Smithsonian Institution (for each pair of images, dorsal habitus left, labels right): **(a, b)** *Gnoma luzonica bilaras* Dillon & Dillon, 1951, Holotype; **(c, d)** *Gnoma luzonica transiens* Kriesche, 1936, Holotype; **(e, f)** *Myromeus luzonicus* Fisher, 1925, Holotype; **(g, h)** *Enes luzonicus* Fisher, 1925, Holotype; **(i, j)** *Charoides lycimnia* Dillon & Dillon, 1945, Holotype; **(k, l)** *Methia lycoides* Chemsak & Linsley, 1971, Holotype; **(m, n)** *Calliclytus macoris* Lingafelter, 2011, Holotype; **(o, p)** *Leptura (Dorcasina) macrocera* Casey, 1913, Lectotype; **(q, r)** *Sulenus macrophthalmus* Breuning, 1954, Holotype; **(s, t)** *Stronglaspis macrotomoides* Tippmann, 1953, Holotype; **(u, v)** *Crossidius maculicollis* Casey, 1912, Lectotype; **(w, x)** *Ectatosia maculosa* Fisher, 1935, Holotype.

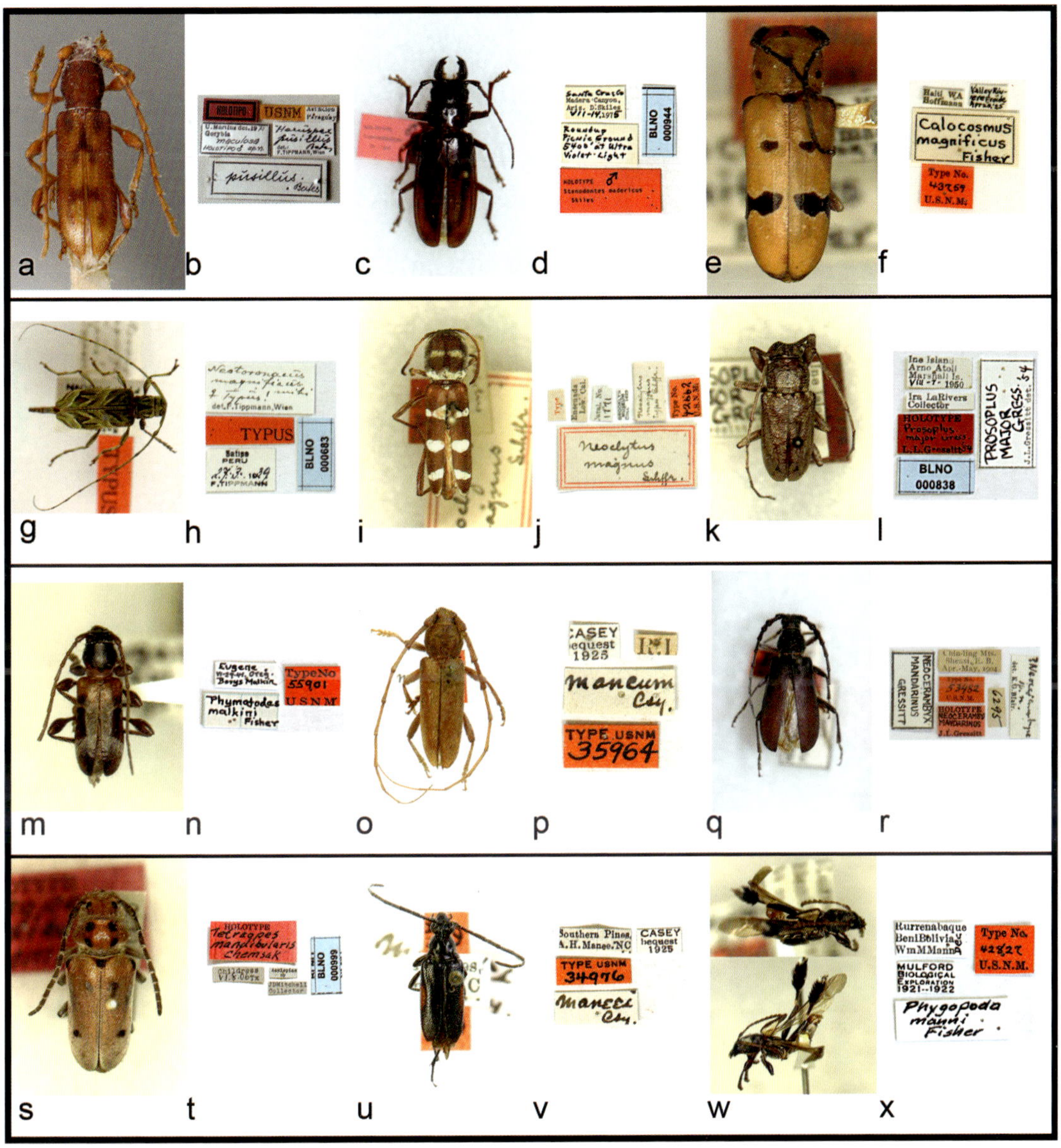

Figure 104. Primary types of Cerambycidae of the Smithsonian Institution (for each pair of images, dorsal habitus left, labels right): **(a, b)** *Gorybia maculosa* Martins, 1976, Holotype; **(c, d)** *Stenodontes (Mallodon) madericus* Skiles, 1978, Holotype; **(e, f)** *Calocosmus magnificus* Fisher, 1932, Holotype; **(g, h)** *Neotoronaeus magnificus* Tippmann, 1953, Holotype; **(i, j)** *Neoclytus magnus* Schaeffer, 1904, Holotype; **(k, l)** *Prosoplus major* Gressitt, 1956, Holotype; **(m, n)** *Phymatodes malkini* Fisher, 1942, Holotype; **(o, p)** *Romaleum mancum* Casey, 1912, Holotype; **(q, r)** *Neocerambyx mandarinus* Gressitt, 1939, Holotype; **(s, t)** *Tetraopes mandibularis* Chemsak, 1963, Holotype; **(u, v)** *Strangalia maneei* Casey, 1914, Holotype; **(w, x)** *Phygopoda manni* Fisher, 1930, Holotype.

Figure 105. Primary types of Cerambycidae of the Smithsonian Institution (for each pair of images, dorsal habitus left, labels right): **(a, b)** *Eburia marginalis* Fisher, 1947, Holotype; **(c, d)** *Necydalis marginalis* Ohbayashi, 1948, Holotype; **(e, f)** *Stenocorus marginellus* Casey, 1924, Holotype; **(g, h)** *Necydalis marginipennis* Gressitt, 1948, Holotype; **(i, j)** *Baryssinus marianae* Martins & Monné, 1974, Holotype; **(k, l)** *Lochmaeocles marmoratus* Casey, 1913, Lectotype; **(m, n)** *Eupogonius marmoratus* Fisher, 1925, Holotype; **(o, p)** *Enes marmoratus* Fisher, 1925, Holotype; **(q, r)** *Dihammus marmoratus* Fisher, 1935, Holotype; **(s, t)** *Presates marmoratus* Gressitt, 1940, Holotype; **(u, v)** *Rhytiphora (Rhytiphora) marmorea* Breuning, 1942, Holotype; **(w, x)** *Oopsis marshallensis* Gressitt, 1956, Holotype.

Figure 106. Primary types of Cerambycidae of the Smithsonian Institution (for each pair of images, dorsal habitus left, labels right): **(a, b)** *Aegocidnus (Miaena) masbatensis* Breuning, 1957, Holotype; **(c, d)** *Euseboides matsudai* Gressitt, 1938, Holotype; **(e, f)** *Demonax matsushitai* Gressitt, 1936, Holotype; **(g, h)** *Pseudaethomerus maximus* Tippmann, 1953, Holotype; **(i, j)** *Elaphidion mayesae* Ivie, 2007, Holotype; **(k, l)** *Saperda mecasoides* Casey, 1913, Holotype; **(m, n)** *Melanauster medenbachi* var. *sumatrensis* Fisher, 1934, Holotype; **(o, p)** *Hypermallus medialis* Casey, 1912, Holotype; **(q, r)** *Tylosis medialis* Casey, 1912, Lectotype; **(s, t)** *Tylosis medialis bifasciata* Casey, 1912, Holotype; **(u, v)** *Hypermallus medialis canadensis* Casey, 1912, Holotype; **(w, x)** *Hypermallus medialis densicollis* Casey, 1912, Holotype.

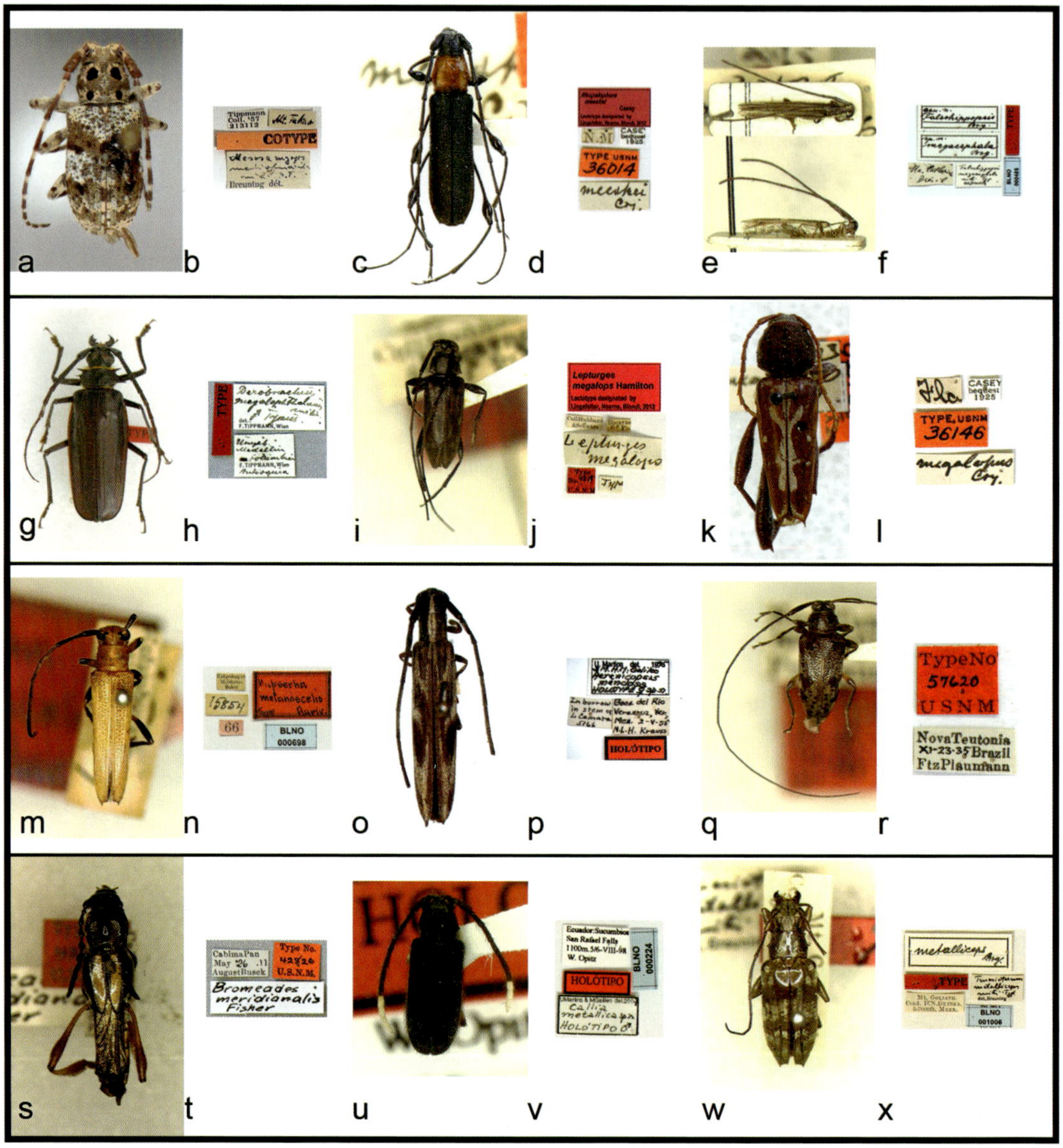

Figure 107. Primary types of Cerambycidae of the Smithsonian Institution (for each pair of images, dorsal habitus left, labels right): **(a, b)** *Mesosa mediofasciata* Breuning, 1942, Holotype; **(c, d)** *Rhopalophora meeskei* Casey, 1891, Lectotype; **(e, f)** *Falsohippopsis megacephala* Breuning, 1940, Holotype; **(g, h)** *Derobrachus megalophthalmus* Tippmann, 1953, Holotype; **(i, j)** *Lepturges megalops* Hamilton, 1896, Lectotype; **(k, l)** *Euryscelis megalopus* Casey, 1924, Holotype; **(m, n)** *Nupserha melanoscelis* Aurivillius, 1922, Holotype; **(o, p)** *Aerenicopsis mendosa* Martins & Galileo, 1998, Holotype; **(q, r)** *Lophopoeum meridianum* Fisher, 1938, Holotype; **(s, t)** *Bromiades meridionalis* Fisher, 1930, Holotype; **(u, v)** *Callia metallica* Galileo & Martins, 2008, Holotype; **(w, x)** *Tmesisternus metalliceps* Breuning, 1940, Holotype.

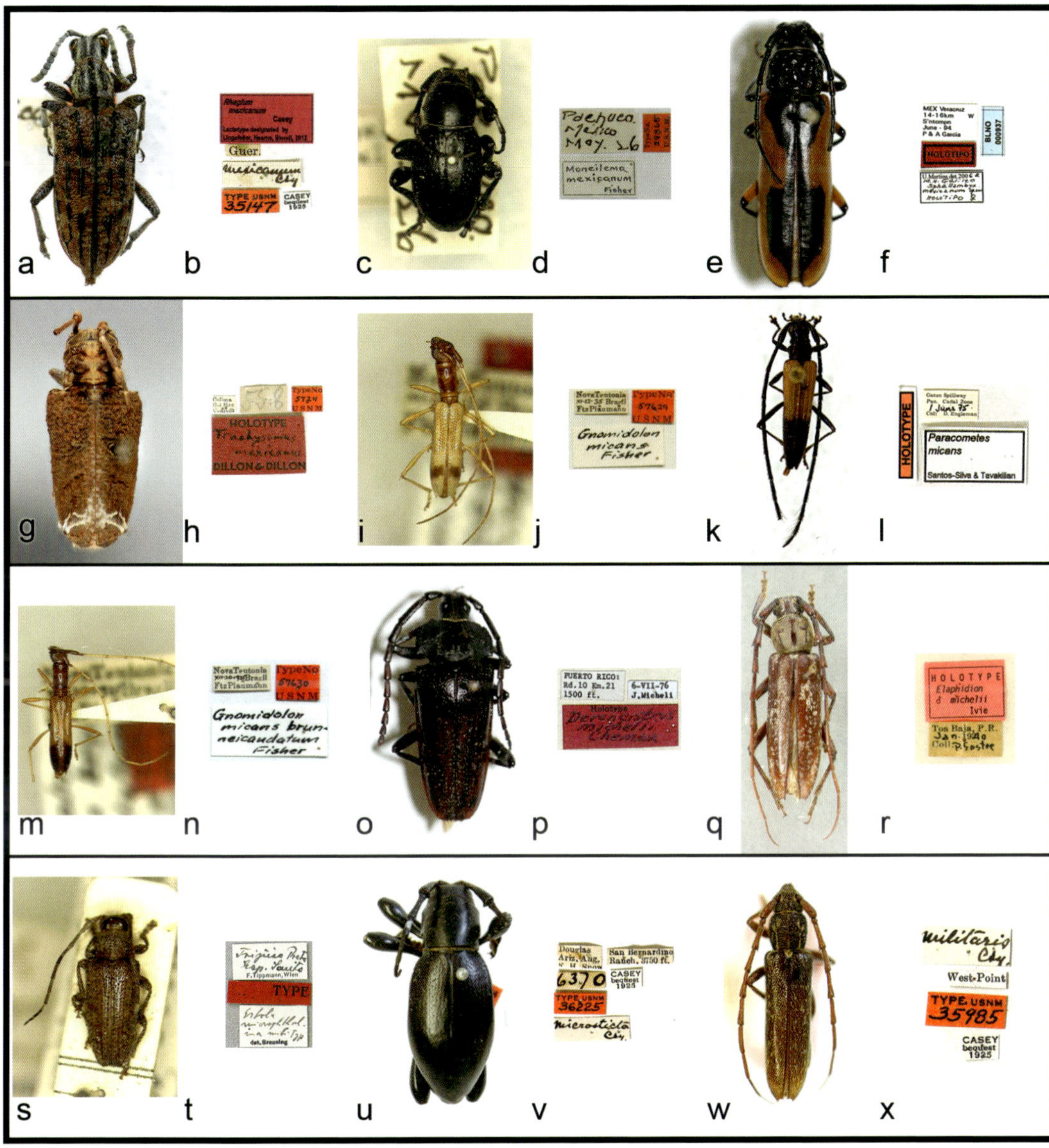

Figure 108. Primary types of Cerambycidae of the Smithsonian Institution (for each pair of images, dorsal habitus left, labels right): **(a, b)** *Rhagium (Hargium) mexicanum* Casey, 1913, Lectotype; **(c, d)** *Moneilema (Collapteryx) mexicanum* Fisher, 1927, Holotype; **(e, f)** *Sphallambyx mexicanum* Galileo & Martins, 2007, Holotype; **(g, h)** *Trachysomus mexicanus* Dillon & Dillon, 1946, Holotype; **(i, j)** *Gnomidolon micans* Fisher, 1937, Holotype; **(k, l)** *Paracometes micans* Santos-Silva & Tavakilian, 2009, Holotype; **(m, n)** *Gnomidolon micans* v. *brunneicaudatum* Fisher, 1937, Holotype; **(o, p)** *Derancistrus michelii* Chemsak, 1979, Holotype; **(q, r)** *Elaphidion michelii* Ivie, 2007, Holotype; **(s, t)** *Estola microphthalma* Breuning, 1942, Holotype; **(u, v)** *Moneilema (Moneilema) microsticta* Casey, 1913, Holotype; **(w, x)** *Hypermallus militaris* Casey, 1914, Holotype.

Figure 109. Primary types of Cerambycidae of the Smithsonian Institution (for each pair of images, dorsal habitus left, labels right): **(a, b)** *Leptostylus milleri* Fisher, 1932, Holotype; **(c, d)** *Oreodera mimetica* Lane, 1970, Holotype; **(e, f)** *Elaphidion mimeticum* Schaeffer, 1905, Lectotype; **(g, h)** *Leiopus mimeticus* Casey, 1891, Holotype; **(i, j)** *Oreodera mimica* Tippmann, 1960, Holotype; **(k, l)** *Exocentrus mindanaoensis* Fisher, 1925, Holotype; **(m, n)** *Noemia mindanaoensis* Gressitt, 1935, Holotype; **(o, p)** *Xylotrechus mindanaonis* Aurivillius, 1928, Syntype; **(q, r)** *Aegocidnus (Miaena) mindanaonis* Breuning, 1957, Holotype; **(s, t)** *Exocentrus (Pseudocentrus) mindoroanus* Breuning, 1957, Holotype; **(u, v)** *Leptacmaeops minima* Casey, 1913, Holotype; **(w, x)** *Miriochrus minimus* Galileo & Martins, 2012, Holotype.

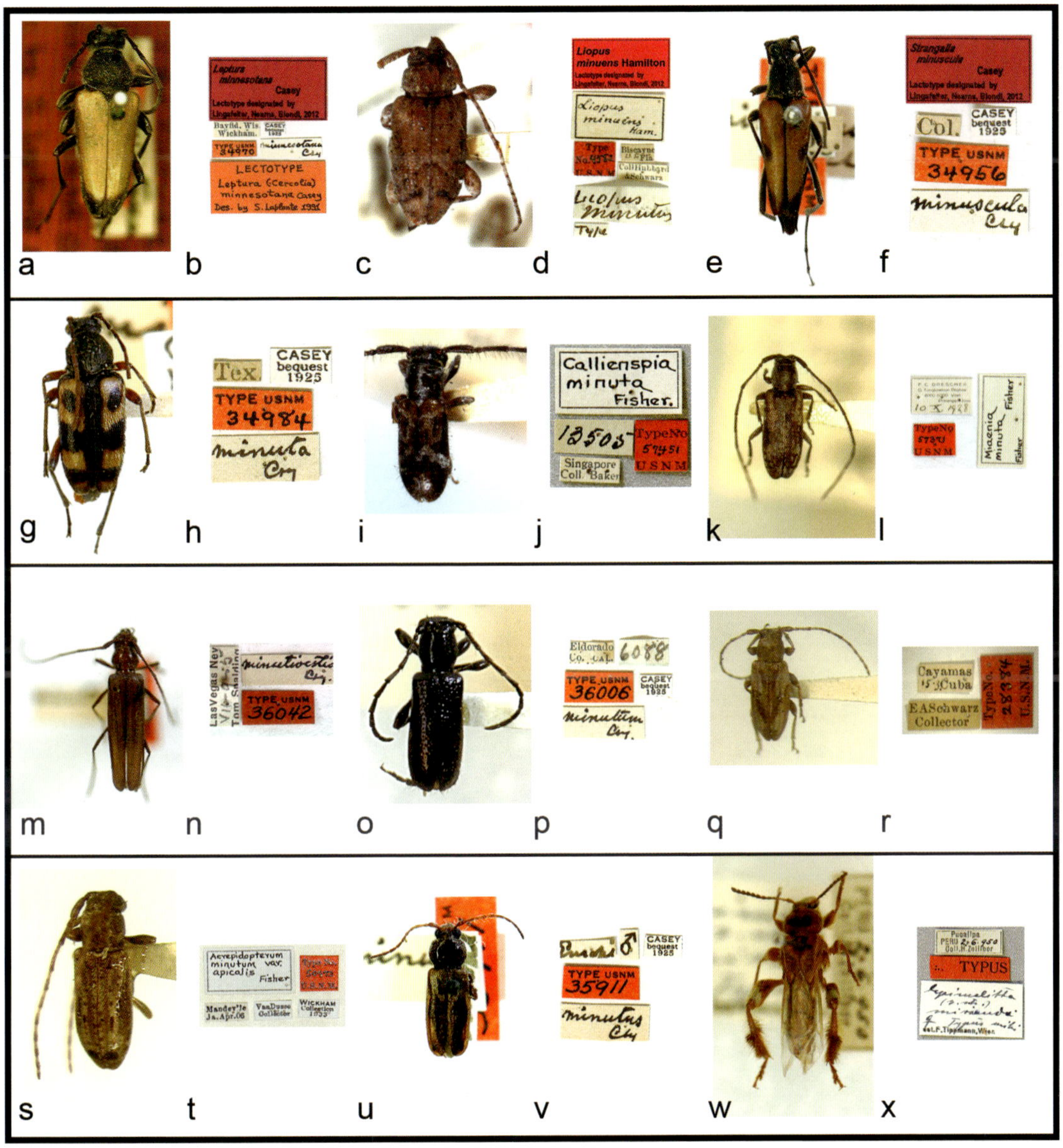

Figure 110. Primary types of Cerambycidae of the Smithsonian Institution (for each pair of images, dorsal habitus left, labels right): **(a, b)** *Leptura (Cercolia) minnesotana* Casey, 1913, Lectotype; **(c, d)** *Liopus minuens* Hamilton in Leng & Hamilton, 1896, Lectotype; **(e, f)** *Strangalia minuscula* Casey, 1913, Lectotype; **(g, h)** *Judolia minuta* Casey, 1924, Holotype; **(i, j)** *Callienispia minuta* Fisher, 1925, Holotype; **(k, l)** *Miaenia minuta* Fisher, 1936, Holotype; **(m, n)** *Oeme minutivestis* Casey, 1924, Holotype; **(o, p)** *Poecilobrium minutum* Casey, 1924, Holotype; **(q, r)** *Acrepidopterum minutum* Fisher, 1926, Holotype; **(s, t)** *Acrepidopterum minutum* var. *apicalis* Fisher, 1935, Holotype; **(u, v)** *Hylotrupes minutus* Casey, 1924, Holotype; **(w, x)** *Epimelitta miranda* Tippmann, 1960, Holotype.

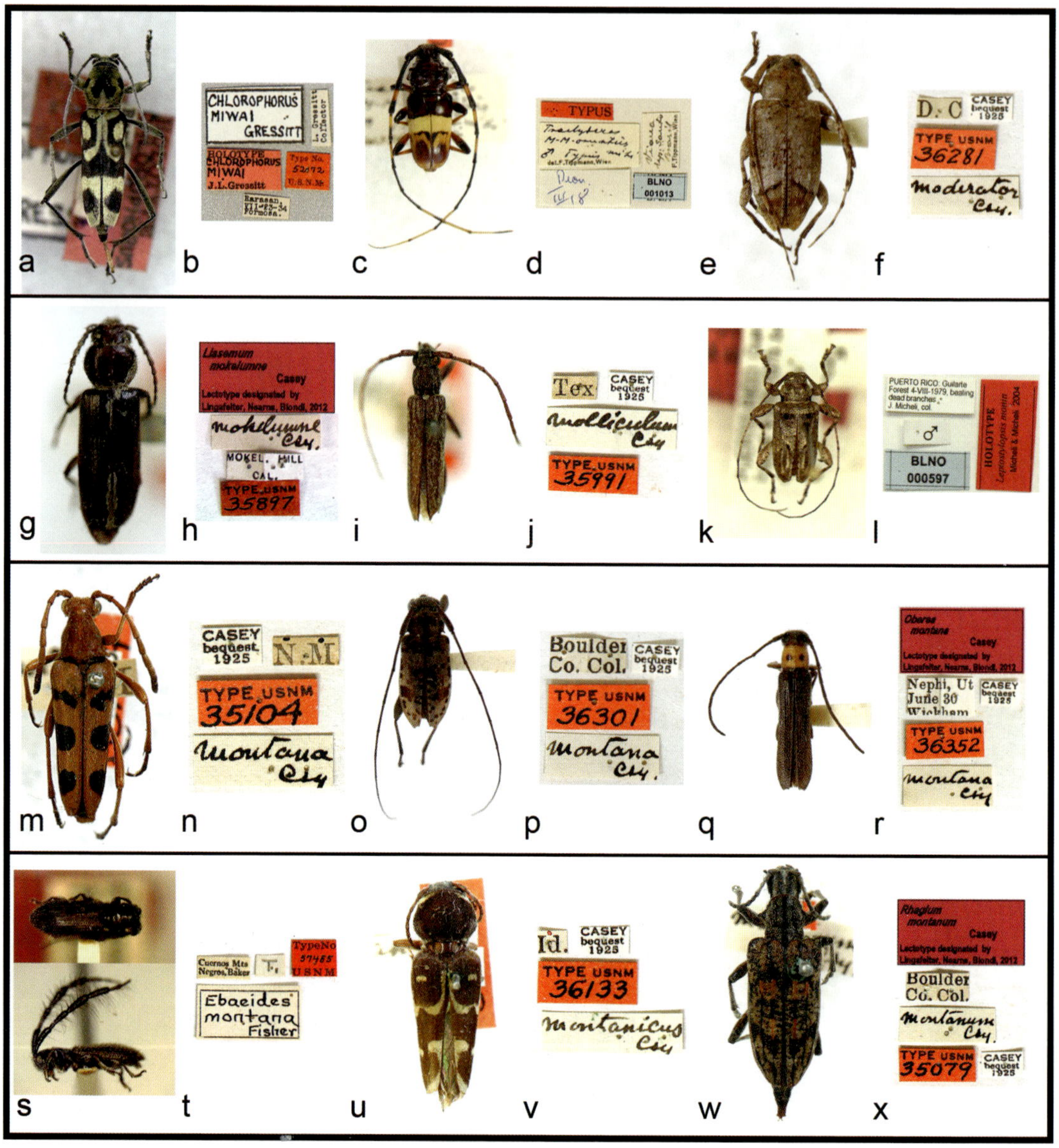

Figure 111. Primary types of Cerambycidae of the Smithsonian Institution (for each pair of images, dorsal habitus left, labels right): **(a, b)** *Chlorophorus miwai* Gressitt, 1936, Holotype; **(c, d)** *Trachyderes mm-ornata* Tippmann, 1960, Holotype; **(e, f)** *Leiopus moderator* Casey, 1913, Holotype; **(g, h)** *Liasemum mokelumne* Casey, 1912, Lectotype; **(i, j)** *Hypermallus molliculus* Casey, 1912, Holotype; **(k, l)** *Leptostylopsis monin* Micheli & Micheli, 2004, Holotype; **(m, n)** *Strangalia montana* Casey, 1891, Holotype; **(o, p)** *Hyperplatys montana* Casey, 1913, Holotype; **(q, r)** *Oberea montana* Casey, 1913, Lectotype; **(s, t)** *Ebaeides montana* Fisher, 1925, Holotype; **(u, v)** *Xylotrechus montanicus* Casey, 1912, Holotype; **(w, x)** *Rhagium (Hargium) montanum* Casey, 1913, Lectotype.

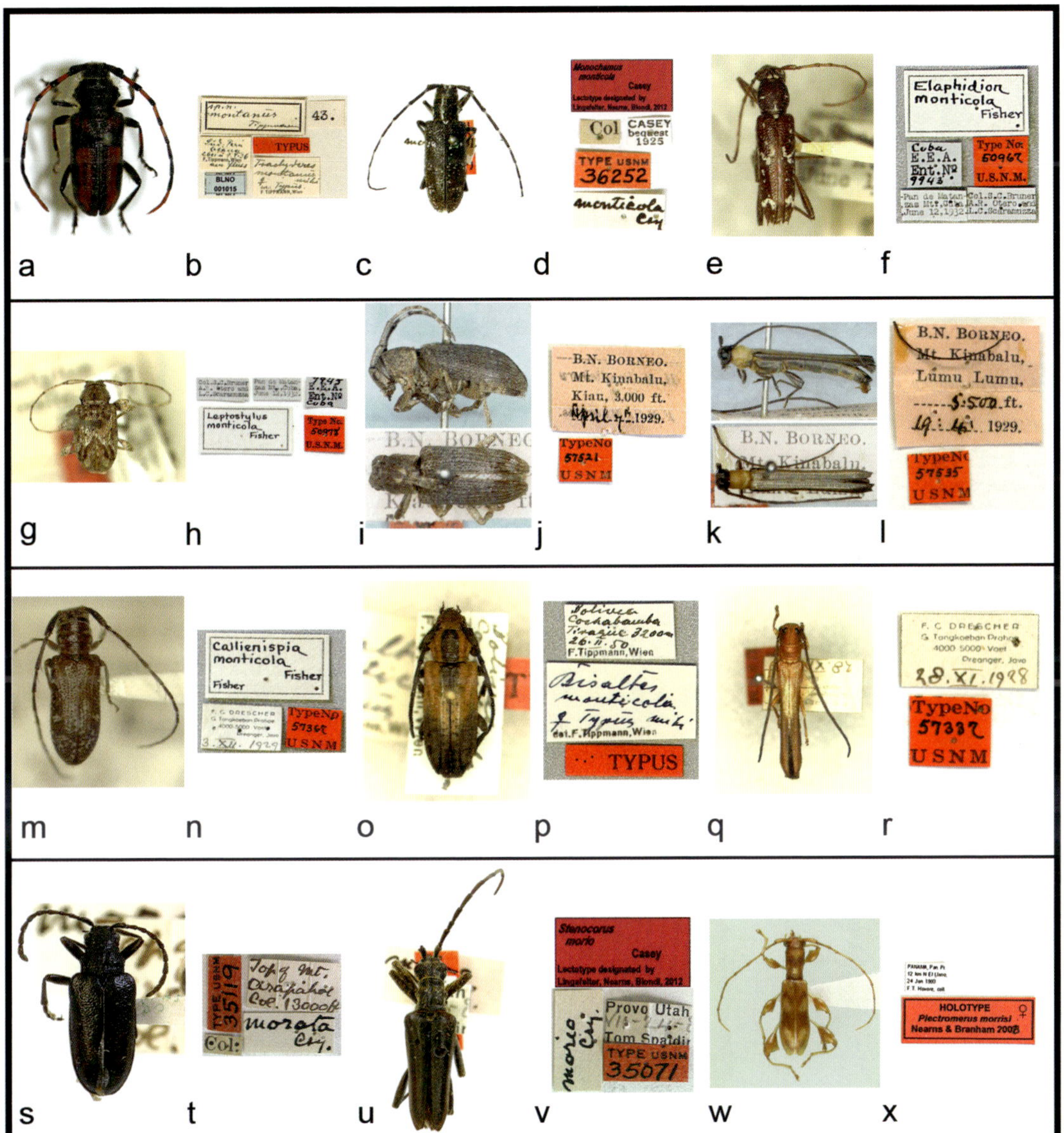

Figure 112. Primary types of Cerambycidae of the Smithsonian Institution (for each pair of images, dorsal habitus left, labels right): **(a, b)** *Trachyderes montanus* Tippmann, 1953, Holotype; **(c, d)** *Monochamus monticola* Casey, 1913, Lectotype; **(e, f)** *Elaphidion monticola* Fisher, 1935, Holotype; **(g, h)** *Leptostylus monticola* Fisher, 1935, Holotype; **(i, j)** *Pterolophia monticola* Fisher, 1935, Holotype; **(k, l)** *Oberea monticola* Fisher, 1935, Holotype; **(m, n)** *Callienispia monticola* Fisher, 1936, Holotype; **(o, p)** *Bisaltes monticola* Tippmann, 1960, Holotype; **(q, r)** *Oberea montivagans* Fisher, 1937, Holotype; **(s, t)** *Brachysomida morata* Casey, 1913, Holotype; **(u, v)** *Stenocorus morio* Casey, 1924, Lectotype; **(w, x)** *Plectromerus morrisi* Nearns & Branham, 2008, Holotype.

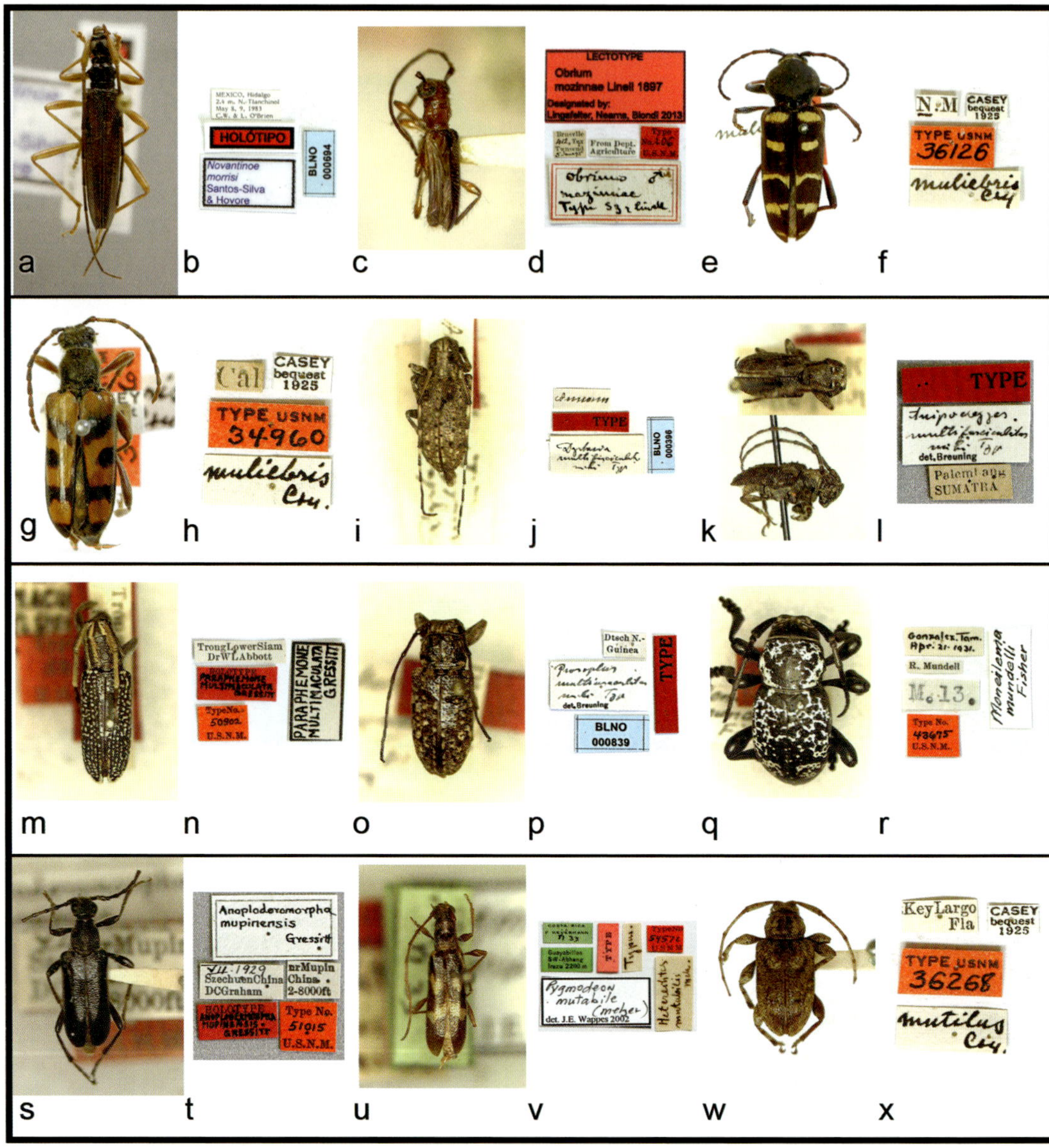

Figure 113. Primary types of Cerambycidae of the Smithsonian Institution (for each pair of images, dorsal habitus left, labels right): **(a, b)** *Novantinoe morrisi* Santos-Silva & Hovore, 2007, Holotype; **(c, d)** *Obrium mozinnae* Linell, 1897, Lectotype; **(e, f)** *Calloides muliebris* Casey, 1912, Holotype; **(g, h)** *Xestoleptura muliebris* Casey, 1913, Holotype; **(i, j)** *Dystasia multifasciculata* Breuning, 1943, Holotype; **(k, l)** *Anipocregyes multifasciculatus* Breuning, 1939, Holotype; **(m, n)** *Paraphemone multimaculata* Gressitt, 1935, Holotype; **(o, p)** *Prosoplus multimaculatus* Breuning, 1943, Holotype; **(q, r)** *Moneilema (Collapteryx) mundelli* Fisher, 1931, Holotype; **(s, t)** *Anoploderamorpha mupinensis* Gressitt, 1935, Holotype; **(u, v)** *Heterachthes mutabile* Melzer, 1935, Holotype; **(w, x)** *Leptostylus mutilus* Casey, 1913, Holotype.

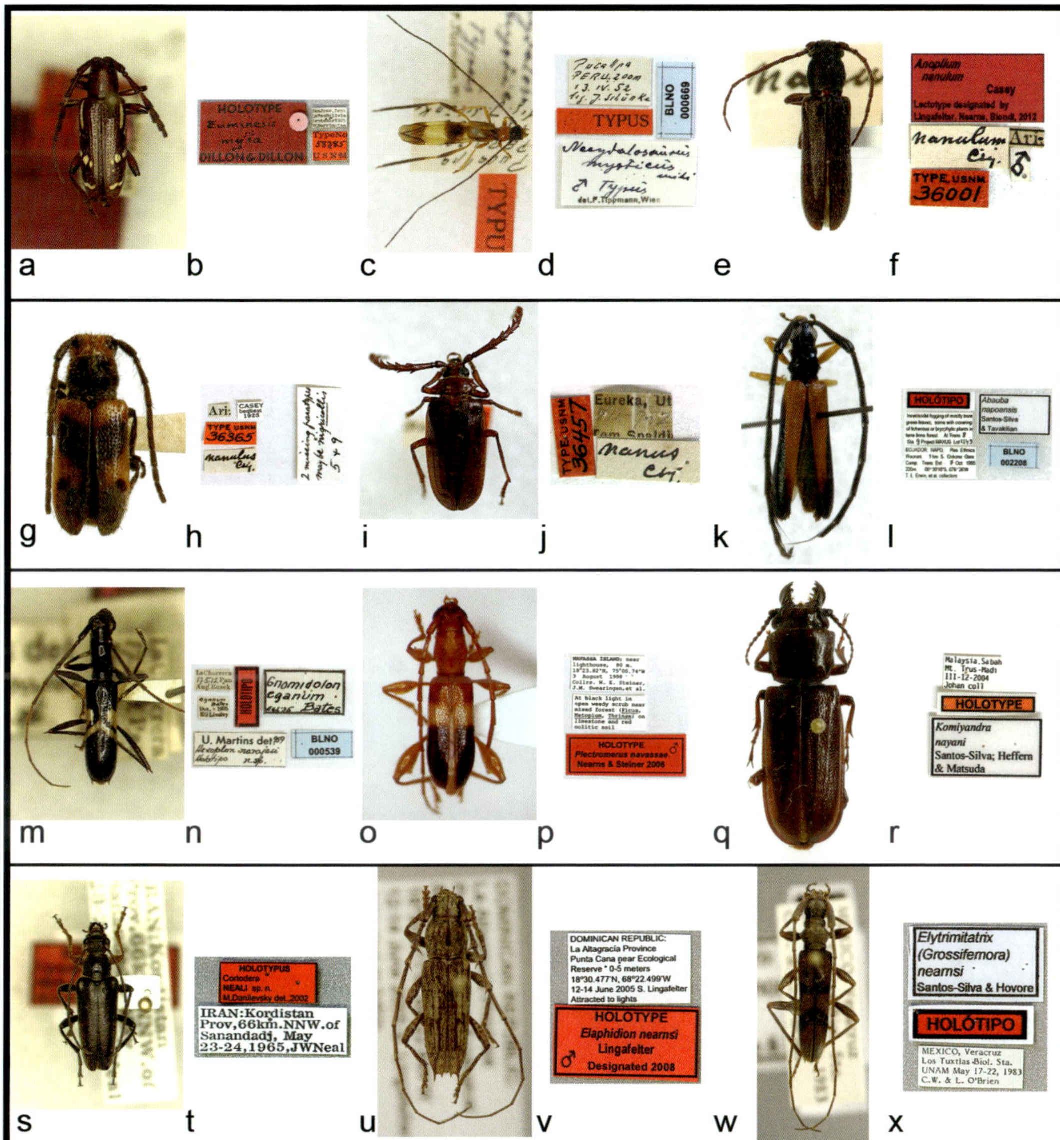

Figure 114. Primary types of Cerambycidae of the Smithsonian Institution (for each pair of images, dorsal habitus left, labels right): **(a, b)** *Eumimesis myia* Dillon & Dillon, 1946, Holotype; **(c, d)** *Necydalosaurus mysticus* Tippmann, 1960, Holotype; **(e, f)** *Anoplium nanulum* Casey, 1924, Lectotype; **(g, h)** *Tetraopes nanulus* Casey, 1913, Holotype; **(i, j)** *Prionus (Prionus) nanus* Casey, 1924, Holotype; **(k, l)** *Abauba napoensis* Santos-Silva & Tavakilian, 2009, Holotype; **(m, n)** *Hexoplon navajasi* Martins, 1959, Holotype; **(o, p)** *Plectromerus navassae* Nearns & Steiner, 2006, Holotype; **(q, r)** *Komiyandra nayani* Santos-Silva, Heffern & Matsuda, 2010, Holotype; **(s, t)** *Cortodera neali* Danilevsky, 2004, Holotype; **(u, v)** *Elaphidion nearnsi* Lingafelter, 2008, Holotype; **(w, x)** *Elytrimitatrix (Grossifemora) nearnsi* Santos-Silva & Hovore, 2008, Holotype.

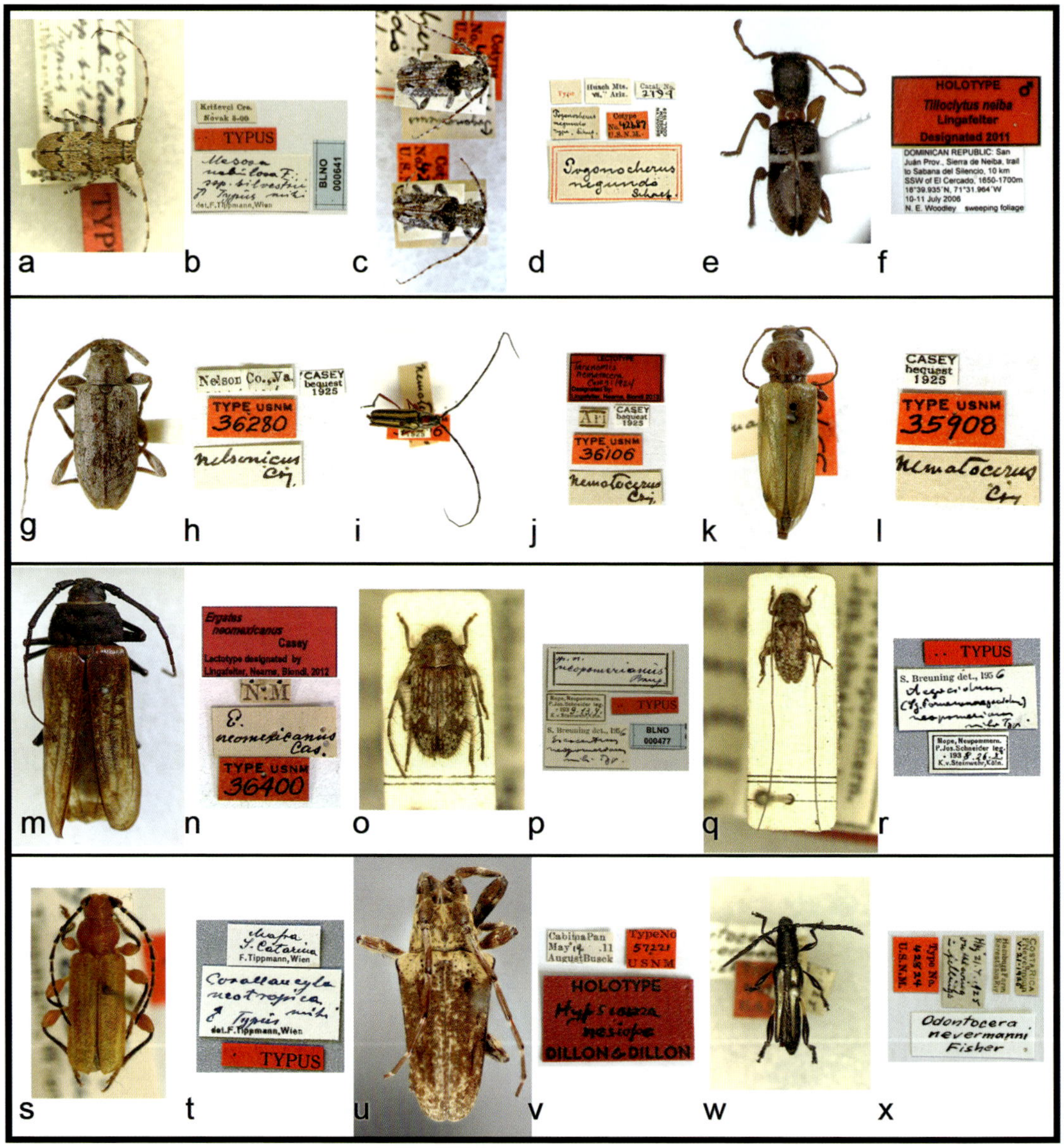

Figure 115. Primary types of Cerambycidae of the Smithsonian Institution (for each pair of images, dorsal habitus left, labels right): **(a, b)** *Mesosa nebulosa silvestrii* Tippmann, 1956, Holotype; **(c, d)** *Pogonocherus negundo* Schaeffer, 1905, Syntype; **(e, f)** *Tilloclytus neiba* Lingafelter, 2011, Holotype; **(g, h)** *Leiopus nelsonicus* Casey, 1924, Holotype; **(i, j)** *Taranomis nematocera* Casey, 1924, Lectotype; **(k, l)** *Hylotrupes nematocerus* Casey, 1924, Holotype; **(m, n)** *Ergates (Trichocnemis) neomexicanus* Casey, 1890, Lectotype; **(o, p)** *Exocentrus (Camptomyme) neopomerianus* Breuning, 1957, Holotype; **(q, r)** *Aegocidnus (Pomeranaegocidnus) neopomerianus* Breuning, 1957, Holotype; **(s, t)** *Corallancyla neotropica* Tippmann, 1960, Holotype; **(u, v)** *Hypsioma nesiope* Dillon & Dillon, 1945, Holotype; **(w, x)** *Odontocera nevermanni* Fisher, 1930, Holotype.

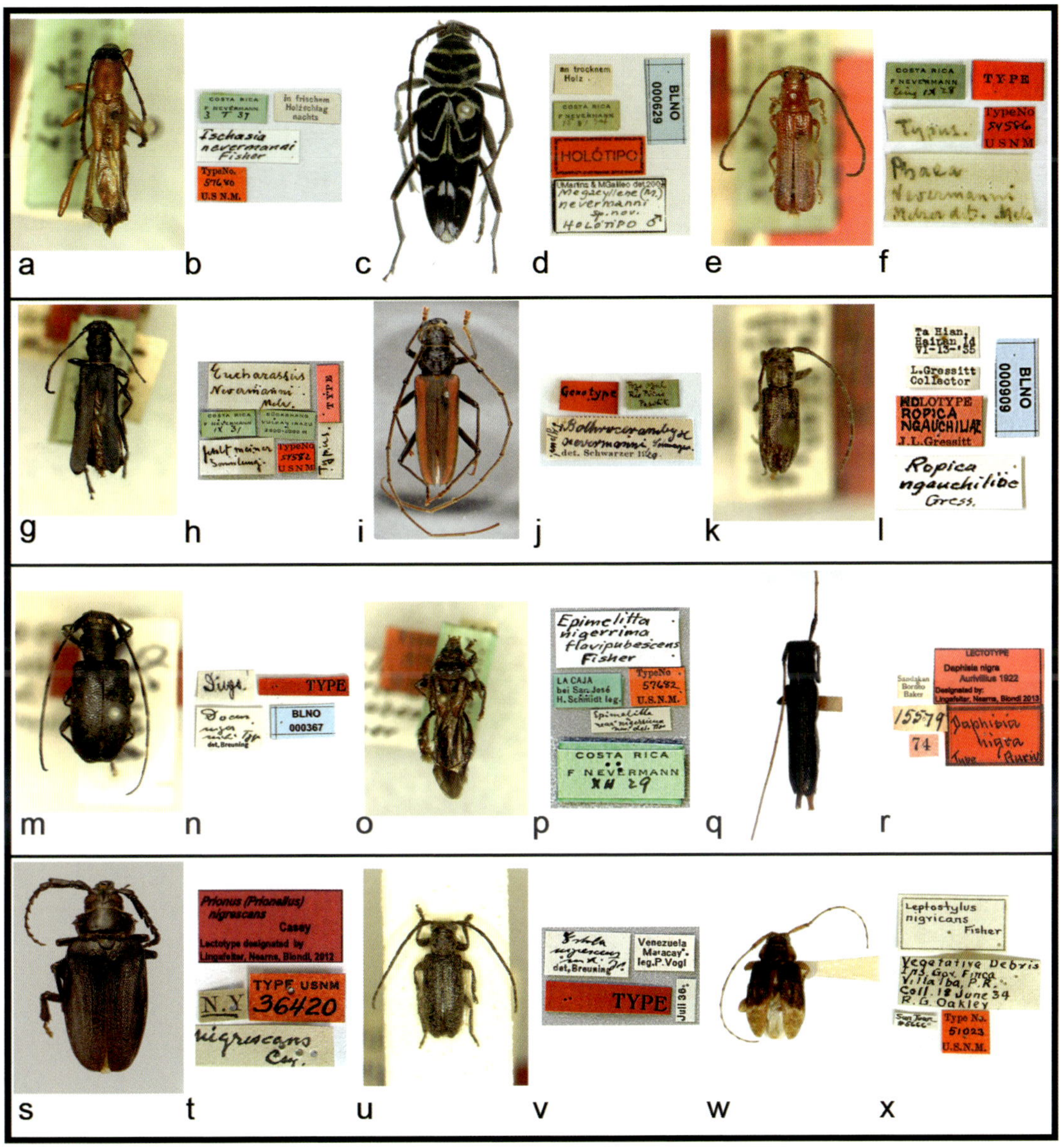

Figure 116. Primary types of Cerambycidae of the Smithsonian Institution (for each pair of images, dorsal habitus left, labels right): **(a, b)** *Ischasia nevermanni* Fisher, 1947, Holotype; **(c, d)** *Megacyllene (Megacyllene) nevermanni* Martins & Galileo, 2008, Holotype; **(e, f)** *Phaea nevermanni* Melzer, 1932, Holotype; **(g, h)** *Eucharassus nevermanni* Melzer, 1934, Holotype; **(i, j)** *Bothrocerambyx nevermanni* Schwarzer, 1929, Holotype; **(k, l)** *Ropica ngauchiliae* Gressitt, 1940, Holotype; **(m, n)** *Docus niger* Breuning, 1942, Holotype; **(o, p)** *Epimelitta nigerrima flavipurescens* Fisher, 1947, Holotype; **(q, r)** *Daphisia nigra* Aurivillius, 1922, Lectotype; **(s, t)** *Prionus (Prionellus) nigrescans* Casey, 1924, Lectotype; **(u, v)** *Estola nigrescens* Breuning, 1943, Holotype; **(w, x)** *Leptostylus nigricans* Fisher, 1935, Holotype.

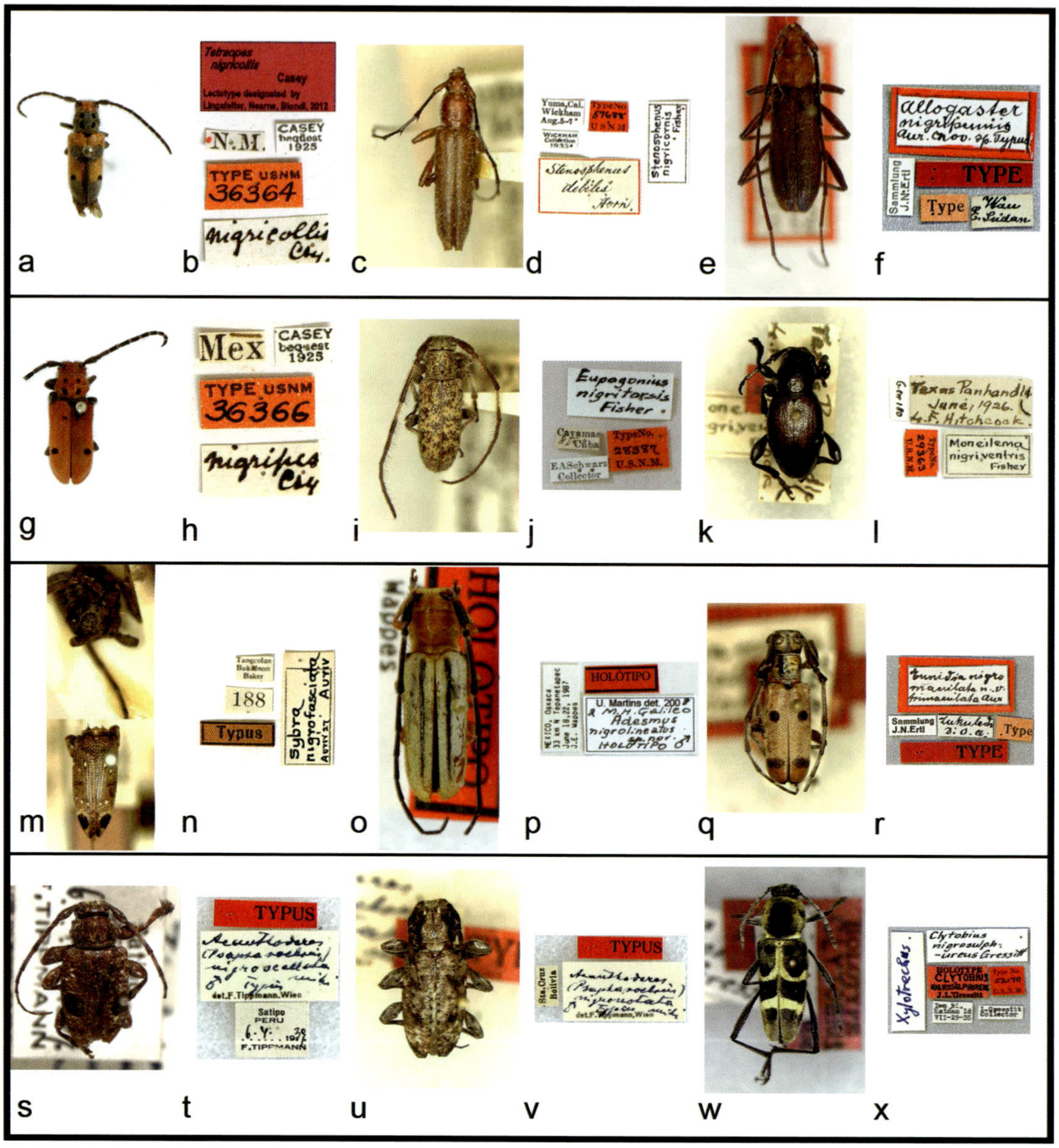

Figure 117. Primary types of Cerambycidae of the Smithsonian Institution (for each pair of images, dorsal habitus left, labels right): **(a, b)** *Tetraopes nigricollis* Casey, 1913, Lectotype; **(c, d)** *Stenosphenus nigricornis* Fisher, 1946, Holotype; **(e, f)** *Allogaster nigripennis* Aurivillius, 1915, Holotype; **(g, h)** *Tetraopes nigripes* Casey, 1913, Holotype; **(i, j)** *Eupogonius nigritarsis* Fisher, 1926, Holotype; **(k, l)** *Moneilema (Moneilema) nigriventris* Fisher, 1927, Holotype; **(m, n)** *Sybra nigrofasciata* Aurivillius, 1927, Syntype; **(o, p)** *Adesmus nigrolineatus* Martins & Galileo, 2008, Holotype; **(q, r)** *Eunidia nigromaculata* var. *trimaculata* Aurivillius, 1915, Holotype; **(s, t)** *Acanthoderes (Psapharochrus) nigroocellata* Tippmann, 1960, Holotype; **(u, v)** *Acanthoderes (Psapharochrus) nigropunctata* Tippmann, 1960, Holotype; **(w, x)** *Xylotrechus nigrosulphureus* Gressitt, 1940, Holotype.

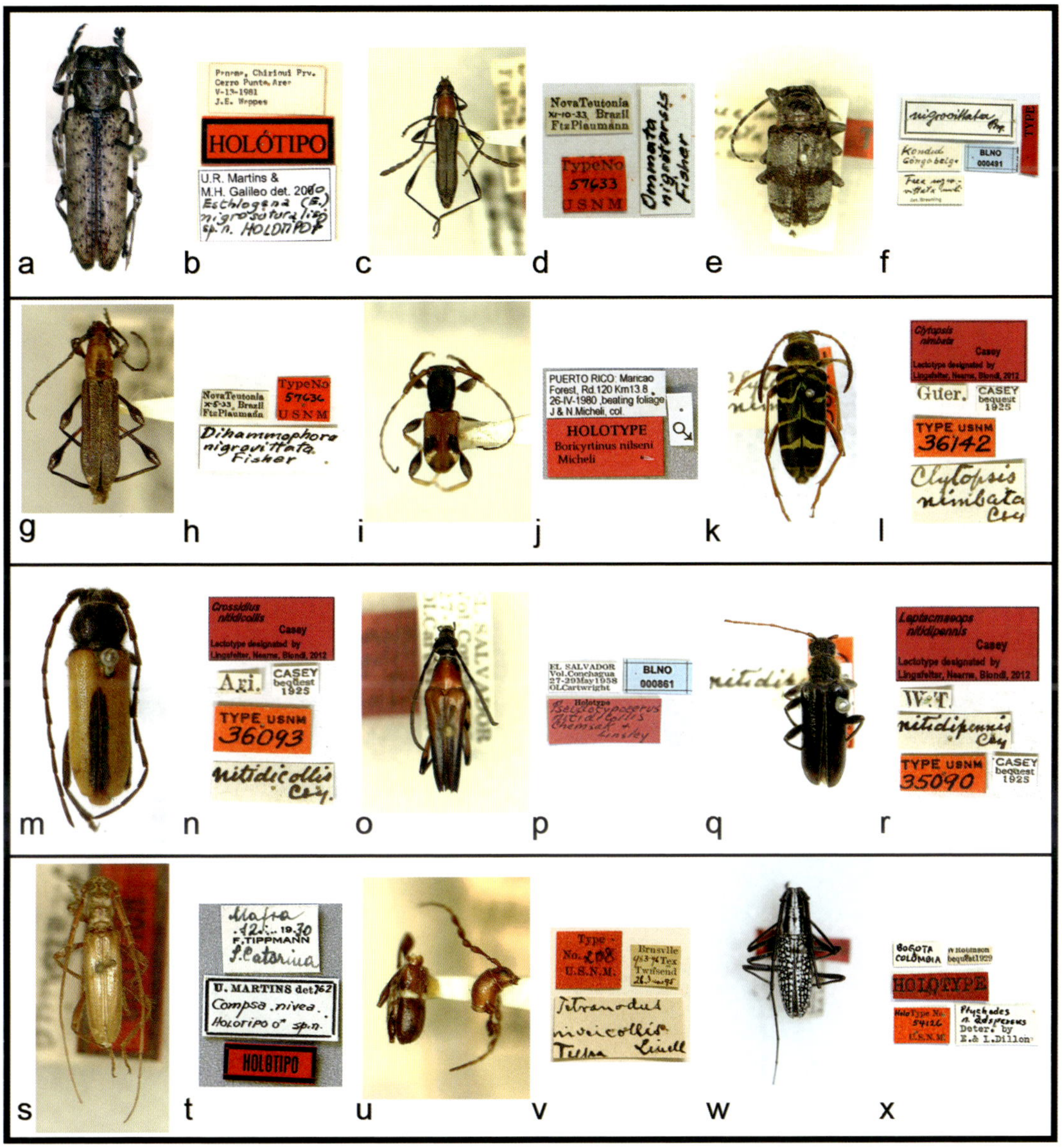

Figure 118. Primary types of Cerambycidae of the Smithsonian Institution (for each pair of images, dorsal habitus left, labels right): **(a, b)** *Esthlogena nigrosuturalis* Galileo & Martins, 2011, Holotype; **(c, d)** *Ommata (Rhopalessa) nigrotarsis* Fisher, 1937, Holotype; **(e, f)** *Frea nigrovittata* Breuning, 1935, Holotype; **(g, h)** *Dihammophora nigrovittata* Fisher, 1937, Holotype; **(i, j)** *Boricyrtinus nilseni* Micheli, 2003, Holotype; **(k, l)** *Clytopsis nimbata* Casey, 1912, Lectotype; **(m, n)** *Crossidius nitidicollis* Casey, 1891, Lectotype; **(o, p)** *Pseudotypocerus nitidicollis* Chemsak & Linsley, 1976, Holotype; **(q, r)** *Leptacmaeops (Leptacmaeops) nitidipennis* Casey, 1913, Lectotype; **(s, t)** *Compsa nivea* Martins, 1962, Holotype; **(u, v)** *Tetranodus niveicollis* Linell, 1897, Holotype; **(w, x)** *Ptychodes niveisparsis adspersus* Dillon & Dillon, 1941, Holotype.

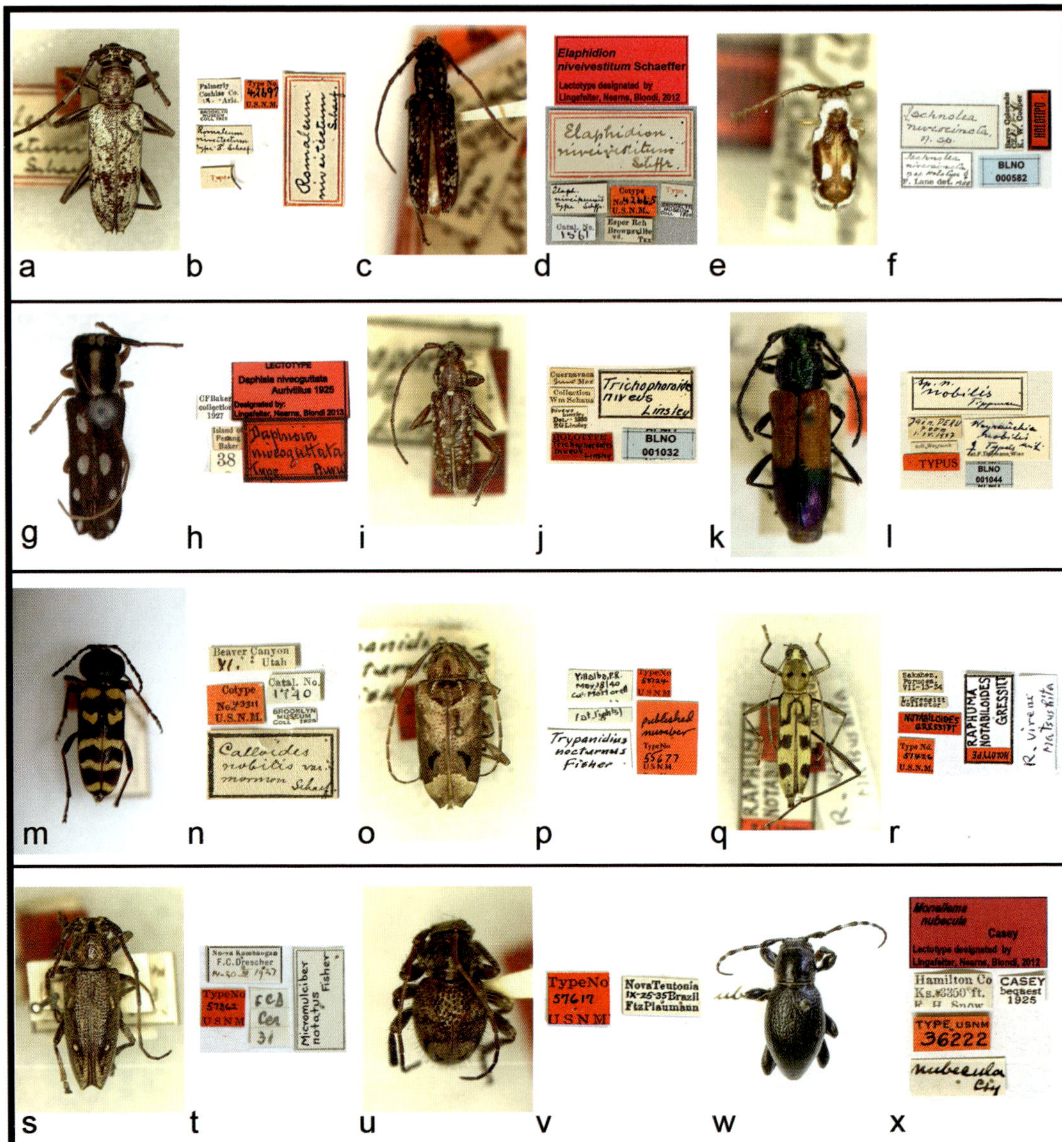

Figure 119. Primary types of Cerambycidae of the Smithsonian Institution (for each pair of images, dorsal habitus left, labels right): **(a, b)** *Romaleum niveitectum* Schaeffer, 1905, Syntype; **(c, d)** *Elaphidion niveivestitum* Schaeffer, 1905, Lectotype; **(e, f)** *Ischnolea niveocincta* Lane, 1959, Holotype; **(g, h)** *Daphisia niveoguttata* Aurivillius, 1925, Lectotype; **(i, j)** *Trichophoroides niveus* Linsley, 1935, Holotype; **(k, l)** *Weyrauchia nobilis* Tippmann, 1960, Holotype; **(m, n)** *Calloides nobilis* var. *mormonus* Schaeffer, 1911, Holotype; **(o, p)** *Trypanidius nocturnus* Fisher, 1942, Holotype; **(q, r)** *Raphuma notabiloides* Gressitt, 1936, Holotype; **(s, t)** *Micromulciber notatus* Fisher, 1936, Holotype; **(u, v)** *Plaumanniella novateutoniae* Fisher, 1938, Holotype; **(w, x)** *Moneilema (Moneilema) nubecula* Casey, 1913, Lectotype.

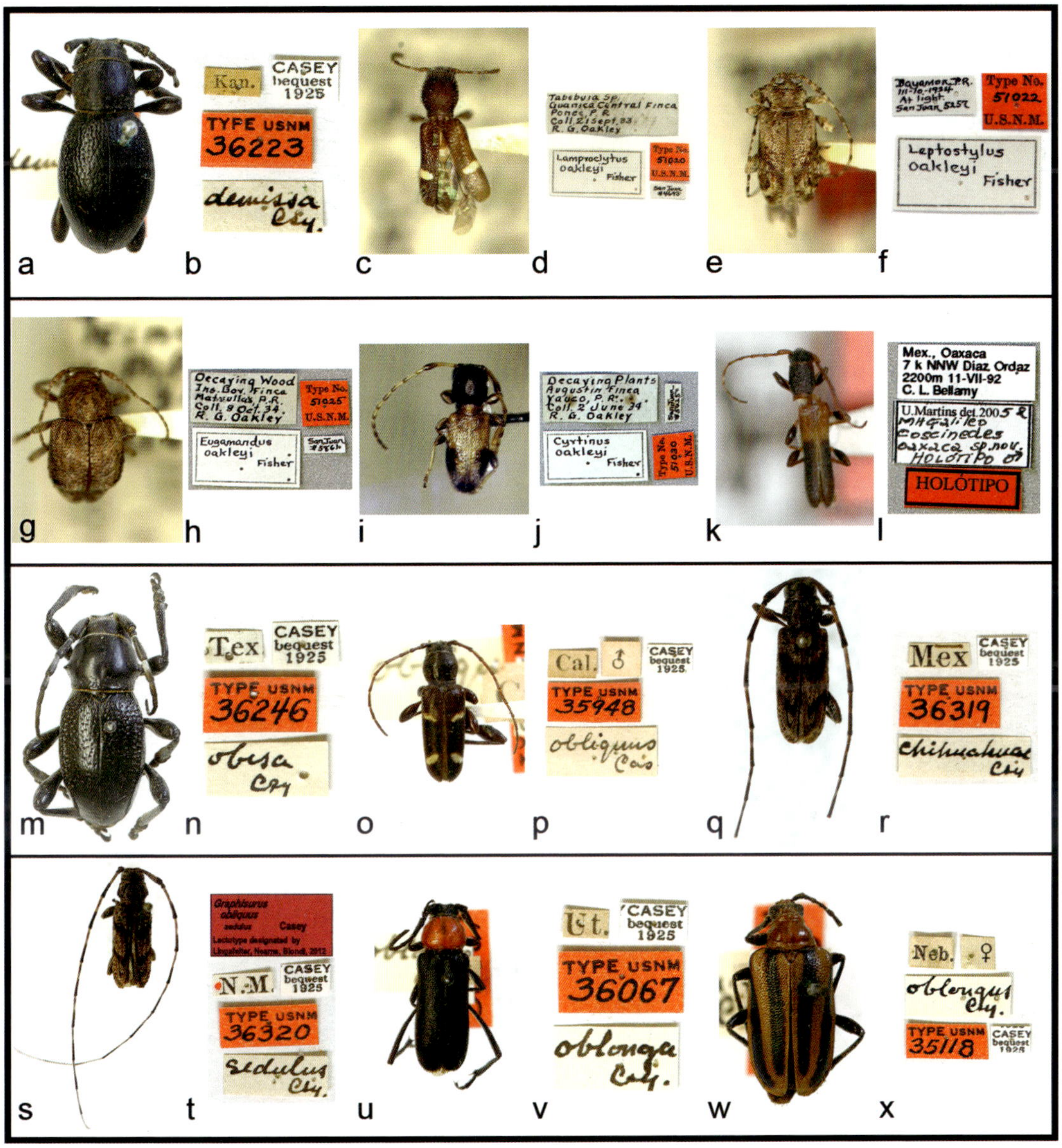

Figure 120. Primary types of Cerambycidae of the Smithsonian Institution (for each pair of images, dorsal habitus left, labels right): **(a, b)** *Moneilema (Moneilema) nubecula demissa* Casey, 1913, Holotype; **(c, d)** *Lamproclytus oakleyi* Fisher, 1935, Holotype; **(e, f)** *Leptostylus oakleyi* Fisher, 1935, Holotype; **(g, h)** *Eugamandus oakleyi* Fisher, 1935, Holotype; **(i, j)** *Cyrtinus oakleyi* Fisher, 1935, Holotype; **(k, l)** *Coscinedes oaxaca* Martins & Galileo, 2006, Holotype; **(m, n)** *Moneilema (Collapteryx) obesa* Casey, 1924, Holotype; **(o, p)** *Phymatodes obliquus* Casey, 1891, Holotype; **(q, r)** *Graphisurus (Graphisurus) obliquus chihuahuae* Casey, 1913, Holotype; **(s, t)** *Graphisurus (Graphisurus) obliquus sedulus* Casey, 1913, Lectotype; **(u, v)** *Batyle oblonga* Casey, 1912, Holotype; **(w, x)** *Gaurotes oblongus* Casey, 1913, Holotype.

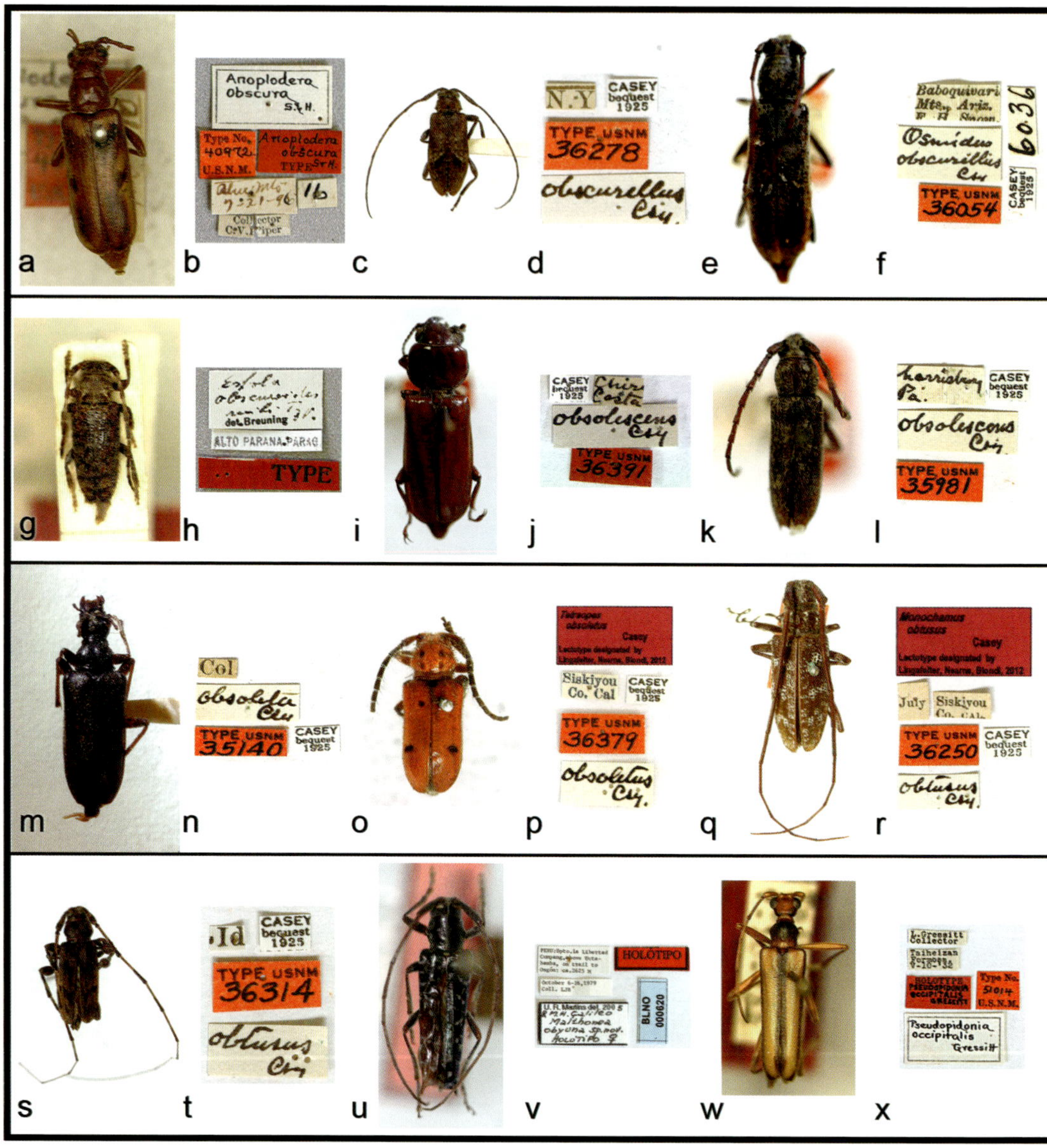

Figure 121. Primary types of Cerambycidae of the Smithsonian Institution (for each pair of images, dorsal habitus left, labels right): **(a, b)** *Anoplodera obscura* Swaine & Hopping, 1928, Holotype; **(c, d)** *Leiopus obscurellus* Casey, 1913, Holotype; **(e, f)** *Osmidus obscurellus* Casey, 1924, Holotype; **(g, h)** *Estola obscuroides* Breuning, 1942, Holotype; **(i, j)** *Parandra obsolescens* Casey, 1912, Holotype; **(k, l)** *Hypermallus obsolescens* Casey, 1924, Holotype; **(m, n)** *Acmaeops obsoleta* Casey, 1913, Holotype; **(o, p)** *Tetraopes obsoletus* Casey, 1913, Lectotype; **(q, r)** *Monochamus obtusus* Casey, 1891, Lectotype; **(s, t)** *Graphisurus obtusus* Casey, 1924, Holotype; **(u, v)** *Malthonea obyuna* Martins & Galileo, 2005, Holotype; **(w, x)** *Pseudopidonia occipitalis* Gressitt, 1935, Holotype.

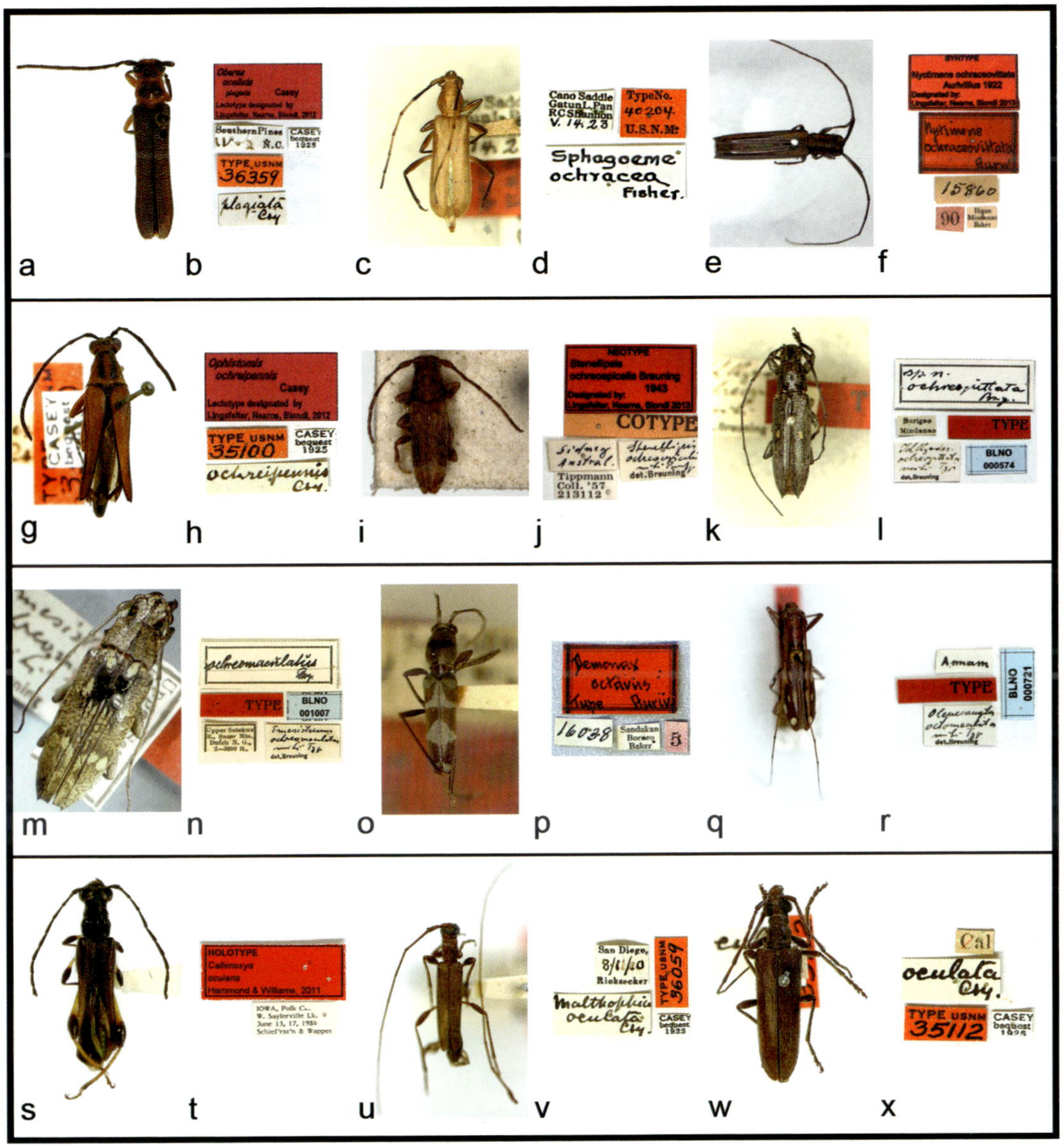

Figure 122. Primary types of Cerambycidae of the Smithsonian Institution (for each pair of images, dorsal habitus left, labels right): **(a, b)** *Oberea ocellata plagiata* Casey, 1913, Lectotype; **(c, d)** *Sphagoeme ochracea* Fisher, 1927, Holotype; **(e, f)** *Nyctimene ochraceovittata* Aurivillius, 1922, Syntype; **(g, h)** *Ophistomis ochreipennis* Casey, 1913, Lectotype; **(i, j)** *Stenellipsis ochreoapicalis* Breuning, 1943, Neotype; **(k, l)** *Ichthyodes (Ichthyodes) ochreoguttata* Breuning, 1942, Holotype; **(m, n)** *Tmesisternus (Arrhenotus) ochreomaculatus* Breuning, 1942, Holotype; **(o, p)** *Demonax octavus* Aurivillius, 1922, Holotype; **(q, r)** *Olenecamptus octomaculatus* Breuning, 1940, Holotype; **(s, t)** *Callimoxys ocularis* Hammond & Williams, 2011, Holotype; **(u, v)** *Malthophia oculata* Casey, 1912, Holotype; **(w, x)** *Centrodera oculata* Casey, 1913, Holotype.

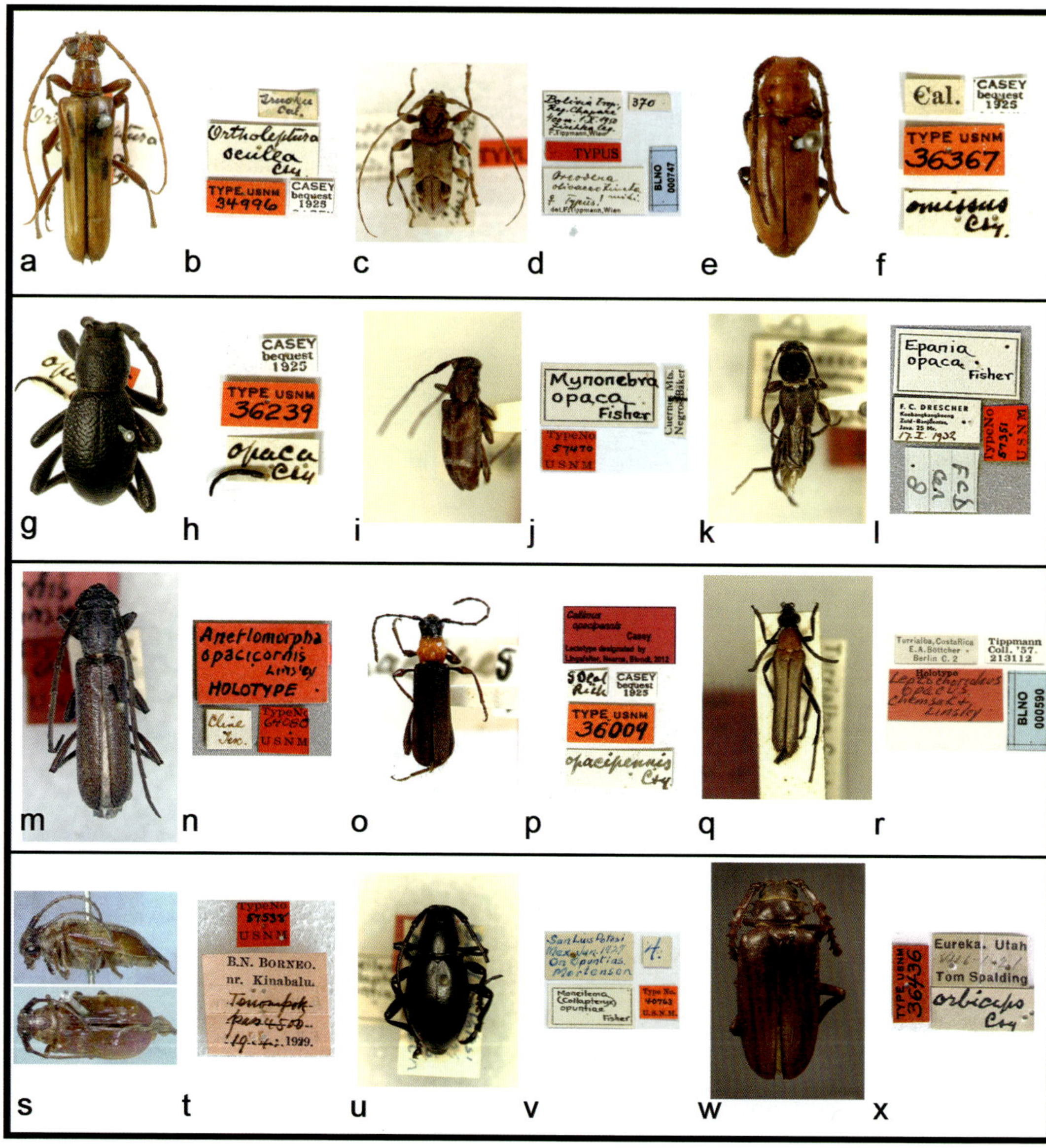

Figure 123. Primary types of Cerambycidae of the Smithsonian Institution (for each pair of images, dorsal habitus left, labels right): **(a, b)** *Ortholeptura oculea* Casey, 1913, Holotype; **(c, d)** *Oreodera olivaceotincta* Tippmann, 1953, Holotype; **(e, f)** *Tetraopes omissus* Casey, 1913, Holotype; **(g, h)** *Moneilema (Collapteryx) opaca* Casey, 1913, Holotype; **(i, j)** *Mynonebra opaca* Fisher, 1925, Holotype; **(k, l)** *Epania opaca* Fisher, 1936, Holotype; **(m, n)** *Aneflomorpha opacicornis* Linsley, 1957, Holotype; **(o, p)** *Callimus opacipennis* Casey, 1912, Lectotype; **(q, r)** *Leptochoriolaus opacus* Chemsak & Linsley, 1976, Holotype; **(s, t)** *Astathes opalescens* Fisher, 1935, Holotype; **(u, v)** *Moneilema (Collapteryx) opuntiae* Fisher, 1928, Holotype; **(w, x)** *Prionus (Prionus) orbiceps* Casey, 1924, Holotype.

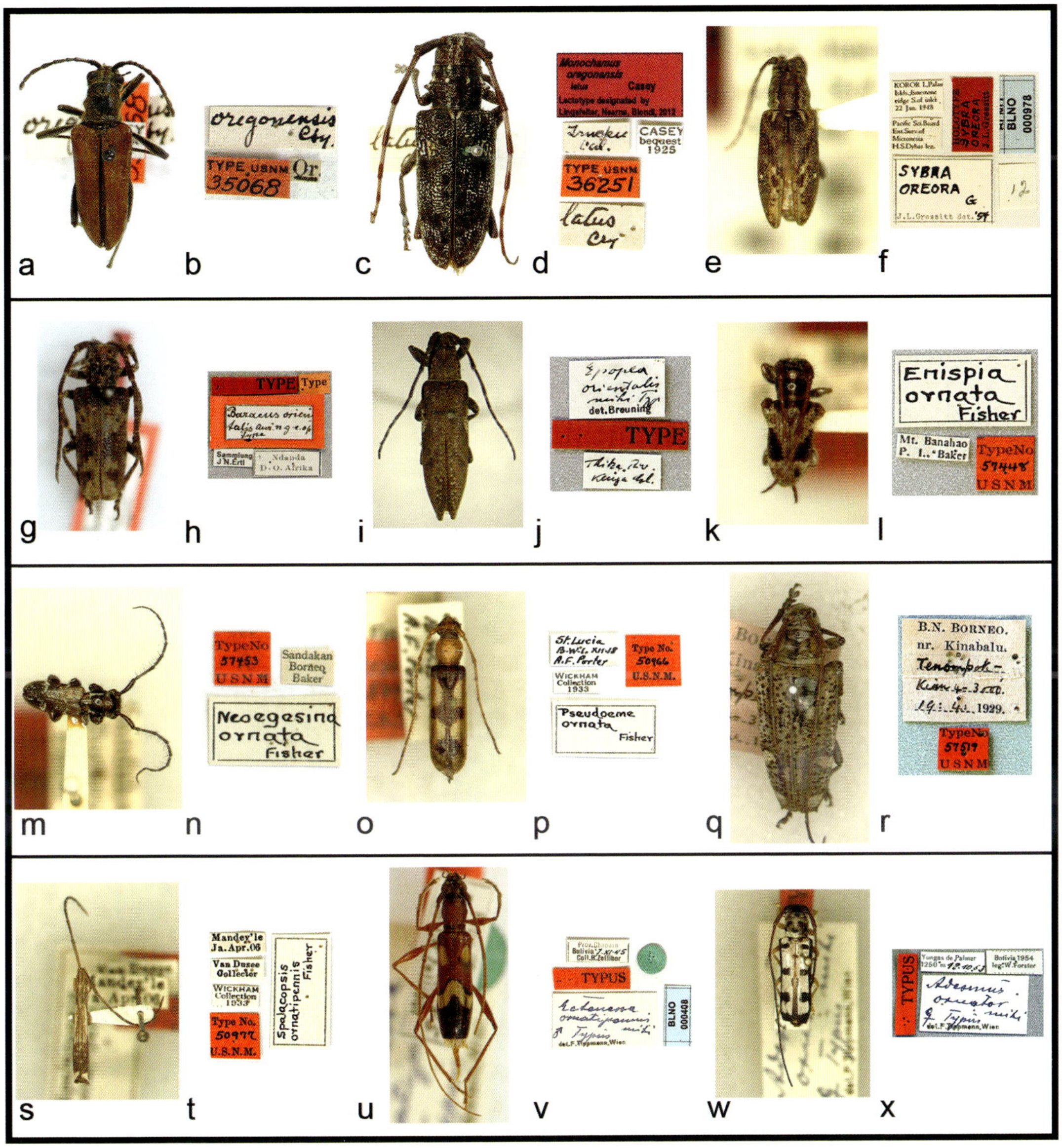

Figure 124. Primary types of Cerambycidae of the Smithsonian Institution (for each pair of images, dorsal habitus left, labels right): **(a, b)** *Stenocorus (Stenocorus) oregonensis* Casey, 1913, Holotype; **(c, d)** *Monochamus oregonensis latus* Casey, 1924, Lectotype; **(e, f)** *Sybra oreora* Gressitt, 1956, Holotype; **(g, h)** *Baraeus orientalis* Aurivillius, 1907, Syntype; **(i, j)** *Epopea orientalis* Breuning, 1940, Holotype; **(k, l)** *Enispia ornata* Fisher, 1925, Holotype; **(m, n)** *Neoegesina ornata* Fisher, 1925, Holotype; **(o, p)** *Pseudoeme ornata* Fisher, 1935, Holotype; **(q, r)** *Combe ornata* Fisher, 1935, Holotype; **(s, t)** *Spalacopsis ornatipennis* Fisher, 1935, Holotype; **(u, v)** *Ectenessa ornatipennis* Tippmann, 1960, Holotype; **(w, x)** *Adesmus ornator* Tippmann, 1960, Holotype.

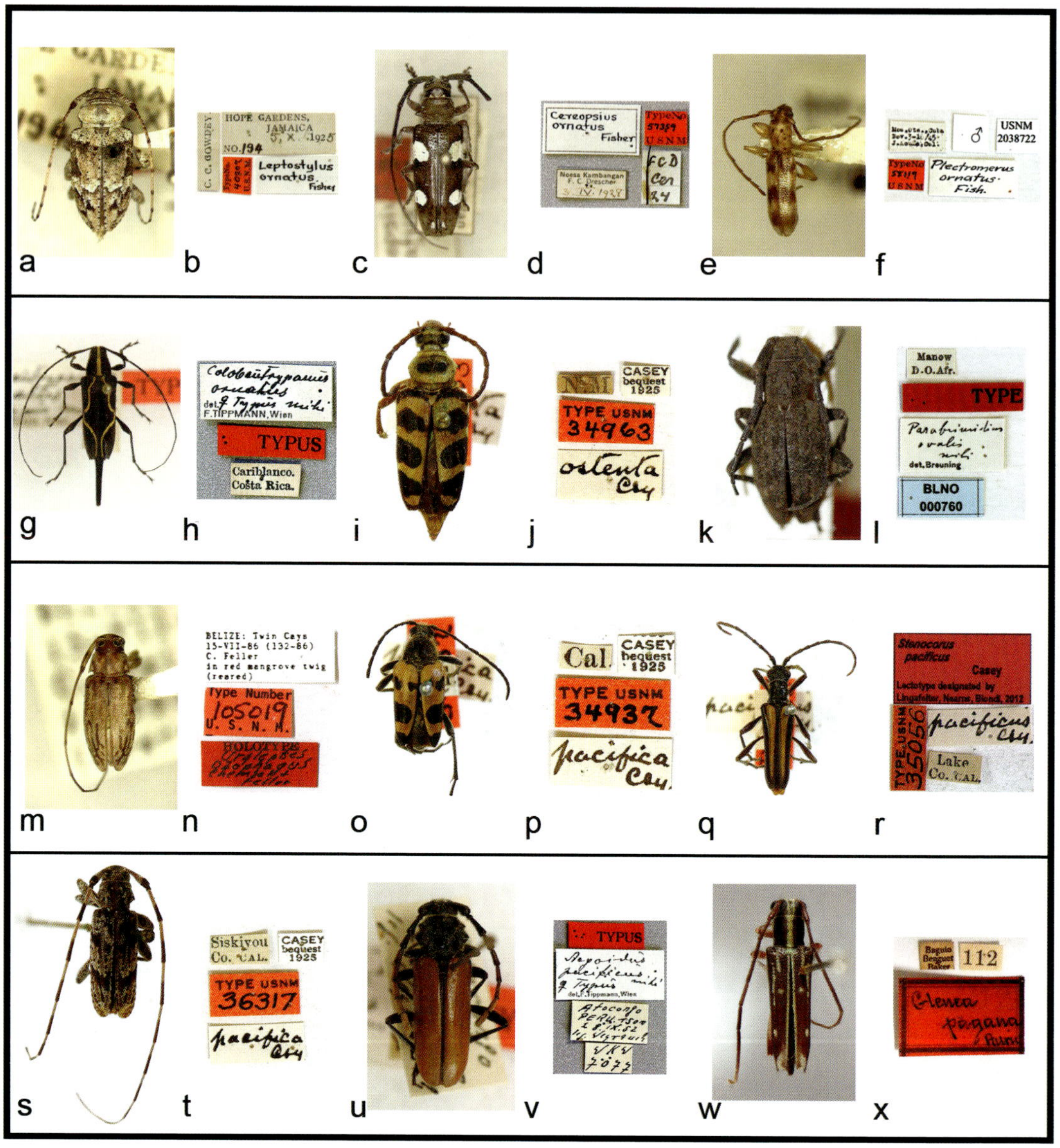

Figure 125. Primary types of Cerambycidae of the Smithsonian Institution (for each pair of images, dorsal habitus left, labels right): **(a, b)** *Leptostylus ornatus* Fisher, 1928, Holotype; **(c, d)** *Cereopsius ornatus* Fisher, 1936, Holotype; **(e, f)** *Plectromerus ornatus* Fisher, 1947, Holotype; **(g, h)** *Colobeutrypanus ornatus* Tippmann, 1953, Holotype; **(i, j)** *Strophiona ostenta* Casey, 1913, Holotype; **(k, l)** *Parabrimidius ovalis* Breuning, 1938, Holotype; **(m, n)** *Urgleptes ozophagus* Chemsak & Feller, 1988, Holotype; **(o, p)** *Judolia pacifica* Casey, 1913, Holotype; **(q, r)** *Stenocorus (Stenocorus) pacificus* Casey, 1913, Lectotype; **(s, t)** *Graphisurus (Graphisurus) pacificus* Casey, 1913, Holotype; **(u, v)** *Aegoidus pacificus* Tippmann, 1960, Holotype; **(w, x)** *Glenea (Glenea) pagana* Aurivillius, 1926, Holotype.

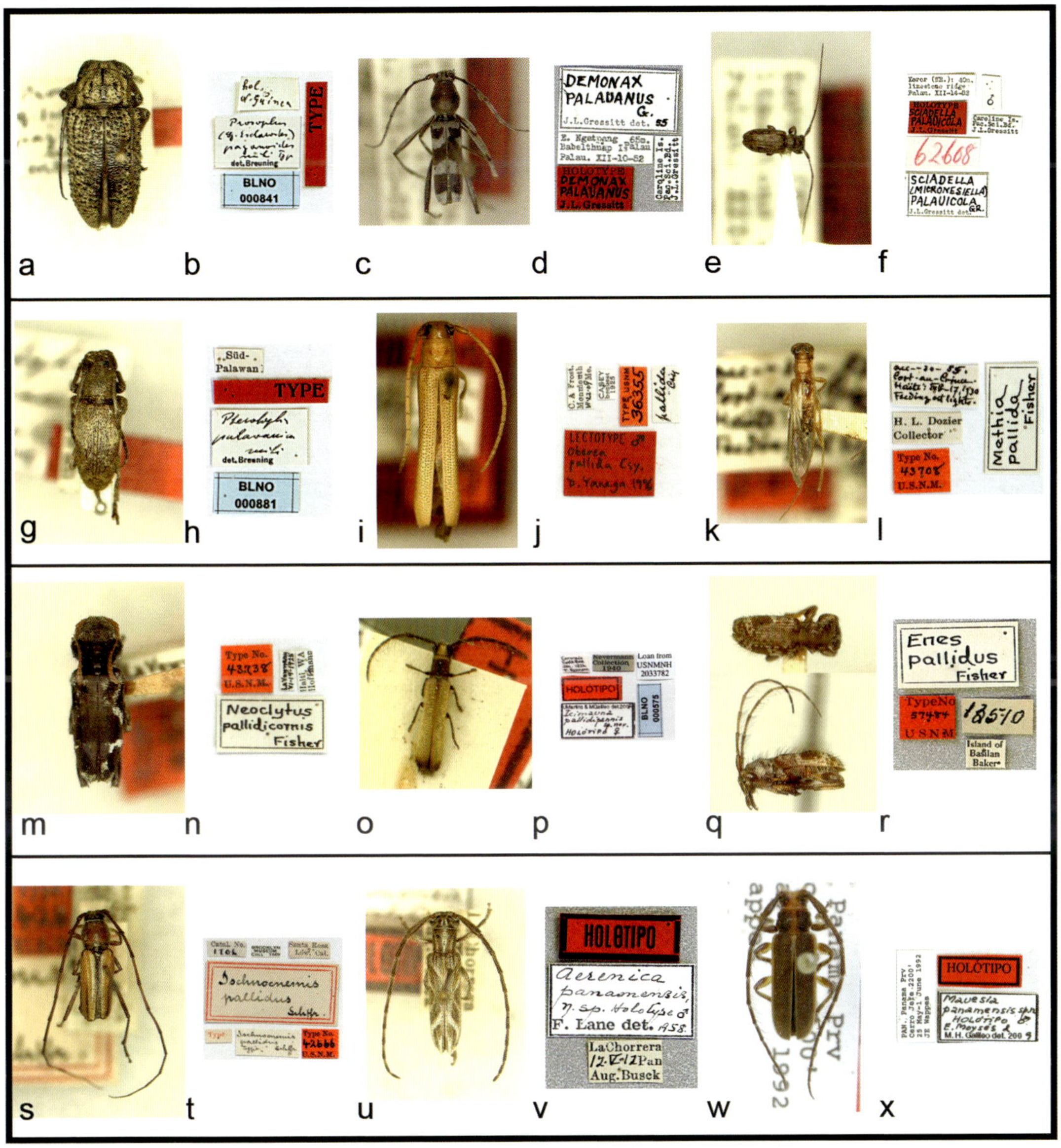

Figure 126. Primary types of Cerambycidae of the Smithsonian Institution (for each pair of images, dorsal habitus left, labels right): **(a, b)** *Prosoplus (Escharodes) paganoides* Breuning, 1940, Holotype; **(c, d)** *Demonax palauanus* Gressitt, 1956, Holotype; **(e, f)** *Sciadella (Micronesiella) palauicola* Gressitt, 1956, Holotype; **(g, h)** *Pterolophia palawanica* Breuning, 1938, Holotype; **(i, j)** *Oberea pallida* Casey, 1913, Lectotype; **(k, l)** *Methia pallida* Fisher, 1932, Holotype; **(m, n)** *Neoclytus pallidicornis* Fisher, 1932, Holotype; **(o, p)** *Icimauna pallidipennis* Martins & Galileo, 2007, Holotype; **(q, r)** *Enes pallidus* Fisher, 1925, Holotype; **(s, t)** *Ischnocuemis pallidus* Schaeffer, 1905, Syntype; **(u, v)** *Aerenica panamensis* Lane, 1959, Holotype; **(w, x)** *Mauesia panamensis* Moyses & Galileo, 2009, Holotype.

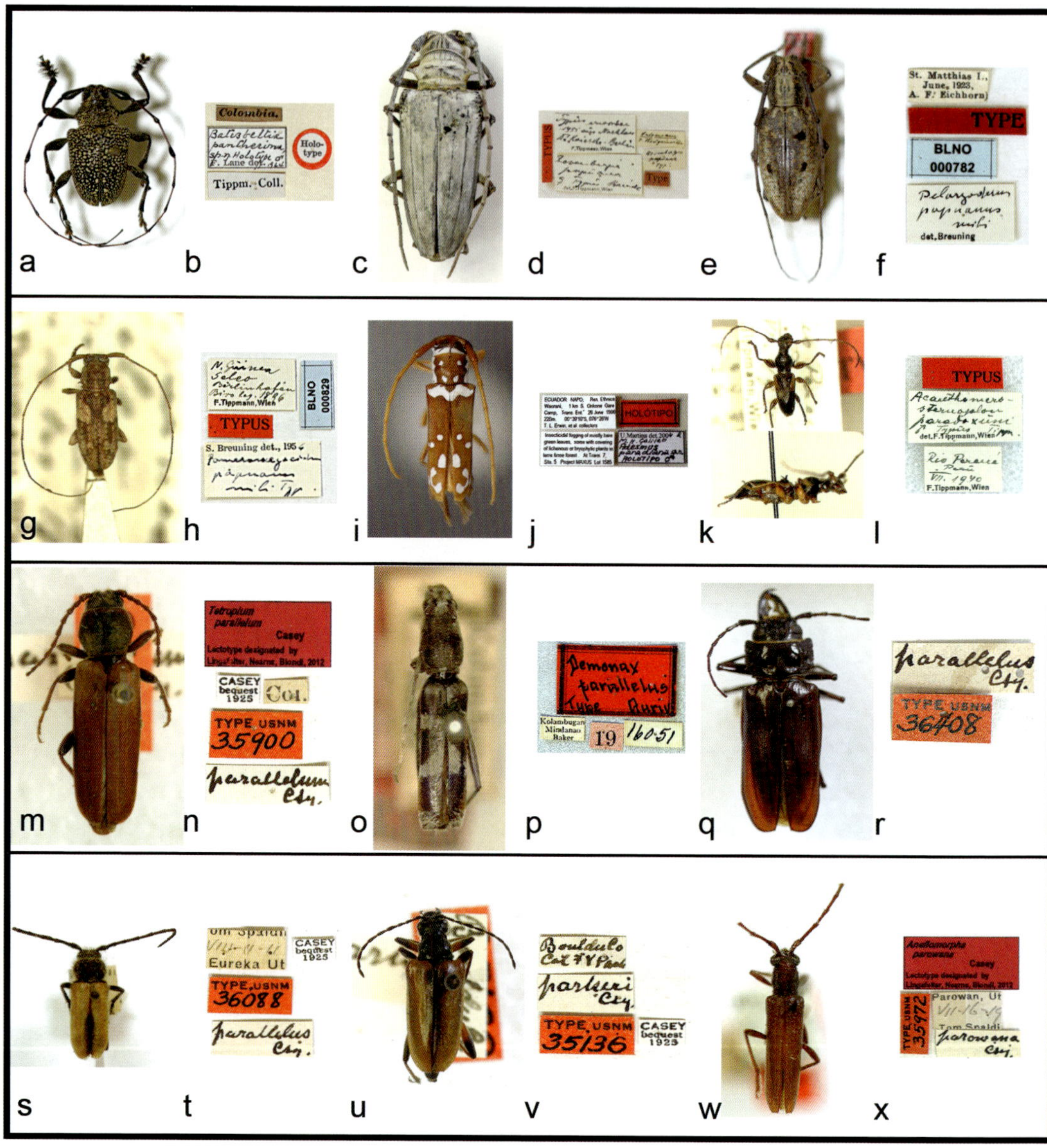

Figure 127. Primary types of Cerambycidae of the Smithsonian Institution (for each pair of images, dorsal habitus left, labels right): **(a, b)** *Batesbeltia pantherina* Lane, 1964, Holotype; **(c, d)** *Rosenbergia papuana* Kriesche, 1920, Holotype; **(e, f)** *Pelargoderus papuanus* Breuning, 1936, Holotype; **(g, h)** *Aegocidnus (Pomeranaegocidnus) papuanus* Breuning, 1957, Holotype; **(i, j)** *Adesmus paradiana* Galileo & Martins, 2004, Holotype; **(k, l)** *Acanthomerosternoplon paradoxum* Tippmann, 1955, Holotype; **(m, n)** *Tetropium parallelum* Casey, 1891, Lectotype; **(o, p)** *Demonax parallelus* Aurivillius, 1922, Holotype; **(q, r)** *Stenodontes parallelus* Casey, 1912, Holotype; **(s, t)** *Crossidius parallelus* Casey, 1924, Holotype; **(u, v)** *Acmaeops parkeri* Casey, 1913, Holotype; **(w, x)** *Aneflomorpha parowana* Casey, 1924, Lectotype.

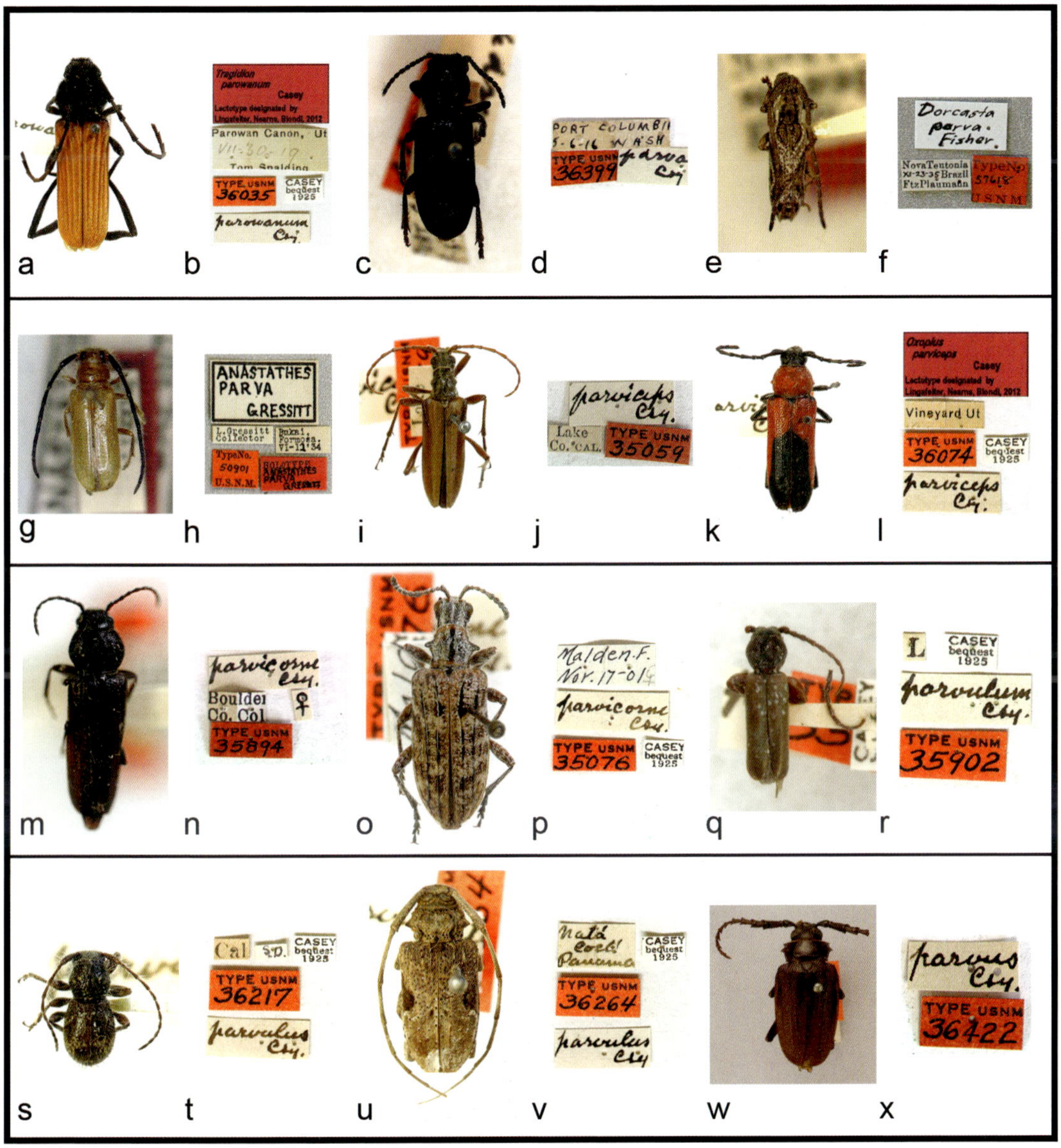

Figure 128. Primary types of Cerambycidae of the Smithsonian Institution (for each pair of images, dorsal habitus left, labels right): **(a, b)** *Tragidion parowanum* Casey, 1924, Lectotype; **(c, d)** *Spondylis parva* Casey, 1924, Holotype; **(e, f)** *Dorcasta parva* Fisher, 1938, Holotype; **(g, h)** *Anastathes parva* Gressitt, 1935, Holotype; **(i, j)** *Stenocorus (Stenocorus) parviceps* Casey, 1913, Holotype; **(k, l)** *Oxoplus parviceps* Casey, 1924, Lectotype; **(m, n)** *Asemum parvicorne* Casey, 1912, Holotype; **(o, p)** *Rhagium (Hargium) parvicorne* Casey, 1913, Holotype; **(q, r)** *Tetropium parvulum* Casey, 1891, Holotype; **(s, t)** *Ipochus parvulus* Casey, 1913, Holotype; **(u, v)** *Lagochirus parvulus* Casey, 1913, Holotype; **(w, x)** *Prionus (Prionus) parvus* Casey, 1912, Holotype.

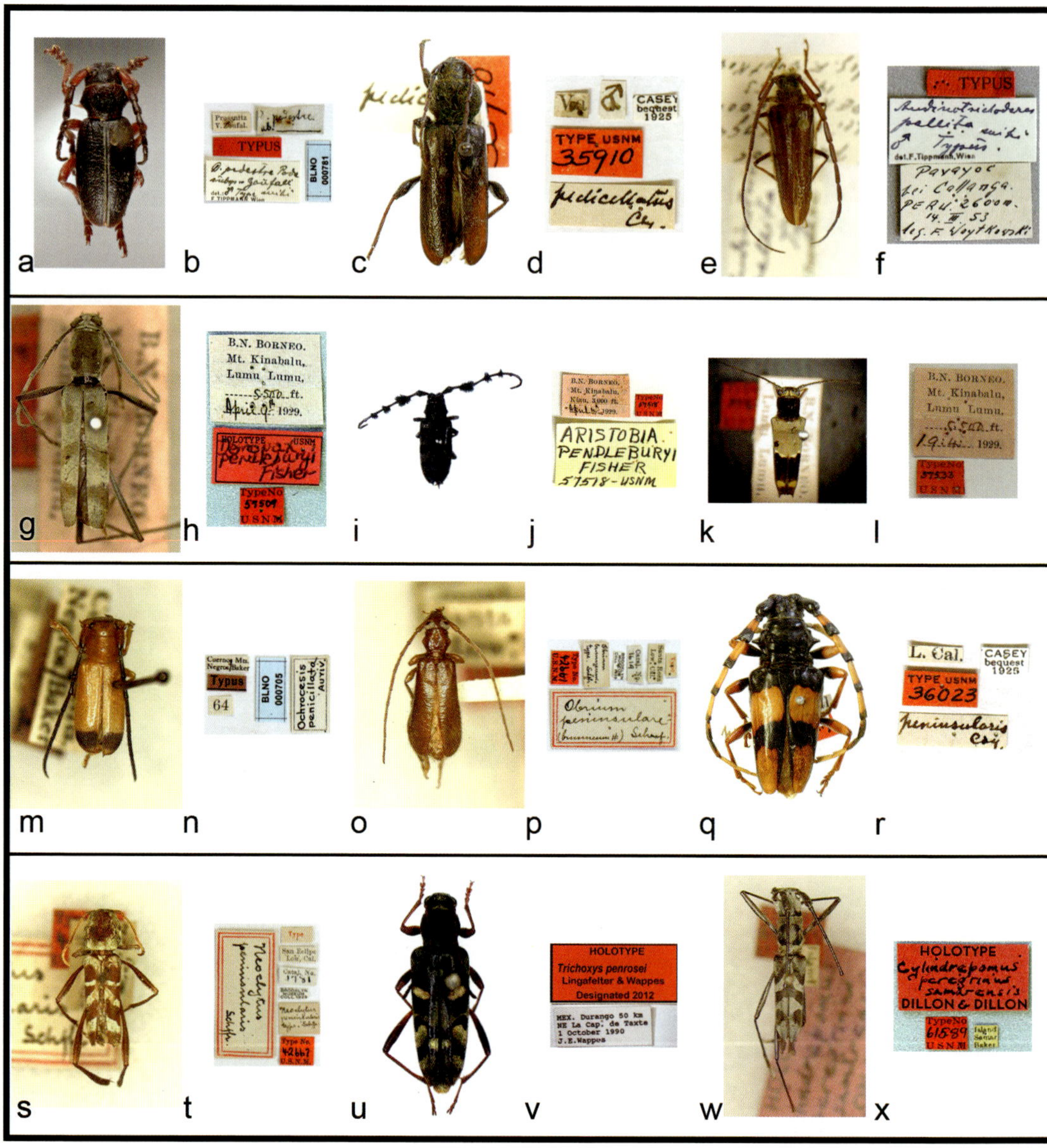

Figure 129. Primary types of Cerambycidae of the Smithsonian Institution (for each pair of images, dorsal habitus left, labels right): **(a, b)** *Pedestredorcadion pedestre zoufali* Tippmann, 1958, Holotype; **(c, d)** *Hylotrupes pedicellatus* Casey, 1924, Holotype; **(e, f)** *Andinotrichoderes pellita* Tippmann, 1960, Holotype; **(g, h)** *Demonax pendleburyi* Fisher, 1935, Holotype; **(i, j)** *Aristobia pendleburyi* Fisher, 1935, Holotype; **(k, l)** *Glenea pendleburyi* Fisher, 1935, Holotype; **(m, n)** *Ochrocesis penicillata* Aurivillius, 1927, Holotype; **(o, p)** *Obrium peninsulare* Schaeffer, 1908, Holotype; **(q, r)** *Dendrobias peninsularis* Casey, 1912, Holotype; **(s, t)** *Neoclytus peninsularis* Schaeffer, 1905, Holotype; **(u, v)** *Trichoxys penrosei* Lingafelter & Wappes, 2012, Holotype; **(w, x)** *Cylindrepomus peregrinus samarensis* Dillon & Dillon, 1948, Holotype.

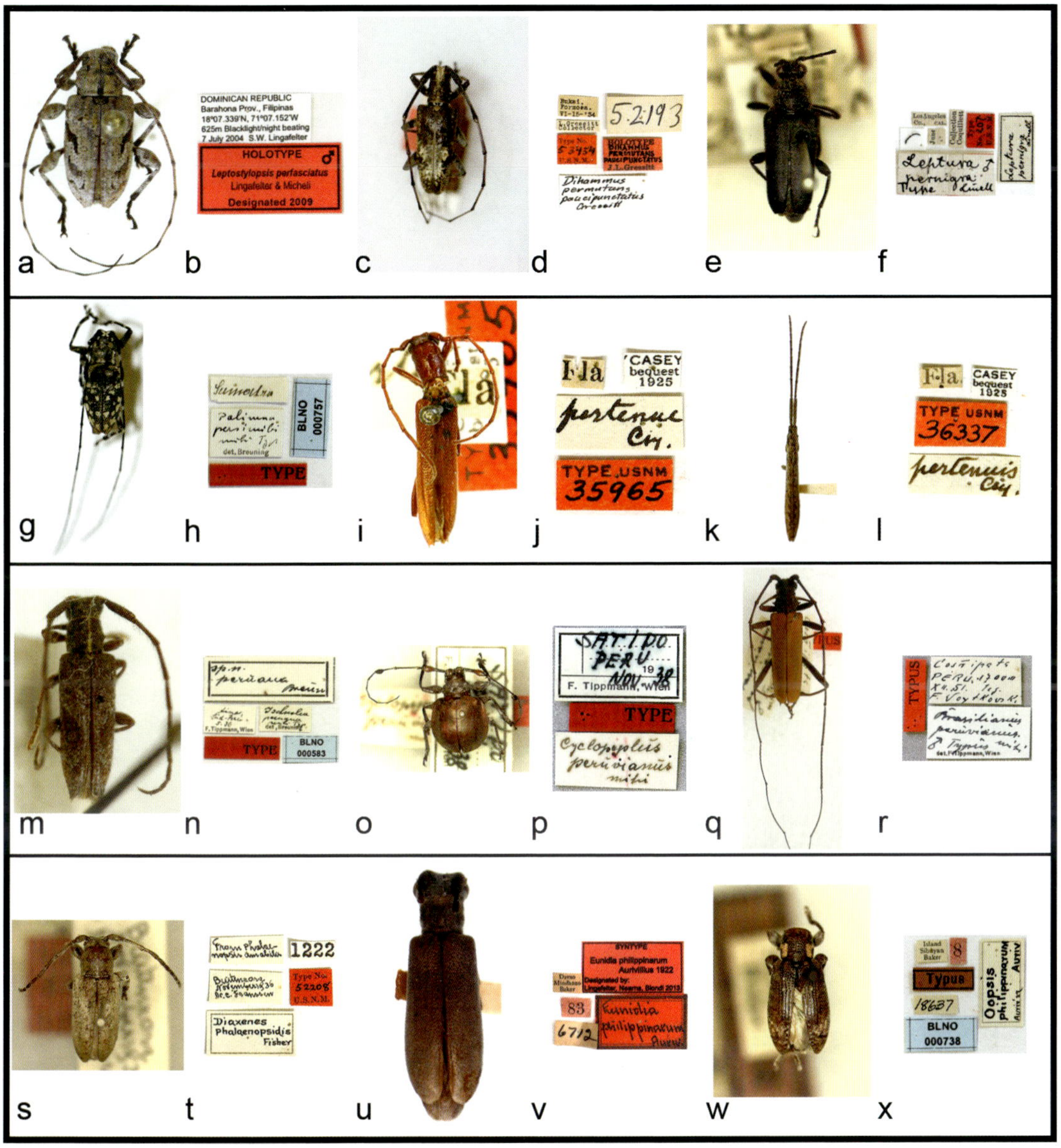

Figure 130. Primary types of Cerambycidae of the Smithsonian Institution (for each pair of images, dorsal habitus left, labels right): **(a, b)** *Leptostylopsis perfasciatus* Lingafelter & Micheli, 2009, Holotype; **(c, d)** *Dihammus permutans paucipunctatus* Gressitt, 1938, Holotype; **(e, f)** *Leptura pernigra* Linell, 1897, Holotype; **(g, h)** *Palimna persimilis* Breuning, 1938, Holotype; **(i, j)** *Pseudibidion pertenue* Casey, 1924, Holotype; **(k, l)** *Spalacopsis (Spalacopsis) pertenuis* Casey, 1913, Holotype; **(m, n)** *Ischnolea peruana* Breuning, 1943, Holotype; **(o, p)** *Cyclopeplus peruvianus* Tippmann, 1939, Holotype; **(q, r)** *Brasilianus peruvianus* Tippmann, 1960, Holotype; **(s, t)** *Diaxenes phalaenopsidis* Fisher, 1937, Holotype; **(u, v)** *Eunidia philippinarum* Aurivillius, 1922, Syntype; **(w, x)** *Oopsis philippinarum* Aurivillius, 1927, Syntype.

Figure 131. Primary types of Cerambycidae of the Smithsonian Institution (for each pair of images, dorsal habitus left, labels right): **(a, b)** *Ipochira philippinarum* Aurivillius, 1927, Syntype; **(c, d)** *Exocentrus philippinus* Fisher, 1925, Holotype; **(e, f)** *Camptomyne philippinus* Fisher, 1925, Holotype; **(g, h)** *Exocentrus (Camptomyme) philippinus transeuns* Breuning, 1957, Holotype; **(i, j)** *Camptomyne philippinus* var. *unicolor* Fisher, 1925, Holotype; **(k, l)** *Eucrossus phoenicis* Casey, 1924, Holotype; **(m, n)** *Phidola picea* Fisher, 1936, Holotype; **(o, p)** *Octoplon piceicollis* Linsley, 1935, Holotype; **(q, r)** *Tyrinthia picticornis* Martins & Galileo, 1991, Holotype; **(s, t)** *Xestoleptura pictipennis* Casey, 1924, Lectotype; **(u, v)** *Physocnemum pictum* Casey, 1924, Holotype; **(w, x)** *Psapharochrus pictus* Galileo & Martins, 2012, Holotype.

Figure 132. Primary types of Cerambycidae of the Smithsonian Institution (for each pair of images, dorsal habitus left, labels right): **(a, b)** *Adesmus pilatus* Galileo & Martins, 2005, Holotype; **(c, d)** *Eburodacrys pilicornis* Fisher, 1944, Holotype; **(e, f)** *Tragosoma pilosicornis* Casey, 1890, Holotype; **(g, h)** *Aneflus (Protaneflus) pilosicornis* Chemsak & Linsley, 1965, Holotype; **(i, j)** *Ebaeides pilosicornis* Fisher, 1925, Holotype; **(k, l)** *Pterolophia pilosipennis* Breuning, 1943, Holotype; **(m, n)** *Acrepidopterum pilosum* Fisher, 1932, Holotype; **(o, p)** *Xylotrechus pilosus* Aurivillius, 1927, Syntype; **(q, r)** *Moneilema (Collapteryx) pimalis* Casey, 1913, Holotype; **(s, t)** *Leptostylus pini* Schaeffer, 1905, Holotype; **(u, v)** *Lepturgoides pini* Schaeffer, 1905, Syntype; **(w, x)** *Ipochus pinicola* Casey, 1913, Lectotype.

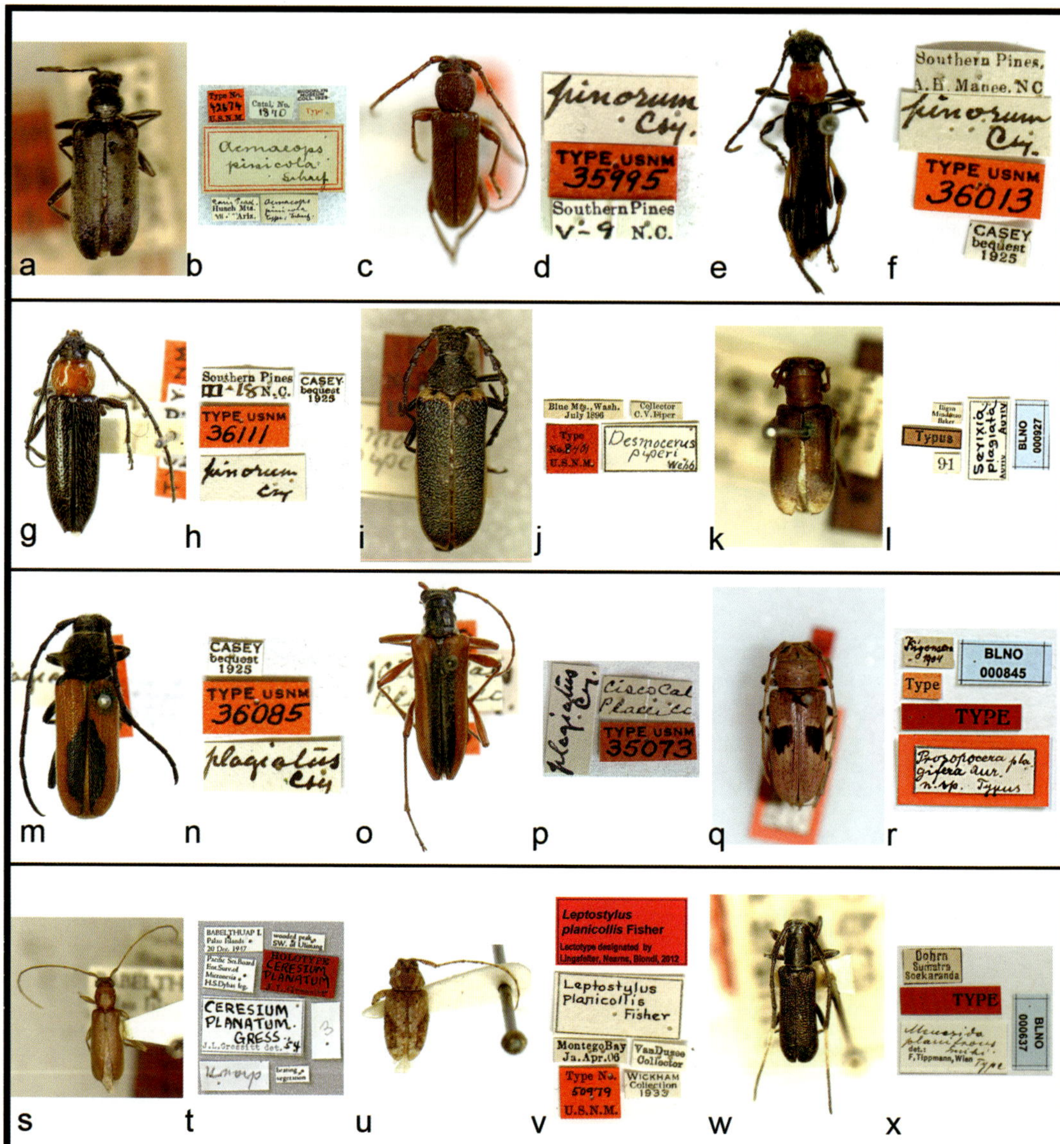

Figure 133. Primary types of Cerambycidae of the Smithsonian Institution (for each pair of images, dorsal habitus left, labels right): **(a, b)** *Acmaeops pinicola* Schaeffer, 1908, Holotype; **(c, d)** *Anoplium pinorum* Casey, 1914, Holotype; **(e, f)** *Callimoxys pinorum* Casey, 1924, Holotype; **(g, h)** *Stenosphenus pinorum* Casey, 1924, Holotype; **(i, j)** *Desmocerus piperi* Webb, 1905, Holotype; **(k, l)** *Serixia plagiata* Aurivillius, 1927, Syntype; **(m, n)** *Crossidius plagiatus* Casey, 1912, Holotype; **(o, p)** *Stenocorus plagiatus* Casey, 1924, Holotype; **(q, r)** *Prosopocera plagifera* Aurivillius, 1907, Holotype; **(s, t)** *Ceresium (Ceresiellum) planatum* Gressitt, 1956, Holotype; **(u, v)** *Leptostylus planicollis* Fisher, 1935, Lectotype; **(w, x)** *Menesida planifrons* Tippmann, 1951, Holotype.

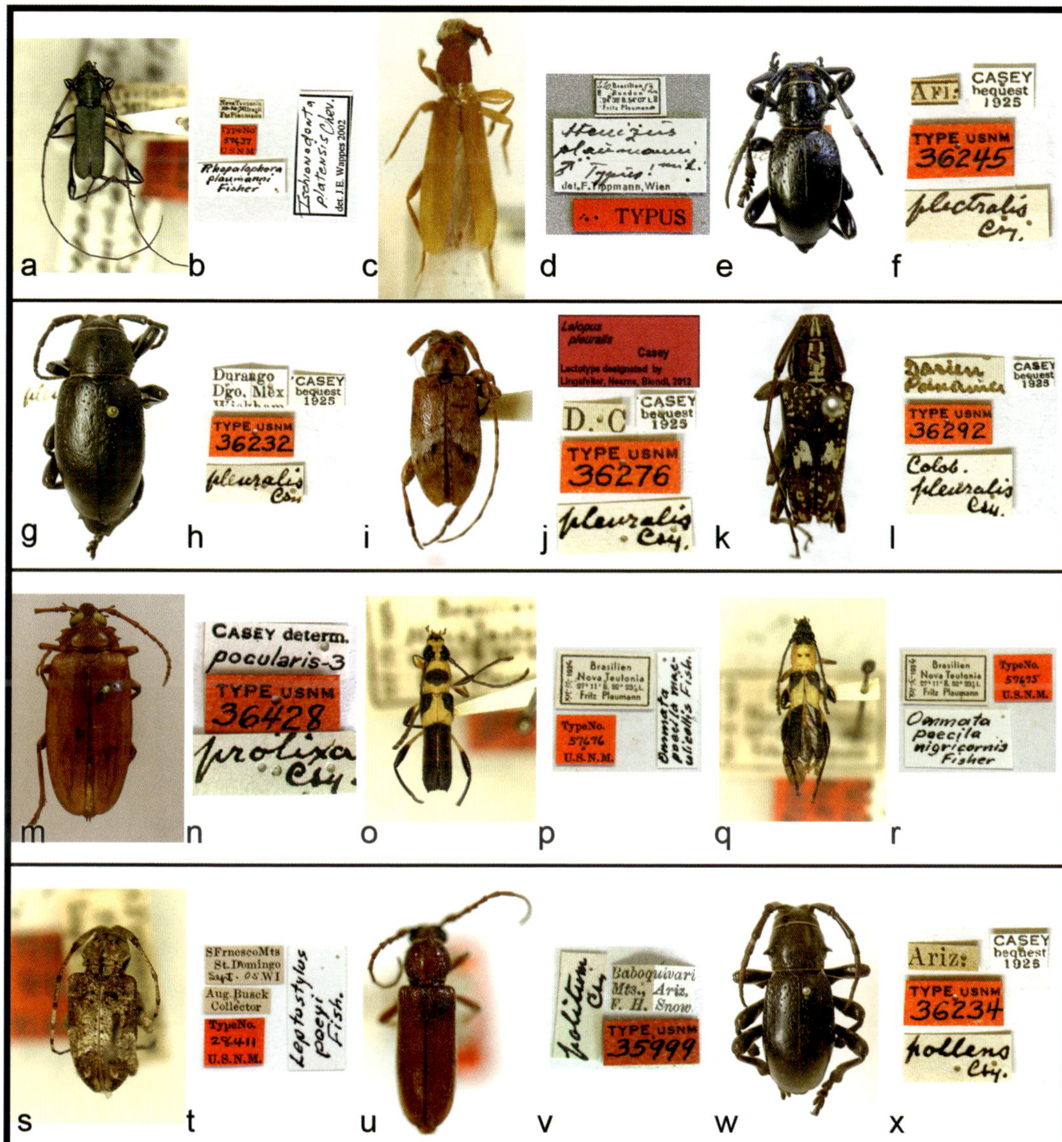

Figure 134. Primary types of Cerambycidae of the Smithsonian Institution (for each pair of images, dorsal habitus left, labels right): **(a, b)** *Rhopalophora (Ischionodonta) plaumanni* Fisher, 1937, Holotype; **(c, d)** *Atenizus plaumanni* Tippmann, 1960, Holotype; **(e, f)** *Moneilema (Collapteryx) plectralis* Casey, 1924, Holotype; **(g, h)** *Moneilema (Collapteryx) pleuralis* Casey, 1913, Holotype; **(i, j)** *Leiopus pleuralis* Casey, 1913, Lectotype; **(k, l)** *Colobothea pleuralis* Casey, 1913, Holotype; **(m, n)** *Prionus (Prionus) pocularis prolixus* Casey, 1912, Holotype; **(o, p)** *Ommata (Eclipta) poecila* v. *maculicollis* Fisher, 1947, Holotype; **(q, r)** *Ommata (Eclipta) poecila* v. *nigricornis* Fisher, 1947, Holotype; **(s, t)** *Leptostylus poeyi* Fisher, 1925, Holotype; **(u, v)** *Anoplium politum* Casey, 1924, Holotype; **(w, x)** *Moneilema (Collapteryx) pollens* Casey, 1913, Holotype.

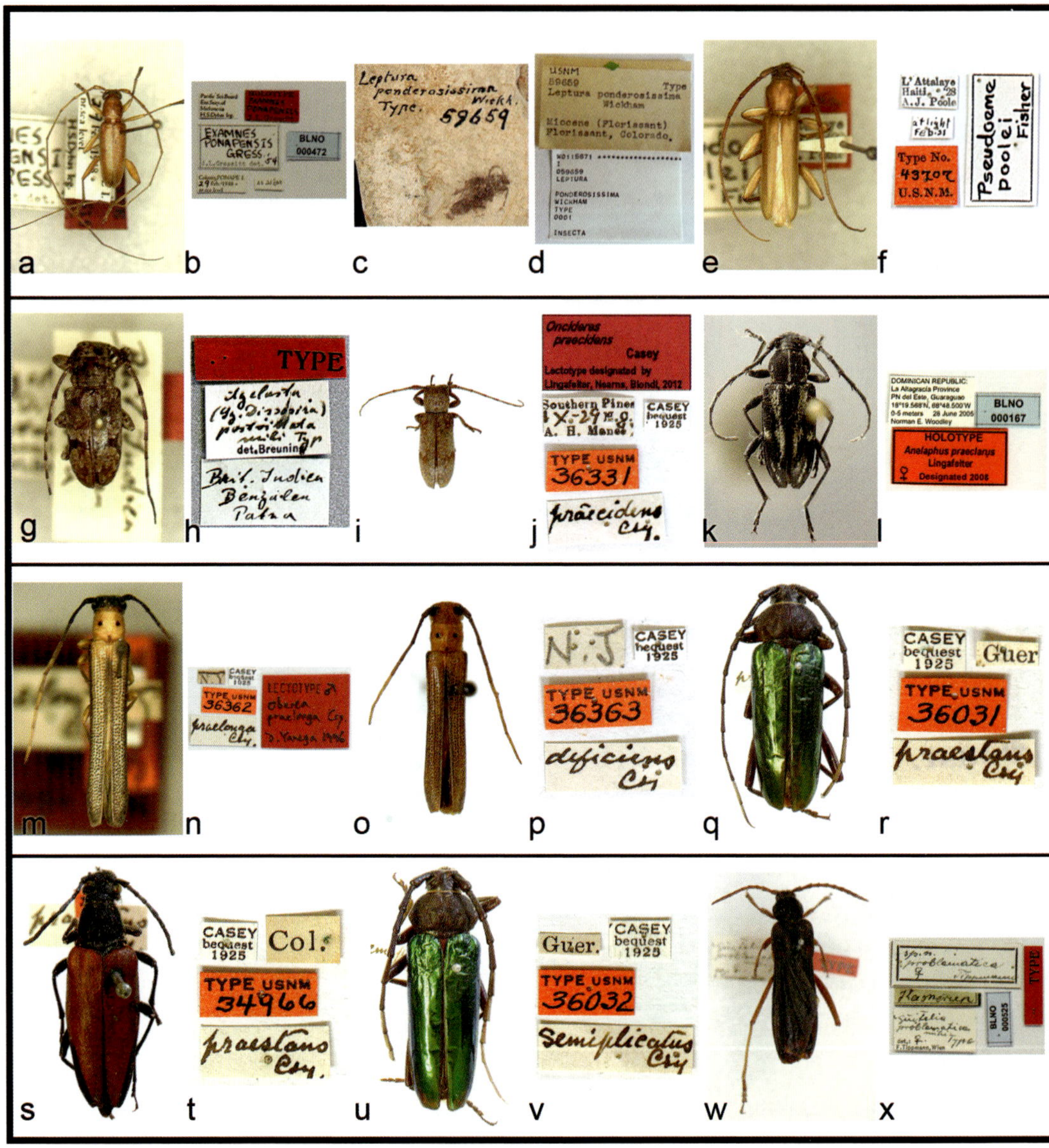

Figure 135. Primary types of Cerambycidae of the Smithsonian Institution (for each pair of images, dorsal habitus left, labels right): **(a, b)** *Examnes ponapensis* Gressitt, 1956, Holotype; **(c, d)** *Leptura ponderosissima* Wickham, 1913, Holotype; **(e, f)** *Pseudoeme poolei* Fisher, 1932, Holotype; **(g, h)** *Agelasta (Dissosira) postvittata* Breuning, 1939, Holotype; **(i, j)** *Oncideres praecidens* Casey, 1913, Lectotype; **(k, l)** *Anelaphus praeclarus* Lingafelter, 2008, Holotype; **(m, n)** *Oberea praelonga* Casey, 1913, Lectotype; **(o, p)** *Oberea praelonga deficiens* Casey, 1924, Holotype; **(q, r)** *Crioprosopus praestans* Casey, 1912, Holotype; **(s, t)** *Leptura (Cosmosalia) praestans* Casey, 1913, Holotype; **(u, v)** *Crioprosopus praestans semiplicatus* Casey, 1912, Holotype; **(w, x)** *Guitelia problematica* Tippmann, 1951, Holotype.

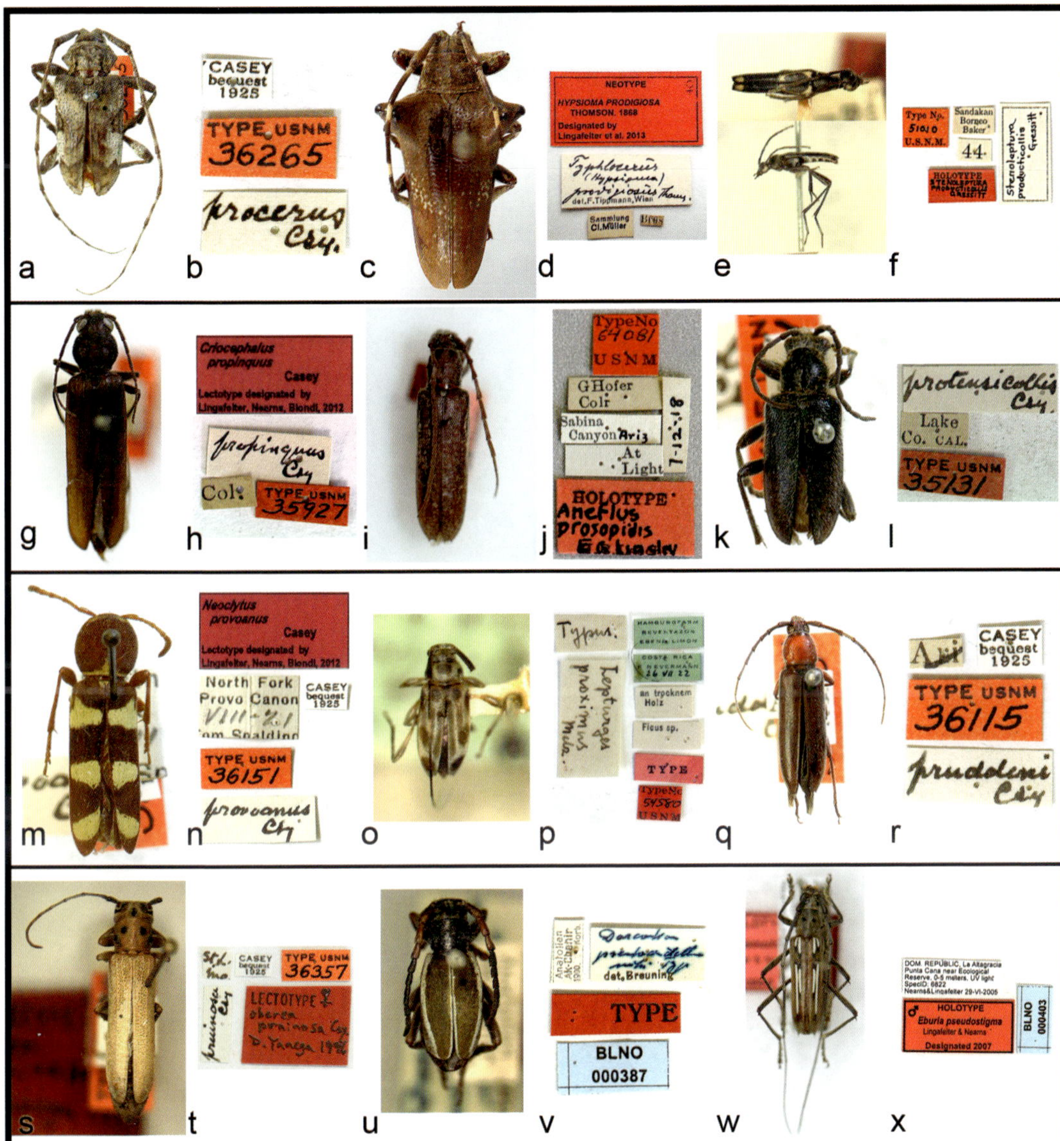

Figure 136. Primary types of Cerambycidae of the Smithsonian Institution (for each pair of images, dorsal habitus left, labels right): **(a, b)** *Lagochirus procerus* Casey, 1913, Holotype; **(c, d)** *Hypsioma prodigiosa* Thomson, 1868, Neotype; **(e, f)** *Stenoleptura producticollis* Gressitt, 1935, Holotype; **(g, h)** *Criocephalus propinquus* Casey, 1924, Lectotype; **(i, j)** *Aneflus prosopidis* Linsley, 1957, Holotype; **(k, l)** *Brachysomida protensicollis* Casey, 1913, Holotype; **(m, n)** *Neoclytus provoanus* Casey, 1924, Lectotype; **(o, p)** *Lepturges proximus* Melzer, 1934, Holotype; **(q, r)** *Stenosphenus pruddeni* Casey, 1912, Holotype; **(s, t)** *Oberea pruinosa* Casey, 1913, Lectotype; **(u, v)** *Dorcadion pseudocinctellum* Breuning, 1943, Holotype; **(w, x)** *Eburia pseudostigma* Lingafelter & Nearns, 2007, Holotype.

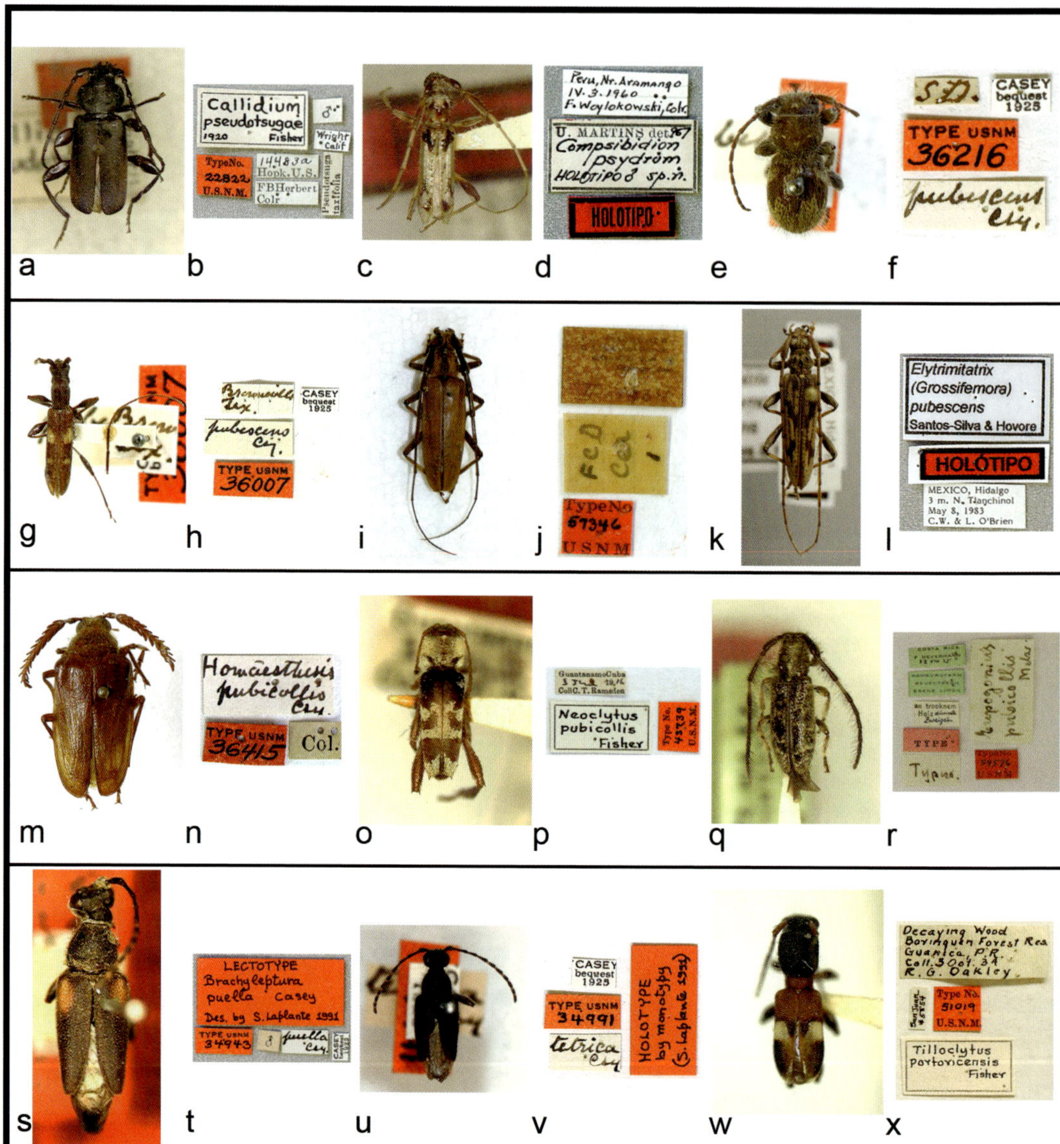

Figure 137. Primary types of Cerambycidae of the Smithsonian Institution (for each pair of images, dorsal habitus left, labels right): **(a, b)** *Callidium pseudotsugae* Fisher, 1920, Holotype; **(c, d)** *Compsibidion psydrum* Martins, 1969, Holotype; **(e, f)** *Ipochus pubescens* Casey, 1891, Holotype; **(g, h)** *Ibidion pubescens* Casey, 1924, Holotype; **(i, j)** *Neocerambyx pubescens* Fisher, 1936, Holotype; **(k, l)** *Elytrimitatrix (Grossifemora) pubescens* Santos-Silva & Hovore, 2008, Holotype; **(m, n)** *Homaesthesis pubicollis* Casey, 1912, Holotype; **(o, p)** *Neoclytus pubicollis* Fisher, 1932, Holotype; **(q, r)** *Eupogonius pubicollis* Melzer, 1933, Holotype; **(s, t)** *Brachyleptura puella* Casey, 1913, Lectotype; **(u, v)** *Brachyleptura puella tetrica* Casey, 1913, Holotype; **(w, x)** *Tilloclytus puertoricensis* Fisher, 1935, Holotype.

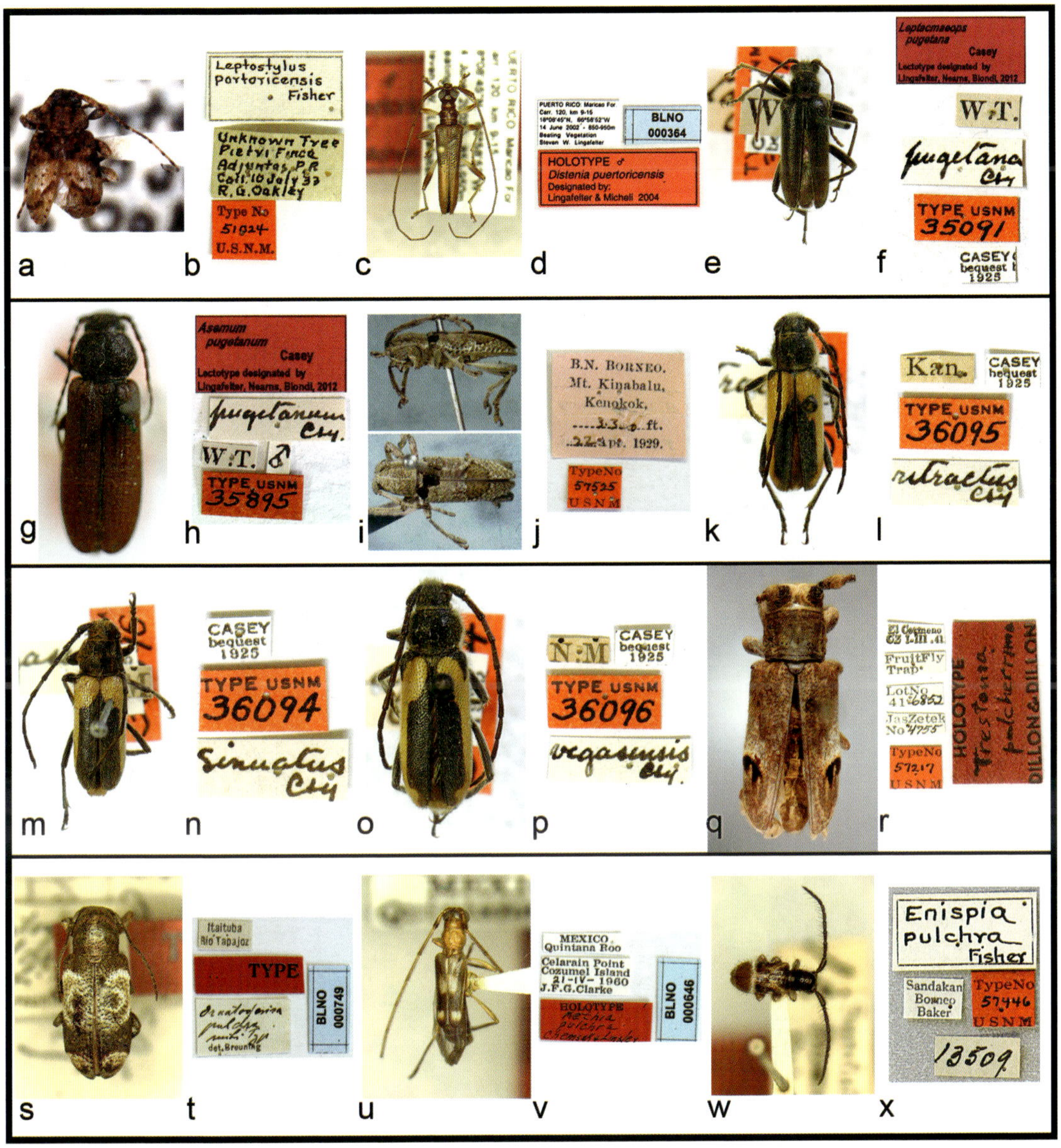

Figure 138. Primary types of Cerambycidae of the Smithsonian Institution (for each pair of images, dorsal habitus left, labels right): **(a, b)** *Leptostylus puertoricensis* Fisher, 1935, Holotype; **(c, d)** *Distenia puertoricensis* Lingafelter & Micheli, 2004, Holotype; **(e, f)** *Leptacmaeops (Leptacmaeops) pugetana* Casey, 1913, Lectotype; **(g, h)** *Asemum pugetanum* Casey, 1912, Lectotype; **(i, j)** *Neoepaphra pulchella* Fisher, 1935, Holotype; **(k, l)** *Crossidius pulchellus retractus* Casey, 1912, Holotype; **(m, n)** *Crossidius pulchellus sinuosus* Casey, 1912, Holotype; **(o, p)** *Crossidius pulchellus vegasensis* Casey, 1912, Holotype; **(q, r)** *Trestonia pulcherrima* Dillon & Dillon, 1946, Holotype; **(s, t)** *Ornatodesisa pulchra* Breuning, 1961, Holotype; **(u, v)** *Methia pulchra* Chemsak & Linsley, 1964, Holotype; **(w, x)** *Enispia pulchra* Fisher, 1925, Holotype.

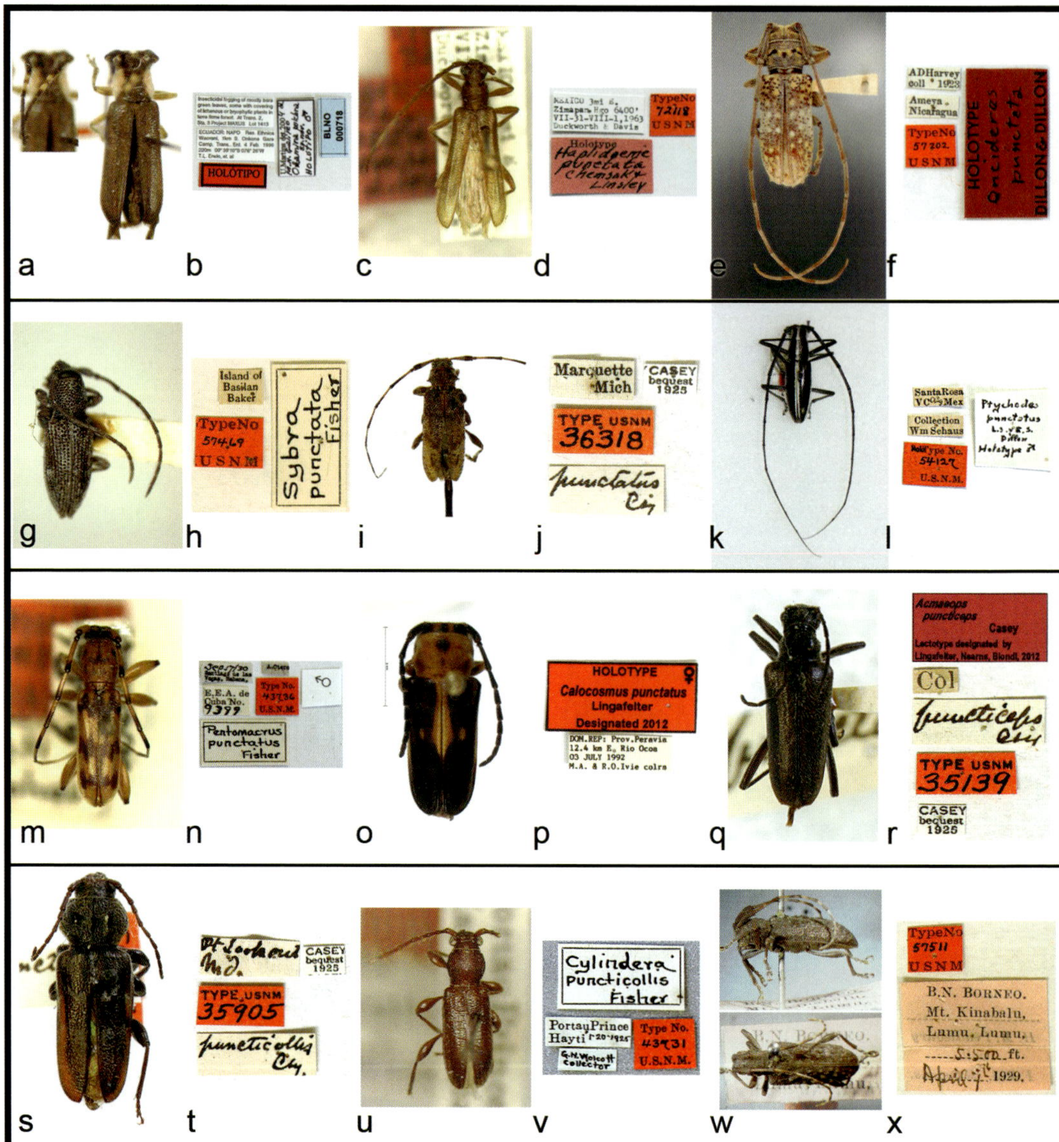

Figure 139. Primary types of Cerambycidae of the Smithsonian Institution (for each pair of images, dorsal habitus left, labels right): **(a, b)** *Okamira pulchra* Galileo & Martins, 2005, Holotype; **(c, d)** *Haplidoeme punctata* Chemsak & Linsley, 1971, Holotype; **(e, f)** *Oncideres punctata* Dillon & Dillon, 1946, Holotype; **(g, h)** *Sybra punctata* Fisher, 1925, Holotype; **(i, j)** *Graphisurus punctatus* Casey, 1924, Holotype; **(k, l)** *Ptychodes punctatus* Dillon & Dillon, 1941, Holotype; **(m, n)** *Pentomacrus punctatus* Fisher, 1932, Holotype; **(o, p)** *Calocosmus punctatus* Lingafelter, 2013, Holotype; **(q, r)** *Acmaeops puncticeps* Casey, 1913, Lectotype; **(s, t)** *Hylotrupes puncticollis* Casey, 1924, Holotype; **(u, v)** *Cylindera puncticollis* Fisher, 1932, Holotype; **(w, x)** *Trachystola puncticollis* Fisher, 1935, Holotype.

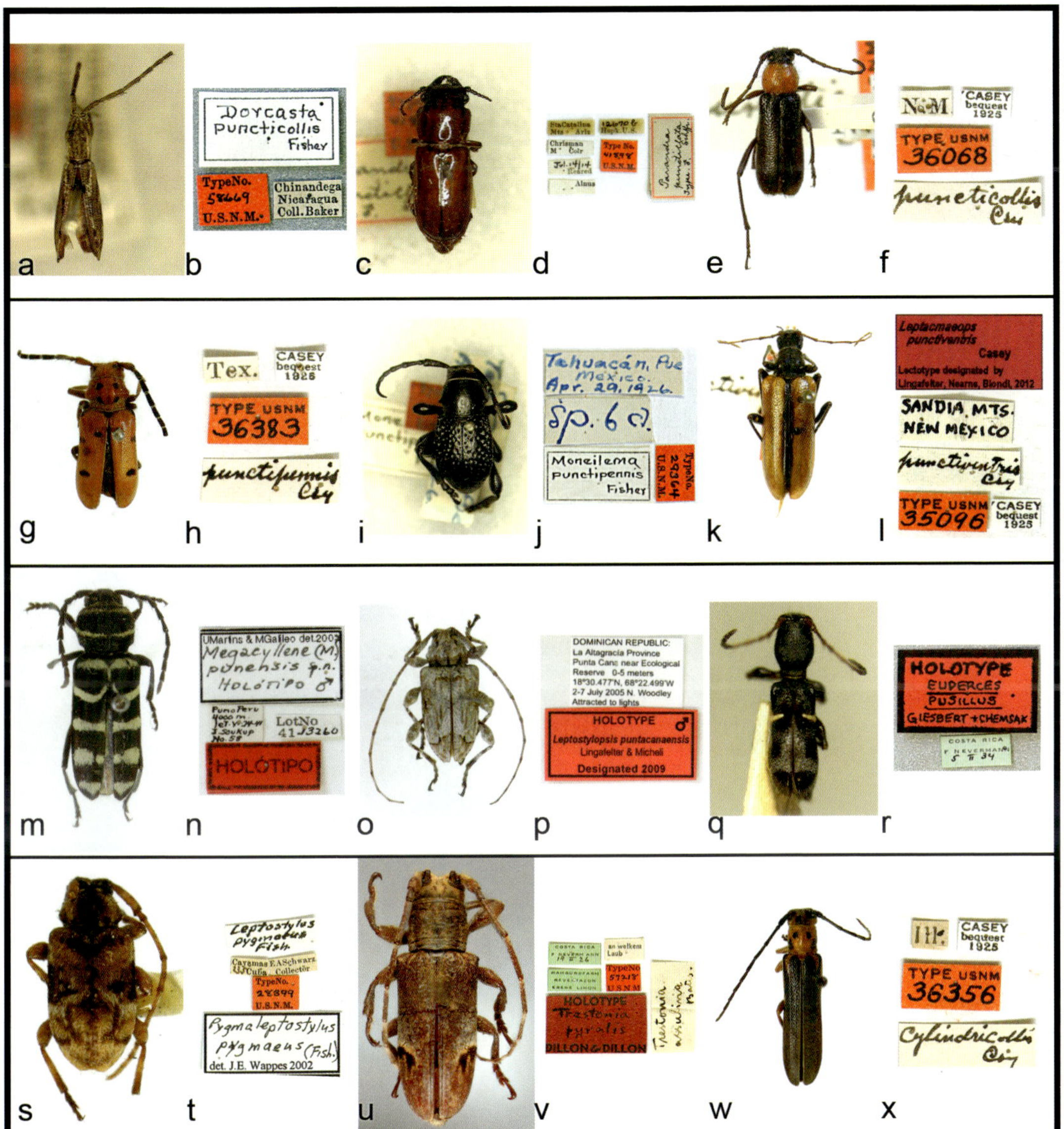

Figure 140. Primary types of Cerambycidae of the Smithsonian Institution (for each pair of images, dorsal habitus left, labels right): **(a, b)** *Dorcasta puncticollis* Fisher, 1947, Holotype; **(c, d)** *Parandra punctillata* Schaeffer, 1929, Holotype; **(e, f)** *Batyle punctipennis* Casey, 1912, Holotype; **(g, h)** *Tetraopes punctipennis* Casey, 1913, Holotype; **(i, j)** *Moneilema (Collapteryx) punctipennis* Fisher, 1927, Holotype; **(k, l)** *Leptacmaeops (Leptacmaeops) punctiventris* Casey, 1913, Lectotype; **(m, n)** *Megacyllene (Megacyllene) punensis* Martins & Galileo, 2008, Holotype; **(o, p)** *Leptostylopsis puntacanaensis* Lingafelter & Micheli, 2009, Holotype; **(q, r)** *Euderces pusillus* Giesbert & Chemsak, 1997, Holotype; **(s, t)** *Leptostylus pygmaeus* Fisher, 1926, Holotype; **(u, v)** *Trestonia pyralis* Dillon & Dillon, 1946, Holotype; **(w, x)** *Oberea quadricallosa cylindricollis* Casey, 1924, Holotype.

Figure 141. Primary types of Cerambycidae of the Smithsonian Institution (for each pair of images, dorsal habitus left, labels right): **(a, b)** *Leptacmaeops quadriceps* Casey, 1913, Lectotype; **(c, d)** *Anoplium quadricolle* Casey, 1924, Holotype; **(e, f)** *Eburia quadricollis* Casey, 1924, Holotype; **(g, h)** *Mimocorus quadricristatus* Breuning, 1942, Holotype; **(i, j)** *Sybra quadriguttata* Aurivillius, 1927, Syntype; **(k, l)** *Opsimus quadrilineatus biplectralis* Casey, 1924, Holotype; **(m, n)** *Leptacmaeops quadrinotata* Casey, 1913, Holotype; **(o, p)** *Serixia quadriplagiata* Aurivillius, 1927, Syntype; **(q, r)** *Cremnosterna quadriplagiata* Breuning, 1940, Holotype; **(s, t)** *Metopocoilus quadrispinosus* var. *basimaculatus* Tippmann, 1953, Holotype; **(u, v)** *Planodes quaternaria evanescnes* Kriesche, 1936, Holotype; **(w, x)** *Xylotrechus quercus* Schaeffer, 1905, Syntype.

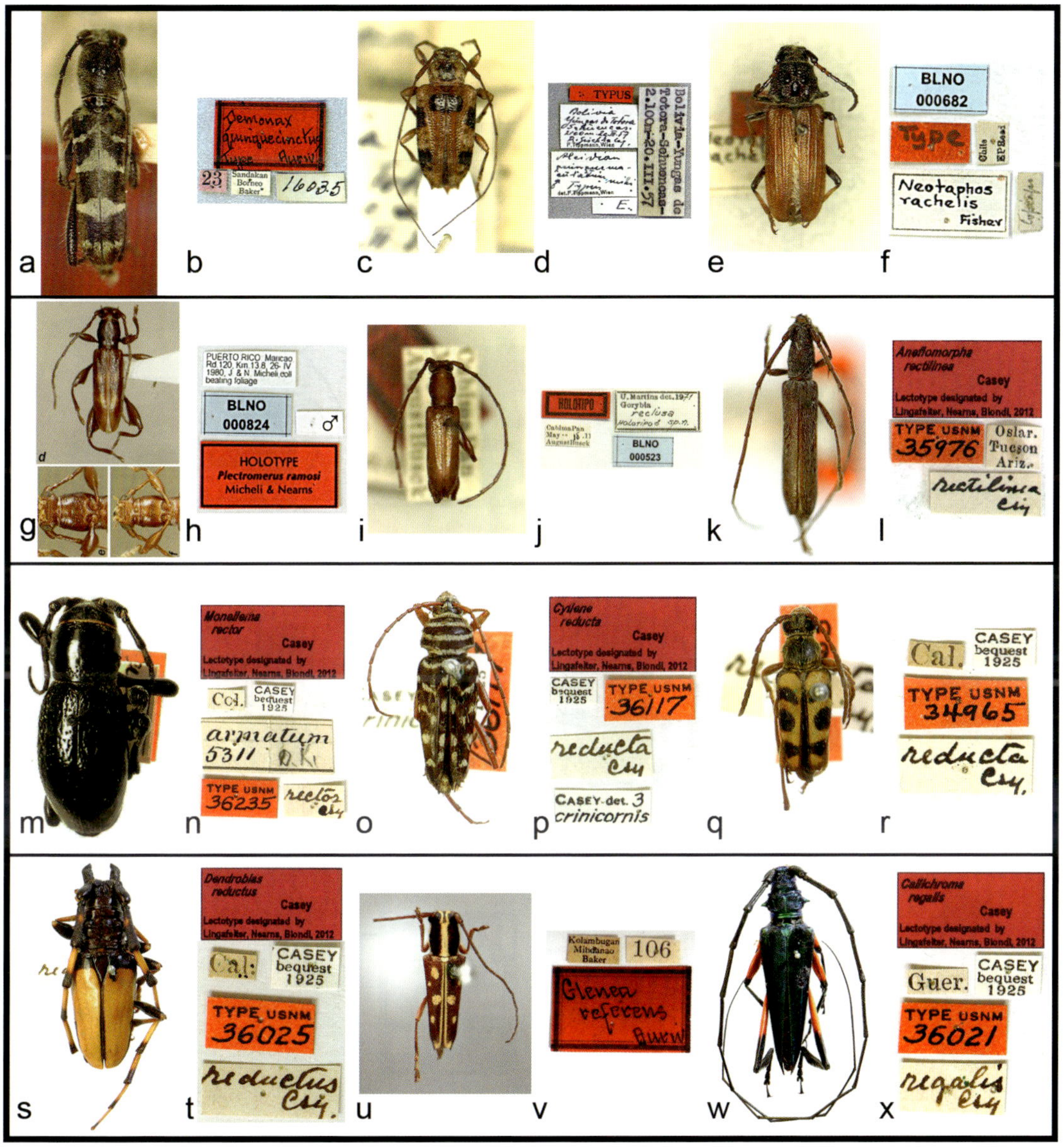

Figure 142. Primary types of Cerambycidae of the Smithsonian Institution (for each pair of images, dorsal habitus left, labels right): **(a, b)** *Demonax quinquecinctus* Aurivillius, 1922, Holotype; **(c, d)** *Alcidion quinquemaculatum* Tippmann, 1960, Holotype; **(e, f)** *Neotaphos rachelis* Fisher, 1937, Holotype; **(g, h)** *Plectromerus ramosi* Micheli & Nearns, 2005, Holotype; **(i, j)** *Gorybia reclusa* Martins, 1976, Holotype; **(k, l)** *Aneflomorpha rectilinea* Casey, 1924, Lectotype; **(m, n)** *Moneilema (Collapteryx) rector* Casey, 1913, Lectotype; **(o, p)** *Cyllene reducta* Casey, 1912, Lectotype; **(q, r)** *Strophiona reducta* Casey, 1913, Holotype; **(s, t)** *Dendrobias reductus* Casey, 1912, Lectotype; **(u, v)** *Glenea (Glenea) referens* Aurivillius, 1926, Holotype; **(w, x)** *Callichroma regalis* Casey, 1912, Lectotype.

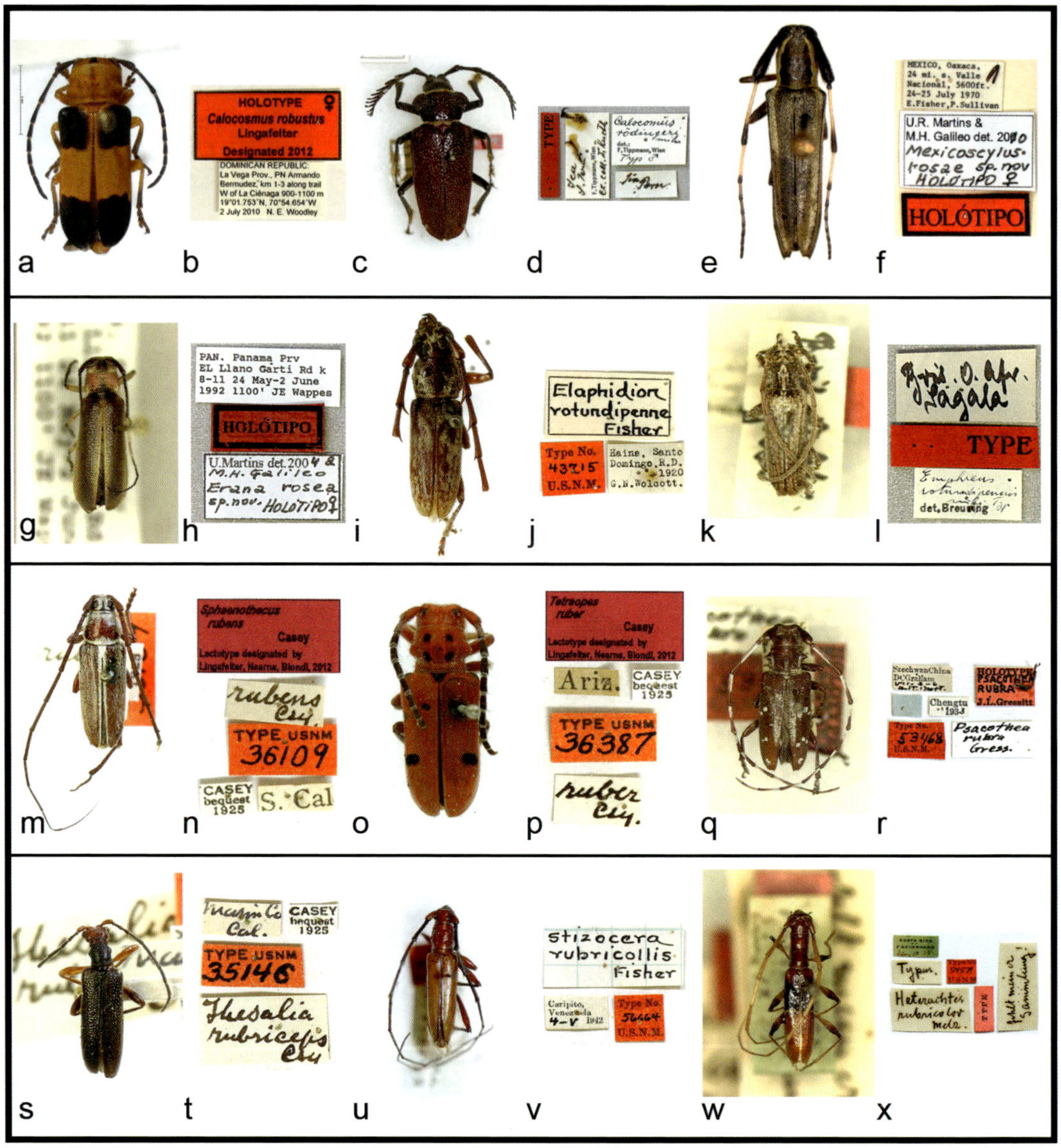

Figure 145. Primary types of Cerambycidae of the Smithsonian Institution (for each pair of images, dorsal habitus left, labels right): **(a, b)** *Calocosmus robustus* Lingafelter, 2013, Holotype; **(c, d)** *Calocomus roedingeri* Tippmann, 1951, Holotype; **(e, f)** *Mexicoscylus rosae* Martins & Galileo, 2011, Holotype; **(g, h)** *Erana rosea* Galileo & Martins, 2004, Holotype; **(i, j)** *Elaphidion rotundipenne* Fisher, 1932, Holotype; **(k, l)** *Emphreus rotundipennis* Breuning, 1950, Holotype; **(m, n)** *Sphaenothecus rubens* Casey, 1891, Lectotype; **(o, p)** *Tetraopes ruber* Casey, 1913, Lectotype; **(q, r)** *Psacothea rubra* Gressitt, 1938, Holotype; **(s, t)** *Thesalia rubriceps* Casey, 1913, Holotype; **(u, v)** *Stizocera rubricollis* Fisher, 1944, Holotype; **(w, x)** *Heterachthes rubricolor* Melzer, 1935, Holotype.

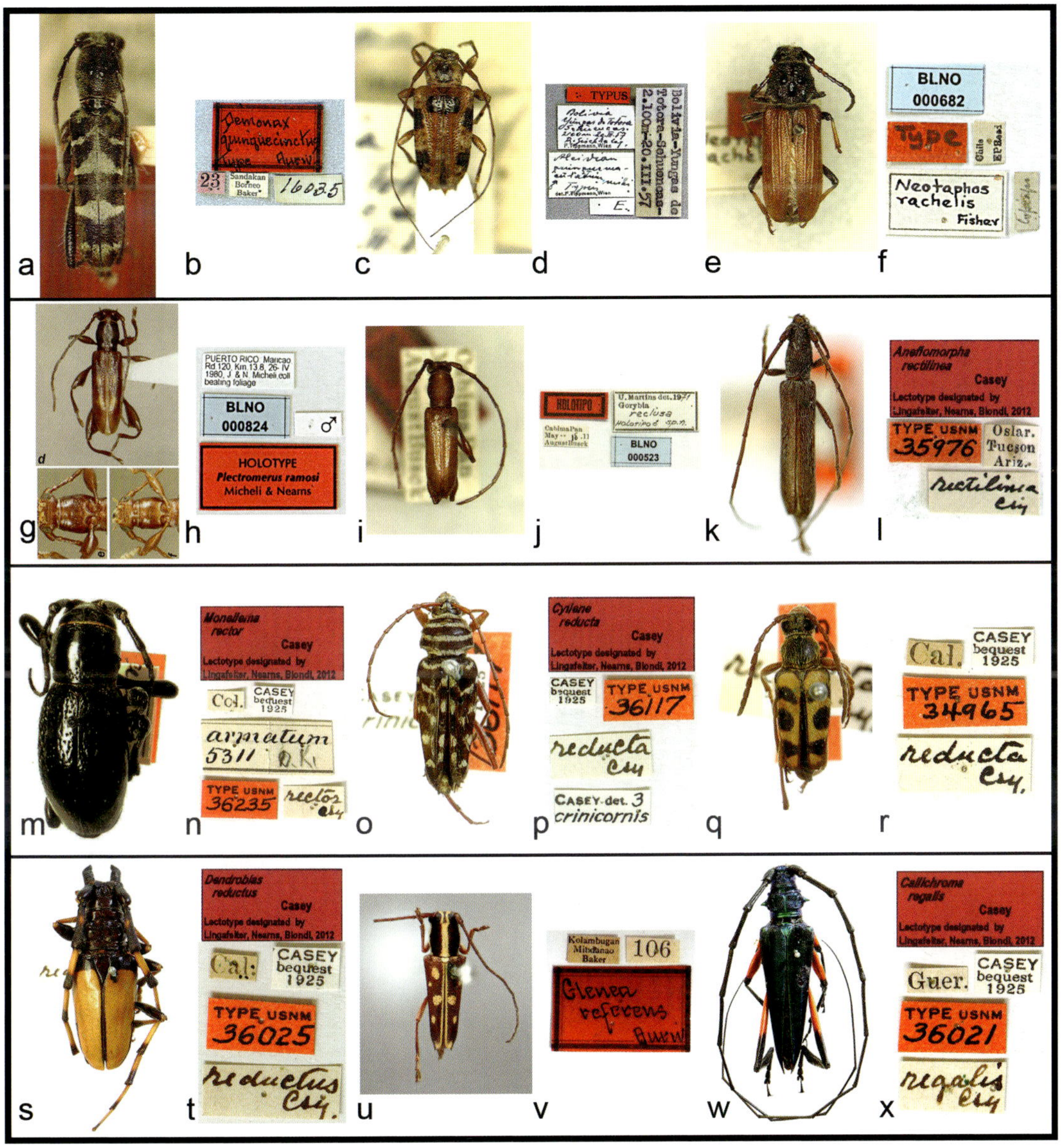

Figure 142. Primary types of Cerambycidae of the Smithsonian Institution (for each pair of images, dorsal habitus left, labels right): **(a, b)** *Demonax quinquecinctus* Aurivillius, 1922, Holotype; **(c, d)** *Alcidion quinquemaculatum* Tippmann, 1960, Holotype; **(e, f)** *Neotaphos rachelis* Fisher, 1937, Holotype; **(g, h)** *Plectromerus ramosi* Micheli & Nearns, 2005, Holotype; **(i, j)** *Gorybia reclusa* Martins, 1976, Holotype; **(k, l)** *Aneflomorpha rectilinea* Casey, 1924, Lectotype; **(m, n)** *Moneilema (Collapteryx) rector* Casey, 1913, Lectotype; **(o, p)** *Cyllene reducta* Casey, 1912, Lectotype; **(q, r)** *Strophiona reducta* Casey, 1913, Holotype; **(s, t)** *Dendrobias reductus* Casey, 1912, Lectotype; **(u, v)** *Glenea (Glenea) referens* Aurivillius, 1926, Holotype; **(w, x)** *Callichroma regalis* Casey, 1912, Lectotype.

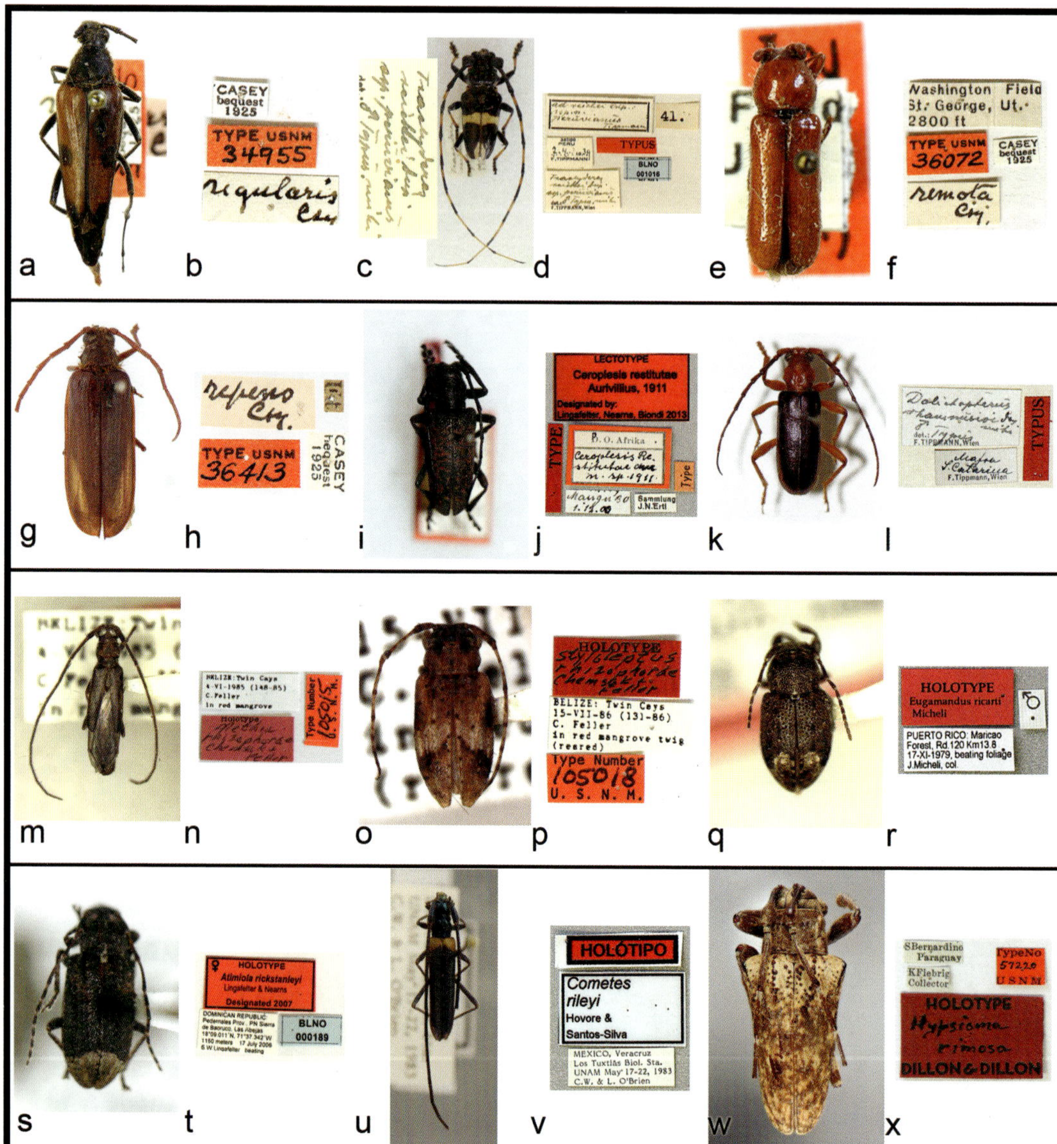

Figure 143. Primary types of Cerambycidae of the Smithsonian Institution (for each pair of images, dorsal habitus left, labels right): **(a, b)** *Strangalia regularis* Casey, 1913, Holotype; **(c, d)** *Trachyderes reichei peruvianus* Tippmann, 1953, Holotype; **(e, f)** *Batyleoma remota* Casey, 1924, Holotype; **(g, h)** *Tragosoma repens* Casey, 1924, Holotype; **(i, j)** *Ceroplesis restitutae* Aurivillius, 1911, Lectotype; **(k, l)** *Dolichopterus rhamnusioides* Tippmann, 1953, Holotype; **(m, n)** *Methia rhizophorae* Chemsak & Feller, 1988, Holotype; **(o, p)** *Styloleptus rhizophorae* Chemsak & Feller, 1988, Holotype; **(q, r)** *Eugamandus ricarti* Micheli, 2003, Holotype; **(s, t)** *Atimiola rickstanleyi* Lingafelter & Nearns, 2007, Holotype; **(u, v)** *Cometes rileyi* Hovore & Santos-Silva, 2007, Holotype; **(w, x)** *Hypsioma rimosa* Dillon & Dillon, 1945, Holotype.

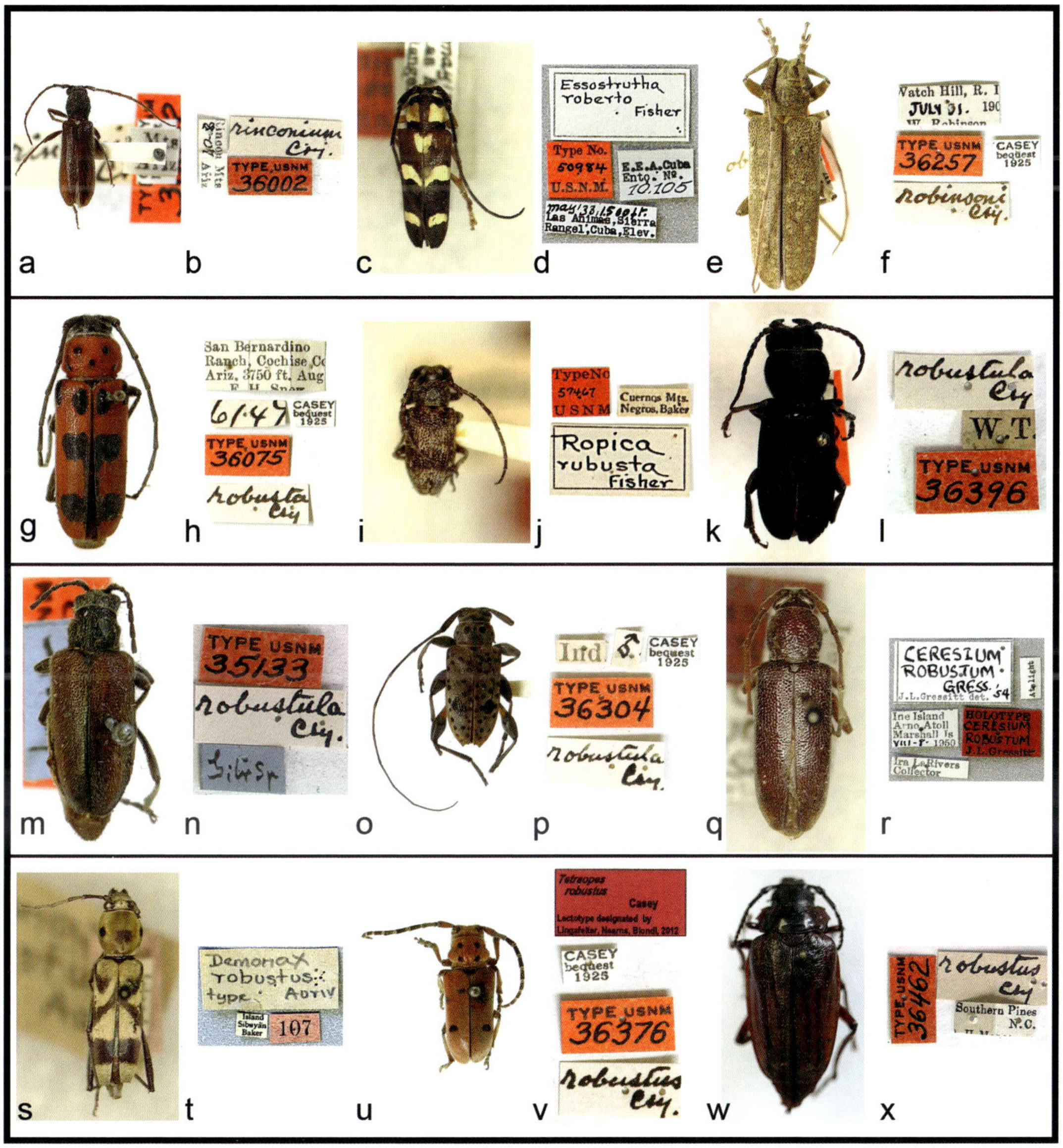

Figure 144. Primary types of Cerambycidae of the Smithsonian Institution (for each pair of images, dorsal habitus left, labels right): **(a, b)** *Anoplium rinconium* Casey, 1924, Holotype; **(c, d)** *Essostrutha roberto* Fisher, 1935, Holotype; **(e, f)** *Goes robinsoni* Casey, 1914, Holotype; **(g, h)** *Tylosis robusta* Casey, 1912, Holotype; **(i, j)** *Ropica robusta* Fisher, 1925, Holotype; **(k, l)** *Spondylis robustula* Casey, 1912, Holotype; **(m, n)** *Brachysomida robustula* Casey, 1913, Holotype; **(o, p)** *Hyperplatys robustula* Casey, 1913, Holotype; **(q, r)** *Ceresium (Ceresium) robustum* Gressitt, 1956, Holotype; **(s, t)** *Demonax robustus* Aurivillius, 1928, Holotype; **(u, v)** *Tetraopes robustus* Casey, 1913, Lectotype; **(w, x)** *Prionus (Neopolyarthron) robustus* Casey, 1924, Holotype.

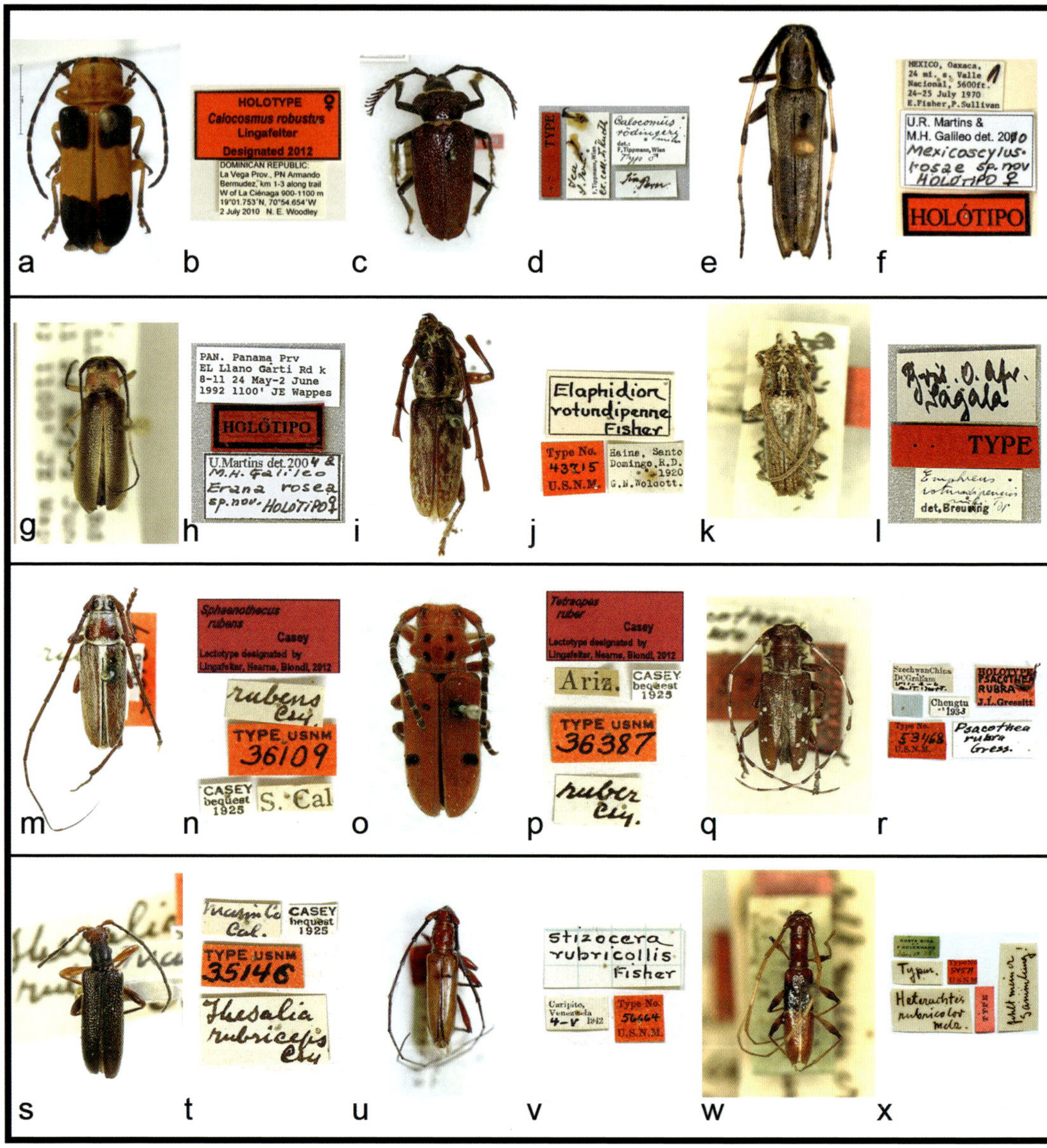

Figure 145. Primary types of Cerambycidae of the Smithsonian Institution (for each pair of images, dorsal habitus left, labels right): **(a, b)** *Calocosmus robustus* Lingafelter, 2013, Holotype; **(c, d)** *Calocomus roedingeri* Tippmann, 1951, Holotype; **(e, f)** *Mexicoscylus rosae* Martins & Galileo, 2011, Holotype; **(g, h)** *Erana rosea* Galileo & Martins, 2004, Holotype; **(i, j)** *Elaphidion rotundipenne* Fisher, 1932, Holotype; **(k, l)** *Emphreus rotundipennis* Breuning, 1950, Holotype; **(m, n)** *Sphaenothecus rubens* Casey, 1891, Lectotype; **(o, p)** *Tetraopes ruber* Casey, 1913, Lectotype; **(q, r)** *Psacothea rubra* Gressitt, 1938, Holotype; **(s, t)** *Thesalia rubriceps* Casey, 1913, Holotype; **(u, v)** *Stizocera rubricollis* Fisher, 1944, Holotype; **(w, x)** *Heterachthes rubricolor* Melzer, 1935, Holotype.

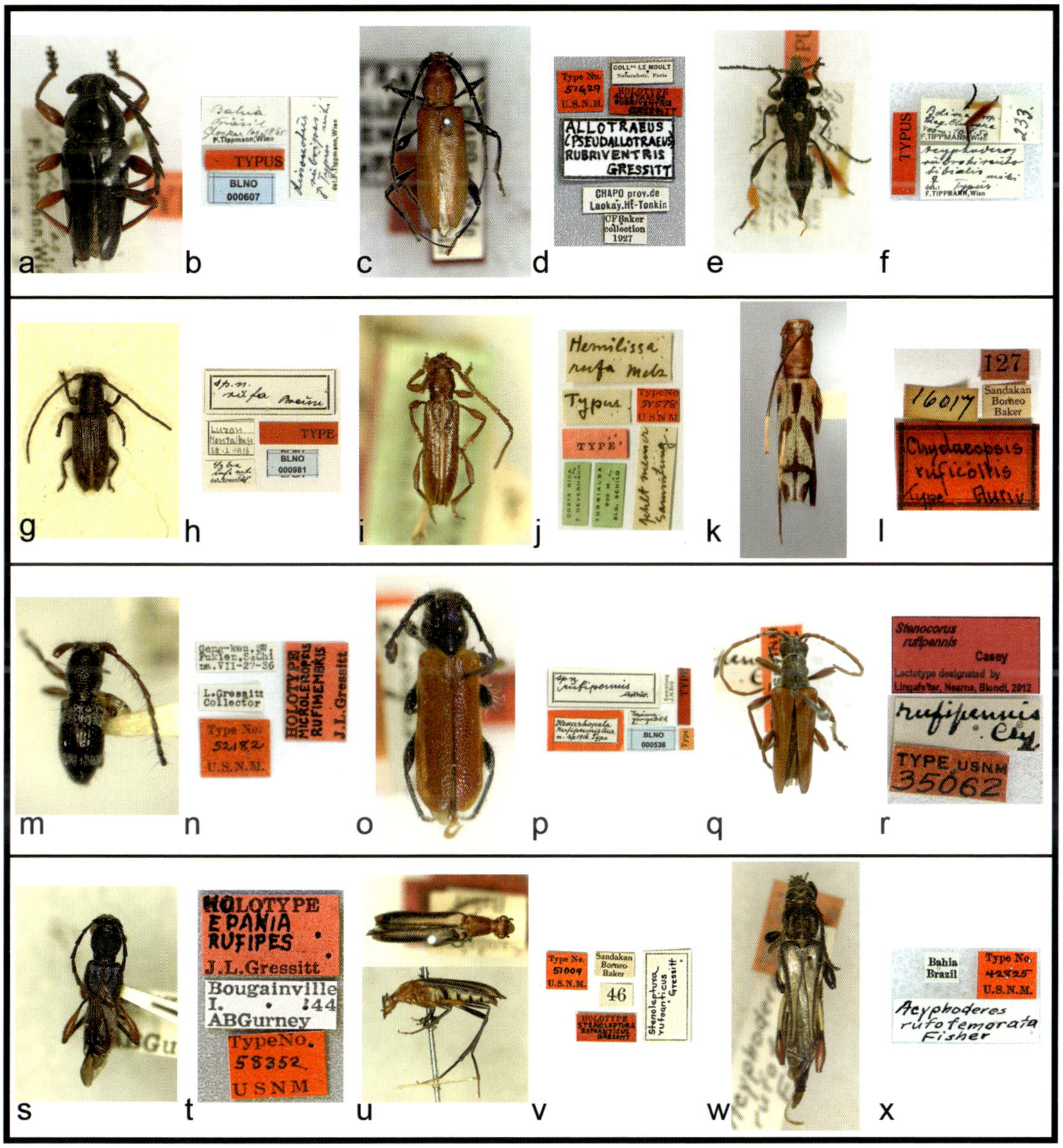

Figure 146. Primary types of Cerambycidae of the Smithsonian Institution (for each pair of images, dorsal habitus left, labels right): **(a, b)** *Lissonotus rubripes* Tippmann, 1960, Holotype; **(c, d)** *Allotraeus (Pseudallotraeus) rubriventris* Gressitt, 1937, Holotype; **(e, f)** *Acyphoderes rubrohirsutotibialis* Tippmann, 1953, Holotype; **(g, h)** *Sybra rufa* Breuning, 1943, Holotype; **(i, j)** *Hemilissa rufa* Melzer, 1934, Holotype; **(k, l)** *Chydaeopsis ruficollis* Aurivillius, 1922, Holotype; **(m, n)** *Microleropsis rufimembris* Gressitt, 1937, Holotype; **(o, p)** *Hexarhopala rufipennis* Aurivillius, 1916, Holotype; **(q, r)** *Stenocorus (Stenocorus) rufipennis* Casey, 1913, Lectotype; **(s, t)** *Epania rufipes* Gressitt, 1951, Holotype; **(u, v)** *Stenoleptura rufoanticus* Gressitt, 1935, Holotype; **(w, x)** *Acyphoderes rufofemorata* Fisher, 1930, Holotype.

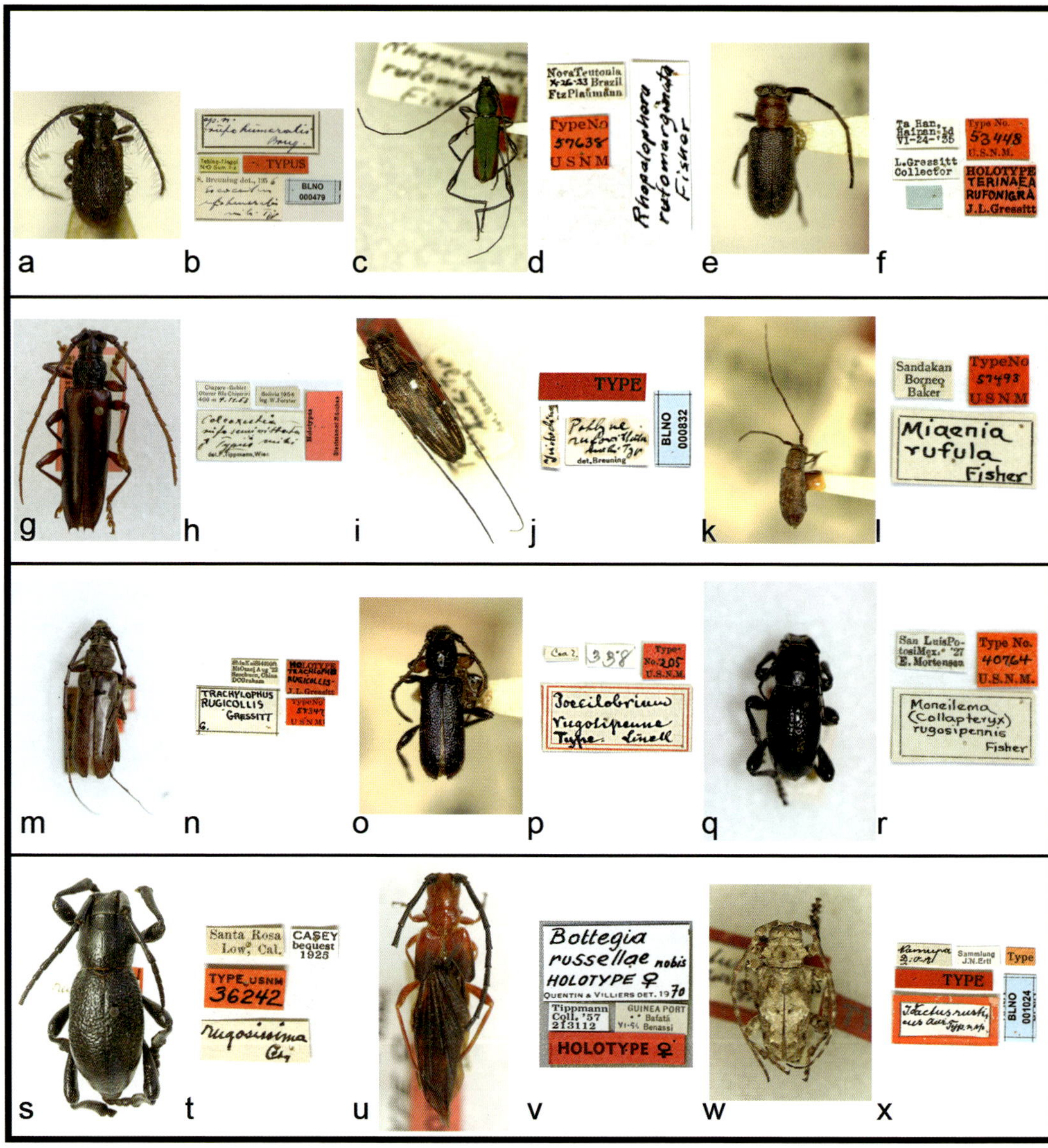

Figure 147. Primary types of Cerambycidae of the Smithsonian Institution (for each pair of images, dorsal habitus left, labels right): **(a, b)** *Exocentrus (Pseudocentrus) rufohumeralis* Breuning, 1957, Holotype; **(c, d)** *Rhopalophora (Ischionodonta) rufomarginata* Fisher, 1937, Holotype; **(e, f)** *Terinaea rufonigra* Gressitt, 1940, Holotype; **(g, h)** *Coleoxestia rufosemivittata* Tippmann, 1960, Holotype; **(i, j)** *Pothyne rufovittata* Breuning, 1943, Holotype; **(k, l)** *Miaenia rufula* Fisher, 1925, Holotype; **(m, n)** *Trachylophus rugicollis* Gressitt, 1948, Holotype; **(o, p)** *Poecilobrium rugosipenne* Linell, 1897, Holotype; **(q, r)** *Moneilema (Collapteryx) rugosipennis* Fisher, 1928, Holotype; **(s, t)** *Moneilema (Collapteryx) rugosissima* Casey, 1924, Holotype; **(u, v)** *Bottegia russellae* Quentin & Villiers, 1971, Holotype; **(w, x)** *Idactus rusticus* Aurivillius, 1916, Syntype.

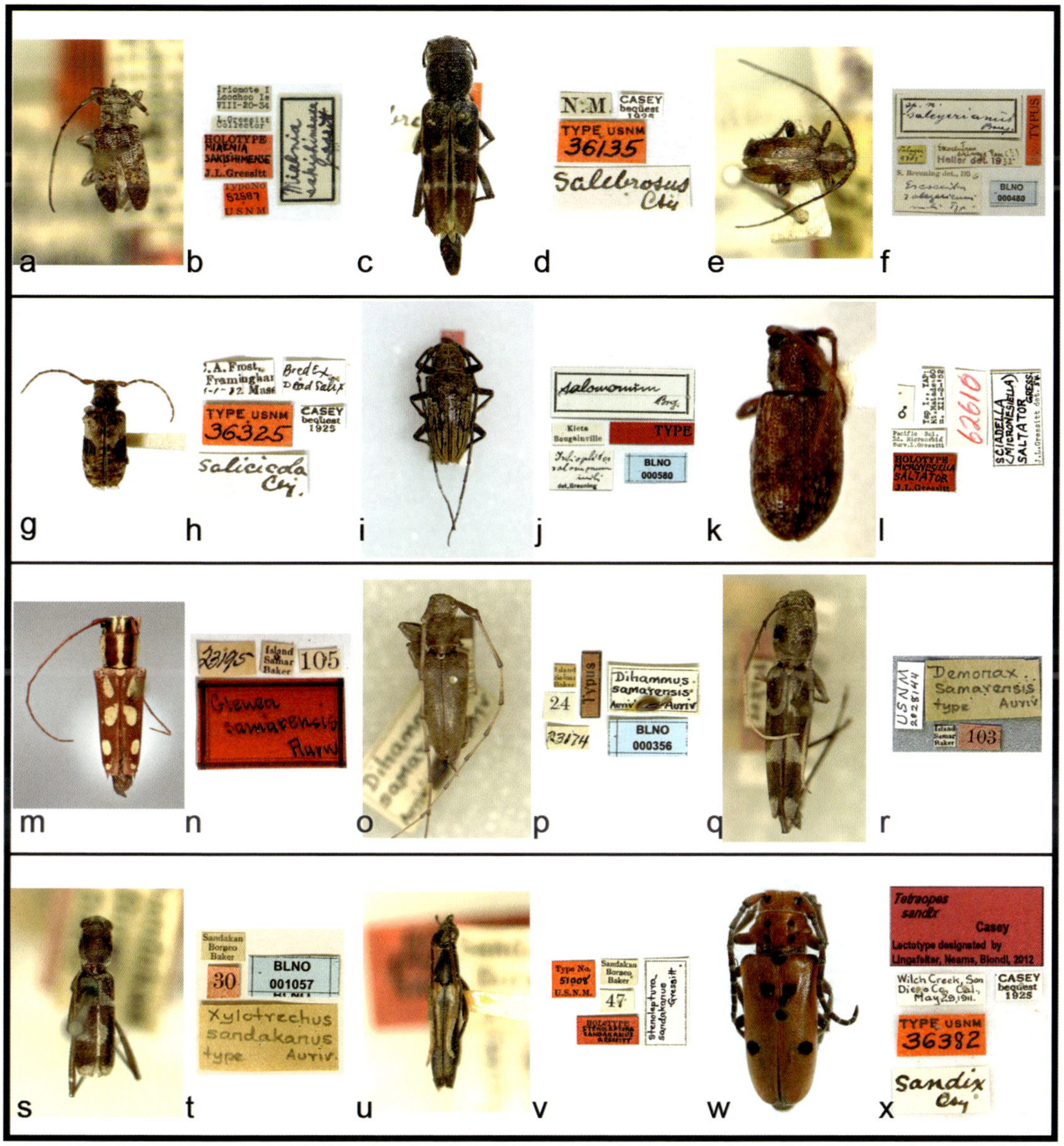

Figure 148. Primary types of Cerambycidae of the Smithsonian Institution (for each pair of images, dorsal habitus left, labels right): **(a, b)** *Miaenia sakishimana* Gressitt, 1951, Holotype; **(c, d)** *Xylotrechus salebrosus* Casey, 1912, Holotype; **(e, f)** *Exocentrus (Pseudocentrus) saleyerianus* Breuning, 1957, Holotype; **(g, h)** *Pogonocherus salicola* Casey, 1913, Holotype; **(i, j)** *Ischioplites salomonum* Breuning, 1938, Holotype; **(k, l)** *Sciadella (Micronesiella) saltator* Gressitt, 1956, Holotype; **(m, n)** *Glenea (Glenea) samarensis* Aurivillius, 1926, Holotype; **(o, p)** *Dihammus samarensis* Aurivillius, 1927, Holotype; **(q, r)** *Demonax samarensis* Aurivillius, 1928, Holotype; **(s, t)** *Xylotrechus sandakanus* Aurivillius, 1927, Syntype; **(u, v)** *Stenoleptura sandakanus* Gressitt, 1935, Holotype; **(w, x)** *Tetraopes sandix* Casey, 1914, Lectotype.

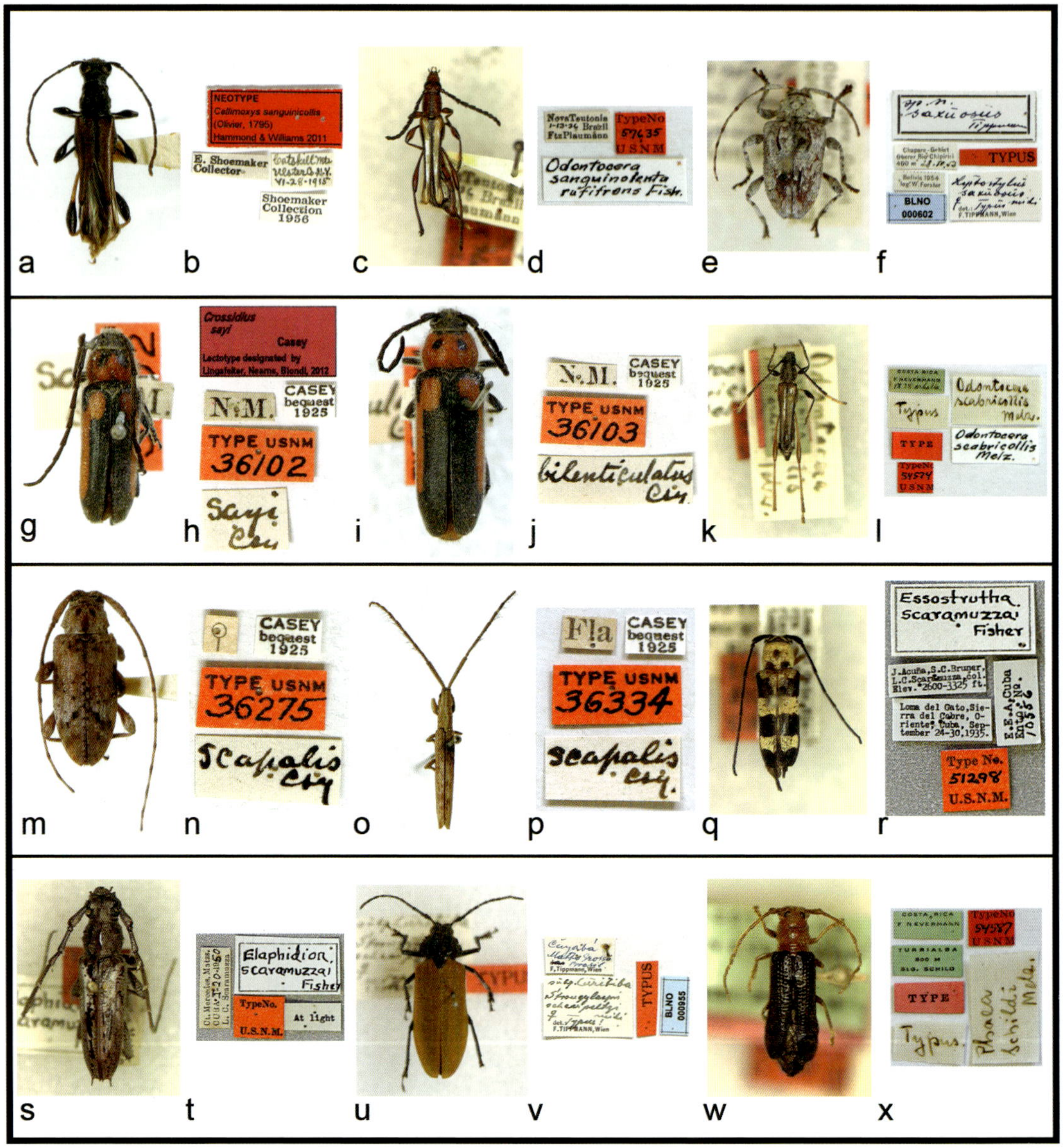

Figure 149. Primary types of Cerambycidae of the Smithsonian Institution (for each pair of images, dorsal habitus left, labels right): **(a, b)** *Necydalis sanguinicollis* Olivier, 1795, Neotype; **(c, d)** *Odontocera sanguinolenta* v. *rufifrons* Fisher, 1937, Holotype; **(e, f)** *Leptostylus saxuosus* Tippmann, 1960, Holotype; **(g, h)** *Crossidius sayi* Casey, 1912, Lectotype; **(i, j)** *Crossidius sayi bilenticulatus* Casey, 1912, Holotype; **(k, l)** *Odontocera scabricollis* Melzer, 1935, Holotype; **(m, n)** *Leiopus scapalis* Casey, 1913, Holotype; **(o, p)** *Spalacopsis scapalis* Casey, 1913, Holotype; **(q, r)** *Essostrutha scaramuzzai* Fisher, 1936, Holotype; **(s, t)** *Elaphidion scaramuzzai* Fisher, 1951, Holotype; **(u, v)** *Strongylaspis (Curitiba) scheerpeltzi* Tippmann, 1953, Holotype; **(w, x)** *Phaea schildi* Melzer, 1935, Holotype.

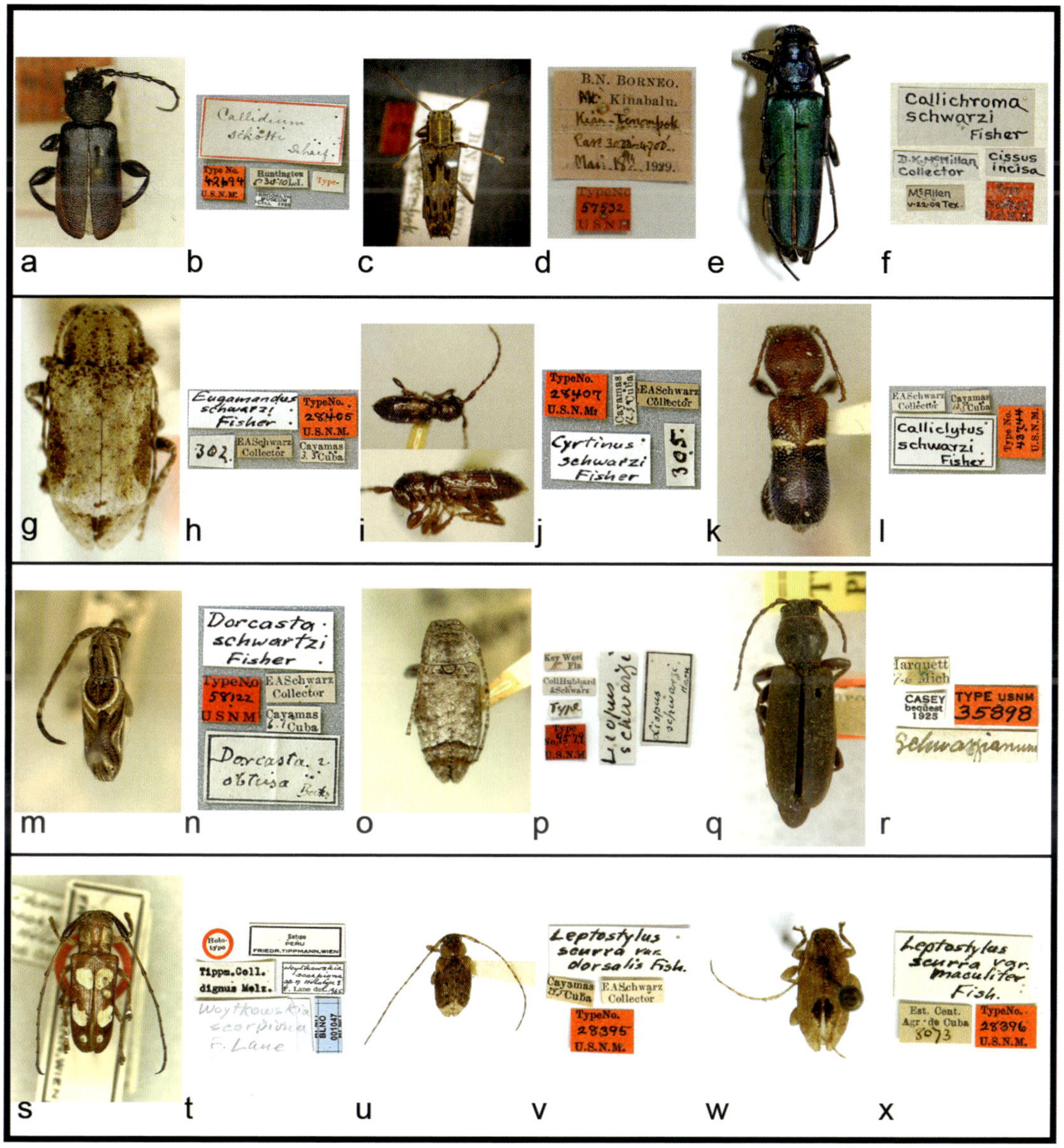

Figure 150. Primary types of Cerambycidae of the Smithsonian Institution (for each pair of images, dorsal habitus left, labels right): **(a, b)** *Callidium schotti* Schaeffer, 1917, Holotype; **(c, d)** *Glenea schwarzeri* Fisher, 1935, Holotype; **(e, f)** *Callichroma schwarzi* Fisher, 1914, Holotype; **(g, h)** *Eugamandus schwarzi* Fisher, 1926, Holotype; **(i, j)** *Cyrtinus schwarzi* Fisher, 1926, Holotype; **(k, l)** *Calliclytus schwarzi* Fisher, 1932, Holotype; **(m, n)** *Dorcasta schwarzi* Fisher, 1947, Holotype; **(o, p)** *Leiopus schwarzi* Hamilton in Leng & Hamilton, 1896, Holotype; **(q, r)** *Tetropium schwarzianum* Casey, 1891, Holotype; **(s, t)** *Woytkowskia scorpiona* Lane, 1966, Holotype; **(u, v)** *Leptostylus scurra* var. *dorsalis* Fisher, 1926, Holotype; **(w, x)** *Leptostylus scurra* var. *maculifer* Fisher, 1926, Holotype.

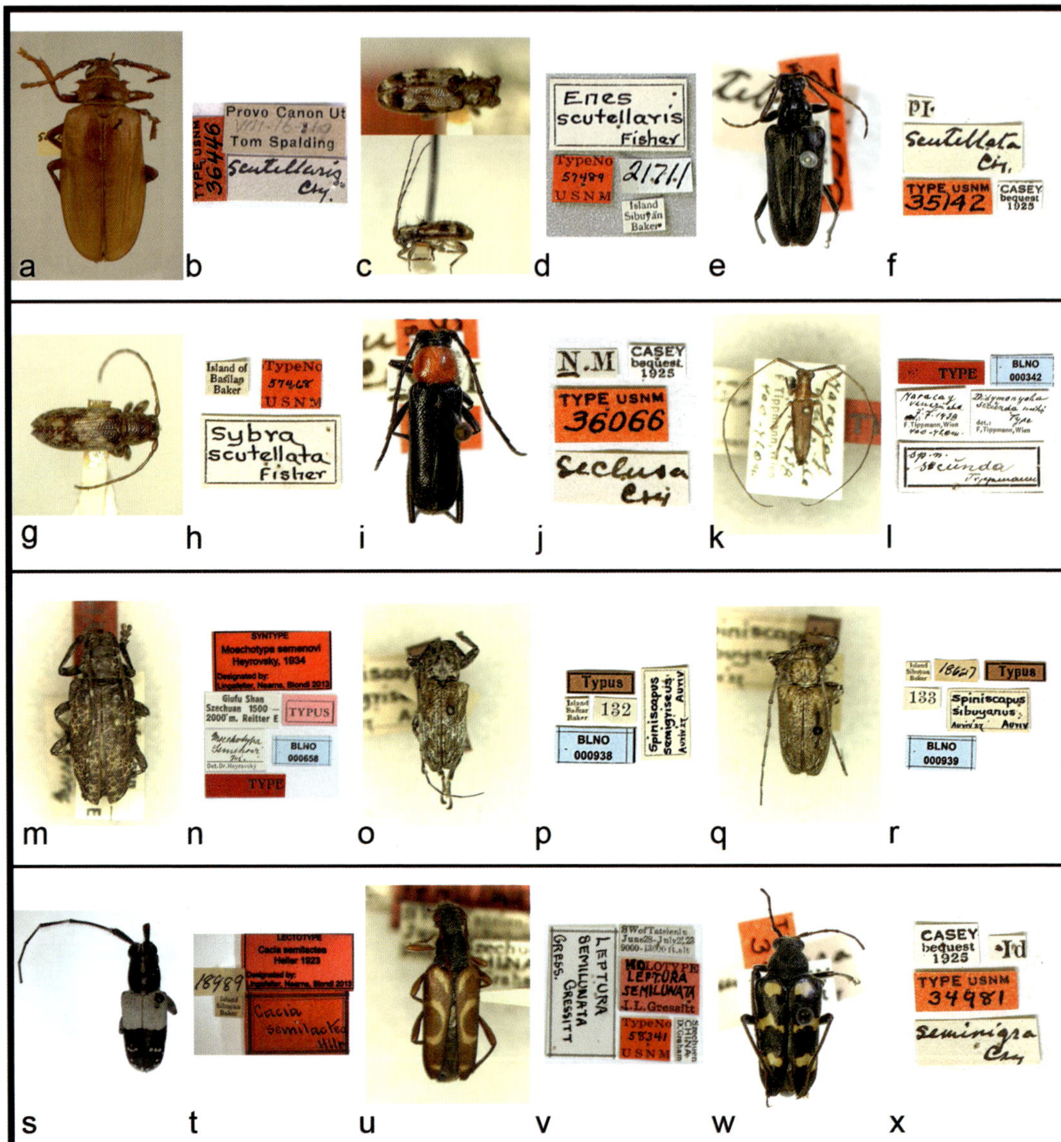

Figure 151. Primary types of Cerambycidae of the Smithsonian Institution (for each pair of images, dorsal habitus left, labels right): **(a, b)** *Prionus (Prionus) scutellaris* Casey, 1924, Holotype; **(c, d)** *Enes scutellaris* Fisher, 1925, Holotype; **(e, f)** *Acmaeops scutellata* Casey, 1924, Holotype; **(g, h)** *Sybra scutellata* Fisher, 1925, Holotype; **(i, j)** *Batyle seclusa* Casey, 1912, Holotype; **(k, l)** *Didymonycha secunda* Tippmann, 1951, Holotype; **(m, n)** *Moechotypa semenovi* Heyrovsky, 1934, Syntype; **(o, p)** *Spiniscapus semigriseus* Aurivillius, 1927, Syntype; **(q, r)** *Spiniscapus semigriseus sibuyanus* Aurivillius, 1927, Syntype; **(s, t)** *Cacia semilactea* Heller, 1923, Lectotype; **(u, v)** *Leptura semilunata* Gressitt, 1948, Holotype; **(w, x)** *Judolia seminigra* Casey, 1924, Holotype.

Figure 152. Primary types of Cerambycidae of the Smithsonian Institution (for each pair of images, dorsal habitus left, labels right): **(a, b)** *Aneflomorpha seminuda* Casey, 1912, Holotype; **(c, d)** *Chreonoma seminuda* Fisher, 1935, Holotype; **(e, f)** *Oreodera semiporosa* Tippmann, 1960, Holotype; **(g, h)** *Aneflomorpha semirufa* Linsley, 1935, Holotype; **(i, j)** *Tropidion semirufum* Martins, 1968, Holotype; **(k, l)** *Erana septuosa* Galileo & Martins, 2004, Holotype; **(m, n)** *Callidium sequarium* Fisher, 1920, Holotype; **(o, p)** *Noctileptura seriata* Chemsak & Linsley, 1984, Holotype; **(q, r)** *Demonax seriatopunctatus* Aurivillius, 1922, Holotype; **(s, t)** *Strongylaspis sericans* Tippmann, 1953, Holotype; **(u, v)** *Stenocorus (Stenocorus) sericatus* Casey, 1913, Holotype; **(w, x)** *Stenhomalus sericeus* Aurivillius, 1922, Syntype.

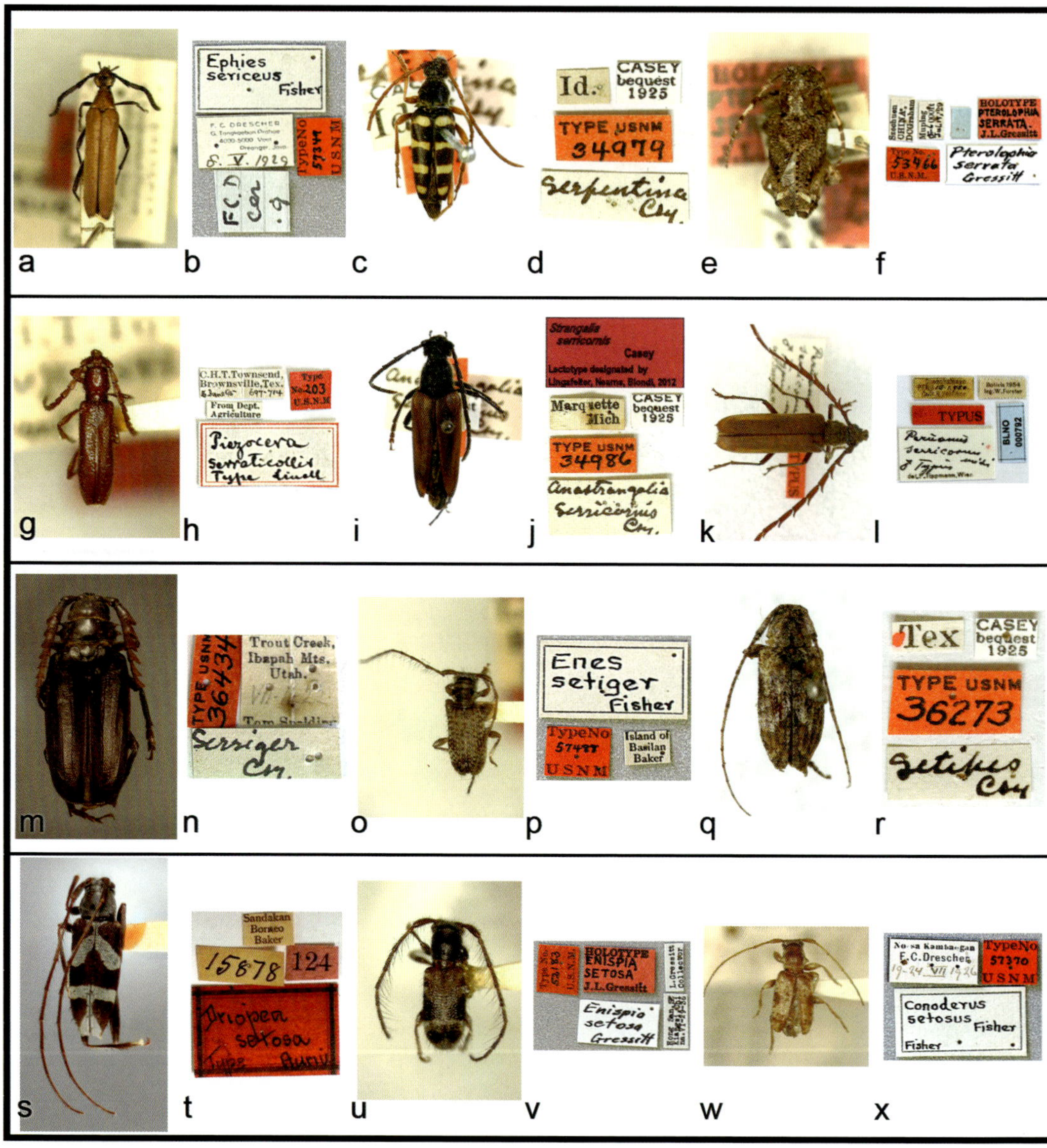

Figure 153. Primary types of Cerambycidae of the Smithsonian Institution (for each pair of images, dorsal habitus left, labels right): **(a, b)** *Ephies sericeus* Fisher, 1936, Holotype; **(c, d)** *Leptura serpentina* Casey, 1891, Holotype; **(e, f)** *Pterolophia serrata* Gressitt, 1938, Holotype; **(g, h)** *Piezocera serraticollis* Linell, 1897, Holotype; **(i, j)** *Strangalia (Anastrangalia) serricornis* Casey, 1924, Lectotype; **(k, l)** *Peruanus serricornis* Tippmann, 1960, Holotype; **(m, n)** *Prionus (Prionus) serriger* Casey, 1924, Holotype; **(o, p)** *Enes setiger* Fisher, 1925, Holotype; **(q, r)** *Leiopus setipes* Casey, 1891, Holotype; **(s, t)** *Driopea setosa* Aurivillius, 1922, Holotype; **(u, v)** *Enispia setosa* Gressitt, 1937, Holotype; **(w, x)** *Controderus setosus* Fisher, 1936, Holotype.

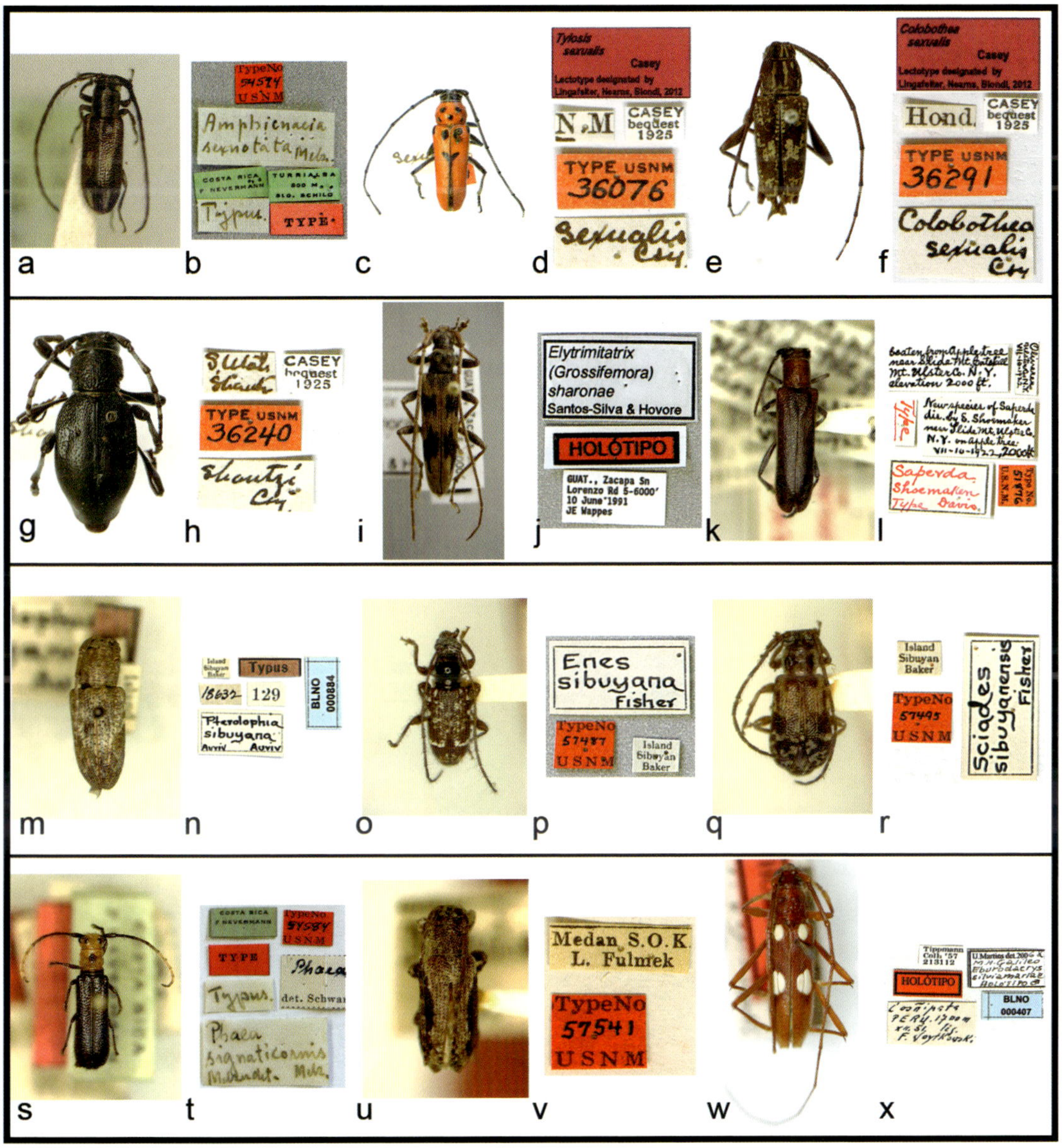

Figure 154. Primary types of Cerambycidae of the Smithsonian Institution (for each pair of images, dorsal habitus left, labels right): **(a, b)** *Amphicnaeia sexnotata* Melzer, 1933, Holotype; **(c, d)** *Tylosis sexualis* Casey, 1912, Lectotype; **(e, f)** *Colobothea sexualis* Casey, 1913, Lectotype; **(g, h)** *Moneilema (Collapteryx) shantzi* Casey, 1924, Holotype; **(i, j)** *Elytrimitatrix (Grossifemora) sharonae* Santos-Silva & Hovore, 2008, Holotype; **(k, l)** *Saperda shoemakeri* Davis, 1923, Holotype; **(m, n)** *Pterolophia sibuyana* Aurivillius, 1927, Syntype; **(o, p)** *Enes sibuyana* Fisher, 1925, Holotype; **(q, r)** *Sciades sibuyanensis* Fisher, 1925, Holotype; **(s, t)** *Phaea signaticornis* Melzer, 1932, Holotype; **(u, v)** *Sybra signatipennis* Fisher, 1927, Holotype; **(w, x)** *Eburodacrys silviamariae* Galileo & Martins, 2006, Holotype.

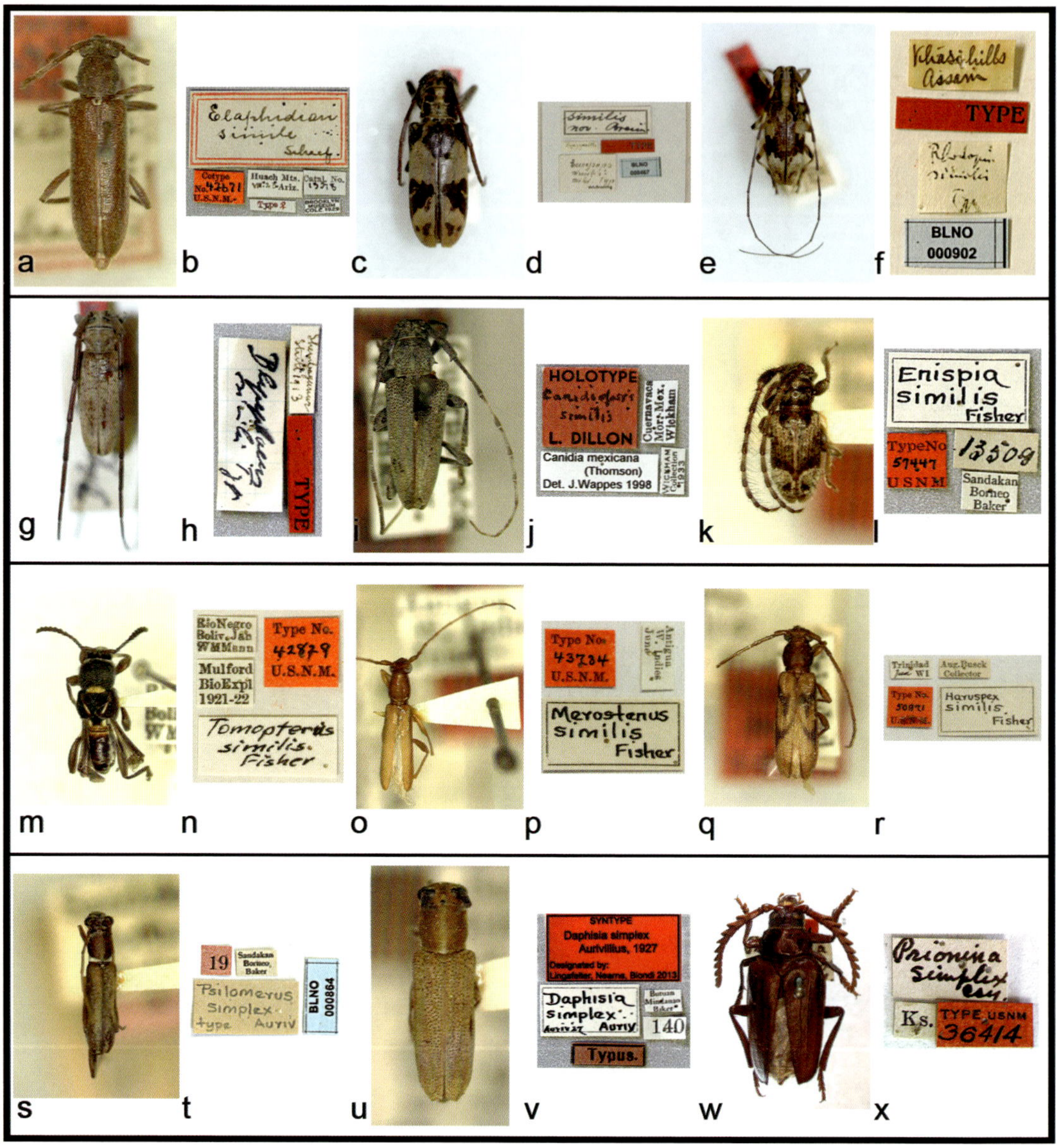

Figure 155. Primary types of Cerambycidae of the Smithsonian Institution (for each pair of images, dorsal habitus left, labels right): **(a, b)** *Elaphidion simile* Schaeffer, 1908, Syntype; **(c, d)** *Eurysops similis* Breuning, 1937, Holotype; **(e, f)** *Rhodopis similis* Breuning, 1940, Holotype; **(g, h)** *Blepephaeus similis* Breuning, 1942, Holotype; **(i, j)** *Canidiopsis similis* Dillon, 1955, Holotype; **(k, l)** *Enispia similis* Fisher, 1925, Holotype; **(m, n)** *Tomopterus similis* Fisher, 1930, Holotype; **(o, p)** *Merostenus similis* Fisher, 1932, Holotype; **(q, r)** *Haruspex similis* Fisher, 1935, Holotype; **(s, t)** *Psilomerus simplex* Aurivillius, 1924, Holotype; **(u, v)** *Daphisia simplex* Aurivillius, 1927, Syntype; **(w, x)** *Prionina simplex* Casey, 1912, Holotype.

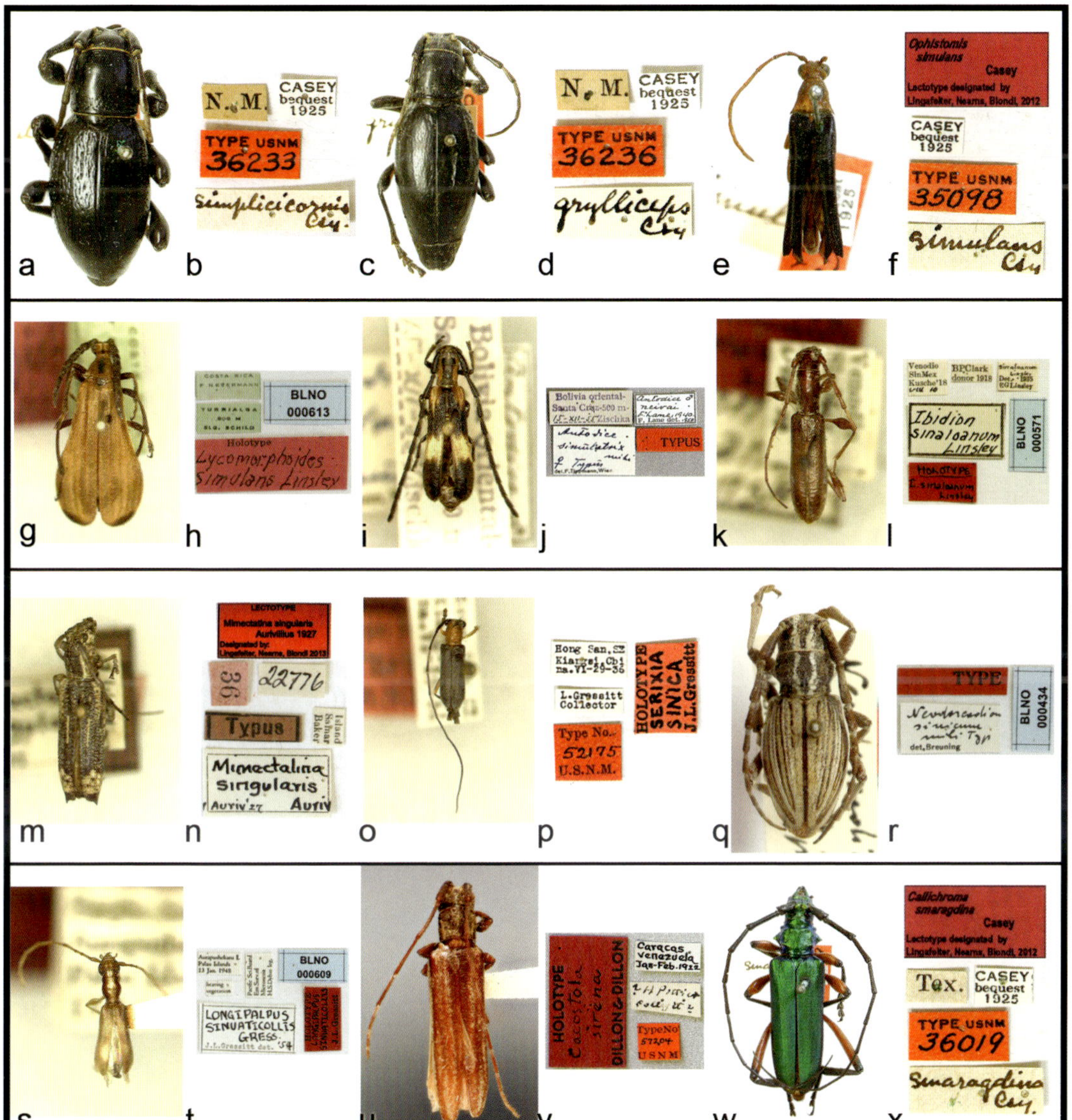

Figure 156. Primary types of Cerambycidae of the Smithsonian Institution (for each pair of images, dorsal habitus left, labels right): **(a, b)** *Moneilema (Collapteryx) simplicicornis* Casey, 1913, Holotype; **(c, d)** *Moneilema (Collapteryx) simplicicornis grylliceps* Casey, 1913, Holotype; **(e, f)** *Ophistomis simulans* Casey, 1913, Lectotype; **(g, h)** *Lycomorphoides simulans* Linsley, 1970, Holotype; **(i, j)** *Antodice simulatrix* Tippmann, 1960, Holotype; **(k, l)** *Ibidion sinaloanum* Linsley, 1935, Holotype; **(m, n)** *Mimectatina singularis* Aurivillius, 1927, Lectotype; **(o, p)** *Serixia sinica* Gressitt, 1937, Holotype; **(q, r)** *Eodorcadion (Ornatodorcadion) sinicum* Breuning, 1948, Holotype; **(s, t)** *Longipalpus sinuaticollis* Gressitt, 1956, Holotype; **(u, v)** *Cacostola sirena* Dillon & Dillon, 1946, Holotype; **(w, x)** *Callichroma smaragdina* Casey, 1912, Lectotype.

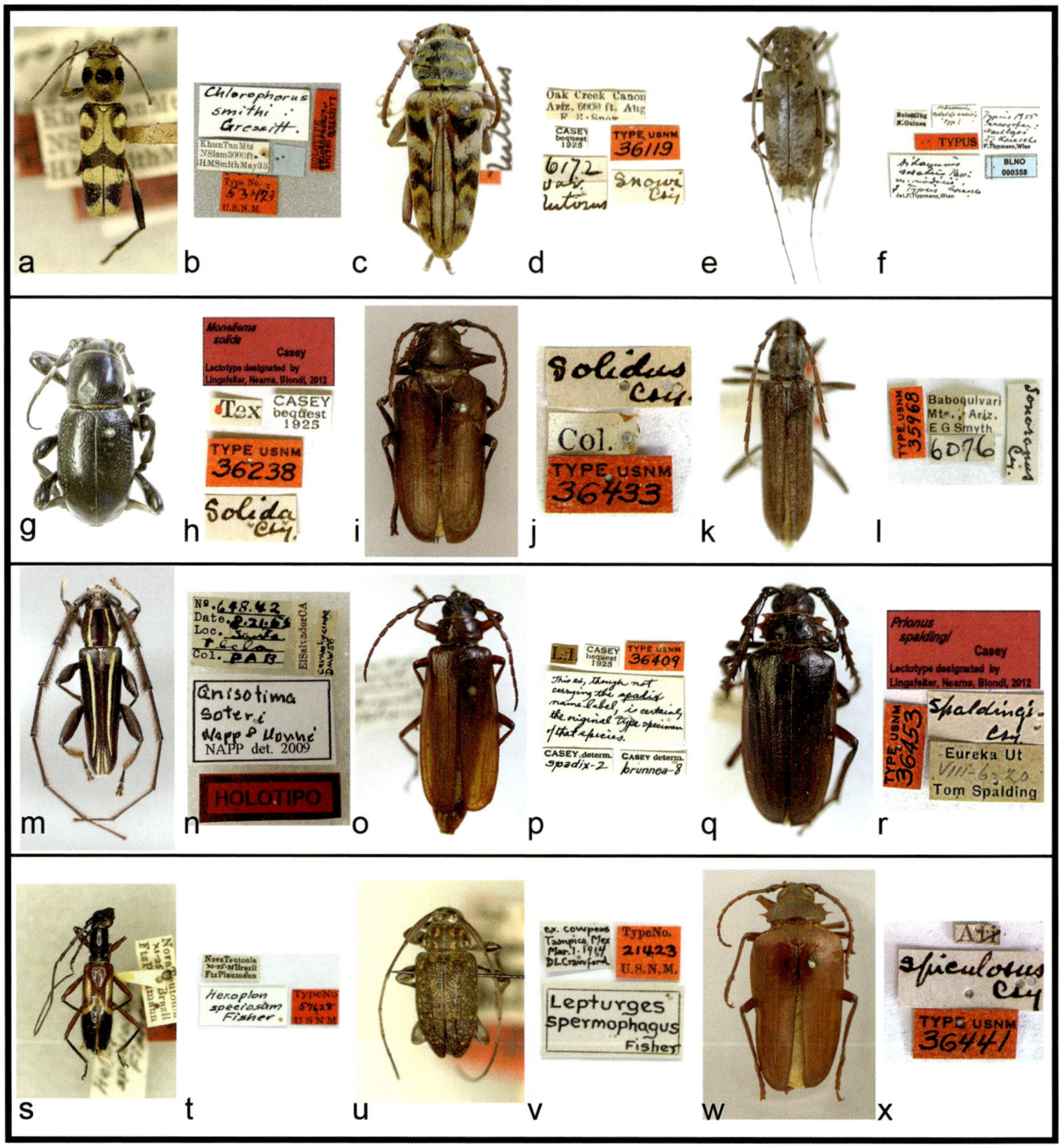

Figure 157. Primary types of Cerambycidae of the Smithsonian Institution (for each pair of images, dorsal habitus left, labels right): **(a, b)** *Chlorophorus smithi* Gressitt, 1941, Holotype; **(c, d)** *Cyllene snowi* Casey, 1912, Holotype; **(e, f)** *Dihammus solatus nodius* Kriesche, 1936, Holotype; **(g, h)** *Moneilema (Collapteryx) solida* Casey, 1913, Lectotype; **(i, j)** *Prionus (Prionus) solidus* Casey, 1912, Holotype; **(k, l)** *Aneflus sonoranus* Casey, 1924, Holotype; **(m, n)** *Anisotyma soteri* Napp & Monné, 2009, Holotype; **(o, p)** *Orthosoma spadix* Casey, 1912, Holotype; **(q, r)** *Prionus (Prionus) spaldingi* Casey, 1924, Lectotype; **(s, t)** *Hexoplon speciosum* Fisher, 1937, Holotype; **(u, v)** *Lepturges spermophagus* Fisher, 1918, Holotype; **(w, x)** *Prionus (Prionus) spiculosus* Casey, 1912, Holotype.

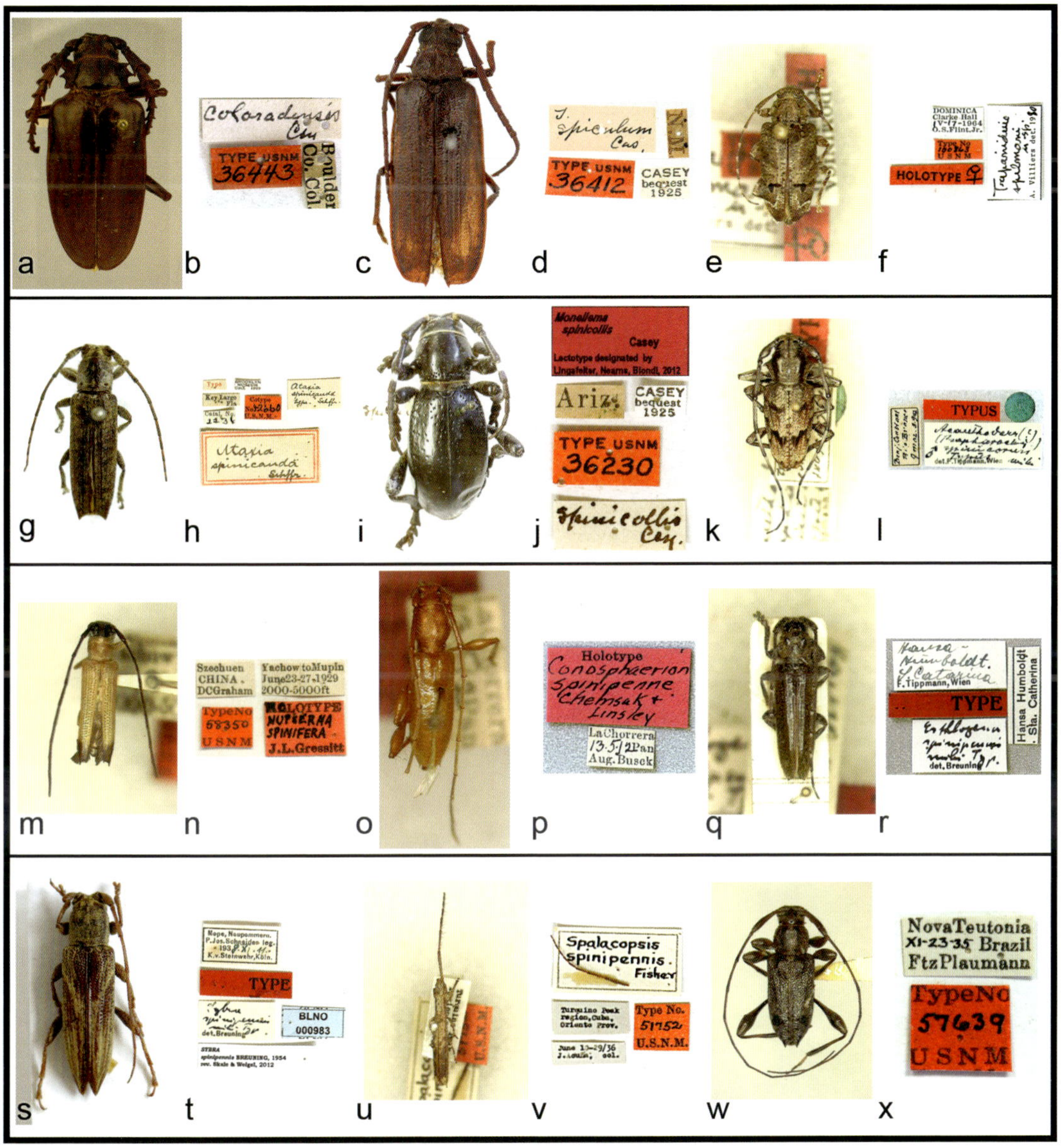

Figure 158. Primary types of Cerambycidae of the Smithsonian Institution (for each pair of images, dorsal habitus left, labels right): **(a, b)** *Prionus (Prionus) spiculosus coloradensis* Casey, 1924, Holotype; **(c, d)** *Tragosoma spiculum* Casey, 1890, Holotype; **(e, f)** *Trypanidius spilmani* Villiers, 1980, Holotype; **(g, h)** *Ataxia spinicauda* Schaeffer, 1904, Syntype; **(i, j)** *Moneilema spinicollis* Casey, 1891, Lectotype; **(k, l)** *Acanthoderes (Psapharochrus) spinicornis* Tippmann, 1960, Holotype; **(m, n)** *Nupserha spinifera* Gressitt, 1948, Holotype; **(o, p)** *Conosphaerion spinipenne* Chemsak & Linsley, 1967, Holotype; **(q, r)** *Esthlogena spinipennis* Breuning, 1942, Holotype; **(s, t)** *Sybra spinipennis* Breuning, 1954, Holotype; **(u, v)** *Spalacopsis spinipennis* Fisher, 1936, Holotype; **(w, x)** *Hyperplatys spinipennis* Fisher, 1938, Holotype.

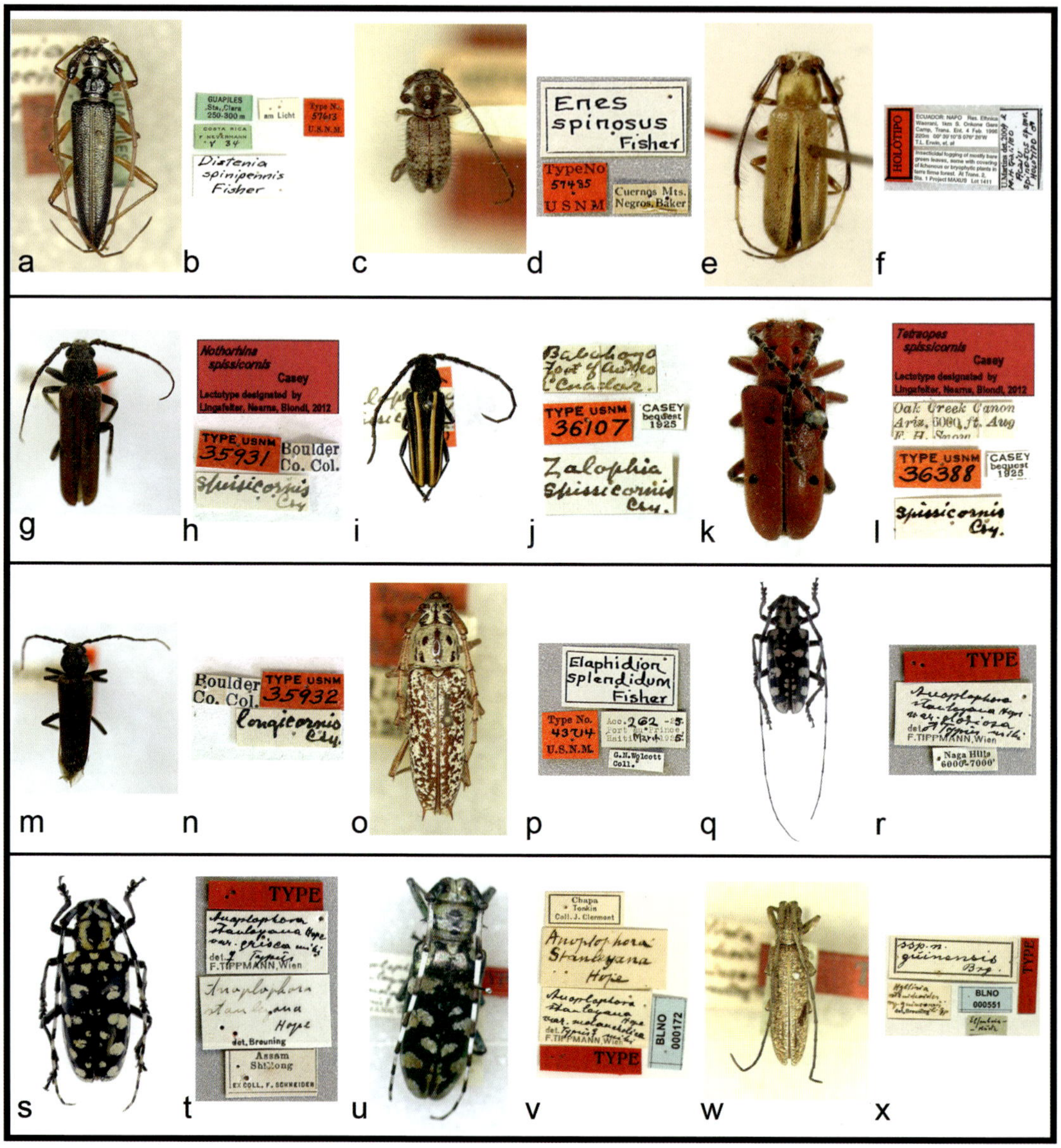

Figure 159. Primary types of Cerambycidae of the Smithsonian Institution (for each pair of images, dorsal habitus left, labels right): **(a, b)** *Distenia spinipennis* Fisher, 1946, Holotype; **(c, d)** *Enes spinosus* Fisher, 1925, Holotype; **(e, f)** *Acaiu spinosus* Galileo & Martins, 2005, Holotype; **(g, h)** *Nothorhina spissicornis* Casey, 1912, Lectotype; **(i, j)** *Zalophia spissicornis* Casey, 1912, Holotype; **(k, l)** *Tetraopes spissicornis* Casey, 1913, Lectotype; **(m, n)** *Nothorhina spissicornis longicornis* Casey, 1912, Holotype; **(o, p)** *Elaphidion splendidum* Fisher, 1932, Holotype; **(q, r)** *Anoplophora stanleyana* var. *gloriosa* Tippmann, 1953, Holotype; **(s, t)** *Anoplophora stanleyana* var. *grisea* Tippmann, 1953, Holotype; **(u, v)** *Anoplophora stanleyana* var. *melancholica* Tippmann, 1953, Holotype; **(w, x)** *Hyllisia stenideoides guineensis* Breuning, 1957, Holotype.

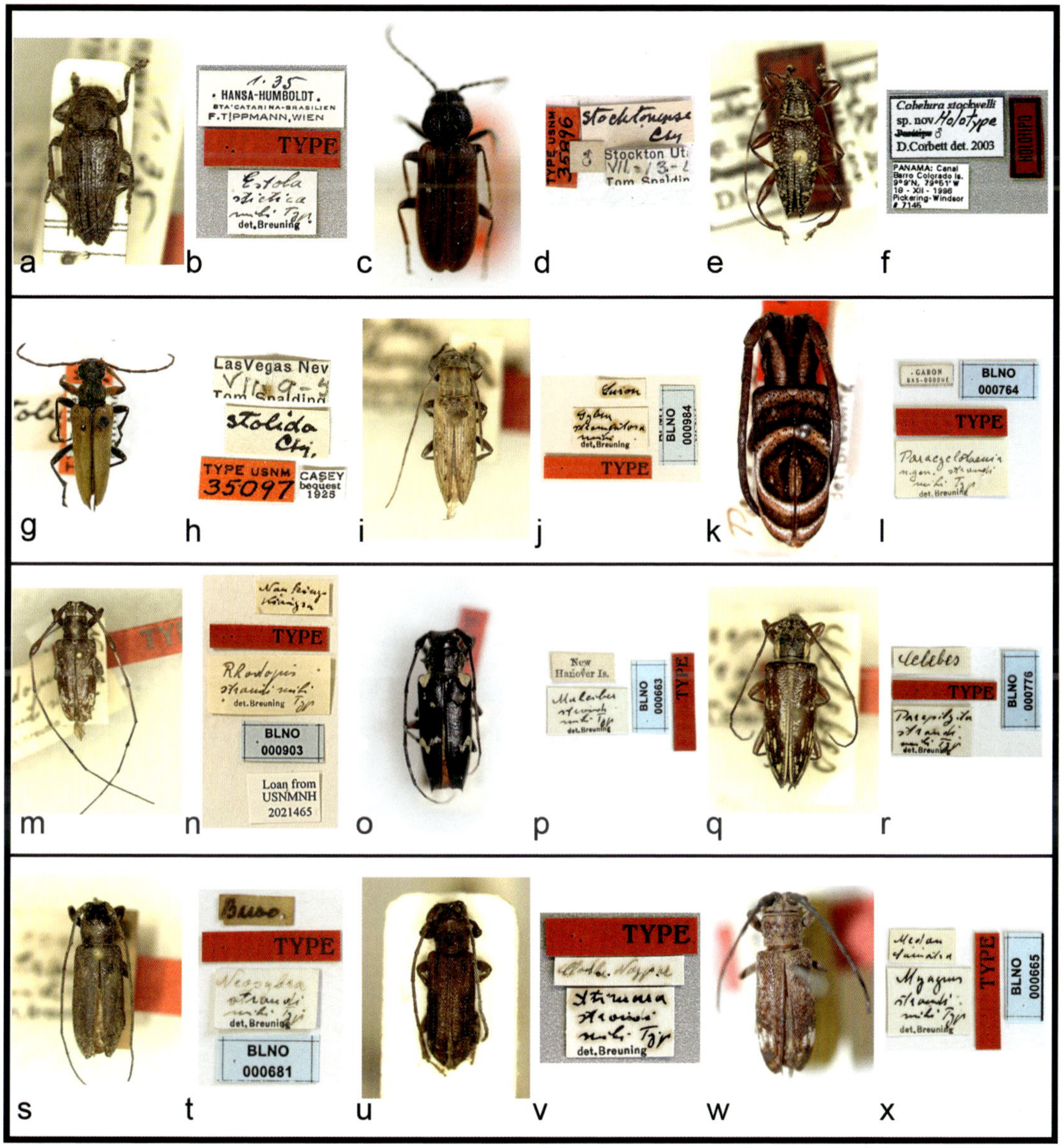

Figure 160. Primary types of Cerambycidae of the Smithsonian Institution (for each pair of images, dorsal habitus left, labels right): **(a, b)** *Estola stictica* Breuning, 1942, Holotype; **(c, d)** *Asemum stocktonense* Casey, 1924, Holotype; **(e, f)** *Cobelura stockwelli* Corbett, 2004, Holotype; **(g, h)** *Leptacmaeops stolida* Casey, 1924, Holotype; **(i, j)** *Sybra stramentosa* Breuning, 1939, Holotype; **(k, l)** *Paracyclotaenia strandi* Breuning, 1935, Holotype; **(m, n)** *Rhodopis strandi* Breuning, 1935, Holotype; **(o, p)** *Mulciber strandi* Breuning, 1939, Holotype; **(q, r)** *Parepilysta strandi* Breuning, 1939, Holotype; **(s, t)** *Neosybra strandi* Breuning, 1939, Holotype; **(u, v)** *Atimura strandi* Breuning, 1940, Holotype; **(w, x)** *Myagrus strandi* Breuning, 1940, Holotype.

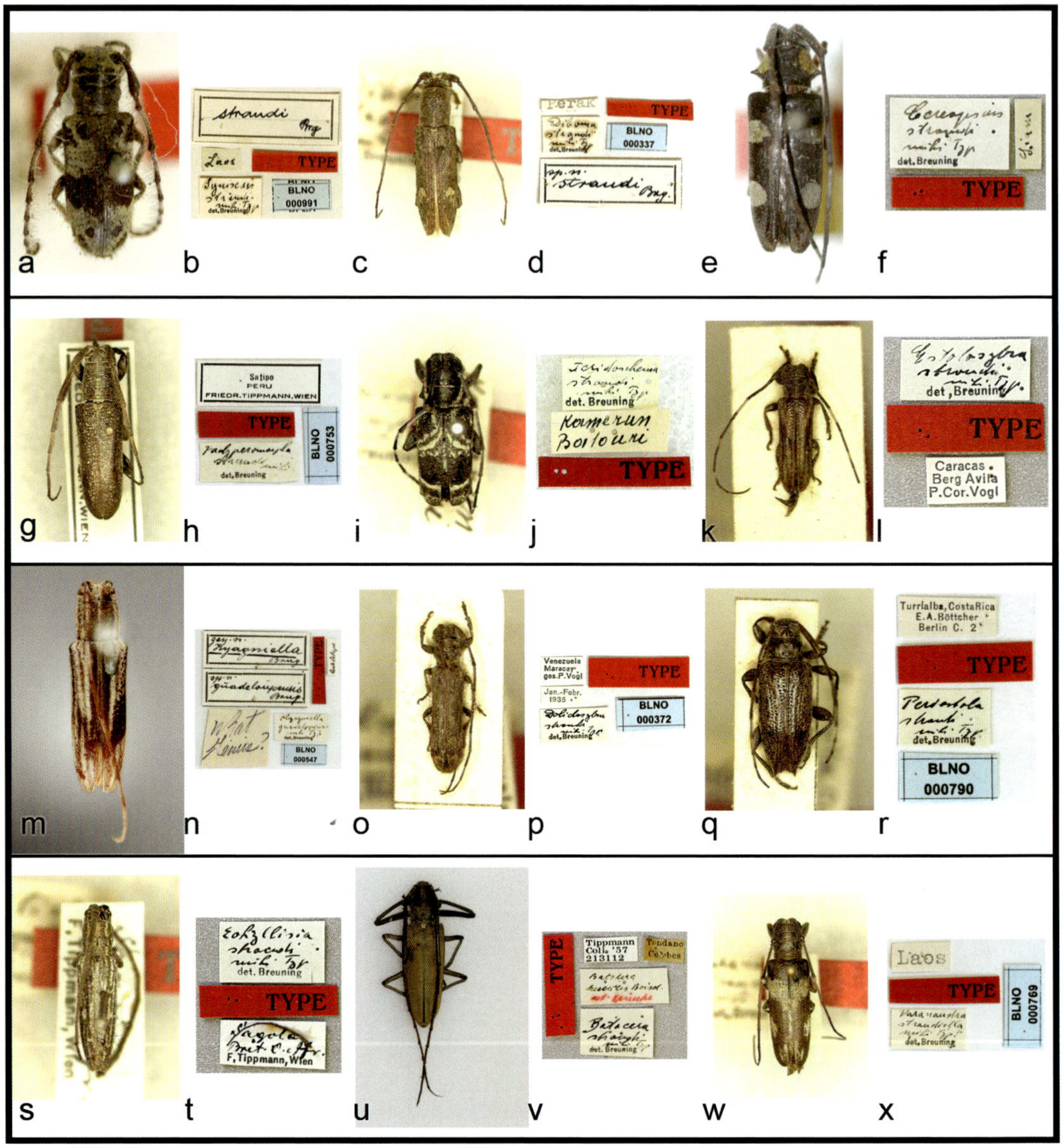

Figure 161. Primary types of Cerambycidae of the Smithsonian Institution (for each pair of images, dorsal habitus left, labels right): **(a, b)** *Synixais strandi* Breuning, 1940, Holotype; **(c, d)** *Diboma strandi* Breuning, 1940, Holotype; **(e, f)** *Cereopsius strandi* Breuning, 1942, Holotype; **(g, h)** *Pachypezomorpha strandi* Breuning, 1942, Holotype; **(i, j)** *Acridoschema strandi* Breuning, 1943, Holotype; **(k, l)** *Estolosybra strandi* Breuning, 1943, Holotype; **(m, n)** *Hyagniellus strandi* Breuning, 1943, Holotype; **(o, p)** *Dolichosybra strandi* Breuning, 1943, Holotype; **(q, r)** *Periestola strandi* Breuning, 1943, Holotype; **(s, t)** *Eohyllisia strandi* Breuning, 1943, Holotype; **(u, v)** *Batocera strandi* Breuning, 1954, Holotype; **(w, x)** *Paranandra strandiella* Breuning, 1940, Holotype.

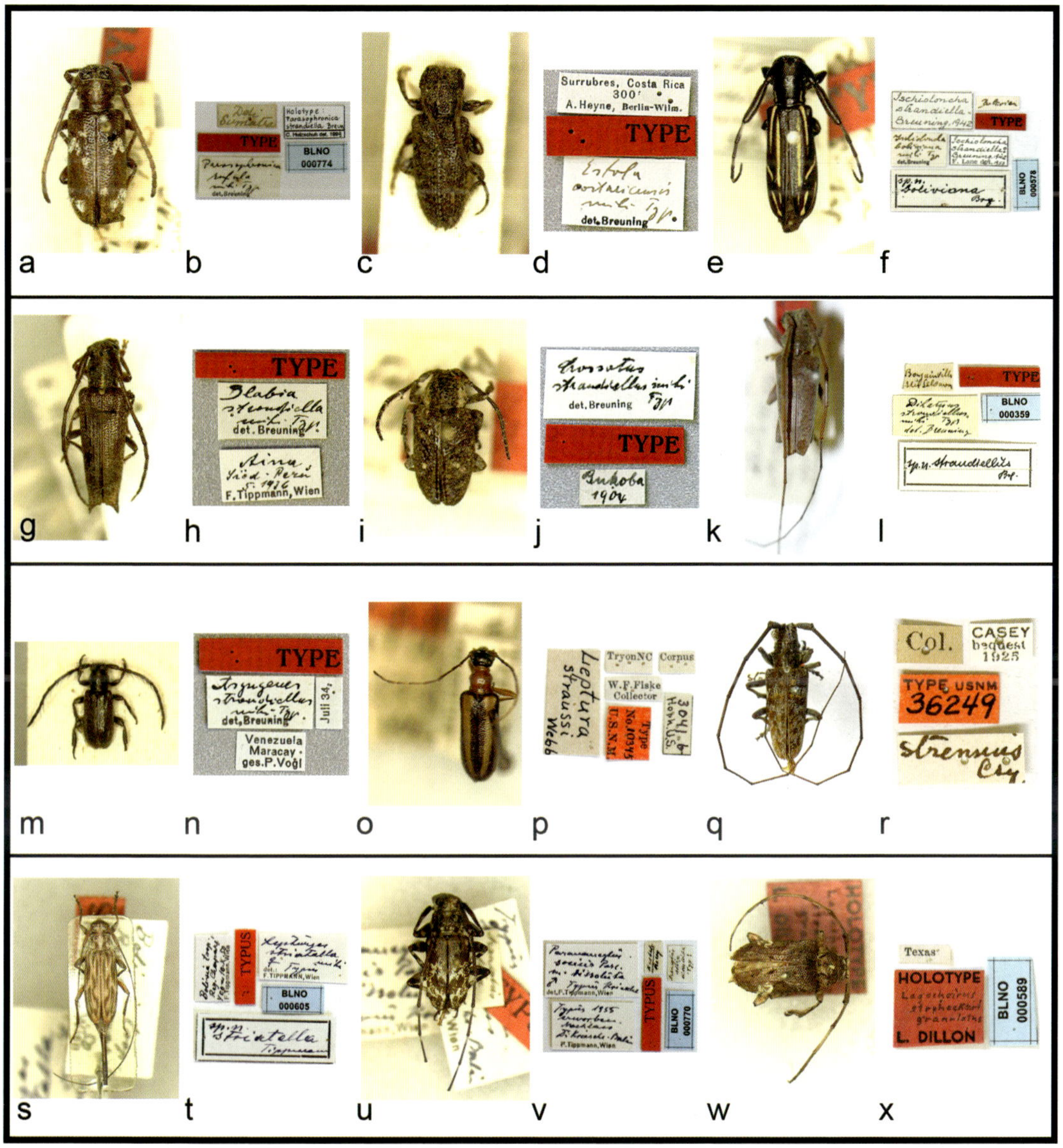

Figure 162. Primary types of Cerambycidae of the Smithsonian Institution (for each pair of images, dorsal habitus left, labels right): **(a, b)** *Parasophronica strandiella* Breuning, 1940, Holotype; **(c, d)** *Estola strandiella* Breuning, 1942, Holotype; **(e, f)** *Ischioloncha strandiella* Breuning, 1942, Holotype; **(g, h)** *Blabia strandiella* Breuning, 1943, Holotype; **(i, j)** *Crossotus strandiellus* Breuning, 1935, Holotype; **(k, l)** *Dihammus strandiellus* Breuning, 1935, Holotype; **(m, n)** *Asyngenes strandiellus* Breuning, 1943, Holotype; **(o, p)** *Leptura straussi* Webb, 1908, Holotype; **(q, r)** *Monochamus strenuus* Casey, 1913, Holotype; **(s, t)** *Lepturges striatella* Tippmann, 1960, Holotype; **(u, v)** *Anancylus strix dissolutus* Kriesche, 1936, Holotype; **(w, x)** *Lagocheirus stroheckeri granulatus* Dillon, 1956, Holotype.

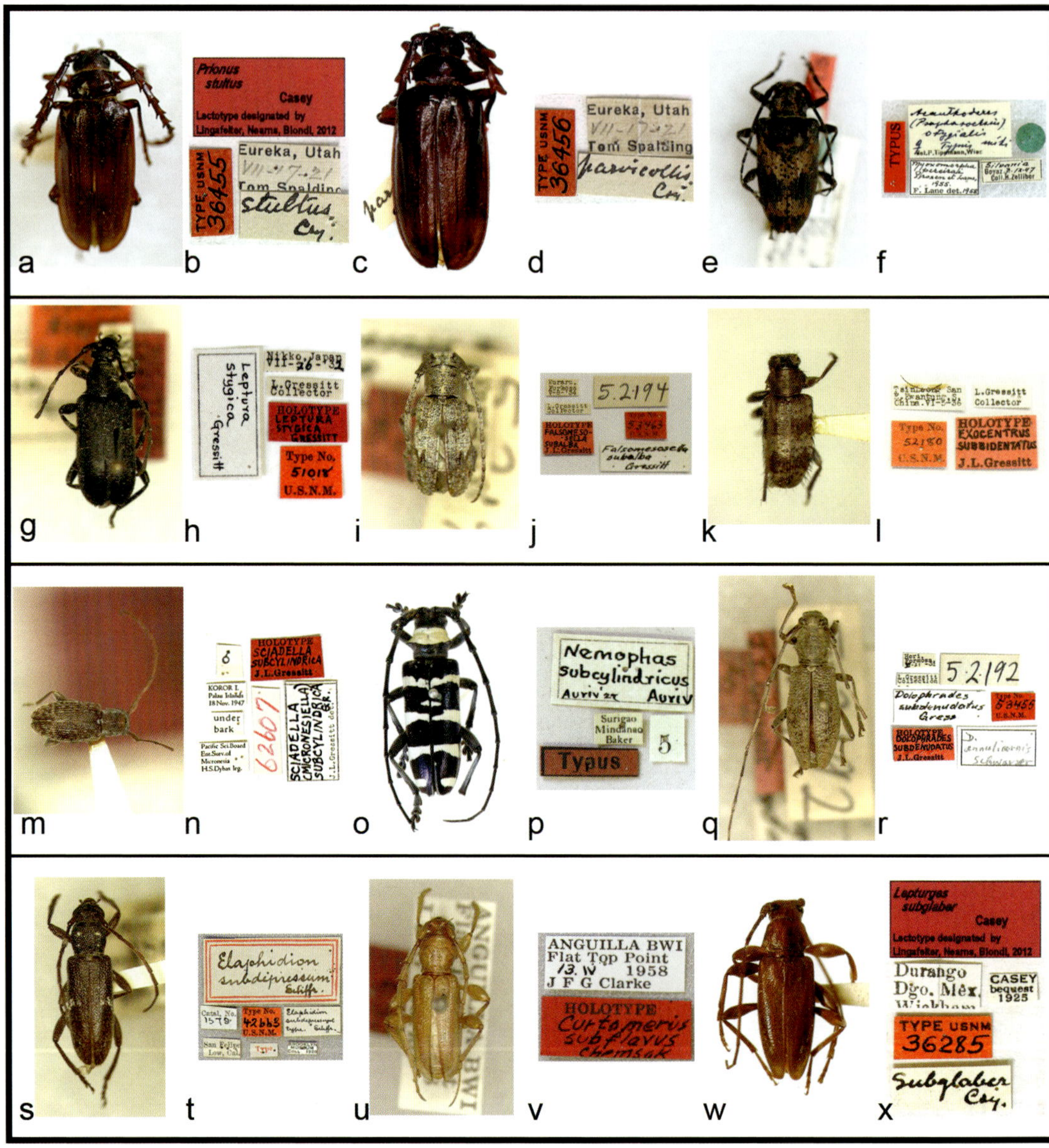

Figure 163. Primary types of Cerambycidae of the Smithsonian Institution (for each pair of images, dorsal habitus left, labels right): **(a, b)** *Prionus (Prionus) stultus* Casey, 1924, Lectotype; **(c, d)** *Prionus (Prionus) stultus parvicollis* Casey, 1924, Holotype; **(e, f)** *Acanthoderes (Psapharochrus) stygialis* Tippmann, 1960, Holotype; **(g, h)** *Leptura stygica* Gressitt, 1935, Holotype; **(i, j)** *Falsomesosella subalba* Gressitt, 1938, Holotype; **(k, l)** *Exocentrus subbidentatus* Gressitt, 1937, Holotype; **(m, n)** *Sciadella (Micronesiella) subcylindrica* Gressitt, 1956, Holotype; **(o, p)** *Nemophas subcylindricus* Aurivillius, 1927, Holotype; **(q, r)** *Dolophrades subdenudatus* Gressitt, 1938, Holotype; **(s, t)** *Elaphidion subdepressum* Schaeffer, 1904, Holotype; **(u, v)** *Curtomerus subflavus* Chemsak, 1966, Holotype; **(w, x)** *Lepturges subglaber* Casey, 1913, Lectotype.

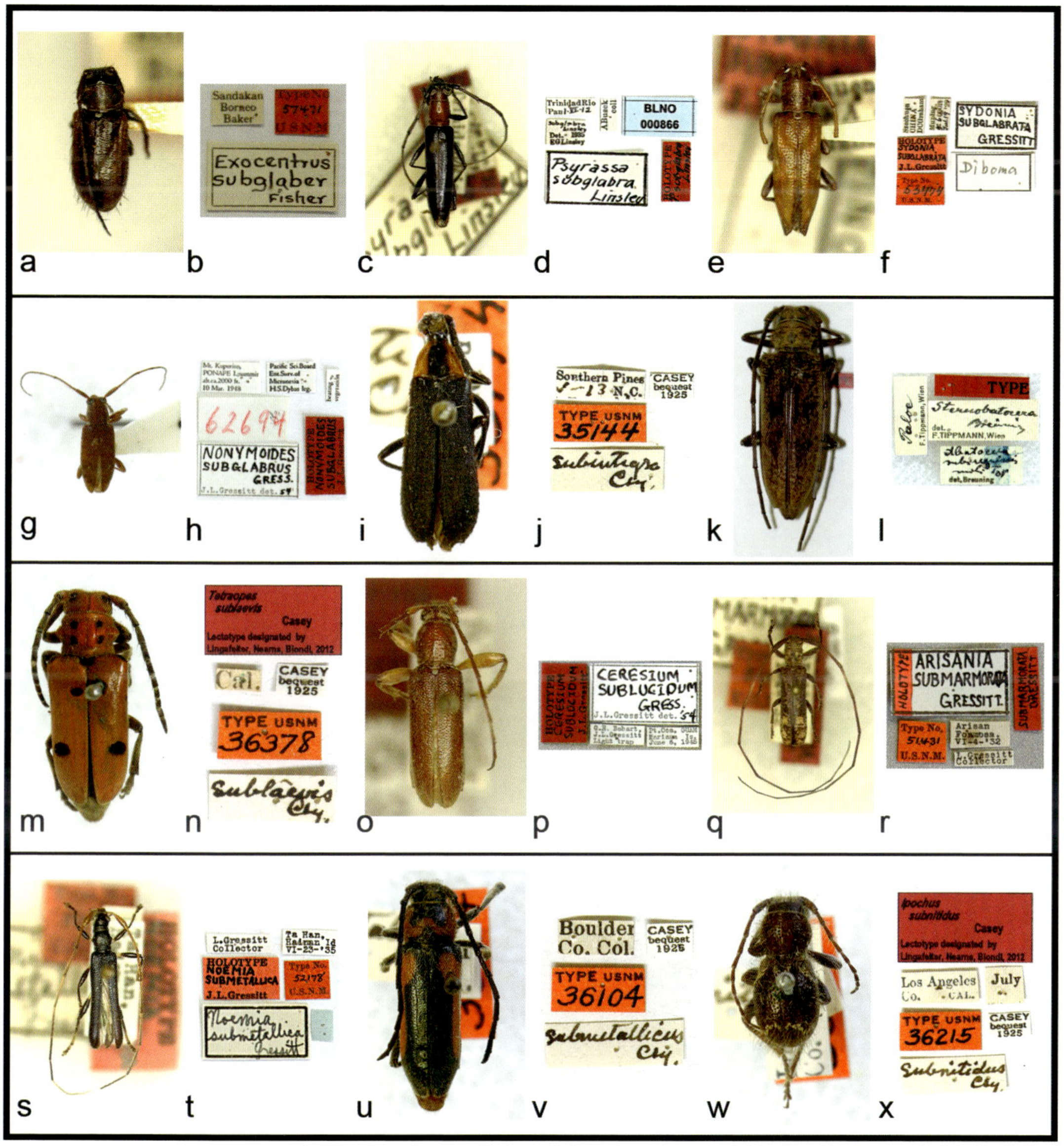

Figure 164. Primary types of Cerambycidae of the Smithsonian Institution (for each pair of images, dorsal habitus left, labels right): **(a, b)** *Exocentrus subglaber* Fisher, 1925, Holotype; **(c, d)** *Psyrassa subglabra* Linsley, 1935, Holotype; **(e, f)** *Sydonia subglabrata* Gressitt, 1938, Holotype; **(g, h)** *Nonymoides subglabrus* Gressitt, 1956, Holotype; **(i, j)** *Euryptera subintegra* Casey, 1924, Holotype; **(k, l)** *Abatocera (Sternobatocera) subirregularis* Breuning, 1954, Holotype; **(m, n)** *Tetraopes sublaevis* Casey, 1913, Lectotype; **(o, p)** *Ceresium (Ceresium) sublucidum* Gressitt, 1956, Holotype; **(q, r)** *Arisania submarmorata* Gressitt, 1936, Holotype; **(s, t)** *Noemia submetallica* Gressitt, 1940, Holotype; **(u, v)** *Crossidius submetallicus* Casey, 1912, Holotype; **(w, x)** *Ipochus subnitidus* Casey, 1891, Lectotype.

Figure 165. Primary types of Cerambycidae of the Smithsonian Institution (for each pair of images, dorsal habitus left, labels right): **(a, b)** *Oeme subobsoleta* Casey, 1924, Lectotype; **(c, d)** *Cyrtinus subopacus* Fisher, 1935, Holotype; **(e, f)** *Stenocorus (Stenocorus) subpinguis* Casey, 1913, Holotype; **(g, h)** *Spondylis subpubescens* Casey, 1912, Lectotype; **(i, j)** *Brachyleptura subquadrata* Casey, 1913, Holotype; **(k, l)** *Nemophas subterrubens* Heller, 1924, Holotype; **(m, n)** *Eupogonius subtessellatus* Melzer, 1933, Holotype; **(o, p)** *Eczemotellus subtilipictus* Heller, 1924, Lectotype; **(q, r)** *Euryscelis subtropica* Casey, 1924, Holotype; **(s, t)** *Anoplium subtropicum* Casey, 1924, Lectotype; **(u, v)** *Oncideres subtropicus* Casey, 1913, Holotype; **(w, x)** *Tylosis subtruncata* Casey, 1912, Holotype.

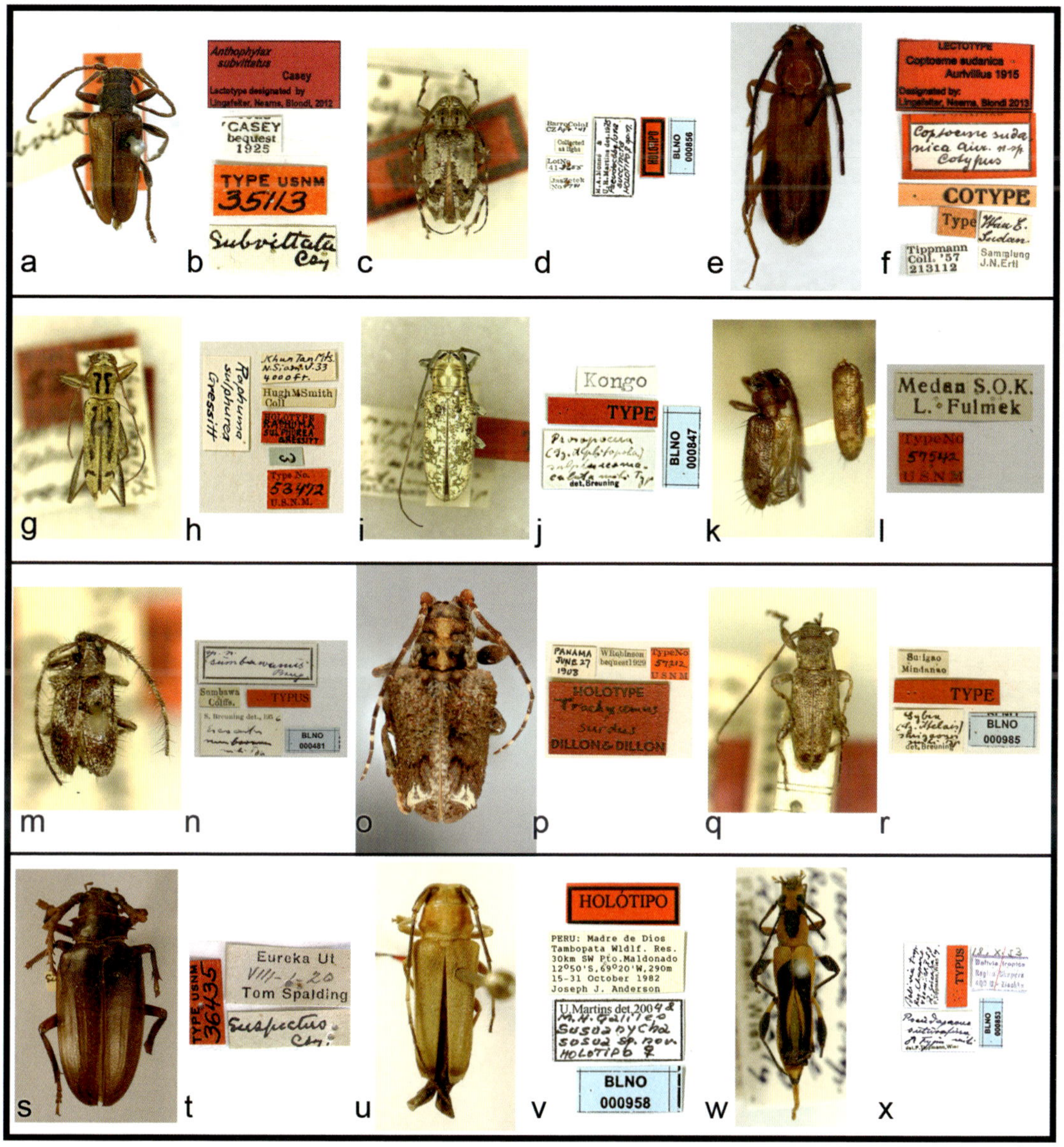

Figure 166. Primary types of Cerambycidae of the Smithsonian Institution (for each pair of images, dorsal habitus left, labels right): **(a, b)** *Anthophilax subvittata* Casey, 1891, Lectotype; **(c, d)** *Pseudocobelura succincta* Monné & Martins, 1976, Holotype; **(e, f)** *Coptoeme sudanica* Aurivillius, 1915, Lectotype; **(g, h)** *Rhaphuma sulphurea* Gressitt, 1941, Holotype; **(i, j)** *Prosopocera (Alphitopola) sulphureomarmorata* Breuning, 1939, Holotype; **(k, l)** *Exocentrus sumatrensis* Fisher, 1927, Holotype; **(m, n)** *Exocentrus (Pseudocentrus) sumbawanus* Breuning, 1957, Holotype; **(o, p)** *Trachysomus surdus* Dillon & Dillon, 1946, Holotype; **(q, r)** *Sybra (Atelais) surigaonis* Breuning, 1943, Holotype; **(s, t)** *Prionus (Prionus) suspectus* Casey, 1924, Holotype; **(u, v)** *Susuanycha susua* Galileo & Martins, 2005, Holotype; **(w, x)** *Pseudagaone suturafissa* Tippmann, 1960, Holotype.

Figure 167. Primary types of Cerambycidae of the Smithsonian Institution (for each pair of images, dorsal habitus left, labels right): **(a, b)** *Leptacmaeops suturalis* Casey, 1924, Lectotype; **(c, d)** *Megopis (Aegosoma) suturalis* Fisher, 1935, Holotype; **(e, f)** *Spalacopsis suturalis* Hamilton in Leng & Hamilton, 1896, Lectotype; **(g, h)** *Amphionycha suturalis* Linell, 1897, Holotype; **(i, j)** *Dihammus szechuanus* Gressitt, 1938, Holotype; **(k, l)** *Flabelloprionus szechuanus* Heyrovsky, 1933, Syntype; **(m, n)** *Leptostylus taeniatus* Casey, 1913, Lectotype; **(o, p)** *Styloleptus taino* Lingafelter & Micheli, 2004, Holotype; **(q, r)** *Metopoplectus taiwanensis* Gressitt, 1936, Holotype; **(s, t)** *Leptura tatsienlua* Gressitt, 1948, Holotype; **(u, v)** *Dihammus tavura* Kriesche, 1936, Holotype; **(w, x)** *Exocentrus tectonae* Fisher, 1934, Holotype.

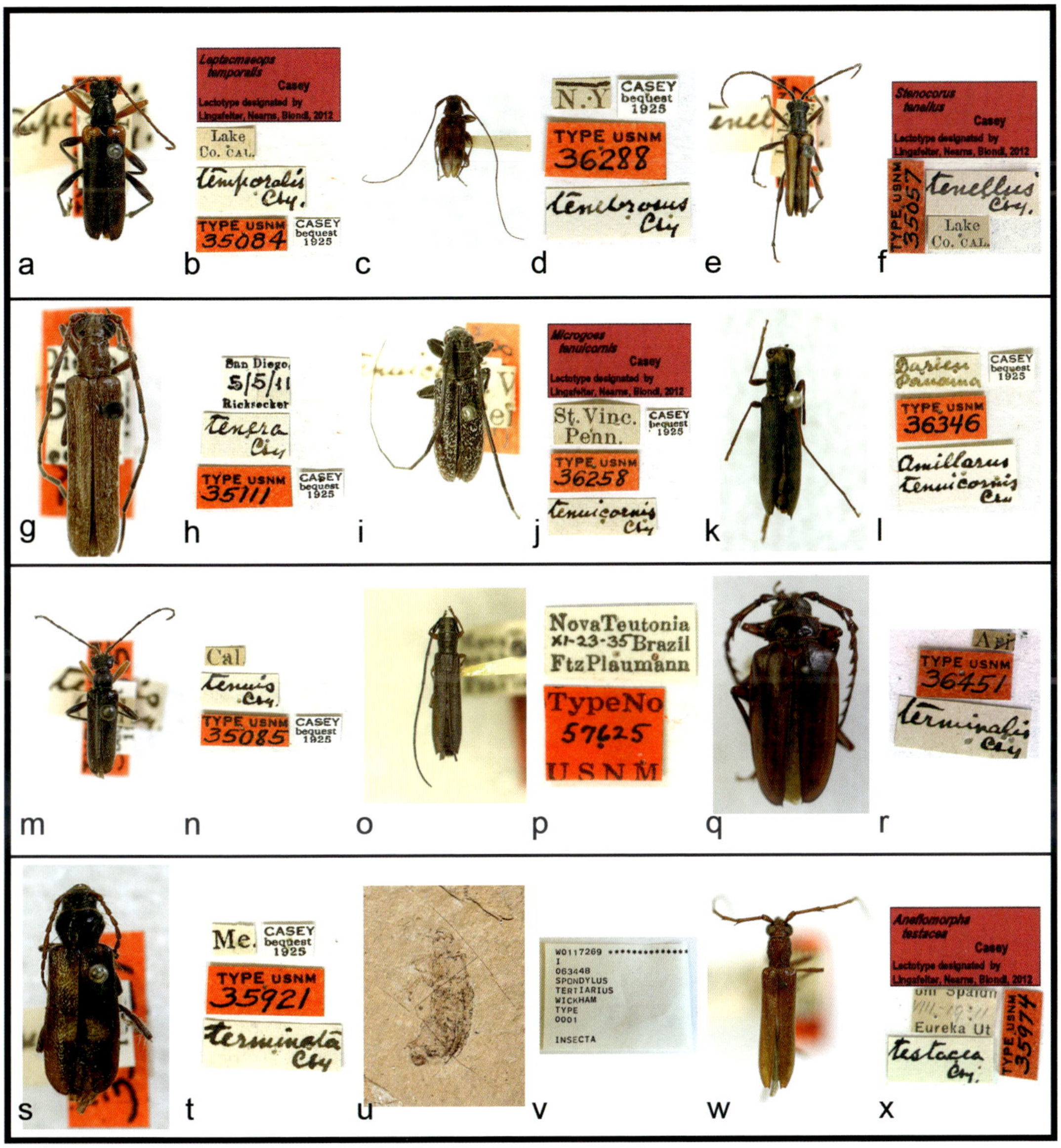

Figure 168. Primary types of Cerambycidae of the Smithsonian Institution (for each pair of images, dorsal habitus left, labels right): **(a, b)** *Leptacmaeops (Leptacmaeops) temporalis* Casey, 1913, Lectotype; **(c, d)** *Lepturges tenebrosus* Casey, 1913, Holotype; **(e, f)** *Stenocorus (Stenocorus) tenellus* Casey, 1913, Lectotype; **(g, h)** *Centrodera tenera* Casey, 1913, Holotype; **(i, j)** *Microgoes tenuicornis* Casey, 1913, Lectotype; **(k, l)** *Amillarus tenuicornis* Casey, 1913, Holotype; **(m, n)** *Leptacmaeops (Leptacmaeops) tenuis* Casey, 1913, Holotype; **(o, p)** *Obereoides tenuis* Fisher, 1938, Holotype; **(q, r)** *Prionus (Prionus) terminalis* Casey, 1912, Holotype; **(s, t)** *Anocomis terminata* Casey, 1912, Holotype; **(u, v)** *Spondylis tertiarius* Wickham, 1917, Holotype; **(w, x)** *Aneflomorpha testacea* Casey, 1924, Lectotype.

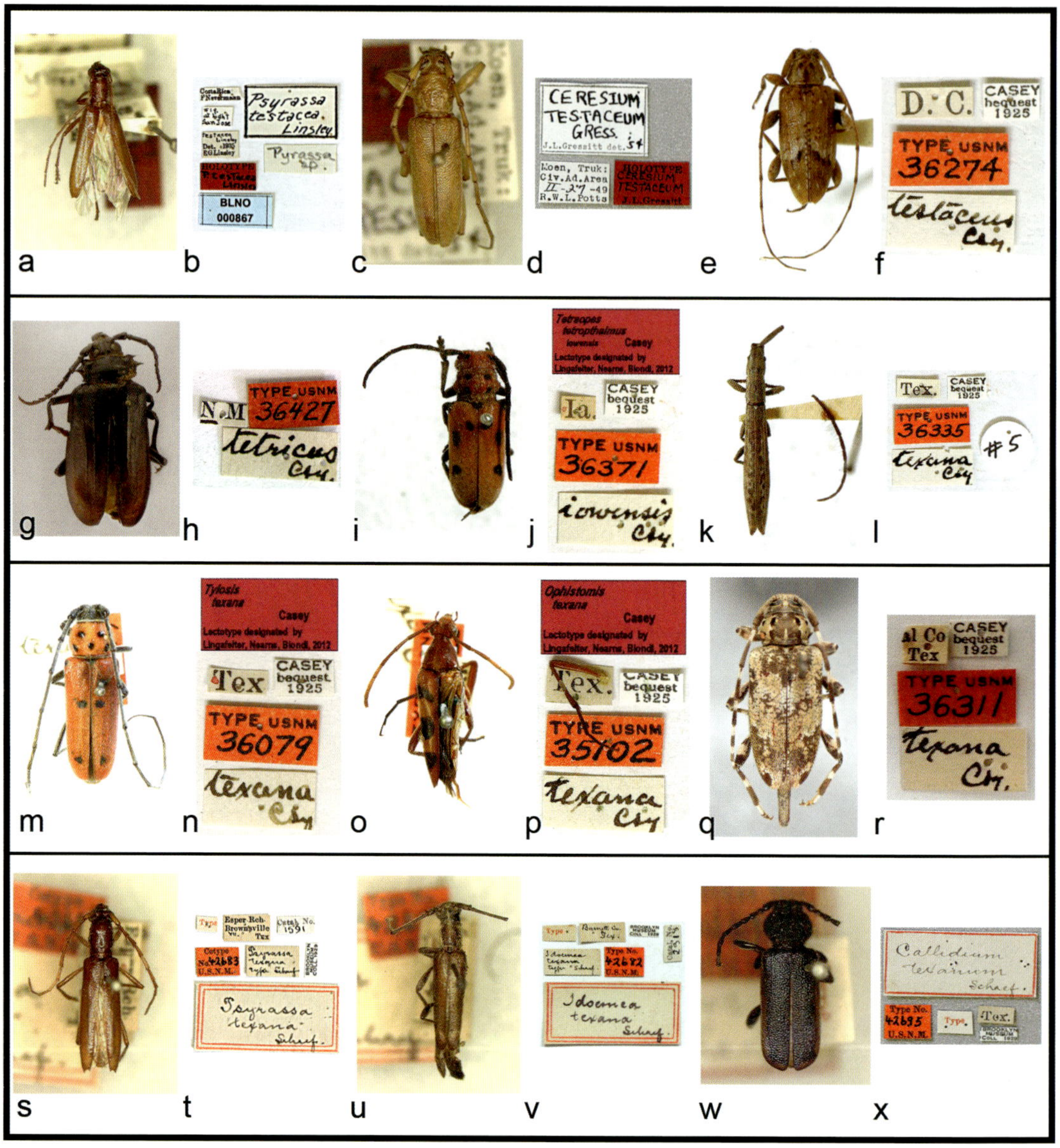

Figure 169. Primary types of Cerambycidae of the Smithsonian Institution (for each pair of images, dorsal habitus left, labels right): **(a, b)** *Psyrassa testacea* Linsley, 1935, Holotype; **(c, d)** *Ceresium (Ceresium) testaceum* Gressitt, 1956, Holotype; **(e, f)** *Leiopus testaceus* Casey, 1913, Holotype; **(g, h)** *Prionus (Prionus) tetricus* Casey, 1912, Holotype; **(i, j)** *Tetraopes tetropthalmus iowensis* Casey, 1913, Lectotype; **(k, l)** *Spalacopsis texana* Casey, 1891, Holotype; **(m, n)** *Tylosis texana* Casey, 1912, Holotype; **(o, p)** *Ophistomis texana* Casey, 1913, Lectotype; **(q, r)** *Urographis texana* Casey, 1924, Holotype; **(s, t)** *Psyrassa texana* Schaeffer, 1905, Holotype; **(u, v)** *Idoemea texana* Schaeffer, 1908, Holotype; **(w, x)** *Callidium texanum* Schaeffer, 1917, Holotype.

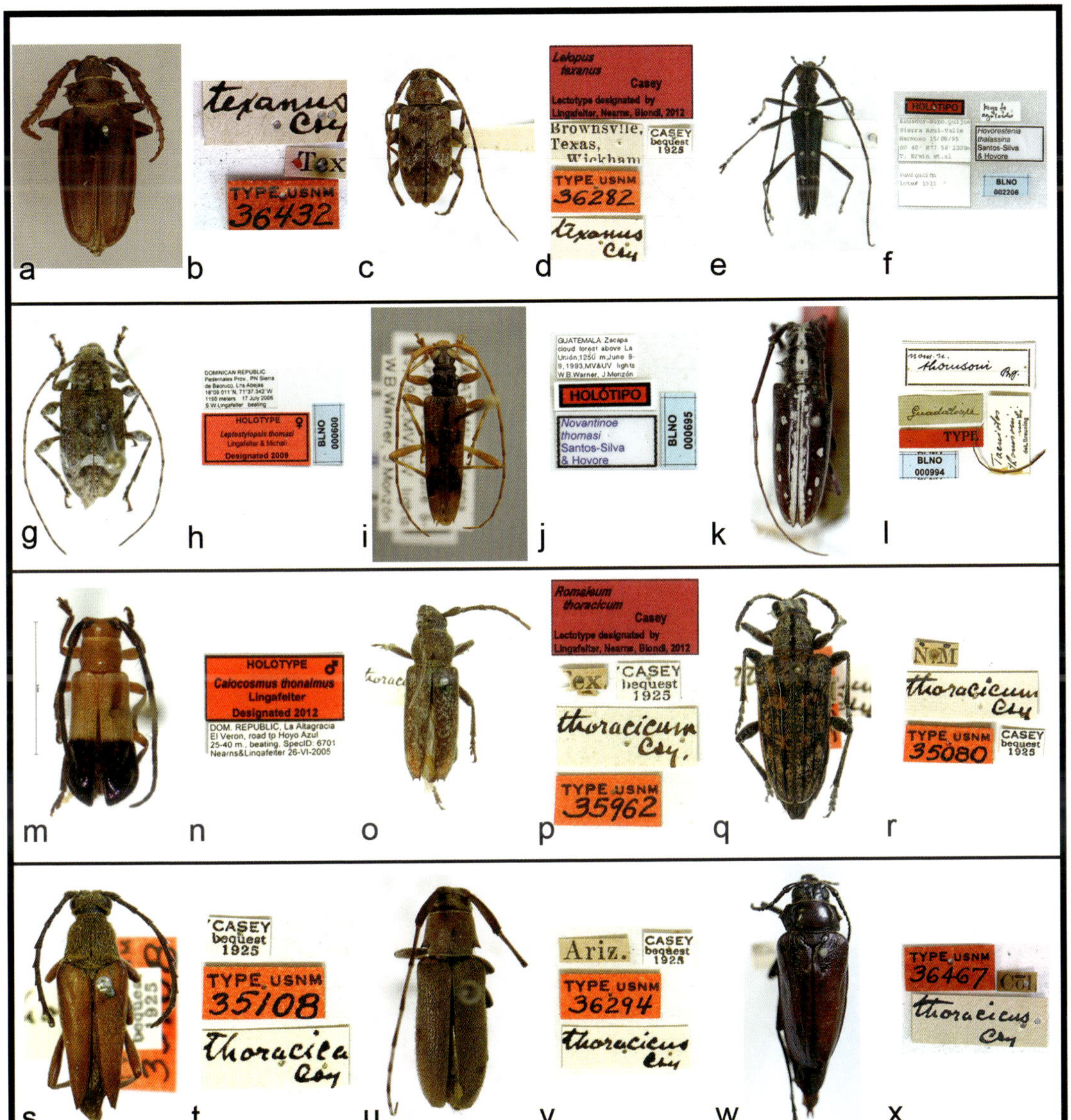

Figure 170. Primary types of Cerambycidae of the Smithsonian Institution (for each pair of images, dorsal habitus left, labels right): **(a, b)** *Prionus (Prionus) texanus* Casey, 1912, Holotype; **(c, d)** *Leiopus texanus* Casey, 1913, Lectotype; **(e, f)** *Hovorestenia thalassina* Santos-Silva & Hovore, 2008, Holotype; **(g, h)** *Leptostylopsis thomasi* Lingafelter & Micheli, 2009, Holotype; **(i, j)** *Novantinoe thomasi* Santos-Silva & Hovore, 2007, Holotype; **(k, l)** *Taeniotes thomsoni* Breuning, 1943, Holotype; **(m, n)** *Calocosmus thonalmus* Lingafelter, 2013, Holotype; **(o, p)** *Romaleum thoracicum* Casey, 1912, Lectotype; **(q, r)** *Rhagium (Hargium) thoracicum* Casey, 1913, Holotype; **(s, t)** *Typocerus thoracicus* Casey, 1913, Holotype; **(u, v)** *Dectes thoracicus* Casey, 1913, Holotype; **(w, x)** *Prionus (Antennalia) thoracicus* Casey, 1924, Holotype.

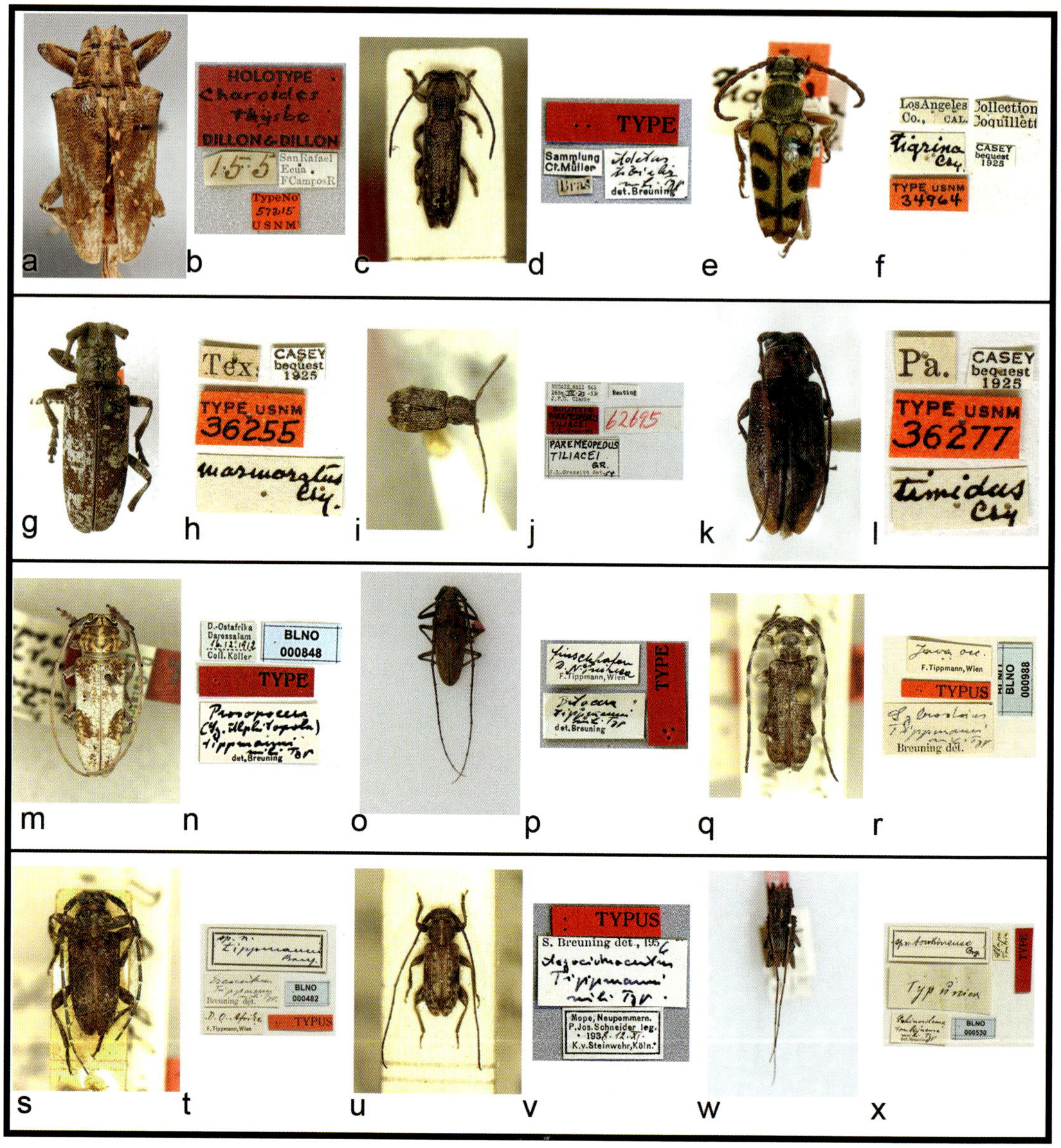

Figure 171. Primary types of Cerambycidae of the Smithsonian Institution (for each pair of images, dorsal habitus left, labels right): **(a, b)** *Charoides thysbe* Dillon & Dillon, 1945, Holotype; **(c, d)** *Adetus tibialis* Breuning, 1943, Holotype; **(e, f)** *Strophiona tigrina* Casey, 1913, Holotype; **(g, h)** *Goes tigrinis marmoratus* Casey, 1913, Holotype; **(i, j)** *Paremeopedus tiliacei* Gressitt, 1956, Holotype; **(k, l)** *Leiopus timidus* Casey, 1913, Holotype; **(m, n)** *Prosopocera (Alphitopola) tippmanni* Breuning, 1942, Holotype; **(o, p)** *Batocera tippmanni* Breuning, 1954, Holotype; **(q, r)** *Sybrodoius tippmanni* Breuning, 1957, Holotype; **(s, t)** *Exocentrus (Pseudocentrus) tippmanni* Breuning, 1957, Holotype; **(u, v)** *Aegocidnexocentrus tippmanni* Breuning, 1957, Holotype; **(w, x)** *Eshinoschema tonkinense* Breuning, 1938, Holotype.

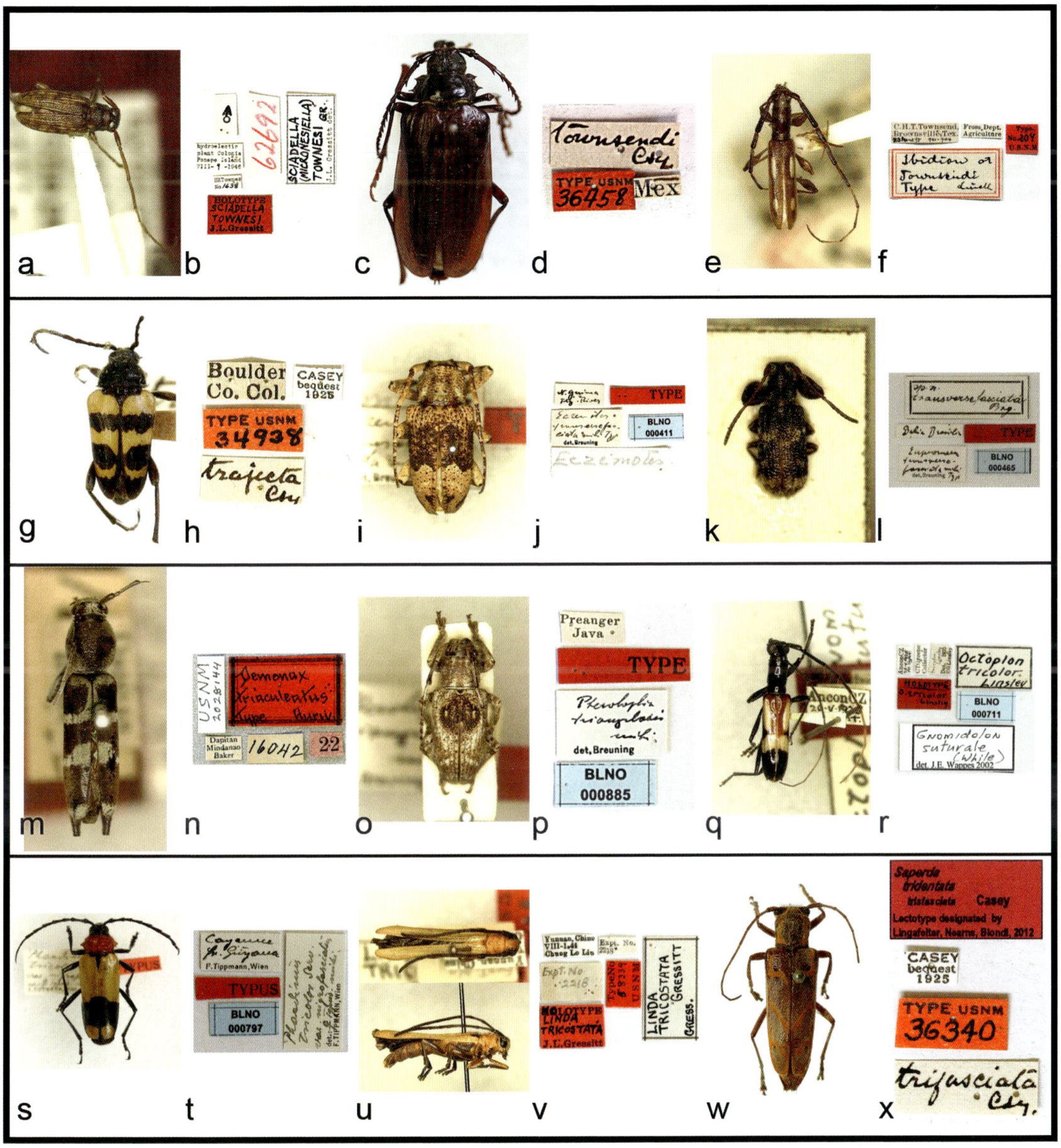

Figure 172. Primary types of Cerambycidae of the Smithsonian Institution (for each pair of images, dorsal habitus left, labels right): **(a, b)** *Sciadella (Micronesiella) townesi* Gressitt, 1956, Holotype; **(c, d)** *Prionus (Riponus) townsendi* Casey, 1912, Holotype; **(e, f)** *Ibidion townsendi* Linell, 1897, Holotype; **(g, h)** *Judolia trajecta* Casey, 1913, Holotype; **(i, j)** *Eczemotes transversefasciata* Breuning, 1940, Holotype; **(k, l)** *Eupromera transversefasciata* Breuning, 1943, Holotype; **(m, n)** *Demonax triaculeatus* Aurivillius, 1922, Holotype; **(o, p)** *Pterolophia triangularis* Breuning, 1938, Holotype; **(q, r)** *Octoplon tricolor* Linsley, 1935, Holotype; **(s, t)** *Phaedinus tricolor* var. *superbus* Tippmann, 1954, Holotype; **(u, v)** *Linda tricostata* Gressitt, 1947, Holotype; **(w, x)** *Saperda tridentata trifasciata* Casey, 1913, Lectotype.

Figure 173. Primary types of Cerambycidae of the Smithsonian Institution (for each pair of images, dorsal habitus left, labels right): **(a, b)** *Driopeomima trifasciata* Aurivillius, 1927, Syntype; **(c, d)** *Demonax trifasciatus* Aurivillius, 1928, Holotype; **(e, f)** *Brachysomida trinitatis* Casey, 1913, Lectotype; **(g, h)** *Oncideres trinodatus* Casey, 1913, Lectotype; **(i, j)** *Odontocera triplaris* Fisher, 1930, Holotype; **(k, l)** *Oberea tripunctata appalachiana* Casey, 1913, Holotype; **(m, n)** *Oberea tripunctata intermedia* Casey, 1913, Lectotype; **(o, p)** *Psenocerus tristis* Casey, 1891, Holotype; **(q, r)** *Prionus (Prionus) tristis* Casey, 1912, Holotype; **(s, t)** *Clytoleptus tristis* Casey, 1912, Holotype; **(u, v)** *Lepturges tristis* Casey, 1913, Holotype; **(w, x)** *Pterolophia trivittata* Breuning, 1940, Holotype.

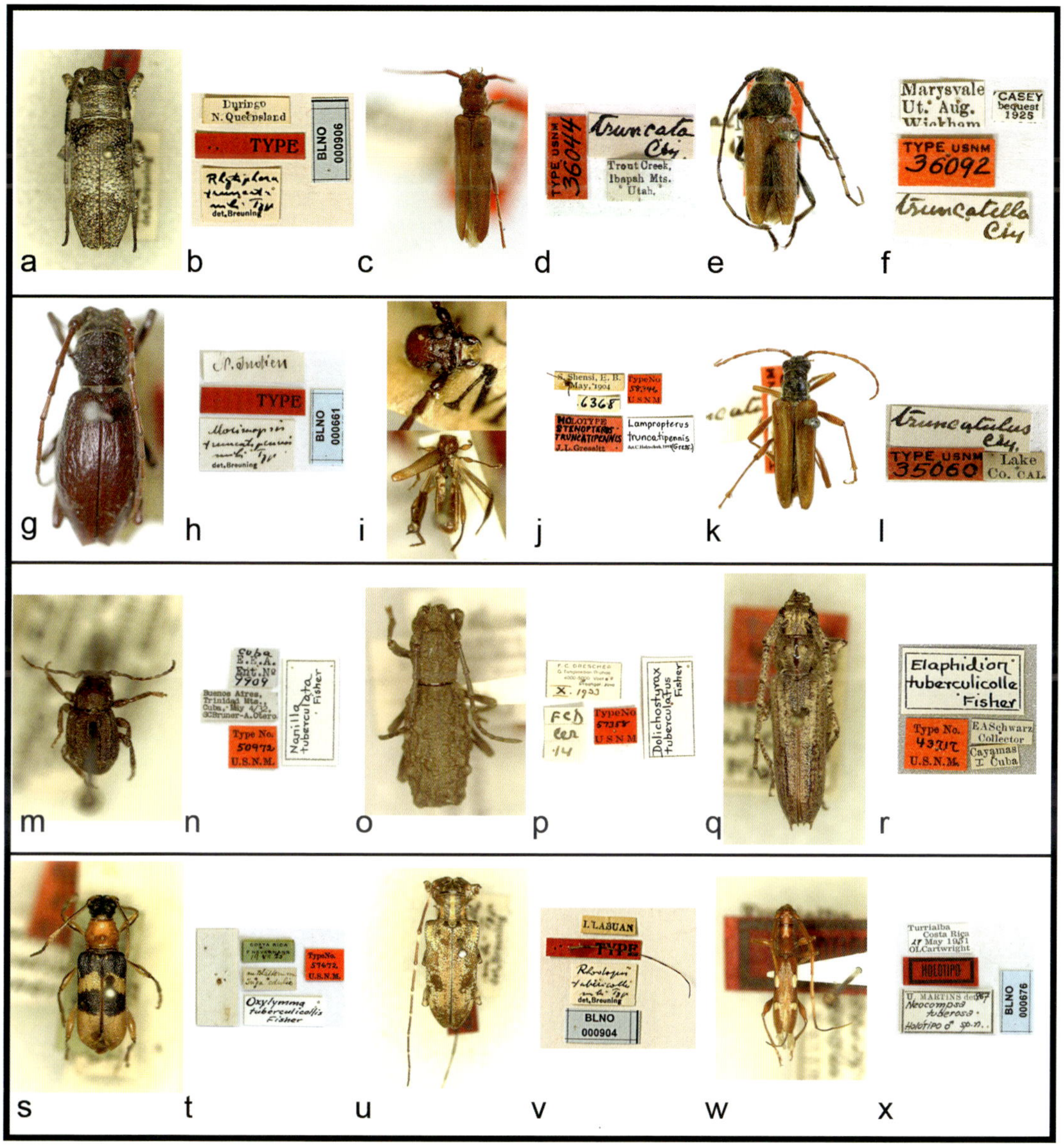

Figure 174. Primary types of Cerambycidae of the Smithsonian Institution (for each pair of images, dorsal habitus left, labels right): **(a, b)** *Rhytiphora truncata* Breuning, 1940, Holotype; **(c, d)** *Oeme truncata* Casey, 1924, Holotype; **(e, f)** *Crossidius truncatellus* Casey, 1912, Holotype; **(g, h)** *Morimopsis truncatipennis* Breuning, 1940, Holotype; **(i, j)** *Stenopterus truncatipennis* Gressitt, 1948, Holotype; **(k, l)** *Stenocorus (Stenocorus) truncatulus* Casey, 1913, Holotype; **(m, n)** *Nanilla tuberculata* Fisher, 1935, Holotype; **(o, p)** *Dolichostyrax tuberculatus* Fisher, 1936, Holotype; **(q, r)** *Elaphidion tuberculicolle* Fisher, 1932, Holotype; **(s, t)** *Oxylymma tuberculicollis* Fisher, 1947, Holotype; **(u, v)** *Rhodopis tubericollis* Breuning, 1943, Holotype; **(w, x)** *Neocompsa tuberosa* Martins, 1970, Holotype.

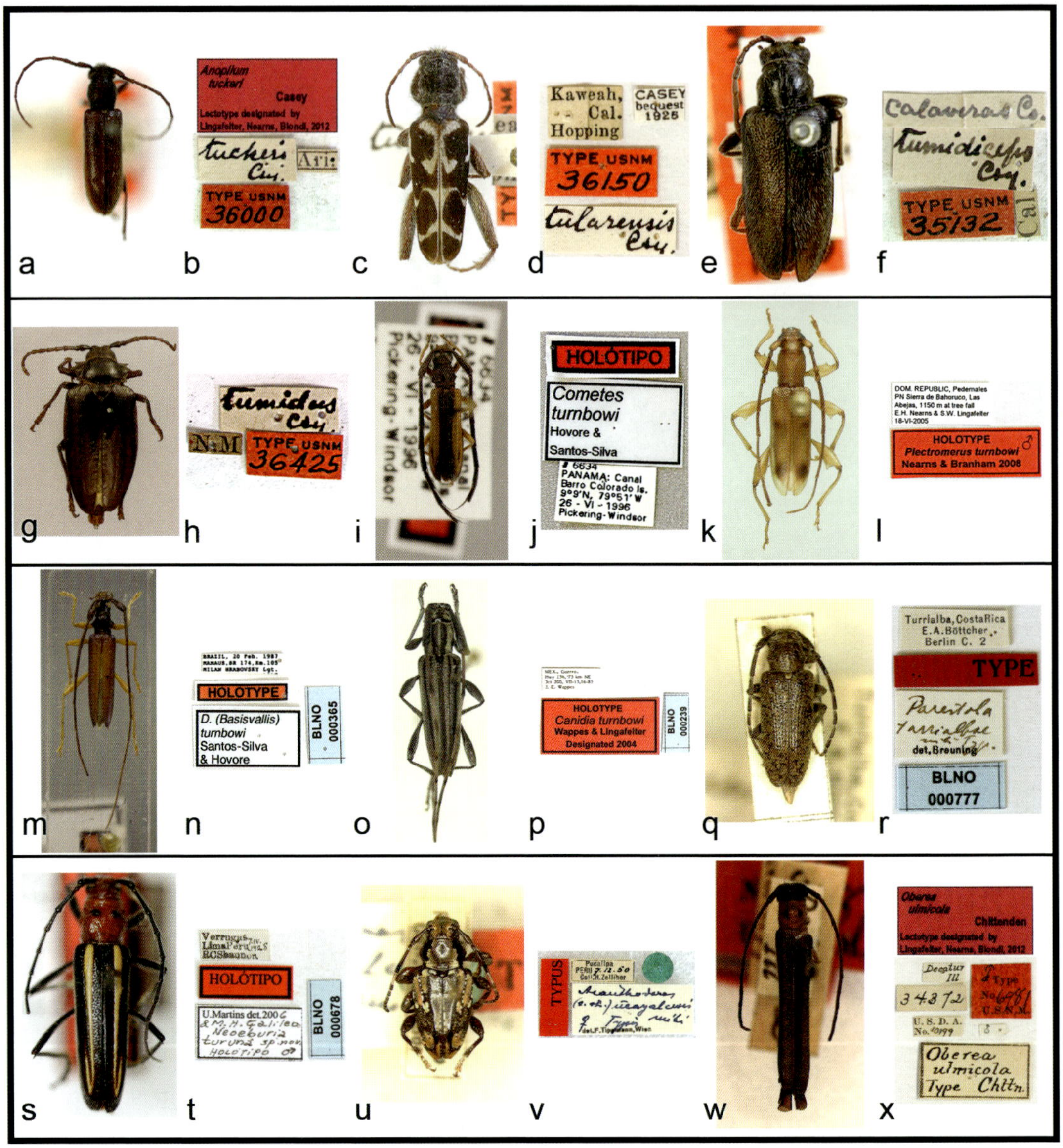

Figure 175. Primary types of Cerambycidae of the Smithsonian Institution (for each pair of images, dorsal habitus left, labels right): **(a, b)** *Anoplium tuckeri* Casey, 1924, Lectotype; **(c, d)** *Neoclytus tularensis* Casey, 1912, Holotype; **(e, f)** *Brachysomida tumidiceps* Casey, 1913, Holotype; **(g, h)** *Prionus (Prionus) tumidus* Casey, 1912, Holotype; **(i, j)** *Cometes turnbowi* Hovore & Santos-Silva, 2007, Holotype; **(k, l)** *Plectromerus turnbowi* Nearns & Branham, 2008, Holotype; **(m, n)** *Distenia (Basisvallis) turnbowi* Santos-Silva & Hovore, 2007, Holotype; **(o, p)** *Canidia turnbowi* Wappes & Lingafelter, 2005, Holotype; **(q, r)** *Estoloides (Estoloides) turrialbae* Breuning, 1943, Holotype; **(s, t)** *Neoeburia turuna* Galileo & Martins, 2006, Holotype; **(u, v)** *Acanthoderes ucayalensis* Tippmann, 1960, Holotype; **(w, x)** *Oberea ulmicola* Chittenden in Webster, 1904, Lectotype.

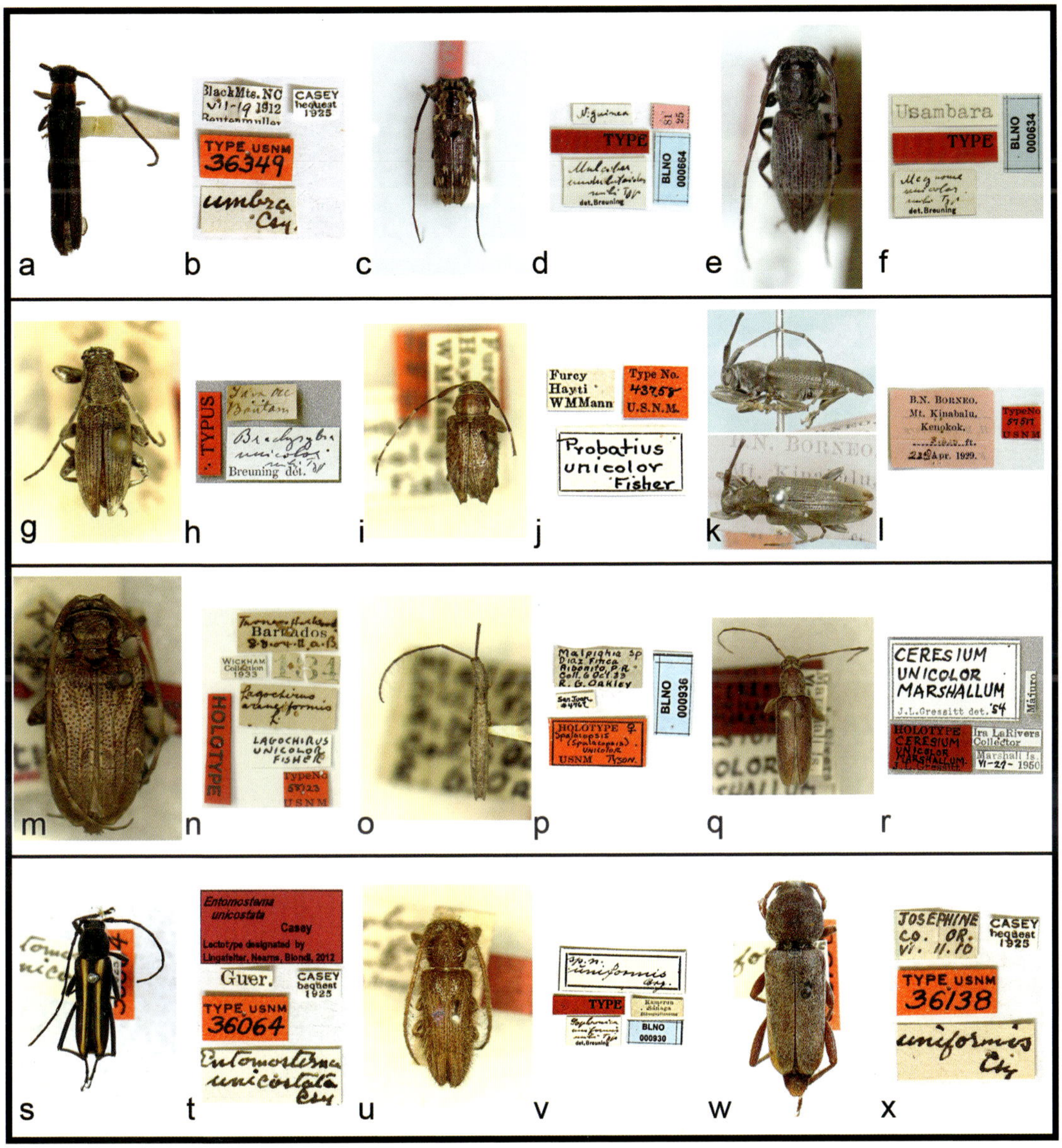

Figure 176. Primary types of Cerambycidae of the Smithsonian Institution (for each pair of images, dorsal habitus left, labels right): **(a, b)** *Oberea umbra* Casey, 1914, Holotype; **(c, d)** *Mulciber undulatoides* Breuning, 1940, Holotype; **(e, f)** *Mecynome (Niphoparmena) unicolor* Breuning, 1940, Holotype; **(g, h)** *Brachysybra unicolor* Breuning, 1957, Holotype; **(i, j)** *Probatius unicolor* Fisher, 1932, Holotype; **(k, l)** *Orsidis unicolor* Fisher, 1935, Holotype; **(m, n)** *Lagochirus unicolor* Fisher, 1947, Holotype; **(o, p)** *Spalacopsis unicolor* Tyson, 1973, Holotype; **(q, r)** *Ceresium (Ceresium) unicolor marshallum* Gressitt, 1956, Holotype; **(s, t)** *Entomosterna unicostata* Casey, 1912, Lectotype; **(u, v)** *Sophronica uniformis* Breuning, 1942, Holotype; **(w, x)** *Xylotrechus uniformis* Casey, 1924, Holotype.

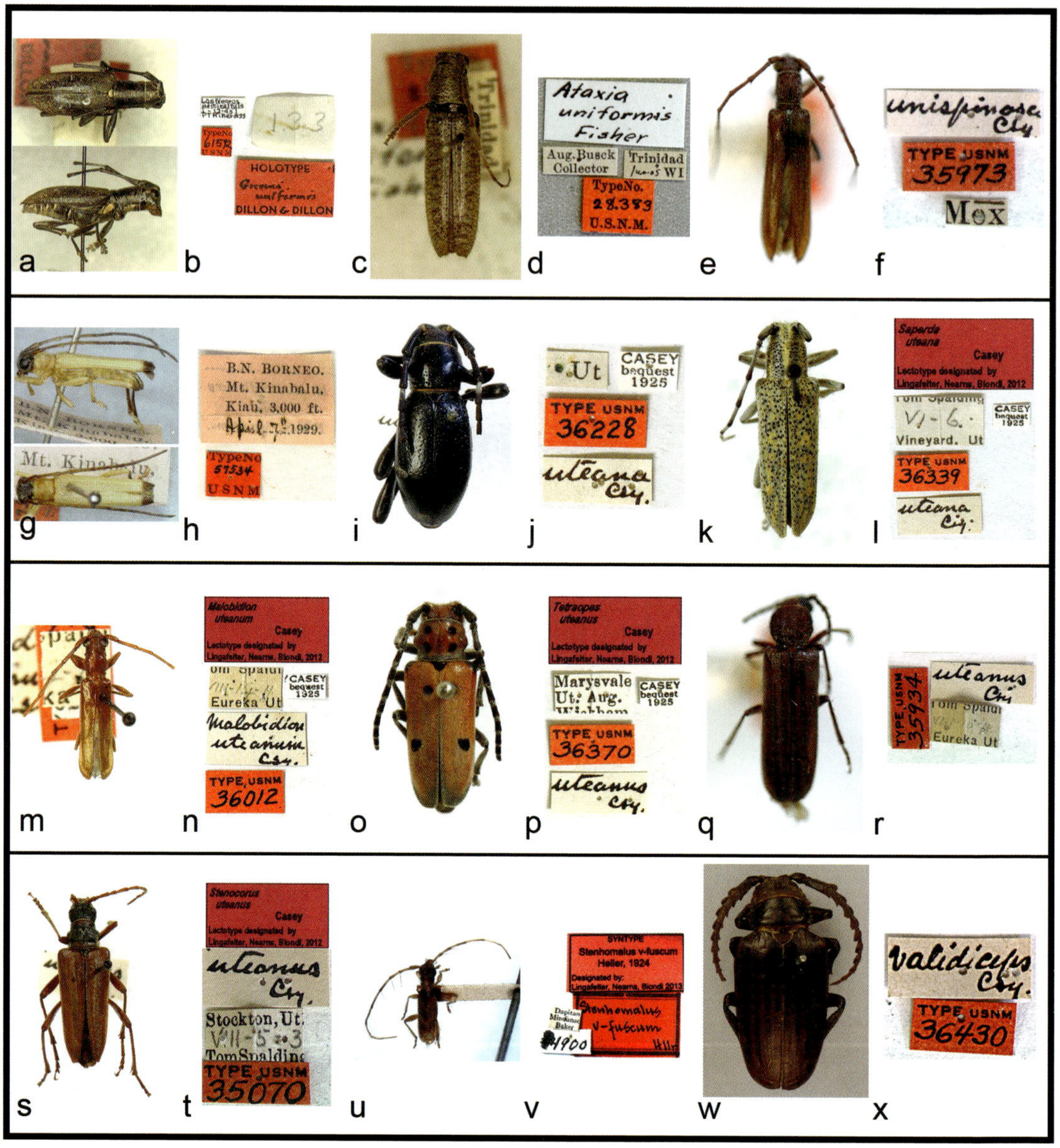

Figure 177. Primary types of Cerambycidae of the Smithsonian Institution (for each pair of images, dorsal habitus left, labels right): **(a, b)** *Gnoma uniformis* Dillon & Dillon, 1951, Holotype; **(c, d)** *Ataxia uniformis* Fisher, 1926, Holotype; **(e, f)** *Aneflomorpha unispinosa* Casey, 1912, Holotype; **(g, h)** *Nupserha ustulata* var. *nigricornis* Fisher, 1935, Holotype; **(i, j)** *Moneilema (Collapteryx) uteana* Casey, 1913, Holotype; **(k, l)** *Saperda uteana* Casey, 1924, Lectotype; **(m, n)** *Malobidion uteanum* Casey, 1924, Lectotype; **(o, p)** *Tetraopes uteanus* Casey, 1913, Lectotype; **(q, r)** *Criocephalus uteanus* Casey, 1924, Holotype; **(s, t)** *Stenocorus uteanus* Casey, 1924, Lectotype; **(u, v)** *Stenhomalus v-fuscum* Heller, 1924, Syntype; **(w, x)** *Prionus (Prionus) validiceps* Casey, 1912, Holotype.

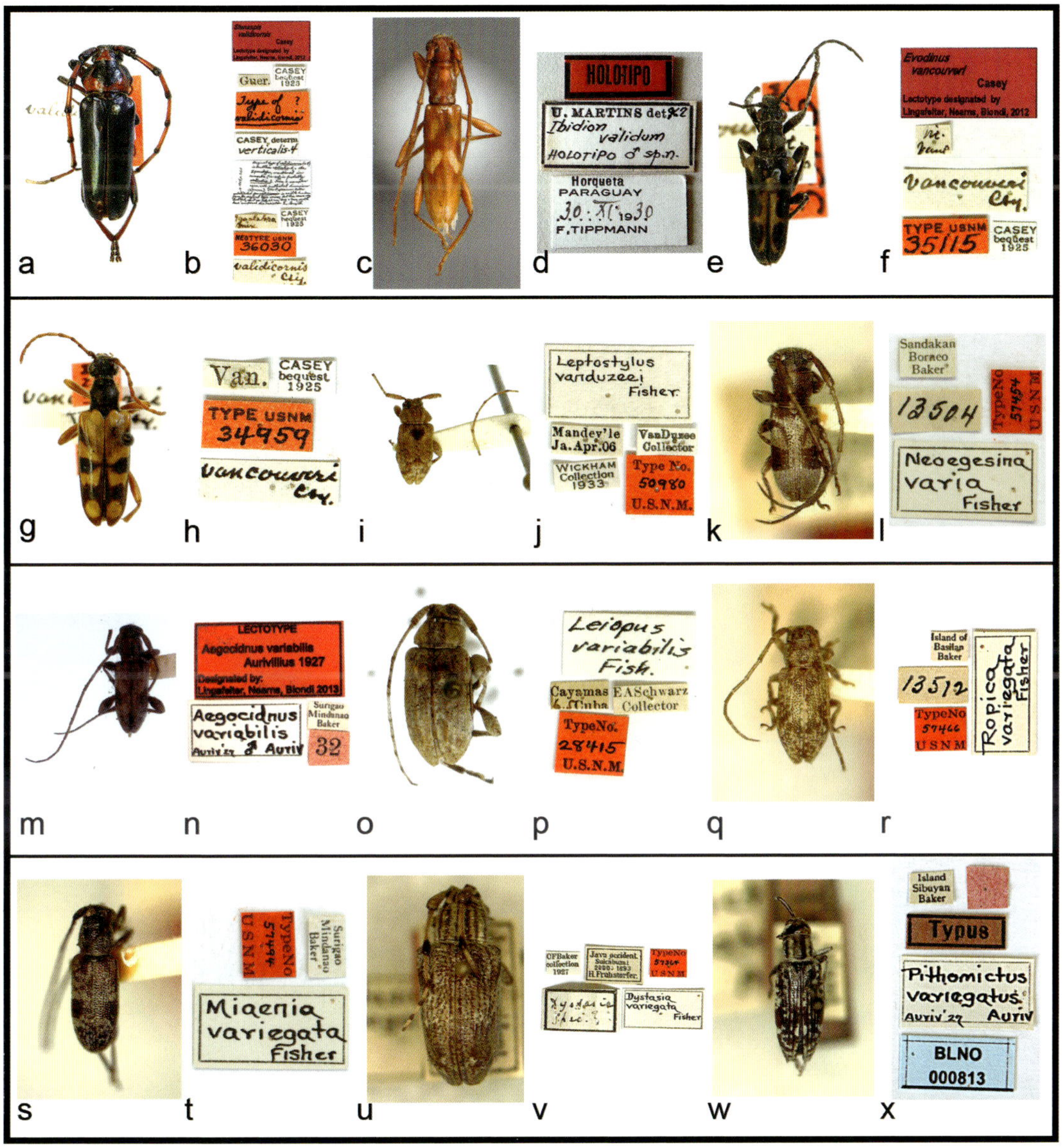

Figure 178. Primary types of Cerambycidae of the Smithsonian Institution (for each pair of images, dorsal habitus left, labels right): **(a, b)** *Stenaspis validicornis* Casey, 1912, Lectotype; **(c, d)** *Ibidion validum* Martins, 1962, Holotype; **(e, f)** *Evodinus vancouveri* Casey, 1913, Lectotype; **(g, h)** *Xestoleptura vancouveri* Casey, 1913, Holotype; **(i, j)** *Leptostylus vanduzeei* Fisher, 1935, Holotype; **(k, l)** *Neoegesina varia* Fisher, 1925, Holotype; **(m, n)** *Aegocidnus variabilis* Aurivillius, 1927, Lectotype; **(o, p)** *Leiopus variabilis* Fisher, 1925, Holotype; **(q, r)** *Ropica variegata* Fisher, 1925, Holotype; **(s, t)** *Miaenia variegata* Fisher, 1925, Holotype; **(u, v)** *Dystasia variegata* Fisher, 1936, Holotype; **(w, x)** *Pithomictus ? variegatus* Aurivillius, 1927, Syntype.

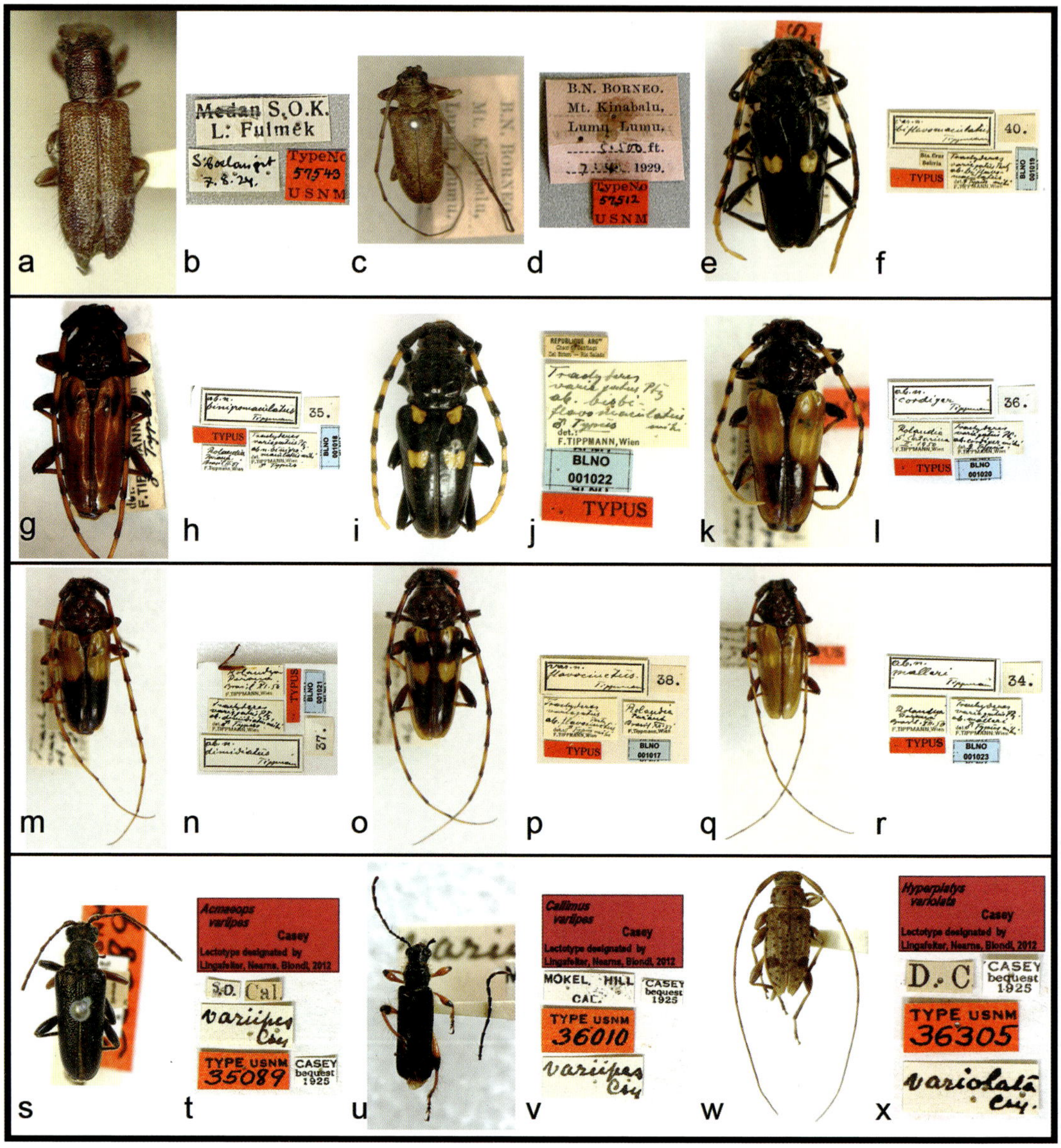

Figure 179. Primary types of Cerambycidae of the Smithsonian Institution (for each pair of images, dorsal habitus left, labels right): **(a, b)** *Emeopedus variegatus* Fisher, 1927, Holotype; **(c, d)** *Anhammus variegatus* Fisher, 1935, Holotype; **(e, f)** *Trachyderes variegatus* var. *biflavomaculatus* Tippmann, 1953, Holotype; **(g, h)** *Trachyderes variegatus* var. *binigromaculatus* Tippmann, 1953, Holotype; **(i, j)** *Trachyderes variegatus* var. *bisbiflavomaculatus* Tippmann, 1953, Holotype; **(k, l)** *Trachyderes variegatus* var. *cordiger* Tippmann, 1953, Holotype; **(m, n)** *Trachyderes variegatus* var. *dimidiatus* Tippmann, 1953, Holotype; **(o, p)** *Trachyderes variegatus* var. *flavocinctus* Tippmann, 1953, Holotype; **(q, r)** *Trachyderes variegatus* var. *malleri* Tippmann, 1953, Holotype; **(s, t)** *Acmaeops variipes* Casey, 1891, Lectotype; **(u, v)** *Callimus variipes* Casey, 1912, Lectotype; **(w, x)** *Hyperplatys variolata* Casey, 1913, Lectotype.

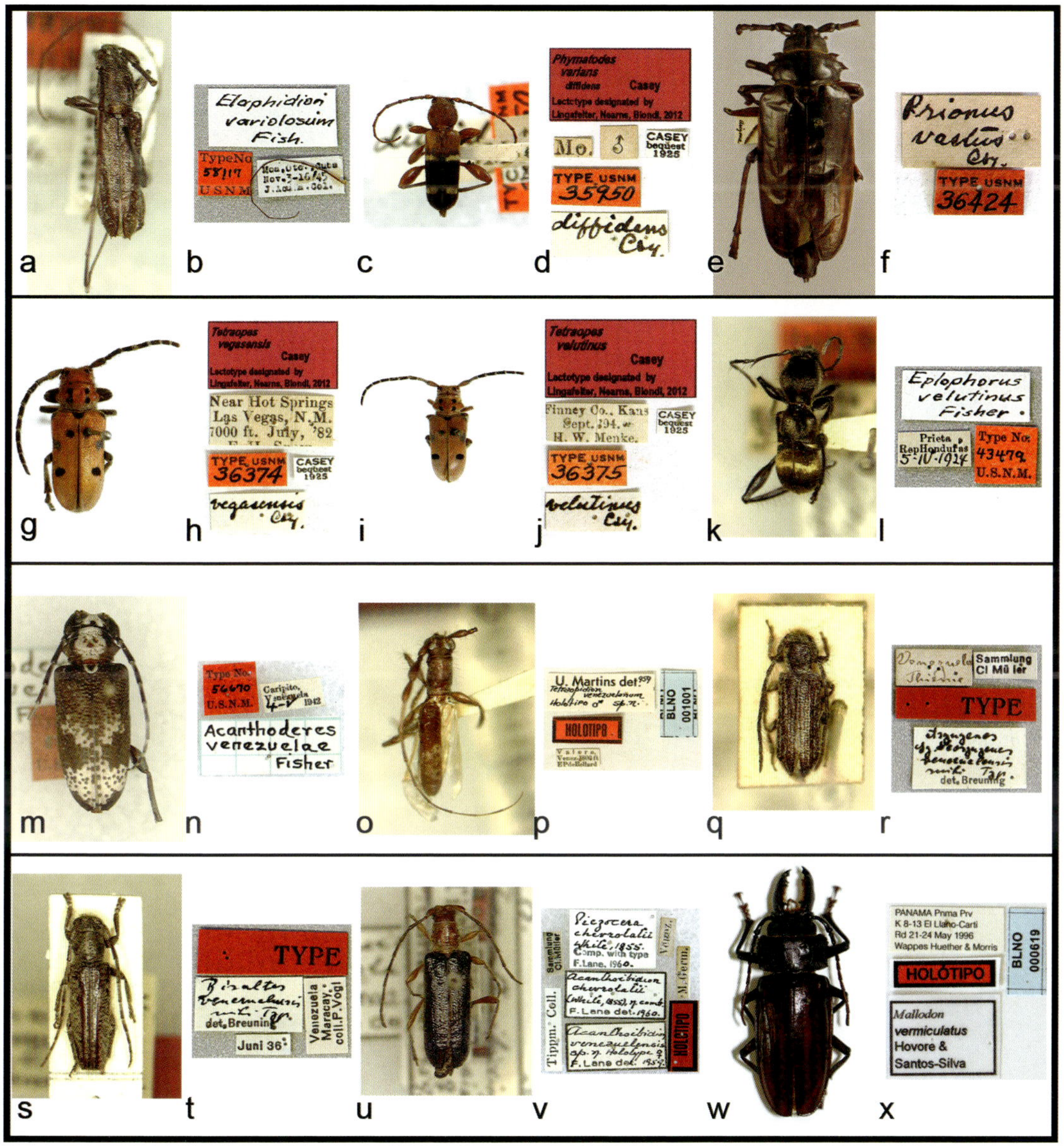

Figure 180. Primary types of Cerambycidae of the Smithsonian Institution (for each pair of images, dorsal habitus left, labels right): **(a, b)** *Elaphidion variolosum* Fisher, 1947, Holotype; **(c, d)** *Phymatodes (Poecilium) varius diffidens* Casey, 1912, Lectotype; **(e, f)** *Prionus (Prionus) vastus* Casey, 1912, Holotype; **(g, h)** *Tetraopes vegasensis* Casey, 1913, Lectotype; **(i, j)** *Tetraopes velutinus* Casey, 1913, Lectotype; **(k, l)** *Eplophorus velutinus* Fisher, 1931, Holotype; **(m, n)** *Acanthoderes venezuelae* Fisher, 1944, Holotype; **(o, p)** *Tetraopidion venezuelanum* Martins, 1960, Holotype; **(q, r)** *Asyngenes (Neosyngenes) venezuelensis* Breuning, 1943, Holotype; **(s, t)** *Bisaltes venezuelensis* Breuning, 1943, Holotype; **(u, v)** *Acanthoibidion venezuelensis* Lane, 1959, Holotype; **(w, x)** *Mallodon vermiculatus* Hovore & Santos-Silva, 2004, Holotype.

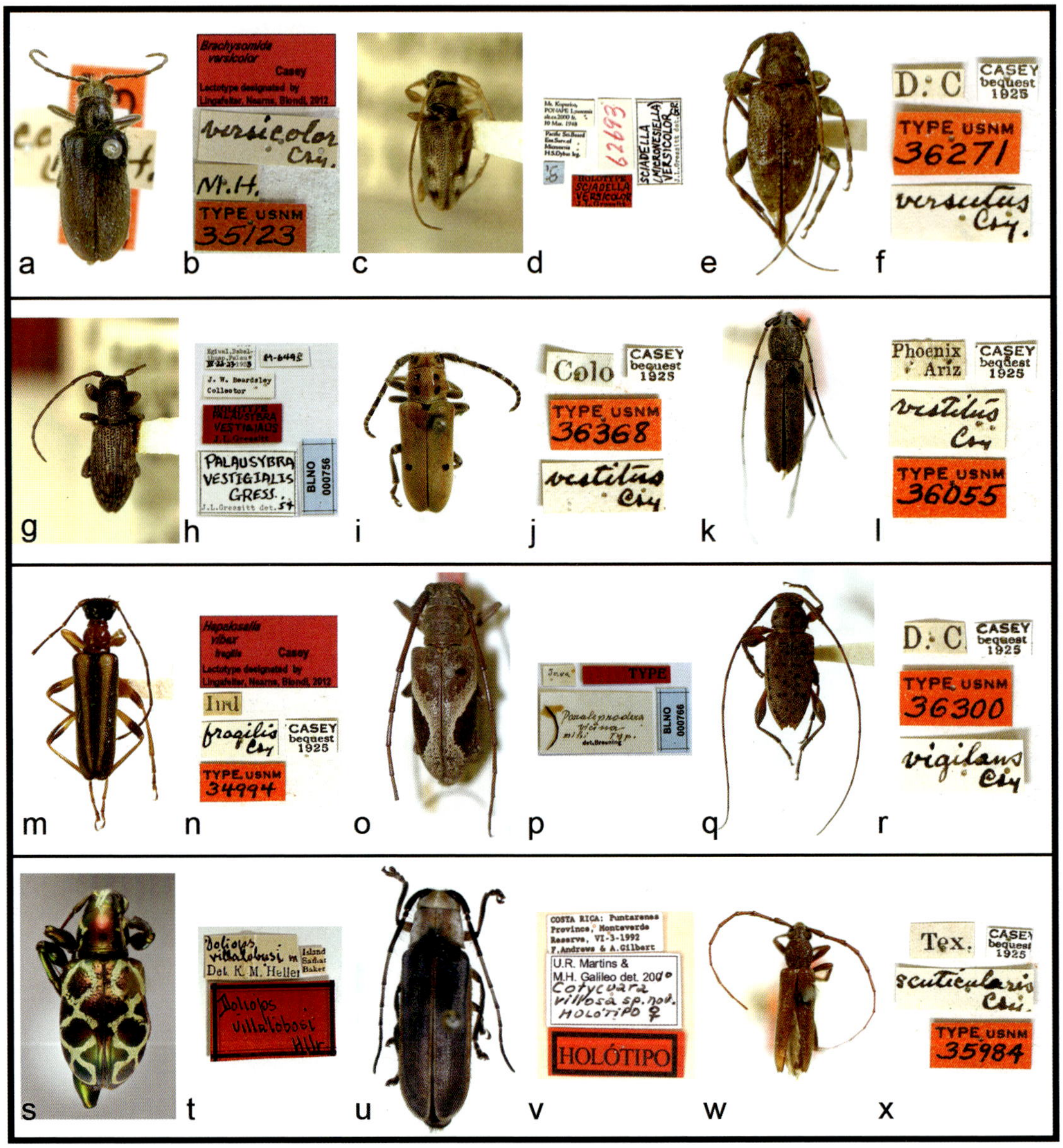

Figure 181. Primary types of Cerambycidae of the Smithsonian Institution (for each pair of images, dorsal habitus left, labels right): **(a, b)** *Brachysomida versicolor* Casey, 1913, Lectotype; **(c, d)** *Sciadella (Micronesiella) versicolor* Gressitt, 1956, Holotype; **(e, f)** *Astylidius versutus* Casey, 1913, Holotype; **(g, h)** *Palausybra vestigialis* Gressitt, 1956, Holotype; **(i, j)** *Tetraopes vestitus* Casey, 1913, Holotype; **(k, l)** *Osmidus vestitus* Casey, 1924, Holotype; **(m, n)** *Hapalosalia vibex fragilis* Casey, 1913, Lectotype; **(o, p)** *Paraleprodera vicina* Breuning, 1940, Holotype; **(q, r)** *Hyperplatys vigilans* Casey, 1913, Holotype; **(s, t)** *Doliops villalobosi* Heller, 1926, Holotype; **(u, v)** *Cotycuara villosa* Martins & Galileo, 2011, Holotype; **(w, x)** *Hypermallus villosus scuticularis* Casey, 1912, Holotype.

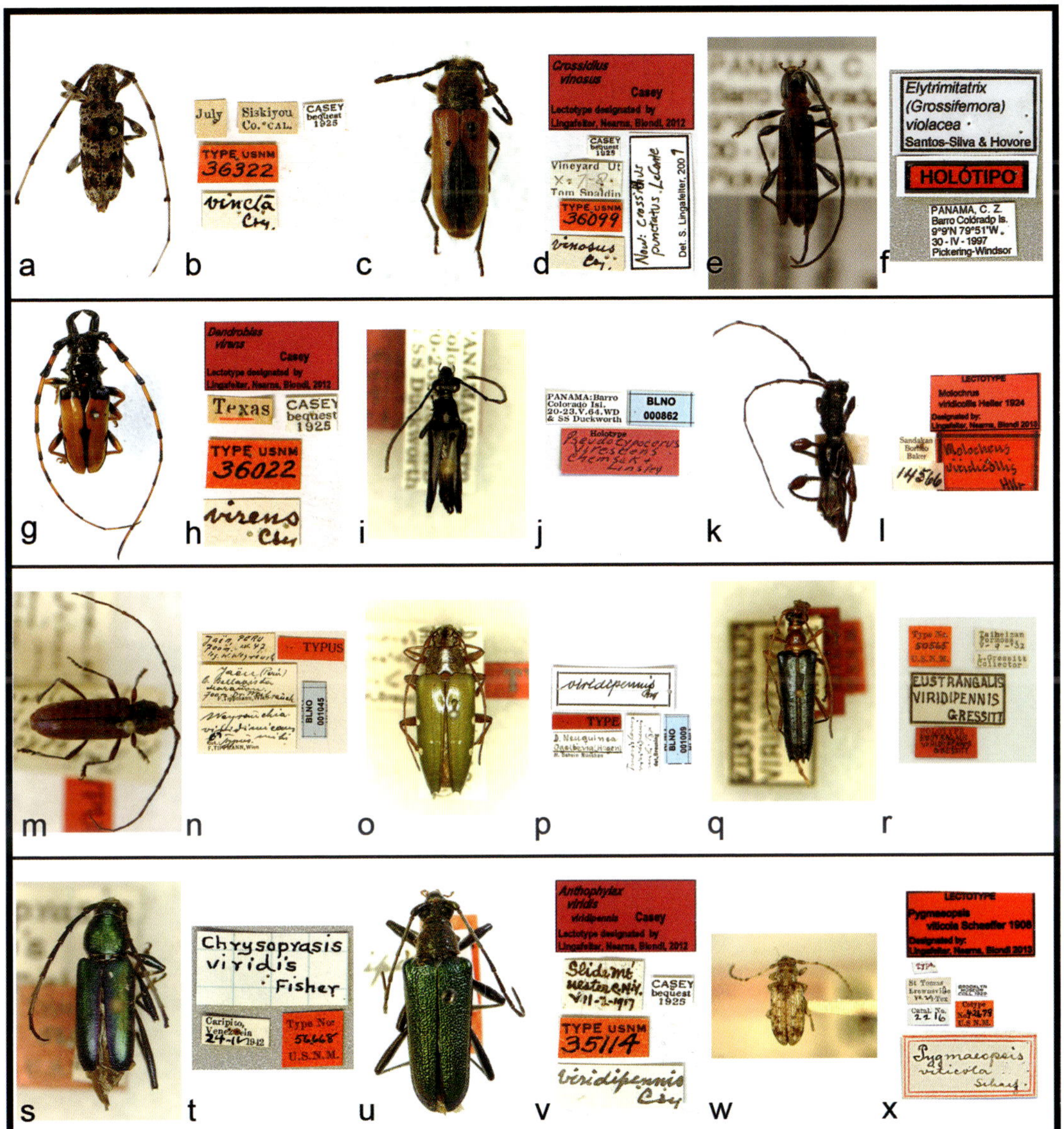

Figure 182. Primary types of Cerambycidae of the Smithsonian Institution (for each pair of images, dorsal habitus left, labels right): **(a, b)** *Graphisurus (Canonura) vinctus* Casey, 1913, Holotype; **(c, d)** *Crossidius vinosus* Casey, 1924, Lectotype; **(e, f)** *Elytrimitatrix (Grossifemora) violacea* Santos-Silva & Hovore, 2008, Holotype; **(g, h)** *Dendrobias virens* Casey, 1912, Lectotype; **(i, j)** *Pseudotypocerus virescens* Chemsak & Linsley, 1976, Holotype; **(k, l)** *Molochrus viridicollis* Heller, 1924, Lectotype; **(m, n)** *Weyrauchia viridimicans* Tippmann, 1953, Holotype; **(o, p)** *Tmesisternus viridipennis* Breuning, 1940, Holotype; **(q, r)** *Eustrangalis viridipennis* Gressitt, 1935, Holotype; **(s, t)** *Chrysoprasis viridis* Fisher, 1944, Holotype; **(u, v)** *Anthophilax viridis viridipennis* Casey, 1918, Lectotype; **(w, x)** *Pygmaeopsis viticola* Schaeffer, 1908, Lectotype.

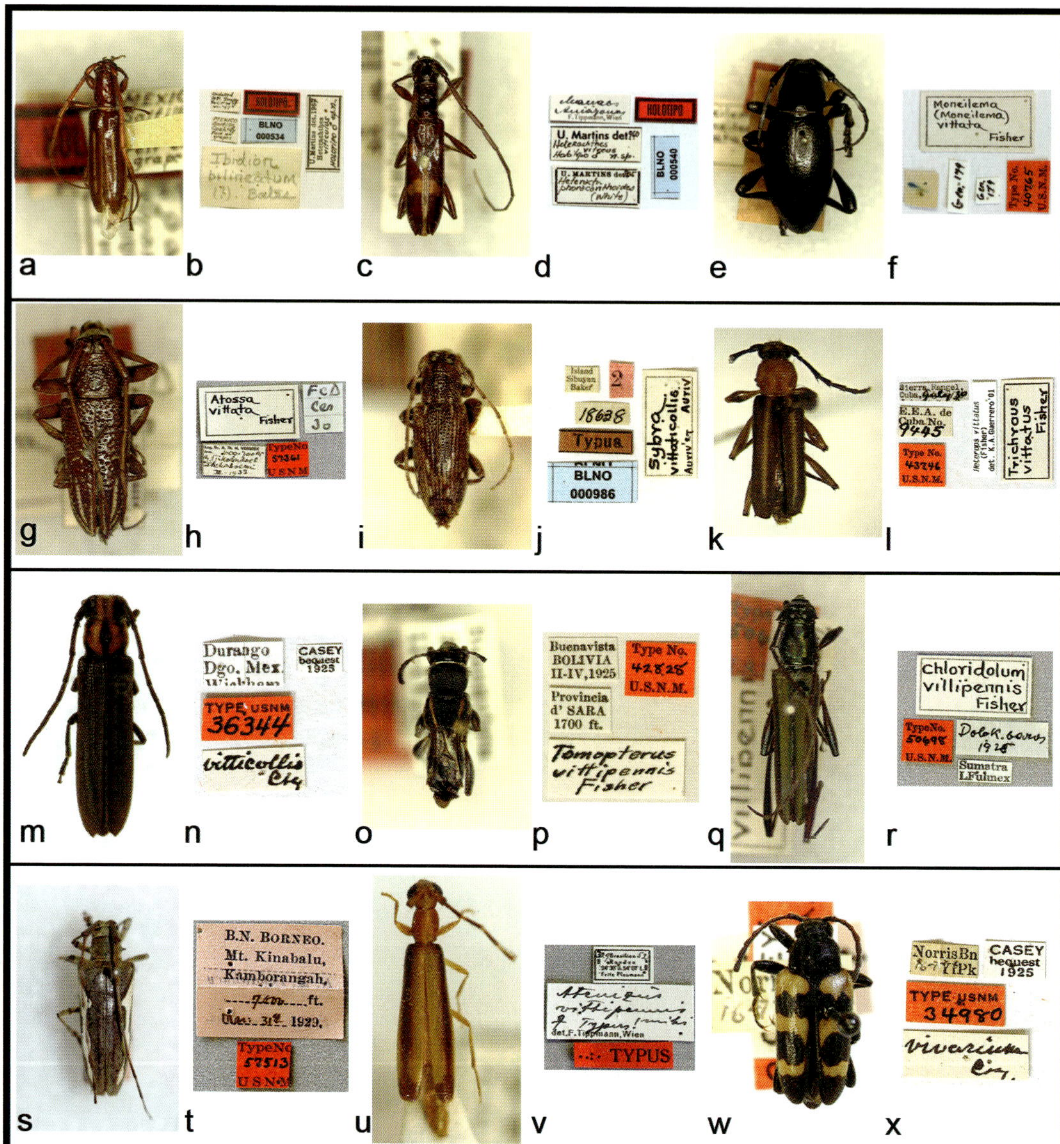

Figure 183. Primary types of Cerambycidae of the Smithsonian Institution (for each pair of images, dorsal habitus left, labels right): **(a, b)** *Heterachthes viticulus* Martins, 1970, Holotype; **(c, d)** *Heterachthes vitreus* Martins, 1962, Holotype; **(e, f)** *Moneilema (Moneilema) vittata* Fisher, 1928, Holotype; **(g, h)** *Atossa vittata* Fisher, 1936, Holotype; **(i, j)** *Sybra vittaticollis* Aurivillius, 1927, Syntype; **(k, l)** *Trichrous vittatus* Fisher, 1932, Holotype; **(m, n)** *Mecas vitticollis* Casey, 1913, Holotype; **(o, p)** *Tomopterus vittipennis* Fisher, 1930, Holotype; **(q, r)** *Chloridolum vittipennis* Fisher, 1934, Holotype; **(s, t)** *Epepeotes vittipennis* Fisher, 1935, Holotype; **(u, v)** *Atenizus vittipennis* Tippmann, 1960, Holotype; **(w, x)** *Judolia vivarium* Casey, 1924, Holotype.

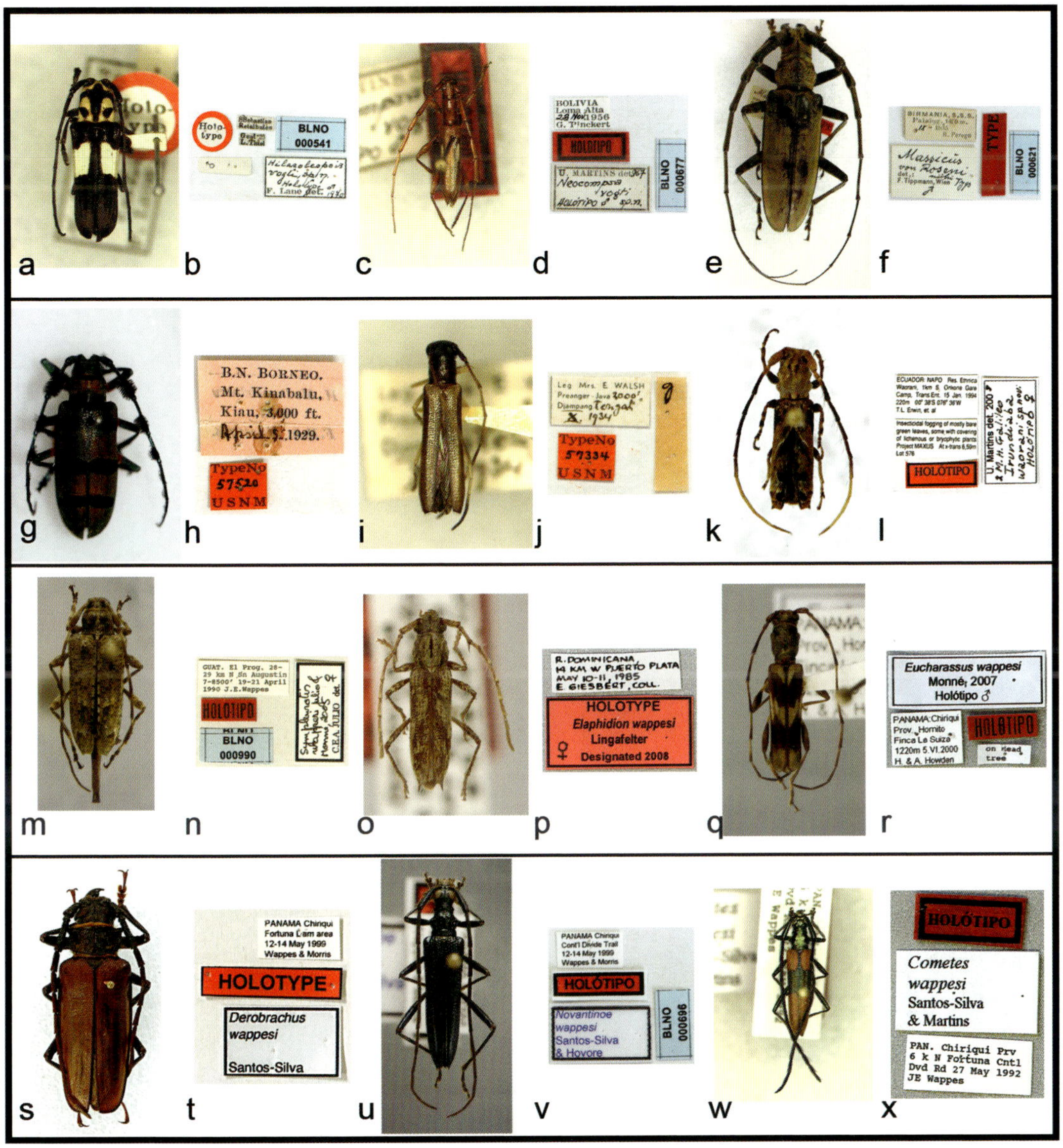

Figure 184. Primary types of Cerambycidae of the Smithsonian Institution (for each pair of images, dorsal habitus left, labels right): **(a, b)** *Hilaroleopsis vogti* Lane, 1970, Holotype; **(c, d)** *Neocompsa vogti* Martins, 1970, Holotype; **(e, f)** *Massicus vonroseni* Tippmann, 1949, Holotype; **(g, h)** *Diastocera wallichi* var. *insularis* Fisher, 1935, Holotype; **(i, j)** *Oberea walshae* Fisher, 1937, Holotype; **(k, l)** *Irundiaba waorani* Martins & Galileo, 2008, Holotype; **(m, n)** *Sympleurotis wappesi* Julio & Monné, 2005, Holotype; **(o, p)** *Elaphidion wappesi* Lingafelter, 2008, Holotype; **(q, r)** *Eucharassus wappesi* Monné, 2007, Holotype; **(s, t)** *Derobrachus wappesi* Santos-Silva, 2007, Holotype; **(u, v)** *Novantinoe wappesi* Santos-Silva & Hovore, 2007, Holotype; **(w, x)** *Cometes wappesi* Santos-Silva & Martins, 2004, Holotype.

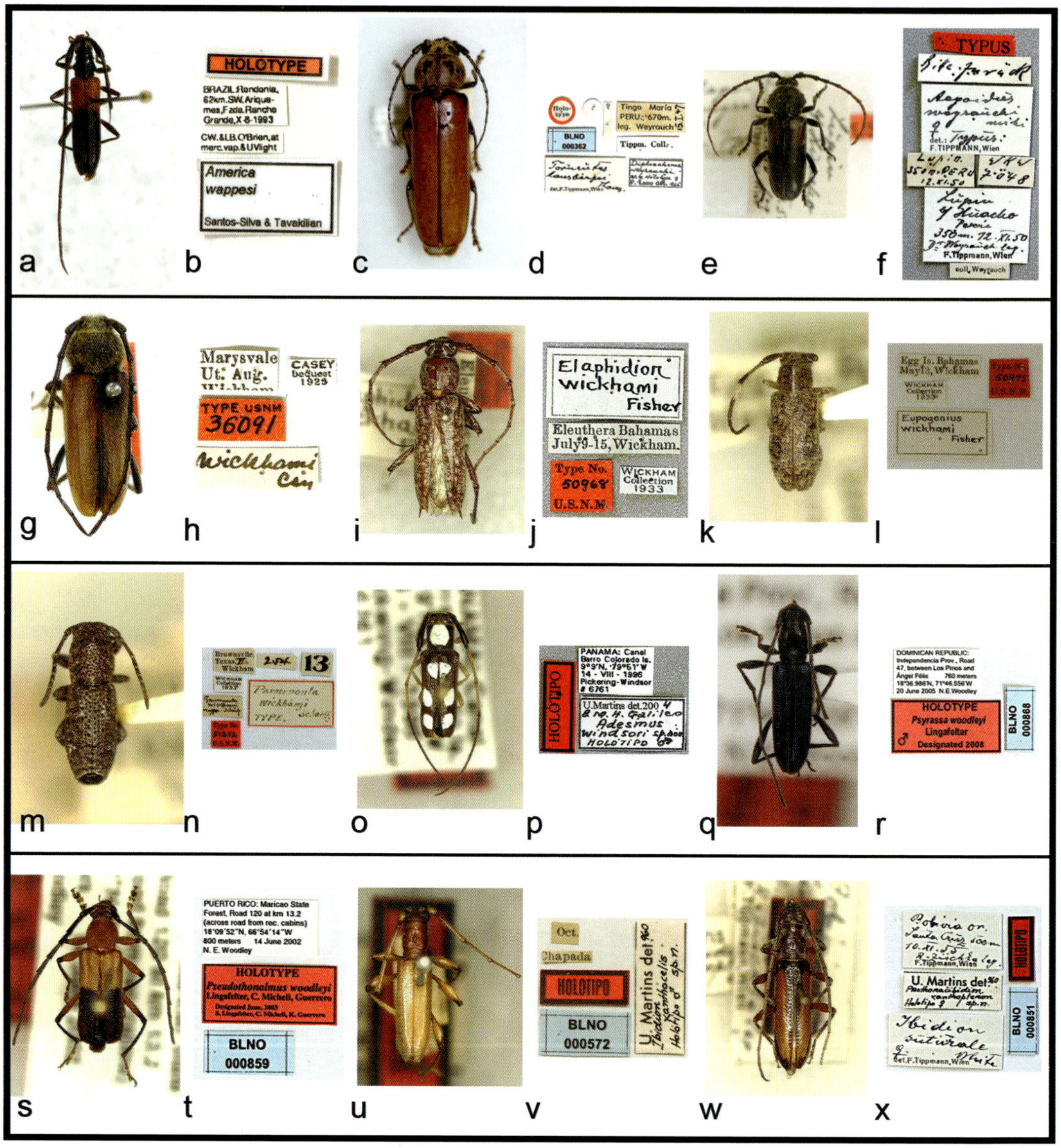

Figure 185. Primary types of Cerambycidae of the Smithsonian Institution (for each pair of images, dorsal habitus left, labels right): **(a, b)** *America wappesi* Santos-Silva & Tavakilian, 2009, Holotype; **(c, d)** *Diploschema weyrauchi* Lane, 1966, Holotype; **(e, f)** *Aegoidus weyrauchi* Tippmann, 1953, Holotype; **(g, h)** *Crossidius wickhami* Casey, 1912, Holotype; **(i, j)** *Elaphidion wickhami* Fisher, 1935, Holotype; **(k, l)** *Eupogonius wickhami* Fisher, 1935, Holotype; **(m, n)** *Parmenonta wickhami* Schaeffer, 1908, Holotype; **(o, p)** *Adesmus windsori* Martins & Galileo, 2004, Holotype; **(q, r)** *Psyrassa woodleyi* Lingafelter, 2008, Holotype; **(s, t)** *Pseudothonalmus woodleyi* Lingafelter, Micheli & Guerrero, 2004, Holotype; **(u, v)** *Ibidion xanthocelis* Martins, 1962, Holotype; **(w, x)** *Prothoracibidion xanthopterum* Martins, 1962, Holotype.

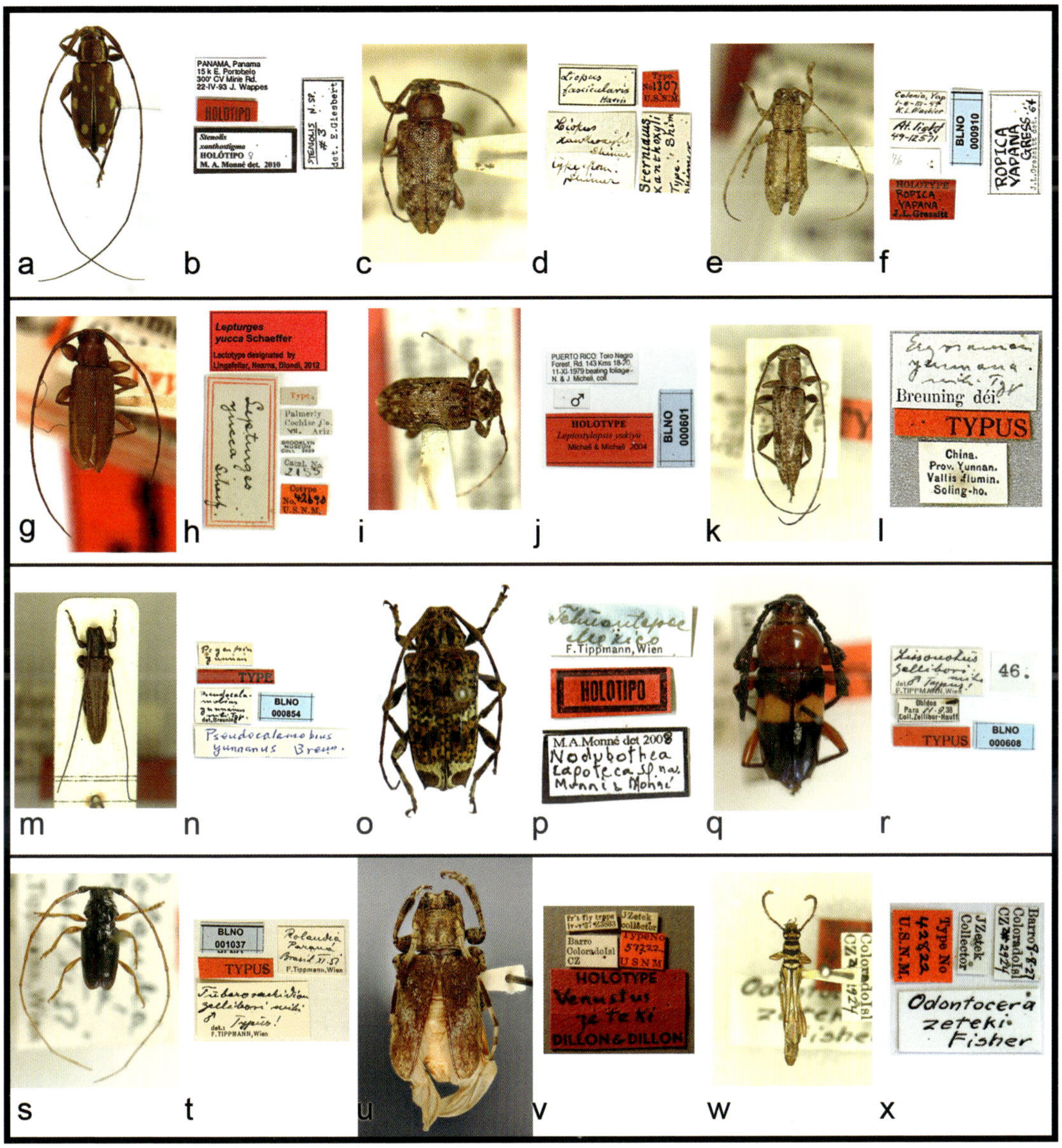

Figure 186. Primary types of Cerambycidae of the Smithsonian Institution (for each pair of images, dorsal habitus left, labels right): **(a, b)** *Stenolis xanthostigma* Monné, 2011, Holotype; **(c, d)** *Liopus xanthoxyli* Shimer, 1868, Holotype; **(e, f)** *Ropica yapana* Gressitt, 1956, Holotype; **(g, h)** *Lepturges yucca* Schaeffer, 1905, Lectotype; **(i, j)** *Leptostylopsis yukiyu* Micheli & Micheli, 2004, Holotype; **(k, l)** *Eryssamena yunnana* Breuning, 1957, Holotype; **(m, n)** *Pseudocalamobius yunnanus* Breuning, 1942, Holotype; **(o, p)** *Nodubothea zapoteca* Monné & Monné, 2008, Holotype; **(q, r)** *Lissonotus zellibori* Tippmann, 1953, Holotype; **(s, t)** *Tuberorachidion zellibori* Tippmann, 1953, Holotype; **(u, v)** *Venustus zeteki* Dillon & Dillon, 1945, Holotype; **(w, x)** *Odontocera zeteki* Fisher, 1930, Holotype.

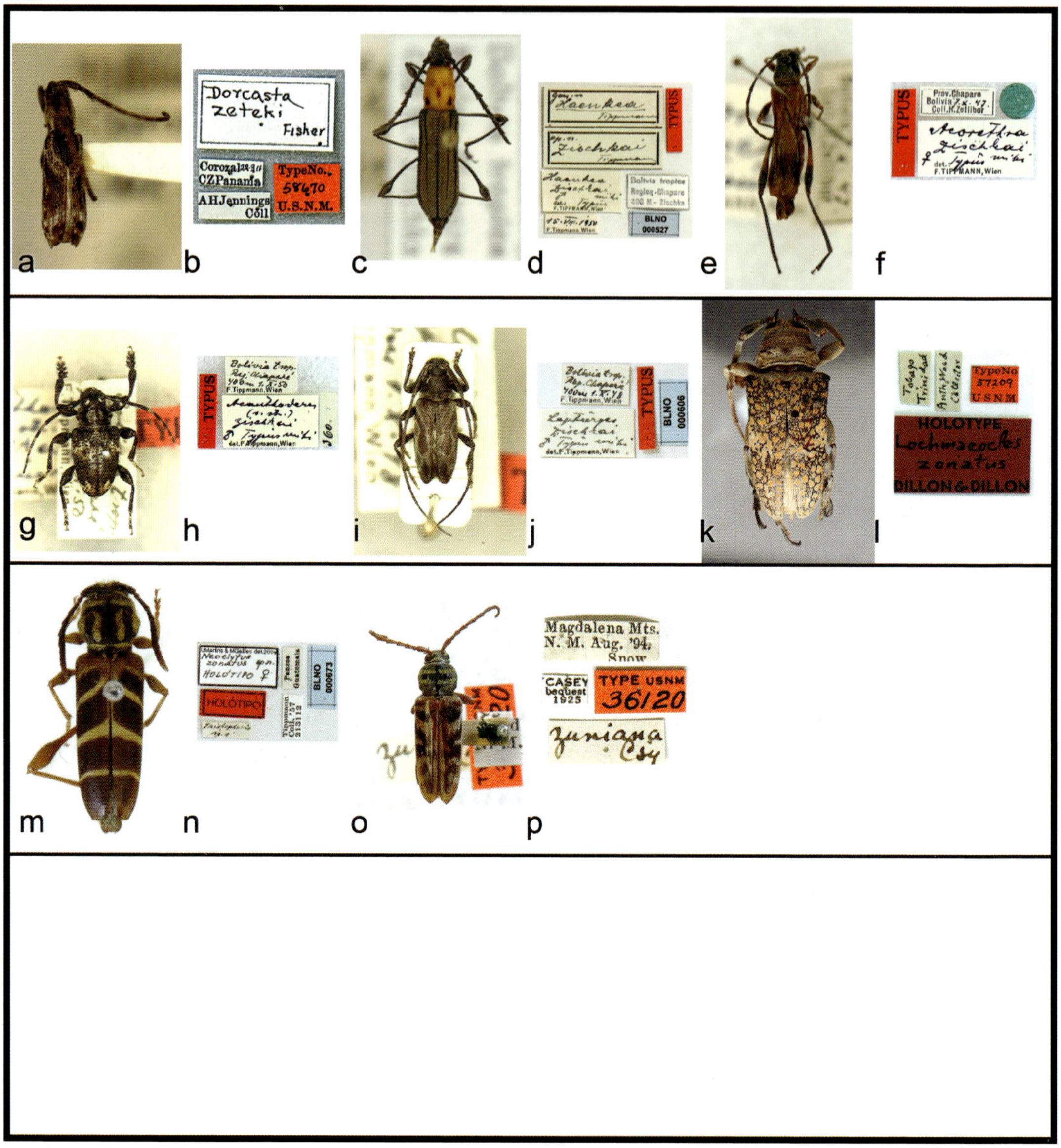

Figure 187. Primary types of Cerambycidae of the Smithsonian Institution (for each pair of images, dorsal habitus left, labels right): **(a, b)** *Dorcasta zeteki* Fisher, 1947, Holotype; **(c, d)** *Haenkea zischkai* Tippmann, 1953, Holotype; **(e, f)** *Acorethra zischkai* Tippmann, 1960, Holotype; **(g, h)** *Acanthoderes zischkai* Tippmann, 1960, Holotype; **(i, j)** *Lepturges zischkai* Tippmann, 1960, Holotype; **(k, l)** *Lochmaeocles zonatus* Dillon & Dillon, 1946, Holotype; **(m, n)** *Neoclytus zonatus* Martins & Galileo, 2008, Holotype; **(o, p)** *Cyllene zuniana* Casey, 1912, Holotype.

ochracea **Fisher, 1927c**: 23 (Fig. 122c, d), Holotype
Original combination: *Sphagoeme ochracea* Fisher
Current name: same
Current placement: Cerambycidae: Cerambycinae: Oemini
Type locality: Panama: Panama Province, Cano Saddle, Gatun Lake

ochraceovittata **Aurivillius, 1922**: 423 (Fig. 122e, f), Syntype
Original combination: *Nyctimene ochraceovittata* Aurivillius
Current name: *Nyctimenius ochraceovittata* (Aurivillius, 1922)
Current placement: Cerambycidae: Lamiinae: Nyctimeniini
Type locality: Philippines: Mindanao, Iligan
Remarks: Other syntype(s) are present in the Naturhistoriska Riksmuseet, Stockholm, Sweden.

ochreipennis **Casey, 1913**: 278 (Fig. 122g, h), Lectotype
Original combination: *Ophistomis ochreipennis* Casey
Current name: *Strangalia famelica* Newman, 1841
Current placement: Cerambycidae: Lepturinae: Lepturini
Type locality: No locality label or published data.
Remarks: This specimen is herein designated as the lectotype in order to stabilize the taxonomy and facilitate further identifications of this species.

ochreoapicalis **Breuning, 1943c**: 12 (Fig. 122i, j), Neotype
Original combination: *Stenellipsis ochreoapicalis* Breuning
Current name: same
Current placement: Cerambycidae: Lamiinae: Acanthocinini
Type locality: Australia: Sydney
Remarks: According to Tavakilian and Chevillotte (2013), the holotype from the Hamburg Museum was probably destroyed in WWII. Since this specimen is identical to the original description, it is here designated as the neotype.

ochreoguttata **Breuning, 1942a**: 125 (Fig. 122k, l), Holotype
Original combination: *Ichthyodes (Ichthyodes) ochreoguttata* Breuning
Current name: *Ichthyodes ochreoguttata* Breuning, 1942a
Current placement: Cerambycidae: Lamiinae: Apomecynini
Type locality: Philippines: Mindanao, Surigao

ochreomaculatus **Breuning, 1942a**: 123 (Fig. 122m, n), Holotype
Original combination: *Tmesisternus (Arrhenotus) ochreomaculatus* Breuning
Current name: *Tmesisternus ochreomaculatus* Breuning, 1942a
Current placement: Cerambycidae: Lamiinae: Tmesisternini
Type locality: Indonesia: Papua, Irian Jaya, Upper Setekwa, Snow Mtns., 2000–3000 ft.

octavus **Aurivillius, 1922**: 420 (Fig. 122o, p), Holotype
Original combination: *Demonax octavus* Aurivillius
Current name: same
Current placement: Cerambycidae: Cerambycinae: Clytini
Type locality: Malaysia: Borneo, Sandakan

octomaculatus **Breuning, 1940a**: 545 (Fig. 122q, r), Holotype
Original combination: *Olenecamptus octomaculatus* Breuning
Current name: same
Current placement: Cerambycidae: Lamiinae: Dorcaschematini
Type locality: Vietnam: Annam

ocularis **Hammond & Williams, 2011**: 280 (Fig. 122s, t), Holotype
Original combination: *Callimoxys ocularis* Hammond & Williams
Current name: same
Current placement: Cerambycidae: Cerambycinae: Stenopterini
Type locality: USA: Iowa, Polk Co., W. Saylorville Lake

oculata **Casey, 1912**: 309 (Fig. 122u, v), Holotype
Original combination: *Malthophia oculata* Casey
Current name: *Styloxus fulleri californicus* (Fall, 1901)
Current placement: Cerambycidae: Cerambycinae: Methiini
Type locality: USA: California, San Diego

oculata **Casey, 1913**: 202 (Fig. 122w, x), Holotype
Original combination: *Centrodera oculata* Casey
Current name: same
Current placement: Cerambycidae: Lepturinae: Rhagiini
Type locality: USA: California, Mt. Diablo, Contra Costa Co.

oculea **Casey, 1913**: 204 (Fig. 123a, b), Holotype
Original combination: *Ortholeptura oculea* Casey
Current name: *Ortholeptura valida* (LeConte, 1857)
Current placement: Cerambycidae: Lepturinae: Lepturini
Type locality: USA: California, Truckee

olivaceotincta **Tippmann, 1953d**: 343 (Fig. 123c, d), Holotype
Original combination: *Oreodera olivaceotincta* Tippmann
Current name: same
Current placement: Cerambycidae: Lamiinae: Acanthoderini
Type locality: Bolivia: Cochabamba, Reg. Chapare 400 m.

omissus **Casey, 1913**: 377 (Fig. 123e, f), Holotype
Original combination: *Tetraopes omissus* Casey
Current name: *Tetraopes basalis* LeConte, 1852
Current placement: Cerambycidae: Lamiinae: Tetraopini
Type locality: USA: California

opaca **Casey, 1913**: 289 (Fig. 123g, h), Holotype
Original combination: *Moneilema (Collapteryx) opaca* Casey
Current name: *Moneilema subrugosum* Bland, 1862
Current placement: Cerambycidae: Lamiinae: Moneilemini
Type locality: USA

opaca **Fisher, 1925a**: 236 (Fig. 123i, j), Holotype
Original combination: *Mynonebra opaca* Fisher
Current name: same
Current placement: Cerambycidae: Lamiinae: Rhodopini
Type locality: Philippines: Negros Oriental, Cuernos Mountains

opaca **Fisher, 1936a**: 175 (Fig. 123k, l), Holotype
Original combination: *Epania opaca* Fisher
Current name: *Molorchoepania opaca* (Fisher, 1936a)
Current placement: Cerambycidae: Cerambycinae: Molorchini
Type locality: Indonesia: Java, Koebangkang Koeng, Zuid-Banjoemas, 25 M

opacicornis **Linsley, 1957c**: 285 (Fig. 123m, n), Holotype
Original combination: *Aneflomorpha opacicornis* Linsley
Current name: same
Current placement: Cerambycidae: Cerambycinae: Elaphidiini
Type locality: USA: Texas, Cline

opacipennis **Casey, 1912**: 311 (Fig. 123o, p), Lectotype
Original combination: *Callimus opacipennis* Casey
Current name: *Lampropterus ruficollis* (LeConte, 1873)
Current placement: Cerambycidae: Cerambycinae: Hyboderini
Type locality: USA: California, San Diego
Remarks: This specimen is herein designated as the lectotype in order to stabilize the taxonomy and facilitate further identifications of this species.

opacus **Chemsak & Linsley, 1976b**: 47 (Fig. 123q, r), Holotype
Original combination: *Leptochoriolaus opacus* Chemsak & Linsley
Current name: same
Current placement: Cerambycidae: Lepturinae: Lepturini
Type locality: Costa Rica: Cartago, Turrialba

opalescens **Fisher, 1935c**: 630 (Fig. 123s, t), Holotype
Original combination: *Astathes opalescens* Fisher
Current name: *Astathes (Tetraophthalmus) japonica* (Thomson, 1857)
Current placement: Cerambycidae: Lamiinae: Astathini
Type locality: Malaysia: Borneo, Sabah, near Mt. Kinabalu, Tenompok Pan, 4500 ft.

opuntiae **Fisher, 1928b**: 1 (Fig. 123u, v), Holotype
Original combination: *Moneilema (Collapteryx) opuntiae* Fisher
Current name: *Moneilema opuntiae* Fisher, 1928b
Current placement: Cerambycidae: Lamiinae: Moneilemini
Type locality: Mexico: San Luis Potosí

orbiceps **Casey, 1924**: 216 (Fig. 123w, x), Holotype
Original combination: *Prionus (Prionus) orbiceps* Casey
Current name: *Prionus californicus* Motschulsky, 1845
Current placement: Cerambycidae: Prioninae: Prionini
Type locality: USA: Utah, Eureka

oregonensis **Casey, 1913**: 214 (Fig. 124a, b), Holotype
Original combination: *Stenocorus (Stenocorus) oregonensis* Casey
Current name: *Stenocorus vestitus* (Haldeman, 1847)
Current placement: Cerambycidae: Lepturinae: Rhagiini
Type locality: USA: Oregon, Clackamas Co. (Big Sandy Camp), Bull Run

oregonensis latus **Casey, 1924**: 290 (Fig. 124c, d), Lectotype
Original combination: *Monochamus oregonensis latus* Casey
Current name: *Monochamus clamator latus* Casey, 1924
Current placement: Cerambycidae: Lamiinae: Monochamini
Type locality: USA: California, Truckee
Remarks: This specimen is herein designated as the lectotype in order to stabilize the taxonomy and facilitate further identifications of this species.

oreora **Gressitt, 1956**: 142 (Fig. 124e, f), Holotype
Original combination: *Sybra oreora* Gressitt
Current name: same
Current placement: Cerambycidae: Lamiinae: Apomecynini
Type locality: Palau: Koror Island, limestone ridge south of inlet

orientalis **Aurivillius, 1907**: 119 (Fig. 124g, h), Syntype
Original combination: *Baraeus orientalis* Aurivillius
Current name: same
Current placement: Cerambycidae: Lamiinae: Pteropliini
Type locality: Tanzania: Ndanda
Remarks: Tavakilian and Chevillotte (2013) state that syntypes are present in the Naturhistoriska Riksmuseet, Stockholm, Sweden.

***orientalis* Breuning, 1940c**: 148 (Fig. 124i, j), Holotype
Original combination: *Epopea orientalis* Breuning
Current name: same
Current placement: Cerambycidae: Lamiinae: Pteropliini
Type locality: Kenya: Thika River Colony
Remarks: Although Tavakilian and Chevillotte (2013) indicate the holotype is in Basel, Switzerland, the Smithsonian specimen matches the original description and data.

***ornata* Fisher, 1925a**: 208 (Fig. 124k, l), Holotype
Original combination: *Enispia ornata* Fisher
Current name: *Athylia ornata* (Fisher, 1925a)
Current placement: Cerambycidae: Lamiinae: Pteropliini
Type locality: Philippines: Luzon, Mt. Banahao

***ornata* Fisher, 1925a**: 216 (Fig. 124m, n), Holotype
Original combination: *Neoegesina ornata* Fisher
Current name: *Egesina ornata* (Fisher, 1925a)
Current placement: Cerambycidae: Lamiinae: Pteropliini
Type locality: Malaysia: Borneo, Sandakan

***ornata* Fisher, 1935b**: 189 (Fig. 124o, p), Holotype
Original combination: *Pseudoeme ornata* Fisher
Current name: *Ochrus ornatus* (Fisher, 1935b)
Current placement: Cerambycidae: Cerambycinae: Hesperophanini
Type locality: Saint Lucia

***ornata* Fisher, 1935c**: 607 (Fig. 124q, r), Holotype
Original combination: *Combe ornata* Fisher
Current name: *Cereopsius guttulatus* Aurivillius, 1924
Current placement: Cerambycidae: Lamiinae: Lamiini
Type locality: Malaysia: Borneo, Sabah, near Mt. Kinabalu, Tenompok-Kian 4–3500 (sic) ft.

***ornatipennis* Fisher, 1935b**: 200 (Fig. 124s, t), Holotype
Original combination: *Spalacopsis ornatipennis* Fisher
Current name: same
Current placement: Cerambycidae: Lamiinae: Agapanthiini
Type locality: Jamaica: Mandeville

***ornatipennis* Tippmann, 1960**: 109 (Fig. 124u, v), Holotype
Original combination: *Ectenessa ornatipennis* Tippmann
Current name: same
Current placement: Cerambycidae: Cerambycinae: Ectenessini
Type locality: Bolivia: Cochabamba, Reg. Chapare

***ornator* Tippmann, 1960**: 204 (Fig. 124w, x), Holotype
Original combination: *Adesmus ornator* Tippmann
Current name: *Phoebe ornator* (Tippmann, 1960)
Current placement: Cerambycidae: Lamiinae: Hemilophini
Type locality: Bolivia: Yungas del Palmar, 1250 m.

***ornatus* Fisher, 1928b**: 3 (Fig. 125a, b), Holotype
Original combination: *Leptostylus ornatus* Fisher
Current name: *Leptostylopsis ornatus* (Fisher, 1928b)
Current placement: Cerambycidae: Lamiinae: Acanthocinini
Type locality: Jamaica: Hope Gardens

***ornatus* Fisher, 1936a**: 182 (Fig. 125c, d), Holotype
Original combination: *Cereopsius ornatus* Fisher
Current name: *Cereopsius javanicus* Breuning, 1936b
Current placement: Cerambycidae: Lamiinae: Lamiini
Type locality: Indonesia: Java, Noesa Kambangan

***ornatus* Fisher, 1947c**: 34 (Fig. 125e, f), Holotype
Original combination: *Plectromerus ornatus* Fisher
Current name: same
Current placement: Cerambycidae: Cerambycinae: Plectromerini
Type locality: Cuba: Santiago de Cuba, Moa, Oriente

***ornatus* Tippmann, 1953d**: 354 (Fig. 125g, h), Holotype
Original combination: *Colobeutrypanus ornatus* Tippmann
Current name: same
Current placement: Cerambycidae: Lamiinae: Acanthocinini
Type locality: Costa Rica: Cariblanco

***ostenta* Casey, 1913**: 265 (Fig. 125i, j), Holotype
Original combination: *Strophiona ostenta* Casey
Current name: *Strophiona laeta* (LeConte, 1857)
Current placement: Cerambycidae: Lepturinae: Lepturini
Type locality: USA: New Mexico

***ovalis* Breuning, 1938d**: 31 (Fig. 125k, l), Holotype
Original combination: *Parabrimidius ovalis* Breuning
Current name: same
Current placement: Cerambycidae: Lamiinae: Parmenini
Type locality: Tanzania: Manow
Remarks: Tavakilian and Chevillotte (2013) suggest the holotype is present in the Institut Royal des Sciences Naturelles de Belgique, Belgium, but this specimen and its labels match the original description.

***ozophagus* Chemsak & Feller, 1988**: 188 (Fig. 125m, n), Holotype
Original combination: *Urgleptes ozophagus* Chemsak & Feller
Current name: same
Current placement: Cerambycidae: Lamiinae: Acanthocinini
Type locality: Belize: Twin Cays

***pacifica* Casey, 1913**: 249 (Fig. 125o, p), Holotype
Original combination: *Judolia pacifica* Casey
Current name: *Judolia instabilis* (Haldeman, 1847)
Current placement: Cerambycidae: Lepturinae: Lepturini
Type locality: USA: California

***pacificus* Casey, 1913**: 209 (Fig. 125q, r), Lectotype
Original combination: *Stenocorus (Stenocorus) pacificus* Casey
Current name: *Stenocorus nubifer* (LeConte, 1859)
Current placement: Cerambycidae: Lepturinae: Rhagiini
Type locality: USA: California, Lake Co.
Remarks: This specimen is herein designated as the lectotype in order to stabilize the taxonomy and facilitate further identifications of this species.

***pacificus* Casey, 1913**: 338 (Fig. 125s, t), Holotype
Original combination: *Graphisurus (Graphisurus) pacificus* Casey
Current name: *Acanthocinus obliquus* (LeConte, 1862)
Current placement: Cerambycidae: Lamiinae: Acanthocinini
Type locality: USA: California, Siskiyou Co.

***pacificus* Tippmann, 1960**: 142 (Fig. 125u, v), Holotype
Original combination: *Aegoidus pacificus* Tippmann
Current name: same
Current placement: Cerambycidae: Cerambycinae: Trachyderini
Type locality: Peru: Atocongo, 150 m.

***pagana* Aurivillius, 1926**: 93 (Fig. 125w, x), Holotype
Original combination: *Glenea (Glenea) pagana* Aurivillius
Current name: *Glenea pagana* Aurivillius, 1926
Current placement: Cerambycidae: Lamiinae: Saperdini
Type locality: Philippines: Luzon, Benguet, Baguio

***paganoides* Breuning, 1940d**: 423 (Fig. 126a, b), Holotype
Original combination: *Prosoplus (Escharodes) paganoides* Breuning
Current name: same
Current placement: Cerambycidae: Lamiinae: Pterolophini
Type locality: Papua New Guinea

***palauanus* Gressitt, 1956**: 108 (Fig. 126c, d), Holotype
Original combination: *Demonax palauanus* Gressitt
Current name: same
Current placement: Cerambycidae: Cerambycinae: Clytini
Type locality: Palau: East Ngatpang, 65 m, Babelthuap Island

***palauicola* Gressitt, 1956**: 158 (Fig. 126e, f), Holotype
Original combination: *Sciadella (Micronesiella) palauicola* Gressitt
Current name: *Sciades (Micronesiella) palauicola* (Gressitt, 1956)
Current placement: Cerambycidae: Lamiinae: Acanthocinini
Type locality: Palau: Caroline Island, Koror, 40 m., limestone ridge

***palawanica* Breuning, 1938b**: 292 (Fig. 126g, h), Holotype
Original combination: *Pterolophia palawanica* Breuning
Current name: *Pterolophia (Ale) palawanica* Breuning, 1938b
Current placement: Cerambycidae: Lamiinae: Pteropliini
Type locality: Philippines: Palawan
Remarks: Tavakilian and Chevillotte (2013) suggest the holotype is present in the Museum national d'Histoire naturelle, Paris, France, however, this specimen and its labels match the original description.

***pallida* Casey, 1913**: 366 (Fig. 126i, j), Lectotype
Original combination: *Oberea pallida* Casey
Current name: same
Current placement: Cerambycidae: Lamiinae: Phytoeciini
Type locality: USA: Maine, Monmouth
Remarks: Lectotype designated by Yanega 1996.

***pallida* Fisher, 1932**: 6 (Fig. 126k, l), Holotype
Original combination: *Methia pallida* Fisher
Current name: *Methia necydalea* (Fabricius, 1798)
Current placement: Cerambycidae: Cerambycinae: Methiini
Type locality: Haiti: Port-au-Prince

***pallidicornis* Fisher, 1932**: 58 (Fig. 126m, n), Holotype
Original combination: *Neoclytus pallidicornis* Fisher
Current name: same
Current placement: Cerambycidae: Cerambycinae: Clytini
Type locality: Haiti: La Vanneau

***pallidipennis* Martins & Galileo, 2007c**: 1145 (Fig. 126o, p), Holotype
Original combination: *Icimauna pallidipennis* Martins & Galileo
Current name: same
Current placement: Cerambycidae: Lamiinae: Hemilophini
Type locality: Costa Rica: Coronado

***pallidus* Fisher, 1925a**: 256 (Fig. 126q, r), Holotype
Original combination: *Enes pallidus* Fisher
Current name: *Enes (Sibuyenes) pallidus* Fisher, 1925a
Current placement: Cerambycidae: Lamiinae: Acanthocinini
Type locality: Philippines: Basilan
Remarks: Allotype: same data

***pallidus* Schaeffer, 1905a**: 132 (Fig. 126s, t), Syntype
Original combination: *Ischnocuemis pallidus* Schaeffer
Current name: *Perarthrus pallidus* (Schaeffer, 1905a)
Current placement: Cerambycidae: Cerambycinae: Trachyderini
Type locality: Mexico: Baja California Sur, Santa Rosa
Remarks: Although the labels have the genus names spelled as "*Ischnocnemis*", it was published as "*Ischnocuemis*"

***panamensis* Lane, 1959a**: 263 (Fig. 126u, v), Holotype
Original combination: *Aerenica panamensis* Lane
Current name: *Recchia hirsuta* (Bates, 1881)
Current placement: Cerambycidae: Lamiinae: Aerenicini
Type locality: Panama: La Chorrera

***panamensis* Moyses & Galileo, 2009**: 498 (Fig. 126w, x), Holotype
Original combination: *Mauesia panamensis* Moyses & Galileo
Current name: same
Current placement: Cerambycidae: Cerambycinae: Mauesini
Type locality: Panama: Panama Province, Cerro Jefe, 2200 ft.

***pantherina* Lane, 1964**: 185 (Fig. 127a, b), Holotype
Original combination: *Batesbeltia pantherina* Lane
Current name: same
Current placement: Cerambycidae: Lamiinae: Anisocerini
Type locality: Colombia

***papuana* Kriesche, 1920**: 197 (Fig. 127c, d), Holotype
Original combination: *Rosenbergia papuana* Kriesche
Current name: *Rosenbergia lactiflua* (Fairmaire, 1883)
Current placement: Cerambycidae: Lamiinae: Batocerini
Type locality: Papua New Guinea: Bougainville Island (Solomon Islands)
Remarks: Tippmann labeled this specimen as having been obtained from the Kriesche Estate in 1955.

***papuanus* Breuning, 1936b**: 287 (Fig. 127e, f), Holotype
Original combination: *Pelargoderus papuanus* Breuning
Current name: same
Current placement: Cerambycidae: Lamiinae: Monochamini
Type locality: Papua New Guinea: New Ireland, St. Matthias Island
Remarks: Tavakilian and Chevillotte (2013) suggest the holotype is present in the Paris Museum. The locality data does not match the published description, so this specimen is probably not the holotype.

***papuanus* Breuning, 1957a**: 15 (Fig. 127g, h), Holotype
Original combination: *Aegocidnus (Pomeranaegocidnus) papuanus* Breuning
Current name: *Sciades (Pomeranaegocidnus) papuanus* (Breuning, 1957a)
Current placement: Cerambycidae: Lamiinae: Acanthocinini
Type locality: Papua New Guinea, Berlinhafen, Saleo

***paradiana* Galileo & Martins, 2004b**: 384 (Fig. 127i, j), Holotype
Original combination: *Adesmus paradiana* Galileo & Martins
Current name: same
Current placement: Cerambycidae: Lamiinae: Hemilophini
Type locality: Ecuador: Napo, Res. Ethnica Waorani, 1km S. Onkone Gare Camp, 220 m., 00°39′10S 076°26′W

***paradoxum* Tippmann, 1955b**: 12 (Fig. 127k, l), Holotype
Original combination: *Acanthomerosternoplon paradoxum* Tippmann
Current name: *Omosarotes paradoxum* (Tippmann, 1955b)
Current placement: Cerambycidae: Lamiinae: Cyrtinini
Type locality: Peru: Rio Perené

***parallelum* Casey, 1891**: 23 (Fig. 127m, n), Lectotype
Original combination: *Tetropium parallelum* Casey
Current name: same
Current placement: Cerambycidae: Spondylidinae: Asemini
Type locality: USA: Colorado
Remarks: This specimen is herein designated as the lectotype in order to stabilize the taxonomy and facilitate further identifications of this species.

***parallelus* Aurivillius, 1922**: 415 (Fig. 127o, p), Holotype
Original combination: *Demonax parallelus* Aurivillius
Current name: same
Current placement: Cerambycidae: Cerambycinae: Clytini
Type locality: Philippines: Mindanao, Kolambugan

***parallelus* Casey, 1912**: 225 (Fig. 127q, r), Holotype
Original combination: *Stenodontes parallelus* Casey
Current name: *Stenodontes chevrolati* Gahan, 1890
Current placement: Cerambycidae: Prioninae: Macrotomini
Type locality: USA: Florida

***parallelus* Casey, 1924**: 265 (Fig. 127s, t), Holotype
Original combination: *Crossidius parallelus* Casey
Current name: *Crossidius hirtipes wickhami* Casey, 1912
Current placement: Cerambycidae: Cerambycinae: Trachyderini
Type locality: USA: Eureka

***parkeri* Casey, 1913**: 242 (Fig. 127u, v), Holotype
Original combination: *Acmaeops parkeri* Casey
Current name: *Acmaeops proteus* (Kirby, 1837)
Current placement: Cerambycidae: Lepturinae: Rhagiini
Type locality: USA: Colorado, Boulder Co.

parowana **Casey, 1924**: 242 (Fig. 127w, x), Lectotype
Original combination: *Aneflomorpha parowana* Casey
Current name: same
Current placement: Cerambycidae: Cerambycinae: Elaphidiini
Type locality: USA: Utah, Parowan
Remarks: This specimen is herein designated as the lectotype in order to stabilize the taxonomy and facilitate further identifications of this species.

parowanum **Casey, 1924**: 263 (Fig. 128a, b), Lectotype
Original combination: *Tragidion parowanum* Casey
Current name: *Tragidion auripenne* Casey, 1893
Current placement: Cerambycidae: Cerambycinae: Trachyderini
Type locality: USA: Utah, Parowan Canyon
Remarks: This specimen is herein designated as the lectotype in order to stabilize the taxonomy and facilitate further identifications of this species.

parva **Casey, 1924**: 226 (Fig. 128c, d), Holotype
Original combination: *Spondylis parva* Casey
Current name: *Neospondylis upiformis* (Mannerheim, 1843)
Current placement: Cerambycidae: Spondylidinae: Spondylidini
Type locality: USA: Washington, Port Columbia

parva **Fisher, 1938**: 138 (Fig. 128e, f), Holotype
Original combination: *Dorcasta parva* Fisher
Current name: *Bebelis parva* (Fisher, 1938)
Current placement: Cerambycidae: Lamiinae: Apomecynini
Type locality: Brazil: Santa Catarina, Nova Teutonia

parva **Gressitt, 1935e**: 193 (Fig. 128g, h), Holotype
Original combination: *Anastathes parva* Gressitt
Current name: same
Current placement: Cerambycidae: Lamiinae: Astathini
Type locality: China: Taiwan, Bukai

parviceps **Casey, 1913**: 210 (Fig. 128i, j), Holotype
Original combination: *Stenocorus (Stenocorus) parviceps* Casey
Current name: *Stenocorus nubifer* (LeConte, 1859)
Current placement: Cerambycidae: Lepturinae: Rhagiini
Type locality: USA: California, Lake Co.

parviceps **Casey, 1924**: 264 (Fig. 128k, l), Lectotype
Original combination: *Oxoplus parviceps* Casey
Current name: *Crossidius coralinus parviceps* (Casey, 1924)
Current placement: Cerambycidae: Cerambycinae: Trachyderini
Type locality: USA: Utah, Vineyard
Remarks: This specimen is herein designated as the lectotype in order to stabilize the taxonomy and facilitate further identifications of this species.

parvicorne **Casey, 1912**: 260 (Fig. 128m, n), Holotype
Original combination: *Asemum parvicorne* Casey
Current name: *Asemum striatum* (Linnaeus, 1758)
Current placement: Cerambycidae: Spondylidinae: Asemini
Type locality: USA: Colorado, Boulder Co.

parvicorne **Casey, 1913**: 195 (Fig. 128o, p), Holotype
Original combination: *Rhagium (Hargium) parvicorne* Casey
Current name: *Rhagium inquisitor* (Linnaeus, 1758)
Current placement: Cerambycidae: Lepturinae: Rhagiini
Type locality: USA: Massachusetts, Malden

parvulum **Casey, 1891**: 24 (Fig. 128q, r), Holotype
Original combination: *Tetropium parvulum* Casey
Current name: same
Current placement: Cerambycidae: Spondylidinae: Asemini
Type locality: USA: Indiana

parvulus **Casey, 1913**: 280 (Fig. 128s, t), Holotype
Original combination: *Ipochus parvulus* Casey
Current name: *Ipochus fasciatus* LeConte, 1852
Current placement: Cerambycidae: Lamiinae: Parmenini
Type locality: USA: California, San Diego

parvulus **Casey, 1913**: 304 (Fig. 128u, v), Holotype
Original combination: *Lagochirus parvulus* Casey
Current name: *Lagocheirus araneiformis ypsilon* (Voet, 1778)
Current placement: Cerambycidae: Lamiinae: Acanthocinini
Type locality: Panama: Cocle, Natá

parvus **Casey, 1912**: 234 (Fig. 128w, x), Holotype
Original combination: *Prionus (Prionus) parvus* Casey
Current name: *Prionus laticollis* (Drury, 1773)
Current placement: Cerambycidae: Prioninae: Prionini
Type locality: No locality label or published information.

pedestre zoufali **Tippmann, 1958**: 156 (Fig. 129a, b), Holotype
Original combination: *Pedestredorcadion pedestre zoufali* Tippmann
Current name: *Dorcadion (Cribridorcadion) pedestre* (Poda, 1761)
Current placement: Cerambycidae: Lamiinae: Dorcadionini
Type locality: Austria: Prossnitz

pedicellatus **Casey, 1924**: 235 (Fig. 129c, d), Holotype
Original combination: *Hylotrupes pedicellatus* Casey
Current name: *Hylotrupes bajulus* (Linnaeus, 1758)
Current placement: Cerambycidae: Cerambycinae: Hylotrupini
Type locality: USA: Virginia

pellita **Tippmann, 1960**: 98 (Fig. 129e, f), Holotype
Original combination: *Andinotrichoderes pellita* Tippmann
Current name: same
Current placement: Cerambycidae: Prioninae: Meroscelisini
Type locality: Peru: Pavayoc, Callanga, 2600 m.

pendleburyi **Fisher, 1935c**: 593 (Fig. 129g, h), Holotype
Original combination: *Demonax pendleburyi* Fisher
Current name: same
Current placement: Cerambycidae: Cerambycinae: Clytini
Type locality: Malaysia: Borneo, Sabah, Mt. Kinabalu, 5500 ft., Lumu Lumu

pendleburyi **Fisher, 1935c**: 606 (Fig. 129i, j), Holotype
Original combination: *Aristobia pendleburyi* Fisher
Current name: same
Current placement: Cerambycidae: Lamiinae: Monochamini
Type locality: Malaysia: Borneo, Sabah, Mt. Kinabalu, Kiau, 3000 ft.

pendleburyi **Fisher, 1935c**: 626 (Fig. 129k, l), Holotype
Original combination: *Glenea pendleburyi* Fisher
Current name: same
Current placement: Cerambycidae: Lamiinae: Saperdini
Type locality: Malaysia: Borneo, Sabah, Mt. Kinabalu, 5500 ft., Lumu Lumu

penicillata **Aurivillius, 1927b**: 588 (Fig. 129m, n), Holotype
Original combination: *Ochrocesis penicillata* Aurivillius
Current name: *Bacchisa penicillata* (Aurivillius, 1927b)
Current placement: Cerambycidae: Lamiinae: Tetraopini
Type locality: Philippines: Negros Oriental, Cuernos Mountains

peninsulare **Schaeffer, 1908**: 338 (Fig. 129o, p), Holotype
Original combination: *Obrium peninsulare* Schaeffer
Current name: same
Current placement: Cerambycidae: Cerambycinae: Obriini
Type locality: Mexico: Baja California Sur, Santa Rosa

peninsularis **Casey, 1912**: 316 (Fig. 129q, r), Holotype
Original combination: *Dendrobias peninsularis* Casey
Current name: *Trachyderes (Dendrobias) mandibularis* (Dupont, 1834)
Current placement: Cerambycidae: Cerambycinae: Trachyderini
Type locality: Mexico: Baja California Sur

peninsularis **Schaeffer, 1905a**: 133 (Fig. 129s, t), Holotype
Original combination: *Neoclytus peninsularis* Schaeffer
Current name: same
Current placement: Cerambycidae: Cerambycinae: Clytini
Type locality: Mexico: Baja California Sur, San Felipe

penrosei **Lingafelter & Wappes, 2012**: 158 (Fig. 129u, v), Holotype
Original combination: *Trichoxys penrosei* Lingafelter & Wappes
Current name: same
Current placement: Cerambycidae: Cerambycinae: Clytini
Type locality: Mexico: Durango, 50 km NE La Cap. de Taxte

peregrinus samarensis **Dillon & Dillon, 1948**: 264 (Fig. 129w, x), Holotype
Original combination: *Cylindrepomus peregrinus samarensis* Dillon & Dillon
Current name: *Cylindrepomus peregrinus* Pascoe, 1858
Current placement: Cerambycidae: Lamiinae: Gnomiini
Type locality: Philippines: Samar

perfasciatus **Lingafelter & Micheli, 2009**: 43 (Fig. 130a, b), Holotype
Original combination: *Leptostylopsis perfasciatus* Lingafelter & Micheli
Current name: same
Current placement: Cerambycidae: Lamiinae: Acanthocinini
Type locality: Dominican Republic: Barahona Province, Filipinas, 18°07.339′N, 71°07.152′W, 625 m.

permutans paucipunctatus **Gressitt, 1938c**: 157 (Fig. 130c, d), Holotype
Original combination: *Dihammus permutans paucipunctatus* Gressitt
Current name: *Acalolepta permutans paucipunctata* (Gressitt, 1938c)
Current placement: Cerambycidae: Lamiinae: Monochamini
Type locality: China: Taiwan, Bukai

pernigra **Linell, 1897a**: 397 (Fig. 130e, f), Holotype
Original combination: *Leptura pernigra* Linell
Current name: *Brachyleptura pernigra* (Linell, 1897a)
Current placement: Cerambycidae: Lepturinae: Lepturini
Type locality: USA: California, Los Angeles County

persimilis **Breuning, 1938a**: 327 (Fig. 130g, h), Holotype
Original combination: *Palimna persimilis* Breuning
Current name: same
Current placement: Cerambycidae: Lamiinae: Ancylonotini
Type locality: Indonesia: Sumatra

pertenue **Casey, 1924**: 248 (Fig. 130i, j), Holotype
Original combination: *Pseudibidion pertenue* Casey
Current name: *Psyrassa pertenuis* (Casey, 1924)
Current placement: Cerambycidae: Cerambycinae: Elaphidiini
Type locality: USA: Florida

***pertenuis* Casey, 1913**: 357 (Fig. 130k, l), Holotype
Original combination: *Spalacopsis (Spalacopsis) pertenuis* Casey
Current name: *Spalacopsis stolata* Newman, 1842
Current placement: Cerambycidae: Lamiinae: Agapanthiini
Type locality: USA: Florida, Palm Beach

***peruana* Breuning, 1943c**: 48 (Fig. 130m, n), Holotype
Original combination: *Ischnolea peruana* Breuning
Current name: same
Current placement: Cerambycidae: Lamiinae: Desmiphorini
Type locality: Peru: Ayacucho, Aina

***peruvianus* Tippmann, 1939**: 77 (Fig. 130o, p), Holotype
Original combination: *Cyclopeplus peruvianus* Tippmann
Current name: same
Current placement: Cerambycidae: Lamiinae: Anisocerini
Type locality: Peru: Satipo, Junin
Remarks: A paratype is present in the Staatliches Museum für Tierkunde, Dresden, Germany.

***peruvianus* Tippmann, 1960**: 111 (Fig. 130q, r), Holotype
Original combination: *Brasilianus peruvianus* Tippmann
Current name: *Jupoata peruviana* (Tippmann, 1960)
Current placement: Cerambycidae: Cerambycinae: Cerambycini
Type locality: Peru: Cosñipata, 1700 m.

***phalaenopsidis* Fisher, 1937b**: 53 (Fig. 130s, t), Holotype
Original combination: *Diaxenes phalaenopsidis* Fisher
Current name: same
Current placement: Cerambycidae: Lamiinae: Apomecynini
Type locality: Indonesia: Java, Buitenzorg

***philippinarum* Aurivillius, 1922**: 425 (Fig. 130u, v), Syntype
Original combination: *Eunidia philippinarum* Aurivillius
Current name: same
Current placement: Cerambycidae: Lamiinae: Eunidiini
Type locality: Philippines: Mindanao, Davao
Remarks: Other syntype(s) are present in the Naturhistoriska Riksmuseet, Stockholm, Sweden.

***philippinarum* Aurivillius, 1927b**: 574 (Fig. 130w, x), Syntype
Original combination: *Oopsis philippinarum* Aurivillius
Current name: *Sybra binotata* Gahan, 1907
Current placement: Cerambycidae: Lamiinae: Apomecynini
Type locality: Philippines: Sibuyan, Romblon
Remarks: Tavakilian and Chevillotte (2013) suggests another syntype is present in the Naturhistoriska Riksmuseet, Stockholm, Sweden.

***philippinarum* Aurivillius, 1927b**: 576 (Fig. 131a, b), Syntype
Original combination: *Ipochira philippinarum* Aurivillius
Current name: same
Current placement: Cerambycidae: Lamiinae: Acanthocinini
Type locality: Philippines: Negros Oriental, Cuernos Mountains
Remarks: Other syntype(s) are present in the Naturhistoriska Riksmuseet, Stockholm, Sweden.

***philippinus* Fisher, 1925a**: 246 (Fig. 131c, d), Holotype
Original combination: *Exocentrus philippinus* Fisher
Current name: same
Current placement: Cerambycidae: Lamiinae: Acanthocinini
Type locality: Philippines: Sibuyan, Romblon

***philippinus* Fisher, 1925a**: 271 (Fig. 131e, f), Holotype
Original combination: *Camptomyne philippinus* Fisher
Current name: *Exocentrus (Camptomyne) philippinensis* Breuning, 1958
Current placement: Cerambycidae: Lamiinae: Acanthocinini
Type locality: Philippines: Luzon, Mt. Banahao

***philippinus transeuns* Breuning, 1957a**: 12 (Fig. 131g, h), Holotype
Original combination: *Exocentrus (Camptomyme) philippinus transeuns* Breuning
Current name: *Exocentrus (Camptomyne) philippinensis* Breuning, 1958
Current placement: Cerambycidae: Lamiinae: Acanthocinini
Type locality: Philippines: Pabentoe

***philippinus* var. *unicolor* Fisher, 1925a**: 272 (Fig. 131i, j), Holotype
Original combination: *Camptomyne philippinus* var. *unicolor* Fisher
Current name: *Exocentrus (Camptomyne) philippinensis* Breuning, 1958
Current placement: Cerambycidae: Lamiinae: Acanthocinini
Type locality: Philippines: Mindanao, Davao
Remarks: The name following the binomen was not expressly given infrasubspecific rank and the work did not indicate unambiguously that the name was infrasubspecific, therefore, it is deemed a subspecies and available.

***phoenicis* Casey, 1924**: 249 (Fig. 131k, l), Holotype
Original combination: *Eucrossus phoenicis* Casey
Current name: *Eucrossus villicornis* LeConte, 1873
Current placement: Cerambycidae: Cerambycinae: Hesperophanini
Type locality: USA: Arizona, Phoenix

***picea* Fisher, 1936d**: 348 (Fig. 131m, n), Holotype
Original combination: *Phidola picea* Fisher
Current name: *Eupogonius piceus* (Fisher, 1936d)
Current placement: Cerambycidae: Lamiinae: Desmiphorini
Type locality: Cuba: Santiago de Cuba, Loma del Gato, Sierra del Cobre, Oriente

***piceicollis* Linsley, 1935b**: 481 (Fig. 131o, p), Holotype
Original combination: *Octoplon piceicollis* Linsley
Current name: *Gnomidolon insulicola* Bates, 1885
Current placement: Cerambycidae: Cerambycinae: Hexoplini
Type locality: Panama: Canal Zone, Paraiso

***picticornis* Martins & Galileo, 1991**: 816 (Fig. 131q, r), Holotype
Original combination: *Tyrinthia picticornis* Martins & Galileo
Current name: same
Current placement: Cerambycidae: Lamiinae: Hemilophini
Type locality: Panama: Panama Province, Trinidad River

***pictipennis* Casey, 1924**: 285 (Fig. 131s, t), Lectotype
Original combination: *Xestoleptura pictipennis* Casey
Current name: *Xestoleptura tibialis* (LeConte, 1850)
Current placement: Cerambycidae: Lepturinae: Lepturini
Type locality: USA: Idaho, Bonner Co., Priest Lake
Remarks: This specimen is herein designated as the lectotype in order to stabilize the taxonomy and facilitate further identifications of this species.

***pictum* Casey, 1924**: 239 (Fig. 131u, v), Holotype
Original combination: *Physocnemum pictum* Casey
Current name: *Physocnemum brevilineum* (Say, 1824)
Current placement: Cerambycidae: Cerambycinae: Callidiini
Type locality: USA: North or South Dakota

***pictus* Galileo & Martins, 2012a**: 65 (Fig. 131w, x), Holotype
Original combination: *Psapharochrus pictus* Galileo & Martins
Current name: same
Current placement: Cerambycidae: Lamiinae: Acanthoderini
Type locality: Panama: Panama Province, Cerro Campana, 800 m.

***pilatus* Galileo & Martins, 2005a**: 276 (Fig. 132a, b), Holotype
Original combination: *Adesmus pilatus* Galileo & Martins
Current name: same
Current placement: Cerambycidae: Lamiinae: Hemilophini
Type locality: Panama: Panama Province, Cerro Campana, 800 m.

***pilicornis* Fisher, 1944a**: 5 (Fig. 132c, d), Holotype
Original combination: *Eburodacrys pilicornis* Fisher
Current name: same
Current placement: Cerambycidae: Cerambycinae: Eburiini
Type locality: Venezuela: Monagas, Caripito

***pilosicornis* Casey, 1890**: 492 (Fig. 132e, f), Holotype
Original combination: *Tragosoma pilosicornis* Casey
Current name: same
Current placement: Cerambycidae: Prioninae: Meroscelisini
Type locality: USA: California, Mount Diablo

***pilosicornis* Chemsak & Linsley, 1965**: 26 (Fig. 132g, h), Holotype
Original combination: *Aneflus (Protaneflus) pilosicornis* Chemsak & Linsley
Current name: *Aneflus pilosicornis* Chemsak & Linsley, 1965
Current placement: Cerambycidae: Cerambycinae: Elaphidiini
Type locality: Costa Rica: San Jose, 1000–2000 m.

***pilosicornis* Fisher, 1925a**: 228 (Fig. 132i, j), Holotype
Original combination: *Ebaeides pilosicornis* Fisher
Current name: same
Current placement: Cerambycidae: Lamiinae: Acanthocinini
Type locality: Philippines: Sibuyan, Romblon

***pilosipennis* Breuning, 1943c**: 29 (Fig. 132k, l), Holotype
Original combination: *Pterolophia pilosipennis* Breuning
Current name: *Pterolophia (Trichopraonetha) pilosipennis* Breuning, 1943c
Current placement: Cerambycidae: Lamiinae: Pteropliini
Type locality: Indonesia: Palang

***pilosum* Fisher, 1932**: 75 (Fig. 132m, n), Holotype
Original combination: *Acrepidopterum pilosum* Fisher
Current name: same
Current placement: Cerambycidae: Lamiinae: Apomecynini
Type locality: Dominican Republic: North of San Cristóbal, Blanton mine

***pilosus* Aurivillius, 1927b**: 550 (Fig. 132o, p), Syntype
Original combination: *Xylotrechus pilosus* Aurivillius
Current name: same
Current placement: Cerambycidae: Cerambycinae: Clytini
Type locality: Malaysia: Borneo, Sandakan
Remarks: Tavakilian and Chevillotte (2013) state that syntypes are present in the Naturhistoriska Riksmuseet, Stockholm, Sweden as well as the USNM.

***pimalis* Casey, 1913**: 285 (Fig. 132q, r), Holotype
Original combination: *Moneilema (Collapteryx) pimalis* Casey
Current name: *Moneilema gigas* LeConte, 1873
Current placement: Cerambycidae: Lamiinae: Moneilemini
Type locality: USA: Arizona, Tuscon

***pini* Schaeffer, 1905b**: 165 (Fig. 132s, t), Holotype
Original combination: *Leptostylus pini* Schaeffer
Current name: *Pseudastylopsis pini* (Schaeffer, 1905b)
Current placement: Cerambycidae: Lamiinae: Acanthocinini
Type locality: USA: Arizona, Huachuca Mts., Carrs Peak, 9000 ft.

***pini* Schaeffer, 1905b**: 166 (Fig. 132u, v), Syntype
Original combination: *Lepturgoides pini* Schaeffer
Current name: *Eutrichillus pini* (Schaeffer, 1905b)
Current placement: Cerambycidae: Lamiinae: Acanthocinini
Type locality: USA: Arizona, Huachuca Mts., Carrs Peak, 9000 ft.

***pinicola* Casey, 1913**: 280 (Fig. 132w, x), Lectotype
Original combination: *Ipochus pinicola* Casey
Current name: *Ipochus fasciatus* LeConte, 1852
Current placement: Cerambycidae: Lamiinae: Parmenini
Type locality: USA: California, Monterey
Remarks: This specimen is herein designated as the lectotype in order to stabilize the taxonomy and facilitate further identifications of this species.

***pinicola* Schaeffer, 1908**: 341 (Fig. 133a, b), Holotype
Original combination: *Acmaeops pinicola* Schaeffer
Current name: *Acmaeops proteus* (Kirby, 1837)
Current placement: Cerambycidae: Lepturinae: Rhagiini
Type locality: USA: Arizona, Carr Peak

***pinorum* Casey, 1914**: 365 (Fig. 133c, d), Holotype
Original combination: *Anoplium pinorum* Casey
Current name: *Anelaphus moestus pinorum* Casey, 1914
Current placement: Cerambycidae: Cerambycinae: Elaphidiini
Type locality: USA: North Carolina, Southern Pines

***pinorum* Casey, 1924**: 261 (Fig. 133e, f), Holotype
Original combination: *Callimoxys pinorum* Casey
Current name: *Callimoxys sanguinicollis* (Olivier, 1795)
Current placement: Cerambycidae: Cerambycinae: Stenopterini
Type locality: USA: North Carolina, Southern Pines

***pinorum* Casey, 1924**: 268 (Fig. 133g, h), Holotype
Original combination: *Stenosphenus pinorum* Casey
Current name: *Stenosphenus notatus* (Olivier, 1795)
Current placement: Cerambycidae: Cerambycinae: Elaphidiini
Type locality: USA: North Carolina, Southern Pines

***piperi* Webb, 1905**: 104 (Fig. 133i, j), Holotype
Original combination: *Desmocerus piperi* Webb
Current name: *Desmocerus aureipennis piperi* Webb, 1905
Current placement: Cerambycidae: Lepturinae: Desmocerini
Type locality: USA: Washington, Blue Mountains

***plagiata* Aurivillius, 1927b**: 584 (Fig. 133k, l), Syntype
Original combination: *Serixia plagiata* Aurivillius
Current name: same
Current placement: Cerambycidae: Lamiinae: Saperdini
Type locality: Philippines: Mindanao, Iligan
Remarks: Tavakilian and Chevillotte (2013) state that syntypes are present in the Naturhistoriska Riksmuseet, Stockholm, Sweden as well as the USNM.

***plagiatus* Casey, 1912**: 336 (Fig. 133m, n), Holotype
Original combination: *Crossidius plagiatus* Casey
Current name: *Crossidius punctatus* LeConte, 1873
Current placement: Cerambycidae: Cerambycinae: Trachyderini
Type locality: No published locality data or label associated with the specimen.

***plagiatus* Casey, 1924**: 273 (Fig. 133o, p), Holotype
Original combination: *Stenocorus plagiatus* Casey
Current name: *Stenocorus nubifer* (LeConte, 1859)
Current placement: Cerambycidae: Lepturinae: Rhagiini
Type locality: USA: California, Placer Co., Cisco

***plagifera* Aurivillius, 1907**: 111 (Fig. 133q, r), Holotype
Original combination: *Prosopocera plagifera* Aurivillius
Current name: *Prosopocera (Hierogyna) plagifera* Aurivillius, 1907
Current placement: Cerambycidae: Lamiinae: Prosopocerini
Type locality: Tanzania: Kigonsera
Remarks: Tavakilian and Chevillotte (2013) suggest the holotype is present in the Zoologische Staatssammlung des Bayerischen Staates, Munich, Germany, however, this specimen and its labels match the original description.

planatum **Gressitt, 1956**: 81 (Fig. 133s, t), Holotype
Original combination: *Ceresium (Ceresiellum) planatum* Gressitt
Current name: same
Current placement: Cerambycidae: Cerambycinae: Callidiopini
Type locality: Palau: Babelthuap Island, southwest of Ulimang

planicollis **Fisher, 1935b**: 203 (Fig. 133u, v), Lectotype
Original combination: *Leptostylus planicollis* Fisher
Current name: *Styloleptus planicollis* (Fisher, 1935b)
Current placement: Cerambycidae: Lamiinae: Acanthocinini
Type locality: Jamaica: Montego Bay

planifrons **Tippmann, 1951**: 318 (Fig. 133w, x), Holotype
Original combination: *Menesida planifrons* Tippmann
Current name: same
Current placement: Cerambycidae: Lamiinae: Saperdini
Type locality: Indonesia: Sumatra, Soekaranda

plaumanni **Fisher, 1937d**: 152 (Fig. 134a, b), Holotype
Original combination: *Rhopalophora (Ischionodonta) plaumanni* Fisher
Current name: *Ischionodonta platensis* (Chevrolat, 1859)
Current placement: Cerambycidae: Cerambycinae: Rhopalophorini
Type locality: Brazil: Santa Catarina, Nova Teutonia

plaumanni **Tippmann, 1960**: 105 (Fig. 134c, d), Holotype
Original combination: *Atenizus plaumanni* Tippmann
Current name: same
Current placement: Cerambycidae: Cerambycinae: Oemini
Type locality: Brazil: Paraná, Rondon, 500 m Rondon, 24°38, 54°07

plectralis **Casey, 1924**: 286 (Fig. 134e, f), Holotype
Original combination: *Moneilema (Collapteryx) plectralis* Casey
Current name: *Moneilema gigas* LeConte, 1873
Current placement: Cerambycidae: Lamiinae: Moneilemini
Type locality: USA: Arizona

pleuralis **Casey, 1913**: 287 (Fig. 134g, h), Holotype
Original combination: *Moneilema (Collapteryx) pleuralis* Casey
Current name: *Moneilema blapsides ulkei* Horn, 1885
Current placement: Cerambycidae: Lamiinae: Moneilemini
Type locality: Mexico: Durango, Ciudad de Durango

pleuralis **Casey, 1913**: 312 (Fig. 134i, j), Lectotype
Original combination: *Leiopus pleuralis* Casey
Current name: *Sternidius alpha* (Say, 1827)
Current placement: Cerambycidae: Lamiinae: Acanthocinini
Type locality: USA: District of Columbia
Remarks: This specimen is herein designated as the lectotype in order to stabilize the taxonomy and facilitate further identifications of this species.

pleuralis **Casey, 1913**: 344 (Fig. 134k, l), Holotype
Original combination: *Colobothea pleuralis* Casey
Current name: same
Current placement: Cerambycidae: Lamiinae: Colobotheini
Type locality: Panama: Darién

pocularis prolixus **Casey, 1912**: 239 (Fig. 134m, n), Holotype
Original combination: *Prionus (Prionus) pocularis prolixus* Casey
Current name: *Prionus pocularis* Dalman in Schoenherr, 1817
Current placement: Cerambycidae: Prioninae: Prionini
Type locality: No published locality data or label associated with the specimen.

poecila **v. *maculicollis*** **Fisher, 1947a**: 51 (Fig. 134o, p), Holotype
Original combination: *Ommata (Eclipta) poecila* v. *maculicollis* Fisher
Current name: *Stultutragus poecilus* (Bates, 1873)
Current placement: Cerambycidae: Cerambycinae: Rhinotragini
Type locality: Brazil: Santa Catarina, Nova Teutonia, 27°11′B, 52°23′L
Remarks: The name following the binomen was not expressly given infrasubspecific rank and the work did not indicate unambiguously that the name was infrasubspecific, therefore, it is deemed a subspecies and available.

poecila **v. *nigricornis*** **Fisher, 1947a**: 51 (Fig. 134q, r), Holotype
Original combination: *Ommata (Eclipta) poecila* v. *nigricornis* Fisher
Current name: *Stultutragus nigricornis* (Fisher, 1947a)
Current placement: Cerambycidae: Cerambycinae: Rhinotragini
Type locality: Brazil: Santa Catarina, Nova Teutonia, 27°11′B, 52°23′L
Remarks: The name following the binomen was not expressly given infrasubspecific rank and the work did not indicate unambiguously that the name was infrasubspecific, therefore, it is deemed a subspecies and available.

poeyi **Fisher, 1925b**: 4 (Fig. 134s, t), Holotype
Original combination: *Leptostylus poeyi* Fisher
Current name: *Leptostylopsis poeyi* (Fisher, 1925b)
Current placement: Cerambycidae: Lamiinae: Acanthocinini
Type locality: Dominican Republic: Distrito Nacional, San Francisco Mts.

politum **Casey, 1924**: 246 (Fig. 134u, v), Holotype
Original combination: *Anoplium politum* Casey
Current name: *Eustromula validum* (LeConte, 1858)
Current placement: Cerambycidae: Cerambycinae: Elaphidiini
Type locality: USA: Arizona, Baboquivari Mts.

pollens **Casey, 1913**: 286 (Fig. 134w, x), Holotype
Original combination: *Moneilema (Collapteryx) pollens* Casey
Current name: *Moneilema gigas* LeConte, 1873
Current placement: Cerambycidae: Lamiinae: Moneilemini
Type locality: USA: Arizona

ponapensis **Gressitt, 1956**: 96 (Fig. 135a, b), Holotype
Original combination: *Examnes ponapensis* Gressitt
Current name: same
Current placement: Cerambycidae: Cerambycinae: Callidiopini
Type locality: Micronesia: Pohnpei (formerly Ponape Island)

ponderosissima **Wickham, 1913**: 295 (Fig. 135c, d), Holotype
Original combination: *Leptura ponderosissima* Wickham
Current name: same
Current placement: Cerambycidae: Lepturinae: Lepturini
Type locality: USA: Colorado, Florissant
Remarks: This is one of four fossil specimens at the USNM and is housed in the Paleobiology (Paleoentomology) section.

poolei **Fisher, 1932**: 5 (Fig. 135e, f), Holotype
Original combination: *Pseudoeme poolei* Fisher
Current name: *Enosmaeus cubanus* Thomson, 1878
Current placement: Cerambycidae: Cerambycinae: Achrysonini
Type locality: Haiti: St. Michele de l'Atalaye

postvittata **Breuning, 1939b**: 520 (Fig. 135g, h), Holotype
Original combination: *Agelasta (Dissosira) postvittata* Breuning
Current name: same
Current placement: Cerambycidae: Lamiinae: Mesosini
Type locality: India: Bengalen, Patna

praecidens **Casey, 1913**: 354 (Fig. 135i, j), Lectotype
Original combination: *Oncideres praecidens* Casey
Current name: *Oncideres c. cingulata* (Say, 1827)
Current placement: Cerambycidae: Lamiinae: Onciderini
Type locality: USA: North Carolina, Southern Pines
Remarks: This specimen is herein designated as the lectotype in order to stabilize the taxonomy and facilitate further identifications of this species.

praeclarus **Lingafelter, 2008**: 359 (Fig. 135k, l), Holotype
Original combination: *Anelaphus praeclarus* Lingafelter
Current name: same
Current placement: Cerambycidae: Cerambycinae: Elaphidiini
Type locality: Dominican Republic: La Altagracia Province, PN del Este, Guaraguao, 0–5 meters, 18°19.568′N, 68°48.500′W

praelonga **Casey, 1913**: 368 (Fig. 135m, n), Lectotype
Original combination: *Oberea praelonga* Casey
Current name: same
Current placement: Cerambycidae: Lamiinae: Phytoeciini
Type locality: USA: New York
Remarks: Lectotype designated by Yanega (1996)

praelonga deficiens **Casey, 1924**: 296 (Fig. 135o, p), Holotype
Original combination: *Oberea praelonga deficiens* Casey
Current name: *Oberea deficiens* Casey, 1924
Current placement: Cerambycidae: Lamiinae: Phytoeciini
Type locality: USA: New Jersey

praestans **Casey, 1912**: 320 (Fig. 135q, r), Holotype
Original combination: *Crioprosopus praestans* Casey
Current name: *Callona praestans* (Casey, 1912)
Current placement: Cerambycidae: Cerambycinae: Trachyderini
Type locality: Mexico: Guerrero

praestans **Casey, 1913**: 267 (Fig. 135s, t), Holotype
Original combination: *Leptura (Cosmosalia) praestans* Casey
Current name: *Pygoleptura nigrella* (Say, 1827)
Current placement: Cerambycidae: Lepturinae: Lepturini
Type locality: USA: Colorado

praestans semiplicatus **Casey, 1912**: 320 (Fig. 135u, v), Holotype
Original combination: *Crioprosopus praestans semiplicatus* Casey
Current name: *Callona praestans* (Casey, 1912)
Current placement: Cerambycidae: Cerambycinae: Trachyderini
Type locality: Mexico: Guerrero

problematica **Tippmann, 1951**: 326 (Fig. 135w, x), Holotype
Original combination: *Guitelia problematica* Tippmann
Current name: *Paraguitelia problematica* (Tippmann, 1951)
Current placement: Cerambycidae: Cerambycinae: Callichromatini
Type locality: Cameroon

procerus **Casey, 1913**: 304 (Fig. 136a, b), Holotype
Original combination: *Lagochirus procerus* Casey
Current name: *Lagocheirus procerus* Casey, 1913
Current placement: Cerambycidae: Lamiinae: Acanthocinini
Type locality: Mexico: Baja California Sur

prodigiosa **Thomson, 1868**: 45 (Fig. 136c, d), Neotype
Original combination: *Hypsioma prodigiosa* Thomson
Current name: *Typhlocerus prodigiosus* (Thomson, 1868)
Current placement: Cerambycidae: Lamiinae: Onciderini
Type locality: Brazil
Remarks: Thomson's material is supposed to be deposited in the Muséum national d'Histoire naturelle in Paris (MNHN), however, we were unable to locate the holotype for this species. Since this specimen matches the original description, it is here designated as the neotype.

producticollis **Gressitt, 1935g**: 272 (Fig. 136e, f), Holotype
Original combination: *Stenoleptura producticollis* Gressitt
Current name: same
Current placement: Cerambycidae: Lepturinae: Lepturini
Type locality: Malaysia: Borneo, Sandakan

propinquus **Casey, 1924**: 230 (Fig. 136g, h), Lectotype
Original combination: *Criocephalus propinquus* Casey
Current name: *Arhopalus productus* (LeConte, 1850)
Current placement: Cerambycidae: Spondylidinae: Asemini
Type locality: USA: Colorado
Remarks: This specimen is herein designated as the lectotype in order to stabilize the taxonomy and facilitate further identifications of this species.

prosopidis **Linsley, 1957a**: 20 (Fig. 136i, j), Holotype
Original combination: *Aneflus prosopidis* Linsley
Current name: *Aneflus paracalvatus* Knull, 1955
Current placement: Cerambycidae: Cerambycinae: Elaphidiini
Type locality: USA: Arizona, Sabino Canyon

protensicollis **Casey, 1913**: 228 (Fig. 136k, l), Holotype
Original combination: *Brachysomida protensicollis* Casey
Current name: *Brachysomida californica* (LeConte, 1851)
Current placement: Cerambycidae: Lepturinae: Rhagiini
Type locality: USA: California, Lake Co.

provoanus **Casey, 1924**: 272 (Fig. 136m, n), Lectotype
Original combination: *Neoclytus provoanus* Casey
Current name: same
Current placement: Cerambycidae: Cerambycinae: Clytini
Type locality: USA: Utah, North Fork Provo Canyon
Remarks: This specimen is herein designated as the lectotype in order to stabilize the taxonomy and facilitate further identifications of this species.

proximus **Melzer, 1934**: 100 (Fig. 136o, p), Holotype
Original combination: *Lepturges proximus* Melzer
Current name: same
Current placement: Cerambycidae: Lamiinae: Acanthocinini
Type locality: Costa Rica: Limón, Hamburg Farm, Reventazón, Ebene

pruddeni **Casey, 1912**: 346 (Fig. 136q, r), Holotype
Original combination: *Stenosphenus pruddeni* Casey
Current name: *Stenosphenus debilis* Horn, 1885
Current placement: Cerambycidae: Cerambycinae: Elaphidiini
Type locality: USA: Arizona, Grand Canyon

pruinosa **Casey, 1913**: 365 (Fig. 136s, t), Lectotype
Original combination: *Oberea pruinosa* Casey
Current name: same
Current placement: Cerambycidae: Lamiinae: Phytoeciini
Type locality: USA: Missouri, St. Louis
Remarks: Lectotype designated by Yanega (1996).

pseudocinctellum **Breuning, 1943b**: 90 (Fig. 136u, v), Holotype
Original combination: *Dorcadion pseudocinctellum* Breuning
Current name: *Dorcadion (Cribridorcadion) pseudocinctellum* Breuning, 1943b
Current placement: Cerambycidae: Lamiinae: Dorcadionini
Type locality: Turkey: Anatolia, Ak-Chehir

pseudostigma **Lingafelter & Nearns, 2007**: 182 (Fig. 136w, x), Holotype
Original combination: *Eburia pseudostigma* Lingafelter & Nearns
Current name: same
Current placement: Cerambycidae: Cerambycinae: Eburiini
Type locality: Dominican Republic: La Altagracia Province, Punta Cana near Ecological Reserve, 0–5 m.

pseudotsugae **Fisher, 1920**: 155 (Fig. 137a, b), Holotype
Original combination: *Callidium pseudotsugae* Fisher
Current name: same
Current placement: Cerambycidae: Cerambycinae: Callidiini
Type locality: USA: California, Wright, on *Pseudotsuga taxifolia*

psydrum **Martins, 1969**: 806 (Fig. 137c, d), Holotype
Original combination: *Compsibidion psydrum* Martins
Current name: same
Current placement: Cerambycidae: Cerambycinae: Neoibidionini
Type locality: Peru: Loreto, near Aramango

pubescens **Casey, 1891**: 45 (Fig. 137e, f), Holotype
Original combination: *Ipochus pubescens* Casey
Current name: *Ipochus fasciatus* LeConte, 1852
Current placement: Cerambycidae: Lamiinae: Parmenini
Type locality: USA: California, San Diego

pubescens **Casey, 1924**: 260 (Fig. 137g, h), Holotype
Original combination: *Ibidion pubescens* Casey
Current name: *Neocompsa quadriplagiata* (LeConte, 1873)
Current placement: Cerambycidae: Cerambycinae: Neoibidionini
Type locality: USA: Texas, Brownsville

pubescens **Fisher, 1936a**: 169 (Fig. 137i, j), Holotype
Original combination: *Neocerambyx pubescens* Fisher
Current name: same
Current placement: Cerambycidae: Cerambycinae: Cerambycini
Type locality: Indonesia: Java, Mt. Tangkoeban Prahoe, Preanger 4000–5000 ft.

pubescens **Santos-Silva & Hovore, 2008a**: 12 (Fig. 137k, l), Holotype
Original combination: *Elytrimitatrix (Grossifemora) pubescens* Santos-Silva & Hovore
Current name: same
Current placement: Disteniidae: Disteniinae: Disteniini
Type locality: Mexico, Hidalgo, 3 miles north of Tlanchinol

pubicollis **Casey, 1912**: 254 (Fig. 137m, n), Holotype
Original combination: *Homaesthesis pubicollis* Casey
Current name: *Prionus (Homaesthesis) emarginatus* Say, 1824
Current placement: Cerambycidae: Prioninae: Prionini
Type locality: USA: Colorado

pubicollis **Fisher, 1932**: 59 (Fig. 137o, p), Holotype
Original combination: *Neoclytus pubicollis* Fisher
Current name: same
Current placement: Cerambycidae: Cerambycinae: Clytini
Type locality: Cuba: Oriente, Guantanamo

pubicollis **Melzer, 1933**: 378 (Fig. 137q, r), Holotype
Original combination: *Eupogonius pubicollis* Melzer
Current name: same
Current placement: Cerambycidae: Lamiinae: Desmiphorini
Type locality: Costa Rica: Limón, Hamburg Farm, Reventazón, Ebene

puella **Casey, 1913**: 254 (Fig. 137s, t), Lectotype
Original combination: *Brachyleptura puella* Casey
Current name: *Brachyleptura vagans* (Olivier, 1795)
Current placement: Cerambycidae: Lepturinae: Lepturini
Type locality: There is no locality data associated with this specimen.
Remarks: A lectotype label was placed on this specimen by Laplante in 1991, however, it was not formally designated until now. This specimen is herein designated as the lectotype in order to stabilize the taxonomy and facilitate further identifications of this species.

puella tetrica **Casey, 1913**: 255 (Fig. 137u, v), Holotype
Original combination: *Brachyleptura puella tetrica* Casey
Current name: *Brachyleptura vagans* (Olivier, 1795)
Current placement: Cerambycidae: Lepturinae: Lepturini
Type locality: There is no locality data associated with this specimen.

puertoricensis **Fisher, 1935a**: 51 (Fig. 137w, x), Holotype
Original combination: *Tilloclytus puertoricensis* Fisher
Current name: *Tilloclytus minutus* Fisher, 1932
Current placement: Cerambycidae: Cerambycinae: Anaglyptini
Type locality: USA: Puerto Rico, Guánica, Borinquen Forest Reserve

puertoricensis **Fisher, 1935a**: 56 (Fig. 138a, b), Holotype
Original combination: *Leptostylus puertoricensis* Fisher
Current name: *Styloleptus nigricans* (Fisher, 1935a)
Current placement: Cerambycidae: Lamiinae: Acanthocinini
Type locality: USA: Puerto Rico, Adjuntas, Pietri Finca
Remarks: The specific epithet is misspelled on the label.

puertoricensis **Lingafelter & Micheli, 2004**: 46 (Fig. 138c, d), Holotype
Original combination: *Distenia puertoricensis* Lingafelter & Micheli
Current name: *Novantinoe puertoricensis* (Lingafelter & Micheli, 2004)
Current placement: Disteniidae: Disteniinae: Disteniini
Type locality: USA: Puerto Rico, Maricao Forest, Carreterra 120, km 9–15, 18°08′45″N, 66°58′52″W, 850–950 m

***pugetana* Casey, 1913**: 237 (Fig. 138e, f), Lectotype
Original combination: *Leptacmaeops (Leptacmaeops) pugetana* Casey
Current name: *Cortodera subpilosa* (LeConte, 1850)
Current placement: Cerambycidae: Lepturinae: Rhagiini
Type locality: USA: Washington, Seattle
Remarks: This specimen is herein designated as the lectotype in order to stabilize the taxonomy and facilitate further identifications of this species

***pugetanum* Casey, 1912**: 261 (Fig. 138g, h), Lectotype
Original combination: *Asemum pugetanum* Casey
Current name: *Asemum striatum* (Linnaeus, 1758)
Current placement: Cerambycidae: Spondylidinae: Asemini
Type locality: USA: Washington, Seattle
Remarks: This specimen is herein designated as the lectotype in order to stabilize the taxonomy and facilitate further identifications of this species.

***pulchella* Fisher, 1935c**: 616 (Fig. 138i, j), Holotype
Original combination: *Neoepaphra pulchella* Fisher
Current name: same
Current placement: Cerambycidae: Lamiinae: Rhodopini
Type locality: Malaysia: Borneo, Sabah, Mt. Kinabalu, Kenokok, 3300 ft.

***pulchellus retractus* Casey, 1912**: 339 (Fig. 138k, l), Holotype
Original combination: *Crossidius pulchellus retractus* Casey
Current name: *Crossidius pulchellus* LeConte, 1861
Current placement: Cerambycidae: Cerambycinae: Trachyderini
Type locality: USA: Kansas

***pulchellus sinuosus* Casey, 1912**: 339 (Fig. 138m, n), Holotype
Original combination: *Crossidius pulchellus sinuosus* Casey
Current name: *Crossidius pulchellus* LeConte, 1861
Current placement: Cerambycidae: Cerambycinae: Trachyderini
Type locality: No published locality data or label associated with the specimen.

***pulchellus vegasensis* Casey, 1912**: 339 (Fig. 138o, p), Holotype
Original combination: *Crossidius pulchellus vegasensis* Casey
Current name: *Crossidius pulchellus* LeConte, 1861
Current placement: Cerambycidae: Cerambycinae: Trachyderini
Type locality: USA: New Mexico, Las Vegas

***pulcherrima* Dillon & Dillon, 1946b**: 285 (Fig. 138q, r), Holotype
Original combination: *Trestonia pulcherrima* Dillon & Dillon
Current name: same
Current placement: Cerambycidae: Lamiinae: Onciderini
Type locality: Panama: Canal Zone, El Cermeno

***pulchra* Breuning, 1961a**: 19 (Fig. 138s, t), Holotype
Original combination: *Ornatodesisa pulchra* Breuning
Current name: *Cryptocranium cazieri* Lane, 1958
Current placement: Cerambycidae: Lamiinae: Pteropliini
Type locality: Brazil: Itaituba, Río Tapajós
Remarks: According to Breuning the holotype is in the Museum national d'Histoire naturelle, Paris, France, however this specimen and its labels match the original description.

***pulchra* Chemsak & Linsley, 1964b**: 159 (Fig. 138u, v), Holotype
Original combination: *Methia pulchra* Chemsak & Linsley
Current name: *Methia necydalea* (Fabricius, 1798)
Current placement: Cerambycidae: Cerambycinae: Methiini
Type locality: Mexico: Quintana Roo, Cozumel Island, Celarain Point

***pulchra* Fisher, 1925a**: 206 (Fig. 138w, x), Holotype
Original combination: *Enispia pulchra* Fisher
Current name: *Athylia pulchra* (Fisher, 1925a)
Current placement: Cerambycidae: Lamiinae: Pteropliini
Type locality: Malaysia: Borneo, Sandakan

***pulchra* Galileo & Martins, 2005b**: 65 (Fig. 139a, b), Holotype
Original combination: *Okamira pulchra* Galileo & Martins
Current name: same
Current placement: Cerambycidae: Lamiinae: Hemilophini
Type locality: Ecuador: Napo, Reserva Étnica Waoranic (00°39′10S, 76°26′W), 1 km south of Onkone Gare Camp, 220 m.

***punctata* Chemsak & Linsley, 1971**: 118 (Fig. 139c, d), Holotype
Original combination: *Haplidoeme punctata* Chemsak & Linsley
Current name: same
Current placement: Cerambycidae: Cerambycinae: Oemini
Type locality: Mexico: Hidalgo, 3 miles east of Zimapan, 6400 ft.

***punctata* Dillon & Dillon, 1946b**: 390 (Fig. 139e, f), Holotype
Original combination: *Oncideres punctata* Dillon & Dillon
Current name: same
Current placement: Cerambycidae: Lamiinae: Onciderini
Type locality: Nicaragua: Ameya

***punctata* Fisher, 1925a**: 235 (Fig. 139g, h), Holotype
Original combination: *Sybra punctata* Fisher
Current name: same
Current placement: Cerambycidae: Lamiinae: Apomecynini
Type locality: Philippines: Basilan

***punctatus* Casey, 1924**: 293 (Fig. 139i, j), Holotype
Original combination: *Graphisurus punctatus* Casey
Current name: *Acanthocinus obsoletus* (Olivier, 1795)
Current placement: Cerambycidae: Lamiinae: Acanthocinini
Type locality: USA: Michigan, Marquette

***punctatus* Dillon & Dillon, 1941**: 53 (Fig. 139k, l), Holotype
Original combination: *Ptychodes punctatus* Dillon & Dillon
Current name: same
Current placement: Cerambycidae: Lamiinae: Monochamini
Type locality: Mexico: Veracruz, Santa Rosa

***punctatus* Fisher, 1932**: 55 (Fig. 139m, n), Holotype
Original combination: *Pentomacrus punctatus* Fisher
Current name: *Curius punctatus* (Fisher, 1932)
Current placement: Cerambycidae: Cerambycinae: Curiini
Type locality: Cuba: Santiago de Las Vegas, Habana

***punctatus* Lingafelter, 2013**: 65 (Fig. 139o, p), Holotype
Original combination: *Calocosmus punctatus* Lingafelter
Current name: same
Current placement: Cerambycidae: Lamiinae: Hemilophini
Type locality: Dominican Republic: Peravia Province, 12.4 km east of Río Ocoa

***puncticeps* Casey, 1913**: 243 (Fig. 139q, r), Lectotype
Original combination: *Acmaeops puncticeps* Casey
Current name: *Acmaeops proteus* (Kirby, 1837)
Current placement: Cerambycidae: Lepturinae: Lepturini
Type locality: USA: Colorado
Remarks: This specimen is herein designated as the lectotype in order to stabilize the taxonomy and facilitate further identifications of this species

***puncticollis* Casey, 1924**: 234 (Fig. 139s, t), Holotype
Original combination: *Hylotrupes puncticollis* Casey
Current name: *Hylotrupes bajulus* (Linnaeus, 1758)
Current placement: Cerambycidae: Cerambycinae: Hylotrupini
Type locality: USA: Maryland, Point Lookout

***puncticollis* Fisher, 1932**: 50 (Fig. 139u, v), Holotype
Original combination: *Cylindera puncticollis* Fisher
Current name: *Curtomerus puncticollis* (Fisher, 1932)
Current placement: Cerambycidae: Cerambycinae: Elaphidiini
Type locality: Haiti: Port au Prince

***puncticollis* Fisher, 1935c**: 596 (Fig. 139w, x), Holotype
Original combination: *Trachystola puncticollis* Fisher
Current name: *Acalolepta puncticollis* (Fisher, 1935c)
Current placement: Cerambycidae: Lamiinae: Monochamini
Type locality: Malaysia: Borneo, Sabah, Mt. Kinabalu, 5500 ft., Lumu Lumu

***puncticollis* Fisher, 1947b**: 178 (Fig. 140a, b), Holotype
Original combination: *Dorcasta puncticollis* Fisher
Current name: *Bebelis puncticollis* (Fisher, 1947b)
Current placement: Cerambycidae: Lamiinae: Apomecynini
Type locality: Nicaragua: Chinandega

***punctillata* Schaeffer, 1929**: 40 (Fig. 140c, d), Holotype
Original combination: *Parandra punctillata* Schaeffer
Current name: *Neandra marginicollis* (Schaeffer, 1929)
Current placement: Cerambycidae: Parandrinae: Parandrini
Type locality: USA: Arizona, Santa Catalina Mts.

***punctipennis* Casey, 1912**: 328 (Fig. 140e, f), Holotype
Original combination: *Batyle punctipennis* Casey
Current name: *Batyle ignicollis ignicollis* (Say, 1824)
Current placement: Cerambycidae: Cerambycinae: Trachyderini
Type locality: USA: New Mexico

***punctipennis* Casey, 1913**: 383 (Fig. 140g, h), Holotype
Original combination: *Tetraopes punctipennis* Casey
Current name: *Tetraopes femoratus* LeConte, 1847
Current placement: Cerambycidae: Lamiinae: Tetraopini
Type locality: USA: Texas

***punctipennis* Fisher, 1927b**: 216 (Fig. 140i, j), Holotype
Original combination: *Moneilema (Collapteryx) punctipennis* Fisher
Current name: *Moneilema punctipennis* Fisher, 1926
Current placement: Cerambycidae: Lamiinae: Moneilemini
Type locality: Mexico: Puebla, Tehuacán

***punctiventris* Casey, 1913**: 239 (Fig. 140k, l), Lectotype
Original combination: *Leptacmaeops (Leptacmaeops) punctiventris* Casey
Current name: *Cortodera longicornis* (Kirby, 1837)
Current placement: Cerambycidae: Lepturinae: Rhagiini
Type locality: USA: New Mexico, Sandia Mts.
Remarks: This specimen is herein designated as the lectotype in order to stabilize the taxonomy and facilitate further identifications of this species.

punensis **Martins & Galileo, 2008a**: 273 (Fig. 140m, n), Holotype
Original combination: *Megacyllene (Megacyllene) punensis* Martins & Galileo
Current name: *Megacyllene punensis* Martins & Galileo, 2008a
Current placement: Cerambycidae: Cerambycinae: Clytini
Type locality: Peru: Puno

puntacanaensis **Lingafelter & Micheli, 2009**: 47 (Fig. 140o, p), Holotype
Original combination: *Leptostylopsis puntacanaensis* Lingafelter & Micheli
Current name: same
Current placement: Cerambycidae: Lamiinae: Acanthocinini
Type locality: Dominican Republic: La Altagracia Province, Punta Cana, near Ecological Reserve, 0–5 m, 18°30.477′N, 68°22.499′W.

pusillus **Giesbert & Chemsak, 1997**: 269 (Fig. 140q, r), Holotype
Original combination: *Euderces pusillus* Giesbert & Chemsak
Current name: same
Current placement: Cerambycidae: Cerambycinae: Tillomorphini
Type locality: Costa Rica: El Congo

pygmaeus **Fisher, 1926**: 25 (Fig. 140s, t), Holotype
Original combination: *Leptostylus pygmaeus* Fisher
Current name: *Pygmaleptostylus pygmaeus* (Fisher, 1926)
Current placement: Cerambycidae: Lamiinae: Acanthocinini
Type locality: Cuba: Cayamas

pyralis **Dillon & Dillon, 1946b**: 283 (Fig. 140u, v), Holotype
Original combination: *Trestonia pyralis* Dillon & Dillon
Current name: same
Current placement: Cerambycidae: Lamiinae: Onciderini
Type locality: Costa Rica: Limón, Hamburg Farm, Reventazón, Ebene

quadricallosa cylindricollis **Casey, 1924**: 295 (Fig. 140w, x), Holotype
Original combination: *Oberea quadricallosa cylindricollis* Casey
Current name: *Oberea pruinosa* Casey, 1913
Current placement: Cerambycidae: Lamiinae: Phytoeciini
Type locality: USA: Illinois

quadriceps **Casey, 1913**: 237 (Fig. 141a, b), Lectotype
Original combination: *Leptacmaeops quadriceps* Casey
Current name: *Cortodera subpilosa* (LeConte, 1850)
Current placement: Cerambycidae: Lepturinae: Rhagiini
Type locality: USA: Utah
Remarks: This specimen is herein designated as the lectotype in order to stabilize the taxonomy and facilitate further identifications of this species.

quadricolle **Casey, 1924**: 246 (Fig. 141c, d), Holotype
Original combination: *Anoplium quadricolle* Casey
Current name: *Anelaphus brevidens* (Schaeffer, 1908)
Current placement: Cerambycidae: Cerambycinae: Elaphidiini
Type locality: USA: Arizona, Tucson

quadricollis **Casey, 1924**: 259 (Fig. 141e, f), Holotype
Original combination: *Eburia quadricollis* Casey
Current name: *Eburia ovicollis* LeConte, 1873
Current placement: Cerambycidae: Cerambycinae: Eburiini
Type locality: USA: Texas, Comal Co.

quadricristatus **Breuning, 1942b**: 15 (Fig. 141g, h), Holotype
Original combination: *Mimocorus quadricristatus* Breuning
Current name: same
Current placement: Cerambycidae: Lamiinae: Crossotini
Type locality: Tanzania: Usambara (Tanganyika)

quadriguttata **Aurivillius, 1927b**: 573 (Fig. 141i, j), Syntype
Original combination: *Sybra quadriguttata* Aurivillius
Current name: same
Current placement: Cerambycidae: Lamiinae: Apomecynini
Type locality: Philippines: Mindanao, Surigao
Remarks: Tavakilian and Chevillotte (2013) state that syntypes are present in the Naturhistoriska Riksmuseet, Stockholm, Sweden as well as the USNM.

quadrilineatus biplectralis **Casey, 1924**: 229 (Fig. 141k, l), Holotype
Original combination: *Opsimus quadrilineatus biplectralis* Casey
Current name: *Opsimus quadrilineatus* Mannerheim, 1843
Current placement: Cerambycidae: Cerambycinae: Opsimini
Type locality: USA: California, Del Norte Co.

quadrinotata **Casey, 1913**: 233 (Fig. 141m, n), Holotype
Original combination: *Leptacmaeops quadrinotata* Casey
Current name: *Cortodera cubitalis* (LeConte, 1861)
Current placement: Cerambycidae: Lepturinae: Rhagiini
Type locality: USA: California, Santa Cruz Mts.

***quadriplagiata* Aurivillius, 1927b**: 582 (Fig. 141o, p), Syntype
Original combination: *Serixia quadriplagiata* Aurivillius
Current name: same
Current placement: Cerambycidae: Lamiinae: Saperdini
Type locality: Philippines: Sibuyan, Romblon
Remarks: Tavakilian and Chevillotte (2013) state that syntypes are present in the Naturhistoriska Riksmuseet, Stockholm, Sweden as well as the USNM.

***quadriplagiata* Breuning, 1940d**: 413 (Fig. 141q, r), Holotype
Original combination: *Cremnosterna quadriplagiata* Breuning
Current name: same
Current placement: Cerambycidae: Lamiinae: Monochamini
Type locality: India: Karnataka, Mysore

***quadrispinosus* var. *basimaculatus* Tippmann, 1953c**: 208 (Fig. 141s, t), Holotype
Original combination: *Metopocoilus quadrispinosus* var. *basimaculatus* Tippmann
Current name: *Metopocoilus quadrispinosus* (Buquet, 1860)
Current placement: Cerambycidae: Cerambycinae: Trachyderini
Type locality: Brazil: Santa Catarina, Nova Teutonia
Remarks: The name following the binomen was not expressly given infrasubspecific rank and the work did not indicate unambiguously that the name was infrasubspecific, therefore, it is deemed a subspecies and available.

***quaternaria evanescnes* Kriesche, 1936**: 69 (Fig. 141u, v), Holotype
Original combination: *Planodes quaternaria evanescnes* Kriesche
Current name: *Planodes quaternaria* Newman, 1842
Current placement: Cerambycidae: Lamiinae: Mesosini
Type locality: Philippines: Masbate, Karbala, Aroroy
Remarks: Tippmann labeled this specimen as having been obtained from the Kriesche Estate in 1955.

***quercus* Schaeffer, 1905b**: 163 (Fig. 141w, x), Syntype
Original combination: *Xylotrechus quercus* Schaeffer
Current name: same
Current placement: Cerambycidae: Cerambycinae: Clytini
Type locality: USA: Arizona, Huachuca Mts.

***quinquecinctus* Aurivillius, 1922**: 418 (Fig. 142a, b), Holotype
Original combination: *Demonax quinquecinctus* Aurivillius
Current name: same
Current placement: Cerambycidae: Cerambycinae: Clytini
Type locality: Malaysia: Borneo, Sandakan

***quinquemaculatum* Tippmann, 1960**: 179 (Fig. 142c, d), Holotype
Original combination: *Alcidion quinquemaculatum* Tippmann
Current name: *Nealcidion quinquemaculatum* (Tippmann, 1960)
Current placement: Cerambycidae: Lamiinae: Acanthocinini
Type locality: Bolivia: Yungas de Totora, Sehuencas, 2100 m.

***rachelis* Fisher, 1937a**: 325 (Fig. 142e, f), Holotype
Original combination: *Neotaphos rachelis* Fisher
Current name: same
Current placement: Cerambycidae: Cerambycinae: Trachyderini
Type locality: Chile: Copiapó, Valle de Elqui

***ramosi* Micheli & Nearns, 2005**: 30 (Fig. 142g, h), Holotype
Original combination: *Plectromerus ramosi* Micheli & Nearns
Current name: same
Current placement: Cerambycidae: Cerambycinae: Plectromerini
Type locality: USA: Puerto Rico, Maricao, Rd. 120, Km. 13.8

***reclusa* Martins, 1976**: 341 (Fig. 142i, j), Holotype
Original combination: *Gorybia reclusa* Martins
Current name: same
Current placement: Cerambycidae: Cerambycinae: Piezocerini
Type locality: Panama: Canal Zone, Cabima

***rectilinea* Casey, 1924**: 243 (Fig. 142k, l), Lectotype
Original combination: *Aneflomorpha rectilinea* Casey
Current name: same
Current placement: Cerambycidae: Cerambycinae: Elaphidiini
Type locality: USA: Arizona, Tucson
Remarks: This specimen is herein designated as the lectotype in order to stabilize the taxonomy and facilitate further identifications of this species.

***rector* Casey, 1913**: 287 (Fig. 142m, n), Lectotype
Original combination: *Moneilema (Collapteryx) rector* Casey
Current name: *Moneilema armatum* LeConte, 1853
Current placement: Cerambycidae: Lamiinae: Moneilemini
Type locality: USA: Colorado
Remarks: This specimen is herein designated as the lectotype in order to stabilize the taxonomy and facilitate further identifications of this species.

***reducta* Casey, 1912**: 351 (Fig. 142o, p), Lectotype
Original combination: *Cyllene reducta* Casey
Current name: *Placosternus difficilis* (Chevrolat, 1862)
Current placement: Cerambycidae: Cerambycinae: Clytini
Type locality: USA: Texas, Brownsville (?)
Remarks: This specimen is herein designated as the lectotype in order to stabilize the taxonomy and facilitate further identifications of this species

***reducta* Casey, 1913**: 266 (Fig. 142q, r), Holotype
Original combination: *Strophiona reducta* Casey
Current name: *Strophiona tigrina* Casey, 1913
Current placement: Cerambycidae: Lepturinae: Lepturini
Type locality: USA: California

***reductus* Casey, 1912**: 316 (Fig. 142s, t), Lectotype
Original combination: *Dendrobias reductus* Casey
Current name: *Trachyderes (Dendrobias) mandibularis reductus* Casey, 1912
Current placement: Cerambycidae: Cerambycinae: Trachyderini
Type locality: USA: California
Remarks: This specimen is herein designated as the lectotype in order to stabilize the taxonomy and facilitate further identifications of this species.

***referens* Aurivillius, 1926**: 95 (Fig. 142u, v), Holotype
Original combination: *Glenea (Glenea) referens* Aurivillius
Current name: *Glenea referens* Aurivillius, 1926
Current placement: Cerambycidae: Lamiinae: Saperdini
Type locality: Philippines: Mindanao, Kolambugan

***regalis* Casey, 1912**: 314 (Fig. 142w, x), Lectotype
Original combination: *Callichroma regalis* Casey
Current name: *Callichroma holochlorum* Bates, 1872
Current placement: Cerambycidae: Cerambycinae: Callichromatini
Type locality: Mexico: Guerrero
Remarks: This specimen is herein designated as the lectotype in order to stabilize the taxonomy and facilitate further identifications of this species.

***regularis* Casey, 1913**: 259 (Fig. 143a, b), Holotype
Original combination: *Strangalia regularis* Casey
Current name: *Leptura propinqua* Bland, 1865
Current placement: Cerambycidae: Lepturinae: Lepturini
Type locality: No locality label or published data.

***reichei peruvianus* Tippmann, 1953d**: 323 (Fig. 143c, d), Holotype
Original combination: *Trachyderes reichei peruvianus* Tippmann
Current name: *Trachyderes cingulatus* Klug, 1825
Current placement: Cerambycidae: Cerambycinae: Trachyderini
Type locality: Peru: Junin, Satipo

***remota* Casey, 1924**: 264 (Fig. 143e, f), Holotype
Original combination: *Batyleoma remota* Casey
Current name: *Batyle suturalis remota* Casey, 1924
Current placement: Cerambycidae: Cerambycinae: Trachyderini
Type locality: USA: Utah, Washington Field, St. George, Ut. 2800 ft.

***repens* Casey, 1924**: 226 (Fig. 143g, h), Holotype
Original combination: *Tragosoma repens* Casey
Current name: *Tragosoma depsarium* (Linnaeus, 1767)
Current placement: Cerambycidae: Prioninae: Meroscelisini
Type locality: USA: Utah

***restitutae* Aurivillius, 1911**: 206 (Fig. 143i, j), Lectotype
Original combination: *Ceroplesis restitutae* Aurivillius
Current name: *Ceroplesis quinquefasciata* (Fabricius, 1793)
Current placement: Cerambycidae: Lamiinae: Ceroplesini
Type locality: Kenya: Nangubo
Remarks: Tavakilian and Chevillotte (2013) indicates that syntypes are in Stockholm, Sweden. This specimen matches the original description and data and is herein designated as the lectotype in order to stabilize the taxonomy and facilitate further identifications of this species.

***rhamnusioides* Tippmann, 1953d**: 337 (Fig. 143k, l), Holotype
Original combination: *Dolichopterus rhamnusioides* Tippmann
Current name: *Tippmannia rhamnusioides* (Tippmann, 1953d)
Current placement: Cerambycidae: Cerambycinae: Hesperophanini
Type locality: Brazil: Santa Catarina, Mafra

***rhizophorae* Chemsak & Feller, 1988**: 181 (Fig. 143m, n), Holotype
Original combination: *Methia rhizophorae* Chemsak & Feller
Current name: *Methia necydalea* (Fabricius, 1798)
Current placement: Cerambycidae: Cerambycinae: Methiini
Type locality: Belize: Twin Cays

rhizophorae **Chemsak & Feller, 1988**: 185 (Fig. 143o, p), Holotype
Original combination: *Styloleptus rhizophorae* Chemsak & Feller
Current name: same
Current placement: Cerambycidae: Lamiinae: Acanthocinini
Type locality: Belize: Twin Cays

ricarti **Micheli, 2003**: 194 (Fig. 143q, r), Holotype
Original combination: *Eugamandus ricarti* Micheli
Current name: same
Current placement: Cerambycidae: Lamiinae: Acanthocinini
Type locality: USA: Puerto Rico, Maricao Forest, Road 120, km 13.8, beating foliage

rickstanleyi **Lingafelter & Nearns, 2007**: 185 (Fig. 143s, t), Holotype
Original combination: *Atimiola rickstanleyi* Lingafelter & Nearns
Current name: same
Current placement: Cerambycidae: Lamiinae: Desmiphorini
Type locality: Dominican Republic: Pedernales Province, PN Sierra de Baoruco, Las Abejas, 1150 m, 18°09.011′N, 71°37.342′W

rileyi **Hovore & Santos-Silva, 2007**: 88 (Fig. 143u, v), Holotype
Original combination: *Cometes rileyi* Hovore & Santos-Silva
Current name: *Arietocometes rileyi* (Hovore & Santos-Silva, 2007)
Current placement: Disteniidae: Disteniinae: Disteniini
Type locality: Mexico: Veracruz, Los Tuxtlas, Biological Station.

rimosa **Dillon & Dillon, 1945**: 24 (Fig. 143w, x), Holotype
Original combination: *Hypsioma rimosa* Dillon & Dillon
Current name: same
Current placement: Cerambycidae: Lamiinae: Onciderini
Type locality: Paraguay: San Bernardino

rinconium **Casey, 1924**: 248 (Fig. 144a, b), Holotype
Original combination: *Anoplium rinconium* Casey
Current name: *Anopliomorpha rinconium* (Casey, 1924)
Current placement: Cerambycidae: Cerambycinae: Elaphidiini
Type locality: USA: Arizona, Rincon Mts.

roberto **Fisher, 1935b**: 209 (Fig. 144c, d), Holotype
Original combination: *Essostrutha roberto* Fisher
Current name: *Oedudes roberto* (Fisher, 1935b)
Current placement: Cerambycidae: Lamiinae: Hemilophini
Type locality: Cuba: Las Animas, 1500 ft., Sierra Rangel

robinsoni **Casey, 1914**: 368 (Fig. 144e, f), Holotype
Original combination: *Goes robinsoni* Casey
Current name: *Goes tesselatus* (Haldeman, 1847)
Current placement: Cerambycidae: Lamiinae: Monochamini
Type locality: USA: Rhode Island, Watch Hill

robusta **Casey, 1912**: 342 (Fig. 144g, h), Holotype
Original combination: *Tylosis robusta* Casey
Current name: *Tylosis maculatus* LeConte, 1850
Current placement: Cerambycidae: Cerambycinae: Trachyderini
Type locality: USA: Arizona, San Bernardino Ranch, Cochise Co.

robusta **Fisher, 1925a**: 233 (Fig. 144i, j), Holotype
Original combination: *Ropica robusta* Fisher
Current name: same
Current placement: Cerambycidae: Lamiinae: Apomecynini
Type locality: Philippines: Negros Oriental, Cuernos Mountains

robustula **Casey, 1912**: 219 (Fig. 144k, l), Holotype
Original combination: *Spondylis robustula* Casey
Current name: *Neospondylis upiformis* (Mannerheim, 1843)
Current placement: Cerambycidae: Spondylidinae: Spondylidini
Type locality: USA: Washington

robustula **Casey, 1913**: 228 (Fig. 144m, n), Holotype
Original combination: *Brachysomida robustula* Casey
Current name: *Brachysomida californica* (LeConte, 1851)
Current placement: Cerambycidae: Lepturinae: Rhagiini
Type locality: USA: California, Gilroy Springs

robustula **Casey, 1913**: 326 (Fig. 144o, p), Holotype
Original combination: *Hyperplatys robustula* Casey
Current name: *Hyperplatys maculata* Haldeman, 1847
Current placement: Cerambycidae: Lamiinae: Acanthocinini
Type locality: USA: Indiana

robustum **Gressitt, 1956**: 88 (Fig. 144q, r), Holotype
Original combination: *Ceresium (Ceresium) robustum* Gressitt
Current name: same
Current placement: Cerambycidae: Cerambycinae: Callidiopini
Type locality: Marshall Islands: Ine Island, Arno Atoll

robustus **Aurivillius, 1928**: 319 (Fig. 144s, t), Holotype
Original combination: *Demonax robustus* Aurivillius
Current name: same
Current placement: Cerambycidae: Cerambycinae: Clytini
Type locality: Philippines: Sibuyan, Romblon

***robustus* Casey, 1913**: 381 (Fig. 144u, v), Lectotype
Original combination: *Tetraopes robustus* Casey
Current name: *Tetraopes femoratus* LeConte, 1847
Current placement: Cerambycidae: Lamiinae: Tetraopini
Type locality: USA: Colorado (?)
Remarks: This specimen is herein designated as the lectotype in order to stabilize the taxonomy and facilitate further identifications of this species.

***robustus* Casey, 1924**: 222 (Fig. 144w, x), Holotype
Original combination: *Prionus (Neopolyarthron) robustus* Casey
Current name: *Prionus (Neopolyarthron) imbricornis* Linnaeus, 1767
Current placement: Cerambycidae: Prioninae: Prionini
Type locality: USA: North Carolina, Southern Pines

***robustus* Lingafelter, 2013**: 69 (Fig. 145a, b), Holotype
Original combination: *Calocosmus robustus* Lingafelter
Current name: same
Current placement: Cerambycidae: Lamiinae: Hemilophini
Type locality: Dominican Republic: La Vega Province, Parque Nacional Armando Bermudez, km 1–3 along trail W of La Ciénaga, 900–1100 m., 19°01.753′N, 70°54.654′W

***roedingeri* Tippmann, 1951**: 304 (Fig. 145c, d), Holotype
Original combination: *Calocomus roedingeri* Tippmann
Current name: *Calocomus rodingeri* Tippmann, 1951
Current placement: Cerambycidae: Prioninae: Calocomini
Type locality: Peru: Ica
Remarks: Per ICZN art. 32.5.2.1, the diacritic in a German word before 1985 changed to "oe".

***rosae* Martins & Galileo, 2011**: 357 (Fig. 145e, f), Holotype
Original combination: *Mexicoscylus rosae* Martins & Galileo
Current name: same
Current placement: Cerambycidae: Lamiinae: Hemilophini
Type locality: Mexico: Oaxaca, 24 miles south Valle Nacional, 5600 ft.

***rosea* Galileo & Martins, 2004a**: 249 (Fig. 145g, h), Holotype
Original combination: *Erana rosea* Galileo & Martins
Current name: *Eranina rosea* (Galileo & Martins, 2004a)
Current placement: Cerambycidae: Lamiinae: Hemilophini
Type locality: Panama: Panama Province, El Llano Carti Rd, km 8–11, 1100 feet

***rotundipenne* Fisher, 1932**: 23 (Fig. 145i, j), Holotype
Original combination: *Elaphidion rotundipenne* Fisher
Current name: same
Current placement: Cerambycidae: Cerambycinae: Elaphidiini
Type locality: Dominican Republic: San Cristóbal, Santo Domingo, Haina

***rotundipennis* Breuning, 1950a**: 306 (Fig. 145k, l), Holotype
Original combination: *Emphreus rotundipennis* Breuning
Current name: same
Current placement: Cerambycidae: Lamiinae: Stenobiini
Type locality: Kenya: Sagala

***rubens* Casey, 1891**: 34 (Fig. 145m, n), Lectotype
Original combination: *Sphaenothecus rubens* Casey
Current name: *Plionoma rubens* (Casey, 1891)
Current placement: Cerambycidae: Cerambycinae: Trachyderini
Type locality: USA: California
Remarks: This specimen is herein designated as the lectotype in order to stabilize the taxonomy and facilitate further identifications of this species.

***ruber* Casey, 1913**: 384 (Fig. 145o, p), Lectotype
Original combination: *Tetraopes ruber* Casey
Current name: *Tetraopes femoratus* LeConte, 1847
Current placement: Cerambycidae: Lamiinae: Tetraopini
Type locality: USA: Arizona
Remarks: This specimen is herein designated as the lectotype in order to stabilize the taxonomy and facilitate further identifications of this species.

***rubra* Gressitt, 1938a**: 49 (Fig. 145q, r), Holotype
Original combination: *Psacothea rubra* Gressitt
Current name: same
Current placement: Cerambycidae: Lamiinae: Monochamini
Type locality: China: Sichuan Province, Chengdu, 1700 ft.

***rubriceps* Casey, 1913**: 198 (Fig. 145s, t), Holotype
Original combination: *Thesalia rubriceps* Casey
Current name: *Pidonia (Thesalia) gnathoides* (LeConte, 1873)
Current placement: Cerambycidae: Lepturinae: Rhagiini
Type locality: USA: California, Marin Co.

***rubricollis* Fisher, 1944a**: 6 (Fig. 145u, v), Holotype
Original combination: *Stizocera rubricollis* Fisher
Current name: *Parastizocera procera* (Erichson, 1848)
Current placement: Cerambycidae: Cerambycinae: Elaphidiini
Type locality: Venezuela: Monagas, Caripito

***rubricolor* Melzer, 1935b**: 180 (Fig. 145w, x), Holotype
Original combination: *Heterachthes rubricolor* Melzer
Current name: same
Current placement: Cerambycidae: Cerambycinae: Neoibidionini
Type locality: Costa Rica: Cartago, Turrialba

***rubripes* Tippmann, 1960**: 150 (Fig. 146a, b), Holotype
Original combination: *Lissonotus rubripes* Tippmann
Current name: same
Current placement: Cerambycidae: Cerambycinae: Lissonotini
Type locality: Brazil: Bahia

***rubriventris* Gressitt, 1937a**: 90 (Fig. 146c, d), Holotype
Original combination: *Allotraeus (Pseudallotraeus) rubriventris* Gressitt
Current name: *Allotraeus (Nysina) rubriventris* Gressitt, 1937a
Current placement: Cerambycidae: Cerambycinae: Phoracanthini
Type locality: Vietnam: Laokay Province, Tonkin, Chapo

***rubrohirsutotibialis* Tippmann, 1953c**: 210 (Fig. 146e, f), Holotype
Original combination: *Acyphoderes rubrohirsutotibialis* Tippmann
Current name: same
Current placement: Cerambycidae: Cerambycinae: Rhinotragini
Type locality: Bolivia: Cochabamba, Reg. Chapare 400 m.

***rufa* Breuning, 1943c**: 45 (Fig. 146g, h), Holotype
Original combination: *Sybra rufa* Breuning
Current name: same
Current placement: Cerambycidae: Lamiinae: Apomecynini
Type locality: Philippines: Luzon, Montalban

***rufa* Melzer, 1934**: 73 (Fig. 146i, j), Holotype
Original combination: *Hemilissa rufa* Melzer
Current name: same
Current placement: Cerambycidae: Cerambycinae: Piezocerini
Type locality: Costa Rica: Cartago, Turrialba, 800 m.

***ruficollis* Aurivillius, 1922**: 428 (Fig. 146k, l), Holotype
Original combination: *Chydaeopsis ruficollis* Aurivillius
Current name: same
Current placement: Cerambycidae: Lamiinae: Acanthocinini
Type locality: Malaysia: Borneo, Sandakan

***rufimembris* Gressitt, 1937b**: 608 (Fig. 146m, n), Holotype
Original combination: *Microleropsis rufimembris* Gressitt
Current name: same
Current placement: Cerambycidae: Lamiinae: Gyartini
Type locality: China: Fujian Province, Gang-ken, southwest Fukien

***rufipennis* Aurivillius, 1916**: 344 (Fig. 146o, p), Holotype
Original combination: *Hexarhopala rufipennis* Aurivillius
Current name: *Hexarrhopala rufipennis* Aurivillius, 1916
Current placement: Cerambycidae: Cerambycinae: Cleomenini
Type locality: Tanzania: Tasamaganga
Remarks: The genus is misspelled in the original publication. Tavakilian and Chevillotte (2013) suggest the holotype is present in the Zoologische Staatssammlung des Bayerischen Staates, Munich, Germany, however, this specimen and its labels match the original description. Apparently only Ertl's later material was donated to Munich.

***rufipennis* Casey, 1913**: 212 (Fig. 146q, r), Lectotype
Original combination: *Stenocorus (Stenocorus) rufipennis* Casey
Current name: *Stenocorus nubifer* (LeConte, 1859)
Current placement: Cerambycidae: Lepturinae: Rhagiini
Type locality: USA: California
Remarks: This specimen is herein designated as the lectotype in order to stabilize the taxonomy and facilitate further identifications of this species.

***rufipes* Gressitt, 1951a**: 29 (Fig. 146s, t), Holotype
Original combination: *Epania rufipes* Gressitt
Current name: same
Current placement: Cerambycidae: Cerambycinae: Molorchini
Type locality: Papua New Guinea: Bougainville Island (Solomon Islands)

***rufoanticus* Gressitt, 1935g**: 271 (Fig. 146u, v), Holotype
Original combination: *Stenoleptura rufoanticus* Gressitt
Current name: same
Current placement: Cerambycidae: Lepturinae: Lepturini
Type locality: Malaysia: Borneo, Sandakan

***rufofemorata* Fisher, 1930a**: 12 (Fig. 146w, x), Holotype
Original combination: *Acyphoderes rufofemorata* Fisher
Current name: *Acyphoderes hirtipes* (Klug, 1825)
Current placement: Cerambycidae: Cerambycinae: Rhinotragini
Type locality: Brazil: Bahia

***rufohumeralis* Breuning, 1957a**: 13 (Fig. 147a, b), Holotype
Original combination: *Exocentrus (Pseudocentrus) rufohumeralis* Breuning
Current name: same
Current placement: Cerambycidae: Lamiinae: Acanthocinini
Type locality: Indonesia: Sumatra, Tebing-Tinggi

***rufomarginata* Fisher, 1937d**: 153 (Fig. 147c, d), Holotype
Original combination: *Rhopalophora (Ischionodonta) rufomarginata* Fisher
Current name: *Ischionodonta rufomarginata* (Fisher, 1937d)
Current placement: Cerambycidae: Cerambycinae: Rhopalophorini
Type locality: Brazil: Santa Catarina, Nova Teutonia

***rufonigra* Gressitt, 1940**: 175 (Fig. 147e, f), Holotype
Original combination: *Terinaea rufonigra* Gressitt
Current name: same
Current placement: Cerambycidae: Lamiinae: Desmiphorini
Type locality: China: Hainan Province, Hainan Island, Ta Han

***rufosemivittata* Tippmann, 1960**: 113 (Fig. 147g, h), Holotype
Original combination: *Coleoxestia rufosemivittata* Tippmann
Current name: same
Current placement: Cerambycidae: Cerambycinae: Cerambycini
Type locality: Bolivia: Cochabamba, Chapare - Gebiet, Obere, Río Chipiriri, 400 m.
Remarks: This holotype may belong to the Zoologische Staatssammlung, Munich, Germany.

***rufovittata* Breuning, 1943c**: 59 (Fig. 147i, j), Holotype
Original combination: *Pothyne rufovittata* Breuning
Current name: same
Current placement: Cerambycidae: Lamiinae: Agapanthiini
Type locality: Vietnam (Indochina)

***rufula* Fisher, 1925a**: 267 (Fig. 147k, l), Holotype
Original combination: *Miaenia rufula* Fisher
Current name: *Sciades (Miaenia) rufulus* (Fisher, 1925a)
Current placement: Cerambycidae: Lamiinae: Acanthocinini
Type locality: Malaysia: Borneo, Sandakan

***rugicollis* Gressitt, 1948a**: 48 (Fig. 147m, n), Holotype
Original combination: *Trachylophus rugicollis* Gressitt
Current name: same
Current placement: Cerambycidae: Cerambycinae: Cerambycini
Type locality: China: Sichuan Province, Shin Kai Si, Mt. Omei, 4,400 ft.

***rugosipenne* Linell, 1897a**: 395 (Fig. 147o, p), Holotype
Original combination: *Poecilobrium rugosipenne* Linell
Current name: *Poecilobrium chalybeum* (LeConte, 1873)
Current placement: Cerambycidae: Cerambycinae: Eumichthini
Type locality: USA: California

***rugosipennis* Fisher, 1928b**: 2 (Fig. 147q, r), Holotype
Original combination: *Moneilema (Collapteryx) rugosipennis* Fisher
Current name: *Moneilema armatum* LeConte, 1853
Current placement: Cerambycidae: Lamiinae: Moneilemini
Type locality: Mexico: San Luis Potosí

***rugosissima* Casey, 1924**: 288 (Fig. 147s, t), Holotype
Original combination: *Moneilema (Collapteryx) rugosissima* Casey
Current name: *Moneilema rugosissimum* Casey, 1924
Current placement: Cerambycidae: Lamiinae: Moneilemini
Type locality: Mexico: Baja California Sur, Santa Rosa

***russellae* Quentin & Villiers, 1971**: 16 (Fig. 147u, v), Holotype
Original combination: *Bottegia russellae* Quentin & Villiers
Current name: same
Current placement: Cerambycidae: Cerambycinae: Psebiini
Type locality: Guinea-Bissau: Bafatá

***rusticus* Aurivillius, 1916**: 348 (Fig. 147w, x), Syntype
Original combination: *Idactus rusticus* Aurivillius
Current name: same
Current placement: Cerambycidae: Lamiinae: Ancylonotini
Type locality: Tanzania: Namupa
Remarks: Tavakilian and Chevillotte (2013) state that syntypes are present in the Naturhistoriska Riksmuseet, Stockholm, Sweden, however, this is one of the type specimens.

***sakishimana* Gressitt, 1951c**: 227 (Fig. 148a, b), Holotype
Original combination: *Miaenia sakishimana* Gressitt
Current name: *Sciades (Estoliops) sakishimanus* (Gressitt, 1951c)
Current placement: Cerambycidae: Lamiinae: Acanthocinini
Type locality: Japan: Ryukyu Islands (formerly Loochoo), Iriomote Island

***salebrosus* Casey, 1912**: 364 (Fig. 148c, d), Holotype
Original combination: *Xylotrechus salebrosus* Casey
Current name: *Xylotrechus sagitatus* (Germar, 1821)
Current placement: Cerambycidae: Cerambycinae: Clytini
Type locality: USA: New Mexico

***saleyerianus* Breuning, 1957a**: 13 (Fig. 148e, f), Holotype
Original combination: *Exocentrus (Pseudocentrus) saleyerianus* Breuning
Current name: same
Current placement: Cerambycidae: Lamiinae: Acanthocinini
Type locality: Indonesia: Sulawesi, Salayar

***salicola* Casey, 1913**: 347 (Fig. 148g, h), Holotype
Original combination: *Pogonocherus salicola* Casey
Current name: *Pogonocherus parvulus* LeConte, 1852
Current placement: Cerambycidae: Lamiinae: Pogonocherini
Type locality: USA: Massachusetts, Framingham
Remarks: The specific epithet is misspelled on the label.

***salomonum* Breuning, 1938d**: 41 (Fig. 148i, j), Holotype
Original combination: *Ischioplites salomonum* Breuning
Current name: same
Current placement: Cerambycidae: Lamiinae: Pteropliini
Type locality: Papua New Guinea: Bougainville Island (Solomon Islands), Kieta
Remarks: Although Tavakilian and Chevillotte (2013) indicate the holotype is in Belgium, this specimen matches the original description and data.

***saltator* Gressitt, 1956**: 159 (Fig. 148k, l), Holotype
Original combination: *Sciadella (Micronesiella) saltator* Gressitt
Current name: *Sciades (Micronesiella) saltator* (Gressitt, 1956)
Current placement: Cerambycidae: Lamiinae: Acanthocinini
Type locality: Micronesia: Yap Island, Mt. Matade, 60 m.

***samarensis* Aurivillius, 1926**: 95 (Fig. 148m, n), Holotype
Original combination: *Glenea (Glenea) samarensis* Aurivillius
Current name: *Glenea samarensis* Aurivillius, 1926
Current placement: Cerambycidae: Lamiinae: Saperdini
Type locality: Philippines: Samar

***samarensis* Aurivillius, 1927b**: 552 (Fig. 148o, p), Holotype
Original combination: *Dihammus samarensis* Aurivillius
Current name: *Acalolepta samarensis* (Aurivillius, 1927b)
Current placement: Cerambycidae: Lamiinae: Monochamini
Type locality: Philippines: Samar

***samarensis* Aurivillius, 1928**: 321 (Fig. 148q, r), Holotype
Original combination: *Demonax samarensis* Aurivillius
Current name: same
Current placement: Cerambycidae: Cerambycinae: Clytini
Type locality: Philippines: Samar

***sandakanus* Aurivillius, 1927b**: 550 (Fig. 148s, t), Syntype
Original combination: *Xylotrechus sandakanus* Aurivillius
Current name: same
Current placement: Cerambycidae: Cerambycinae: Clytini
Type locality: Malaysia: Borneo, Sandakan
Remarks: Tavakilian and Chevillotte (2013) state that syntypes are present in the Naturhistoriska Riksmuseet, Stockholm, Sweden as well as the USNM.

***sandakanus* Gressitt, 1935g**: 270 (Fig. 148u, v), Holotype
Original combination: *Stenoleptura sandakanus* Gressitt
Current name: same
Current placement: Cerambycidae: Lepturinae: Lepturini
Type locality: Malaysia: Borneo, Sandakan

***sandix* Casey, 1914**: 370 (Fig. 148w, x), Lectotype
Original combination: *Tetraopes sandix* Casey
Current name: *Tetraopes sublaevis* Casey, 1913
Current placement: Cerambycidae: Lamiinae: Tetraopini
Type locality: USA: California, San Diego Co., Witch Creek
Remarks: This specimen is herein designated as the lectotype in order to stabilize the taxonomy and facilitate further identifications of this species.

***sanguinicollis* Olivier, 1795**: 9 (Fig. 149a, b), Neotype
Original combination: *Necydalis sanguinicollis* Olivier
Current name: *Callimoxys sanguinicollis* (Olivier, 1795)
Current placement: Cerambycidae: Cerambycinae: Stenopterini
Type locality: USA: New York, Ulster Co., Catskill Mts.
Remarks: Hammond and Williams (2011) should have indicated *Necydalis*, the original combination, on the label when they designated this specimen as the neotype.

***sanguinolenta* v. *rufifrons* Fisher, 1937d**: 151 (Fig. 149c, d), Holotype
Original combination: *Odontocera sanguinolenta* v. *rufifrons* Fisher
Current name: *Odontocera sanguinolenta* Bates, 1873
Current placement: Cerambycidae: Cerambycinae: Rhinotragini
Type locality: Brazil: Santa Catarina, Nova Teutonia
Remarks: The name following the binomen was not expressly given infrasubspecific rank and the work did not indicate unambiguously that the name was infrasubspecific, therefore, it is deemed a subspecies and available.

saxuosus **Tippmann, 1960**: 189 (Fig. 149e, f), Holotype
Original combination: *Leptostylus saxuosus* Tippmann
Current name: *Erphaea pumicosa* Erichson, 1847
Current placement: Cerambycidae: Lamiinae: Acanthocinini
Type locality: Bolivia: Cochabamba, Chapare-Gebiet Oberer, 400 m., Río Chipiriri
Remarks: Tavakilian and Chevillotte (2013) suggest the holotype is present in the Zoologische Staatssammlung des Bayerischen Staates, Munich, Germany, however, this specimen and its labels match the original description.

sayi **Casey, 1912**: 340 (Fig. 149g, h), Lectotype
Original combination: *Crossidius sayi* Casey
Current name: *Crossidius discoideus sayi* Casey, 1912
Current placement: Disteniidae: Disteniinae: Trachyderini
Type locality: USA: New Mexico
Remarks: This specimen is herein designated as the lectotype in order to stabilize the taxonomy and facilitate further identifications of this species.

sayi bilenticulatus **Casey, 1912**: 340 (Fig. 149i, j), Holotype
Original combination: *Crossidius sayi bilenticulatus* Casey
Current name: *Crossidius discoideus sayi* Casey, 1912
Current placement: Cerambycidae: Cerambycinae: Trachyderini
Type locality: USA: New Mexico

scabricollis **Melzer, 1935b**: 222 (Fig. 149k, l), Holotype
Original combination: *Odontocera scabricollis* Melzer
Current name: same
Current placement: Cerambycidae: Cerambycinae: Rhinotragini
Type locality: Costa Rica: Cartago, Turrialba

scapalis **Casey, 1913**: 312 (Fig. 149m, n), Holotype
Original combination: *Leiopus scapalis* Casey
Current name: *Sternidius alpha* (Say, 1827)
Current placement: Cerambycidae: Lamiinae: Acanthocinini
Type locality: USA: Indiana

scapalis **Casey, 1913**: 355 (Fig. 149o, p), Holotype
Original combination: *Spalacopsis scapalis* Casey
Current name: *Spalacopsis filum costulatum* Casey, 1913
Current placement: Cerambycidae: Lamiinae: Agapanthiini
Type locality: USA: Florida, Palm Beach

scaramuzzai **Fisher, 1936c**: 271 (Fig. 149q, r), Holotype
Original combination: *Essostrutha scaramuzzai* Fisher
Current name: *Oedudes scaramuzzai* (Fisher, 1936c)
Current placement: Cerambycidae: Lamiinae: Hemilophini
Type locality: Cuba: Santiago de Cuba, Loma del Gato, 2600–3325 ft, Sierra del Cobre, Oriente

scaramuzzai **Fisher, 1951**: 77 (Fig. 149s, t), Holotype
Original combination: *Elaphidion scaramuzzai* Fisher
Current name: same
Current placement: Cerambycidae: Cerambycinae: Elaphidiini
Type locality: Cuba: Matanzas, central Mercedes

scheerpeltzi **Tippmann, 1953c**: 182 (Fig. 149u, v), Holotype
Original combination: *Strongylaspis (Curitiba) scheerpeltzi* Tippmann
Current name: *Curitiba brunni* (Lameere, 1903)
Current placement: Cerambycidae: Prioninae: Macrotomini
Type locality: Brazil: Mato Grosso, Cuyabá

schildi **Melzer, 1935a**: 240 (Fig. 149w, x), Holotype
Original combination: *Phaea schildi* Melzer
Current name: *Phaea scuticollis* Bates, 1872
Current placement: Cerambycidae: Lamiinae: Tetraopini
Type locality: Costa Rica: Cartago, Turrialba, 800 m.

schotti **Schaeffer, 1917**: 184 (Fig. 150a, b), Holotype
Original combination: *Callidium schotti* Schaeffer
Current name: same
Current placement: Cerambycidae: Cerambycinae: Callidiini
Type locality: USA: New York, Huntington, Long Island

schwarzeri **Fisher, 1935c**: 624 (Fig. 150c, d), Holotype
Original combination: *Glenea schwarzeri* Fisher
Current name: same
Current placement: Cerambycidae: Lamiinae: Saperdini
Type locality: Malaysia: Borneo, Sabah, Mt. Kinabalu, 3000 ft., Kiau-Tenompok Pass

schwarzi **Fisher, 1914**: 97 (Fig. 150e, f), Holotype
Original combination: *Callichroma schwarzi* Fisher
Current name: *Plinthocoelium schwarzi* (Fisher, 1914)
Current placement: Cerambycidae: Cerambycinae: Callichromatini
Type locality: USA: Texas, McAllen

schwarzi **Fisher, 1926**: 34 (Fig. 150g, h), Holotype
Original combination: *Eugamandus schwarzi* Fisher
Current name: same
Current placement: Cerambycidae: Lamiinae: Acanthocinini
Type locality: Cuba: Cayamas

schwarzi **Fisher, 1926**: 36 (Fig. 150i, j), Holotype
Original combination: *Cyrtinus schwarzi* Fisher
Current name: same
Current placement: Cerambycidae: Lamiinae: Cyrtinini
Type locality: Cuba: Cayamas

schwarzi **Fisher, 1932**: 66 (Fig. 150k, l), Holotype
Original combination: *Calliclytus schwarzi* Fisher
Current name: same
Current placement: Cerambycidae: Cerambycinae: Tillomorphini
Type locality: Cuba: Cayamas

schwarzi **Fisher, 1947c**: 37 (Fig. 150m, n), Holotype
Original combination: *Dorcasta schwarzi* Fisher
Current name: *Bebelis schwarzi* (Fisher, 1947c)
Current placement: Cerambycidae: Lamiinae: Apomecynini
Type locality: Cuba: Cayamas
Remarks: Specific epithet in original publication is *schwarzi* (without the t).

schwarzi **Hamilton in Leng & Hamilton, 1896**: 124 (Fig. 150o, p), Holotype
Original combination: *Leiopus schwarzi* Hamilton in Leng & Hamilton
Current name: *Sternidius mimeticus* (Casey, 1891)
Current placement: Cerambycidae: Lamiinae: Acanthocinini
Type locality: USA: Florida, Key West

schwarzianum **Casey, 1891**: 24 (Fig. 150q, r), Holotype
Original combination: *Tetropium schwarzianum* Casey
Current name: same
Current placement: Cerambycidae: Spondylidinae: Asemini
Type locality: USA: Michigan, Marquette

scorpiona **Lane, 1966b**: 282 (Fig. 150s, t), Holotype
Original combination: *Woytkowskia scorpiona* Lane
Current name: same
Current placement: Cerambycidae: Lamiinae: Hemilophini
Type locality: Peru: Junin, Satipo

scurra **var.** *dorsalis* **Fisher, 1926**: 21 (Fig. 150u, v), Holotype
Original combination: *Leptostylus scurra* var. *dorsalis* Fisher
Current name: *Styloleptus biustus dorsalis* (Fisher, 1926)
Current placement: Cerambycidae: Lamiinae: Acanthocinini
Type locality: Cuba: Cayamas
Remarks: The name following the binomen was not expressly given infrasubspecific rank and the work did not indicate unambiguously that the name was infrasubspecific, therefore, it is deemed a subspecies and available.

scurra **var.** *maculifer* **Fisher, 1926**: 21 (Fig. 150w, x), Holotype
Original combination: *Leptostylus scurra* var. *maculifer* Fisher
Current name: *Styloleptus scurra* (Chevrolat, 1862)
Current placement: Cerambycidae: Lamiinae: Acanthocinini
Type locality: Cuba: Estacion Central Agricultural de Cuba
Remarks: The name following the binomen was not expressly given infrasubspecific rank and the work did not indicate unambiguously that the name was infrasubspecific, therefore, it is deemed a subspecies and available.

scutellaris **Casey, 1924**: 219 (Fig. 151a, b), Holotype
Original combination: *Prionus (Prionus) scutellaris* Casey
Current name: *Prionus californicus* Motschulsky, 1845
Current placement: Cerambycidae: Prioninae: Prionini
Type locality: USA: Utah, North Fork, Provo Canyon

scutellaris **Fisher, 1925a**: 262 (Fig. 151c, d), Holotype
Original combination: *Enes scutellaris* Fisher
Current name: *Enes (Sibuyenes) scutellaris* Fisher, 1925a
Current placement: Cerambycidae: Lamiinae: Acanthocinini
Type locality: Philippines: Sibuyan, Romblon

scutellata **Casey, 1924**: 276 (Fig. 151e, f), Holotype
Original combination: *Acmaeops scutellata* Casey
Current name: *Acmaeops proteus* (Kirby, 1837)
Current placement: Cerambycidae: Lepturinae: Rhagiini
Type locality: USA: Idaho, Priest Lake, Bonner Co.

scutellata **Fisher, 1925a**: 234 (Fig. 151g, h), Holotype
Original combination: *Sybra scutellata* Fisher
Current name: same
Current placement: Cerambycidae: Lamiinae: Apomecynini
Type locality: Philippines: Basilan

seclusa **Casey, 1912**: 327 (Fig. 151i, j), Holotype
Original combination: *Batyle seclusa* Casey
Current name: *Batyle ignicollis* (Say, 1824)
Current placement: Cerambycidae: Cerambycinae: Trachyderini
Type locality: USA: New Mexico, Jemez Springs

secunda **Tippmann, 1951**: 301 (Fig. 151k, l), Holotype
Original combination: *Didymonycha secunda* Tippmann
Current name: *Amillarus secundus* (Tippmann, 1951)
Current placement: Cerambycidae: Lamiinae: Agapanthiini
Type locality: Venezuela: Maracay, 400–420 m, Aragua

***semenovi* Heyrovsky, 1934**: 172 (Fig. 151m, n), Syntype
Original combination: *Moechotypa semenovi* Heyrovsky
Current name: same
Current placement: Cerambycidae: Lamiinae: Crossotini
Type locality: China: Sichuan Province, Giufu Shan, 1500–2000 m.
Remarks: The original description indicates that there were 15 syntypes in the Reitter and Heyrovsky collections.

***semigriseus* Aurivillius, 1927b**: 562 (Fig. 151o, p), Syntype
Original combination: *Spiniscapus semigriseus* Aurivillius
Current name: *Dystasia semicana* Pascoe, 1864
Current placement: Cerambycidae: Lamiinae: Pteropliini
Type locality: Philippines: Sibuyan, Romblon
Remarks: Tavakilian and Chevillotte (2013) state that syntypes are present in the Naturhistoriska Riksmuseet, Stockholm, Sweden as well as the USNM.

***semigriseus sibuyanus* Aurivillius, 1927b**: 563 (Fig. 151q, r), Syntype
Original combination: *Spiniscapus semigriseus sibuyanus* Aurivillius
Current name: *Dystasia sibuyana* (Aurivillius, 1927b)
Current placement: Cerambycidae: Lamiinae: Pteropliini
Type locality: Philippines: Sibuyan, Romblon
Remarks: Tavakilian and Chevillotte (2013) state that syntypes are present in the Naturhistoriska Riksmuseet, Stockholm, Sweden as well as the USNM.

***semilactea* Heller, 1923b**: 19 (Fig. 151s, t), Lectotype
Original combination: *Cacia semilactea* Heller
Current name: *Cacia (Ipocregyes) semilactea* Heller, 1923b
Current placement: Cerambycidae: Lamiinae: Mesosini
Type locality: Philippines: Sibuyan
Remarks: This specimen is herein designated as the lectotype in order to stabilize the taxonomy and facilitate further identifications of this species.

***semilunata* Gressitt, 1948a**: 44 (Fig. 151u, v), Holotype
Original combination: *Leptura semilunata* Gressitt
Current name: *Leptura arcifera* (Blanchard, 1871)
Current placement: Cerambycidae: Lepturinae: Lepturini
Type locality: China: Sichuan Province, southwest of Tatsienlu, 9,000–13,000 ft.

***seminigra* Casey, 1924**: 283 (Fig. 151w, x), Holotype
Original combination: *Judolia seminigra* Casey
Current name: *Judolia montivagans* (Couper, 1864)
Current placement: Cerambycidae: Lepturinae: Lepturini
Type locality: USA: Idaho, Priest Lake, Bonner Co.

***seminuda* Casey, 1912**: 294 (Fig. 152a, b), Holotype
Original combination: *Aneflomorpha seminuda* Casey
Current name: same
Current placement: Cerambycidae: Cerambycinae: Elaphidiini
Type locality: USA: Texas

***seminuda* Fisher, 1935c**: 628 (Fig. 152c, d), Holotype
Original combination: *Chreonoma seminuda* Fisher
Current name: *Bacchisa testacea* (Thomson, 1857)
Current placement: Cerambycidae: Lamiinae: Tetraopini
Type locality: Malaysia: Borneo, Sabah, Mt. Kinabalu, Kiau, 3000 ft.

***semiporosa* Tippmann, 1960**: 160 (Fig. 152e, f), Holotype
Original combination: *Oreodera semiporosa* Tippmann
Current name: same
Current placement: Cerambycidae: Lamiinae: Acanthoderini
Type locality: Bolivia: Cochabamba, Reg. Chapare 400 m.

***semirufa* Linsley, 1935a**: 146 (Fig. 152g, h), Holotype
Original combination: *Aneflomorpha semirufa* Linsley
Current name: same
Current placement: Cerambycidae: Cerambycinae: Elaphidiini
Type locality: Mexico: Oaxaca, Salina Cruz

***semirufum* Martins, 1968**: 422 (Fig. 152i, j), Holotype
Original combination: *Tropidion semirufum* Martins
Current name: same
Current placement: Cerambycidae: Cerambycinae: Neoibidionini
Type locality: Brazil

***septuosa* Galileo & Martins, 2004a**: 249 (Fig. 152k, l), Holotype
Original combination: *Erana septuosa* Galileo & Martins
Current name: *Eranina septuosa* (Galileo & Martins, 2004a)
Current placement: Cerambycidae: Lamiinae: Hemilophini
Type locality: Panama: Chiriquí, Continental Divide Trail, 3600 ft.

***sequarium* Fisher, 1920**: 156 (Fig. 152m, n), Holotype
Original combination: *Callidium sequarium* Fisher
Current name: *Callidium sequoiarium* Fisher, 1920
Current placement: Cerambycidae: Cerambycinae: Callidiini
Type locality: USA: California, Giant Forest, Sequoia

seriata **Chemsak & Linsley, 1984**: 283 (Fig. 152o, p), Holotype
Original combination: *Noctileptura seriata* Chemsak & Linsley
Current name: same
Current placement: Cerambycidae: Lepturinae: Lepturini
Type locality: Guatemala: Livingston

seriatopunctatus **Aurivillius, 1922**: 416 (Fig. 152q, r), Holotype
Original combination: *Demonax seriatopunctatus* Aurivillius
Current name: same
Current placement: Cerambycidae: Cerambycinae: Clytini
Type locality: Philippines: Luzon, Mt. Banahao

sericans **Tippmann, 1953c**: 184 (Fig. 152s, t), Holotype
Original combination: *Strongylaspis sericans* Tippmann
Current name: same
Current placement: Cerambycidae: Prioninae: Macrotomini
Type locality: Colombia: Bogotá

sericatus **Casey, 1913**: 213 (Fig. 152u, v), Holotype
Original combination: *Stenocorus (Stenocorus) sericatus* Casey
Current name: *Stenocorus vestitus* (Haldeman, 1847)
Current placement: Cerambycidae: Lepturinae: Rhagiini
Type locality: USA: California

sericeus **Aurivillius, 1922**: 407 (Fig. 152w, x), Syntype
Original combination: *Stenhomalus sericeus* Aurivillius
Current name: same
Current placement: Cerambycidae: Cerambycinae: Obriini
Type locality: Malaysia: Borneo, Sandakan
Remarks: Tavakilian and Chevillotte (2013) state that syntypes are present in the Naturhistoriska Riksmuseet, Stockholm, Sweden as well as the USNM.

sericeus **Fisher, 1936a**: 173 (Fig. 153a, b), Holotype
Original combination: *Ephies sericeus* Fisher
Current name: same
Current placement: Cerambycidae: Lepturinae: Lepturini
Type locality: Indonesia: Java, Mt. Tangkoeban Prahoe, Preanger, 4000–5000 ft.

serpentina **Casey, 1891**: 41 (Fig. 153c, d), Holotype
Original combination: *Leptura serpentina* Casey
Current name: *Stenostrophia tribalteata serpentina* (Casey, 1891)
Current placement: Cerambycidae: Lepturinae: Lepturini
Type locality: USA: Idaho

serrata **Gressitt, 1938a**: 52 (Fig. 153e, f), Holotype
Original combination: *Pterolophia serrata* Gressitt
Current name: same
Current placement: Cerambycidae: Lamiinae: Pteropliini
Type locality: China: Sichuan Province, Muping, 5000–6000 ft.

serraticollis **Linell, 1897a**: 394 (Fig. 153g, h), Holotype
Original combination: *Piezocera serraticollis* Linell
Current name: same
Current placement: Cerambycidae: Cerambycinae: Piezocerini
Type locality: USA: Texas, Brownsville

serricornis **Casey, 1924**: 279 (Fig. 153i, j), Lectotype
Original combination: *Strangalia (Anastrangalia) serricornis* Casey
Current name: *Pygoleptura nigrella* (Say, 1827)
Current placement: Cerambycidae: Lepturinae: Lepturini
Type locality: USA: Michigan, Marquette
Remarks: This specimen is herein designated as the lectotype in order to stabilize the taxonomy and facilitate further identifications of this species.

serricornis **Tippmann, 1960**: 111 (Fig. 153k, l), Holotype
Original combination: *Peruanus serricornis* Tippmann
Current name: same
Current placement: Cerambycidae: Cerambycinae: Cerambycini
Type locality: Peru: Junin, Chanchamayo (it is unclear why another label indicating Bolivia is present.)

serriger **Casey, 1924**: 215 (Fig. 153m, n), Holotype
Original combination: *Prionus (Prionus) serriger* Casey
Current name: *Prionus californicus* Motschulsky, 1845
Current placement: Cerambycidae: Prioninae: Prionini
Type locality: USA: Utah, Ibapah Mts., Trout Creek

setiger **Fisher, 1925a**: 261 (Fig. 153o, p), Holotype
Original combination: *Enes setiger* Fisher
Current name: *Enes (Sibuyenes) setiger* Fisher, 1925a
Current placement: Cerambycidae: Lamiinae: Acanthocinini
Type locality: Philippines: Basilan

setipes **Casey, 1891**: 48 (Fig. 153q, r), Holotype
Original combination: *Leiopus setipes* Casey
Current name: *Trichastylopsis albidus* (LeConte, 1852)
Current placement: Cerambycidae: Lamiinae: Acanthocinini
Type locality: USA: Texas, El Paso

setosa **Aurivillius, 1922**: 427 (Fig. 153s, t), Holotype
Original combination: *Driopea setosa* Aurivillius
Current name: same
Current placement: Cerambycidae: Lamiinae: Acanthocinini
Type locality: Malaysia: Borneo, Sandakan

***setosa* Gressitt, 1937b**: 603 (Fig. 153u, v), Holotype
Original combination: *Enispia setosa* Gressitt
Current name: *Egesina setosa* (Gressitt, 1937b)
Current placement: Cerambycidae: Lamiinae: Pteropliini
Type locality: China: Jiangxi Province, Hong Shan

***setosus* Fisher, 1936a**: 196 (Fig. 153w, x), Holotype
Original combination: *Controderus setosus* Fisher
Current name: *Trichocontoderes setosus* (Fisher, 1936a)
Current placement: Cerambycidae: Lamiinae: Acanthocinini
Type locality: Indonesia: Java, Noesa Kambangan
Remarks: The genus name is *Controderus* in the original publication, however, it is misspelled on the label as *Conoderus*.

***sexnotata* Melzer, 1933**: 374 (Fig. 154a, b), Holotype
Original combination: *Amphicnaeia sexnotata* Melzer
Current name: same
Current placement: Cerambycidae: Lamiinae: Apomecynini
Type locality: Costa Rica: Cartago, Turrialba, 800 m.

***sexualis* Casey, 1912**: 343 (Fig. 154c, d), Lectotype
Original combination: *Tylosis sexualis* Casey
Current name: *Tylosis maculatus* LeConte, 1850
Current placement: Cerambycidae: Cerambycinae: Trachyderini
Type locality: USA: New Mexico, Jemez Springs
Remarks: This specimen is herein designated as the lectotype in order to stabilize the taxonomy and facilitate further identifications of this species.

***sexualis* Casey, 1913**: 344 (Fig. 154e, f), Lectotype
Original combination: *Colobothea sexualis* Casey
Current name: same
Current placement: Cerambycidae: Lamiinae: Colobotheini
Type locality: Honduras
Remarks: This specimen is herein designated as the lectotype in order to stabilize the taxonomy and facilitate further identifications of this species.

***shantzi* Casey, 1924**: 286 (Fig. 154g, h), Holotype
Original combination: *Moneilema (Collapteryx) shantzi* Casey
Current name: *Moneilema semipunctatum* LeConte, 1852
Current placement: Cerambycidae: Lamiinae: Moneilemini
Type locality: USA: New Mexico

***sharonae* Santos-Silva & Hovore, 2008a**: 16 (Fig. 154i, j), Holotype
Original combination: *Elytrimitatrix (Grossifemora) sharonae* Santos-Silva & Hovore
Current name: same
Current placement: Disteniidae: Disteniinae: Disteniini
Type locality: Guatemala: Zacapa, San Lorenzo, Rd 5–6000 ft.

***shoemakeri* Davis, 1923**: 97 (Fig. 154k, l), Holotype
Original combination: *Saperda shoemakeri* Davis
Current name: *Saperda fayi* Bland, 1863
Current placement: Cerambycidae: Lamiinae: Saperdini
Type locality: USA: New York, Ulster Co., Oliverca, elevation 2000 ft.

***sibuyana* Aurivillius, 1927b**: 560 (Fig. 154m, n), Syntype
Original combination: *Pterolophia sibuyana* Aurivillius
Current name: same
Current placement: Cerambycidae: Lamiinae: Pteropliini
Type locality: Philippines: Sibuyan, Romblon
Remarks: Tavakilian and Chevillotte (2013) state that syntypes are present in the Naturhistoriska Riksmuseet, Stockholm, Sweden as well as the USNM.

***sibuyana* Fisher, 1925a**: 260 (Fig. 154o, p), Holotype
Original combination: *Enes sibuyana* Fisher
Current name: *Enes (Sibuyenes) sibuyana* Fisher, 1925a
Current placement: Cerambycidae: Lamiinae: Acanthocinini
Type locality: Philippines: Sibuyan, Romblon

***sibuyanensis* Fisher, 1925a**: 269 (Fig. 154q, r), Holotype
Original combination: *Sciades sibuyanensis* Fisher
Current name: *Sciades (Sibuyomiaenia) sibuyanensis* Fisher, 1925a
Current placement: Cerambycidae: Lamiinae: Acanthocinini
Type locality: Philippines: Sibuyan, Romblon

***signaticornis* Melzer, 1932**: 235 (Fig. 154s, t), Holotype
Original combination: *Phaea signaticornis* Melzer
Current name: same
Current placement: Cerambycidae: Lamiinae: Tetraopini
Type locality: Costa Rica: San José, 1000–1200 m.

***signatipennis* Fisher, 1927a**: 83 (Fig. 154u, v), Holotype
Original combination: *Sybra signatipennis* Fisher
Current name: same
Current placement: Cerambycidae: Lamiinae: Apomecynini
Type locality: Indonesia: Sumatra, Medan

***silviamariae* Galileo & Martins, 2006c**: 181 (Fig. 154w, x), Holotype
Original combination: *Eburodacrys silviamariae* Galileo & Martins
Current name: same
Current placement: Cerambycidae: Cerambycinae: Eburiini
Type locality: Peru: Cusco, Cosñipata (1700 m.)

***simile* Schaeffer, 1908**: 334 (Fig. 155a, b), Syntype
Original combination: *Elaphidion simile* Schaeffer
Current name: *Anelaphus simile* (Schaeffer, 1908)
Current placement: Cerambycidae: Cerambycinae: Elaphidiini
Type locality: USA: Arizona, Huachuca Mts.

***similis* Breuning, 1937**: 277 (Fig. 155c, d), Holotype
Original combination: *Eurysops similis* Breuning
Current name: same
Current placement: Cerambycidae: Lamiinae: Phrynetini
Type locality: Congo: Brazzaville

***similis* Breuning, 1940d**: 414 (Fig. 155e, f), Holotype
Original combination: *Rhodopis similis* Breuning
Current name: *Rhodopina similis* (Breuning, 1940d)
Current placement: Cerambycidae: Lamiinae: Rhodopini
Type locality: India: Assam, Khasi Hills

***similis* Breuning, 1942a**: 119 (Fig. 155g, h), Holotype
Original combination: *Blepephaeus similis* Breuning
Current name: same
Current placement: Cerambycidae: Lamiinae: Monochamini
Type locality: India: Shembaganur, 5 km before Kodaikanal

***similis* Dillon, 1955**: 182 (Fig. 155i, j), Holotype
Original combination: *Canidiopsis similis* Dillon
Current name: *Canidia mexicana* Thomson, 1860
Current placement: Cerambycidae: Lamiinae: Acanthocinini
Type locality: Mexico: Morelos, Cuernavaca

***similis* Fisher, 1925a**: 207 (Fig. 155k, l), Holotype
Original combination: *Enispia similis* Fisher
Current name: *Athylia similis* (Fisher, 1925a)
Current placement: Cerambycidae: Lamiinae: Pteropliini
Type locality: Malaysia: Borneo, Sandakan

***similis* Fisher, 1930a**: 19 (Fig. 155m, n), Holotype
Original combination: *Tomopterus similis* Fisher
Current name: same
Current placement: Cerambycidae: Cerambycinae: Rhinotragini
Type locality: Bolivia: Río Negro (Department unknown)

***similis* Fisher, 1932**: 53 (Fig. 155o, p), Holotype
Original combination: *Merostenus similis* Fisher
Current name: *Caribbomerus similis* (Fisher, 1932)
Current placement: Cerambycidae: Cerambycinae: Graciliini
Type locality: Antigua

***similis* Fisher, 1935b**: 196 (Fig. 155q, r), Holotype
Original combination: *Haruspex similis* Fisher
Current name: *Haruspex inscriptus* Gahan, 1895
Current placement: Cerambycidae: Cerambycinae: Piezocerini
Type locality: Trinidad and Tobago: Trinidad

***simplex* Aurivillius, 1924**: 445 (Fig. 155s, t), Holotype
Original combination: *Psilomerus simplex* Aurivillius
Current name: same
Current placement: Cerambycidae: Cerambycinae: Clytini
Type locality: Malaysia: Borneo, Sandakan

***simplex* Aurivillius, 1927b**: 588 (Fig. 155u, v), Syntype
Original combination: *Daphisia simplex* Aurivillius
Current name: *Serixia simplex* (Aurivillius, 1927b)
Current placement: Cerambycidae: Lamiinae: Saperdini
Type locality: Philippines: Mindanao, Butuan
Remarks: Other syntype(s) are present in the Naturhistoriska Riksmuseet, Stockholm, Sweden.

***simplex* Casey, 1912**: 253 (Fig. 155w, x), Holotype
Original combination: *Prionina simplex* Casey
Current name: *Prionus (Homaesthesis) simplex* (Casey, 1912)
Current placement: Cerambycidae: Prioninae: Prionini
Type locality: USA: Kansas

***simplicicornis* Casey, 1913**: 286 (Fig. 156a, b), Holotype
Original combination: *Moneilema (Collapteryx) simplicicornis* Casey
Current name: *Moneilema armatum* LeConte, 1853
Current placement: Cerambycidae: Lamiinae: Moneilemini
Type locality: USA: New Mexico

***simplicicornis grylliceps* Casey, 1913**: 287 (Fig. 156c, d), Holotype
Original combination: *Moneilema (Collapteryx) simplicicornis grylliceps* Casey
Current name: *Moneilema armatum* LeConte, 1853
Current placement: Cerambycidae: Lamiinae: Moneilemini
Type locality: USA: New Mexico

***simulans* Casey, 1913**: 278 (Fig. 156e, f), Lectotype
Original combination: *Ophistomis simulans* Casey
Current name: *Strangalia biclor* (Swederus, 1787)
Current placement: Cerambycidae: Lepturinae: Lepturini
Type locality: USA: Colorado
Remarks: This specimen is herein designated as the lectotype in order to stabilize the taxonomy and facilitate further identifications of this species.

***simulans* Linsley, 1970**: 133 (Fig. 156g, h), Holotype
Original combination: *Lycomorphoides simulans* Linsley
Current name: same
Current placement: Cerambycidae: Lepturinae: Lepturini
Type locality: Costa Rica: Cartago, Turrialba, 800 m.

***simulatrix* Tippmann, 1960**: 207 (Fig. 156i, j), Holotype
Original combination: *Antodice simulatrix* Tippmann
Current name: *Antodice neivai* Lane, 1940
Current placement: Cerambycidae: Lamiinae: Aerenicini
Type locality: Bolivia: Santa Cruz 500 m.

***sinaloanum* Linsley, 1935b**: 484 (Fig. 156k, l), Holotype
Original combination: *Ibidion sinaloanum* Linsley
Current name: *Neocompsa sinaloana* (Linsley, 1935b)
Current placement: Cerambycidae: Cerambycinae: Neoibidionini
Type locality: Mexico: Sinaloa, El Venedillo

***singularis* Aurivillius, 1927b**: 575 (Fig. 156m, n), Lectotype
Original combination: *Mimectatina singularis* Aurivillius
Current name: same
Current placement: Cerambycidae: Lamiinae: Rhodopini
Type locality: Philippines: Samar
Remarks: This specimen is herein designated as the lectotype in order to stabilize the taxonomy and facilitate further identifications of this species. Two additional paralectotypes are present in the USNM.

***sinica* Gressitt, 1937b**: 616 (Fig. 156o, p), Holotype
Original combination: *Serixia sinica* Gressitt
Current name: same
Current placement: Cerambycidae: Lamiinae: Saperdini
Type locality: China: Jiangxi Province, Hong Shan

***sinicum* Breuning, 1948a**: 57 (Fig. 156q, r), Holotype
Original combination: *Eodorcadion (Ornatodorcadion) sinicum* Breuning
Current name: *Eodorcadion sinicum* Breuning, 1948a
Current placement: Cerambycidae: Lamiinae: Dorcadionini
Type locality: China (no specific locality data)

***sinuaticollis* Gressitt, 1956**: 98 (Fig. 156s, t), Holotype
Original combination: *Longipalpus sinuaticollis* Gressitt
Current name: same
Current placement: Cerambycidae: Cerambycinae: Obriini
Type locality: Palau: Aurapushekaru Island

***sirena* Dillon & Dillon, 1946b**: 267 (Fig. 156u, v), Holotype
Original combination: *Cacostola sirena* Dillon & Dillon
Current name: same
Current placement: Cerambycidae: Lamiinae: Onciderini
Type locality: Venezuela: Distrito Federal, Caracas

***smaragdina* Casey, 1912**: 314 (Fig. 156w, x), Lectotype
Original combination: *Callichroma smaragdina* Casey
Current name: *Plinthocoelium suaveolens plicatum* (LeConte, 1853)
Current placement: Cerambycidae: Cerambycinae: Callichromatini
Type locality: USA: Texas
Remarks: This specimen is herein designated as the lectotype in order to stabilize the taxonomy and facilitate further identifications of this species.

***smithi* Gressitt, 1941**: 334 (Fig. 157a, b), Holotype
Original combination: *Chlorophorus smithi* Gressitt
Current name: same
Current placement: Cerambycidae: Cerambycinae: Clytini
Type locality: Thailand: Khun Tan Mts., 3000 ft.

***snowi* Casey, 1912**: 353 (Fig. 157c, d), Holotype
Original combination: *Cyllene snowi* Casey
Current name: *Megacyllene snowi* (Casey, 1912)
Current placement: Cerambycidae: Cerambycinae: Clytini
Type locality: USA: Arizona, Oak Creek Canyon, 6000 ft.

***solatus nodius* Kriesche, 1936**: 67 (Fig. 157e, f), Holotype
Original combination: *Dihammus solatus nodius* Kriesche
Current name: *Acalolepta bolanica* (Aurivillius, 1925)
Current placement: Cerambycidae: Lamiinae: Monochamini
Type locality: Papua New Guinea: Bolau-Gebirge
Remarks: Tippmann labeled this specimen as having been obtained from the Kriesche Estate in 1955.

***solida* Casey, 1913**: 288 (Fig. 157g, h), Lectotype
Original combination: *Moneilema (Collapteryx) solida* Casey
Current name: *Moneilema armatum* LeConte, 1853
Current placement: Cerambycidae: Lamiinae: Moneilemini
Type locality: USA: Texas
Remarks: This specimen is herein designated as the lectotype in order to stabilize the taxonomy and facilitate further identifications of this species.

***solidus* Casey, 1912**: 238 (Fig. 157i, j), Holotype
Original combination: *Prionus (Prionus) solidus* Casey
Current name: *Prionus californicus* Motschulsky, 1845
Current placement: Cerambycidae: Prioninae: Prionini
Type locality: USA: Colorado

***sonoranus* Casey, 1924**: 241 (Fig. 157k, l), Holotype
Original combination: *Aneflus sonoranus* Casey
Current name: same
Current placement: Cerambycidae: Cerambycinae: Elaphidiini
Type locality: USA: Arizona, Baboquivari Mts.

soteri **Napp & Monné, 2009**: 30 (Fig. 157m, n), Holotype
Original combination: *Anisotyma soteri* Napp & Monné
Current name: same
Current placement: Cerambycidae: Cerambycinae: Dryobiini
Type locality: El Salvador: La Liberdad, Santa Tecla

spadix **Casey, 1912**: 231 (Fig. 157o, p), Holotype
Original combination: *Orthosoma spadix* Casey
Current name: *Orthosoma brunneum* (Forster, 1771)
Current placement: Cerambycidae: Prioninae: Prionini
Type locality: USA: New York, Long Island, Willets Point

spaldingi **Casey, 1924**: 221 (Fig. 157q, r), Lectotype
Original combination: *Prionus (Prionus) spaldingi* Casey
Current name: *Prionus californicus* Motschulsky, 1845
Current placement: Cerambycidae: Prioninae: Prionini
Type locality: USA: Utah, Eureka
Remarks: This specimen is herein designated as the lectotype in order to stabilize the taxonomy and facilitate further identifications of this species.

speciosum **Fisher, 1937d**: 145 (Fig. 157s, t), Holotype
Original combination: *Hexoplon speciosum* Fisher
Current name: same
Current placement: Cerambycidae: Cerambycinae: Hexoplini
Type locality: Brazil: Santa Catarina, Nova Teutonia

spermophagus **Fisher, 1918**: 173 (Fig. 157u, v), Holotype
Original combination: *Lepturges spermophagus* Fisher
Current name: *Atrypanius spermophagus* (Fisher, 1918)
Current placement: Cerambycidae: Lamiinae: Acanthocinini
Type locality: Mexico: Tamaulipas, Tampico

spiculosus **Casey, 1912**: 240 (Fig. 157w, x), Holotype
Original combination: *Prionus (Prionus) spiculosus* Casey
Current name: *Prionus californicus* Motschulsky, 1845
Current placement: Cerambycidae: Prioninae: Prionini
Type locality: USA: Arizona, Canyon of Colorado River

spiculosus coloradensis **Casey, 1924**: 218 (Fig. 158a, b), Holotype
Original combination: *Prionus (Prionus) spiculosus coloradensis* Casey
Current name: *Prionus californicus* Motschulsky, 1845
Current placement: Cerambycidae: Prioninae: Prionini
Type locality: USA: Colorado, Boulder Co.

spiculum **Casey, 1890**: 492 (Fig. 158c, d), Holotype
Original combination: *Tragosoma spiculum* Casey
Current name: *Tragosoma depsarium* (Linnaeus, 1767)
Current placement: Cerambycidae: Prioninae: Meroscelisini
Type locality: USA: New Mexico, Las Vegas

spilmani **Villiers, 1980**: 95 (Fig. 158e, f), Holotype
Original combination: *Trypanidius spilmani* Villiers
Current name: same
Current placement: Cerambycidae: Lamiinae: Acanthocinini
Type locality: Dominica: Clarke Hall

spinicauda **Schaeffer, 1904**: 224 (Fig. 158g, h), Syntype
Original combination: *Ataxia spinicauda* Schaeffer
Current name: same
Current placement: Cerambycidae: Lamiinae: Pteropliini
Type locality: USA: Florida, Key Largo
Remarks: There are 2 syntypes present in USNM.

spinicollis **Casey, 1891**: 45 (Fig. 158i, j), Lectotype
Original combination: *Moneilema spinicollis* Casey
Current name: *Moneilema gigas* LeConte, 1873
Current placement: Cerambycidae: Lamiinae: Moneilemini
Type locality: USA: Arizona
Remarks: This specimen is herein designated as the lectotype in order to stabilize the taxonomy and facilitate further identifications of this species.

spinicornis **Tippmann, 1960**: 172 (Fig. 158k, l), Holotype
Original combination: *Acanthoderes (Psapharochrus) spinicornis* Tippmann
Current name: *Catuana spinicornis* (Tippmann, 1960)
Current placement: Cerambycidae: Lamiinae: Acanthoderini
Type locality: Brazil: Amazonas, Benjamin Constant, Rio Branco

spinifera **Gressitt, 1948b**: 57 (Fig. 158m, n), Holotype
Original combination: *Nupserha spinifera* Gressitt
Current name: same
Current placement: Cerambycidae: Lamiinae: Phytoeciini
Type locality: China: Sichuan Province, Yachow to Mupin, 2,000–5,000 ft.

spinipenne **Chemsak & Linsley, 1967**: 289 (Fig. 158o, p), Holotype
Original combination: *Conosphaerion spinipenne* Chemsak & Linsley
Current name: same
Current placement: Cerambycidae: Cerambycinae: Elaphidiini
Type locality: Panama: La Chorrera

spinipennis **Breuning, 1942a**: 137 (Fig. 158q, r), Holotype
Original combination: *Esthlogena spinipennis* Breuning
Current name: *Esthlogena (Esthlogena) spinipennis* Breuning, 1942a
Current placement: Cerambycidae: Lamiinae: Pteropliini
Type locality: Brazil: Santa Catarina, Corupá (formerly Hansa-Humboldt)

***spinipennis* Breuning, 1954**: 22 (Fig. 158s, t), Holotype
Original combination: *Sybra spinipennis* Breuning
Current name: *Sybra spinipennis* Breuning, 1954
Current placement: Cerambycidae: Lamiinae: Apomecynini
Type locality: Papua New Guinea: East New Britain, Mope

***spinipennis* Fisher, 1936d**: 347 (Fig. 158u, v), Holotype
Original combination: *Spalacopsis spinipennis* Fisher
Current name: same
Current placement: Cerambycidae: Lamiinae: Agapanthiini
Type locality: Cuba: Santiago de Cuba, Oriente Province, Turquino Peak region

***spinipennis* Fisher, 1938**: 146 (Fig. 158w, x), Holotype
Original combination: *Hyperplatys spinipennis* Fisher
Current name: *Hyperplatys cana* (Bates, 1863)
Current placement: Cerambycidae: Lamiinae: Acanthocinini
Type locality: Brazil: Santa Catarina, Nova Teutonia

***spinipennis* Fisher, 1946a**: 330 (Fig. 159a, b), Holotype
Original combination: *Distenia spinipennis* Fisher
Current name: *Distenia (Basisvallis) spinipennis* Fisher, 1946a
Current placement: Disteniidae: Disteniinae: Disteniini
Type locality: Costa Rica: Guapiles, Santa Clara, 250–300 m.

***spinosus* Fisher, 1925a**: 257 (Fig. 159c, d), Holotype
Original combination: *Enes spinosus* Fisher
Current name: *Enes (Sibuyenes) spinosus* Fisher, 1925a
Current placement: Cerambycidae: Lamiinae: Acanthocinini
Type locality: Philippines: Negros Oriental, Cuernos Mountains

***spinosus* Galileo & Martins, 2005b**: 64 (Fig. 159e, f), Holotype
Original combination: *Acaiu spinosus* Galileo & Martins
Current name: same
Current placement: Cerambycidae: Lamiinae: Hemilophini
Type locality: Ecuador: Napo Reserve, Ethnica Waorani, 1km south of Onkone Gare Camp, Trans. Ent., 220 m., 00°39´10S, 76°26´W

***spissicornis* Casey, 1912**: 264 (Fig. 159g, h), Lectotype
Original combination: *Nothorhina spissicornis* Casey
Current name: *Arhopalus rusticus montanus* (LeConte, 1873)
Current placement: Cerambycidae: Spondylidinae: Asemini
Type locality: USA: Colorado, Boulder Co.
Remarks: This specimen is herein designated as the lectotype in order to stabilize the taxonomy and facilitate further identifications of this species.

***spissicornis* Casey, 1912**: 336 (Fig. 159i, j), Holotype
Original combination: *Zalophia spissicornis* Casey
Current name: *Zalophia funebris* (Bates, 1880)
Current placement: Cerambycidae: Cerambycinae: Trachyderini
Type locality: Ecuador: Babahoyo, foot of Andes

***spissicornis* Casey, 1913**: 384 (Fig. 159k, l), Lectotype
Original combination: *Tetraopes spissicornis* Casey
Current name: *Tetraopes femoratus* LeConte, 1847
Current placement: Cerambycidae: Lamiinae: Tetraopini
Type locality: USA: Arizona, Oak Creek Canyon, 6000 ft.
Remarks: This specimen is herein designated as the lectotype in order to stabilize the taxonomy and facilitate further identifications of this species.

***spissicornis longicornis* Casey, 1912**: 264 (Fig. 159m, n), Holotype
Original combination: *Nothorhina spissicornis longicornis* Casey
Current name: *Arhopalus rusticus montanus* (LeConte, 1873)
Current placement: Cerambycidae: Spondylidinae: Asemini
Type locality: USA: Colorado, Boulder Co.

***splendidum* Fisher, 1932**: 19 (Fig. 159o, p), Holotype
Original combination: *Elaphidion splendidum* Fisher
Current name: same
Current placement: Cerambycidae: Cerambycinae: Elaphidiini
Type locality: Haiti: Port-au-Prince

***stanleyana* var. *gloriosa* Tippmann, 1953b**: 152 (Fig. 159q, r), Holotype
Original combination: *Anoplophora stanleyana* var. *gloriosa* Tippmann
Current name: *Anoplophora stanleyana* Hope, 1839
Current placement: Cerambycidae: Lamiinae: Monochamini
Type locality: India: Naga Hills, 6000–7000 ft.
Remarks: The name following the binomen was not expressly given infrasubspecific rank and the work did not indicate unambiguously that the name was infrasubspecific, therefore, it is deemed a subspecies and available.

stanleyana **var.** ***grisea*** **Tippmann, 1953b**: 153 (Fig. 159s, t), Holotype
Original combination: *Anoplophora stanleyana* var. *grisea* Tippmann
Current name: *Anoplophora birmanica* Hüdepohl, 1990
Current placement: Cerambycidae: Lamiinae: Monochamini
Type locality: India: Assam, Shillong
Remarks: Although Tippmann used both "ab." and "var." on his labels, the published work uses "var.". The name following the binomen was not expressly given infrasubspecific rank and the work did not indicate unambiguously that the name was infrasubspecific, therefore, it is deemed a subspecies and available.

stanleyana **var.** ***melancholica*** **Tippmann, 1953b**: 153 (Fig. 159u, v), Holotype
Original combination: *Anoplophora stanleyana* var. *melancholica* Tippmann
Current name: *Anoplophora stanleyana* Hope, 1839
Current placement: Cerambycidae: Lamiinae: Monochamini
Type locality: Vietnam: Tonkin, Chapa
Remarks: The name following the binomen was not expressly given infrasubspecific rank and the work did not indicate unambiguously that the name was infrasubspecific, therefore, it is deemed a subspecies and available.

stenideoides guineensis **Breuning, 1957b**: 818 (Fig. 159w, x), Holotype
Original combination: *Hyllisia stenideoides guineensis* Breuning
Current name: same
Current placement: Cerambycidae: Lamiinae: Agapanthiini
Type locality: Ivory Coast

stictica **Breuning, 1942a**: 163 (Fig. 160a, b), Holotype
Original combination: *Estola stictica* Breuning
Current name: *Drycothaea truncatipennis* Tavakilian, 1997
Current placement: Cerambycidae: Lamiinae: Caliini
Type locality: Brazil: Santa Catarina, Corupá (formerly Hansa-Humboldt)

stocktonense **Casey, 1924**: 227 (Fig. 160c, d), Holotype
Original combination: *Asemum stocktonense* Casey
Current name: *Asemum striatum* (Linnaeus, 1758)
Current placement: Cerambycidae: Spondylidinae: Asemini
Type locality: USA: Utah, Stockton

stockwelli **Corbett, 2004**: 278 (Fig. 160e, f), Holotype
Original combination: *Cobelura stockwelli* Corbett
Current name: same
Current placement: Cerambycidae: Lamiinae: Acanthocinini
Type locality: Panama: Canal Zone, Barro Colorado Island

stolida **Casey, 1924**: 277 (Fig. 160g, h), Holotype
Original combination: *Leptacmaeops stolida* Casey
Current name: *Cortodera stolida* (Casey, 1924)
Current placement: Cerambycidae: Lepturinae: Rhagiini
Type locality: USA: Nevada, Las Vegas

stramentosa **Breuning, 1939c**: 262 (Fig. 160i, j), Holotype
Original combination: *Sybra stramentosa* Breuning
Current name: same
Current placement: Cerambycidae: Lamiinae: Apomecynini
Type locality: Philippines: Luzon
Remarks: The original description states the type is in the Museum von Hamburg. Tavakilian and Chevillotte (2013) suggest the holotype was destroyed in World War II, however, it is possible that this is the specimen that was protected in Tippmann's collection.

strandi **Breuning, 1935a**: 170 (Fig. 160k, l), Holotype
Original combination: *Paracyclotaenia strandi* Breuning
Current name: same
Current placement: Cerambycidae: Lamiinae: Petrognathini
Type locality: Gabon: Bas - Ogooue
Remarks: Tavakilian and Chevillotte (2013) suggest the holotype is present in Basel, Switzerland, however, this specimen and its labels match the original description.

strandi **Breuning, 1935a**: 173 (Fig. 160m, n), Holotype
Original combination: *Rhodopis strandi* Breuning
Current name: *Nanohammus sinicus* (Pic, 1926)
Current placement: Cerambycidae: Lamiinae: Rhodopini
Type locality: China: Jiangsu Province, Nanjing

strandi **Breuning, 1939c**: 163 (Fig. 160o, p), Holotype
Original combination: *Mulciber strandi* Breuning
Current name: same
Current placement: Cerambycidae: Lamiinae: Homonoeini
Type locality: Papua New Guinea: New Ireland Province, New Hanover Island
Remarks: The original description states the holotype is in The Natural History Museum, London. However, the BMNH database does not reflect this. The type locality is correct for the USNM specimen, therefore it is determined to be the primary type. Tavakilian and Chevillotte (2013) has the type locality incorrect for this species, stating Sri Lanka.

strandi **Breuning, 1939c**: 243 (Fig. 160q, r), Holotype
Original combination: *Parepilysta strandi* Breuning
Current name: same
Current placement: Cerambycidae: Lamiinae: Apomecynini
Type locality: Indonesia: Sulawesi (formerly Celebes)

***strandi* Breuning, 1939c**: 278 (Fig. 160s, t), Holotype
Original combination: *Neosybra strandi* Breuning
Current name: same
Current placement: Cerambycidae: Lamiinae: Apomecynini
Type locality: Malaysia: Borneo
Remarks: Tavakilian and Chevillotte (2013) suggest the holotype is present in the Naturhistoriska Riksmuseet, Stockholm, Sweden. It is unclear if this is the actual holotype.

***strandi* Breuning, 1940c**: 162 (Fig. 160u, v), Holotype
Original combination: *Atimura strandi* Breuning
Current name: same
Current placement: Cerambycidae: Lamiinae: Apomecynini
Type locality: India: Chota Nagpur
Remarks: Tavakilian and Chevillotte (2013) indicate the holotype is in Basel, Switzerland, however, ours matches the original description and locality data.

***strandi* Breuning, 1940d**: 413 (Fig. 160w, x), Holotype
Original combination: *Myagrus strandi* Breuning
Current name: same
Current placement: Cerambycidae: Lamiinae: Lamiini
Type locality: Indonesia: Sumatra, Medan

***strandi* Breuning, 1940c**: 421 (Fig. 161a, b), Holotype
Original combination: *Synixais strandi* Breuning
Current name: *Synixais strandi* Breuning, 1940
Current placement: Cerambycidae: Lamiinae: Pteropliini
Type locality: Laos

***strandi* Breuning, 1940b**: 79 (Fig. 161c, d), Holotype
Original combination: *Diboma strandi* Breuning
Current name: *Zotalemimon strandi* (Breuning, 1940b)
Current placement: Cerambycidae: Lamiinae: Apomecynini
Type locality: Malaysia: Perak

***strandi* Breuning, 1942a**: 120 (Fig. 161e, f), Holotype
Original combination: *Cereopsius strandi* Breuning
Current name: *Cereopsius strandi* Breuning, 1942
Current placement: Cerambycidae: Lamiinae: Lamiini
Type locality: Thailand

***strandi* Breuning, 1942a**: 166 (Fig. 161g, h), Holotype
Original combination: *Pachypezomorpha strandi* Breuning
Current name: *Helvina strandi* (Breuning, 1942a)
Current placement: Cerambycidae: Lamiinae: Agapanthiini
Type locality: Peru: Junin, Satipo

***strandi* Breuning, 1943c**: 16 (Fig. 161i, j), Holotype
Original combination: *Acridoschema strandi* Breuning
Current name: *Acridoschema ligata* Quedenfeldt, 1882
Current placement: Cerambycidae: Lamiinae: Acmocerini
Type locality: Cameroon: Batouri
Remarks: Original description lists "Kongo" as type locality.

***strandi* Breuning, 1943c**: 41 (Fig. 161k, l), Holotype
Original combination: *Estolosybra strandi* Breuning
Current name: *Cacostola strandi* (Breuning, 1943)
Current placement: Cerambycidae: Lamiinae: Onciderini
Type locality: Venezuela: Distrito Federal, Caracas, Berg Avila

***strandi* Breuning, 1943c**: 42 (Fig. 161m, n), Holotype
Original combination: *Hyagniellus strandi* Breuning
Current name: *Cacostola ornata* Fleutiaux & Sallé, 1889
Current placement: Cerambycidae: Lamiinae: Onciderini
Type locality: Guadeloupe
Remarks: The labels incorrectly state the specific epithet as *guadaloupensis* and the genus name as *Hyagniella* (it was actually published as *Hyagniellus strandi*).

***strandi* Breuning, 1943c**: 44 (Fig. 161o, p), Holotype
Original combination: *Dolichosybra strandi* Breuning
Current name: same
Current placement: Cerambycidae: Lamiinae: Apomecynini
Type locality: Venezuela: Maracay, Aragua

***strandi* Breuning, 1943c**: 54 (Fig. 161q, r), Holotype
Original combination: *Periestola strandi* Breuning
Current name: *Periestola strandi* Breuning, 1943
Current placement: Cerambycidae: Lamiinae: Acanthocinini
Type locality: Costa Rica: Cartago, Turrialba
Remarks: Breuning had another specimen labeled as a cotype with this determination label, however, that specimen is actually conspecific with *Estoloides strandiella*.

***strandi* Breuning, 1943c**: 60 (Fig. 161s, t), Holotype
Original combination: *Eohyllisia strandi* Breuning
Current name: *Eohyllisia strandi* Breuning, 1943c
Current placement: Cerambycidae: Lamiinae: Agapanthiini
Type locality: Kenya: Sagola

***strandi* Breuning, 1954**: 8 (Fig. 161u, v), Holotype
Original combination: *Batocera strandi* Breuning
Current name: same
Current placement: Cerambycidae: Lamiinae: Batocerini
Type locality: Indonesia: Sulawesi, Utara, Tondano

***strandiella* Breuning, 1940c**: 191 (Fig. 161w, x), Holotype
Original combination: *Paranandra strandiella* Breuning
Current name: same
Current placement: Cerambycidae: Lamiinae: Agapanthiini
Type locality: Laos
Remarks: Tavakilian and Chevillotte (2013) suggest the holotype is present in Basel, Switzerland, however, this specimen and its labels match the original description.

***strandiella* Breuning, 1940d**: 430 (Fig. 162a, b), Holotype
Original combination: *Parasophronica strandiella* Breuning
Current name: *Parasophronica strandiella* Breuning, 1940d
Current placement: Cerambycidae: Lamiinae: Rhodopini
Type locality: Indonesia: Sumatra, Deli

***strandiella* Breuning, 1942a**: 163 (Fig. 162c, d), Holotype
Original combination: *Estola strandiella* Breuning
Current name: same
Current placement: Cerambycidae: Lamiinae: Desmiphorini
Type locality: Costa Rica: Surrubres, 300 ft.
Remarks: The specific epithet is incorrect on label.

***strandiella* Breuning, 1942a**: 174 (Fig. 162e, f), Holotype
Original combination: *Ischioloncha strandiella* Breuning
Current name: same
Current placement: Cerambycidae: Lamiinae: Apomecynini
Type locality: Bolivia

***strandiella* Breuning, 1943c**: 53 (Fig. 162g, h), Holotype
Original combination: *Blabia strandiella* Breuning
Current name: same
Current placement: Cerambycidae: Lamiinae: Desmiphorini
Type locality: Peru: Aina, Ayacucho

***strandiellus* Breuning, 1935a**: 167 (Fig. 162i, j), Holotype
Original combination: *Crossotus strandiellus* Breuning
Current name: *Corus strandiellus* (Breuning, 1935a)
Current placement: Cerambycidae: Lamiinae: Crossotini
Type locality: Tanzania: Ziwa Magharibi, Bukoba
Remarks: Although Tavakilian and Chevillotte (2013) indicates that this holotype is in Naturhistoriska Riksmuseet, Stockholm, Sweden, it matches the original description and data.

***strandiellus* Breuning, 1935b**: 253 (Fig. 162k, l), Holotype
Original combination: *Dihammus strandiellus* Breuning
Current name: *Acalolepta strandiella* (Breuning, 1935b)
Current placement: Cerambycidae: Lamiinae: Monochamini
Type locality: Papua New Guinea: Bougainville Island (Solomon Islands)
Remarks: Although Tavakilian and Chevillotte (2013) indicated the holotype is in Basel, Switzerland, this specimen agrees with the original description and data.

***strandiellus* Breuning, 1943c**: 39 (Fig. 162m, n), Holotype
Original combination: *Asyngenes strandiellus* Breuning
Current name: same
Current placement: Cerambycidae: Lamiinae: Apomecynini
Type locality: Venezuela: Maracay, Aragua

***straussi* Webb, 1908**: 41 (Fig. 162o, p), Holotype
Original combination: *Leptura straussi* Webb
Current name: *Pidonia ruficollis* (Say, 1824)
Current placement: Cerambycidae: Lepturinae: Rhagiini
Type locality: USA: North Carolina, Tryon

***strenuus* Casey, 1913**: 293 (Fig. 162q, r), Holotype
Original combination: *Monochamus strenuus* Casey
Current name: *Monochamus clamator* (Leonte, 1852)
Current placement: Cerambycidae: Lamiinae: Monochamini
Type locality: USA: Colorado

***striatella* Tippmann, 1960**: 192 (Fig. 162s, t), Holotype
Original combination: *Lepturges striatella* Tippmann
Current name: *Nyssodrysternum striatellum* (Tippmann, 1960)
Current placement: Cerambycidae: Lamiinae: Acanthocinini
Type locality: Bolivia: Cochabamba, Reg. Chapare 400 m.

***strix dissolutus* Kriesche, 1936**: 69 (Fig. 162u, v), Holotype
Original combination: *Anancylus strix dissolutus* Kriesche
Current name: *Anancylus (Paranancylus) socius* Pascoe, 1865
Current placement: Cerambycidae: Lamiinae: Mesosini
Type locality: Philippines: Masbate, Karbala, Aroroy
Remarks: Tippmann labeled this specimen as having been obtained from the Kriesche Estate in 1955.

***stroheckeri granulatus* Dillon, 1956**: 139 (Fig. 162w, x), Holotype
Original combination: *Lagocheirus stroheckeri granulatus* Dillon
Current name: *Lagocheirus araneiformis ypsilon* (Voet, 1778)
Current placement: Cerambycidae: Lamiinae: Acanthocinini
Type locality: USA: Texas

***stultus* Casey, 1924**: 221 (Fig. 163a, b), Lectotype
Original combination: *Prionus (Prionus) stultus* Casey
Current name: *Prionus californicus* Motschulsky, 1845
Current placement: Cerambycidae: Prioninae: Prionini
Type locality: USA: Utah, Eureka
Remarks: This specimen is herein designated as the lectotype in order to stabilize the taxonomy and facilitate further identifications of this species.

***stultus parvicollis* Casey, 1924**: 222 (Fig. 163c, d), Holotype
Original combination: *Prionus (Prionus) stultus parvicollis* Casey
Current name: *Prionus californicus* Motschulsky, 1845
Current placement: Cerambycidae: Prioninae: Prionini
Type locality: USA: Utah, Eureka

***stygialis* Tippmann, 1960**: 173 (Fig. 163e, f), Holotype
Original combination: *Acanthoderes (Psapharochrus) stygialis* Tippmann
Current name: *Psapharochrus pereirai* (Prosen & Lane, 1955)
Current placement: Cerambycidae: Lamiinae: Acanthoderini
Type locality: Brazil: Goyaz, Silvania

***stygica* Gressitt, 1935a**: 168 (Fig. 163g, h), Holotype
Original combination: *Leptura stygica* Gressitt
Current name: *Judolidia bangi* (Pic, 1901)
Current placement: Cerambycidae: Lepturinae: Lepturini
Type locality: Japan: Nikko

***subalba* Gressitt, 1938c**: 162 (Fig. 163i, j), Holotype
Original combination: *Falsomesosella subalba* Gressitt
Current name: same
Current placement: Cerambycidae: Lamiinae: Mesosini
Type locality: China: Taiwan, Kuraru

***subbidentatus* Gressitt, 1937b**: 615 (Fig. 163k, l), Holotype
Original combination: *Exocentrus subbidentatus* Gressitt
Current name: same
Current placement: Cerambycidae: Lamiinae: Pogonocherini
Type locality: China: Guangdong Province, Tsin Leong Shan

***subcylindrica* Gressitt, 1956**: 158 (Fig. 163m, n), Holotype
Original combination: *Sciadella (Micronesiella) subcylindrica* Gressitt
Current name: *Sciades (Micronesiella) subcylindrus* (Gressitt, 1956)
Current placement: Cerambycidae: Lamiinae: Acanthocinini
Type locality: Palau: Koror Island

***subcylindricus* Aurivillius, 1927b**: 551 (Fig. 163o, p), Holotype
Original combination: *Nemophas subcylindricus* Aurivillius
Current name: same
Current placement: Cerambycidae: Lamiinae: Monochamini
Type locality: Philippines: Surigao, Mindanao

***subdenudatus* Gressitt, 1938c**: 153 (Fig. 163q, r), Holotype
Original combination: *Dolophrades subdenudatus* Gressitt
Current name: *Dolophrades annulicornis* (Schwarzer, 1925)
Current placement: Cerambycidae: Lamiinae: Parmenini
Type locality: China: Taiwan, Hori

***subdepressum* Schaeffer, 1904**: 222 (Fig. 163s, t), Holotype
Original combination: *Elaphidion subdepressum* Schaeffer
Current name: *Anelaphus subdepressum* (Schaeffer, 1904)
Current placement: Cerambycidae: Cerambycinae: Elaphidiini
Type locality: Mexico: Baja California (Sur), San Felipe

***subflavus* Chemsak, 1966**: 213 (Fig. 163u, v), Holotype
Original combination: *Curtomerus subflavus* Chemsak
Current name: *Curtomerus flavus* (Fabricius, 1775)
Current placement: Cerambycidae: Cerambycinae: Elaphidiini
Type locality: Anguilla: Flat Top Point

***subglaber* Casey, 1913**: 319 (Fig. 163w, x), Lectotype
Original combination: *Lepturges subglaber* Casey
Current name: same
Current placement: Cerambycidae: Lamiinae: Acanthocinini
Type locality: Mexico: Durango, Ciudad de Durango
Remarks: This specimen is herein designated as the lectotype in order to stabilize the taxonomy and facilitate further identifications of this species.

***subglaber* Fisher, 1925a**: 238 (Fig. 164a, b), Holotype
Original combination: *Exocentrus subglaber* Fisher
Current name: *Exocentrus (Camptomyne) subglaber* Fisher, 1925a
Current placement: Cerambycidae: Lamiinae: Acanthocinini
Type locality: Malaysia: Borneo, Sandakan

***subglabra* Linsley, 1935a**: 142 (Fig. 164c, d), Holotype
Original combination: *Psyrassa subglabra* Linsley
Current name: same
Current placement: Cerambycidae: Cerambycinae: Elaphidiini
Type locality: Panama: Panama Province, Río Trinidad

***subglabrata* Gressitt, 1938a**: 54 (Fig. 164e, f), Holotype
Original combination: *Sydonia subglabrata* Gressitt
Current name: *Zotalemimon subglabrata* (Gressitt, 1938a)
Current placement: Cerambycidae: Lamiinae: Rhodopini
Type locality: China: Sichuan, Muping 5,000–6,000 ft.

***subglabrus* Gressitt, 1956**: 178 (Fig. 164g, h), Holotype
Original combination: *Nonymoides subglabrus* Gressitt
Current name: *Sciades (Nonymoides) subglabrus* (Gressitt, 1956)
Current placement: Cerambycidae: Lamiinae: Acanthocinini
Type locality: Micronesia: Pohnpei (formerly Ponape Island), Mt. Kupuriso, summit, ca. 2000 ft.

***subintegra* Casey, 1924**: 285 (Fig. 164i, j), Holotype
Original combination: *Euryptera subintegra* Casey
Current name: *Lycochoriolaus lateralis* (Olivier, 1795)
Current placement: Cerambycidae: Lepturinae: Lepturini
Type locality: USA: North Carolina, Southern Pines

***subirregularis* Breuning, 1954**: 8 (Fig. 164k, l), Holotype
Original combination: *Abatocera (Sternobatocera) subirregularis* Breuning
Current name: same
Current placement: Cerambycidae: Lamiinae: Batocerini
Type locality: Indonesia: Celebes, Paloe

***sublaevis* Casey, 1913**: 382 (Fig. 164m, n), Lectotype
Original combination: *Tetraopes sublaevis* Casey
Current name: same
Current placement: Cerambycidae: Lamiinae: Tetraopini
Type locality: USA: California
Remarks: This specimen is herein designated as the lectotype in order to stabilize the taxonomy and facilitate further identifications of this species.

***sublucidum* Gressitt, 1956**: 92 (Fig. 164o, p), Holotype
Original combination: *Ceresium (Ceresium) sublucidum* Gressitt
Current name: same
Current placement: Cerambycidae: Cerambycinae: Callidiopini
Type locality: USA: Guam, Pt. Oca, Mariana Island

***submarmorata* Gressitt, 1936**: 107 (Fig. 164q, r), Holotype
Original combination: *Arisania submarmorata* Gressitt
Current name: *Paruraecha (Arisania) submarmorata* (Gressitt, 1936)
Current placement: Cerambycidae: Lamiinae: Monochamini
Type locality: China: Taiwan, Arisan

***submetallica* Gressitt, 1940**: 29 (Fig. 164s, t), Holotype
Original combination: *Noemia submetallica* Gressitt
Current name: same
Current placement: Disteniidae: Disteniinae: Disteniini
Type locality: China: Hainan Province, Hainan Island, Ta Han

***submetallicus* Casey, 1912**: 341 (Fig. 164u, v), Holotype
Original combination: *Crossidius submetallicus* Casey
Current name: *Crossidius discoideus sayi* Casey, 1912
Current placement: Cerambycidae: Cerambycinae: Trachyderini
Type locality: USA: Colorado, Boulder Co.

***subnitidus* Casey, 1891**: 45 (Fig. 164w, x), Lectotype
Original combination: *Ipochus subnitidus* Casey
Current name: *Ipochus fasciatus* LeConte, 1852
Current placement: Cerambycidae: Lamiinae: Parmenini
Type locality: USA: California, Los Angeles Co.
Remarks: This specimen is herein designated as the lectotype in order to stabilize the taxonomy and facilitate further identifications of this species.

***subobsoleta* Casey, 1924**: 253 (Fig. 165a, b), Lectotype
Original combination: *Oeme subobsoleta* Casey
Current name: *Oeme rigida deserta* Casey, 1924
Current placement: Cerambycidae: Cerambycinae: Oemini
Type locality: USA: Utah, Trout Creek, Ibapah Mts.
Remarks: This specimen is herein designated as the lectotype in order to stabilize the taxonomy and facilitate further identifications of this species.

***subopacus* Fisher, 1935a**: 61 (Fig. 165c, d), Holotype
Original combination: *Cyrtinus subopacus* Fisher
Current name: same
Current placement: Cerambycidae: Lamiinae: Cyrtinini
Type locality: USA: Puerto Rico, Adjuntos

***subpinguis* Casey, 1913**: 213 (Fig. 165e, f), Holotype
Original combination: *Stenocorus (Stenocorus) subpinguis* Casey
Current name: *Stenocorus vestitus* (Haldeman, 1847)
Current placement: Cerambycidae: Lepturinae: Rhagiini
Type locality: USA: Utah

***subpubescens* Casey, 1912**: 219 (Fig. 165g, h), Lectotype
Original combination: *Spondylis subpubescens* Casey
Current name: *Neospondylis upiformis* (Mannerheim, 1843)
Current placement: Cerambycidae: Spondylidinae: Spondylidini
Type locality: USA: California
Remarks: This specimen is herein designated as the lectotype in order to stabilize the taxonomy and facilitate further identifications of this species.

***subquadrata* Casey, 1913**: 251 (Fig. 165i, j), Holotype
Original combination: *Brachyleptura subquadrata* Casey
Current name: *Brachyleptura circumdata* (Olivier, 1795)
Current placement: Cerambycidae: Lepturinae: Lepturini
Type locality: USA: Colorado

***subterrubens* Heller, 1924d**: 164 (Fig. 165k, l), Holotype
Original combination: *Nemophas subterrubens* Heller
Current name: same
Current placement: Cerambycidae: Lamiinae: Monochamini
Type locality: Philippines: Samar

***subtessellatus* Melzer, 1933**: 380 (Fig. 165m, n), Holotype
Original combination: *Eupogonius subtessellatus* Melzer
Current name: same
Current placement: Cerambycidae: Lamiinae: Desmiphorini
Type locality: Costa Rica: San Jose, 1000–1200 m

***subtilipictus* Heller, 1924c**: 432 (Fig. 165o, p), Lectotype
Original combination: *Eczemotellus subtilipictus* Heller
Current name: same
Current placement: Cerambycidae: Lamiinae: Pteropliini
Type locality: Philippines: Mindanao, Davao
Remarks: This specimen is herein designated as the lectotype in order to stabilize the taxonomy and facilitate further identifications of this species. Other paralectotype(s) are present in the Staatliches Museum für Tierkunde, Dresden, Germany.

***subtropica* Casey, 1924**: 271 (Fig. 165q, r), Holotype
Original combination: *Euryscelis subtropica* Casey
Current name: *Euryscelis suturalis* (Olivier, 1795)
Current placement: Cerambycidae: Cerambycinae: Clytini
Type locality: USA: Florida, Key Largo

***subtropicum* Casey, 1924**: 245 (Fig. 165s, t), Lectotype
Original combination: *Anoplium subtropicum* Casey
Current name: *Anelaphus cinereus* (Olivier, 1795)
Current placement: Cerambycidae: Cerambycinae: Elaphidiini
Type locality: USA: Florida, Key West
Remarks: This specimen is herein designated as the lectotype in order to stabilize the taxonomy and facilitate further identifications of this species.

***subtropicus* Casey, 1913**: 353 (Fig. 165u, v), Holotype
Original combination: *Oncideres subtropicus* Casey
Current name: *Oncideres cingulata texana* Horn, 1885
Current placement: Cerambycidae: Lamiinae: Onciderini
Type locality: USA: Texas, Brownsville, Esperanza Ranch

***subtruncata* Casey, 1912**: 343 (Fig. 165w, x), Holotype
Original combination: *Tylosis subtruncata* Casey
Current name: *Tylosis maculatus* LeConte, 1850
Current placement: Cerambycidae: Cerambycinae: Trachyderini
Type locality: USA: New Mexico, Jemez Springs

***subvittata* Casey, 1891**: 37 (Fig. 166a, b), Lectotype
Original combination: *Anthophilax subvittata* Casey
Current name: *Neanthophylax subvittatus* (Casey, 1891)
Current placement: Cerambycidae: Lepturinae: Rhagiini
Type locality: USA: Colorado (?)
Remarks: This specimen is herein designated as the lectotype in order to stabilize the taxonomy and facilitate further identifications of this species.

***succincta* Monné & Martins, 1976**: 84 (Fig. 166c, d), Holotype
Original combination: *Pseudocobelura succincta* Monné & Martins
Current name: same
Current placement: Cerambycidae: Lamiinae: Acanthocinini
Type locality: Panama: Canal Zone, Barro Colorado Island

***sudanica* Aurivillius, 1915**: 319 (Fig. 166e, f), Lectotype
Original combination: *Coptoeme sudanica* Aurivillius
Current name: *Coptoeme depressa* (Klug, 1835)
Current placement: Cerambycidae: Cerambycinae: Oemini
Type locality: Sudan: Wau (east Sudan)
Remarks: This specimen is herein designated as the lectotype in order to stabilize the taxonomy and facilitate further identifications of this species.

***sulphurea* Gressitt, 1941**: 336 (Fig. 166g, h), Holotype
Original combination: *Rhaphuma sulphurea* Gressitt
Current name: same
Current placement: Cerambycidae: Cerambycinae: Clytini
Type locality: Thailand: Khun Tan Mts., 4000 ft.

***sulphureomarmorata* Breuning, 1939a**: 69 (Fig. 166i, j), Holotype
Original combination: *Prosopocera (Alphitopola) sulphureomarmorata* Breuning
Current name: same
Current placement: Cerambycidae: Lamiinae: Prosopocerini
Type locality: Congo

***sumatrensis* Fisher, 1927a**: 83 (Fig. 166k, l), Holotype
Original combination: *Exocentrus sumatrensis* Fisher
Current name: *Exocentrus (Camptomyne) sumatrensis* Fisher, 1927a
Current placement: Cerambycidae: Lamiinae: Acanthocinini
Type locality: Indonesia: Sumatera Utara, Medan

***sumbawanus* Breuning, 1957a**: 13 (Fig. 166m, n), Holotype
Original combination: *Exocentrus (Pseudocentrus) sumbawanus* Breuning
Current name: same
Current placement: Cerambycidae: Lamiinae: Acanthocinini
Type locality: Indonesia: Sumbawa Islands, Nusa Tenggara Barat

***surdus* Dillon & Dillon, 1946b**: 242 (Fig. 166o, p), Holotype
Original combination: *Trachysomus surdus* Dillon & Dillon
Current name: same
Current placement: Cerambycidae: Lamiinae: Onciderini
Type locality: Panama

surigaonis **Breuning, 1943c**: 45 (Fig. 166q, r), Holotype
Original combination: *Sybra (Atelais) surigaonis* Breuning
Current name: *Atelais surigaonis* (Breuning, 1943c)
Current placement: Cerambycidae: Lamiinae: Apomecynini
Type locality: Philippines: Mindanao, Surigao

suspectus **Casey, 1924**: 215 (Fig. 166s, t), Holotype
Original combination: *Prionus (Prionus) suspectus* Casey
Current name: *Prionus californicus* Motschulsky, 1845
Current placement: Cerambycidae: Prioninae: Prionini
Type locality: USA: Utah, Eureka

susua **Galileo & Martins, 2005a**: 274 (Fig. 166u, v), Holotype
Original combination: *Susuanycha susua* Galileo & Martins
Current name: same
Current placement: Cerambycidae: Lamiinae: Hemilophini
Type locality: Peru: Madre de Dios, Reserva Tambopata (30 km southwest of Puerto Maldonado)

suturafissa **Tippmann, 1960**: 126 (Fig. 166w, x), Holotype
Original combination: *Pseudagaone suturafissa* Tippmann
Current name: same
Current placement: Cerambycidae: Cerambycinae: Rhinotragini
Type locality: Bolivia: Cochabamba, Reg. Chapare 400 m.

suturalis **Casey, 1924**: 277 (Fig. 167a, b), Lectotype
Original combination: *Leptacmaeops suturalis* Casey
Current name: *Cortodera longicornis* (Kirby, 1837)
Current placement: Cerambycidae: Lepturinae: Rhagiini
Type locality: USA: California
Remarks: This specimen is herein designated as the lectotype in order to stabilize the taxonomy and facilitate further identifications of this species.

suturalis **Fisher, 1935c**: 581 (Fig. 167c, d), Holotype
Original combination: *Megopis (Aegosoma) suturalis* Fisher
Current name: *Megobaralipton suturale* (Fisher, 1935c)
Current placement: Cerambycidae: Prioninae: Megopini
Type locality: Malaysia: Borneo, Sabah, Mt. Kinabalu, Kiau, 3000 ft.

suturalis **Hamilton in Leng & Hamilton, 1896**: 145 (Fig. 167e, f), Lectotype
Original combination: *Spalacopsis suturalis* Hamilton in Leng & Hamilton
Current name: *Spalacopsis suffusa* Newman, 1842
Current placement: Cerambycidae: Lamiinae: Agapanthiini
Type locality: USA: Florida, Punta Gorda

suturalis **Linell, 1897a**: 398 (Fig. 167g, h), Holotype
Original combination: *Amphionycha suturalis* Linell
Current name: *Hemierana marginata suturalis* (Linell, 1897a)
Current placement: Cerambycidae: Lamiinae: Hemilophini
Type locality: USA: Texas, Brownsville

szechuanus **Gressitt, 1938b**: 155 (Fig. 167i, j), Holotype
Original combination: *Dihammus szechuanus* Gressitt
Current name: *Acalolepta szechuana* (Gressitt, 1938b)
Current placement: Cerambycidae: Lamiinae: Monochamini
Type locality: China: Sichuan Province, Shin Kai Si, Mt. Omei, near Kiating, 4,400 ft.

szechuanus **Heyrovsky, 1933**: 134 (Fig. 167k, l), Syntype
Original combination: *Flabelloprionus szechuanus* Heyrovsky
Current name: *Casiphia (Flabelloprionus) szechuana* (Heyrovsky, 1933)
Current placement: Cerambycidae: Prioninae: Anacolini
Type locality: China: Sichuan Province, Tatsienlu Dsaschi
Remarks: Tavakilian and Chevillotte (2013) indicate that this species is represented by 4 syntypes in Prague and Budapest, however, ours is from the same series and is a syntype as well.

taeniatus **Casey, 1913**: 306 (Fig. 167m, n), Lectotype
Original combination: *Leptostylus taeniatus* Casey
Current name: *Leptostylopsis argentatus* (Jacuelin du Val in Sagra, 1857)
Current placement: Cerambycidae: Lamiinae: Acanthocinini
Type locality: USA: Florida, Lake Worth
Remarks: This specimen is herein designated as the lectotype in order to stabilize the taxonomy and facilitate further identifications of this species.

taino **Lingafelter & Micheli, 2004**: 45 (Fig. 167o, p), Holotype
Original combination: *Styloleptus taino* Lingafelter & Micheli
Current name: same
Current placement: Cerambycidae: Lamiinae: Acanthocinini
Type locality: USA: Puerto Rico: Aguirre State Forest off Rd. 7710, 17°59′N, 65°09′W

taiwanensis **Gressitt, 1936**: 105 (Fig. 167q, r), Holotype
Original combination: *Metopoplectus taiwanensis* Gressitt
Current name: same
Current placement: Cerambycidae: Lamiinae: Agapanthiini
Type locality: China: Taiwan, Sakahen

tatsienlua **Gressitt, 1948a**: 45 (Fig. 167s, t), Holotype
Original combination: *Leptura tatsienlua* Gressitt
Current name: *Leptura guerryi* Pic, 1902
Current placement: Cerambycidae: Lepturinae: Lepturini
Type locality: China: Sichuan Province, near Tatsienlu, 10–12000 ft.

tavura **Kriesche, 1936**: 67 (Fig. 167u, v), Holotype
Original combination: *Dihammus tavura* Kriesche
Current name: *Acalolepta blairi* (Breuning, 1935c)
Current placement: Cerambycidae: Lamiinae: Monochamini
Type locality: Papua New Guinea: Bougainville Island (Solomon Islands)
Remarks: Tippmann labeled this specimen as having been obtained from the Kriesche Estate in 1955.

tectonae **Fisher, 1934b**: 36 (Fig. 167w, x), Holotype
Original combination: *Exocentrus tectonae* Fisher
Current name: *Exocentrus (Pseudocentrus) tectonae* Fisher, 1934b
Current placement: Cerambycidae: Lamiinae: Acanthocinini
Type locality: Indonesia: Java, Semarang, teak-forest

temporalis **Casey, 1913**: 233 (Fig. 168a, b), Lectotype
Original combination: *Leptacmaeops (Leptacmaeops) temporalis* Casey
Current name: *Cortodera cubitalis* (LeConte, 1861)
Current placement: Cerambycidae: Lepturinae: Rhagiini
Type locality: USA: California, Lake Co.
Remarks: This specimen is herein designated as the lectotype in order to stabilize the taxonomy and facilitate further identifications of this species.

tenebrosus **Casey, 1913**: 320 (Fig. 168c, d), Holotype
Original combination: *Lepturges tenebrosus* Casey
Current name: *Urgleptes signatus* (LeConte, 1852)
Current placement: Cerambycidae: Lamiinae: Acanthocinini
Type locality: USA: New York, Bluff Point, Lake Champlain

tenellus **Casey, 1913**: 210 (Fig. 168e, f), Lectotype
Original combination: *Stenocorus (Stenocorus) tenellus* Casey
Current name: *Stenocorus nubifer* (LeConte, 1859)
Current placement: Cerambycidae: Lepturinae: Rhagiini
Type locality: USA: California, Lake Co.
Remarks: This specimen is herein designated as the lectotype in order to stabilize the taxonomy and facilitate further identifications of this species.

tenera **Casey, 1913**: 203 (Fig. 168g, h), Holotype
Original combination: *Centrodera tenera* Casey
Current name: same
Current placement: Cerambycidae: Lepturinae: Rhagiini
Type locality: USA: California, San Diego

tenuicornis **Casey, 1913**: 298 (Fig. 168i, j), Lectotype
Original combination: *Microgoes tenuicornis* Casey
Current name: *Microgoes oculatus* (LeConte, 1862)
Current placement: Cerambycidae: Lamiinae: Monochamini
Type locality: USA: Pennsylvania, Saint Vincent
Remarks: This specimen is herein designated as the lectotype in order to stabilize the taxonomy and facilitate further identifications of this species.

tenuicornis **Casey, 1913**: 363 (Fig. 168k, l), Holotype
Original combination: *Amillarus tenuicornis* Casey
Current name: *Grammopsoides tenuicornis* (Casey, 1913)
Current placement: Cerambycidae: Lamiinae: Agapanthiini
Type locality: Panama: Darién

tenuis **Casey, 1913**: 233 (Fig. 168m, n), Holotype
Original combination: *Leptacmaeops (Leptacmaeops) tenuis* Casey
Current name: *Cortodera cubitalis* (LeConte, 1861)
Current placement: Cerambycidae: Lepturinae: Rhagiini
Type locality: USA: California

tenuis **Fisher, 1938**: 150 (Fig. 168o, p), Holotype
Original combination: *Obereoides tenuis* Fisher
Current name: *Obereoides joergenseni* (Bruch, 1911)
Current placement: Cerambycidae: Lamiinae: Forsteriini
Type locality: Brazil: Santa Catarina, Nova Teutonia

terminalis **Casey, 1912**: 244 (Fig. 168q, r), Holotype
Original combination: *Prionus (Prionus) terminalis* Casey
Current name: *Prionus californicus* Motschulsky, 1845
Current placement: Cerambycidae: Prioninae: Prionini
Type locality: USA: Arizona, Canyon of Colorado River

terminata **Casey, 1912**: 274 (Fig. 168s, t), Holotype
Original combination: *Anocomis terminata* Casey
Current name: *Semanotus litigiosus* (Casey, 1891)
Current placement: Cerambycidae: Cerambycinae: Callidiini
Type locality: USA: Maine

tertiarius **Wickham, 1917**: 469 (Fig. 168u, v), Holotype
Original combination: *Spondylis tertiarius* Wickham
Current name: *Protospondylis florissantensis* (Wickham, 1920)
Current placement: Cerambycidae: Spondylidinae: Spondylidini
Type locality: USA: Colorado, Florissant
Remarks: This is one of four fossil specimens at the USNM and is housed in the Paleobiology (Paleoentomology) section.

***testacea* Casey, 1924**: 243 (Fig. 168w, x), Lectotype
Original combination: *Aneflomorpha testacea* Casey
Current name: *Aneflomorpha parowana* Casey, 1924
Current placement: Cerambycidae: Cerambycinae: Elaphidiini
Type locality: USA: Utah, Eureka
Remarks: This specimen is herein designated as the lectotype in order to stabilize the taxonomy and facilitate further identifications of this species.

***testacea* Linsley, 1935a**: 144 (Fig. 169a, b), Holotype
Original combination: *Psyrassa testacea* Linsley
Current name: same
Current placement: Cerambycidae: Cerambycinae: Elaphidiini
Type locality: Costa Rica: San Jose

***testaceum* Gressitt, 1956**: 87 (Fig. 169c, d), Holotype
Original combination: *Ceresium (Ceresium) testaceum* Gressitt
Current name: same
Current placement: Cerambycidae: Cerambycinae: Callidiopini
Type locality: Micronesia: Moen [Truk] Island, Civil Administration Area

***testaceus* Casey, 1913**: 312 (Fig. 169e, f), Holotype
Original combination: *Leiopus testaceus* Casey
Current name: *Sternidius alpha* (Say, 1827)
Current placement: Cerambycidae: Lamiinae: Acanthocinini
Type locality: USA: District of Columbia

***tetricus* Casey, 1912**: 237 (Fig. 169g, h), Holotype
Original combination: *Prionus (Prionus) tetricus* Casey
Current name: *Prionus heroicus* Semenov, 1907
Current placement: Cerambycidae: Prioninae: Prionini
Type locality: USA: New Mexico, Jemez Springs

***tetropthalmus iowensis* Casey, 1913**: 386 (Fig. 169i, j), Lectotype
Original combination: *Tetraopes tetropthalmus iowensis* Casey
Current name: *Tetraopes tetrophthalmus* (Forster, 1771)
Current placement: Cerambycidae: Lamiinae: Tetraopini
Type locality: USA: Iowa, Keokuk and vicinity
Remarks: This specimen is herein designated as the lectotype in order to stabilize the taxonomy and facilitate further identifications of this species.

***texana* Casey, 1891**: 51 (Fig. 169k, l), Holotype
Original combination: *Spalacopsis texana* Casey
Current name: same
Current placement: Cerambycidae: Lamiinae: Agapanthiini
Type locality: USA: Texas

***texana* Casey, 1912**: 343 (Fig. 169m, n), Holotype
Original combination: *Tylosis texana* Casey
Current name: *Tylosis maculatus* LeConte, 1850
Current placement: Cerambycidae: Cerambycinae: Trachyderini
Type locality: USA: Texas

***texana* Casey, 1913**: 276 (Fig. 169o, p), Lectotype
Original combination: *Ophistomis texana* Casey
Current name: *Strangalia sexnotata* Haldeman, 1847
Current placement: Cerambycidae: Lepturinae: Lepturini
Type locality: USA: Texas
Remarks: This specimen is herein designated as the lectotype in order to stabilize the taxonomy and facilitate further identifications of this species.

***texana* Casey, 1924**: 292 (Fig. 169q, r), Holotype
Original combination: *Urographis texana* Casey
Current name: *Graphisurus triangulifer* (Haldeman, 1847)
Current placement: Cerambycidae: Lamiinae: Acanthocinini
Type locality: USA: Texas, Comal Co.

***texana* Schaeffer, 1905b**: 160 (Fig. 169s, t), Holotype
Original combination: *Psyrassa texana* Schaeffer
Current name: *Psyrassa castanea* Bates, 1880
Current placement: Cerambycidae: Cerambycinae: Elaphidiini
Type locality: USA: Texas, Brownsville, Esperanza Ranch

***texana* Schaeffer, 1908**: 351 (Fig. 169u, v), Holotype
Original combination: *Idoemea texana* Schaeffer
Current name: *Styloxus fulleri fulleri* (Horn, 1880)
Current placement: Cerambycidae: Cerambycinae: Methiini
Type locality: USA: Texas, Burnett Co.

***texanum* Schaeffer, 1917**: 185 (Fig. 169w, x), Holotype
Original combination: *Callidium texanum* Schaeffer
Current name: same
Current placement: Cerambycidae: Cerambycinae: Callidiini
Type locality: USA: Texas

***texanus* Casey, 1912**: 243 (Fig. 170a, b), Holotype
Original combination: *Prionus (Prionus) texanus* Casey
Current name: *Prionus californicus* Motschulsky, 1845
Current placement: Cerambycidae: Prioninae: Prionini
Type locality: USA: Texas, El Paso

***texanus* Casey, 1913**: 315 (Fig. 170c, d), Lectotype
Original combination: *Leiopus texanus* Casey
Current name: *Sternidius mimeticus* (Casey, 1891)
Current placement: Cerambycidae: Lamiinae: Acanthocinini
Type locality: USA: Texas, Brownsville
Remarks: This specimen is herein designated as the lectotype in order to stabilize the taxonomy and facilitate further identifications of this species.

thalassina **Santos-Silva & Hovore, 2008b**: 2 (Fig. 170e, f), Holotype
Original combination: *Hovorestenia thalassina* Santos-Silva & Hovore
Current name: same
Current placement: Disteniidae: Disteniinae: Disteniini
Type locality: Ecuador: Napo, Quijos, Sierra Azul, Valle Hermoso

thomasi **Lingafelter & Micheli, 2009**: 50 (Fig. 170g, h), Holotype
Original combination: *Leptostylopsis thomasi* Lingafelter & Micheli
Current name: same
Current placement: Cerambycidae: Lamiinae: Acanthocinini
Type locality: Dominican Republic: Pedernales Province, Parque Nacional Sierra de Baoruco, Las Abejas, 1150 m, 18°09.011´N, 71°37.342´W.

thomasi **Santos-Silva & Hovore, 2007b**: 584 (Fig. 170i, j), Holotype
Original combination: *Novantinoe thomasi* Santos-Silva & Hovore
Current name: same
Current placement: Disteniidae: Disteniinae: Disteniini
Type locality: Guatemala: Zacapa, cloud forest above La Union, 1250 m.

thomsoni **Breuning, 1943a**: 246 (Fig. 170k, l), Holotype
Original combination: *Taeniotes thomsoni* Breuning
Current name: *Taeniotes insularis* Thomson, 1857
Current placement: Cerambycidae: Lamiinae: Monochamini
Type locality: Guadeloupe

thonalmus **Lingafelter, 2013**: 70 (Fig. 170m, n), Holotype
Original combination: *Calocosmus thonalmus* Lingafelter
Current name: same
Current placement: Cerambycidae: Lamiinae: Hemilophini
Type locality: Dominican Republic: La Altagracia Province, El Veron, road to Hoyo Azul, 25–40 m.

thoracicum **Casey, 1912**: 297 (Fig. 170o, p), Lectotype
Original combination: *Romaleum thoracicum* Casey
Current name: *Enaphalodes atomarius* (Drury, 1773)
Current placement: Cerambycidae: Cerambycinae: Elaphidiini
Type locality: USA: Texas
Remarks: This specimen is herein designated as the lectotype in order to stabilize the taxonomy and facilitate further identifications of this species.

thoracicum **Casey, 1913**: 196 (Fig. 170q, r), Holotype
Original combination: *Rhagium (Hargium) thoracicum* Casey
Current name: *Rhagium inquisitor* (Linnaeus, 1758)
Current placement: Cerambycidae: Lepturinae: Rhagiini
Type locality: USA: New Mexico, Las Vegas

thoracicus **Casey, 1913**: 274 (Fig. 170s, t), Holotype
Original combination: *Typocerus thoracicus* Casey
Current name: *Typocerus acuticauda* Casey, 1913
Current placement: Cerambycidae: Lepturinae: Lepturini
Type locality: No locality label or published data.

thoracicus **Casey, 1913**: 342 (Fig. 170u, v), Holotype
Original combination: *Dectes thoracicus* Casey
Current name: *Dectes texanus* LeConte, 1862
Current placement: Cerambycidae: Lamiinae: Acanthocinini
Type locality: USA: Arizona

thoracicus **Casey, 1924**: 223 (Fig. 170w, x), Holotype
Original combination: *Prionus (Antennalia) thoracicus* Casey
Current name: *Prionus (Antennalia) fissicornis* Haldeman, 1848
Current placement: Cerambycidae: Prioninae: Prionini
Type locality: USA: Colorado, Akron (northeast part of state)

thysbe **Dillon & Dillon, 1945**: 68 (Fig. 171a, b), Holotype
Original combination: *Charoides thysbe* Dillon & Dillon
Current name: *Tulcus thysbe* (Dillon & Dillon, 1945)
Current placement: Cerambycidae: Lamiinae: Onciderini
Type locality: Ecuador: San Rafael

tibialis **Breuning, 1943c**: 37 (Fig. 171c, d), Holotype
Original combination: *Adetus tibialis* Breuning
Current name: same
Current placement: Cerambycidae: Lamiinae: Apomecynini
Type locality: Brazil

tigrina **Casey, 1913**: 266 (Fig. 171e, f), Holotype
Original combination: *Strophiona tigrina* Casey
Current name: same
Current placement: Cerambycidae: Lepturinae: Lepturini
Type locality: USA: California, Los Angeles Co.

tigrinis marmoratus **Casey, 1913**: 296 (Fig. 171g, h), Holotype
Original combination: *Goes tigrinis marmoratus* Casey
Current name: *Goes tigrinus* (Degeer, 1775)
Current placement: Cerambycidae: Lamiinae: Monochamini
Type locality: USA: Texas

***tiliacei* Gressitt, 1956**: 179 (Fig. 171i, j), Holotype
Original combination: *Paremeopedus tiliacei* Gressitt
Current name: same
Current placement: Cerambycidae: Lamiinae: Acanthocinini
Type locality: Micronesia: Kusaie Hill 541, 165 m.

***timidus* Casey, 1913**: 313 (Fig. 171k, l), Holotype
Original combination: *Leiopus timidus* Casey
Current name: *Sternidius alpha* (Say, 1827)
Current placement: Cerambycidae: Lamiinae: Acanthocinini
Type locality: USA: Pennsylvania

***tippmanni* Breuning, 1942d**: 144 (Fig. 171m, n), Holotype
Original combination: *Prosopocera (Alphitopola) tippmanni* Breuning
Current name: same
Current placement: Cerambycidae: Lamiinae: Prosopocerini
Type locality: Tanzania: Tanganyika, Dar-es-Salaam

***tippmanni* Breuning, 1954**: 8 (Fig. 171o, p), Holotype
Original combination: *Batocera tippmanni* Breuning
Current name: same
Current placement: Cerambycidae: Lamiinae: Batocerini
Type locality: Papua New Guinea: Finschhaven

***tippmanni* Breuning, 1957a**: 11 (Fig. 171q, r), Holotype
Original combination: *Sybrodoius tippmanni* Breuning
Current name: same
Current placement: Cerambycidae: Lamiinae: Rhodopini
Type locality: Indonesia: Java

***tippmanni* Breuning, 1957a**: 12 (Fig. 171s, t), Holotype
Original combination: *Exocentrus (Pseudocentrus) tippmanni* Breuning
Current name: *Exocentrus (Pseudocentrus) rhodesianus* Breuning, 1956
Current placement: Cerambycidae: Lamiinae: Acanthocinini
Type locality: Tanzania: Tanganyika

***tippmanni* Breuning, 1957a**: 15 (Fig. 171u, v), Holotype
Original combination: *Aegocidnexocentrus tippmanni* Breuning
Current name: same
Current placement: Cerambycidae: Lamiinae: Acanthocinini
Type locality: Papua New Guinea: East New Britain, Mope

***tonkinense* Breuning, 1938b**: 188 (Fig. 171w, x), Holotype
Original combination: *Eshinoschema tonkinense* Breuning
Current name: *Hechinoschema vitalisi* Pic, 1925
Current placement: Cerambycidae: Lamiinae: Monochamini
Type locality: Vietnam: Tonkin, Chapa

***townesi* Gressitt, 1956**: 158 (Fig. 172a, b), Holotype
Original combination: *Sciadella (Micronesiella) townesi* Gressitt
Current name: *Sciades (Micronesiella) townesi* (Gressitt, 1956)
Current placement: Cerambycidae: Lamiinae: Acanthocinini
Type locality: Micronesia: Ponape Island, Colonia, Hydro-electric plant

***townsendi* Casey, 1912**: 246 (Fig. 172c, d), Holotype
Original combination: *Prionus (Riponus) townsendi* Casey
Current name: same
Current placement: Cerambycidae: Prioninae: Prionini
Type locality: Mexico: Chihuahua, Meadow Valley (6 mi. south Colonia García) (Sierra Madre Mtns.)

***townsendi* Linell, 1897a**: 395 (Fig. 172e, f), Holotype
Original combination: *Ibidion townsendi* Linell
Current name: *Neocompsa mexicana* (Thomson, 1865)
Current placement: Cerambycidae: Cerambycinae: Neoibidionini
Type locality: USA: Texas, Brownsville

***trajecta* Casey, 1913**: 250 (Fig. 172g, h), Holotype
Original combination: *Judolia trajecta* Casey
Current name: *Judolia instabilis* (Haldeman, 1847)
Current placement: Cerambycidae: Lepturinae: Lepturini
Type locality: USA: Colorado, Boulder County

***transversefasciata* Breuning, 1940d**: 419 (Fig. 172i, j), Holotype
Original combination: *Eczemotes transversefasciata* Breuning
Current name: same
Current placement: Cerambycidae: Lamiinae: Niphonini
Type locality: Papua New Guinea: Teg. River

***transversefasciata* Breuning, 1943c**: 58 (Fig. 172k, l), Holotype
Original combination: *Eupromera transversefasciata* Breuning
Current name: *Pibanga transversefasciata* (Breuning, 1943c)
Current placement: Cerambycidae: Lamiinae: Eupromerini
Type locality: Brazil: Bahia, Brasilia

***triaculeatus* Aurivillius, 1922**: 417 (Fig. 172m, n), Holotype
Original combination: *Demonax triaculeatus* Aurivillius
Current name: same
Current placement: Cerambycidae: Cerambycinae: Clytini
Type locality: Philippines: Mindanao, Dapitan Mindanao

***triangularis* Breuning, 1938b**: 280 (Fig. 172o, p), Holotype
Original combination: *Pterolophia triangularis* Breuning
Current name: same
Current placement: Cerambycidae: Lamiinae: Pteropliini
Type locality: Indonesia: Java, Preanger
Remarks: Tavakilian and Chevillotte (2013) suggest the holotype is present in the BMNH, however, this specimen and its labels match the original description.

***tricolor* Linsley, 1935b**: 481 (Fig. 172q, r), Holotype
Original combination: *Octoplon tricolor* Linsley
Current name: *Gnomidolon suturale* (White, 1855)
Current placement: Cerambycidae: Cerambycinae: Hexoplini
Type locality: Panama: Canal Zone, Ancon

***tricolor* var. *superbus* Tippmann, 1954**: 317 (Fig. 172s, t), Holotype
Original combination: *Phaedinus tricolor* var. *superbus* Tippmann
Current name: *Phaedinus tricolor* Dupont, 1834
Current placement: Cerambycidae: Cerambycinae: Trachyderini
Type locality: French Guiana: Cayenne
Remarks: Although the label indicates *Phaedinus tricolor* var. *nigrofasciatus*, it was published as *Phaedinus tricolor* var. *superbus*. The name following the binomen was not expressly given infrasubspecific rank and the work did not indicate unambiguously that the name was infrasubspecific, therefore, it is deemed a subspecies and available.

***tricostata* Gressitt, 1947**: 547 (Fig. 172u, v), Holotype
Original combination: *Linda tricostata* Gressitt
Current name: *Linda semivittata* (Fairmaire, 1887)
Current placement: Cerambycidae: Lamiinae: Saperdini
Type locality: China: Yunnan Province

***tridentata trifasciata* Casey, 1913**: 359 (Fig. 172w, x), Lectotype
Original combination: *Saperda tridentata trifasciata* Casey
Current name: *Saperda tridentata* Olivier, 1795
Current placement: Cerambycidae: Lamiinae: Saperdini
Type locality: USA: Indiana
Remarks: This specimen is herein designated as the lectotype in order to stabilize the taxonomy and facilitate further identifications of this species.

***trifasciata* Aurivillius, 1927b**: 580 (Fig. 173a, b), Syntype
Original combination: *Driopeomima trifasciata* Aurivillius
Current name: *Philicus trifasciatus* (Aurivillius, 1927b)
Current placement: Cerambycidae: Lamiinae: Rhodopini
Type locality: Philippines: Mindanao, Surigao
Remarks: Other syntype(s) are present in the Naturhistoriska Riksmuseet, Stockholm, Sweden.

***trifasciatus* Aurivillius, 1928**: 320 (Fig. 173c, d), Holotype
Original combination: *Demonax trifasciatus* Aurivillius
Current name: same
Current placement: Cerambycidae: Cerambycinae: Clytini
Type locality: Philippines: Negros Oriental, Cuernos Mountains

***trinitatis* Casey, 1913**: 226 (Fig. 173e, f), Lectotype
Original combination: *Brachysomida trinitatis* Casey
Current name: *Brachysomida californica* (LeConte, 1851)
Current placement: Cerambycidae: Lepturinae: Rhagiini
Type locality: USA: California, Hoopa Valley, Trinity River, Humboldt Co (Fort Gaston)
Remarks: This specimen is herein designated as the lectotype in order to stabilize the taxonomy and facilitate further identifications of this species.

***trinodatus* Casey, 1913**: 352 (Fig. 173g, h), Lectotype
Original combination: *Oncideres trinodatus* Casey
Current name: *Oncideres rhodosticta* Bates, 1885
Current placement: Cerambycidae: Lamiinae: Onciderini
Type locality: USA: Texas, El Paso
Remarks: This specimen is herein designated as the lectotype in order to stabilize the taxonomy and facilitate further identifications of this species.

***triplaris* Fisher, 1930a**: 3 (Fig. 173i, j), Holotype
Original combination: *Odontocera triplaris* Fisher
Current name: same
Current placement: Cerambycidae: Cerambycinae: Rhinotragini
Type locality: Colombia: Magdalena, Río Frio

***tripunctata appalachiana* Casey, 1913**: 367 (Fig. 173k, l), Holotype
Original combination: *Oberea tripunctata appalachiana* Casey
Current name: *Oberea tripunctata* (Swederus, 1787)
Current placement: Cerambycidae: Lamiinae: Phytoeciini
Type locality: USA: North Carolina, Asheville

***tripunctata intermedia* Casey, 1913**: 367 (Fig. 173m, n), Lectotype
Original combination: *Oberea tripunctata intermedia* Casey
Current name: *Oberea tripunctata* (Swederus, 1787)
Current placement: Cerambycidae: Lamiinae: Phytoeciini
Type locality: USA
Remarks: This specimen is herein designated as the lectotype in order to stabilize the taxonomy and facilitate further identifications of this species.

***tristis* Casey, 1891**: 46 (Fig. 173o, p), Holotype
Original combination: *Psenocerus tristis* Casey
Current name: *Psenocerus supernotatus* (Say, 1824)
Current placement: Cerambycidae: Lamiinae: Desmiphorini
Type locality: USA: New Mexico

***tristis* Casey, 1912**: 236 (Fig. 173q, r), Holotype
Original combination: *Prionus (Prionus) tristis* Casey
Current name: *Prionus heroicus* Semenov, 1907
Current placement: Cerambycidae: Prioninae: Prionini
Type locality: USA: New Mexico, Jemez Springs

***tristis* Casey, 1912**: 371 (Fig. 173s, t), Holotype
Original combination: *Clytoleptus tristis* Casey
Current name: *Clytoleptus albofasciatus* (Castelnau & Gory, 1841)
Current placement: Cerambycidae: Cerambycinae: Clytini
Type locality: USA: Pennsylvania, Lemoyne

***tristis* Casey, 1913**: 321 (Fig. 173u, v), Holotype
Original combination: *Lepturges tristis* Casey
Current name: *Urgleptes querci* (Fitch, 1858)
Current placement: Cerambycidae: Lamiinae: Acanthocinini
Type locality: USA: Pennsylvania, Franklin Co., Buena Vista Spring

***trivittata* Breuning, 1940c**: 138 (Fig. 173w, x), Holotype
Original combination: *Pterolophia trivittata* Breuning
Current name: same
Current placement: Cerambycidae: Lamiinae: Pteropliini
Type locality: India: Kerala, Travancore

***truncata* Breuning, 1940d**: 424 (Fig. 174a, b), Holotype
Original combination: *Rhytiphora truncata* Breuning
Current name: same
Current placement: Cerambycidae: Lamiinae: Pteropliini
Type locality: Australia: Queensland, Duringo

***truncata* Casey, 1924**: 253 (Fig. 174c, d), Holotype
Original combination: *Oeme truncata* Casey
Current name: *Oeme rigida deserta* Casey, 1924
Current placement: Cerambycidae: Cerambycinae: Oemini
Type locality: USA: Utah, Trout Creek, Ibapah Mts.

***truncatellus* Casey, 1912**: 338 (Fig. 174e, f), Holotype
Original combination: *Crossidius truncatellus* Casey
Current name: *Crossidius hirtipes wickhami* Casey, 1912
Current placement: Cerambycidae: Cerambycinae: Trachyderini
Type locality: USA: Utah, Marysvale

***truncatipennis* Breuning, 1940d**: 410 (Fig. 174g, h), Holotype
Original combination: *Morimopsis truncatipennis* Breuning
Current name: same
Current placement: Cerambycidae: Lamiinae: Morimopsini
Type locality: India: (northern)

***truncatipennis* Gressitt, 1948a**: 52 (Fig. 174i, j), Holotype
Original combination: *Stenopterus truncatipennis* Gressitt
Current name: *Lampropterus (Nathriopterus) truncatipennis* (Gressitt, 1948a)
Current placement: Cerambycidae: Cerambycinae: Stenopterini
Type locality: China: Shaanxi Province

***truncatulus* Casey, 1913**: 211 (Fig. 174k, l), Holotype
Original combination: *Stenocorus (Stenocorus) truncatulus* Casey
Current name: *Stenocorus nubifer* (LeConte, 1859)
Current placement: Cerambycidae: Lepturinae: Lepturini
Type locality: USA: California, Lake Co.
Remarks: Although the original description suggests there may be more than one specimen in the original series, there is only one in the USNM collection.

***tuberculata* Fisher, 1935b**: 196 (Fig. 174m, n), Holotype
Original combination: *Nanilla tuberculata* Fisher
Current name: same
Current placement: Cerambycidae: Lamiinae: Parmenini
Type locality: Cuba: Buenos Aires, Trinidad Mts.

***tuberculatus* Fisher, 1936a**: 181 (Fig. 174o, p), Holotype
Original combination: *Dolichostyrax tuberculatus* Fisher
Current name: same
Current placement: Cerambycidae: Lamiinae: Morimopsini
Type locality: Indonesia: Java, Mt. Tangkoeban Prahoe, Preanger 4000–5000 ft.

***tuberculicolle* Fisher, 1932**: 25 (Fig. 174q, r), Holotype
Original combination: *Elaphidion tuberculicolle* Fisher
Current name: same
Current placement: Cerambycidae: Cerambycinae: Elaphidiini
Type locality: Cuba: Cayamas

***tuberculicollis* Fisher, 1947a**: 48 (Fig. 174s, t), Holotype
Original combination: *Oxylymma tuberculicollis* Fisher
Current name: same
Current placement: Cerambycidae: Cerambycinae: Rhinotragini
Type locality: Costa Rica: Limón, Hamburg Farm, Reventazón, Ebene

tubericollis **Breuning, 1943c**: 16 (Fig. 174u, v), Holotype
Original combination: *Rhodopis tubericollis* Breuning
Current name: *Rhodopina tubericollis* (Breuning, 1943c)
Current placement: Cerambycidae: Lamiinae: Rhodopini
Type locality: Malaysia: Borneo, Sabah, Labuan

tuberosa **Martins, 1970a**: 1098 (Fig. 174w, x), Holotype
Original combination: *Neocompsa tuberosa* Martins
Current name: same
Current placement: Cerambycidae: Cerambycinae: Neoibidionini
Type locality: Costa Rica: Cartago, Turrialba

tuckeri **Casey, 1924**: 247 (Fig. 175a, b), Lectotype
Original combination: *Anoplium tuckeri* Casey
Current name: *Anelaphus tuckeri* (Casey, 1924)
Current placement: Cerambycidae: Cerambycinae: Elaphidiini
Type locality: USA: Arizona, Tucson
Remarks: This specimen is herein designated as the lectotype in order to stabilize the taxonomy and facilitate further identifications of this species.

tularensis **Casey, 1912**: 369 (Fig. 175c, d), Holotype
Original combination: *Neoclytus tularensis* Casey
Current name: *Neoclytus conjunctus* (LeConte, 1857)
Current placement: Cerambycidae: Cerambycinae: Clytini
Type locality: USA: California, Kaweah

tumidiceps **Casey, 1913**: 227 (Fig. 175e, f), Holotype
Original combination: *Brachysomida tumidiceps* Casey
Current name: *Brachysomida californica* (LeConte, 1851)
Current placement: Cerambycidae: Lepturinae: Rhagiini
Type locality: USA: California, Calaveras Co.

tumidus **Casey, 1912**: 235 (Fig. 175g, h), Holotype
Original combination: *Prionus (Prionus) tumidus* Casey
Current name: *Prionus heroicus* Semenov, 1908
Current placement: Cerambycidae: Prioninae: Prionini
Type locality: USA: New Mexico, Fort Wingate

turnbowi **Hovore & Santos-Silva, 2007**: 94 (Fig. 175i, j), Holotype
Original combination: *Cometes turnbowi* Hovore & Santos-Silva
Current name: *Cupecuara turnbowi* (Hovore & Santos-Silva, 2007)
Current placement: Disteniidae: Disteniinae: Disteniini
Type locality: Panama: Canal Zone, Barro Colorado Island, 9°9′N 79°51′W

turnbowi **Nearns & Branham, 2008**: 52 (Fig. 175k, l), Holotype
Original combination: *Plectromerus turnbowi* Nearns & Branham
Current name: same
Current placement: Cerambycidae: Cerambycinae: Plectromerini
Type locality: Dominican Republic: Pedernales, PN Sierra de Bahoruco, Las Abejas, 1150 m., at tree fall

turnbowi **Santos-Silva & Hovore, 2007a**: 12 (Fig. 175m, n), Holotype
Original combination: *Distenia (Basisvallis) turnbowi* Santos-Silva & Hovore
Current name: same
Current placement: Disteniidae: Disteniinae: Disteniini
Type locality: Brazil: Manaus, BR 174, km 105

turnbowi **Wappes & Lingafelter, 2005**: 10 (Fig. 175o, p), Holotype
Original combination: *Canidia turnbowi* Wappes & Lingafelter
Current name: same
Current placement: Cerambycidae: Lamiinae: Acanthocinini
Type locality: Mexico: Guerrero, Hwy 134, 73 km northeast Jct. 200
Remarks: On long term term loan from UNAM (Mexico)

turrialbae **Breuning, 1943c**: 56 (Fig. 175q, r), Holotype
Original combination: *Estoloides (Estoloides) turrialbae* Breuning
Current name: *Drycothaea turrialbae* (Breuning, 1943c)
Current placement: Cerambycidae: Lamiinae: Calliini
Type locality: Costa Rica: Cartago, Turrialba
Remarks: Following Article 32.5.2 (ICZN, 1999), the hyphen was removed from the specific epithet. The holotype has a label with an incorrect genus name, *Parestola*.

turuna **Galileo & Martins, 2006c**: 179 (Fig. 175s, t), Holotype
Original combination: *Neoeburia turuna* Galileo & Martins
Current name: same
Current placement: Cerambycidae: Cerambycinae: Eburiini
Type locality: Peru: Lima, Verrugas

ucayalensis **Tippmann, 1960**: 168 (Fig. 175u, v), Holotype
Original combination: *Acanthoderes ucayalensis* Tippmann
Current name: *Irundisaua ucayalensis* (Tippmann, 1960)
Current placement: Cerambycidae: Lamiinae: Acanthoderini
Type locality: Peru: Pucallpa

***ulmicola* Chittenden in Webster, 1904**: 4 (Fig. 175w, x), Lectotype
Original combination: *Oberea ulmicola* Chittenden in Webster
Current name: same
Current placement: Cerambycidae: Lamiinae: Phytoeciini
Type locality: USA: Illinois, Decatur

***umbra* Casey, 1914**: 370 (Fig. 176a, b), Holotype
Original combination: *Oberea umbra* Casey
Current name: *Oberea perspicillata* Haldeman, 1847
Current placement: Cerambycidae: Lamiinae: Phytoeciini
Type locality: USA: North Carolina, Black Mts.

***undulatoides* Breuning, 1940d**: 415 (Fig. 176c, d), Holotype
Original combination: *Mulciber undulatoides* Breuning
Current name: same
Current placement: Cerambycidae: Lamiinae: Homonoeini
Type locality: Papua New Guinea

***unicolor* Breuning, 1940d**: 409 (Fig. 176e, f), Holotype
Original combination: *Mecynome (Niphoparmena) unicolor* Breuning
Current name: *Niphoparmena unicolor* (Breuning, 1940d)
Current placement: Cerambycidae: Lamiinae: Parmenini
Type locality: Tanzania: Usambara

***unicolor* Breuning, 1957a**: 10 (Fig. 176g, h), Holotype
Original combination: *Brachysybra unicolor* Breuning
Current name: same
Current placement: Cerambycidae: Lamiinae: Apomecynini
Type locality: Indonesia: Java, Bantam

***unicolor* Fisher, 1932**: 87 (Fig. 176i, j), Holotype
Original combination: *Probatius unicolor* Fisher
Current name: *Alcidion unicolor* (Fisher, 1932)
Current placement: Cerambycidae: Lamiinae: Acanthocinini
Type locality: Haiti: Furcy

***unicolor* Fisher, 1935c**: 605 (Fig. 176k, l), Holotype
Original combination: *Orsidis unicolor* Fisher
Current name: *Acalolepta unicolor* (Fisher, 1935c)
Current placement: Cerambycidae: Lamiinae: Monochamini
Type locality: Malaysia: Borneo, Sabah, Mt. Kinabalu, Kenokok, 3300ft.

***unicolor* Fisher, 1947c**: 38 (Fig. 176m, n), Holotype
Original combination: *Lagochirus unicolor* Fisher
Current name: *Lagocheirus unicolor* Fisher, 1947c
Current placement: Cerambycidae: Lamiinae: Acanthocinini
Type locality: Barbados: Turners Hall Woods

***unicolor* Tyson, 1973**: 135 (Fig. 176o, p), Holotype
Original combination: *Spalacopsis unicolor* Tyson
Current name: same
Current placement: Cerambycidae: Lamiinae: Agapanthiini
Type locality: USA: Puerto Rico, Diaz Finca, Aibonito, on Malpighia sp.

***unicolor marshallum* Gressitt, 1956**: 95 (Fig. 176q, r), Holotype
Original combination: *Ceresium (Ceresium) unicolor marshallum* Gressitt
Current name: same
Current placement: Cerambycidae: Cerambycinae: Callidiopini
Type locality: Marshall Islands: Majuro

***unicostata* Casey, 1912**: 332 (Fig. 176s, t), Lectotype
Original combination: *Entomosterna unicostata* Casey
Current name: *Stenobatyle eburata* (Chevrolat, 1862)
Current placement: Cerambycidae: Cerambycinae: Trachyderini
Type locality: Mexico: Guerrero
Remarks: This specimen is herein designated as the lectotype in order to stabilize the taxonomy and facilitate further identifications of this species.

***uniformis* Breuning, 1942a**: 161 (Fig. 176u, v), Holotype
Original combination: *Sophronica uniformis* Breuning
Current name: same
Current placement: Cerambycidae: Lamiinae: Desmiphorini
Type locality: Cameroon: Sanaga, Dibongopflanzung

***uniformis* Casey, 1924**: 270 (Fig. 176w, x), Holotype
Original combination: *Xylotrechus uniformis* Casey
Current name: *Xylotrechus nunenmacheri* Van Dyke, 1920
Current placement: Cerambycidae: Cerambycinae: Clytini
Type locality: USA: Oregon, Josephine Co.

***uniformis* Dillon & Dillon, 1951**: 34 (Fig. 177a, b), Holotype
Original combination: *Gnoma uniformis* Dillon & Dillon
Current name: same
Current placement: Cerambycidae: Lamiinae: Gnomini
Type locality: Papua New Guinea: Manus Province, Los Negros Island (Admiralty Islands)

***uniformis* Fisher, 1926**: 4 (Fig. 177c, d), Holotype
Original combination: *Ataxia uniformis* Fisher
Current name: same
Current placement: Cerambycidae: Lamiinae: Pteropliini
Type locality: Trinidad and Tobago

unispinosa **Casey, 1912**: 295 (Fig. 177e, f), Holotype
Original combination: *Aneflomorpha unispinosa* Casey
Current name: same
Current placement: Cerambycidae: Cerambycinae: Elaphidiini
Type locality: Mexico: Chihuahua, Meadow Valley (6 mi south of Colonia Garcia), Sierra Madre, 7300 ft.

ustulata **var. *nigricornis*** **Fisher, 1935c**: 627 (Fig. 177g, h), Holotype
Original combination: *Nupserha ustulata* var. *nigricornis* Fisher
Current name: *Nupserha nigricornis* Fisher, 1935c
Current placement: Cerambycidae: Lamiinae: Phytoeciini
Type locality: Malaysia: Borneo, Sabah, Mt. Kinabalu, Kiau, 3000 ft.
Remarks: The name following the binomen was not expressly given infrasubspecific rank and the work did not indicate unambiguously that the name was infrasubspecific, therefore, it is deemed a subspecies and available.

uteana **Casey, 1913**: 285 (Fig. 177i, j), Holotype
Original combination: *Moneilema (Collapteryx) uteana* Casey
Current name: *Moneilema gigas* LeConte, 1873
Current placement: Cerambycidae: Lamiinae: Moneilemini
Type locality: USA: Utah

uteana **Casey, 1924**: 294 (Fig. 177k, l), Lectotype
Original combination: *Saperda uteana* Casey
Current name: *Saperda hornii* Joutel, 1902
Current placement: Cerambycidae: Lamiinae: Saperdini
Type locality: USA: Utah, Vineyard
Remarks: This specimen is herein designated as the lectotype in order to stabilize the taxonomy and facilitate further identifications of this species.

uteanum **Casey, 1924**: 258 (Fig. 177m, n), Lectotype
Original combination: *Malobidion uteanum* Casey
Current name: *Malobidion brunneum* Schaeffer, 1908
Current placement: Cerambycidae: Cerambycinae: Hesperophanini
Type locality: USA: Utah, Eureka
Remarks: This specimen is herein designated as the lectotype in order to stabilize the taxonomy and facilitate further identifications of this species.

uteanus **Casey, 1913**: 379 (Fig. 177o, p), Lectotype
Original combination: *Tetraopes uteanus* Casey
Current name: *Tetraopes annulatus* LeConte, 1847
Current placement: Cerambycidae: Lamiinae: Tetraopini
Type locality: USA: Utah, Marysvale
Remarks: This specimen is herein designated as the lectotype in order to stabilize the taxonomy and facilitate further identifications of this species.

uteanus **Casey, 1924**: 231 (Fig. 177q, r), Holotype
Original combination: *Criocephalus uteanus* Casey
Current name: *Megasemum asperum* (LeConte, 1854)
Current placement: Cerambycidae: Spondylidinae: Asemini
Type locality: USA: Utah, Eureka

uteanus **Casey, 1924**: 273 (Fig. 177s, t), Lectotype
Original combination: *Stenocorus uteanus* Casey
Current name: *Stenocorus nubifer* (LeConte, 1859)
Current placement: Cerambycidae: Lepturinae: Rhagiini
Type locality: USA: Utah, Provo
Remarks: This specimen is herein designated as the lectotype in order to stabilize the taxonomy and facilitate further identifications of this species.

v-fuscum **Heller, 1924b**: 196 (Fig. 177u, v), Syntype
Original combination: *Stenhomalus v-fuscum* Heller
Current name: same
Current placement: Cerambycidae: Cerambycinae: Stenhomalini
Type locality: Philippines: Mindanao, Dapitan
Remarks: Other syntype(s) are present in the Staatliches Museum für Tierkunde, Dresden, Germany.

validiceps **Casey, 1912**: 235 (Fig. 177w, x), Holotype
Original combination: *Prionus (Prionus) validiceps* Casey
Current name: *Prionus pocularis* Dalman, 1817
Current placement: Cerambycidae: Prioninae: Prionini
Type locality: No locality label or published data.

validicornis **Casey, 1912**: 319 (Fig. 178a, b), Lectotype
Original combination: *Stenaspis validicornis* Casey
Current name: same
Current placement: Cerambycidae: Cerambycinae: Trachyderini
Type locality: Mexico: Guerrero
Remarks: This specimen is herein designated as the lectotype in order to stabilize the taxonomy and facilitate further identifications of this species.

validum **Martins, 1962b**: 104 (Fig. 178c, d), Holotype
Original combination: *Ibidion validum* Martins
Current name: *Tropidion validum* (Martins, 1962b)
Current placement: Cerambycidae: Cerambycinae: Neoibidionini
Type locality: Paraguay: Concepción, Horqueta

***vancouveri* Casey, 1913**: 216 (Fig. 178e, f), Lectotype
Original combination: *Evodinus vancouveri* Casey
Current name: *Evodinus monticola vancouveri* Casey, 1913
Current placement: Cerambycidae: Lepturinae: Rhagiini
Type locality: Canada: British Columbia, Vancouver Island
Remarks: This specimen is herein designated as the lectotype in order to stabilize the taxonomy and facilitate further identifications of this species.

***vancouveri* Casey, 1913**: 263 (Fig. 178g, h), Holotype
Original combination: *Xestoleptura vancouveri* Casey
Current name: *Xestoleptura crassipes* (LeConte, 1857)
Current placement: Cerambycidae: Lepturinae: Lepturini
Type locality: Canada: British Columbia, Vancouver

***vanduzeei* Fisher, 1935b**: 204 (Fig. 178i, j), Holotype
Original combination: *Leptostylus vanduzeei* Fisher
Current name: *Styloleptus planicollis* (Fisher, 1935b)
Current placement: Cerambycidae: Lamiinae: Acanthocinini
Type locality: Jamaica: Mandeville

***varia* Fisher, 1925a**: 217 (Fig. 178k, l), Holotype
Original combination: *Neoegesina varia* Fisher
Current name: *Egesina varia* (Fisher, 1925a)
Current placement: Cerambycidae: Lamiinae: Pteropliini
Type locality: Malaysia: Borneo, Sandakan

***variabilis* Aurivillius, 1927b**: 579 (Fig. 178m, n), Lectotype
Original combination: *Aegocidnus variabilis* Aurivillius
Current name: *Miaenia (Aegocidnus) variabilis* (Aurivillius, 1927b)
Current placement: Cerambycidae: Lamiinae: Acanthocinini
Type locality: Philippines: Mindanao, Surigao
Remarks: This specimen is herein designated as the lectotype in order to stabilize the taxonomy and facilitate further identifications of this species.

***variabilis* Fisher, 1925b**: 11 (Fig. 178o, p), Holotype
Original combination: *Leiopus variabilis* Fisher
Current name: *Styloleptus variabilis* (Fisher, 1925b)
Current placement: Cerambycidae: Lamiinae: Acanthocinini
Type locality: Cuba: Cayamas

***variegata* Fisher, 1925a**: 232 (Fig. 178q, r), Holotype
Original combination: *Ropica variegata* Fisher
Current name: same
Current placement: Cerambycidae: Lamiinae: Apomecynini
Type locality: Philippines: Basilan

***variegata* Fisher, 1925a**: 268 (Fig. 178s, t), Holotype
Original combination: *Miaenia variegata* Fisher
Current name: *Sciades (Miaenia) variegatus* (Fisher, 1925a)
Current placement: Cerambycidae: Lamiinae: Acanthocinini
Type locality: Philippines: Mindanao, Surigao

***variegata* Fisher, 1936a**: 188 (Fig. 178u, v), Holotype
Original combination: *Dystasia variegata* Fisher
Current name: same
Current placement: Cerambycidae: Lamiinae: Pteropliini
Type locality: Indonesia: Java, Sukabumi 2000 ft.

***variegatus* Aurivillius, 1927b**: 575 (Fig. 178w, x), Syntype
Original combination: *Pithomictus ? variegatus* Aurivillius
Current name: *Pithomictus (Parablattostoma) emandibularis* (Heller, 1924b)
Current placement: Cerambycidae: Lamiinae: Rhodopini
Type locality: Philippines: Sibuyan, Romblon
Remarks: Other syntypes are present in the USNM and the Naturhistoriska Riksmuseet, Stockholm, Sweden.

***variegatus* Fisher, 1927a**: 85 (Fig. 179a, b), Holotype
Original combination: *Emeopedus variegatus* Fisher
Current name: *Emeopedus (Variegatemeopedus) variegatus* Fisher, 1927a
Current placement: Cerambycidae: Lamiinae: Acanthocinini
Type locality: Indonesia: Sumatra, Siboelangit

***variegatus* Fisher, 1935c**: 597 (Fig. 179c, d), Holotype
Original combination: *Anhammus variegatus* Fisher
Current name: *Monochamus fisheri* Breuning, 1944
Current placement: Cerambycidae: Lamiinae: Monochamini
Type locality: Malaysia: Borneo, Mt. Kinabalu, Lumu Lumu, 5500 ft.

***variegatus* var. *biflavomaculatus* Tippmann, 1953d**: 322 (Fig. 179e, f), Holotype
Original combination: *Trachyderes variegatus* var. *biflavomaculatus* Tippmann
Current name: *Andraegoidus variegatus* (Perty, 1832)
Current placement: Cerambycidae: Cerambycinae: Trachyderini
Type locality: Bolivia: Santa Cruz
Remarks: The name following the binomen was not expressly given infrasubspecific rank and the work did not indicate unambiguously that the name was infrasubspecific, therefore, it is deemed a subspecies and available. Note that Tippmann's labels use "ab.", but "var." is used in the publication.

***variegatus* var. *binigromaculatus* Tippmann, 1953d**: 321 (Fig. 179g, h), Holotype
Original combination: *Trachyderes variegatus* var. *binigromaculatus* Tippmann
Current name: *Andraegoidus variegatus* (Perty, 1832)
Current placement: Cerambycidae: Cerambycinae: Trachyderini
Type locality: Brazil: Paraná, Rolandia
Remarks: The name following the binomen was not expressly given infrasubspecific rank and the work did not indicate unambiguously that the name was infrasubspecific, therefore, it is deemed a subspecies and available. Note that Tippmann's labels use both "var." and "ab.", but "var." is used in the publication.

***variegatus* var. *bisbiflavomaculatus* Tippmann, 1953d**: 322 (Fig. 179i, j), Holotype
Original combination: *Trachyderes variegatus* var. *bisbiflavomaculatus* Tippmann
Current name: *Andraegoidus variegatus* (Perty, 1832)
Current placement: Cerambycidae: Cerambycinae: Trachyderini
Type locality: Argentina: Santiago del Estero, Chaco de Santiago del Estero, Rio Salado
Remarks: The name following the binomen was not expressly given infrasubspecific rank and the work did not indicate unambiguously that the name was infrasubspecific, therefore, it is deemed a subspecies and available. Note that Tippmann's labels use "ab.", but "var." is used in the publication.

***variegatus* var. *cordiger* Tippmann, 1953d**: 321 (Fig. 179k, l), Holotype
Original combination: *Trachyderes variegatus* var. *cordiger* Tippmann
Current name: *Andraegoidus variegatus* (Perty, 1832)
Current placement: Cerambycidae: Cerambycinae: Trachyderini
Type locality: Brazil: Santa Catarina, Rolandia
Remarks: The name following the binomen was not expressly given infrasubspecific rank and the work did not indicate unambiguously that the name was infrasubspecific, therefore, it is deemed a subspecies and available. Note that Tippmann's labels use "ab.", but "var." is used in the publication.

***variegatus* var. *dimidiatus* Tippmann, 1953d**: 321 (Fig. 179m, n), Holotype
Original combination: *Trachyderes variegatus* var. *dimidiatus* Tippmann
Current name: *Andraegoidus variegatus* (Perty, 1832)
Current placement: Cerambycidae: Cerambycinae: Trachyderini
Type locality: Brazil: Paraná, Rolandia
Remarks: The name following the binomen was not expressly given infrasubspecific rank and the work did not indicate unambiguously that the name was infrasubspecific, therefore, it is deemed a subspecies and available. Note that Tippmann's labels use "ab.", but "var." is used in the publication.

***variegatus* var. *flavocinctus* Tippmann, 1953d**: 322 (Fig. 179o, p), Holotype
Original combination: *Trachyderes variegatus* var. *flavocinctus* Tippmann
Current name: *Andraegoidus variegatus* (Perty, 1832)
Current placement: Cerambycidae: Cerambycinae: Trachyderini
Type locality: Brazil: Paraná, Rolandia
Remarks: The name following the binomen was not expressly given infrasubspecific rank and the work did not indicate unambiguously that the name was infrasubspecific, therefore, it is deemed a subspecies and available. Note that Tippmann's labels use "ab.", but "var." is used in the publication.

***variegatus* var. *malleri* Tippmann, 1953d**: 320 (Fig. 179q, r), Holotype
Original combination: *Trachyderes variegatus* var. *malleri* Tippmann
Current name: *Andraegoidus variegatus* (Perty, 1832)
Current placement: Cerambycidae: Cerambycinae: Trachyderini
Type locality: Brazil: Paraná, Rolandia
Remarks: The name following the binomen was not expressly given infrasubspecific rank and the work did not indicate unambiguously that the name was infrasubspecific, therefore, it is deemed a subspecies and available. Note that Tippmann's labels use "ab.", but "var." is used in the publication.

***variipes* Casey, 1891**: 38 (Fig. 179s, t), Lectotype
Original combination: *Acmaeops variipes* Casey
Current name: *Cortodera militaris variipes* (Casey, 1891)
Current placement: Cerambycidae: Lepturinae: Rhagiini
Type locality: USA: California
Remarks: This specimen is herein designated as the lectotype in order to stabilize the taxonomy and facilitate further identifications of this species.

***variipes* Casey, 1912**: 311 (Fig. 179u, v), Lectotype
Original combination: *Callimus variipes* Casey
Current name: *Lampropterus cyanipennis* (LeConte, 1873)
Current placement: Cerambycidae: Cerambycinae: Hyboderini
Type locality: USA: California, Mokelumne Hill, Calaveras Co.
Remarks: This specimen is herein designated as the lectotype in order to stabilize the taxonomy and facilitate further identifications of this species.

***variolata* Casey, 1913**: 328 (Fig. 179w, x), Lectotype
Original combination: *Hyperplatys variolata* Casey
Current name: *Hyperplatys maculata* Haldeman, 1847
Current placement: Cerambycidae: Lamiinae: Acanthocinini
Type locality: USA: District of Columbia
Remarks: This specimen is herein designated as the lectotype in order to stabilize the taxonomy and facilitate further identifications of this species.

***variolosum* Fisher, 1947c**: 32 (Fig. 180a, b), Holotype
Original combination: *Elaphidion variolosum* Fisher
Current name: *Trichophoroides variolosum* (Fisher, 1947c)
Current placement: Cerambycidae: Cerambycinae: Elaphidiini
Type locality: Cuba: Santiago de Cuba, Moa Oriente

***varius diffidens* Casey, 1912**: 280 (Fig. 180c, d), Lectotype
Original combination: *Phymatodes (Poecilium) varius diffidens* Casey
Current name: *Phymatodes varius* (Fabricius, 1776)
Current placement: Cerambycidae: Cerambycinae: Callidiini
Type locality: USA: Missouri
Remarks: This specimen is herein designated as the lectotype in order to stabilize the taxonomy and facilitate further identifications of this species.

***vastus* Casey, 1912**: 236 (Fig. 180e, f), Holotype
Original combination: *Prionus (Prionus) vastus* Casey
Current name: *Prionus heroicus* Semenov, 1908
Current placement: Cerambycidae: Prioninae: Prionini
Type locality: USA: Arizona

***vegasensis* Casey, 1913**: 380 (Fig. 180g, h), Lectotype
Original combination: *Tetraopes vegasensis* Casey
Current name: *Tetraopes femoratus* LeConte, 1847
Current placement: Cerambycidae: Lamiinae: Tetraopini
Type locality: USA: New Mexico, Near Hot Springs, Las Vegas, 7000 ft.
Remarks: This specimen is herein designated as the lectotype in order to stabilize the taxonomy and facilitate further identifications of this species.

***velutinus* Casey, 1913**: 380 (Fig. 180i, j), Lectotype
Original combination: *Tetraopes velutinus* Casey
Current name: *Tetraopes femoratus* LeConte, 1847
Current placement: Cerambycidae: Lamiinae: Tetraopini
Type locality: USA: Kansas, Finney Co.
Remarks: This specimen is herein designated as the lectotype in order to stabilize the taxonomy and facilitate further identifications of this species.

***velutinus* Fisher, 1931a**: 100 (Fig. 180k, l), Holotype
Original combination: *Eplophorus velutinus* Fisher
Current name: *Euderces velutinus* (Fisher, 1931a)
Current placement: Cerambycidae: Cerambycinae: Tillomorphini
Type locality: Honduras: Prieta Rep.

***venezuelae* Fisher, 1944a**: 11 (Fig. 180m, n), Holotype
Original combination: *Acanthoderes venezuelae* Fisher
Current name: *Myoxomorpha funesta* (Erichson, 1848)
Current placement: Cerambycidae: Lamiinae: Acanthoderini
Type locality: Venezuela: Monagas, Caripito

***venezuelanum* Martins, 1960a**: 25 (Fig. 180o, p), Holotype
Original combination: *Tetraopidion venezuelanum* Martins
Current name: same
Current placement: Cerambycidae: Cerambycinae: Neoibidionini
Type locality: Venezuela: Trujillo, Valera, 1800 ft.

***venezuelensis* Breuning, 1943c**: 39 (Fig. 180q, r), Holotype
Original combination: *Asyngenes (Neosyngenes) venezuelensis* Breuning
Current name: same
Current placement: Cerambycidae: Lamiinae: Apomecynini
Type locality: Venezuela

***venezuelensis* Breuning, 1943c**: 39 (Fig. 180s, t), Holotype
Original combination: *Bisaltes venezuelensis* Breuning
Current name: same
Current placement: Cerambycidae: Lamiinae: Apomecynini
Type locality: Venezuela: Maracay, Aragua

***venezuelensis* Lane, 1959b**: 18 (Fig. 180u, v), Holotype
Original combination: *Acanthoibidion venezuelensis* Lane
Current name: *Acanthoibidion chevrolatii* (White, 1855)
Current placement: Cerambycidae: Cerambycinae: Eligmodermini
Type locality: Venezuela

vermiculatus **Hovore & Santos-Silva, 2004**: 49 (Fig. 180w, x), Holotype
Original combination: *Mallodon vermiculatus* Hovore & Santos-Silva
Current name: same
Current placement: Cerambycidae: Prioninae: Mallodonini
Type locality: Panama: Panama Province, Rodovia El Llano-Cartí (km 8–13)

versicolor **Casey, 1913**: 229 (Fig. 181a, b), Lectotype
Original combination: *Brachysomida versicolor* Casey
Current name: *Brachysomida californica* (LeConte, 1851)
Current placement: Cerambycidae: Lepturinae: Lepturini
Type locality: USA: California, Mokelumne Hill, Calaveras Co.
Remarks: This specimen is herein designated as the lectotype in order to stabilize the taxonomy and facilitate further identifications of this species.

versicolor **Gressitt, 1956**: 158 (Fig. 181c, d), Holotype
Original combination: *Sciadella (Micronesiella) versicolor* Gressitt
Current name: *Sciades (Micronesiella) versicolor* (Gressitt, 1956)
Current placement: Cerambycidae: Lamiinae: Acanthocinini
Type locality: Micronesia: Ponape Island, Mt. Kupuriso, summit, ca. 2000 ft.

versutus **Casey, 1913**: 308 (Fig. 181e, f), Holotype
Original combination: *Astylidius versutus* Casey
Current name: *Astylidius parvus* (LeConte, 1873)
Current placement: Cerambycidae: Lamiinae: Acanthocinini
Type locality: USA: District of Columbia

vestigialis **Gressitt, 1956**: 153 (Fig. 181g, h), Holotype
Original combination: *Palausybra vestigialis* Gressitt
Current name: same
Current placement: Cerambycidae: Lamiinae: Apomecynini
Type locality: Palau: Babelthuap Island, Ngiwal

vestitus **Casey, 1913**: 378 (Fig. 181i, j), Holotype
Original combination: *Tetraopes vestitus* Casey
Current name: *Tetraopes annulatus* LeConte, 1847
Current placement: Cerambycidae: Lamiinae: Tetraopini
Type locality: USA: Colorado

vestitus **Casey, 1924**: 255 (Fig. 181k, l), Holotype
Original combination: *Osmidus vestitus* Casey
Current name: *Osmidus guttatus* LeConte, 1873
Current placement: Cerambycidae: Cerambycinae: Hesperophanini
Type locality: USA: Arizona, Phoenix

vibex fragilis **Casey, 1913**: 202 (Fig. 181m, n), Lectotype
Original combination: *Hapalosalia vibex fragilis* Casey
Current name: *Pidonia ruficollis* (Say, 1824)
Current placement: Cerambycidae: Lepturinae: Rhagiini
Type locality: USA: Indiana
Remarks: This specimen is herein designated as the lectotype in order to stabilize the taxonomy and facilitate further identifications of this species.

vicina **Breuning, 1940d**: 413 (Fig. 181o, p), Holotype
Original combination: *Paraleprodera vicina* Breuning
Current name: same
Current placement: Cerambycidae: Lamiinae: Monochamini
Type locality: Indonesia: Java

vigilans **Casey, 1913**: 329 (Fig. 181q, r), Holotype
Original combination: *Hyperplatys vigilans* Casey
Current name: *Hyperplatys aspersa* (Say, 1824)
Current placement: Cerambycidae: Lamiinae: Acanthocinini
Type locality: USA: District of Columbia

villalobosi **Heller, 1926**: 43 (Fig. 181s, t), Holotype
Original combination: *Doliops villalobosi* Heller
Current name: same
Current placement: Cerambycidae: Lamiinae: Apomecynini
Type locality: Philippines: Samar

villosa **Martins & Galileo, 2011**: 358 (Fig. 181u, v), Holotype
Original combination: *Cotycuara villosa* Martins & Galileo
Current name: same
Current placement: Cerambycidae: Lamiinae: Hemilophini
Type locality: Costa Rica: Puntarenas Province, Monteverde Reserve

villosus scuticularis **Casey, 1912**: 302 (Fig. 181w, x), Holotype
Original combination: *Hypermallus villosus scuticularis* Casey
Current name: *Anelaphus villosus* (Fabricius, 1793)
Current placement: Cerambycidae: Cerambycinae: Elaphidiini
Type locality: USA: Texas

vinctus **Casey, 1913**: 340 (Fig. 182a, b), Holotype
Original combination: *Graphisurus (Canonura) vinctus* Casey
Current name: *Acanthocinus princeps* (Walker in Lord, 1866)
Current placement: Cerambycidae: Lamiinae: Acanthocinini
Type locality: USA: California, Siskiyou Co.

vinosus **Casey, 1924**: 266 (Fig. 182c, d), Lectotype
Original combination: *Crossidius vinosus* Casey
Current name: *Crossidius punctatus* LeConte, 1873
Current placement: Cerambycidae: Cerambycinae: Trachyderini
Type locality: USA: Utah, Vineyard
Remarks: This specimen is herein designated as the lectotype in order to stabilize the taxonomy and facilitate further identifications of this species.

violacea **Santos-Silva & Hovore, 2008a**: 4 (Fig. 182e, f), Holotype
Original combination: *Elytrimitatrix (Grossifemora) violacea* Santos-Silva & Hovore
Current name: same
Current placement: Disteniidae: Disteniinae: Disteniini
Type locality: Panama: Canal Zone, Barro Colorado Island, 9°9′N 79°51′W

virens **Casey, 1912**: 316 (Fig. 182g, h), Lectotype
Original combination: *Dendrobias virens* Casey
Current name: *Trachyderes (Dendrobias) mandibularis virens* Casey, 1912
Current placement: Cerambycidae: Cerambycinae: Trachyderini
Type locality: USA: Texas
Remarks: This specimen is herein designated as the lectotype in order to stabilize the taxonomy and facilitate further identifications of this species.

virescens **Chemsak & Linsley, 1976a**: 173 (Fig. 182i, j), Holotype
Original combination: *Pseudotypocerus virescens* Chemsak & Linsley
Current name: same
Current placement: Cerambycidae: Lepturinae: Lepturini
Type locality: Panama: Canal Zone, Barro Colorado Island

viridicollis **Heller, 1924a**: 27 (Fig. 182k, l), Lectotype
Original combination: *Molochrus viridicollis* Heller
Current name: same
Current placement: Cerambycidae: Cerambycinae: Molorchini
Type locality: Malaysia: Borneo, Sandakan
Remarks: This specimen is herein designated as the lectotype in order to stabilize the taxonomy and facilitate further identifications of this species. Other paralectotype(s) are present in the Staatliches Museum für Tierkunde, Dresden, Germany.

viridimicans **Tippmann, 1953c**: 227 (Fig. 182m, n), Holotype
Original combination: *Weyrauchia viridimicans* Tippmann
Current name: same
Current placement: Cerambycidae: Cerambycinae: Trachyderini
Type locality: Peru: Cajamarca, Jaen, Bellavista Marañon, 700 m.

viridipennis **Breuning, 1940c**: 131 (Fig. 182o, p), Holotype
Original combination: *Tmesisternus viridipennis* Breuning
Current name: *Tmesisternus viridipennis* Breuning, 1940c
Current placement: Cerambycidae: Lamiinae: Tmesisternini
Type locality: Indonesia: Papua, Irian Jaya, Ogelbeng (Hagen)
Remarks: Tavakilian and Chevillotte (2013) suggest the holotype is present in the Zoologische Staatssammlung des Bayerischen Staates, Munich, Germany, however, this specimen and its labels match the original description.

viridipennis **Gressitt, 1935d**: 380 (Fig. 182q, r), Holotype
Original combination: *Eustrangalis viridipennis* Gressitt
Current name: same
Current placement: Cerambycidae: Lepturinae: Lepturini
Type locality: China: Taiwan, Taiheizan

viridis **Fisher, 1944a**: 8 (Fig. 182s, t), Holotype
Original combination: *Chrysoprasis viridis* Fisher
Current name: same
Current placement: Cerambycidae: Cerambycinae: Heteropsini
Type locality: Venezuela: Monagas, Caripito

viridis viridipennis **Casey, 1918**: 416 (Fig. 182u, v), Lectotype
Original combination: *Anthophilax viridis viridipennis* Casey
Current name: *Anthophylax viridis* LeConte, 1850
Current placement: Cerambycidae: Lepturinae: Rhagiini
Type locality: USA: New York, Slide Mt., Ulster C.
Remarks: This specimen is herein designated as the lectotype in order to stabilize the taxonomy and facilitate further identifications of this species.

viticola **Schaeffer, 1908**: 348 (Fig. 182w, x), Lectotype
Original combination: *Pygmaeopsis viticola* Schaeffer
Current name: same
Current placement: Cerambycidae: Lamiinae: Pogonocherini
Type locality: USA: Texas, Brownsville, St. Tomas
Remarks: This specimen is herein designated as the lectotype in order to stabilize the taxonomy and facilitate further identifications of this species.

***viticulus* Martins, 1970b**: 1332 (Fig. 183a, b), Holotype
Original combination: *Heterachthes viticulus* Martins
Current name: same
Current placement: Cerambycidae: Cerambycinae: Neoibidionini
Type locality: Mexico: Coahuila de Zaragoza, Saltillo

***vitreus* Martins, 1962a**: 293 (Fig. 183c, d), Holotype
Original combination: *Heterachthes vitreus* Martins
Current name: *Heterachthes laesicollis* (Germar, 1824)
Current placement: Cerambycidae: Cerambycinae: Neoibidionini
Type locality: Brazil: Amazonas, Manaus

***vittata* Fisher, 1928b**: 3 (Fig. 183e, f), Holotype
Original combination: *Moneilema (Moneilema) vittata* Fisher
Current name: *Moneilema appressum* LeConte, 1852
Current placement: Cerambycidae: Lamiinae: Moneilemini
Type locality: Mexico: Aguascalientes

***vittata* Fisher, 1936a**: 185 (Fig. 183g, h), Holotype
Original combination: *Atossa vittata* Fisher
Current name: *Grammoechus bipartitus* (Ritsema, 1890)
Current placement: Cerambycidae: Lamiinae: Pteropliini
Type locality: Indonesia: Java, G. Tjikoendoel, Sockaloemi, 600–700 m.

***vittaticollis* Aurivillius, 1927b**: 568 (Fig. 183i, j), Syntype
Original combination: *Sybra vittaticollis* Aurivillius
Current name: same
Current placement: Cerambycidae: Lamiinae: Apomecynini
Type locality: Philippines: Sibuyan, Romblon
Remarks: Tavakilian and Chevillotte (2013) state that syntypes are present in the Naturhistoriska Riksmuseet, Stockholm, Sweden as well as the USNM.

***vittatus* Fisher, 1932**: 69 (Fig. 183k, l), Holotype
Original combination: *Trichrous vittatus* Fisher
Current name: same
Current placement: Cerambycidae: Cerambycinae: Heteropsini
Type locality: Cuba: Sierra Rangel

***vitticollis* Casey, 1913**: 363 (Fig. 183m, n), Holotype
Original combination: *Mecas vitticollis* Casey
Current name: *Mecas (Dylobolus) rotundicollis* (Thomson, 1868)
Current placement: Cerambycidae: Lamiinae: Phytoeciini
Type locality: Mexico: Durango, Ciudad de Durango

***vittipennis* Fisher, 1930a**: 17 (Fig. 183o, p), Holotype
Original combination: *Tomopterus vittipennis* Fisher
Current name: *Tomopterus larroides* White, 1855
Current placement: Cerambycidae: Cerambycinae: Rhinotragini
Type locality: Bolivia: Buenavista, Dept. Santa Cruz, Provincia Sara, 1700 ft.

***vittipennis* Fisher, 1934a**: 2 (Fig. 183q, r), Holotype
Original combination: *Chloridolum vittipennis* Fisher
Current name: same
Current placement: Cerambycidae: Cerambycinae: Callichromatini
Type locality: Indonesia: Sumatra, Dolok Baros

***vittipennis* Fisher, 1935c**: 599 (Fig. 183s, t), Holotype
Original combination: *Epepeotes vittipennis* Fisher
Current name: *Cereopsius vittipennis* (Fisher, 1935c)
Current placement: Cerambycidae: Lamiinae: Monochamini
Type locality: Malaysia: Borneo, Sabah, Mt. Kinabalu, Kamborangah, 7000 ft.

***vittipennis* Tippmann, 1960**: 106 (Fig. 183u, v), Holotype
Original combination: *Atenizus vittipennis* Tippmann
Current name: *Atenizus taunayi* Melzer, 1920
Current placement: Cerambycidae: Cerambycinae: Oemini
Type locality: Brazil: Paraná, Rondon, 24°38′B, 54°07′L, 500 m.

***vivarium* Casey, 1924**: 282 (Fig. 183w, x), Holotype
Original combination: *Judolia vivarium* Casey
Current name: *Judolia gaurotoides* (Casey, 1893)
Current placement: Cerambycidae: Lepturinae: Lepturini
Type locality: USA: Wyoming, Yellowstone National Park, Norris Geyser Basin

***vogti* Lane, 1970**: 406 (Fig. 184a, b), Holotype
Original combination: *Hilaroleopsis vogti* Lane
Current name: same
Current placement: Cerambycidae: Lamiinae: Hemilophini
Type locality: Guatemala: Retalhuleu, San Sebastian

***vogti* Martins, 1970a**: 1126 (Fig. 184c, d), Holotype
Original combination: *Neocompsa vogti* Martins
Current name: same
Current placement: Cerambycidae: Cerambycinae: Neoibidionini
Type locality: Bolivia: Pando Dept. [possibly], Loma Alta

vonroseni **Tippmann, 1949**: 293 (Fig. 184e, f), Holotype
Original combination: *Massicus vonroseni* Tippmann
Current name: *Massicus unicolor* Gahan, 1906
Current placement: Cerambycidae: Cerambycinae: Cerambycini
Type locality: Myanmar: Palaing, 1450 m.

wallichi **var.** ***insularis*** **Fisher, 1935c**: 609 (Fig. 184g, h), Holotype
Original combination: *Diastocera wallichi* var. *insularis* Fisher
Current name: *Diastocera wallichi tricincta* (Duncan, 1835)
Current placement: Cerambycidae: Lamiinae: Ceroplesini
Type locality: Malaysia: Borneo, Sabah, Mt. Kinabalu, 3000 ft., Kiau
Remarks: The name following the binomen was not expressly given infrasubspecific rank and the work did not indicate unambiguously that the name was infrasubspecific, therefore, it is deemed a subspecies and available.

walshae **Fisher, 1937c**: 199 (Fig. 184i, j), Holotype
Original combination: *Oberea walshae* Fisher
Current name: *Obereopsis walshae* (Fisher, 1937c)
Current placement: Cerambycidae: Lamiinae: Phytoeciini
Type locality: Indonesia: Java, Djampang, Tongah 2000 ft.

waorani **Martins & Galileo, 2008c**: 279 (Fig. 184k, l), Holotype
Original combination: *Irundiaba waorani* Martins & Galileo
Current name: same
Current placement: Cerambycidae: Lamiinae: Apomecynini
Type locality: Ecuador: Napo, Res. Ethnica Waorani, 1km S. Onkone Gare Camp, 220 m., 00°39′10S 076°26′W

wappesi **Julio & Monné, 2005**: 395 (Fig. 184m, n), Holotype
Original combination: *Sympleurotis wappesi* Julio & Monné
Current name: same
Current placement: Cerambycidae: Lamiinae: Colobotheini
Type locality: Guatemala: El Progreso, 28–29 km north San Augustin, 7–8500 ft.

wappesi **Lingafelter, 2008**: 371 (Fig. 184o, p), Holotype
Original combination: *Elaphidion wappesi* Lingafelter
Current name: same
Current placement: Cerambycidae: Cerambycinae: Elaphidiini
Type locality: Dominican Republic: Puerto Plata, 14 km west of Puerto Plata
Remarks: On long term loan from FSCA, pending revision of *Elaphidion* by Lingafelter, in prep.

wappesi **Monné, 2007**: 209 (Fig. 184q, r), Holotype
Original combination: *Eucharassus wappesi* Monné
Current name: same
Current placement: Cerambycidae: Cerambycinae: Necydalopsini
Type locality: Panama: Chiriquí Province, Hornito Finca La Suiza, 1220 m.

wappesi **Santos-Silva, 2007**: 32 (Fig. 184s, t), Holotype
Original combination: *Derobrachus wappesi* Santos-Silva
Current name: same
Current placement: Cerambycidae: Prioninae: Prioninae
Type locality: Panama: Chiriquí Province, Fortuna Dam

wappesi **Santos-Silva & Hovore, 2007b**: 581 (Fig. 184u, v), Holotype
Original combination: *Novantinoe wappesi* Santos-Silva & Hovore
Current name: same
Current placement: Disteniidae: Disteniinae: Disteniini
Type locality: Panama: Chiriquí Province, Continental Divide Trail

wappesi **Santos-Silva & Martins, 2004**: 150 (Fig. 184w, x), Holotype
Original combination: *Cometes wappesi* Santos-Silva & Martins
Current name: *Arietocometes wappesi* (Santos-Silva & Martins, 2004)
Current placement: Disteniidae: Disteniinae: Disteniini
Type locality: Panama: Chiriquí Province, 6 km north of Fortuna, Continental Divide Road

wappesi **Santos-Silva & Tavakilian, 2009**: 34 (Fig. 185a, b), Holotype
Original combination: *America wappesi* Santos-Silva & Tavakilian
Current name: same
Current placement: Disteniidae: Disteniinae: Disteniini
Type locality: Brazil: Rondonia, 62 km., southwest Ariquemes, fazenda Rancho Grande

weyrauchi **Lane, 1966c**: 80 (Fig. 185c, d), Holotype
Original combination: *Diploschema weyrauchi* Lane
Current name: same
Current placement: Cerambycidae: Cerambycinae: Torneutini
Type locality: Peru: Huanuco, Tingo Maria, 670 m.

weyrauchi **Tippmann, 1953d**: 316 (Fig. 185e, f), Holotype
Original combination: *Aegoidus weyrauchi* Tippmann
Current name: same
Current placement: Cerambycidae: Cerambycinae: Trachyderini
Type locality: Peru: Lupin, 350m, Huacho

wickhami **Casey, 1912**: 337 (Fig. 185g, h), Holotype
Original combination: *Crossidius wickhami* Casey
Current name: *Crossidius hirtipes wickhami* Casey, 1912
Current placement: Cerambycidae: Cerambycinae: Trachyderini
Type locality: USA: Utah, Marysvale

wickhami **Fisher, 1935b**: 192 (Fig. 185i, j), Holotype
Original combination: *Elaphidion wickhami* Fisher
Current name: *Elaphidion manni* Fisher, 1932
Current placement: Cerambycidae: Cerambycinae: Elaphidiini
Type locality: Bahamas: Eleuthera

wickhami **Fisher, 1935b**: 199 (Fig. 185k, l), Holotype
Original combination: *Eupogonius wickhami* Fisher
Current name: same
Current placement: Cerambycidae: Lamiinae: Desmiphorini
Type locality: Bahamas: Egg Island

wickhami **Schaeffer, 1908**: 350 (Fig. 185m, n), Holotype
Original combination: *Parmenonta wickhami* Schaeffer
Current name: same
Current placement: Cerambycidae: Lamiinae: Apomecynini
Type locality: USA: Texas, Brownsville

windsori **Martins & Galileo, 2004**: 538 (Fig. 185o, p), Holotype
Original combination: *Adesmus windsori* Martins & Galileo
Current name: same
Current placement: Cerambycidae: Lamiinae: Hemilophini
Type locality: Panama: Canal Zone, Barro Colorado Island, 9°9′N, 79°51′W

woodleyi **Lingafelter, 2008**: 363 (Fig. 185q, r), Holotype
Original combination: *Psyrassa woodleyi* Lingafelter
Current name: same
Current placement: Cerambycidae: Cerambycinae: Elaphidiini
Type locality: Dominican Republic: Independencia Province, Road 47, between Los Pinos and Angel Felix, 760 meters, 18°36.986′N, 71°46.556′W

woodleyi **Lingafelter, Micheli & Guerrero, 2004**: 44 (Fig. 185s, t), Holotype
Original combination: *Pseudothonalmus woodleyi* Lingafelter, Micheli & Guerrero
Current name: same
Current placement: Cerambycidae: Cerambycinae: Heteropsini
Type locality: USA: Puerto Rico, Maricao State Forest, Road 120, km 13.2 (18°09′52N,66°54′14W) 800 meters

xanthocelis **Martins, 1962a**: 305 (Fig. 185u, v), Holotype
Original combination: *Ibidion xanthocelis* Martins
Current name: *Tropidion xanthocele* (Martins, 1962a)
Current placement: Cerambycidae: Cerambycinae: Neoibidionini
Type locality: Brazil: Mato Grosso, Chapada

xanthopterum **Martins, 1962a**: 281 (Fig. 185w, x), Holotype
Original combination: *Prothoracibidion xanthopterum* Martins
Current name: same
Current placement: Cerambycidae: Cerambycinae: Neoibidionini
Type locality: Bolivia: Santa Cruz, 500 m.

xanthostigma **Monné, 2011**: 40 (Fig. 186a, b), Holotype
Original combination: *Stenolis xanthostigma* Monné
Current name: same
Current placement: Cerambycidae: Cerambycinae: Acanthocinini
Type locality: Panama: Panama Province, 15 km E. Portobelo, 300 ft., CV Mine Rd.

xanthoxyli **Shimer, 1868**: vii (Fig. 186c, d), Holotype
Original combination: *Liopus xanthoxyli* Shimer
Current name: *Sternidius alpha* (Say, 1827)
Current placement: Cerambycidae: Lamiinae: Acanthocinini
Type locality: USA: Illinois, 1 mile West of Mount Carroll, on prickly ash

yapana **Gressitt, 1956**: 138 (Fig. 186e, f), Holotype
Original combination: *Ropica yapana* Gressitt
Current name: same
Current placement: Cerambycidae: Lamiinae: Apomecynini
Type locality: Micronesia: Yap Island, Colonia

yucca **Schaeffer, 1905b**: 167 (Fig. 186g, h), Lectotype
Original combination: *Lepturges yucca* Schaeffer
Current name: same
Current placement: Cerambycidae: Lamiinae: Acanthocinini
Type locality: USA: Arizona, Cochise County, Palmerly
Remarks: This specimen is herein designated as the lectotype in order to stabilize the taxonomy and facilitate further identifications of this species.

yukiyu **Micheli & Micheli, 2004**: 25 (Fig. 186i, j), Holotype
Original combination: *Leptostylopsis yukiyu* Micheli & Micheli
Current name: *Leptostylopsis yukiyu* Micheli & Micheli, 2004
Current placement: Cerambycidae: Lamiinae: Acanthocinini
Type locality: USA: Puerto Rico, Toro Negro Forest, Rd. 143 Kms 18–20

yunnana **Breuning, 1957a**: 11 (Fig. 186k, l), Holotype
Original combination: *Eryssamena yunnana* Breuning
Current name: *Rondibilis yunnana* (Breuning, 1957a)
Current placement: Cerambycidae: Lamiinae: Acanthocinini
Type locality: China: Yunnan Province, Vallis flumin, Soling-ho
Remarks: From Hauser Collection

yunnanus **Breuning, 1942a**: 172 (Fig. 186m, n), Holotype
Original combination: *Pseudocalamobius yunnanus* Breuning
Current name: same
Current placement: Cerambycidae: Lamiinae: Agapanthiini
Type locality: China: Yunnan Province, Pe-yen-tsin

zapoteca **Monné & Monné, 2008**: 504 (Fig. 186o, p), Holotype
Original combination: *Nodubothea zapoteca* Monné & Monné
Current name: same
Current placement: Cerambycidae: Cerambycinae: Colobotheini
Type locality: Mexico: Tehuantepec

zellibori **Tippmann, 1953d**: 331 (Fig. 186q, r), Holotype
Original combination: *Lissonotus zellibori* Tippmann
Current name: same
Current placement: Cerambycidae: Cerambycinae: Lissonotini
Type locality: Brazil: Pará, Obidos

zellibori **Tippmann, 1953d**: 334 (Fig. 186s, t), Holotype
Original combination: *Tuberorachidion zellibori* Tippmann
Current name: *Tuberorachidion pumilio* (Gounelle, 1911)
Current placement: Cerambycidae: Cerambycinae: Trachyderini
Type locality: Brazil: Paraná, Rolandia

zeteki **Dillon & Dillon, 1945**: 116 (Fig. 186u, v), Holotype
Original combination: *Venustus zeteki* Dillon & Dillon
Current name: same
Current placement: Cerambycidae: Lamiinae: Onciderini
Type locality: Panama: Panama Province, Barro Colorado Island

zeteki **Fisher, 1930a**: 6 (Fig. 186w, x), Holotype
Original combination: *Odontocera zeteki* Fisher
Current name: same
Current placement: Cerambycidae: Cerambycinae: Rhinotragini
Type locality: Panama: Barro Colorado Island

zeteki **Fisher, 1947b**: 179 (Fig. 187a, b), Holotype
Original combination: *Dorcasta zeteki* Fisher
Current name: *Bebelis modesta* (Belon, 1903)
Current placement: Cerambycidae: Lamiinae: Apomecynini
Type locality: Panama: Canal Zone, Corozal

zischkai **Tippmann, 1953c**: 222 (Fig. 187c, d), Holotype
Original combination: *Haenkea zischkai* Tippmann
Current name: same
Current placement: Cerambycidae: Cerambycinae: Rhopalophorini
Type locality: Bolivia: Cochabamba, Reg. Chapare 400 m.

zischkai **Tippmann, 1960**: 130 (Fig. 187e, f), Holotype
Original combination: *Acorethra zischkai* Tippmann
Current name: *Pseudacorethra zischkai* (Tippmann, 1960)
Current placement: Cerambycidae: Cerambycinae: Rhinotragini
Type locality: Bolivia: Chapare, Cochabamba

zischkai **Tippmann, 1960**: 165 (Fig. 187g, h), Holotype
Original combination: *Acanthoderes zischkai* Tippmann
Current name: same
Current placement: Cerambycidae: Lamiinae: Acanthoderini
Type locality: Bolivia: Cochabamba, Reg. Chapare 400 m.

zischkai **Tippmann, 1960**: 192 (Fig. 187i, j), Holotype
Original combination: *Lepturges zischkai* Tippmann
Current name: *Lepturges curvilinea* Gilmour, 1959
Current placement: Cerambycidae: Lamiinae: Acanthocinini
Type locality: Bolivia: Cochabamba, Reg. Chapare 400 m.

zonatus **Dillon & Dillon, 1946b**: 214 (Fig. 187k, l), Holotype
Original combination: *Lochmaeocles zonatus* Dillon & Dillon
Current name: same
Current placement: Cerambycidae: Lamiinae: Onciderini
Type locality: Trinidad and Tobago: Tobago

zonatus **Martins & Galileo, 2008a**: 273 (Fig. 187m, n), Holotype
Original combination: *Neoclytus zonatus* Martins & Galileo
Current name: *Unaiuba zonata* (Martins & Galileo, 2008)
Current placement: Cerambycidae: Cerambycinae: Clytini
Type locality: Guatemala: Panzoa

zuniana **Casey, 1912**: 353 (Fig. 187o, p), Holotype
Original combination: *Cyllene zuniana* Casey
Current name: *Megacyllene snowi zuniana* (Casey, 1912)
Current placement: Cerambycidae: Cerambycinae: Clytini
Type locality: USA: New Mexico, Magdalena Mts.

Appendix 1

Taxa Originally Included as Primary Types at the NMNH but Determined as Infrasubspecific and Therefore Unavailable

Infrasubspecific determinations follow the provisions of Article 45.6 (1999) of the International Commission on Zoological Nomenclature.

Acmaeops (*Dinoptera*) *collaris* ab. *marginaticollis* Tippmann 1951: 316
Aegoidus pacificus ab. *corallithorax* Tippmann 1960: 144
Alphus senilis ab. *uniformis* Tippmann 1960: 165
Anamera strandi ab. *flavescens* Breuning 1935: 65
Aromia moschata ab. *steinmanni* Tippmann 1956: 486
Batocera davidis m. *yunnana* Breuning 1948: 14
Batocera hector m. *menelaus* Breuning 1948: 14
Batocera nebulosa ab. *diversimaculata* Breuning 1948: 14
Batocera wyliei m. *granulipennis* Breuning 1948: 15
Batus barbicornis itaitubensis ab. *picticornis* Tippmann 1953c: 228
Batus hirticornis ab. *melancholicus* Tippmann 1953c: 228
Bisaltes monticola ab. *bistriata* Tippmann 1960: 153
Cacia semiluctuosa m. *reducta* Breuning 1939b: 459.
Calocomus desmaresti ab. *bimaculatus* fma. *ruficollis* Tippmann 1960: 96
Calocomus desmaresti ab. *bimaculatus* Tippmann 1960: 96
Calocomus desmaresti ab. *conjunctus* Tippmann 1960: 97
Calocomus desmaresti ab. *conjunctus*, fma. *ruficollis* Tippmann 1960: 97
Calocomus desmaresti ab. *melancholicus* fma. *ruficollis* Tippmann 1960: 96
Calocomus desmaresti ab. *melancholicus* Tippmann 1960: 96
Cereopsius quaestor var. *connecta* Fisher 1934: 2
Cyclopeplus peruvianus ab. *coeruleometallescens* Tippmann 1939: 77
Cyclopeplus peruvianus ab. *submetallicus* Tippmann 1939: 77
Cyclopeplus peruvianus ab. *thoracicus* Tippmann 1939: 77
Cyclopeplus peruvianus ab. *viridimetallescens* Tippmann 1939: 77
Cyriocrates elegans ab. *albonotata* var. *coloremutabilis* Tippmann 1953: 152
Dichostathes kuntzeni ab. *ochreicollis* Breuning 1962: 438
Dichostathes lobatus m. *albomaculata* Breuning 1942: 52

Dichostathes lobatus m. *ochraceomaculaus* Breuning 1942: 52
Didymonycha secunda ab. *infuscata* Tippmann 1951: 302
Dorcadion (*Carinatorcadion*) *aethiops* ab. *brunneipenne* Tippmann 1958: 152
Dorcadion (*Carinatorcadion*) *aethiops* ab. *rufosuturatum* Tippmann 1958: 153
Dorcadion (*Carinatorcadion*) *aethiops* m. n. *lucidum* Tippmann 1958: 151
Dorcadion fulvum ab. *nigripenne* m. *bozsinkae* Tippmann 1958: 162
Dorcadion fulvum ab. *nigripenne* m. *breuningi* Tippmann 1958: 162
Dorcadion fulvum ab. *nigripenne* m. *gitulinkae* Tippmann 1958: 162
Dorcadion fulvum ab. *uniformebrunneum* Tippmann 1958: 160
Dorcadion fulvum rugipenne ab. *atrum* Tippmann 1958: 155
Dorcadion fulvum rugipenne ab. *rubrobrunneum* Tippmann 1958: 155
Dorcadion (*Maculatodorcadion*) *triste* m. *postmedioreductum* Breuning 1946: 97
Eburia octoguttata ab. *basibiguttata* Tippmann 1960: 115
Eburia octoguttata ab. *inguttata* Tippmann 1960: 115
Eburia octoguttata ab. *melanescens* Tippmann 1960: 115
Eburia octoguttata ab. *quadricostata* Tippmann 1960: 115
Eburia octoguttata ab. *basiquadriguttata* Tippmann 1960: 115
Epepeotes luscus m. *interruptus* Breuning 1943: 223
Erosida formosa ab. *conjunctula* Tippmann 1960: 116
Eurymerus eburioides ab. *nigricans* Tippmann 1960: 109
Evodinus breiti ab. *cincticollis* Tippmann 1946: 126
Evodinus breiti ab. *testaceipennis* Tippmann 1946: 126
Gaurotes excellens ab. *korbeli* Sekera 1935: 139
Gaurotes virginea ab. *nigronotaticollis* Tippmann 1951: 310
Gaurotes virginea ab. *occipitalis* Tippmann 1951: 311
Glenea lobata ab. *biguttulata* Aurivillius 1926
Glenea (*Glenea*) *beccarii* m. *discoinvittata* Breuning 1956: 878
Glenea (*Glenea*) *discomaculata* m. *trimaculicollis* Breuning 1958: 243
Isthmiade ichneumoniformis ab. *rufithorax* Tippmann 1960: 127
Leptura scutellata var. *stichai* n. m. Heyrovsky 1931: 120
Lycaneptia amicta ab. *nigrobasalis* Tippmann 1960: 203
Lycidola palliata ab. *nigrohumeralis* Tippmann 1960: 203
Megacyllene (*Sierracyllene*) *horioni* ab. *bifasciata* Tippmann 1960: 134
Megacyllene (*Sierracyllene*) *horioni* ab. *parvidecorata* Tippmann 1960: 134
Megacyllene (*Sierracyllene*) *horioni* ab. *suturaalba* Tippmann 1960: 134
Megacyllene mellyi ab. *modesta* Tippmann 1960: 135
Megaderus stigma ab. *bimaculata* Tippmann 1960: 151
Megaderus stigma ab. *interruptefasciata* Tippmann 1960: 151
Oberea infantula ab. *suensoni* Heyrovsky 1938: 29
Odontocera lineatocollis ab. crucifera Tippmann 1960: 126
Olenecamptus similis m. *congoanus* Dillon & Dillon 1948: 213
Pedestredorcadion pusillum m. *humeratum* Podany 1953: 33
Pedestredorcadion pusillum m. *romanicum* Podany 1953: 33
Penichroa fasciata ab. *lanceolata* Tippmann 1956: 479
Penichroa fasciata ab. *latefasciata* Tippmann 1956: 478
Penichroa fasciata ab. *moussardi* Tippmann 1956: 480
Penichroa fasciata ab. *ornata* Tippmann 1956: 479

Periaptodes testator m. *trilobata* Breuning 1944: 508
Phosphorus virescens ab. *albidipennis* Breuning 1942d: 144
Phosphorus virescens grisescens Breuning 1934: 91
Plagionotus arcuatus ab. *tippmanni* Tippmann 1952: 146
Plagionotus detritus ab. *curvatofasciata* Tippmann 1952: 141
Plagionotus detritus ab. *freyi* Tippmann 1952: 140
Plagionotus detritus ab. *kulzeri* Tippmann 1952: 141
Plagionotus detritus ab. *reithofferi* Tippmann 1952: 142
Plagionotus detritus m. *causicola*, ab. *lepesmei* Tippmann 1952: 142
Prosopocera blairi m. *immaculata* Breuning 1942d: 144
Prosopocera (*Dalterus*) *peeli* m. *confluentina* Breuning 1940: 214
Protaxis var. *fulviventris* m. *nigripes* Tippmann 1956: 475
Psenocerus supernotatus var. *funebris* Casey 1914: 367
Pseudocyriocrates strandi m. *ochraceomaculatus* Breuning 1944: 297
Pteroplatus anchora ab. *fereniger* Tippmann 1960: 140
Pteroplatus anchora ab. *latefasciatus* Tippmann 1960: 140
Pteroplatus anchora ab. *latesuturalis* Tippmann 1960: 140
Pteroplatus anchora ab. *medioelargatosignatus* Tippmann 1960: 140
Pteroplatus anchora ab. *parvodelineatus* Tippmann 1960: 139
Pteroplatus anchora ab. *zischkai* Tippmann 1960: 140
Rhamnusium bicolor var. *de maggii* f. *aeneomicans* Tippmann 1956: 484
Rhinotragus bizonatus ab. *rufithorax* Tippmann 1960: 121
Rhodocharis anacoloides ab. *bicolor* Tippmann 1953: 204
Rondibilis parvula ab. *immaculipennis* Breuning 1957: 11
Rosalia batesi ab. *adjectomaculata* Tippmann 1956: 487
Rosalia batesi ab. *chujoi* Tippmann 1956: 487
Rosalia batesi ab. *nigricollis* Tippmann 1948: 37
Rosalia batesi ab. *thoracebasiimaculata* Tippmann 1956: 487
Semanotus japonicus ab. *fasciatus* Tippmann 1956: 492
Semanotus japonicus ab. *immaculatus* Tippmann 1956: 492
Semanotus japonicus ab. *posticeinterruptus* Tippmann 1956: 492
Semanotus japonicus ab. *trimaculatus* Tippmann 1956: 492
Strangalia arcuata ab. *heyrovskyella* Balthasar 1935: 11
Strangalia arcuata ab. *latenotata* Balthasar 1935: 11
Strangalia quadrifasciata ab. *dichisa* Jesatko 1935: 186
Sybra alternans m. *discomaculata* Breuning 1950c: 268
Tragocephala nobilis ab. *azurea* Breuning 1934: 78
Weyrauchia viridimicans ab. *rubra* Tippmann 1960: 141
Xylosteus Bartoni ab. *Purkynei* Obenberger & Maran 1933: 130

Appendix 2

Names of Taxa Originally Published with the Qualifier "var." or "v." but Determined to Be Available Species-Group Names

The status of all names as either subspecific or infrasubspecific had to be determined by examination of the original description according to Article 45.6.4 (ICZN, 1999) and using Lingafelter & Nearns (2013) since all were published before 1961.

Achryson immaculipenne var. *humeromaculatum* Tippmann 1960: 107
Achryson lineolatum var. *andinum* Tippmann 1960: 108
Achryson lineolatum var. *nigricans* Tippmann 1960: 109
Acrepidopterum minutum var. *apicalis* Fisher 1935: 197
Anoplophora stanleyana var. *gloriosa* Tippmann 1953: 152
Anoplophora stanleyana var. *grisea* Tippmann 1953: 153
Anoplophora stanleyana var. *melancholica* Tippmann 1953: 153
Calloides nobilis var. *mormonus* Schaeffer 1911: 124
Camptomyne philippinus var. *unicolor* Fisher 1925: 272
Chion cinctus var. *sonorensis* Schaeffer 1908: 332
Coloborhombus drescheri var. *flavipes* Fisher 1936: 177
Daphisia bakeri var. *semisignata* Aurivillius 1926: 114
Daphisia bakeri var. *vittulata* Aurivillius 1926: 114
Daphisia bakeri var. *immaculata* Aurivillius 1926: 114
Dectes (*Canidia*) *balteata* var. *inapicalis* Tippmann 1960: 190
Demonax angusticollis var. *sibuyanus* Aurivillius 1928: 319
Diastocera wallichi var. *insularis* Fisher 1935: 609
Dihammaphora aurovittata var. *debilis* Melzer 1935: 184
Ecyrus dasycerus var. *texanus* Schaeffer 1908: 347
Ethemon basale var. *inbasale* Tippmann 1960: 133
Eunidia nigromaculata var. *trimaculata* Aurivillius 1915: 330
Gnomidolon micans v. *brunneicaudatum* Fisher 1937: 147
Hexoplon armatum var. *integrum* Tippmann 1960: 118
Isthmiade laevicollis var. *rubrosignata* Tippmann 1953: 215
Leptostylus biustus var. *cubanus* Fisher 1926: 22
Leptostylus scurra var. *dorsalis* Fisher 1926: 21

Leptostylus scurra var. *maculifer* Fisher 1926: 21
Liopus alpha var. *floridanus* Hamilton 1896: 125
Melanauster medenbachi var. *sumatrensis* Fisher 1934: 3
Metopocoilus quadrispinosus var. *basimaculatus* Tippmann 1953: 208
Nupserha ustulata var. *nigricornis* Fisher 1935: 627
Nyssodrys calligramma var. *consobrina* Melzer 1934: 109
Odontocera sanguinolenta v. *rufifrons* Fisher 1937: 151
Ommata (*Eclipta*) *brevipennis* v. *flavofemorata* Fisher 1937: 150
Ommata (*Eclipta*) *brevipennis* v. *fulvipes* Fisher 1947: 50
Ommata (*Eclipta*) *brevipennis* v. *sanguinicollis* Fisher 1947: 50
Ommata (*Eclipta*) *poecila* v. *maculicollis* Fisher 1947: 51
Ommata (*Eclipta*) *poecila* v. *nigricornis* Fisher 1947: 51
Oreodera glauca var. *jamaicensis* Fisher 1942: 26
Pantonyssus erichsonii var. *flavipes* Fisher 1944: 6
Phaea auricapilla var. *vittulata* Melzer 1935: 240
Phaedinus tricolor var. *superbus* Tippmann 1954: 317
Rhamnusium bicolor var. *demaggii* Tippmann 1956: 484
Trachyderes variegatus var. *bisbiflavomaculatus* Tippmann 1953: 322
Trachyderes variegatus var. *binigromaculatus* Tippmann 1953: 321
Trachyderes variegatus var. *cordiger* Tippmann 1953: 321
Trachyderes variegatus var. *dimidiatus* Tippmann 1953: 321
Trachyderes variegatus var. *malleri* Tippmann 1953: 320
Trachyderes variegatus var. *biflavomaculatus* Tippmann 1953: 322
Trachyderes variegatus var. *flavocinctus* Tippmann 1953: 322
Tragidion annulatum var. *peninsulare* Schaeffer 1908: 339

Appendix 3

Unpublished Taxa Originally Labeled and Included as Primary Types in the NMNH Collection

The following were determined as *nomina nuda* because there is no evidence that they were published. Even if it is later determined that some were published, the aberrations are nevertheless unavailable following the provisions of ICZN Article 45.6 (1999).

Acestrilla laterifusca Breuning
Acridoschema varians m. *oblongomaculatum* Breuning
Aerenea longeantennata Breuning
Amblymora grisescens Breuning
Amphicnaeia crustuloides Breuning
Anaespogonius brevepilis Gressitt
Anelaphus niveisparsus Linsley
Batus barbicornis ab. *tomentosus* Tippmann
Brachychilus literatus albomarmoratus Breuning
Calocomus hamatiferus ab. *ruficollis* Tippmann
Chloridolum loochooanum annammanum Gressitt
Chrysaperda mediocincta Lane
Clytus tropicus ab. *heyrovskyi* Tippmann
Cylindrepomus grammicus m. *latefasciata* Breuning
Desisella bituberosa Breuning
Desmiphora pulchra Breuning
Dihammus semenovi Breuning
Dorcadion mniszechi tippmanni Breuning
Dorcadion pusillum ab. *humeratum* Podany
Dorcadion pusillum ab. *romanicum* Podany
Eupogonius vestitus Breuning
Eupromera alboscutellaris Breuning
Evodinus variabilis ab. *subconstrictus* Plavilstchikov
Falsodihammus nigrovittatus Breuning
Glenea verduta [?] *pseudoaccessoria* Breuning
Glenea indistinctevittata m. *viridivittata* Breuning
Glenea longipes Breuning
Glenea viridicincta rufofemorata Breuning

Glenea nobilor Breuning
Glenea collaroides Breuning
Glenea anaglica [?] m. *quadripunctata* Breuning
Glenea kuraruana Breuning
Glenea indistinctevittata Breuning
Glenea apicomaculata Breuning
Hyagnis congoanus Breuning
Hypsioma picticornis columbianum Breuning
Hypsioma picticornis peruana Breuning
Hypsioma (Hypsioma) persimilis Breuning
Hypsioma (Hypsioma) tubericollis Breuning
Hypsioma (Hypsioma) costipennis Breuning
Hypsioma (Hypsioma) gigas Breuning
Hypsioma (Hypsioma) gissen Breuning
Hypsioma (Hypsioma) gusanalia Breuning
Hypsioma (Hypsiomoida) rubrovarinota Breuning
Hypsioma (Hypsiomoides) triangularis Breuning
Hypsioma (Pseudohypsioma) obliguenachta Breuning
Ischioplites rubigineus Tippmann
Ischnolea albostictica Breuning
Lachaerus eudesmoides Breuning
Lypsimena bicristata Breuning
Metagnoma affinis Breuning
Mimosophronica kuldshensis Breuning
Monochamus plumbeus m. *dissoluta* Breuning
Oberea holoxantha clarior Breuning
Oberea holoxantha ruficornis Breuning
Obrium sibiricum Gressitt
Oncideres compacta Breuning
Oncideres (Ishciocentra) costaricensis Breuning
Oncideres (Ishciocentra) fuscovittata Breuning
Oncideres (Ishciocentra) similis Breuning
Oncideres (Oncideres) albosterna Breuning
Oncideres (Oncideres) proscina Breuning
Oncideres (Paraghaloud) nigropunctata Breuning
Oncideres (Paraglypthaga) plerodomiana Breuning
Oncideres (Pseudoncideres) forticornis Breuning
Oncideres (Pseudotrestonia) fuscoapicata Breuning
Pachypezoides brunneus Breuning
Paracylicasta ceroplesoides Breuning
Paracylicasta incognaria Breuning
Paracylicasta marmorata Breuning
Paranaches simplex m. *inveoplagiatus* Breuning
Paranca laevipennis Breuning
Paraplundia albofasciata var. *albida* Breuning
Parindnia strandiella Breuning
Parmenonta brunnea Breuning

Parobages quadriplagiatus Breuning
Pericycos strandi Breuning
Phantasis gorgo var. *amerewensis* Schwarzer
Phantasis rhodesiana Schwarzer
Phantasis roseni Schwarzer
Phantasis sansibarica var. *montana* Schwarzer
Phantasis zorgoides Schwarzer
Plagionotus detritus ab. *lanceolata* Tippmann
Protilema tuberculifera Breuning
Purpuricenus budensis ab. *trimaculicollis* Tippmann
Sternotomis runsoriensis ab. *ochracea* Breuning
Sybra conicollis m. *maculata* Breuning
Sybra pseudosolida Breuning
Sybrasoma parmeniforme Breuning
Taeniotes subocellatus ab. *albomaculatus* Tippmann
Timoreticus dejeani ab. *pseudocornifrons* Breuning
Trichodocus unicolor Breuning
Trichohyllisia unquicalaris Breuning
Trichopothyne antennalis Breuning

Appendix 4

Specimens in the NMNH Labeled as Primary Types but Due to Inconsistencies with the Locality Labels or Other Information Determined to Not Be Primary Types

Adetus venezuelensis Breuning 1942. The holotype is deposited in the Muséum national d'Histoire naturelle, Paris, France.

Apomecyna arabica Breuning 1938. The holotype is deposited in the Natural History Museum, London.

Cleptometopus simulimus Breuning 1947. The holotype is deposited in the Naturhistoriska Riksmuseet, Stockholm, Sweden and the locality data do not match exactly the published description.

Coptops japonica Breuning 1936. The holotype is deposited in the Muséum national d'Histoire naturelle, Paris, France and the original publication states the holotype is from Osaka, Japan. The NMNH specimen is likely a paratype.

Crossotus parallelus Breuning 1935. The NMNH specimen does not match the published description.

Dihammus elongatus Breuning 1935. The NMNH specimen does not match the published description. The holotype is deposited in Basel, Switzerland.

Dihammus griseofasciatus Breuning 1935. The NMNH specimen does not match the published description. The holotype is deposited in the British Museum, London.

Dihammus neopomeranius Breuning 1938. The NMNH specimen does not match the published description. The holotype is deposited in the British Museum, London.

Dihammus pseudotincturatus Breuning 1935. The NMNH specimen does not match the published description. The holotype is deposited in Basel, Switzerland.

Dihammus salomonum Breuning 1938. The NMNH specimen does not match the published description. The holotype is deposited in the Brussells, Belgium.

Dolichosybra tubericollis Breuning 1942. The NMNH specimen does not match the published description. The holotype is deposited in the Muséum national d'Histoire naturelle, Paris, France.

Eunidia laterialba Breuning 1948b. The NMNH specimen does not match the published description. The holotype is deposited in the British Museum, London.

Frea (Frea) albomarmorata Breuning 1935. The NMNH specimen does not match the published description.

Frea (Frea) unicolor Breuning 1938. The NMNH specimen does not match the published description.

Glenea (*Glenea*) *didymoides* Breuning 1956. The NMNH specimen does not match the published description.

Glenea (*Glenea*) *rufobasalis* Breuning. The NMNH specimen is not a primary type. The original publication says the holotype is in the Royal Museum in Tervuren, Belgium.

Mispila curvifascia Breuning 1938. The NMNH specimen probably is not a type since the label data do not match the original description. The holotype is deposited in the Muséum national d'Histoire naturelle, Paris, France.

Mulciber maculosus Breuning 1939. The NMNH specimen does not match the published description. The holotype is deposited in the British Museum, London.

Pelargoderus papuanus Breuning 1936. Tavakilian and Chevillotte (2013) suggest the holotype is present in the Muséum national d'Histoire naturelle, Paris, France and the locality data do not match the published description, so this specimen probably is not the holotype.

Prosopcera luteomarmorata Breuning 1936. The NMNH specimen probably is not a primary type since the original description states the type locality as Elisabethville, Belgian Congo and the holotype is deposited in Tervuren, Belgium.

Prosopocera multiguttata Breuning 1936. The NMNH specimen probably is not a primary type since the original description states the type locality as Elisabethville, Belgian Congo, and that the holotype is deposited in London.

Prosopocera prasina Breuning 1936. Since the original description states the holotype is deposited in Tervuren, Belgium and the label data does not exactly match, it is presumed that this specimen is not a primary type.

Pterolophia fuscoscutellata Breuning 1938. The NMNH specimen probably is not the primary type since Tavakilian and Chevillotte (2013) state that the holotype is present in the Naturhistoriska Riksmuseet, Stockholm, Sweden and the locality data do not match exactly the published description.

Pterolophia pluricarinata Breuning 1940. The NMNH specimen probably is not a type since the label data does not match the original description. The holotype is deposited in the Muséum national d'Histoire naturelle, Paris, France.

Pterolophia tubericollis Breuning 1938. Since the original description states the holotype is a male from Darjeeling, India and the NMNH specimen is from Sumatra, it is not a primary type.

Quercivir zikani Melzer 1919. The NMNH specimen is a paralectotype. The specimen in MZSP is the lectotype.

Sodus papuanus Breuning 1940. Tavakilian and Chevillotte (2013) suggest the holotype is present in the Paris Museum. The locality data do not match the published description, so this specimen probably is not the holotype.

Sybra formosana Breuning 1964. The NMNH specimen is actually a paratype of a species whose holotype is deposited in Senckenberg Naturhistorische Sammlungen, Dresden, Germany.

Taeniotes affinis Breuning 1935c. Since Breuning identified one specimen as the "Typ" in his paper, a male from Peru, that is a holotype designation. Therefore there are no syntypes. The holotype, in the Hamburg Museum, was likely destroyed in World War II. The NMNH does have two other female specimens, labeled as paratypes by Breuning, both from Ecuador. One of these specimens could be chosen as the neotype if it is certain the holotype is lost.

Appendix 5

Primary Types That Probably Belong to the NMNH but Are Presumed Lost or Wrongly Deposited at Another Institution

Note that this list may be incomplete as it is very difficult to determine the history of all published taxa to determine their proper deposition.

Acestrilla strandiella Breuning 1942
Agelasta (Dissosira) tonkinea omeishana Gressitt 1951b
Amphicnaia strandi Breuning 1942
Anaespogonius omeimontis Gressitt 1938
Astathes bakeri Aurivillius 1927
Brachyolene sereimaculata Breuning 1942
Cleptometopus biapicatus Breuning 1942
Cnemolia marmorata Breuning 1942
Cyanomethia pseudothonalmus Philips & Ivie 1998
[*note*: this specimen found during book production]
Dolichosybra strandiella Breuning 1942
Dorcadion (Carinatodorcadion) fulvum freyi Tippmann 1958
Dorcadion (Carinatodorcadion) fulvum kulzeri Tippmann 1958
Eburia bahamicae Fisher 1932
Eranina hovorei Galileo and Martins 2008
Linda macilenta Gressitt 1947
Methia mormona Linell 1896
Necydalis diversicollis Schaeffer 1932
Neoclytus abbreviatus Schaeffer 1908
Oberea javaensis Fisher 1937
Obrium brunneum Schaeffer 1904
Oemospiloides bengalensis Fisher 1940
Oncideres sparsemaculatus Martins & Galileo 2010
Ozineus nyssodroides Tippmann 1960
Pachydissus xyliae Fisher 1940
Tragosoma parvicollis Casey 1899
Tragosoma sodalis Casey 1899

Parauxa strandiella Breuning 1942
Parazygocera flavolineata Breuning 1943
Parischnia strandiella Breuning 1943
Parobages strandi Breuning 1940
Penessada brunneoscutellaris Tippmann 1960
Phryneta pulchra Tippmann 1958
Saperda concolor var. *unicolor* Felt & Joutel 1904
Saperda imitans Felt & Joutel 1904
Saperda lateralis var. *connecta* Felt & Joutel 1904
Sybra incanoides Breuning 1942a
Sybra strandiella Breuning 1942a
Trichopothyne strandiella Breuning 1942a

Appendix 6

Neotype Designations from NMNH Specimens

Entries marked with an asterisk are neotypes designated in this publication.

Elaphidion fullonium Newman 1841: 110
Hypsioma prodigiosa Thomson 1868: 45*
Necydalis sanguinicollis Olivier 1795: 9
Stenellipsis ochreoapicalis Breuning 1943: 12*
Tragidion agave Swift & Ray 2008: 17*
Tragidion deceptum Swift & Ray 2008: 17*

Appendix 7

New Lectotype Designations from NMNH Syntypes

Achryson galapagoensis Linell 1899: 259
Acmaeops cavicollis Casey 1913: 244
Acmaeops coloradensis Casey 1913: 244
Acmaeops puncticeps Casey 1913: 243
Acmaeops variipes Casey 1891: 38
Aegocidnus variabilis Aurivillius 1927b: 579
Aneflomorpha parowana Casey 1924: 242
Aneflomorpha rectilinea Casey 1924: 243
Aneflomorpha testacea Casey 1924: 243
Anocomis ampla conformis Casey 1924: 236
Anocomis ampla Casey 1912: 273
Anocomis ampla terracensis Casey 1924: 237
Anocomis canadensis Casey 1924: 238
Anoplium laterale Casey 1914: 365
Anoplium linelli Casey 1924: 246
Anoplium nanulum Casey 1924: 247
Anoplium subtropicum Casey 1924: 245
Anoplium tuckeri Casey 1924: 247
Anthophilax subvittata Casey 1891: 37
Anthophilax viridis viridipennis Casey 1918: 416
Apriona bicolor Kriesche 1920: 193
Asemum carolinum Casey 1924: 227
Asemum pugetanum Casey 1912: 261
Batus barbicornis itaitubensis Tippmann 1953: 227
Batyle acomana Casey 1912: 328
Brachyleptura (Stictoleptura) cribripennis divisa Casey 1924: 281
Brachysomida caerulea Casey 1913: 227
Brachysomida californica proxima Casey 1913: 225
Brachysomida celestina Casey 1913: 222
Brachysomida hirsuta Casey 1913: 228
Brachysomida lepidula Casey 1913: 229
Brachysomida trinitatis Casey 1913: 226
Brachysomida versicolor Casey 1913: 229

Cacia semilactea Heller 1923: 19
Callichroma regalis Casey 1912: 314
Callichroma smaragdina Casey 1912: 314
Callidium antennatum hesperum Casey 1912: 288
Callimus dehiscens Casey 1912: 312
Callimus opacipennis Casey 1912: 311
Callimus variipes Casey 1912: 311
Calloides biformis Casey 1912: 358
Calloides coloradensis Casey 1912: 358
Centrodera decolorata lacustris Casey 1918: 416
Ceresium femoratum Aurivillius 1927: 529
Ceresium flavicorne Aurivillius 1927: 529
Ceroplesis restitutae Aurivillius 1911: 206
Chion cinctus var. *sonorensis* Schaeffer 1908: 332
Clytoleptus gracilicornis Casey 1912: 371
Clytopsis nimbata Casey 1912: 373
Colobothea sexualis Casey 1913: 344
Coptoeme sudanica Aurivillius 1915: 319
Criocephalus elongatus Casey 1924: 230
Criocephalus propinquus Casey 1924: 230
Crossidius blandi Casey 1893: 589
Crossidius longipennis Casey 1891: 31
Crossidius maculicollis Casey 1912: 337
Crossidius nitidicollis Casey 1891: 33
Crossidius sayi Casey 1912: 340
Crossidius vinosus Casey 1924: 266
Cyllene chara kansana Casey 1912: 355
Cyllene chara solida Casey 1912: 355
Cyllene reducta Casey 1912: 351
Cyrtocris fulvicornis Aurivillius 1903: 322
Daphisia bakeri Aurivillius 1926: 113
Daphisia bakeri var. *vittulata* Aurivillius 1926: 114
Daphisia bakeri var. *semisignata* Aurivillius 1926: 114
Daphisia bakeri var. *immaculata* Aurivillius 1926: 114
Daphisia fasciolata Aurivillius 1922: 431
Daphisia nigra Aurivillius 1922: 430
Daphisia niveoguttata Aurivillius 1925: 522
Dectes alticola Casey 1913: 342
Dectes aridus Casey 1913: 343
Dectes brevis Casey 1913: 342
Dectes brevisetosus Casey 1913: 343
Demonax angusticollis var. *sibuyanus* Aurivillius 1928: 319
Demonax includens Aurivillius 1928: 322
Dendrobias debilicollis Casey 1912: 316
Dendrobias reductus Casey 1912: 316
Dendrobias virens Casey 1912: 316
Dihammus dentifer Aurivillius 1927: 551

Dihammus fuscopunctatus Aurivillius 1927: 552
Doliops conspersa Aurivillius 1927: 563
Driopea cyrtomera Aurivillius 1922: 428
Eburia bauri Linell 1899: 260
Eburia densa Casey 1924: 259
Ecyrus dasycerus var. *texanus* Schaeffer 1908: 347
Eczemotellus subtilipictus Heller 1924: 432
Elaphidion arizonense Casey 1891: 28
Elaphidion levettei Casey 1891: 29
Elaphidion mimeticum Schaeffer 1905: 132
Elaphidion niveivestitum Schaeffer 1905: 132
Entomosterna unicostata Casey 1912: 332
Ergates (Trichocnemis) neomexicanus Casey 1890: 491
Euclea gloriosa Schultze 1922: 571
Evodinus carolinensis Casey 1924: 275
Evodinus vancouveri Casey 1913: 216
Glenea benguetana Aurivillius 1926: 99
Glenea lobata Aurivillius 1926: 103
Graphisurus (Graphisurus) floridanus Casey 1913: 336
Graphisurus (Graphisurus) obliquus sedulus Casey 1913: 337
Hapalosalia vibex fragilis Casey 1913: 202
Haploparmena angolana Aurivillius 1913: 239
Heterachthes exilissimus Casey 1912: 307
Heterachthes floridanus Casey 1912: 307
Hylotrupes inaequalis Casey 1924: 233
Hylotrupes litigiosus Casey 1891: 25
Hyperplatys californica Casey 1891: 51
Hyperplatys cryptica Casey 1913: 327
Hyperplatys frigida Casey 1913: 328
Hyperplatys variolata Casey 1913: 328
Hypocrites aureovittatus Aurivillius 1903: 316
Ipochus pinicola Casey 1913: 280
Ipochus subnitidus Casey 1891: 45
Judolia convolvens Casey 1913: 250
Leiopus houstoni Casey 1913: 315
Leiopus pleuralis Casey 1913: 312
Leiopus texanus Casey 1913: 315
Leptacmaeops quadriceps Casey 1913: 237
Leptacmaeops suturalis Casey 1924: 277
Leptacmaeops (Leptacmaeops) auricollis Casey 1913: 232
Leptacmaeops (Leptacmaeops) basalis dichroma Casey 1913: 231
Leptacmaeops (Leptacmaeops) nitidipennis Casey 1913: 236
Leptacmaeops (Leptacmaeops) pugetana Casey 1913: 237
Leptacmaeops (Leptacmaeops) punctiventris Casey 1913: 239
Leptacmaeops (Leptacmaeops) temporalis Casey 1913: 233
Leptostylus taeniatus Casey 1913: 306
Leptura haldemani Casey 1891: 42

Leptura lacustris Casey 1891: 43
Leptura (*Cercolia*) *minnesotana* Casey 1913: 269
Leptura (*Cosmosalia*) *aureola* Casey 1913: 268
Leptura (*Dorcasina*) *macrocera* Casey 1913: 270
Lepturges megalops Hamilton 1896: 127
Lepturges subglaber Casey 1913: 319
Lepturges yucca Schaeffer 1905: 167
Liasemum mokelumne Casey 1912: 262
Liopus alpha var. *floridanus* Hamilton 1896: 125
Liopus foveatocollis Hamilton in Leng & Hamilton 1896: 125
Liopus minuens Hamilton in Leng & Hamilton 1896: 123
Lochmaeocles marmoratus Casey 1913: 351
Mallodon castaneum Casey 1924: 225
Malobidion uteanum Casey 1924: 258
Mecas cineracea Casey 1913: 361
Merionoeda atricollis Heller 1924: 31
Microgoes tenuicornis Casey 1913: 298
Mimectatina singularis Aurivillius 1927: 575
Molochrus viridicollis Heller 1924: 27
Moneilema spinicollis Casey 1891: 45
Moneilema (*Collapteryx*) *rector* Casey 1913: 287
Moneilema (*Collapteryx*) *solida* Casey 1913: 288
Moneilema (*Moneilema*) *nubecula* Casey 1913: 282
Monochamus monticola Casey 1913: 293
Monochamus obtusus Casey 1891: 47
Monochamus oregonensis latus Casey 1924: 290
Neoclytus conjunctus occiduus Casey 1912: 369
Neoclytus provoanus Casey 1924: 272
Nothorhina aspera impressa Casey 1912: 264
Nothorhina spissicornis Casey 1912: 264
Oberea balineae Heller 1915: 43
Oberea exilis Casey 1913: 368
Oberea insignis Casey 1913: 370
Oberea iowensis Casey 1913: 370
Oberea montana Casey 1913: 369
Oberea ocellata plagiata Casey 1913: 371
Oberea tripunctata intermedia Casey 1913: 367
Obrium constricticolle Schaeffer 1908: 338
Obrium mozinnae Linell 1897: 395
Oeme subobsoleta Casey 1924: 253
Oncideres praecidens Casey 1913: 354
Oncideres trinodatus Casey 1913: 352
Ophistomis eversa Casey 1913: 277
Ophistomis ochreipennis Casey 1913: 278
Ophistomis simulans Casey 1913: 278
Ophistomis texana Casey 1913: 276
Oxoplus coccineus Casey 1893: 588

Oxoplus parviceps Casey 1924: 264
Paraclytus brevitarsis Casey 1912: 375
Paramallus arizonicus Casey 1912: 228
Parandra ampliceps Casey 1912: 216
Parandra gravidula Casey 1912: 216
Peritragus ertli Aurivillius 1913: 248
Phymatodes frosti Casey 1924: 240
Phymatodes (*Microcallidium*) *amoenus hudsonicus* Casey 1912: 284
Phymatodes (*Microcallidium*) *densipennis* Casey 1912: 284
Phymatodes (*Poecilium*) *harfordi* Casey 1912: 279
Phymatodes (*Poecilium*) *latipennis* Casey 1912: 280
Phymatodes (*Poecilium*) *varius diffidens* Casey 1912: 280
Pogonocherus carinatus Casey 1913: 346
Prionus debilis Casey 1891: 21
Prionus (*Prionellus*) *frosti* Casey 1924: 210
Prionus (*Prionellus*) *nigrescans* Casey 1924: 211
Prionus (*Prionus*) *angustulus* Casey 1912: 241
Prionus (*Prionus*) *californicus compar* Casey 1924: 217
Prionus (*Prionus*) *ineptus uintanus* Casey 1924: 220
Prionus (*Prionus*) *ineptus ambiguus* Casey 1924: 219
Prionus (*Prionus*) *spaldingi* Casey 1924: 221
Prionus (*Prionus*) *stultus* Casey 1924: 221
Prionus (*Riponus*) *aztecus* Casey 1912: 246
Prionus (*Riponus*) *diversus* Casey 1912: 247
Prionus (*Riponus*) *diversus cuneatus* Casey 1912: 247
Pygmaeopsis viticola Schaeffer 1908: 348
Pyrodes cervicalis Casey 1912: 256
Rhagium (*Hargium*) *californicum* Casey 1913: 195
Rhagium (*Hargium*) *cariniventre* Casey 1913: 196
Rhagium (*Hargium*) *mexicanum* Casey 1913: 197
Rhagium (*Hargium*) *montanum* Casey 1913: 197
Rhopalophora meeskei Casey 1891: 30
Romaleum thoracicum Casey 1912: 297
Rosalia funebris mandibularis Casey 1912: 313
Saperda tridentata trifasciata Casey 1913: 359
Saperda uteana Casey 1924: 294
Spalacopsis (*Euthuorus*) *costulata* Casey 1913: 355
Sphaenothecus rubens Casey 1891: 34
Spondylis erebea Casey 1912: 220
Spondylis subpubescens Casey 1912: 219
Stenaspis lugubris Casey 1912: 318
Stenaspis validicornis Casey 1912: 319
Stenocorus morio Casey 1924: 274
Stenocorus uteanus Casey 1924: 273
Stenocorus (*Stenocorus*) *brevicollis* Casey 1913: 214
Stenocorus (*Stenocorus*) *pacificus* Casey 1913: 209
Stenocorus (*Stenocorus*) *rufipennis* Casey 1913: 212

Stenocorus (Stenocorus) tenellus Casey 1913: 210
Stenodontes cubensis Casey 1912: 224
Strangalia idahoensis Casey 1913: 259
Strangalia minuscula Casey 1913: 260
Strangalia (Anastrangalia) apicata Casey 1924: 280
Strangalia (Anastrangalia) serricornis Casey 1924: 279
Tetraopes atrisetosus Casey 1913: 383
Tetraopes brevisetosus Casey 1913: 381
Tetraopes canescens fontinalis Casey 1913: 378
Tetraopes coccineus Casey 1913: 382
Tetraopes femoratus amnicola Casey 1913: 383
Tetraopes latior Casey 1924: 296
Tetraopes nigricollis Casey 1913: 376
Tetraopes obsoletus Casey 1913: 382
Tetraopes robustus Casey 1913: 381
Tetraopes ruber Casey 1913: 384
Tetraopes sandix Casey 1914: 370
Tetraopes spissicornis Casey 1913: 384
Tetraopes sublaevis Casey 1913: 382
Tetraopes tetropthalmus iowensis Casey 1913: 386
Tetraopes uteanus Casey 1913: 379
Tetraopes vegasensis Casey 1913: 380
Tetraopes velutinus Casey 1913: 380
Tetropium parallelum Casey 1891: 23
Tragidion auripenne Casey 1893: 587
Tragidion coquus filicorne Casey 1912: 323
Tragidion parowanum Casey 1924: 263
Tylosis medialis Casey 1912: 344
Tylosis sexualis Casey 1912: 343
Typocerus acuticauda Casey 1913: 274
Typocerus arapahoe Casey 1924: 278
Typocerus caligans Casey 1913: 273
Typocerus confluens Casey 1913: 272
Urographis fasciata reducta Casey 1913: 333
Urographis hebes Casey 1913: 333
Xestoleptura corusca Casey 1913: 261
Xestoleptura pictipennis Casey 1924: 285
Xylotrechus incongruens Casey 1912: 366
Xylotrechus longitarsis Casey 1912: 362

Literature Cited

Alten, R. L., G. R. Alten, and T. S. Ramey. 2009. A new genus and species of Cerambycidae from Utah (Coleoptera: Cerambycidae: Cerambycinae). *Pan-Pacific Entomologist*, 85(1): 1–7.

Aurivillius, C. 1903. Neue Oder Wenig Bekannte Coleoptera Longicornia. 8. *Arkiv för Zoologi*, 1: 313–328.

Aurivillius, C. 1907. Neue Oder Wenig Bekannte Coleoptera Longicornia. 9. *Arkiv för Zoologi*, 3(18): 93–131.

Aurivillius, C. 1911. Neue Oder Wenig Bekannte Coleoptera Longicornia. 12. *Arkiv för Zoologi*, 7(19): 187–228.

Aurivillius, C. 1913. Neue Oder Wenig Bekannte Coleoptera Longicornia. 13. *Arkiv för Zoologi*, 8(22): 229–263.

Aurivillius, C. 1915. Neue Oder Wenig Bekannte Coleoptera Longicornia. 15. *Arkiv för Zoologi*, 9(8) [1914]: 319–333.

Aurivillius, C. 1916. Neue Oder Wenig Bekannte Coleoptera Longicornia. 16. *Arkiv för Zoologi*, 10(19): 335–360.

Aurivillius, C. 1922. Neue Oder Wenig Bekannte Coleoptera Longicornia. 18. *Arkiv för Zoologi*, 14(18): 405–436.

Aurivillius, C. 1924. Neue Oder Wenig Bekannte Coleoptera Longicornia. 19. *Arkiv för Zoologi*, 15(25) [1923]: 437–479.

Aurivillius, C. 1925. Neue Oder Wenig Bekannte Coleoptera Longicornia. 21. *Arkiv för Zoologi*, 18A(9): 503–524.

Aurivillius, C. 1926. Revision of the Philippine Species of the Gleneini (Coleoptera, Longicornia). *The Philippine Journal of Science*, 30(1): 89–115.

Aurivillius, C. 1927a. Neue Oder Wenig Bekannte Coleoptera Longicornia. 22. *Arkiv för Zoologi*, 19A(17): 525–547.

Aurivillius, C. 1927b. Neue Oder Wenig Bekannte Coleoptera Longicornia. 23. *Arkiv för Zoologi*, 19A(23): 549–589.

Aurivillius, C. 1928. Revision of the Philippine Species of the Clytini (Coleoptera, Longicornia). *Philippine Journal of Science*, 36(3): 307–326.

Bezark, L. G., and M. A. Monné. 2013. Checklist of the Oxypeltidae, Vesperidae, Disteniidae and Cerambycidae (Coleoptera) of the Western Hemisphere. Available from: https://apps2.cdfa.ca.gov/publicApps/plant/bycidDB/wdefault.asp (accessed on 2/06/2014).

Bouchard, P., Y. Bousquet, A. E. Davies, M. A. Alonso-Zarazaga, J. F. Lawrence, C. Lyal, C. A. M. Reid, M. Schmitt, S. A. Ślipiński, and A. B. T. Smith. 2011. Family-group names in Coleoptera (Insecta). *ZooKeys*, 88: 1–972.

Bousquet, Y. 2007. Nomenclature and bibliographical notes on Cerambycidae (Coleoptera). *The Coleopterists Bulletin*, 61(4): 616–631.

Bousquet Y., D. J. Heffern, P. Bouchard, and E. H. Nearns. 2009. Catalogue of family-group names in Cerambycidae (Coleoptera). *Zootaxa*, 2321: 1–80.

Blackwelder, R. E. 1950. The Casey Room: Memorial to a Coleopterist. *The Coleopterists Bulletin*, 4(5): 65–80.

Breuning, S. 1935a. Novae species Cerambycidarum. *Folia Zoologica et Hydrobiologica*, 7(2): 153–174.

Breuning, S. 1935b. Novae species Cerambycidarum. II. *Folia Zoologica et Hydrobiologica*, 7(2): 241–254.

Breuning, S. 1935c. Novae species Cerambycidarum. III. *Folia Zoologica et Hydrobiologica*, 8: 51–71.

Breuning, S. 1936a. Études sur les Lamiaires (Coléop., Cerambycidae). Troisième Tribu: Prosopocerini Thomson. *Novitates Entomologicae*, 3ème supplément (21–29): 157–230.

Breuning, S. 1936b. Novae species Cerambycidarum. V. *Festschrift zum 60. Geburtstage von Professor Dr. Embrik Strand*, 1: 274–325.

Breuning, S. 1937. Études sur les Lamiaires (Coléop., Cerambycidae). Cinquième Tribu: Phrynetini Thomson. *Novitates Entomologiques*, 3éme Supplement (46–49): 271–302.

Breuning, S. 1938a. Études sur les Lamiaires (Coléop., Cerambycidae). VII. Ancylonotini. *Novitates Entomologicae*, 3ème supplément (41–46): 319–364.

Breuning, S. 1938b. Novae species Cerambycidarum. VI. *Festschrift zum 60. Geburtstage von Professor Dr. Embrik Strand*, 4 [1937]: 180–392.

Breuning, S. 1938c. Quelques nouveaux Lamiides du Musée du Congo Belge. *Revue de Zoologie et Botanique Africaine*, 30: 469–475.

Breuning, S. 1938d. Études sur les Lamiaires (Coléop., Cerambycidae). IX. *Novitates Entomologicae*: 30–63.

Breuning, S. 1939a. Nouveaux Cerambycidae (Col.) (2me note). *Novitates Entomologicae*, 9: 65–74.

Breuning, S. 1939b. Études sur les Lamiaires (Coléop., Cerambycidae). Huitième tribu: Mesosini Thomson. *Novitates Entomologicae*, 3ème supplément (50–66): 393–526.

Breuning, S. 1939c. Novae species Cerambycidarum. VII. *Festschrift zum 60. Geburtstage von Professor Dr. Embrik Strand*, 5: 144–290.

Breuning, S. 1940a. Études sur les Lamiaires (Coléop. Cerambycidae). Neuvième Tribu: Dorcaschematini Thoms. *Novitates Entomologicae*, 3ème supplément (67–71): 527–568.

Breuning, S. 1940b. Novae species Cerambycidarum. VIII. *Folia Zoologica et Hydrobiologica*, 10(1): 37–85.

Breuning, S. 1940c. Novae species Cerambycidarum. IX. *Folia Zoologica et Hydrobiologica*, 10(1): 115–214.

Breuning, S. 1940d. Novae species Cerambycidarum. X. *Folia Zoologica et Hydrobiologica*, 10(2): 407–437.

Breuning, S. 1942a. Novae species Cerambycidarum. XI. *Folia Zoologica et Hydrobiologica*, 11: 113–175.

Breuning, S. 1942b. Études sur les Lamiaires (Coléop., Cerambycidae). Dixième tribu: Crossotini Thoms. *Novitates Entomologicae*, 3ème supplément (73–84): 8–101.

Breuning, S. 1942c. Études sur les Lamiaires (Coléop., Cerambycidae). Onzième tribu: Phrissomini Lac. *Novitates Entomologicae*, 3ème supplément (84–88): 102–136.

Breuning, S. 1942d. Nouveaux Cerambycidae (3e note). *Novitates Entomologicae*, 12(19): 142–148.

Breuning, S. 1943a. Études sur les Lamiaires (Coleop. Cerambycidae). Douzième tribu: Agniini Thomson. *Novitates Entomologicae*, 3ème supplément (89–106): 137–280.

Breuning, S. 1943b. Nouveaux Cérambycides paléarctiques (1re note). *Miscellanea Entomologica*, 40(9): 89–104.

Breuning, S. 1943c. Novae species Cerambycidarum. XII. *Folia Zoologica et Hydrobiologica*, 12: 12–66.

Breuning, S. 1946. Nouveaux Cérambycides paléarctiques (Col.) (3ème note). *Miscellanea Entomologica*, 43(3): 21–24.

Breuning, S. 1948a. Nouvelles formes de *Dorcadion* (Col., Cerambycidae) (3e note). *Miscellanea Entomologica*, 45: 57–64.

Breuning, S. 1948b. Nouvelles formes de Lamiaires (deuxième partie). *Bulletin du Musée Royal d'Histoire Naturelle de Belgique*, 24(47): 1–47.

Breuning, S. 1950a. Révision des Stenobiini. *Longicornia*, I: 305–315, Paul Lechevalier, Paris.

Breuning, S. 1950b. Révision des Homonoeini. *Longicornia*, I: 317–377, Paul Lechevalier, Paris.

Breuning, S. 1950c. Quelques nouveaux Lamiaires (Coleoptera, Cerambycidae) du Riksmuseum. *Arkiv för Zoologi*, Uppsala (series 2) 1(19): 265–274.

Breuning, S. 1954. Nouvelles formes de Lamiaires (sixième partie). *Bulletin de l'Institut Royal des Sciences Naturelles de Belgique*, 30(28): 1–23.

Breuning, S. 1956. Revision der Gattung *Glenea* Newm. (1. Fortsetzung). *Entomologischen Arbeiten aus der Museum G. Frey*, 7(2): 671–893.

Breuning, S. 1957a. Neue Lamiinae aus der Sammlung F. Tippmann (Cerambycidae, Coleoptera). *Entomologisches Nachrichtenblatt für Österreichischer und Schweizer Entomologen*, 8(3): 10–16.

Breuning, S. 1957b. Révision des Agapanthiini africains (Cerambycidae, Col.). *Bulletin de lInstitut Français dAfrique Noire*, 19(3), série A: 792–869.

Breuning, S. 1961a. Révision des Pteropliini (Col., Cerambycidae). *Pesquisas, Zoologia*, 5(9): 5–61.

Breuning, S. 1961b. Catalogue des Lamiaires du Monde (Col. Céramb.). *Verlag des Museums G. Frey, Tutzing bei München*, (4): 183–284.

Buchanan, L. L. 1935. Thomas Lincoln Casey and the Casey Collection of Coleoptera. *Smithsonian Miscellaneous Collections*, 94(3): 1–15.

Carpenter, M. M. 1945. Bibliography of Biographies of Entomologists. *The American Midland Naturalist*, 33(1): 1–116.

Casey, T. L. 1890. VII. – Coleopterological Notices. II. *Annals of the New York Academy of Sciences*, 5: 307–504.

Casey, T. L. 1891. II. – Coleopterological Notices. III. *Annals of the New York Academy of Sciences*, 6: 9–214.

Casey, T. L. 1893. III. – Coleopterological Notices. V. *Annals of the New York Academy of Science*, 7: 281–606.

Casey, T. L. 1912. III – Studies in the Longicornia of North America. *Memoirs on the Coleoptera*, 3: 215–376.

Casey, T. L. 1913. II – Further Studies among the American Longicornia. *Memoirs on the Coleoptera*, 4: 193–388.

Casey, T. L. 1914. Miscellaneous notes and new species. *Memoirs on the Coleoptera*, 5: 355–378.

Casey, T. L. 1918. Miscellaneous notes and corrections. *Memoirs on the Coleoptera*, 8: 413–416.

Casey, T. L. 1924. I – Additions to the known Coleoptera of North America. *Memoirs on the Coleoptera*, 11: 1–347.

Chemsak, J. A. 1963. Taxonomy and bionomics of the genus *Tetraopes* (Cerambycidae: Coleoptera). *University of California Publications in Entomology*, 30(1): 1–90.

Chemsak, J. A. 1966. Descriptions and records of West Indian Cerambycidae (Coleoptera). *Proceedings of the United States National Museum*, 118(3526): 209–220.

Chemsak, J. A. 1969. Records and descriptions of Mexican and Central American Tillomorphini (Coleoptera: Cerambycidae). *The Pan-Pacific Entomologist*, 45(4): 303–317.

Chemsak, J. A. 1977. Records and descriptions of some Mexican species of the genus *Phaea* Newman (Coleoptera: Cerambycidae). *The Pan-Pacific Entomologist*, 53(4): 269–276.

Chemsak, J. A. 1979. New species of neotropical Prioninae (Coleoptera: Cerambycidae). *The Coleopterists Bulletin*, 33(1): 125–128.

Chemsak, J. A. 1983. A new Central American species of *Derancistrus* (Coleoptera: Cerambycidae). *Proceedings of the Entomological Society of Washington*, 85(4): 714–716.

Chemsak, J. A. 1997. A new Central American Rhinotragine (Coleoptera: Cerambycidae). *Occasional Papers of the Consortium Coleopterorum*, 1(1): 15–16.

Chemsak, J. A. 2000. Revision of the genus *Phaea* Newman (Coleoptera, Cerambycidae). *Occasional Papers of the Consortium Coleopterorum*, 3(1): 36–101.

Chemsak, J. A., and C. Feller. 1988. New species of Cerambycidae from Twin Cays, Belize (Coleoptera). *Proceedings of the Entomological Society of Washington*, 90(2): 179–188.

Chemsak, J. A., and E. G. Linsley. 1959. Some new species and subspecies of *Crossidius* from Western North America (Coleoptera: Cerambycidae). *Journal of the Kansas Entomological Society*, 32(4): 176–183.

Chemsak, J. A., and E. G. Linsley. 1964a. Descriptions and records of Hesperophanini, with keys to the species of *Anatinomma* and *Amphelictus* (Coleoptera, Cerambycidae). *Journal of the Kansas Entomological Society*, 37(3): 216–226.

Chemsak, J. A., and E. G. Linsley. 1964b. Descriptions and records of Mexican Methiini (Coleoptera: Cerambycidae). *The Pan-Pacific Entomologist*, 40(3): 158–161.

Chemsak, J. A., and E. G. Linsley. 1965. The Elaphidionine subgenus *Protaneflus* (Coleoptera: Cerambycidae). *Proceedings of the Entomological Society of Washington*, 67(1): 24–28.

Chemsak, J. A., and E. G. Linsley. 1967. Notes on the genera *Pseudoperiboeum* Linsley and *Conosphaerion* Linsley (Coleoptera: Cerambycidae). *Journal of the Kansas Entomological Society*, 40(3): 285–290.

Chemsak, J. A., and E. G. Linsley. 1971. New Neotropical Methiine Cerambycidae (Coleoptera). *The Pan-Pacific Entomologist*, 47(2): 117–122.

Chemsak, J. A., and E. G. Linsley. 1974. Notes and descriptions of some Lycid-like Neotropical lepturine Cerambycidae (Coleoptera). *The Pan-Pacific Entomologist*, 50(1): 35–43.

Chemsak, J. A., and E. G. Linsley. 1976a. The Lepturine genus *Pseudotypocerus* (Coleoptera: Cerambycidae). *The Coleopterists Bulletin*, 30(2): 171–175.

Chemsak, J. A., and E. G. Linsley. 1976b. Further studies among the Mexican and Central American lepturine Cerambycidae related to *Euryptera* and *Choriolaus* (Coleoptera). *The Pan-Pacific Entomologist*, 52(1): 38–49.

Chemsak, J. A., and E. G. Linsley. 1979. Review of the Rhinotragini of Mexico (Coleoptera: Cerambycidae). *Proceedings of the California Academy of Sciences*, 42(3): 69–85.

Chemsak, J. A., and E. G. Linsley. 1984. New genera and species of neotropical Cerambycidae (Coleoptera). *The Pan-Pacific Entomologist*, 60(4): 279–288.

Chittenden, F. H. 1897. Martin Larsson Linell, Obituary. *Entomological News*, 8: 159–160.

Chittenden, F. H. 1904. Studies of the life history, habits, and taxonomic relations of a new species of *Oberea* (*Oberea ulmicola* Chittenden). *Bulletin of the Illinois State Laboratory of Natural History*, 7: 2–5

Chittenden, F. H., and M. L. Linell. 1896. A pod-inhabiting longicorn found at the Columbian Exposition. *Proceedings of the Entomological Society of Washington*, 4: 42–43.

Corbett, D. 2004. Sinopse do gênero *Cobelura* (Coleoptera, Cerambycidae, Lamiinae, Acanthocinini), com descrição de três espécies novas. *Iheringia (Série Zoologia)*, 94(3): 277–280.

Cushman, H. G., C. F. W. Muesebeck, and G. B. Vogt. 1972. Warren Samuel Fisher 1878–1971. *Proceedings of the Entomological Society of Washington*, 74(3): 344–352.

Cushman, R. A. 1929. Bringing to America the Baker collection of Malayan Insects. Exploration and Field Work of the Smithsonian Institute, 1928: 90–100.

Daniel, K., and J. Daniel. 1898. Zwanzig neue Arten aus dem paläarktischen Faunengebiet. Coleopteren- Studien 2(7): 61–82. Munich: Druck C. Wolf & Sohn.

Danilevsky, M. L. 2004. *Cortodera neali* sp. n. from Iran and *Dorcadion shirvanicum azerbajdzhanicum* Plavilstshikov, 1937 stat. n. from Azerbajdzhan (Coleoptera, Cerambycidae). *Les Cahiers Magellanes*, 32: 1–7.

Danilevsky M. L. 2010. Additions and corrections to the new Catalogue of Palaearctic Cerambycidae (Coleoptera), ed. I. Löbl and A. Smetana, 2010. *Russian Entomological Journal*, 19(3): 215–239.

Danilevsky M. L. 2011. Additions and corrections to the new Catalogue of Palaearctic Cerambycidae (Coleoptera), ed. I. Löbl and A. Smetana, 2010. Part. II. *Russian Entomological Journal*, 19(4)[2010]: 313–324.

Danilevsky M. L. 2012a. Additions and corrections to the new Catalogue of Palaearctic Cerambycidae (Coleoptera), ed. I. Löbl and A. Smetana, 2010. Part. III. *Munis Entomology & Zoolog*, 7(1): 109–173.

Danilevsky M. L. 2012b. Additions and corrections to the new Catalogue of Palaearctic Cerambycidae (Coleoptera), ed. I. Löbl and A. Smetana, 2010. Part IV, Humanity Space. *International Almanac*, 1(1): 86–136.

Danilevsky M. L. 2012c. Additions and corrections to the new Catalogue of Palaearctic Cerambycidae (Coleoptera), ed. I. Löbl and A. Smetana, 2010. Part V, Humanity Space. *International Almanac*, 1(3): 695–741.

Davis, W. T. 1923. A New Longicorn Beetle from the Catskill Mountains. *Bulletin of the Brooklyn Entomological Society*, 18: 97–98.

Demets, Y. 1976. Notes sur les Callichromatini (Coleoptera, Cerambycidae). IV. Étude preliminaire des genres *Schwarzerion* Schmidt, 1924 et *Xenochroma* Schmidt, 1924. *Papéis Avulsos de Zoologia*, 29(17): 121–140.

Dillon, L. S. 1955. Revision of the Neotropical Acanthocinini (Coleoptera: Cerambycidae). I. The genus *Canidia* and its allies. *Entomological News*, 66(6–7): 141–149, 176–187.

Dillon, L. S. 1956. The Nearctic components of the tribe Acanthocinini (Coleoptera: Cerambycidae). Part I. *Annals of the Entomological Society of America*, 49(2): 134–167.

Dillon, L. S. 1957. Revision of the Neotropical Acanthocinini (Coleoptera: Cerambycidae). 2. The genus *Lagocheirus*. *Bulletin of the British Museum of Natural History (Entomology)*, 6(6): 137–166.

Dillon, L. S., and E. S. Dillon. 1941. The tribe Monochamini in the Western Hemisphere (Coleoptera: Cerambycidae). *Scientific Publications of the Reading Public Museum*, 1: 1–135.

Dillon, L. S., and E. S. Dillon. 1945. The tribe Onciderini (Coleoptera: Cerambycidae). Part I. *Scientific Publications of the Reading Public Museum*, 5: 1–186.

Dillon, L. S., and E. S. Dillon. 1946a. A review of the tribe Gryllicini (Coleoptera, Cerambycidae). *Livro de homenagem a Romualdo Ferreira d'Almeida*, 14: 155–164.

Dillon, L. S., and E. S. Dillon. 1946b. The tribe Onciderini (Coleoptera: Cerambycidae). Part II. *Scientific Publications of the Reading Public Museum*, 6: 189–413.

Dillon, L. S., and E. S. Dillon. 1948. The Tribe Dorcaschematini (Coleoptera: Cerambycidae). *Transactions of the American Entomological Society*, 73: 173–298.

Dillon, L. S., and E. S. Dillon. 1951. The Lamiinae Tribe Gnomini (Coleoptera, Cerambycidae). *The Philippine Journal of Science*, 79(1): 11–57.

Dillon, L. S., and E. S. Dillon. 1959. The Monochamini (Cerambycidae) of the Ethiopian Faunistic Region I. Subtribe Monochamidi. Pseudhammus and Allies. *Scientific Publications Reading Public Museum and Art Gallery*, 9: 1–72.

Fisher, W. S. 1914. A new species of *Callichroma* from Texas. *Proceedings of the Entomological Society of Washington*, 16(3): 97–98.

Fisher, W. S. 1915. One new genus and two new species of Cerambycidae. *Proceedings of the Entomological Society of Washington*, 17(2): 77–79.

Fisher, W. S. 1917. A new species of *Xylotrechus*. *Proceedings of the Entomological Society of Washington*, 18(2) [1916]: 214–216.

Fisher, W. S. 1918. A new species of longhorn beetle infesting cowpeas from Mexico. *Proceedings of the Entomological Society of Washington*, 19(1–4) [1917]: 173–174.

Fisher, W. S. 1919. A new genus and species of Cerambycidae from Colorado. (Coleo.) *Proceedings of the Entomological Society of Washington*, 21(2): 38–40.

Fisher, W. S. 1920. A new genus and several new species of Cerambycidae (Coleoptera). *Proceedings of the Entomological Society of Washington*, 22(7): 153–159.

Fisher, W. S. 1921. A new cerambycid beetle from California. *Proceedings of the Entomological Society of Washington*, 23(9): 206–208.

Fisher, W. S. 1922. A new cerambycid beetle from Santo Domingo (Col.). *Entomological News*, 33(2): 52.

Fisher, W.S. 1924. A new species of *Ataxia* from the United States. (Coleoptera: Cerambycidae). *The Canadian Entomologist*, 56: 253–254.

Fisher, W. S. 1925a. New Malaysian Cerambycidae, subfamily Lamiinae. *The Philippine Journal of Science*, 28(2): 205–275.

Fisher, W. S. 1925b. New West Indian Cerambycidae (Coleoptera). Subfamily Lamiinae. *American Museum Novitates*, 174: 1–16.

Fisher, W. S. 1925c. Two new Mexican Cerambycidae (Coleoptera). *Proceedings of the Entomological Society of Washington*, 27(1): 15–16.

Fisher, W. S. 1925d. A new species of *Leptostylus* from the United States. (Coleoptera: Cerambycidae). *Proceedings of the Entomological Society of Washington*, 27(5): 103–105.

Fisher, W. S. 1926. Descriptions of new West Indian longicorn beetles of the subfamily Lamiinae. *Proceedings of the United States National Museum*, 68(22): 1–40.

Fisher, W. S. 1927a. Fauna sumatrensis. (Beitrag Nr. 37). Cerambycidae (Col.). *Supplementa Entomologica*, 15: 80–86.

Fisher, W. S. 1927b. New cactus beetles. *Proceedings of the Entomological Society of Washington*, 28(9) [1926]: 214–217.

Fisher, W.S. 1927c. A new cerambycid beetle from Colombia and Central America (Coleoptera). *Proceedings of the Entomological Society of Washington*, 29(1): 23–24.

Fisher, W. S. 1928a. Appendix: New species of Jamaican Buprestidae and Cerambycidae. In C. G. Gowdey, Catalogus insectorum jamaicensis. *Entomological Bulletin of the Department of Agriculture*, 4(3): 16–18.

Fisher, W. S. 1928b. New cactus beetles. II. *Proceedings of the Entomological Society of Washington*, 30(1): 1–7.

Fisher, W. S. 1930a. Notes on the rhinotragine beetles of the family Cerambycidae, with descriptions of new species. *Proceedings of the United States National Museum*, 77(19): 1–20.

Fisher, W. S. 1930b. New West Indian Buprestidae and Cerambycidae. *The Canadian Entomologist*, 62(1): 7–11.

Fisher, W. S. 1931a. A new ant-like cerambycid beetle from Honduras. *Psyche*, 38(2–3): 99–101.

Fisher, W. S. 1931b. A new longhorn beetle from Costa Rica (Coleoptera: Cerambycidae). *Journal of the Washington Academy of Sciences*, 21(2): 23–24.

Fisher, W. S. 1931c. New cactus beetles. III. *Proceedings of the Entomological Society of Washington*, 33(8): 197–201.

Fisher, W. S. 1932. New West Indian cerambycid beetles. *Proceedings of the United States National Museum*, 80(2922) 22: 1–93.

Fisher, W. S. 1934a. Cerambycidae from Sumatra, with descriptions of new species. *Miscellanea Zoologica Sumatrana*, 88: 1–7.

Fisher, W. S. 1934b. New Species of Cerambycidae (Col.) from Java. *Stylops*, 3(2): 35–42.

Fisher, W. S. 1934c. Malayan Cerambycidae (Coleopt.). *Bulletin du Musée Royal dHistoire Naturelle de Belgique*, 10(15): 1–6.

Fisher, W. S. 1935a. New cerambycid beetles from Puerto Rico. *Journal of Agriculture of the University of Puerto Rico*, 19(2): 51–63.

Fisher, W. S. 1935b. New West Indian cerambycid beetles. *Proceedings of the United States National Museum*, 83(2979): 189–210.

Fisher, W. S. 1935c. Cerambycidae from Mount Kinabalu. *Journal of the Federated Malay States Museum*, 17(30): 581–631.

Fisher, W. S. 1936a. Fauna Javanica. New Cerambycidae from Java. *Tijdschrift voor Entomologie*, 79: 169–198.

Fisher, W. S. 1936b. New cactus beetle from Argentine Republic (Cerambycidae). *Proceedings of the Entomological Society of Washington*, 38(1): 7–8.

Fisher, W. S. 1936c. New cerambycid and buprestid beetles from Cuba. Memorias *de la Sociedad Cubana de Historia Natural Felipe Poey*, 9(4): 271–273.

Fisher, W. S. 1936d. New buprestid and cerambycid beetles from Cuba. *Memorias de la Sociedad Cubana de Historia Natural Felipe Poey*, 10(5): 343–348.

Fisher, W. S. 1937a. A new cerambycid beetle from Chile. *Revista Chilena de Historia Natural*, 40 [1936]: 324–326.

Fisher, W. S. 1937b. A New Cerambycid Beetle Injurious to Orchids in Java. *Entomologische Mededeelingen van Nederlandsch-Indie*, 3(4): 53–54.

Fisher, W. S. 1937c. New Cerambycid beetles belonging to the genus *Oberea* from Java. *Treubia*, 16(2): 197–203.

Fisher, W. S. 1937d. New neotropical Cerambycidae (Col.). *Revista de Entomologia*, 7(2–3): 145–154.

Fisher, W.S. 1938. New Neotropical Cerambycidae (Coleoptera). II. *Revista de Entomologia*, 8(1–2): 135–153.

Fisher, W. S. 1942a. New West Indian cerambycid beetles. III. *Torreia*, 10: 1–43.

Fisher, W. S. 1942b. New Coleoptera from Puerto Rico. *Journal of Agriculture of the University of Puerto Rico*, 25[1941]: 37–39.

Fisher, W. S. 1942c. A new Cerambycid beetle from Oregon (Coleoptera). *The Canadian Entomologist*, 74(3): 51.

Fisher, W. S. 1943. A new cerambycid beetle from the Canal Zone. *Boletín del Departamento de Sanidad Vegetal*, 3(1): 12–14.

Fisher, W. S. 1944a. Cerambycidae (Coleoptera) of Caripito, Venezuela. *Zoologica*, 29(1): 3–12.

Fisher, W. S. 1944b. New American Cerambycidae (Coleoptera) from British Guiana and Costa Rica. *Zoologica*, 29(1): 1–2.

Fisher, W. S. 1945. A new species of *Oberea* from Canada. *The Canadian Entomologist*, 77: 56.

Fisher, W. S. 1946a. New Cerambycid Beetles Belonging to the Tribe Disteniini from Central and South America. *Proceedings of the United States National Museum*, 96(3201)[1945]: 329–333.

Fisher, W. S. 1946b. Synopsis of the cerambycid beetles of the genus *Stenosphenus* Haldeman found in America, north of Mexico. *Journal of the Washington Academy of Sciences*, 36(3): 86–94.

Fisher, W. S. 1947a. New cerambycid beetles belonging to the tribe Rhinotragini. *Proceedings of the United States National Museum*, 97(3209): 47–57.

Fisher, W. S. 1947b. New neotropical Cerambycidae, belonging to the genus *Doscasta* Pascoe (Coleoptera). *Revista de Entomologia*, 18(1–2): 173–182.
Fisher, W. S. 1947c. New West Indian cerambycid beetles. IV. *Memorias de la Sociedad Cubana de Historia Natural Felipe Poey*, 19(1): 29–41.
Fisher, W. S. 1951. A new species of *Elaphidion* (family Cerambycidae) from Cuba. *Memorias de la Sociedad Cubana de Historia Natural Felipe Poey*, 20(2): 77–79.
Flint, O. S., Jr., A. S. Menke, F. C. Thompson, and R. A. Ward. 1976. *The United States National Entomological Collections*. Washington, DC: Smithsonian Institution Press. 47 pp.
Galileo, M. H. M., and U. R. Martins. 1996. Revisão do gênero *Isomerida* Bates, 1866 (Coleoptera, Cerambycidae, Lamiinae, Hemilophini). *Papéis Avulsos de Zoologia*, 39(12): 231–248.
Galileo, M. H. M., and U. R. Martins. 2004a. Novos Hemilophini (Coleoptera, Cerambycidae, Lamiinae) da Região Neotropical. *Iheringia (Série Zoologia)*, 94(3): 247–252.
Galileo, M. H. M., and U. R. Martins. 2004b. Novos táxons em Hemilophini (Coleoptera, Cerambycidae) com única carena elitral. *Iheringia (Série Zoologia)*, 94(4): 381–388.
Galileo, M. H. M., and U. R. Martins. 2005a. Novos táxons de Hemilophini (Coleoptera, Cerambycidae) da Região Neotropical. *Iheringia (Série Zoologia)*, 95(3): 269–279.
Galileo, M. H. M., and U. R. Martins. 2005b. Novos táxons de Hemilophini (Coleoptera, Cerambycidae, Lamiinae) sem carenas nos élitros da Região Neotropical. *Revista Brasileira de Entomologia*, 49(1): 63–68.
Galileo, M. H. M., and U. R. Martins. 2006a. Novos táxons de Apomecynini (Coleoptera, Cerambycidae, Lamiinae). *Papéis Avulsos de Zoologia*, 46(2): 11–19.
Galileo, M. H. M., and U. R. Martins. 2006b. Novos taxons em Hippopsini, Desmiphorini, Xenofreini e Acanthoderini (Coleoptera, Cerambycidae, Lamiinae). *Papeis avulsos de Zoologia*, 46(3): 21–29.
Galileo, M. H. M., and U. R. Martins. 2006c. Notas e descrições em Eburiini (Coleoptera, Cerambycidae). *Iheringia (Série Zoologia)*, 96(2): 179–184.
Galileo, M. H. M., and U. R. Martins. 2007a. Description of taxa in Hippopsini, Desmiphorini e Pteroplini (Cerambycidae, Lamiinae). *Papeis Avulsos de Zoologia*, 47(25): 349–357.
Galileo, M. H. M., and U. R. Martins. 2007b. Notas e descrições em Acanthoderini (Coleoptera, Cerambycidae, Lamiinae). II. Espécies novas do gênero *Nesozineus*. Revista *Brasileira de Zoologia*, 24(3): 631–634.
Galileo, M. H. M., and U. R. Martins. 2007c. Novos taxons em Cerambycini, Thyrsiini e Necydalopsini (Coleoptera, Cerambycinae). *Lambillionea*, 106(4): 703–710.
Galileo, M. H. M., and U. R. Martins. 2008a. Novos Taxons de Lamiinae Neotropicais (Coleoptera, Cerambycidae, Lamiinae). *Les Cahiers Magellanes*, 73: 1–9.
Galileo, M. H. M., and U. R. Martins. 2008b. Espécies novas de Calliini e Falsamblesthiini (Coleoptera, Cerambycidae, Lamiinae) da Região Neotropical. *Revista Brasileira de Zoologia*, 52(1): 32–35.
Galileo, M. H. M., and U. R. Martins. 2009. Novas espécies de Hemilophini da América do Sul (Coleoptera, Cerambycidae, Lamiinae). *Les Cahiers Magellanes*, 97: 1–10.
Galileo, M. H. M., and U. R. Martins. 2010. Novas espécies de Prioninae e de Cerambycinae (Cerambycidae) da Região Neotropical. *Revista Brasileira de Entomologia*, 54(1): 32–37.
Galileo, M. H. M., and U. R. Martins. 2011. Novas especies de *Esthlogena* s. str. Thomson (Coleoptera, Cerambycidae, Lamiinae). *Revista Brasileira de Entmologia*, 55(2): 183–186.
Galileo, M. H. M., and U. R. Martins. 2012a. Novos táxon e nova sininímia em Acanthoderini neotropicais (Coleoptera, Cerambycidae, Lamiinae). *Les Cahiers Magellanes (NS)*, 7: 61–69.
Galileo, M. H. M., and U. R. Martins. 2012b. Desmiphorini: dois gêneros novos e espécie nova de *Pseudestola* Breuning, 1940 (Coleoptera, Cerambycidae, Lamiinae). Les *Cahiers Magellanes (NS)*, 8: 106–112.
Galileo, M. H. M., and U. R. Martins. 2012c. Novos táxons e novos registros sobre Cerambycidae (Coleoptera) Neotropicais. *Revista Brasileira entomololgia*, 56(3): 297–303.
Galileo, M. H. M., and U. R. Martins. 2013. Novos gêneros, novas espécies e redescrição em Hemilophini (Coleoptera, Cerambycidae, Lamiinae) das Américas. *Iheringia (Série Zoologia)*, 57(1): 31–39.
Galileo, M. H. M., M. A. Monné, M. L. Monné, S. W. Lingafelter, and E. H. Nearns. 2013. Cerambycidae Holotypes of the Museu de Ciências Naturais, Fundação Zoobotânica do Rio Grande do Sul. Available from: http://www.cerambycids.com/brazil/mcnz/

Giesbert, E. F. 1996. A new genus and species in the subfamily Lamiinae, with comments on *Laticranium mandibulare* Lane in Central America (Coleoptera: Cerambycidae). *Insecta Mundi*, 10(1–4): 317–319.

Giesbert, E. F., and J. A. Chemsak. 1997. A review of the genus *Euderces* LeConte (Coleoptera, Cerambycidae, Tillomorphini). *Proceedings of the California Academy of Sciences*, 49(8): 211–286.

Gilbert, P. 1977 [1978]. *A Compendium of the Biographical Literature on Deceased Entomologists*. London: British Museum (Natural History). 455 pp.

Gressitt, J. L. 1934. New Longicorns from the Japan Empire (Coleopt., Cerambycidae). *Pan-Pacific Entomologist*, 9(4): 163–170.

Gressitt, J. L. 1935a. New Japanese Longicorn Beetles (Coleoptera: Cerambycidae). *Kontyu*, 9(4): 166–179.

Gressitt, J. L. 1935b. New longicorn beetles of the subfamily Lamiinae from Formosa (Coleoptera: Cerambycidae). *Transactions of the Natural History Society of Formosa*, 25(143): 286–292.

Gressitt, J. L. 1935c. New Longicorn Beetles from China (Coleoptera: Cerambycidae). *Lingnan Science Journal*, 14(4): 567–574.

Gressitt, J. L. 1935d. New Longicorn Beetles from the Japanese Empire, II (Coleoptera: Cerambycidae). *The Philippine Journal of Sciences*, 55(4): 379–386.

Gressitt, J. L. 1935e. New species and records of longicorn from Formosa (Coleoptera, Cerambycidae). *The Philippine Journal of Sciences*, 57(2): 181–194.

Gressitt, J. L. 1935f. New longicorn beetles from Formosa, II (Coleoptera, Cerambycidae). *The Philippine Journal of Sciences*, 58(2): 253–266.

Gressitt, J. L. 1935g. New longicorn beetles from the Philippines, Borneo and Siam (Coleoptera, Cerambycidae). *The Philippine Journal of Sciences*, 58(2): 267–280.

Gressitt, J. L. 1936. New Longicorn beetles from Formosa, III (Coleoptera, Cerambycidae). *The Philippine Journal of Sciences*, 61(1): 89–111.

Gressitt, J. L. 1937a. New Longicorn beetles from China, II. *Lingnan Science Journal*, 16(1): 89–94.

Gressitt, J. L. 1937b. New Longicorn beetles from China, IV (Coleoptera, Cerambycidae). *Lingnan Science Journal*, 16(4): 595–621.

Gressitt, J. L. 1938a. New Longicorn beetles from China, V (Coleoptera, Cerambycidae). *Lingnan Science Journal*, 17(1): 45–56.

Gressitt, J. L. 1938b. New Longicorn beetles from China, VI (Coleoptera, Cerambycidae). *Lingnan Science Journal*, 17(2): 151–159.

Gressitt, J. L. 1938c. New Longicorn beetles from Formosa, IV (Coleoptera, Cerambycidae). *The Philippine Journal of Sciences*, 65(3): 147–173.

Gressitt, J. L. 1939a. A study of the longicorn beetles of Kwangtung Province, S. China (Coleoptera: Cerambycidae). *Lingnan Science Journal*, 18(1): 1–122.

Gressitt, J. L. 1939b. New Longicorn Beetles from China, VII (Coleoptera: Cerambycidae). *Lingnan Science Journal*, 18(2): 209–216.

Gressitt, J. L. 1940. The Longicorn beetles of Hainan Island, Coleoptera: Cerambycidae. *The Philippine Journal of Sciences*, 72(1–2): 1–239.

Gressitt, J. L. 1941. A collection of longicorn beetles from Thai, Coleoptera: Cerambycidae. *The Philippines Journal of Sciences*, 74(4): 331–347.

Gressitt, J. L. 1947. Chinese longicorn beetles of the genus *Linda* (Coleoptera: Cerambycidae). *Annals of the Entomological Society of America*, 40(3): 545–555.

Gressitt, J. L. 1948a. New Longicorn Beetles from China, XII (Col.: Ceramb.). *Lingnan Science Journal*, 22(1–4): 43–52.

Gressitt, J. L. 1948b. Chinese longicorn beetles of the genus *Nupserha* (Col., Ceramb.). *Lingnan Science Journal*, 22(1–4): 55–68.

Gressitt, J. L. 1951a. Longicorn Beetles from New Guinea and the South Pacific (Coleoptera: Cerambycidae). Part I. *Annals of the Entomological Society of America*, 44(1): 1–30.

Gressitt, J. L. 1951b. Longicorn beetles of China. *Longicornia*, 2: 1–667.

Gressitt, J. L. 1951c. Longicorn Beetles from the Ryukyu Islands (Coleoptera: Cerambycidae). *The Philippine Journal of Sciences*, 79(2): 193–235.

Gressitt, J. L. 1954. The Longicorne Genus *Thranius* (Coleoptera: Cerambycidae). *Proceedings of the Hawaii Entomological Society*, 15(2): 317–326.

Gressitt, J. L. 1956. Coleoptera: Cerambycidae. *Insects of Micronesia*, 17(2): 61–183.

Gurney, A. B. 1976. A short history of the Entomological Society of Washington. *Proceedings of the Entomological Society of Washington*, 78(3): 225–239.

Hammond, H. E. J., and D. J. Williams. 2011. A Revision of the Genus *Callimoxys* Kraatz (Coleoptera: Cerambycidae) in America North of Mexico and Review of World Species. *The Coleopterists Bulletin*, 65(3): 246–289.

Heller, K. M. 1915. Neue Käfer von den Philippinen: II. *The Philippine Journal of Science*, 10(1) Section D: 19–47.

Heller, K. M. 1916. Philippinische Käfer, gesammelt von Prof. C. Fuller-Baker, Los Baños. *Deutsche entomologische Zeitschrift*, [1916](3–4): 269–311.

Heller, K. M. 1923a. Neue Bockkäfer von den Philippinen und aus Borneo. *Deutsche entomologische Zeitschrift*, 7(4): 414–425.

Heller, K. M. 1923b. Neue philippinische Käfer (XXVII. Beitrag). *Stettiner Entomologische Zeitung*, 84(1): 3–24.

Heller, K. M. 1924a. Bestimmungsschlüssel außereuropäischer Käfer. Cerambicidae, Molorchini: genera *Epania* und *Merionoeda*. *Entomologische Blätter*, 20(1): 26–34.

Heller, K. M. 1924b. Neue, vorwiegend philippinische Bockkäfer. *Entomologische Mitteilungen*, 13(4/5): 195–214.

Heller, K. M. 1924c. Philippinische Bockkäfer. *Deutsche entomologische Zeitschrift*, 1924(5): 424–436.

Heller, K. M. 1924d. Neue Philippinische Bock- und Busselkafer aus der Sammlung von Prof. C. F. Baker. *Wiener Entomologische Zeitung*, 41(4–10): 163–180.

Heller, K. M. 1926. Neue, altweltliche Bockkäfer. *Tijdschrift voor Entomologie*, 69: 19–50.

Heyrovsky, L. 1933. *Flabelloprionus* Gen. Nov. (Col. Cerambycidae). *Sbornik Entom. Odd Nar. Musea v Praze*, 11(91): 133–134.

Heyrovsky, L. 1934. Fünfter Beitrag zur Kenntnis der Cerambyciden aus der chinesischen Provinz Sze- Chuan. (Col., Ceramb.). Časopis České Společnosti Entomologické, 31(4): 171–173.

Hoffman, R. L. 1958. The Subspecies of *Typocerus lunatus*, a Cerambycid Beetle (Coleoptera, Cerambycidae). *Proceedings of the Entomological Society of Washington*, 60(5): 217–221.

Hovore, F. T., and A. Santos-Silva. 2004. Notes and description of Neotropical Macrotomini (Coleoptera, Cerambycidae, Prioninae). *Revista Brasileira de Entomologia*, 48(1): 49–54.

Hovore, F. T., and A. Santos-Silva. 2007. Novas espécies de *Cometes* Audinet-Serville, 1828 (Coleoptera, Cerambycidae). *Papeis Avulsos de Zoologia*, 47(5): 75–96.

Howard, L. O. 1929. Eugene Amandus Schwarz 1844–1928. *Annals of the Entomological Society of America*, 22(1): 142–143.

Howard, L. O., H. S. Barber, and A. Busck. 1928 [1929]. Memorial to E. A. Schwarz. *Proceedings of the Entomological Society of Washington*, 30(9): 154–183.

Howden, H. F. 1960. Two new species of *Cyrtinus* LeConte (Coleoptera, Cerambycidae). *The Canadian Entomologist*, 92(3): 173–177.

ICZN (International Commission of Zoological Nomenclature). 1999. International Code of Zoological Nomenclature. London: The International Trust for Zoological Nomenclature. xxix + 306 pp.

Ivie, M. A., and M. L. Schwengel-Regala. 2007. The *Elaphidion* Audinet-Serville of the Puerto Rican Bank: new species, distributions, taxonomic corrections, and a key to species (Coleoptera: Cerambycidae: Elaphidiini). *Zootaxa*, 1503: 55–68.

Joutel, L. H. 1902. A New Species of *Saperda*. *Entomological News*, 13(2): 33–34.

Joutel, L. H. 1911. Miscellaneous notes. *Phymatodes lengi* Joutel. *Journal of the New York Entomological Society*, 19(3): 200.

Julio, C. E. de A., and M. A. Monné. 2005. Revisão do gênero *Sympleurotis* Bates (Coleoptera, Cerambycidae, Lamiinae, Colobotheini). *Revista Brasileira de Entomologia*, 49(3): 392–396.

Kasatkin, D. G., and A. I. Miroshnikov. 2011. [Some notes to corrections and additions to the new Catalog of Palaearctic timber-beetles (Coleoptera, Cerambycidae)]. Available from: http://www.zin.ru/animalia/coleoptera/rus/corcenew.htm [in Russian].

Knull, J. N. 1934. New Coleoptera (Buprestidae and Cerambycidae). *Ohio Journal of Science*, 34(5): 333–336.
Knull, J. N. 1962. Three new species of long-horned beetles from southern Texas with a key tothe genus *Obrium* (Coleoptera, Cerambycidae). *Ohio Journal of Sciences*, 62(2): 105–107.
Kriesche, R. 1920. Neue Batoceriden (Col. Ceramb.). *Archiv für Naturgeschichte*, 84A(5)[1919]: 192–198.
Kriesche, R. 1936. Neue indoaustralische Lamiinen (Col. Longic.). *Entomologische Blätter*, 32(2): 66–69.
Kriesche, R. 1940. Neue malaio-papuanische Lamiinen (Col. Longic.). *Entomologische Blätter*, 36(4): 115–117.
Lane, F. 1958. A new neotropical genus of Ischiolonchini (Coleoptera, Lamiidae). *American Museum Novitates*, 1889: 1–17.
Lane, F. 1959a. Cerambycoidea Neotropica nova V (Coleoptera). *Papéis Avulsos de Zoologia*, 13(21): 251–264.
Lane, F. 1959b. Three new genera of Ibidionini. (Coleoptera, Cerambycidae). *Journal of the New York Entomological Society*, 67: 13–19.
Lane, F. 1959c. Two new species of *Ischnolea* Thomson, 1860 (Coleoptera, Lamiidae). *Studia Entomologica*, 2(1–4): 219–224.
Lane, F. 1964. Novos gêneros e espécies de Anisocerini (Coleoptera, Lamiidae). *Studia Entomologica*, 7: 179–200.
Lane, F. 1966a. Novas especies do genero *Eulachnesia* Bates, 1872 (Coleoptera, Lamiidae). *Papeis Avulsos de Zoologia*, 18(24): 271–275.
Lane, F. 1966b. Novos gêneros e espécies de Hemilophini II (Coleoptera, Lamiidae). *Papéis Avulsos de Zoologia*, 18(26): 281–289.
Lane, F. 1966c. Novas espécies de Torneutini (Coleoptera, Cerambycidae). *Papéis Avulsos de Zoologia*, 19(7): 79–94.
Lane, F. 1970. Cerambycoidea Neotropica nova. VII (Coleoptera). *Studia Entomologica*, 13(1–4): 369–428.
Lane, F. 1972. Cerambycoidea Neotropica nova (Coleoptera). VIII. *Studia Entomologica*, 15(1–4): 352–382.
Lane, F. 1977. Uma nova espécie de *Oncideres* Serville, 1835 (Coleoptera, Lamiidae). *Revista de Agricultura*, 52(4): 233–236.
Lane, F. 1978. Novas espécies do gênero *Polyrhaphis* Serville, 1835 (Coleoptera). *Studia Entomologica*, 20(1–4): 63–74.
Ledesma Arias, J., J. L. Aramayo, J. E. Wappes, S. W. Lingafelter, M. A. Monné, and E. H. Nearns. 2008–2012. Cerambycidae Holotypes of the Museo de Historia Natural Noel Kempff Mercado, Santa Cruz, Bolivia (MNKM). Available from: http://www.cerambycids.com/bolivia/mnkm/
Leng, C. W., and J. Hamilton. 1896. The Lamiinae of North America. *Transactions of the American Entomological Society*, 23: 101–178.
Linell, M. L. 1896. List of Coleoptera Collected on the Tana River and on the Jombene Range, East Africa, by Mr. William Astor Chanler and Lieutenant Ludwig von Höhnel, with Descriptions of new Genera and Species. *Proceedings of the United States National Museum*, 18(1094)[1895]: 687–716.
Linell, M. L. 1897a. Descriptions of new species of North American Coleoptera in the families Cerambycidae and Scarabaeidae. *Proceedings of the United States National Museum*, 19(1113): 393–401.
Linell, M. L. 1897b. On the Insects Collected by Dr. Abbott on the Seychelles, Aldabra, Glorioso and Providence Islands, with Descriptions of Nine New Species of Coleoptera. *Proceedings of the United States National Museum*, 19(1119): 695–706.
Linell, M. L. 1899. On the coleopterous insects of the Galápagos Islands. *Proceedings of the United States National Museum*, 21[1898](1143): 249–268.
Lingafelter, S. W. 2008. Seven New Species of Elaphidiini (Coleoptera: Cerambycidae) from the Dominican Republic with Taxonomic Notes, New Country Records, and a Key to *Elaphidion* Audinet- Serville from Hispaniola. *The Coleopterists Bulletin*, 62(3): 353–379.
Lingafelter, S.W. 2010. Methiini and Oemini (Coleoptera: Cerambycidae: Cerambycinae) of Hispaniola. *The Coleopterists Bulletin*, 64(3): 265–269.
Lingafelter, S.W. 2011. New myrmecomorphous longhorned beetles from Haiti and the Dominican Republic with a key to Anaglyptini and Tillomorphini of Hispaniola (Coleoptera, Cerambycidae, Cerambycinae). *ZooKeys*, 106: 55–75.
Lingafelter, S.W. 2013. Hispaniolan Hemilophini (Coleoptera, Cerambycidae, Lamiinae). *ZooKeys*, 258: 53–83.

Lingafelter, S. W., and J. A. Chemsak. 2002. A new species of *Enaphalodes* Haldeman from Florida (Coleoptera: Cerambycidae) with review of genus, synonymies, and key to species. *The Coleopterists Bulletin*, 56(4): 569–581.

Lingafelter, S. W., A. Garzón-Moreno, and E. H. Nearns. 2013. Primary types of Chinese longhorned woodboring beetles (Coleoptera: Cerambycidae and Disteniidae) of the Smithsonian Institution. In *In Memory of Mr. Wenhsin Lin*, ed. M.-Y. Lin and C. Chen, pp. 117–146. Taiwan: Formosa Ecological Company.

Lingafelter, S. W., and M. E. Ivie. 2013. A new genus of mimetic longhorned beetle from St. Lucia, Lesser Antilles (Coleoptera: Cerambycidae: Rhinotragini). *Proceedings of the Entomological Society of Washington*, 115(3): 268–273.

Lingafelter, S. W., and C. J. Micheli. 2004. New species of Cerambycidae (Coleoptera) from Puerto Rico with records and notes for other species. *Journal of the New York Entomological Society*, 112(1): 37–55.

Lingafelter, S. W., and C. J. Micheli. 2009. The genus *Leptostylopsis* of Hispaniola (Coleoptera: Cerambycidae: Acanthocinini). *ZooKeys*, 17: 1–55.

Lingafelter, S. W., M. A. Monné, and E. H. Nearns. 2004–2014. Online Image Database of Cerambycidae Primary Types of the Smithsonian Institution. http://smithsoniancerambycidae.com

Lingafelter, S. W., and E. H. Nearns. 2007. Five new species of longhorned beetles (Coleoptera: Cerambycidae) from Dominican Republic in genera *Ataxia* Haldeman, *Atimiola* Bates, *Drycothaea* Thomson, *Eburia* Lepeletier & Audinet-Serville, and *Hormathus* Gahan. *The Coleopterists Bulletin*, 61(2): 177–191.

Lingafelter, S. W., and E. H. Nearns. 2013. Elucidating Article 45.6 of the International Code of Zoological Nomenclature: A dichotomous key for the determination of subspecific or infrasubspecific rank. *Zootaxa*, 3709(6): 597–600.

Lingafelter, S. W., and J. E. Wappes. 2012. A new species of *Trichoxys* Chevrolat (Cerambycidae: Cerambycinae: Clytini) from Mexico, with a key to known species. *The Pan-Pacific Entomologist*, 88(2): 154–162.

Lingafelter, S. W., and N. E. Woodley. 2007. A new species of *Derancistrus* Audinet-Serville (Coleoptera: Cerambycidae: Prioninae) from the Dominican Republic with notes on other species of Prioninae from Hispaniola. *The Coleopterists Bulletin*, 61(2): 165–175.

Linsley, E. G. 1935a. Notes and descriptions of new or little known Neotropical Sphaerionini (Coleoptera, Cerambycidae). *Revista de Entomologia*, 5(2): 139–149.

Linsley, E. G. 1935b. Notes and descriptions of new or little known neotropical Ibidionini (Col., Cerambycidae). *Revista de Entomologia*, 5(4): 479–486.

Linsley, E. G. 1957a. Descriptive and synonymical notes on some North American Cerambycidae (Coleoptera). *American Museum Novitates*, 1828: 1–21.

Linsley, E. G. 1957b. New longicorn beetles from Texas (Coleoptera: Cerambycidae). *Entomological News*, 68(6): 159–161.

Linsley, E. G. 1957c. Some new genera and species of North American Cerambycidae (Coleoptera). *The Canadian Entomologist*, 89(6): 283–287.

Linsley, E. G. 1970. New genera and species in the lepturine complex related to *Euryptera* and *Choriolaus* (Coleoptera: Cerambycidae). *The Pan-Pacific Entomologist*, 46(2):128–141.

Löbl, I., and A. Smetana. 2009. *Catalogue of Palaearctic Coleoptera*, Vol. 6. Stenstrup, Denmark: Apollo Books. 373pp.

Löbl, I., and A. Smetana. 2011. Errata for Vol. 6, pp. 35–61. In *Catalogue of Palaearctic Coleoptera*, Vol. 7, ed. I. Löbl and A. Smetana. Stenstrup, Denmark: Apollo Books. 373pp.

Mallis, A. 1971. *American Entomologists*. New Brunswick, NJ: Rutgers University Press. 549 pp.

Mandl, K. 1978 [1979]. Friedrich F. Tippmann, Ein Nachruf. *Zeitschrift der Arbeitsgemeinschaft Arbeitsgemeinschaft* Österreichischer *Entomologen*, 30(3/4): 137–140.

Martins, U. R. 1959. Ibidionini (Coleoptera, Cerambycinae) IV. Comentários, redescrição e espécies novas do gênero *Hexoplon* Thomson. *Papéis Avulsos de Zoologia*, 13(30): 339–349.

Martins, U. R. 1960a. Ibidionini (Coleoptera, Cerambycinae) VI. Gêneros de corpo parcialmente sericeo-piloso. *Papéis Avulsos de Zoologia*, 14(3): 17–29.

Martins, U. R. 1960b. Ibidionini (Coleoptera, Cerambycinae) IX. Novas espécies do gênero *Octoplon*. *Papéis Avulsos de Zoologia*, 14(10): 85–92.

Martins, U. R. 1960c. Ibidionini (Coleoptera, Cerambycinae) X. Novas espécies do gênero *Ibidion* Serville. *Papéis Avulsos de Zoologia*, 14(11): 93–98.

Martins, U.R. 1960d. Ibidionini (Coleoptera, Cerambycinae) XI. Gêneros de pronoto enrugado. *Papéis Avulsos de Zoologia*, 14(13): 105–113.

Martins, U. R. 1960e. Ibidionini (Coleoptera, Cerambycinae) XIV. Novas espécies dos gêneros *Heterachthes* e *Octoplon*. *Papéis Avulsos de Zoologia*, 14(19): 173–182.

Martins, U. R. 1962a. Ibidionini (Coleoptera, Cerambycinae) XV. Novas espécies, notas sinonímicas, redescrições. *Papéis Avulsos de Zoologia*, 14(30): 267–310.

Martins, U. R. 1962b. Ibidionini (Coleoptera, Cerambycidae) XX. Descricoes, redescricoes e novas combinacoes. *Papeis Avulsos de Zoologia*, 15(10): 91–110.

Martins, U. R. 1965. Ibidionini (Coleoptera, Cerambycinae). XXIV Divisão do gênero *Compsa* Perty, 1832. *Papéis Avulsos de Zoologia*, 17(9): 89–107.

Martins, U. R. 1968. Monografia da tribo Ibidionini (Coleoptera, Cerambycinae). Parte II. *Arquivos de Zoologia*, 16(2): 321–630.

Martins, U. R. 1969. Monografia da tribo Ibidionini (Coleoptera, Cerambycinae). Parte III. *Arquivos de Zoologia*, 16(3): 631–877.

Martins, U. R. 1970a. Monografia da tribo Ibidionini (Coleoptera, Cerambycinae). Parte IV. *Arquivos de Zoologia*, 16(4): 879–1149.

Martins, U. R. 1970b. Monografia da tribo Ibidionini (Coleoptera, Cerambycinae). Parte V. *Arquivos de Zoologia*, 16(5): 1151–1342.

Martins, U. R. 1976. Sistematica e evolucao da tribo Piezocerini (Coleoptera, Cerambycidae). *Arquivos de Zoologia*, 27(3–4): 165–370.

Martins, U. R. 1977. Transference of the genera *Oemida* Gahan, *Comusia* Thomson and *Hypomares* Thomson from the Methiini to Orbriini (Coleoptera, Cerambycidae). *Papéis Avulsos de Zoologia*, 31(6): 103–118.

Martins, U. R., J. A. Chemsak, and E. G. Linsley. 1966. A generic revision of the tribe Methiini in the Western Hemisphere. *Arquivos de Zoologia*, 14(3): 197–221.

Martins, U. R., and M. H. M. Galileo. 1990. Onciderini (Coleoptera, Cerambycidae, Lamiinae): Sinonímias, novos táxons, chaves e notas. *Papéis Avulsos de Zoologia*, 37(4): 53–95.

Martins, U. R., and M. H. M. Galileo. 1991. O genero *Tyrintha* Bates, 1866 e generos afins (Coleoptera, Cerambycidae, Lamiinae, Hemilophini). *Revista Brasileira de Entomologia*, 35(4): 813–824.

Martins, U. R., and M. H. M. Galileo. 1996. Transferencia de generos de Aerenicini para Hemilophini, novos taxons e chave para as especies de *Apagomerina* Gilmour, 1962 (Coleoptera, Cerambcidae, Lamiinae). *Revista Brasileira de Entomologia*, 40(1): 21–24.

Martins, U. R., and M. H. M. Galileo. 1998. Revisão da Tribo Aerenicini Lacordaire, 1872 (Coleoptera, Cerambycidae, Lamiinae). *Arquivos de Zoologia*, 35(1): 1–133.

Martins, U. R., and M. H. M. Galileo. 2001. Duas espécies novas do genero *Acakyra* Martins & Galileo, 1996 (Coleoptera, Cerambycidae, Lamiinae, Acanthoderini). *Revista Brasileira de Entomologia*, 45(2): 103–105.

Martins, U. R., and M. H. M. Galileo. 2004. Sobre Hemilophini (Coleoptera, Cerambycidae, Lamiinae) da Região Neotropical: espécies novas e novos registros. *Revista Brasileira de Zoologia*, 21(3): 535–541.

Martins, U. R., and M. H. M. Galileo. 2005a. Novas espécies de *Adetus* (Coleoptera, Cerambycidae, Lamiinae, Apomecynini). *Iheringia (Série Zoologia)*, 95(1): 25–28.

Martins, U. R., and M. H. M. Galileo. 2005b. Novos táxons e notas sobre Cerambycinae (Coleoptera, Cerambycidae) da Região Neotropical. *Revista Brasileira de Zoologia*, 22(3): 764–770.

Martins, U. R., and M. H. M. Galileo. 2005c. Contribuição para o conhecimento dos Desmiphorini (Coleoptera, Cerambycidae, Lamiinae) da Região Neotropical. *Papéis Avulsos de Zoologia*, 45(19): 243–248.

Martins, U. R., and M. H. M. Galileo. 2006a. Calliini (Coleoptera, Cerambycidae, Lamiinae): descrições, homonímia, novo registro e chave para as espécies de *Graminea* Thomson, 1864. *Papéis Avulsos de Zoologia*, 46(18): 211–218.

Martins, U. R., and M. H. M. Galileo. 2006b. Novos táxons de Cerambycinae (Coleoptera, Cerambycidae) da Região Neotropical. *Iheringia (Série Zoologia)*, 96(2): 173–178.

Martins, U. R., and M. H. M. Galileo. 2007a. Genero *Hippopsis* (Coleoptera, Cerambycidae): chave para as especies, sinonimia e descricao de especies novas. *Revista Brasileira de Entomologia*, 50(4)[2006]: 475–487.

Martins, U. R., and M. H. M. Galileo. 2007b. Notes and New Species of Onciderini (Coleoptera, Cerambycidae, Lamiinae). *Papéis Avulsos de Zoologia*, 47(9): 127–135.

Martins, U. R., and M. H. M. Galileo. 2007c. New Taxa in Hemilophini (Coleoptera, Cerambycidae, Lamiinae). *Revista Brasileira de Zoologia*, 24(4): 1145–1149.

Martins, U. R., and M. H. M. Galileo. 2008a. Clytini neotropicais II (Coleoptera, Cerambycidae, Cerambycinae). *Revista Brasileira de Entomologia*, 52(2): 272–276.

Martins, U. R., and M. H. M. Galileo. 2008b. Novos taxons e novo registro em Hemilophini (Coleoptera, Cerambycidae, Lamiinae). *Revista Brasileira de Zoologia*, 25(3): 547–554.

Martins, U. R., and M. H. M. Galileo. 2008c. Novos Taxons de Apomecynini (Coleoptera, Cerambycidae, Lamiinae). *Papéis Avulsos Zoologia*, 48(24): 275–280.

Martins, U. R., and M. H. M. Galileo. 2009. Nota e espécies novas de *Cyrtinus* LeConte (Coleoptera, Cerambycidae, Lamiinae, Cyrtinini) da Região Neotropical. *Revsta Brasileira de Entomologia*, 53(1): 65–68

Martins, U. R., and M. H. M. Galileo. 2010. Novas espécies, nova sinonímia e novo registro em Cerambycidae (Coleoptera) do México e da América do Sul. *Papéis Avulsos de Zoologia*, 50(7): 83–90.

Martins, U. R., and M. H. M. Galileo. 2011. Novos táxons de Hemilophini (Coleoptera, Cerambycidae, Lamiinae) das Américas. *Iheringia (Série Zoologia)*, 100(4) [2010]: 356–360.

Martins, U. R., and M. A. Monné. 1974. Contribuição ao Conhecimento dos Acanthocinini, II (Coleoptera, Cerambycidae: Lamiinae). *Studia Entomologica*, 17(1–4): 317–332.

Martins, U. R., M. A. Monné, M. L. Monné, S. W. Lingafelter, C. J. Micheli, and E. H. Nearns. 2008–2012. Cerambycidae Holotypes of the Museu de Zoologia Universidade de São Paulo (MZSP). Available from: http://www.cerambycids.com/brazil/mzsp/

Melzer, J. 1919. Os longicórneos brazileiros da subfamília Prioninae tomando em consideração particular as espécies do Estado de São Paulo. *Revista do Museu Paulista*, 11: 1–208.

Melzer, J. 1932. Vinte espécies novas de cerambycideos neotrópicos, principalmente do Brasil. *Revista de Entomologia*, 2(2): 216–238.

Melzer, J. 1933. Cerambycideos neotrópicos, novos ou pouco conhecidos. *Revista de Entomologia*, 3(3): 367–382.

Melzer, J. 1934. Novos subsidios para o conhecimento dos cerambycideos neotrópicos (Col.). *Revista de Entomologia*, 4(1): 70–110.

Melzer, J. 1935a. Longicórneos Americanos, principalmente do Brasil, novos ou pouco conhecidos (Coleoptera, Cerambycidae) IV. *Arquivos do Instituto Biologico*, 5[1934]: 213–244.

Melzer, J. 1935b. Novos cerambycideos do Brasil, da Argentina e de Costa Rica. *Archivos do Instituto de Biologia Vegetal*, 2(2): 173–205.

Micheli, C. J., and E. H. Nearns. 2005. Two new species of *Plectromerus* Haldeman (Coleoptera: Cerambycidae) from the West Indies. *Zootaxa*, 1028: 23–36.

Micheli, J. 1983. *Curiosa dominicana*, a new genus and species of Curiini (Coleoptera: Cerambycidae) from Dominican Republic. *The Coleopterists Bulletin*, 37(3): 261–266.

Micheli, J. 2003. New longhorns beetles from Puerto Rico (West Indies) (Coleoptera: Cerambycidae). *The Coleopterists Bulletin*, 57(2): 191–204.

Micheli, J., and C. J. Micheli. 2004. Five new Puerto Rican longhorn beetles and other notes on West Indian Acanthocinini (Coleoptera, Cerambycidae, Lamiinae). *Journal of the New York Entomological Society*, 112(1): 18–36.

Miller, D. R., D. C. Ferguson, A. S. Menke, D. A. Nickle, R. L. Smiley, and R. E. White. 1998. National Museum of Natural History Entomological Collection. In *Systematic Collections of the Agricultural Research Service*, Miscellaneous Publication No. 1343, ed. J. R. Lichtenfels, J. H. Kirkbride, Jr., and D. J. Chitwood, pp. 28–42. Washington, DC: USDA. 77 pp.

Miroshnikov, A. I. 2011a. The Longicorn beetles (Cerambycidae) in «Catalogue of Palaearctic Coleoptera. Stenstrup, 2010». Remarks and additions. Available from: http://www.zin.ru/animalia/coleoptera/rus/corcemir.htm [in Russian].

Miroshnikov, A. I. 2011b. The longicorn beetles (Cerambycidae) in "Catalogue of Palaearctic Coleoptera. Stenstrup, 2010." Remarks and additions. Entomologia Kubanica. Supplement No. 1. Krasnodar. 113pp [in Russian with English abstract].

Miroshnikov, A. I. 2011c. [Notes to «Catalogue of Palaearctic Coleoptera. Stenstrup, 2010». 2] [in Russian]. Available from: http://www.zin.ru/animalia/coleoptera/rus/mirosh13.htm

Miroshnikov, A. I. 2011d. [Addition to the future article preparing for publication] [in Russian]. Available from: http://www.zin.ru/animalia/coleoptera/rus/mirosh13.htm

Monné, M. A. 2005a. Catalogue of the Cerambycidae (Coleoptera) of the Neotropical Region. Part I. Subfamily Cerambycinae. *Zootaxa*, 946: 1–765.

Monné, M. A. 2005b. Catalogue of the Cerambycidae (Coleoptera) of the Neotropical Region. Part II. Subfamily Lamiinae. *Zootaxa*, 1023: 1–759.

Monné, M. A. 2006. Catalogue of the Cerambycidae (Coleoptera) of the Neotropical Region. Part III. Subfamilies Parandrinae, Prioninae, Anoplodermatinae, Aseminae, Spondylidinae, Lepturinae, Oxypeltinae, and addenda to the Cerambycinae and Lamiinae. *Zootaxa*, 1212: 1–244.

Monné, M. A. 2008. Gênero *Colobothea* Lepeletier & Audinet-Serville: novas espécies e sinonimia (Coleoptera, Cerambycidae, Lamiinae). *Revista Brasileira de Entomologia*, 49(2): 245–248.

Monné, M. A. 2011. Revision of the genus *Stenolis* Bates (Coleoptera, Cerambycidae, Lamiinae). *Zootaxa*, 2831: 39–53.

Monné, M. A. 2013a. Catalogue of the Cerambycidae (Coleoptera) of the Neotropical Region. Part I. Subfamily Cerambycinae. Electronic-only update to Monné, 2005a.

Monné, M. A. 2013b. Catalogue of the Cerambycidae (Coleoptera) of the Neotropical Region. Part II. Subfamily Lamiinae. Electronic-only update to Monné, 2005b.

Monné, M. A. 2013c. Catalogue of the Cerambycidae (Coleoptera) of the Neotropical Region. Part III. Subfamilies Lepturinae, Necydalinae, Parandrinae, Prioninae, Spondylidinae and Families Oxypeltidae, Vesperidae and Disteniidae. Electronic-only update to Monné, 2006.

Monné, M. A., and U. R. Martins. 1976. Contribuição ao conhecimento dos Acanthocinini (Coleoptera, Cerambycidae, Lamiinae). *Papéis Avulsos de Zoologia*, 30(4): 43–98.

Monné, M. A., and M. L. Monné. 2007. Novas espécies neotropicais de Acanthocinini (Coleoptera, Cerambycidae). *Revista Brasileira de Zoologia*, 24(4): 1038–1042.

Monné, M.A., and M. L. Monné. 2008. Novo genero, sinonimia e novas combinicoes em Colobotheini (Coleoptera: Cerambycidae). *Revista Brasileira de Zoologia*, 25(3): 502–506.

Monné, M. A., and M. L. Monné. 2009. Synopsis of the genus *Sangaris* Dalman in South America (Coleoptera, Cerambycidae), with description of two new species and new combinations. *Zootaxa*, 2230: 29–41.

Monné, M. A., and M. L. Monné. 2011. A new genus and species of Neotropical Anacolini and new distributional records (Insecta, Coleoptera, Cerambycidae). *Journal of Natural History*, 45(29–30): 1859–1866.

Monné, M. A., M. L. Monné, S. W. Lingafelter, C. J. Micheli, and E. H. Nearns. 2008–2012. Cerambycidae Holotypes of the Museu Nacional Rio de Janeiro (MNRJ). Available from: http://www.cerambycids.com/brazil/mnrj/

Monné, M. A., and M. L. Monné. 2009. Three new species and synonymy in *Nealcidion* Monné (Coleoptera, Cerambycidae, Lamiinae). *Revista Brasileira de Entomologia*, 53(3): 361–364.

Monné, M. A., and G. L. Tavakilian. 2011. New species of *Nyssodrysternum* Gilmour (Coleoptera: Cerambycidae, Lamiinae) from Central and South America. *Papeis Avulsos Zoologia*, 51(29): 453–464.

Monné, M. L. 2007. Sinopse do gênero *Eucharassus* Bates (Coleoptera, Cerambycidae). *Revista Brasileira de Entomologia*, 51(2): 205–209.

Morati, J., and M. Huet. 2004. Revision du Genre *Pachyteria*. *Magellanes. Collection Systematique*, 9: 1–148.

Moyses, E., and M. H. M. Galileo. 2009. Duas novas espécies de *Mauesia* (Cerambycidae, Lamiinae, Mauesini) e chave para identificação das espécies do gênero. *Papéis Avulsos de Zoologia*, 49(37): 497–501.

Napp, D. S., and M. A. Monné. 2009. New Neotropical genera and species of Dryobiini and Rhopalophorini (Coleoptera: Cerambycidae: Cerambycinae). *Zootaxa*, 2137: 29–34.

Nearns, E. H., and M. A. Branham. 2008. Revision and phylogeny of the tribes Curiini LeConte and Plectromerini Nearns & Branham, new tribe (Coleoptera: Cerambycidae: Cerambycinae). *Memoirs of the American Entomological Society*, 47: 1–117.

Nearns, E. H., and A. M. Ray. 2006. A new species of *Curius* Newman (Coleoptera: Cerambycidae) from Venezuela with notes on sexual dimorphism within the genus. *Zootaxa*, 1256: 49–57.

Nearns, E. H., and W. E. Steiner, Jr. 2006. A new species of *Plectromerus* Haldeman (Coleoptera: Cerambycidae) from Navassa Island, Greater Antilles. *Zootaxa*, 1163: 61–68.

Ohbayashi, K. 1948. Studies of Longicornia. *The Entomological Review of Japan*, 1(1): 12–14.

Olivier, G. A. 1795. Entomologie, ou Histoire Naturelle des Insectes, avec leurs Caractères Génériques et Spécifiques, leur Description, leur Synonymie et leur Figure Illuminée. *Coléoptères*, 4: 519 pp.

Özdikmen, H. 2011. Additions and corrections to the new Catalogue of Palaearctic Cerambycidae (Coleoptera), ed. I. Löbl and A. Smetana (2010) for Turkish taxa. *Munis Entomology & Zoology*, 6(2): 686–734.

Quentin, R. M., and A. Villiers. 1971. Revision des Psebiini (Col. Cerambycidae, Cerambycinae). *Annales de la Société Entomologique de France*, 7(1): 3–38.

Radovsky, F. J. 1983. J. Linsley Gressitt (1914–1982). *International Journal of Entomology*, 25(1): 1–10.

Rice, M. E., and R. F. Morris. 1992. *Megacyllene comanchei*, a new species of longhorned beetle (Cerambycidae: Coleoptera) from Texas. *Journal of the Kansas Entomological Society*, 65(2): 200–202.

Russo, G. 1930. Contributo alla conoscenza dei Cerambycidae (Coleoptera) della Rep. Dominicana (Antille) com descrizione di una nuova specie. *Bollettino dei Laboratorio di Zoologia Generale e Agraria del Istituto Superiore Agrario di Portici*, 24(14): 140–147.

Sabrosky, C. W. 1964. Taxonomic Entomology in the U.S. Department of Agriculture. *Bulletin of the Entomological Society of America*, 10: 211–220.

Santos-Silva, A. 2007. Revisão do gênero *Derobrachus* Audinet-Serville, 1832 (Coleoptera, Cerambycidae, Prioninae). *Arquivos de Zoologia*, 38(1): 1–94.

Santos-Silva, A., D. Heffern, and K. Matsuda. 2010. Revision of Hawaiian, Australasian, Oriental, and Japanese Parandrinae (Coleoptera, Cerambycidae). *Insecta Mundi*, 0130: 1–119.

Santos-Silva, A., and F. T. Hovore. 2007a. Espécies de *Distenia* (*Basivallis*) Santos-Silva & Hovore (Coleoptera, Cerambycidae, Disteniinae). *Les Cahiers Magellanes*, 71: 1–16.

Santos-Silva, A., and F. T. Hovore. 2007b. Species of *Novantinoe* Santos-Silva & Hovore (Coleoptera, Cerambycidae, Disteniinae). *Lambillionea*, 107(4): 573–610.

Santos-Silva, A., and F. T. Hovore. 2008a. Species of *Elytrimitatrix* (*Grossifemora*) Santos-Silva & Hovore. 2. New Species and key to the subgenus (Coleoptera, Cerambycidae, Disteniinae). *Les Cahiers Magellanes*, 75: 1–25.

Santos-Silva, A., and F. T. Hovore. 2008b. Novas espécies de Disteniinae americanas (Coleoptera, Cerambycidae). *Les Cahiers Magellanes*, 81: 1–16.

Santos-Silva, A., and H. J. Lezama. 2010. Contribuição para o conhecimento dos Parandrini da Costa Rica e do Panamá (Coleoptera, Cerambycidae, Parandrinae). *Revista Brasileira de Entomologia*, 54(3): 399–405.

Santos-Silva, A., and U. R. Martins. 2004. Notas e descrições em Disteniinae (Coleoptera, Cerambycidae). *Revista Brasileira de Zoologia*, 21(1): 145–152.

Santos-Silva, A., and G. Tavakilian. 2009. Demembrement du genre *Cometes* Lepeletier & Audinet-Serville, 1828 1st partie (Coleoptera, Cerambycidae, Distenninae). *Les Cahiers Magellanes*, 102: 1–24.

Schaeffer, C. F. A. 1904. Class I, Hexapoda. Order II, Coleoptera. New genera and species of Coleoptera. *Journal of the New York Entomological Society*, 12(4): 197–236.

Schaeffer, C. F. A. 1905a. Additions to the Coleoptera of the United States with notes on some known species. *Science Bulletin of the Museum of the Brooklyn Institute of Arts and Sciences*, 1(6): 123–140.

Schaeffer, C. F. A. 1905b. Some additional new genera and species of Coleoptera found within the limit of the United States. *Science Bulletin of the Museum of the Brooklyn Institute of Arts and Sciences*, 1(7): 141–179.

Schaeffer, C. F. A. 1908. List of the longicorn Coleoptera collected on the Museum expeditions to Brownsville, Texas and the Huachuca Mts., Arizona, with descriptions of new genera and species and notes on known species. *Science Bulletin of the Museum of the Brooklyn Institute of Arts and Sciences*, 1(12): 325–352.

Schaeffer, C. F. A. 1909. New Coleoptera chiefly from Arizona. *Science Bulletin of the Museum of the Brooklyn Institute of Arts and Sciences*, 1(16): 375–386.

Schaeffer, C. F. A. 1911. New Coleoptera and miscellaneous notes. *Journal of the New York Entomological Society*, 19(2): 113–126.

Schaeffer, C. F. A. 1917. On *Merium* and some blue *Callidium* (Col.). *Journal of the New York Entomological Society*, 25(3): 183–187.

Schaeffer, C. F. A. 1929. The North American species of Parandra (Col. Cerambycidae). *Bulletin of the Brooklyn Entomological Society*, 24(1): 38–40.

Schiefer, T. L. 2001. A new species of *Astylopsis* Casey (Coleoptera: Cerambycidae: Acanthocinini) from the southeastern United States. *The Coleopterists Bulletin*, 54(4): 533–539.

Schultze, W.C.M. 1922. X. Beitrag zur Coleoptera Fauna der Philippinen. *The Philippine Journal of Science*, 21(6): 569–596.

Schwarz, E. A. 1891. Labeling specimens. *Proceedings of the Entomological Society of Washington*, 2(1): 46–50.

Schwarz, E. A. 1926. Condition of the coleopterous collection of the National Museum in 1906. *Proceedings of the Entomological Society of Washington*, 28(4): 71–86.

Schwarz, E. A., and W. M. Mann. 1925. Colonel Thomas Lincoln Casey. *Proceedings of the Entomological Society of Washington*, 27(3): 42–43.

Schwarzer, B. 1925. Sauters Formosa-Ausbeute (Cerambycidae Col.). (Subfamilie Cerambycinae). *Entomologische Blätter*, 21(1): 20–30.

Schwarzer, B. 1929. Neue Cerambyciden des Senckenberg-Museums (Ins., Col.). *Senckenbergiana*, 11(5–6): 353–379.

Sherman, J. D., Jr. 1929. Letters of E. A. Schwarz. *Journal of the New York Entomological Society*, 37: 181–392.

Shimer, H. 1868. Notes on insects bred from the Prickly Ash (*Xanthoxylum americanum*). *Transactions of the American Entomological Society*, 2: vii-viii.

Skiles, D. D. 1978. Taxonomy and description of two prionine Cerambycidae from southern Arizona: a new species of *Stenodontes* and new status for *Neomallodon arizonicus* (Coleoptera). *Proceedings of the Entomological Society of Washington*, 80(3): 407–423.

Solís, Á., S. W. Lingafelter, I. P. Swift, L. G. Bezark, and E. H. Nearns. 2008–2012. Cerambycidae Holotypes of the Instituto Nacional de Biodiversidad (INBio). Available from: http://www.cerambycids.com/costarica/inbio/

Spilman, T. J. 1989. Vignettes of the Presidents of the Entomological Society of America, 1889–1989. *Bulletin of the Entomological Society of America*, 35(3): 33–65.

Swaine, J. M., and R. Hopping. 1928. The Lepturini of America north of Mexico. Part I. *Bulletin of the Canadian National Museum* (Biological series 14) 52: 1–97.

Swift, I. P. 2007. A new species of *Xylotrechus* (Coleoptera: Cerambycidae: Cerambycinae: Clytini) from Southern California, U.S.A. *The Coleopterists Bulletin*, 61(2): 287–291.

Swift, I. P., and A. M. Ray. 2008. A review of the genus *Tragidion* Audinet-Serville, 1834 (Coleoptera: Cerambycidae: Cerambycinae: Trachyderini). *Zootaxa*, 1892: 1–25.

Tavakilian, G., and H. Chevillotte. 2013. Titan: base de données internationales sur les Cerambycidae ou Longicornes. Version 3.0. Available from: http://lully.snv.jussieu.fr/titan/ (accessed on 9/16/2013).

Tippmann, F. F. 1952. Studien uber *Plagionotus detritus* L. und *arcuatus* L. (Coleoptera: Cerambycidae, Subfam. Cerambycinae, Tribus Clytini). *Sonderabdruck aus den Mitteilungen der Munchner Entomologischen Gesellscaft*, 42: 134–147.

Tippmann, F. F. 1939. Eine neue *Cyclopeplus*-Art aus Peru (Coleoptera, Cerambycidae, Lamiinae). Časopis České *Společnosti Entomologické*, 36(2): 77.

Tippmann, F. F. 1946a. Description dun Cérambycide nouveau *Evodinus breiti* sp. n. de la chaîne de montagne Sajan de Sibérie. *Sborník Entomologickébo Oddeliní Národního Musea v Praze*, 24 (316): 125–127.

Tippmann, F. F. 1946b. Studien Uber *Xixuthrus heros* Heer und Seine Nachsten Verwandten, sowie Beschreibung Einer Neuen Gigantischen Art von den Fidschi-Inseln. *Xixuthrus heyrobskyi* nova species. Privately published [1945]: 1–12.

Tippmann, F. F. 1949. Ein neuer, überaus massiger *Massicus* Pasc. Aus Birma. *Massicus von Roseni* sp. n. (Col. Cerambycidae, subfam. Cerambycinae s. str.). *Mitteilungen der* Münchener *entomologischen Gessellschaft*, 35–39: 291–293.

Tippmann, F. F. 1951. Beitrage zur Kenntnis der Cerambyciden (Col.). *Mitteilungen der* Münchener *Entomologische Gesellschaft*, 41: 291–328.

Tippmann, F. F. 1952. Eine neue *Acanthocinus* Steph.-Form aus Dalmatien: *Acanthocinus griseus* Fabr. Navaki, subsp. Nova (Coleoptera: Cerambycidae, Subfam. Lamiinae). *Sonderabdruck aus den Mitteilungen der Munchner Entomologischen Gesellschaft*, 42: 148–154.

Tippmann, F. F. 1953a. Eine neue Batocerini-Gattung und-art Aus Mittel-Celebes (Coleoptera, Cerambycidae, Lamiinae). *Tijdschrift voor Entomologie*, 96(1–2): 101–104.

Tippmann, F. F. 1953b. Lamiinen Mit Interferenz-Farbenspiel (Coleoptera: Cerambycidae, Lamiinae). *Entomologische Blätter für Biologie und Systematik der Käfer*, 49: 148–153.

Tippmann, F. F. 1953c. Studien über neotropische Longicornier - I (Coleoptera: Cerambycidae). *Dusenia*, 4(4–5): 181–228.

Tippmann, F. F. 1953d. Studien über neotropische Longicornier - II (Coleoptera: Cerambycidae). *Dusenia*, 4(5–6): 313–363.

Tippmann, F. F. 1955a. Zur Kenntnis der Cerambycidenfauna Fukiens (Sud-Ost-China). *Koleopterologische Rundschau*, 33(1–6): 88–137.

Tippmann, F. F. 1955b. Eine neue morphologisch einmalige Lamiine aus Ost-Peru (Coleoptera, Cerambycidae, subfam. Lamiinae). *Entomologische Blätter für Biologie und Systematik der Käfer*, 51: 8–14.

Tippmann, F. F. 1956. Uber einige, vorwiegend palaearktische Cerambyciden und Beschreibung neuer Formen. *Bollettino del Laboratorio di Zoologia Generale e Agraria della facolta agraria in Portici*, 33: 473–492.

Tippmann, F. F. 1958. Dorcadion-Belustigungen am Leitha-Gebrige (Austria inferior). *Mitteilungen der Münchener entomologischen Gessellschaft*, 48: 130–165.

Tippmann, F. F. 1960. Studien über neotropische Longicornier - III (Coleoptera, Cerambycidae). *Koleopterologische Rundschau*, 37/38 [1959/60] (1–6): 82–218.

Toledo, V. H., and F. T. Hovore. 2005. Notes on the genus *Lagocheirus* Dejean: records and descriptions (Coleoptera: Cerambycidae: Lamiinae: Acanthocinini). *Zootaxa*, 1021: 29–36.

Tyson, W. H. 1973. The *Spalacopsis* of the West Indies and America, north of Mexico (Coleoptera, Cerambycidae). *The Coleopterists Bulletin*, 27(3): 117–137.

Villiers, A. 1980. Coléoptères Cerambycidae des Petites Antilles. Nouveaux genres, nouvelles espèces. Désignation de types (Troisième note). *Revue Française d'Entomologie (N.S.)*, 2(2): 86–98.

Wappes, J. E., and S. W. Lingafelter. 2005. The genus *Canidia* Thomson, 1857 (Coleoptera: Cerambycidae, Lamiinae, Acanthocinini). *Zootaxa*, 927: 1–27.

Webb, J. L. 1905. Description of a new species of *Desmocerus* with a synoptic of the genus. *Proceedings of the Entomological Society of Washington*, 7(2–3): 104–105.

Webb, J. L. 1908. A new species of *Leptura*. *Proceedings of the Entomological Society of Washington*, 9(1–4): 41.

Wickham, H. F. 1905. New species of Coleoptera from the Western United States. *The Canadian Entomologist*, 37 (5): 165–171.

Wickham, H. F. 1913. Fossil Coleoptera from Florissant in the United States National Museum. *Proceedings of the United States National Museum*, 45(1982): 283–303.

Wickham, H. F. 1917. New species of fossil beetles from Florissant, Colorado. *Proceedings of the United States National Museum*, 52(2189): 463–472.

Yanega, D. 1996. Field Guide to Northeastern Longhorned Beetles (Coleoptera: Cerambycidae). Illinois Natural History Survey Manual 6. Champaign: Illinois Natural History Survey. 184 pp.

About the Authors

Steven W. Lingafelter has been a Research Entomologist with the Systematic Entomology Laboratory, United States Department of Agriculture, based at the Smithsonian Institution's National Museum of Natural History since 1996. He received his Bachelor and Master of Science Degrees in Biology at Midwestern State University (Wichita Falls, Texas) in 1989 and 1991, respectively, advised by Norman Horner. He received his Doctorate in Entomology advised by the late Steve Ashe at the University of Kansas (Lawrence, Kansas) in 1996. He has published over 60 papers and 4 books on beetle systematics and taxonomy, with an emphasis on Neotropical Cerambycidae.

Eugenio H. Nearns is a United States Department of Agriculture Postdoctoral Researcher at the Smithsonian Institution's National Museum of Natural History. He received his Bachelor of Fine Arts and Master of Science degrees from the University of Florida (Gainesville, Florida) in 1996 and 2006, respectively. He received his Doctorate in Biology from the University of New Mexico (Albuquerque, New Mexico) in 2013. He has published 33 papers on beetle systematics and taxonomy, with an emphasis on Neotropical Cerambycidae. He has developed and maintained many websites that can be accessed from http://cerambycids.com/.

Gérard Luc Tavakilian, after studying biochemistry at Paris University, was employed as a Research Entomologist by the Institut de recherche pour le développement (IRD; a French government institute for researchers abroad). He spent four years in the Ivory Coast and 19 years in French Guiana, where he primarily studied Cerambycidae in relation with their host trees. He initiated a database of American species of Cerambycidae in 1996 and decided to enlarge his project to worldwide fauna in 2002 (supported by grants from the European GBIF program 4D for Life). He retired in 2012 but remains active in Entomology (as Attaché au Laboratoire d'Entomologie du Muséum d'Histoire Naturelle de Paris) and continues to publish about French Guiana's fauna and update the database TITAN (http://lully.snv.jussieu.fr/titan). He has published over 70 papers mainly on systematics, taxonomy, and host–plant relations.

Miguel Angel Monné has been a Professor of Entomology in the Museu Nacional, Universidade Federal do Rio de Janeiro, from 1974 to 2008. He retired in 2008 and received the honorary title of Emeritus Researcher of the Museu Nacional. He received his Doctorate in Zoology from the

University of São Paulo, was head of the Department of Entomology for 10 years at the Museu Nacional, and became Associate Director of Research and Teaching and Vice Director at the same institution in 1990. He has been a Research Fellow of the National Council for Scientific and Technological Development of Brazil since 1980. He has published 167 papers and authored or contributed to 35 books, with emphasis on Neotropical Cerambycidae.

Michael Biondi worked as a museum technician at the Smithsonian Institution's National Museum of Natural History between 2009 and 2011. He received his Bachelor of Science degree in Biology from Erskine College and his Master of Science in Entomology with a focus on stream ecology from Clemson University. He currently co-owns and operates a micro-brewery in Charleston, South Carolina.

